PURE AND APPLIED MATHEMATICS

A Program of Monographs, Textbooks, and Lecture Notes

NUMBER THEORY

MONOGRAPHS AND TEXTBOOKS IN
PURE AND APPLIED MATHEMATICS

54. *J. Cronin,* Differential Equations (1980)
55. *C. W. Groetsch,* Elements of Applicable Functional Analysis (1980)
56. *I. Vaisman,* Foundations of Three-Dimensional Euclidean Geometry (1980)
57. *H. I. Freedan,* Deterministic Mathematical Models in Population Ecology (1980)
58. *S. B. Chae,* Lebesgue Integration (1980)
59. *C. S. Rees et al.,* Theory and Applications of Fourier Analysis (1981)
60. *L. Nachbin,* Introduction to Functional Analysis (R. M. Aron, trans.) (1981)
61. *G. Orzech and M. Orzech,* Plane Algebraic Curves (1981)
62. *R. Johnsonbaugh and W. E. Pfaffenberger,* Foundations of Mathematical Analysis (1981)
63. *W. L. Voxman and R. H. Goetschel,* Advanced Calculus (1981)
64. *L. J. Corwin and R. H. Szczarba,* Multivariable Calculus (1982)
65. *V. I. Istrǎţescu,* Introduction to Linear Operator Theory (1981)
66. *R. D. Järvinen,* Finite and Infinite Dimensional Linear Spaces (1981)
67. *J. K. Beem and P. E. Ehrlich,* Global Lorentzian Geometry (1981)
68. *D. L. Armacost,* The Structure of Locally Compact Abelian Groups (1981)
69. *J. W. Brewer and M. K. Smith, eds.,* Emmy Noether: A Tribute (1981)
70. *K. H. Kim,* Boolean Matrix Theory and Applications (1982)
71. *T. W. Wieting,* The Mathematical Theory of Chromatic Plane Ornaments (1982)
72. *D. B.Gauld,* Differential Topology (1982)
73. *R. L. Faber,* Foundations of Euclidean and Non-Euclidean Geometry (1983)
74. *M. Carmeli,* Statistical Theory and Random Matrices (1983)
75. *J. H. Carruth et al.,* The Theory of Topological Semigroups (1983)
76. *R. L. Faber,* Differential Geometry and Relativity Theory (1983)
77. *S. Barnett,* Polynomials and Linear Control Systems (1983)
78. *G. Karpilovsky,* Commutative Group Algebras (1983)
79. *F. Van Oystaeyen and A. Verschoren,* Relative Invariants of Rings (1983)
80. *I. Vaisman,* A First Course in Differential Geometry (1984)
81. *G. W. Swan,* Applications of Optimal Control Theory in Biomedicine (1984)
82. *T. Petrie and J. D. Randall,* Transformation Groups on Manifolds (1984)
83. *K. Goebel and S. Reich,* Uniform Convexity, Hyperbolic Geometry, and Nonexpansive Mappings (1984)
84. *T. Albu and C. Nǎstǎsescu,* Relative Finiteness in Module Theory (1984)
85. *K. Hrbacek and T. Jech,* Introduction to Set Theory: Second Edition (1984)
86. *F. Van Oystaeyen and A. Verschoren,* Relative Invariants of Rings (1984)
87. *B. R. McDonald,* Linear Algebra Over Commutative Rings (1984)
88. *M. Namba,* Geometry of Projective Algebraic Curves (1984)
89. *G. F. Webb,* Theory of Nonlinear Age-Dependent Population Dynamics (1985)
90. *M. R. Bremner et al.,* Tables of Dominant Weight Multiplicities for Representations of Simple Lie Algebras (1985)
91. *A. E. Fekete,* Real Linear Algebra (1985)
92. *S. B. Chae,* Holomorphy and Calculus in Normed Spaces (1985)
93. *A. J. Jerri,* Introduction to Integral Equations with Applications (1985)
94. *G. Karpilovsky,* Projective Representations of Finite Groups (1985)
95. *L. Narici and E. Beckenstein,* Topological Vector Spaces (1985)
96. *J. Weeks,* The Shape of Space (1985)
97. *P. R. Gribik and K. O. Kortanek,* Extremal Methods of Operations Research (1985)
98. *J.-A. Chao and W. A. Woyczynski, eds.,* Probability Theory and Harmonic Analysis (1986)
99. *G. D. Crown et al.,* Abstract Algebra (1986)
100. *J. H. Carruth et al.,* The Theory of Topological Semigroups, Volume 2 (1986)
101. *R. S. Doran and V. A. Belfi,* Characterizations of C*-Algebras (1986)
102. *M. W. Jeter,* Mathematical Programming (1986)
103. *M. Altman,* A Unified Theory of Nonlinear Operator and Evolution Equations with Applications (1986)
104. *A. Verschoren,* Relative Invariants of Sheaves (1987)
105. *R. A. Usmani,* Applied Linear Algebra (1987)
106. *P. Blass and J. Lang,* Zariski Surfaces and Differential Equations in Characteristic $p > 0$ (1987)
107. *J. A. Reneke et al.,* Structured Hereditary Systems (1987)
108. *H. Busemann and B. B. Phadke,* Spaces with Distinguished Geodesics (1987)
109. *R. Harte,* Invertibility and Singularity for Bounded Linear Operators (1988)

Additional Volumes in Preparation

NUMBER THEORY

An Introduction

Don Redmond
Southern Illinois University at Carbondale
Carbondale, Illinois

Marcel Dekker, Inc.   New York • Basel • Hong Kong

Library of Congress Cataloging-in-Publication Data

Redmond, Don.
  Number theory : an introduction / Don Redmond.
    p. cm. — (Monographs and textbooks in pure and applied
  mathematics ; 201)
    Includes bibliographical references (p. –   ) and index.
    ISBN 0-8247-9696-9 (alk. paper)
    1. Number theory. I. Title. II. Series.
  QA241.R36 1996
  512'.72—dc20   96-1657
                                                             CIP

The publisher offers discounts on this book when ordered in bulk quantities.
For more information, write to Special Sales/Professional Marketing at the
address below.

This book is printed on acid-free paper.

MARCEL DEKKER, INC.
270 Madison Avenue, New York, New York 10016

Current printing (last digit):
10 9 8 7 6 5 4 3 2 1

PRINTED IN THE UNITED STATES OF AMERICA

To the memory of my parents
and
To Charlotte, as always

Preface

What is number theory? Number theory is, in the strictest sense, the study of the properties of the natural numbers. It started that way in the days of the Babylonians, Egyptians, et al., as well as all who have lived since. This book is an introduction to this study. As can be seen from the references at the end of this book there are many other books on number theory and many more will be written as long as books (or whatever books turn into) continue to be produced. Number theory is a subject that has fascinated people for a long time and will continue to do so. It is hoped that readers will also discover this fascination as they read this book. This fascination accounts, in large part, for the plethora of books on this subject.

Another reason for interest in number theory is that it starts out being very easy. The beginning problems require very little background knowledge on the part of the reader: some elementary algebraic skills and a knowledge of the logic that goes into proving a theorem. As one goes further into the subject the material gets harder, and in the end number theory has managed to use nearly all branches of mathematics in pursuit of its truths. The most recent example of this universality is Wiles' proof of Fermat's Last Theorem. One of the purposes of this book is to show how some of the rest of mathematics enters into the study of the properties of the integers.

Originally these properties were discovered by trial and error. Indeed, from what we now know, the Babylonians and Egyptians had no concept of mathematical proof, but that did not stop them from using results that they had discovered. Even the Greeks, who eventually (say, with the arrival of Thales) proved results in a reasonably rigorous manner, had to find the results somewhere. Results continue to be discovered by trial and error and the reader is encouraged not to be afraid to make conjectures, collect data, and see if the conjectures can be proved. However, these days one can speed up the trial and error by use of a computer.

This book is designed to be used with or without a computer. For those who do have access to a computer problems have been provided that are directly related to computer use. Sometimes the problems just ask the reader to collect data and see if any patterns emerge. Other computer problems ask the reader to implement some of the procedures ("algorithms") on a computer. Even if a computer is not available many of the computer problems can be tackled on a calculator even if it isn't programmable. The moral of the story is that if you can't see a pattern, then produce more examples.

Since this is not a book on computing, but on number theory, the computer problems have been put at the ends of the chapters in the sections of additional problems. This should not be construed to mean that the computer problems are not considered important, but that they are in addition to the material in the text. For a reader having access to a computer and some expertise in programming, the computer problems should be taken on as a useful part of the course. Indeed, it may be that readers with no computer experience will be tempted to gain some by trying to do the computer problems.

To learn number theory means to do number theory, which is the reason for the large number of problems in this book. Not only are there problems to go with each section, but at the end of each chapter there are additional problems that review and extend the material of the chapter as a whole as well as that of previous chapters. Number theory is also a subject that abounds in techniques. One's familiarity with these techniques is dependent on the amount of time spent doing them. It is usually not possible to watch someone else do them.

However, to get a start on the techniques one must see them for the first time. Since all books are finite an attempt has been made to expand the horizons of this book by giving references throughout to other places where the reader can find extensions or variations of the results that are presented. Many of these references occur in the problem sets where a result is given as a problem with a reference to its source.

There is enough material in the book for a two-semester course. The first semester would cover Chapters 1, 2, 3, and 5 and would then depend on time and one's taste. One possibility would be to cover Chapter 4 and then do Chapter 6, where the material on continued fractions in Chapter 4 could be put to use solving certain Diophantine equations. Another path might be to cover the beginning of Chapter 6 on linear equations and sums of squares and then finish up with parts of Chapter 7 on arithmetic functions.

The second semester would cover Chapters 7, 8, 9, and 10, which is the more advanced material. I have tried to make these chapters as accessible as possible, and therefore self-contained. That means, for example, that in Chapter 8 there is an introduction to Riemann-Stieltjes integration and in Chapter 10 there is a proof of the fundamental theorem of algebra. These results can just be taken for granted without real loss to the continuity of the presentation. Also, in Chapter 7 there is a large section on trigonometric sums which can be omitted as the results are not used elsewhere in the book.

I should also mention how the numbering system works. In each chapter, definitions, examples, equations, and theorems are numbered consecutively in each category. Each is given two numbers: the first refers to the chapter in which it appears and the second refers to its position in the sequence. For example, Theorem 1.18 is the 18th theorem in Chapter 1. Lemmas and corollaries are numbered in relation to the theorems to which they are related. For example, Lemma 1.15.1 refers to the first lemma to Theorem 1.15 and Corollary 1.19.3 refers to the third corollary of Theorem 1.19.

I would like to thank the various secretaries in the mathematics department of Southern Illinois University (all of whom have strangely departed the department by now) who have typed versions of the manuscript as it evolved from lecture notes. I would like to express my thanks to my Math 525 class of the Spring 1995 semester who used a beta version of Chapters 4, 7 ,8 and 9 as the text for the course. Their comments were very helpful. Usually married authors end these thank you's by commending their spouses and children for their patience. In my case I enlisted the aid of the word processing skills of my wife to help in the formatting of the manuscript so as to make it more readable. The good looks are hers. The faults are mine. Last (but not least), I would like to thank my long-suffering editors at Marcel Dekker, Inc., who put up with my lack of speed.

Don Redmond

Contents

NUMBER THEORY

A Historical Introduction

Number theory is one of the oldest branches of mathematics. It might be considered the oldest if you feel that numbers were considered before shapes, but, in any case, it is, at worst, tied with geometry. It is concerned with the study of the properties of the integers:

$$\ldots, -4, -3, -2, -1, 0, 1, 2, 3, 4, \ldots.$$

After working with these numbers long enough one can begin to see patterns emerge and it is the study of these patterns that evolved into number theory.

The first recorded "works" on number theory are old cuneiform tablets from Mesopotamia and later (ca. 2000 BC) various papyri of the Egyptians. These were what we would call work books: concrete problems and only one solution given with no attempt to generalize. Later the Pythagoreans (ca. 550 BC), who considered numbers to be everything, made a more systematic study of the properties of positive integers. The fruit of this labor appeared several centuries later in Nicomachus' book *Introduction to Arithmetic* (ca. 100 AD). In it there is almost as much about the philosophical ramifications of numbers as about the numbers and their properties.

The first systematic written work on the theory of numbers was in Books VII, VIII and IX of Euclid's *Elements* (ca. 300 BC). In here is much of the ordinary rules of arithmetic as well as several very important results: the Euclidean algorithm (see Theorem 1.16) in Book VII and the infinitude of primes (see Theorem 1.13) in Book IX. Also in Book VII is Lemma 1.19.1, which is instrumental in many of the proofs of unique factorization (Theorem 1.19). All of these results are couched in the geometric language then in vogue in Greek mathematics.

Later (ca. 300 AD) Diophantus produced a work, *Arithmetica*, that deals with the subject of solving equations in rational numbers and integers. The exposition is in a symbolic form, as opposed to the geometric language of Euclid, making Diophantus one of the grandfathers of algebra. However, unlike Euclid, Diophantus did not present a systematic treatise. The *Arithmetica* is more like a collection of problems in which general problems are solved using specific examples. In a sense the

Arithmetica is like a sophisticated Rhind papyrus. After Diophantus the theory of numbers went into somewhat of a decline in the West. The Arabs and especially the Hindus kept up the study, but on the whole their work continued along the Diophantine line: doing specific cases rather than a general theory, whether they were doing original work or not. Some of this material is discussed in Chapters 5 and 6.

The great number theory revival began with P. de Fermat (1601-1665). He started with Diophantus' book and between it and his own fertile imagination produced a great variety of results, which, alas, he never published. He left much for later number theorists to prove and, indeed, bequeathed to number theory one problem that may only have just recently been solved. This is the so-called "Fermat's Last Theorem," which states that if $n \geq 3$ is a positive integer, then the equation

$$x^n + y^n = z^n$$

has no solutions in positive integers x, y and z. We will meet many of his other results below, including the famous method of descent, a sort of backwards mathematical induction. In one version one assumes that one has the minimal counterexample to a theorem one is trying to prove and then produces a yet smaller one. We shall see this method in action in Chapter 5.

After Fermat the one who did the most for number theory was probably L. Euler (1707-1783). He began his work in number theory trying to prove many of Fermat's statements, went on to generalize some of these results and finally began to produce original results of his own. He was responsible, in large part, for the introduction of analysis (in his proof of the infinitude of primes as well as his work on partitions) and algebraic numbers (in one of his proofs for the case $n = 3$ of the Fermat conjecture above) into number theory. (See Chapters Eight, Nine and Ten.) While he never wrote a text specifically on the subject, his numerous writings on the various aspects of number theory (some of which appear as parts of his textbooks) form a great body of work that helped to spur on further research.

Two of the people who followed Euler were J. L. Lagrange (1735-1813) and A. M. Legendre (1752-1833). Lagrange, who was a firm believer in the rigorous proof, finished up many of Euler's incomplete proofs as well as making contributions of his own (see, for example, Section 5 of Chapter Four.) Legendre contributed less, but wrote the magnificent treatise, *Essai sur la theorie des nombres*, in 1798, which was the greatest systemization of what was known at that time.

The next giant was a true giant of mathematics: C. F. Gauss (1777-1855).

The shape of number theory today is due, in large part, to Gauss. In his great work, *Disquistiones Arithmeticae* (1801), Gauss started afresh almost in ignorance of Legendre's treatise. Indeed Gauss claims that he did know of Legendre's work until after he had finished his. In any case, his work went beyond that of Legendre's in many areas. He reintroduced, on a firmer basis, the use of algebraic numbers and the theory of quadratic forms. (For a small peek at the latter see Section 6 of Chapter Six.) He also gave the first proof of the, by then well-known, unique factorization of the positive integers. He also used analysis in the study of arithmetic questions, primarily the use of elliptic functions (as in the published work of Jacobi sometime later) and at the age of fourteen conjectured the prime number theorem (Theorem 1.18 and Corollary 9.7.1). Unfortunately he abandoned number theory at a relatively early age and seldom returned to it.

The *Disquisitiones* is a hard work to go through. In an attempt to facilitate the study of Gauss' treatise P. G. L. Dirichlet (1805-1859) wrote a number theory treatise of his own: *Vorlesungen über Zahlentheorie* (1863), wherein he expanded on the work of Gauss adding many of his own contributions including the first great use of analysis in number theory: the proof of the infinitude of primes in arithmetic progressions (Theorems 1.5 and 9.9). This work was later expanded by R. Dedekind (1831-1916) in 1893 in which he added his own work on the foundations of algebraic number theory, which was started by E. E. Kummer (1810-1893) in his attempts to prove Fermat's conjecture.

B. Riemann (1826-1866) finished the introduction of analysis into number theory by his use of complex variable techniques in an attempt to prove the prime number theorem. He wrote only one number theory paper, but it sparked great activity on the part of others to fill in the gaps and in the process complex function theory. Finally, in 1896 C. de la Vallee Poisson (1866-1962) and J. Hadamard (1865-1963) filled in Riemann's argument and independently finished the proof of the prime number theorem. One piece of Riemann's legacy is still open: the famous Riemann Hypothesis about the location of the nontrivial zeros of the zeta function.

In the twentieth century all the tools of modern mathematics have been brought to bear on the study of the theory of numbers often with great success. If there is a golden age of number theory, it is now.

This is a very compact version of the history of number theory. For a fuller account see [25], [86] or [126].

CHAPTER ONE
Primes and Divisibility

1.1. INTRODUCTION

In this chapter we discuss the foundations of number theory. It all begins with the concept of divisibility. As we shall see there is a subset of the integers that will allow us to write all integers as products of them. These numbers are the prime numbers and they are the building blocks of the integers. Much of what we discuss in this chapter is covered in the number theory books of Euclid's *Elements*. One of these results is the concept of greatest common divisor which is defined multiplicatively, but, because of Euclid's Algorithm, can be found through a process than in essence involves only addition and subtraction. With these results in hand we can then prove that integers can be uniquely factored into prime numbers. As we shall see in that section unique factorization is not something that can be taken for granted. Finally, we conclude the chapter by applying some of these results to methods for factoring integers.

There are two results that we will find very useful in this chapter. Since we are at the beginning of our study we do not have many results off of which to build, and so we must use some of the deep properties of the natural numbers. These results are the following.

Proposition 1. (The Principle of Mathematical Induction). Let S be a subset of $\mathbf{N}$. If 1 is an element of S and n being an element of S implies that $n + 1$ is an element of S, then $S = \mathbf{N}$.

Proposition 2. (The Well-Ordering Principle). Any nonempty set of natural numbers contains a least element. In other words, if S is a nonempty set of natural numbers, then there exists some element m, of S such that $m < n$ for all elements n of S.

Before giving some examples to illustrate the use of these two propositions we prove another version of mathematical induction that is sometimes useful.

Theorem 1.1. Let S be a subset of $\mathbf{N}$ such that 1 is in S and $k + 1$ is in S

whenever 1, 2, ... , k are in S. Then $S = \mathbf{N}$.

Proof. Let A consist of the integers of $\mathbf{N}$ that are not in S. If we suppose A is nonempty, then, by Proposition 2, there is a least element of A, say a.

Since 1 is in S we see that $a \neq 1$. Since a is the least element of A we see that 1, 2, ... , $a - 1$ are in S. The hypothesis on S then implies that $(a - 1) + 1 = a$ is in S. This is a contradiction to the definition of A. Therefore we must have that A is empty. Thus $S = \mathbf{N}$. ∎

To help familiarize the reader with these two results we illustrate them with two examples, as well as include some problems related to these two principles.

Example 1.1. Use mathematical induction to show that

$$1 + 2 + \cdots + n = n(n + 1)/2.$$

Solution. If $N = 1$, then we have

$$1 = 1(1 + 1)/2$$

Thus the result is true for $N = 1$.

Suppose that the result is true for $N = n$, that is, assume that

$$1 + 2 + \cdots + n = n(n + 1)/2.$$

Then for $N = n + 1$ we have

$$\begin{aligned}
1 + 2 + \cdots + n + (n+1) &= (1 + 2 + \cdots + n) + (n+1) \\
&= n(n+1)/2 + (n+1) \\
&= (n+1)(n/2 + 1) \\
&= (n+1)(n+2)/2,
\end{aligned}$$

which is the result for $N = n + 1$.

Thus, by Proposition 1, the result is true for all natural numbers. □

Example 1.2. Show that if a and b are any natural numbers, then there exists a natural number n such that $na > b$.

Solution. If $b \leq a$, then we can take $n = 1$ (if $b < a$) or $n = 2$ (if $b = a$).

Assume that $b > a$ and let

$$S = \{b - ka : k \in \mathbf{N}\} \cap \mathbf{N}.$$

Then S is a nonempty subset of the natural numbers since $b - a > 0$. Thus, by Proposition 2, S has a least element, say c. Thus, by the definition of S, there is a natural number m such that

$$c = b - ma.$$

Now $c - a < c$, since $a > 0$, and $c - a$ is an integer. Since c is the least integer in S, we see that we must have $c - a < 0$. Since $c - a = b - (m + 1)a$, we see that

$$b < (m + 1)a.$$

Thus we can take $n = m + 1$ and get our result.   □

This second result is a special case of the more general Archimedes' Principle that holds for the real number system. This result says that if a and b are any positive real numbers, then there exists a natural number n such that $na > b$. The proof is somewhat similar to the one given here, but uses some of the deep properties of the real numbers. (See [6, p. 54].)

Problem Set 1.1

1. Prove that $1^3 + 2^3 + \cdots + n^3 = (n(n + 1)/2)^2$.

2. (a) Prove that $1^2 + 2^2 + \cdots + n^2 = n(n + 1)(2n + 1)/6$.

   (b) P r o v e   t   h   a   t
   $1^2 + 7^2 + 13^2 + \cdots + (6n + 1)^2 = (n + 1)(12n^2 + 12n + 1)$.

3. Prove that $\left(\dfrac{1}{2}+1\right)\left(\dfrac{1}{2}+\dfrac{1}{2}\right)\left(\dfrac{1}{2}+\dfrac{1}{3}\right)\cdots\left(\dfrac{1}{2}+\dfrac{1}{n}\right)=\left(\dfrac{(n+1)(n+2)}{2^{n+1}}\right)$.

4. (a) Prove the following set of inequalities:

$$(n!)^2 2^n \le (2n)! < (n!)^2 4^{n-1}$$

   (b) Prove that $n! \le ((n + 1)/2)^n$.

   (c) Prove that

$$(n!)^2 > n^n$$

5. Show that if $a \ge -1$, then $(a + 1)^n \ge 1 + na$.

6. Show that if n is a natural number, then

$$1 \cdot 1! + 2 \cdot 2! + \cdots + n \cdot n! = (n + 1)! - 1.$$

7. Prove that $1 + 3 + 5 + \cdots + (2n - 1) = n^2$.

8. Prove that if n is a natural number, then

$$a^n - 1 = (a - 1)(a^{n-1} + a^{n-2} + \cdots + a^2 + a + 1).$$

9. (a) Prove the following generalization of Proposition 1. Let a be an integer and let

$$A = \{n \in \mathbf{Z} : n \ge a\}$$

   Let S be a subset of A such that a is in S and whenever k is in S so is

$k + 1$. Then $S = A$. (Hint: make a change of variables and use Proposition 1.)

(b) Prove the analogue of Theorem 1.1 for the set A of part (a).

10. Prove that if $n \geq 4$, then $n! > n^2$ and that if $n \geq 6$, then $n! > n^3$.

11. Show that if $n \geq 0$ and $x > 0$, then

$$e^x > 1 + x + \frac{x^2}{2!} + \cdots + \frac{x^n}{n!}.$$

12. Show that if $0 \leq x < n$ and $n \geq 1$, then

$$0 \leq e^{-x} - \left(1 - \frac{x}{n}\right)^n \leq \frac{x^2 e^{-x}}{n}.$$

13. Show that if $n \geq 264$, then a postage of n cents can be entirely made up using only 12 cent stamps and 25 cent stamps.

14. Show that if $\{v_k\}$ is a sequence defined by $v_0 = 0$, $v_1 = 1$ and $v_n = v_{n-1} - v_{n-2}$ for $n \geq 2$, then for $n \geq 0$ we have $v_{n+3} = -v_n$ and $v_{n+6} = v_n$

15. Define the Fibonacci sequence $\{u_k\}$ by $u_0 = 0$, $u_1 = 1$ and, for $n \geq 1$,

$$u_{n+1} = u_n + u_{n-1}$$

Prove the following properties of the Fibonacci sequence.

(a) Show that

$$u_1 + \cdots + u_n = u_{n+2} - 1.$$

(b) Show that

$$u_1 - u_2 + \cdots + (-1)^{n-1}u_n = 1 + (-1)^{n-1}u_{n-1}.$$

(c) Show that

$$u_{m+n} = u_{m-1}u_n + u_m u_{n+1}.$$

(d) Let $\alpha = \left(1 + \sqrt{5}\right)/2$ and $\beta = \left(1 - \sqrt{5}\right)/2$. Show that

$$(\alpha - \beta)u_n = \alpha^n - \beta^n \text{ and that, for } n \geq 3, u_n > \alpha^{n-2}.$$

Define the Lucas' sequence $\{v_n\}$ by $v_0 = 2$, $v_1 = 1$ and, for $n \geq 1$,

$$v_{n+1} = v_n + v_{n-1}.$$

Try to discover analogues of (a) - (d). (See also Section 10 of Chapter 7.)

16. Prove that Proposition 1 and Proposition 2 are equivalent to one another.

1.2. DIVISIBILITY

Definition 1.1. We say that a nonzero integer a **divides** an integer b, if there exists an integer c such that

$$ac = b.$$

If a divides b, we write $a \mid b$. If a does not divide b, we write $a \nmid b$.

The following theorem gives the key properties of divisibility.

Theorem 1.2.

(a) If a is a nonzero integer, then $a \mid 0$.

(b) If a is an integer, then $1 \mid a$ and, if $a = 0$, then $a \mid a$.

(c) If $a \mid b$ and $b \mid c$, then $a \mid c$.

(d) If $a \mid b$ and c is a nonzero integer, then $ac \mid bc$ and $a \mid bc$.

(e) If $a \mid b$ and $a \mid c$, then for all integers m and n we have $a \mid (mb + nc)$.

(f) If $a \mid b$ and $b \mid a$, then $a = \pm b$.

(g) If $a \mid b$ and a and b are positive integers, then $a < b$.

Proof.

(a) We have $0 = a \cdot 0$.

(b) We have $a = 1 \cdot a$ and $a = a \cdot 1$.

(c) Since $a \mid b$ and $b \mid c$ there exist integers d and e such that

$$b = ad \text{ and } c = be.$$

Thus $c = (ad)e = a(de)$. Since de is an integer the result follows.

(d) Since $a \mid b$ we have that there is some integer d such that

$$b = ad.$$

Multiplying both sides of this equality by c gives

$$bc = (ac)d \text{ and } bc = a(cd).$$

Since c is not zero, we see that ac is not zero and so $ac \mid bc$. The second equality gives $a \mid bc$, whether or not $c = 0$.

(e) If $a \mid b$ and $a \mid c$, then there exist integers d and e such that

$$b = ad \text{ and } c = ae.$$

Thus we have, for any integers m and n,

$$mb + nc = m(ad) + n(ae) = a(md + ne)$$

and the result follows.

(f) If $a \mid b$ and $b \mid a$, then neither a nor b can be zero, by our definition. Also there are integers c and d such that

$$b = ac \text{ and } a = bd.$$

Thus

$$b = (bd)c \text{ and } a = (ac)d.$$

Thus

$$cd = 1,$$

and so

$$c = \pm 1 \text{ and } d = \pm 1.$$

Thus $a = \pm b$.

(g) If $a \mid b$, then there is an integer c such that

$$b = ac.$$

Since a and b are positive we see that c is positive, so that $c \geq 1$. Thus

$$b - a = ac - a = a(c - 1) \geq 0$$

and the result follows.   ■

The next result is called the Division Algorithm. It describes the general situation between two integers whether one divides the other or not.

Theorem 1.3. Let a and b be integers with a positive. Then there exist unique integers q and r such that

$$b = qa + r,$$

where $0 \leq r < a$. If $a \nmid b$, then we have $0 < r < a$.

Proof. Let

$$R = \{b - qa : q \in \mathbf{Z}\} \cap \mathbf{Z}_+.$$

Then by the well-ordering principle, there exists a least non-negative member of R, say r. From the definition of R we see that $r \geq 0$.

If $r \geq a$, then $r = b - qa \geq a$, or $b - (q + 1)a \geq 0$. Since $b - (q + 1)a < b - qa$ we have a contradiction to the minimality of r. Thus $r < a$.

Thus we have $0 \leq r < a$.

Suppose there is another pair q_1 and r_1 satisfying the conditions of the theorem. If $r = r_1$, then $r < r_1$, since r is minimal. Thus $0 < r_1 - r < a$ and, since $r_1 - r = a(q - q_1)$, we have $a \mid (r_1 - r)$. This is a contradiction to (g) of Theorem 1.2. Thus $r = r_1$, and so $q = q_1$. ∎

We can use this theorem and mathematical induction to prove the representability of integers in any base.

Theorem 1.4. Let a and g be positive integers with $g > 1$. Then a can be uniquely represented in the form

$$a = c_0 + c_1 g + \cdots + c_n g^n,$$

where $0 \leq c_m < g$, $m = 0, 1, \ldots, n$.

Proof. We prove the representability by induction on a.

If $a = 1$, we can take $n = 0$ and $c_0 = 1$.

Assume $a > 1$ and assume that the result holds for $m = 1, 2, \ldots, a - 1$. Since $g > 1$ we see that $g^{m+1} > g^m$ for all non-negative integers m. By the Archimedes' Principle (Example 1.2) we see that there exists an integer n such that $g^n \leq a < g^{n+1}$. By the Division Algorithm we can write

$$a = c_n g^n + r,$$

where $0 \leq r < g^n$. Since $c_n g^n = a - r > g^n - g^n = 0$, we see that $c_n > 0$. Since $c_n g^n \leq a < g^{n+1}$, we see that $c_n < g$

If $r = 0$, then

$$a = 0 + 0 \cdot g + \cdots + 0 \cdot g^{n-1} + c_n g^n.$$

If $r > 0$, then the induction hypothesis shows that we can write

$$r = b_0 + b_1 g + \cdots + b_t g^t,$$

where $0 \leq b_m < g$, $m = 0, 1, \ldots, t$. Since $r < g^n$, we see that $t < n$. Thus, in this case,

$$a = b_0 + b_1 g + \cdots + b_t g^t + 0 \cdot g^{t+1} + \cdots + 0 \cdot g^{n-1} + c_n g^n$$

The representability result then follows by Proposition 2.

We now prove uniqueness.

Assume that we have

$$a = c_0 + c_1 g + \cdots + c_n g^n = d_0 + d_1 g + \cdots + d_m g^n,$$

with $n \geq 0$, $c_n > 0$, $0 \leq c_k < g$, $k = 0, 1, \ldots, n$ and $m \geq 0$, $d_m > 0$, $0 \leq$

$d_j < g, j = 0, 1, \ldots, m$. If we subtract, we get

$$0 = e_0 + e_1 g + \cdots + e_s g^s,$$

where $e_k = c_k - d_k$ and s is the largest values of k for which $c_k \neq d_k$. Then $e_s \neq 0$.

If $s = 0$, then we have the contradiction that $e_s = e_0 = 0$. If $s > 0$, then we have

$$|e_k| = |c_k - d_k| \leq g - 1$$

and

$$e_s g^s = -(e_0 + \cdots + e_{s-1} g^{s-1}).$$

Then

$$\begin{aligned}
g^s \leq |e_s g^s| &= |e_0 + \cdots + e_{s-1} g^{s-1}| \\
&\leq |e_0| + \cdots + |e_{s-1}| g^{s-1} \\
&\leq (g-1)(1 + g + \cdots + g^{s-1}) = g^s - 1,
\end{aligned}$$

which is also a contradiction. Thus we must have $n = m$ and $c_k = d_k$, for $k = 0$, $1, \ldots, n$. Thus the representation is unique. ∎

An immediate consequence is the following result.

Corollary 1.4.1. Let a be a given positive integer. Then all integers fall into one of the arithmetic progressions $ak + r$, where $0 \leq r < a$.

This result implies, for example, that all integers can be written in the form $2k$ or $2k + 1$, that is, all integers are either even or odd. Likewise all integers are in one of the three forms: $3k$, $3k + 1$ or $3k + 2$, and so on.

Definition 1.2. We define the **greatest integer in x,** denoted by $[x]$, to be that unique integer n satisfying

$$n \leq x < n + 1.$$

Such an integer exists by the well-ordering principle.

Example 1.3. We have

$$[3] = 3, \ [-5] = -5, \ [3.134] = 3 \ \text{and} \ [-2.1821] = -3.$$

Corollary 1.4.2. The integer q of Theorem 1.4 is $[b/a]$ and the integer $r = b - a[b/a]$.

With this corollary in hand we can now see how to calculate the coefficients in the b-ary expansion of a natural number n. Note that we can write

$$n = a_0 + a_1 b + \cdots + a_k b^k = a_0 + b(a_1 + \cdots + b(a_{k-1} + a_k b)\ldots).$$

Thus, if we write $n = bq + r$, with $0 \le r < b$, then $a_0 = r$. If we let $n_1 = [n/b]$, then, by Corollary 1.4.2, we have $a_0 = n - bn_1$. In general, if we let $n_j = [n_{j-1}/b]$, then we have that $a_{j-1} = n_{j-1} - bn_j$.

We illustrate with an example.

Example 1.4. Find the base 3 expansion of 1234.

Solution. We have

$$1234 = 3 \cdot 411 + 1, \ 411 = 3 \cdot 137 + 0, \ 137 = 3 \cdot 45 + 2, \ 45 = 3 \cdot 15 + 0,$$
$$15 = 3 \cdot 5 + 0, \ 5 = 3 \cdot 1 + 2, \ 1 = 3 \cdot 0 + 1.$$

Thus

$$1234 = 1 + 0 \cdot 3 + 2 \cdot 3^2 + 0 \cdot 3^3 + 0 \cdot 3^4 + 2 \cdot 3^5 + 1 \cdot 3^6$$

Definition 1.3. An integer g is said to be a **common divisor** of the integers a and b if $g \mid a$ and $g \mid b$. The largest one of these common divisors is called the **greatest common divisor** and is denoted by (a, b). We say that a and b are **relatively prime** if $(a, b) = 1$.

We can make an analogous definition for $(a_1, \ldots, a_n)$, the greatest common divisor of the integers $a_1, \ldots, a_n$.

Since each nonzero integer has only a finite number of divisors, by (g) of Theorem 1.2, we see that the greatest common divisor is well-defined and is always positive. Note that if $a = b = 0$, then $(0, 0)$ is undefined. Otherwise we always have $(a, b) \ge 1$. Below it is to be understood that whenever we write (a, b) we are implying that the integers a and b are not both zero.

Theorem 1.5. Let $g = (a, b)$. Then there exist integers x_0 and y_0 such that

$$g = ax_0 + by_0.$$

Proof. Let

$$G = \{ax + by: x, y \in \mathbf{Z}\} \cap \mathbf{N}.$$

Then G has a least element since it is clear that no matter what the signs of a and b there is at least some pair of values of x and y such that $ax + by > 0$. (For example, we could take $x = \mathrm{sgn}(a)$ and $y = \mathrm{sgn}(b)$, at least one of which is nonzero.) Let x_0 and y_0 be a pair that give this least element. Let $d = ax_0 + by_0$ and suppose that $d \nmid a$. Then, by Theorem 1.3, there exist q and r such that $a = dq + r$, with $0 < r < d$. Then

$$r = a - dq = a - q(ax_0 + by_0) = a(1 - qx_0) + b(-qy_0).$$

Thus r is in G and $0 < r < d$, which contradicts the minimality of d. Thus $d \mid a$. In a similar way one can show that $d \mid b$.

Now $g = (a, b)$ implies that for some integers A and B we have $a = gA$ and $b = gB$. Thus

$$d = ax_0 + by_0 = g(Ax_0 + By_0),$$

that is, $g \mid d$. Thus, by (g) of Theorem 1.2, we have $g \leq d$ and, by definition, g is the largest number dividing both a and b. Thus $g \geq d$. Thus $g = d$. ∎

The question arises: short of attempting to find the least positive element of the set G of Theorem 1.5, how can we calculate (a, b) when we are only given a and b. One answer to this question is the Euclidean algorithm, which we will discuss now. Another answer will be found in Corollary 1.19.1.

We begin with a lemma whose proof uses only the definition of the greatest common divisor and Theorem 1.2. We leave the details of the proof to the reader (see Problem 12 below.)

Lemma 1.6.1. If m is an integer, then for all integers a and b, not both zero, we have

$$(a, b) = (b, a) = (a, -b) = (a, b + am).$$

Thus to compute (a,b) we can assume that both a and b are positive since the Lemma tells us that $(a, b) = (a, |b|) = (|a|, |b|)$. Also we have $(a, 0) = a$.

Theorem 1.6 (Euclidean Algorithm). Let a and b be integers with $a > 0$. Applying Theorem 1.3 repeatedly we obtain the following sequence of equations:

$$b = aq_1 + r_1, \qquad 0 < r_1 < a$$
$$a = r_1 q_2 + r_2, \qquad 0 < r_2 < r$$
$$\vdots$$
$$r_{j-2} = r_{j-1} q_j + r_j, \qquad 0 < r_j < r_{j-1}$$
$$r_{j-1} = r_j q_{j+1}.$$

Then $(a, b) = r_j$, the last nonzero remainder. Also the values of x_0 and y_0 such that

$$(a, b) = ax_0 + by_0$$

can be obtained from the $r_{j-1}, \ldots, r_2, r_1$ in this sequence of equations.

Proof. The sequence of equations must stop since there are only a finite number of integers greater than zero and less than a.

By Lemma 1.6.1 we have

$$(b,a) = (b - aq_1, a) = (r_1, a)$$
$$= (r_1, a - r_1 q_2) = (r_1, r_2)$$
$$\vdots$$
$$= (r_{j-2}, r_{j-1}) = (r_{j-2} - r_{j-1} q_j, r_{j-1})$$
$$= (r_j, r_{j-1}) = r_j,$$

since $r_j \mid r_{j-1}$ and $r_j > 0$.

To write (a, b) as a linear combination of a and b we start with

$$r_j = r_{j-2} - r_{j-1} q_j$$
$$r_{j-2} - (r_{j-3} - r_{j-2} q_{j-1}) q_j = r_{j-2}(1 + q_j q_{j-1}) - r_{j-3} q_j$$

and then eliminate r_{j-2}, r_{j-3}, etc. using the sequence of equations backwards. ■

Example 1.5. Find $(963,657)$ and an x and a y such that

$$(963, \ 657) = 963x + 657y.$$

Solution. We have

$$963 = 657 \cdot 1 + 306$$
$$657 = 306 \cdot 2 + 45$$
$$306 = 45 \cdot 6 + 36$$
$$45 = 36 \cdot 1 + 9$$
$$36 = 9 \cdot 4.$$

Thus $(963, 657) = 9$. Working backwards gives

$$9 = 45 - 36$$
$$= 45 - (306 - 45 \cdot 6) = -306 + 7 \cdot 45$$
$$= -306 + 7(657 - 306 \cdot 2) = 7 \cdot 657 - 15 \cdot 306$$
$$= 7 \cdot 657 - 15(963 - 657)$$
$$= 22 \cdot 657 + (-15) \cdot 963,$$

and so we may take $x = 22$ and $y = -15$.   □

Later (in Section 2 of Chapter Six) we shall see that we can produce infinitely many different values of x and y such that

$$9 = 657x + 963y.$$

As can be seen above this is a rather roundabout way to find these values for x and y since it requires us to keep all values of the r's. There is a better way, especially from a computer standpoint, wherein we can use the r's as we calculate with them. This is contained in the following result.

Theorem 1.7. Let a and b be positive integers. Define the sequences $\{s_k\}$ and $\{t_k\}$ by $s_0 = 1$, $t_0 = 0$, $s_1 = 0$, $t_1 = 1$ and, for $k \geq 1$, we define

$$s_{k+1} = s_{k-1} - q_k s_k \quad \text{and} \quad t_{k+1} = t_{k-1} - q_k t_k,$$

where the q's are the numbers given in the Euclidean algorithm. Then, for $k = 2$, ..., n, where r_n is the last nonzero remainder, we have

$$r_k = a s_{k+1} + b t_{k+1}$$

Proof. Note that we have

$$r_{k+1} = r_{k-1} - q_k r_k.$$

Also note that we have

$$r_2 = b - q_1 a = a s_3 + b t_3 \quad \text{and} \quad r_3 = a s_4 + b t_4.$$

If we assume we have that

$$r_j = a s_{j+1} + b t_{j+1}$$

for $j = k - 1$ and k, then we have, as above

$$\begin{aligned}
r_{k+1} &= r_{k-1} - q_k r_k \\
&= a s_k + b t_k - q_k \left(a s_{k+1} + b t_{k+1} \right) \\
&= a(s_k - q_k s_{k+1}) + b(t_k - q_k t_{k+1}) \\
&= a s_{k+2} + b t_{k+2},
\end{aligned}$$

which is the result for $j = k + 1$. The result follows by Proposition 1. ∎

Example 1.6. Find integers x and y such that

$$963x + 657y = (963, \ 657).$$

Solution. From Example 1.5 we have that $(963, 657) = 9$ and

$$q_1 = 1, \ q_2 = 2, \ q_3 = 6 \ \text{and} \ q_4 = 1.$$

This gives us the following sequences

$$s_0 = 1 \qquad\qquad t_0 = 0$$
$$s_1 = 0 \qquad\qquad t_1 = 1$$
$$s_2 = 1 - 1 \cdot 0 = 1 \qquad t_2 = 0 - 1 \cdot 1 = 1$$
$$s_3 = 0 - 1 \cdot 2 = -2 \qquad t_3 = 1 \cdot (-1) - 2 = -3$$
$$s_4 = 1 - (-2) \cdot 6 = 13 \qquad t_4 = -1 - 3 \cdot 6 = -19$$
$$s_5 = -2 - 13 \cdot 1 = -15 \qquad t_5 = 3 - (-19) \cdot 1 = 22.$$

Thus

$$963 \cdot (-15) + 657 \cdot 22 = 9,$$

as before. $\square$

In the exercises (see Additional Problem 53 below) we will give a problem that gives an upper bound on the number of steps required by the Euclidean algorithm.

Now that we know how to compute the greatest common divisor of any pair of integers with relative ease we shall continue to derive further properties of the greatest common divisor.

Theorem 1.8. Let $g = (a, b)$. Then the following two statements are equivalent to the definition of g.

(a)   We have $g = \min\{ax + by > 0\colon x, y \in \mathbf{Z}\}$.

(b)   If $d \mid a$ and $d \mid b$, then $d \mid g$.

Proof. (a) follows immediately from Theorem 1.5.

To prove (b) note that if $d \mid a$ and $d \mid b$, then, by (e) of Theorem 1.2 and by Theorem 1.5, $d \mid g$. By (6) of Theorem 1.2 we see that there cannot be two distinct numbers, g_1 and g_2, with this property.   ∎

The following result generalizes Theorems 1.5 and 1.8. Since the proof is straightforward we omit it.

Theorem 1.9. Let $a_1, \ldots, a_n$ be n integers, not all zero, with greatest common divisor g. Then there exist integers $x_1, \ldots, x_n$ such that

$$g = a_1 x_1 + \cdots + a_n x_n$$

Moreover, g is the least positive value of this linear form and is the common divisor divisible by all other common divisors.

The following result shows that in working with greatest common divisors we essentially only have to consider the case of two integers.

Corollary 1.9.1. We have $(a_1, \ldots, a_n) = ((a_1, \ldots, a_{n-1}), a_n)$ for any integers $a_1, \ldots, a_n$, not all zero, with $n \geq 2$.

Proof. Let

$$g_1 = (a_1, \ldots, a_n) \quad \text{and} \quad g_2 = ((a_1, \ldots, a_{n-1}), a_n).$$

By definition we have $g_1 \mid a_j, j = 1, \ldots, n..$ Thus, by Theorem 1.9, we have $g_1 \mid (a_1, \ldots, a_{n-1})$, and so $g_1 \mid ((a_1, \ldots, a_{n-1}), a_n) = g_2$. Thus, by (g) of Theorem 1.2, we have $g_1 \leq g_2$.

In a similar way $g_2 \mid (a_1, \ldots, a_{n-1})$ and $g_2 \mid a_n$. Thus $g_2 \mid a_j, j = 1, \ldots, n$, and so $g_2 \mid (a_1, \ldots, a_n) = g_1$, by Theorem 1.9. Thus, by (g) of Theorem 1.2, we have $g_2 \leq g_1$.

Thus $g_1 = g_2$.   ■

Theorem 1.10. If m is a positive integer, then

$$(ma, mb) = m(a, b).$$

Proof. By Theorem 1.8 we have

$$\begin{aligned} (ma, \ mb) &= \min\{max + mby > 0 \colon x, y \in \mathbf{Z}\} \\ &= m \cdot \min\{ax + by > 0 \colon x, y \in \mathbf{Z}\} \\ &= m(a, b), \end{aligned}$$

since m is positive.   ■

Corollary 1.10.1. If $d \mid a$ and $d \mid b$, $d > 0$, the $(a/d, b/d) = (a,b)/d$. In particular, if $(a, b) = g$, then $(a/g, b/g) = 1$.

Proof. The second statement follows from the first by letting $d = g$.

The first statement follows from Theorem 1.10 by taking $m = d$ and $a = a/d$ and $b = b/d$.   ■

Corollary 1.10.2. if $c \mid ab$ and $(b, c) = 1$, then $c \mid a$.

Proof. By Theorem 1.10, we have

$$(ab, ac) = a(b, c) = a.$$

Since $c \mid ab$ and $c \mid ac$ we have, by Theorem 1.8, $c \mid a$.   ■

Problem Set 1.2

1. (a) If a and b are natural numbers, then $(a + b)!/a!b!$ is an integer.
   (b) If k is a natural number, then $k! \mid (a + 1)(a + 2) \cdots (a + k)$, for any

integer a.

(c) If k is a natural number and $\binom{2k}{k}$ denotes the binomial coefficient, then

$$2 \mid \binom{2k}{k}.$$

(d) If $a_1, \ldots, a_k$ are natural numbers such that $a_1 + \cdots + a_k = n$, then $n!/a_1! \cdots a_k!$ is an integer.

2. Let k be an integer greater than 2. If $k \mid (q_1 - 1)$, $k \mid (q_2 - 1)$, $\ldots$, $k \mid (q_r - 1)$, then $k \mid (q_1 q_2, \ldots, q_r - 1)$.

3. Let a, b and c be nonzero integers. If $ac \mid bc$, then $a \mid b$.

4. Let a and b be nonzero integers.
   (a) $a \mid b$ if and only if $a^2 \mid b^2$.
   (b) Show that $a^m \mid b^m$ implies that $a \mid b$ if and only if $m \geq n$.

5. Let n be an odd number. Then
   (a) $8 \mid (n^2 - 1)$
   and
   (b) if $3 \nmid n$, then $6 \mid (n^2 - 1)$.

6. Show that if $a \mid b$ and $c \mid d$, then $ac \mid bd$.

7. Show that $8^n \mid (4n)!$ and $16^n \mid (6n)!$ for all natural numbers n. Can you generalize these results?

8. Let n be a natural number. Show that $6 \mid n^3 - n$ and if n is odd, then $24 \mid n^3 - n$.

9. Prove that if n is a natural number, then $2^n \mid (n + 1)(n + 2) \cdots (2n)$.

10. Show that if n is a natural number, then $3 \mid 2^{5n} - 1$ and $17 \mid 2^{8n - 4} - 1$.

11. Let a, b and c be natural numbers. Then $a^2 + b^2 = c^2$ implies that $3 \mid abc$.

12. Give the details of the proof of Lemma 1.6.1.

13. Give the details of the proof of Theorem 1.9.

14. (a) Compute $(187, 221)$, $(6188, 4709)$, $(314, 159)$, $(819, 1430)$, $(227, 659)$, $(584, 1606)$, $(10587, 534)$, $(19800, 180)$ and $(1987645, 675515)$. In each case find the corresponding values of x and y such that $ax + by = (a, b)$.
    (b) Compute $(81719, 52003, 33649, 30107)$.

15. Let m and n be positive integers.
    (a) Let a be a positive integer greater than one and let $d = (m, n)$. Then

$$(a^m - 1, \ a^n - 1) = a^d - 1.$$

(b) Suppose m is odd. Then

$$(2^m - 1, \ 2^n + 1 \ ) = 1.$$

16. (a) Let a, m and n be natural numbers. Show that if

$$(a^{2^m} + 1, \ a^{2^n} + 1) = d,$$

then $d = 1$ if a is even and $d = 2$ if a is odd. (Hint: Show that $(a^{2^n} + 1)$ divides $(a^{2^m} - 1)$, if $m > n$.).

(b) If n is a nonnegative number, let

$$F_n = 2^{2^n} + 1$$

Show that $(F_n, F_m) = 1$, if $m \neq n$.

17. (a) Prove that if $b_1, b_2, \ldots, b_k, \ldots$ is a sequence of integers each greater than 1, then every natural number can be uniquely represented in the form

$$a_0 + a_1 b_1 + a_2 b_1 b_2 + \cdots + a_k b_1 \cdots b_k.$$

where the a_i are integers that satisfy $0 \leq a_i < b_{i+1}$. (Note that this generalizes Theorem 1.6 which is the special case when we have $b_i = g$, for some natural number $g > 1$.)

(b) As a corollary to the result in (a) show that each positive integer n can be uniquely written in the form

$$n = a_m m! + a_{m-1} (m-1)! + \cdots + a_2 2! + a_1 1!,$$

where $0 \leq a_j \leq j, j = 1, \ldots, m$. This expansion is called the Cantor expansion.

18. Let a and b be natural numbers such that $(a, b) = 1$.

(a) Show that if

$$(a + b, a - b) = d,$$

then $d = 1$ if a and b have different parity and $d = 2$ if a and b have the same parity.

(b) Show that

$$(a + b, a - b, ab) = 1.$$

(c) If

$$(a + b, a^2 - ab + b^2) = d,$$

Show that $d = 1$ if $3\!\!\!/(a + b)$ and $d = 3$ if $3 \,|\, (a + b)$.

19. If a, b, x and y are integers, then $(a, b) = (a, b, ax + by)$.

20 (a) If $(a, m) = 1 = (b, m)$, then $(ab, m) = 1$.

    (b) If $b \,|\, a$, $c \,|\, a$ and $(b, c) = 1$, then $bc \,|\, a$.

    (c) If $(a, b) = 1$ and m is any integer, then $(a, m)(b, m) = (ab, m)$.

21. If n is a natural number, then $(a, b)^n = (a^n, b^n)$.

22. Determine $(a^2 + b^2, a + b)$ if $(a, b) = 1$. What can be said if $(a, b) > 1$?

23. If $(m, n) = 1$ and k is any integer, determine $(m + kn, m^3 + n^3)$.

24. Suppose that $(n, n + k) = 1$ for all natural numbers n. Show that $k = \pm 1$.

25. Suppose that a, b, c and d are integers such that $ad - bc = 1$.

    (a) Show that $(a + b, c + d) = 1$.

    (b) If $mn \neq 0$, show that $(m, n) = (am + bn, cm + dn)$.

26. If $(a, 4) = 2$ and $(b, 4) = 2$, show that $(a + b, 4) = 4$.

27. Let a and b be natural numbers such that $(a, b) = 1$.

    (a) If c is a natural number, then there are infinitely many integers m such that

$$(a + bm, c) = 1.$$

    (b) Show that the arithmetic progression $\{a + bk :: k = 0, 1, 2, \ldots\}$ contains an infinite subsequence such that any two members are relatively prime.

28. (a) Let a and N be positive integers. Determine how many positive integers up to N are divisible by a.

    (b) How many multiples of 23 are there between 1732 and 6155?

    (c) How many integers not greater than 1000 are not divisible by 2, 3 or 5?

1.3. PRIME NUMBERS

Definition 1.4. a positive integer p, greater than one, is called a **prime number**, if, whenever $p = ab$, have $p = a$ or $p = b$. An integer greater than one that is not a prime number is called a **composite number**.

For example, if we let p_n denote the nth prime, then we have

$$p_1 = 2, \; p_2 = 3, \; p_3 = 5, \; \ldots, \; p_{100} = 541, \; \ldots$$

The question might be raised about the last set of dots; do they really indicate that the list of primes is infinite? This is indeed true, as we shall see below in Theorem 1.13.

The real importance of prime numbers to number theory is given in the following theorem.

Theorem 1.11. every positive integer greater than one can be written as a product of prime numbers.

Proof. The proof is by induction. For $n = 2$ the result is true since 2 is a prime. Suppose the result holds for all $2 \leq n \leq m - 1$, that is, n is a product of primes. Then either m is a prime or it is composite. If m is a prime, we're done. If not, then $m = ab$ where both a and b are not equal to m. By (g) of Theorem 1.2 we see $a, b \leq m - 1$, and so are themselves a product of primes. Thus m is a product of primes. ∎

Theorem 1.12. If n is composite, then there exists a prime $p \leq \sqrt{n}$ such that $p \mid n$.

Proof. If n is composite, then it is the product of at least two primes (which may not be distinct). If p and q are two of these and $p, q > \sqrt{n}$, then $n \geq pq > n$, by (g) of Theorem 1.2. Since this is a contradiction we see that one of p or q must be less than or equal to $\sqrt{n}$. ☐

Another way of stating this result is as follows: if $p \nmid n$ for all primes $p \leq \sqrt{n}$, then n is a prime. This gives us a test to see if a number is prime or not and is the first of several results that will be found in this text which either gives ways to factor a number or to show that a number is prime. We shall give these results as they occur as applications of the more theoretical results. (See Sections 1.5 and 4.6.)

Theorem 1.12, or rather its contrapositive stated above, is the basis of the "Sieve of Eratoshenes" which can be used to find prime numbers under a given limit. The procedure is simple.

(1) Write down all the numbers from 1 to N, where N is the given limit. (Below we write them in rows of ten only because it makes certain patterns more apparent.)

(2) Cross out all numbers ≥ 4 which are divisible by 2.

(3) Take the first number not crossed out, which in the first case is 3, and cross out all multiples of it.

(4) Continue by picking the first number left after each crossing out.

(5) This process keeps up until you reach $1 + \sqrt{N}$. All the numbers not crossed out will be the primes less than or equal to N.

We illustrate by producing a list of primes less than 50. Since $\sqrt{50} \approx 7.07$ we need only cross out numbers that are multiples of the numbers that are ≤ 8. For clarity we italicize the number crossed out as multiples of 2, the remaining crossed out multiples of 3 are light, the remaining crossed out multiples of 5 are in bold italics, and the remaining crossed out multiples of 7 (which is just 49) are in outline.

1	2	3	4	5	6	7	8	9	10
11	*12*	**13**	*14*	*15*	*16*	**17**	*18*	**19**	*20*
21	*22*	**23**	*24*	*2 5*	*26*	*27*	*28*	*29*	*30*
31	*32*	*33*	*34*	*3 5*	*36*	**37**	*38*	*39*	*40*
41	*42*	**43**	*44*	*45*	*46*	**47**	*48*	49	*50*

Thus the primes less than 50 are

$$2, 3, 5, 7, 11, 13, 17, 19, 23, 29, 31, 37, 41, 43, 47.$$

One can also use this process to find the distinct prime factors of a number. Instead of only crossing out unmarked numbers one crosses out all multiples of a given number and one uses all unmarked numbers whether or not they are less than $\sqrt{n}$. In this way we would find, for example, that 6 has as its distinct prime divisors 2 and 3, since it would be crossed out by both 2 and 3, and that 33 has 3 and 11 as its distinct prime divisors, since 33 would be crossed out by 3 and 11, even though $11 > \sqrt{50}$.

These theorems, especially Theorem 1.11, bring with them several questions. Two of these, which we will be able to answer relatively soon are:

1. Are there only a finite number of primes that generate all the positive integers?

2. How many different ways can number be written as product of primes?

We answer the first question in the negative with the result that follows and the second question shall be answered little later. (See Theorem 1.19.)

Theorem 1.13. There are an infinite number of primes.

Proof. To prove the theorem it will suffice to prove that if $\{p_1, \ldots, p_n\}$ is any finite set of primes, then we can find a prime number that is not in the set.

Let

$$P = p_1 p_2 \cdots p_n + 1.$$

By Theorem 1.11, we know that P is either a prime or is a product of primes. If P is a prime, then the result follows since $P > p_j$, $1 \le j \le n$, and so P cannot be in our given set of primes. If P is composite, then P is divisible by some prime, say Q. Now Q cannot be any of the p_j, $1 \le j \le n$, for if it was, then, by (e) of Theorem 1.2, we would have $p_j \mid 1$, which is impossible. Thus, if P is composite, it must be divisible by some prime Q which is not in $\{p_1, \cdots, p_n\}$.

In either case we have found prime that is not in our given set of primes. ■

The above proof is due to Euclid and will be found in *The Elements*, Book IX, though there it is couched in geometric language. It is one of the simplest of the many proofs of this theorem and perhaps the most beautiful. For collection of other proofs see [89, Chapter 1]. (See also Problem 16 below.)

If we check the proof in the first few cases we obtain the following results:

$$2 \cdot 3 + 1 = 7, \text{ which is prime,}$$

$$2 \cdot 3 \cdot 5 + 1 = 31, \text{ which is prime,}$$

$$2 \cdot 3 \cdot 5 \cdot 7 + 1 = 211, \text{ which is prime,}$$

$$2 \cdot 3 \cdot 5 \cdot 7 \cdot 11 + 1 = 2311, \text{ which is prime.}$$

At that is point we might be tempted to conjecture that we always get primes, but the whole thing breaks down and we find the next cases are all composite. For example,

$$2 \cdot 3 \cdot 5 \cdot 7 \cdot 11 \cdot 13 \cdot 17 + 1 = 510511 = 19 \cdot 97 \cdot 277.$$

At least we see that the proof is correct. One might then ask whether this number we construct in this proof is ever prime again? If it is prime again, is it prime for an infinite sequence of primes? Is it composite for an infinite sequence of primes? The answers to the last two questions are unknown, but as of 1988 the largest known prime for which this number, P, is also prime is $p = 13649$. (See [89, p. 4].)

One may generalize this theorem to a theorem about primes the arithmetic progression $\{an + b: n \in \mathbf{Z}_+\}$. If we take $a = b = 1$, then we have $\mathbf{N}$, the set of natural numbers, and the previous theorem shows that there are infinitely many primes in this arithmetic progression. If we take $a = 2$ and $b = 1$, then we have the set of positive odd numbers and we know it contains infinitely many primes, since all primes greater than 2 must be odd. The general theorem is due to Dirichlet.

Theorem 1.14. Suppose a and b are integers with a positive and with no common divisors except 1. Then there are infinitely many primes of the form $an + b$, where n runs through $\mathbf{Z}_+$.

The proof of this theorem can be found in Corollary 9.9.1. For a completely elementary proof of the cases $b = 1$ or $b = -1$ see [57, pp. 436-446]. We content ourselves with proof for the progression $4n - 1$. This proof is modeled on Euclid's proof of Theorem 1.13. ■

Theorem 1.15. There are infinitely many primes of the form $4n - 1$.

Proof. Let $F = \{3, 7, \ldots, p_n\}$ be list of primes of the form $4n - 1$ and let

$$P = (2^2 \cdot 3 \cdot 7 \cdot \cdots \cdot p_n) - 1.$$

Then P is of the form $4k - 1$ and, as in the proof of Theorem 1.13, we see that P is not divisible by any of the primes in F. By theorem 1.11 P is either prime or product of primes. If P is prime, then we are done. If P is composite, then it cannot be divisible solely by primes of the form $4k + 1$, since the product of two numbers of this form is again of this form. Thus P must be divisible by some prime of the form $4k - 1$ and we're done.   ∎

The corresponding theorem for the progression $4n + 1$ is harder and we shall have to delay it for awhile.

These investigations lead to many interesting questions:

(1) Under "suitable" conditions are there polynomials that generate infinitely many primes besides the linear ones? The conjecture is yes, but no one knows that answer. As to "suitable," it is bit hard to describe, but we give two examples to illustrate.

The polynomial $n^2 + 1$ should yield an infinite number of primes, but no one knows for sure. However, the polynomials $n^2 + n$ and $2n^2 + 4$ do not yield any primes if $n > 1$. For what's known see [39, p. 157] and [89, Chapter 3].

(2) Are there any formulas that give only primes? The answer here is qualified yes. Mills [67] proved that there exists number A such that the largest integer less than or equal to A^{3^n} is a prime for all positive integers n. Unfortunately, no one knows what the value of A is. See [115] for generalizations of this result.

On the other hand, we have the following negative result.

Theorem 1.16. No polynomial in one variable, f with integral coefficients, which is not constant, can be prime for all n or even for all large n.

Proof. Let

$$f(x) = a_m x^m + \cdots + a_0,$$

where the a_i, $1 \le i \le m$, are integers. Without loss of generality we may assume that $a_m > 0$, since, otherwise, $f(x) < 0$ for $x > x_0$, and so can't be prime. Then $f(n) \to +\infty$, as $n \to +\infty$, and so there is an integer N such that if $n > N$, then $f(n) > 1$. If r is a positive integer, then, if $f(n) = M$, we have

$$f(rM + n) = a_m(rM + n)^m + \cdots + a_0$$
$$= M[a_m r^m M^{m-1} + \cdots + a_1 r] + f(n)$$
$$= MT,$$

where T is some positive integer. Since $f(rM + n) \to +\infty$, the result follows. ∎

(3) How many primes are there? This question can be answered reasonably well. Let $\pi(x)$ be the number of primes from 2 up to x, where x is a real variable. For example, we have $\pi(1000) = 168$. Then Theorem 1.5 says that $\pi(x) \to +\infty$ as $x \to +\infty$. So the question here might be rephrased as: how fast does $\pi(x)$ tend to infinity? For this one can give estimates (see Section 9.2), but the following obvious result shows that $\pi(x)$ can be constant for long stretches.

Theorem 1.17. If n is a positive integer greater than one, then all of the numbers $n! + 2, n! + 3, \ldots , n! + n$ are composite.

The truth of the matter is given by the following result.

Theorem 1.18. We have

$$\lim_{x \to +\infty} \pi(x) \log(x) / x = 1.$$

The proof of this result is contained in Corollary 9.7.1. The theorem, which is called the Prime Number Theorem, was first proved in 1896 independently by the French mathematician J. Hadamard and the Belgian mathematician C. de la Vallee-Poussin. (See [38] and [80].) The original proofs used complex function theory and were, to some extent, the crowning achievement of nineteenth century complex function theory. The proof given in Chapter 9 does not use complex function theory, but only real variable integration and some involved estimations.

Problem Set 1.3

1 . (a) Use the Sieve of Eratosthenes to find all primes less than 200.

(b) Find the distinct prime factors of all numbers less than or equal to 200.

2. Show that if $p \nmid n$ for all primes $p \le \sqrt[3]{n}$, then n is either a prime or the product of two primes.

3. Show that if $n^2 - 2$ and $n^2 + 2$ are both primes, then $3 \mid n$.

4. If $a^n + 1$ is prime, then a is even and n is a power of 2.

5. If $2^n - 1$ is a prime, then n is a prime.

6. (a) If p and $p + 2$ are consecutive primes, then 6 divides the composite

number between them for $p > 3$.

(b) If p and $p + 4$ are consecutive primes, then 9 divides the sum of the composite numbers between them.

(c) Generalize the results of (a) and (b) to the sum of the consecutive composite integers between the consecutive primes p and $p + 2k$, $k \geq 3$.

7. Prove that every natural number greater than 11 is the sum of two composite integers.

8. (a) Show that if $n \geq 4$, then $p_{n+1}^2 < p_1 \cdot \cdots \cdot p_n$.

(b) Show that 30 is the largest positive integer N with the property that if k is a natural number less than N and k and N have no common divisors, then k is a prime.

(c) If $n \geq 5$, then $p_{n+1}^3 < p_1 \cdot \cdots \cdot p_n$. (Hint: $p_{j-1} \geq 2j + 2$ for $j \geq 10$.)

9. (a) If $n > 1$, then $n^4 + 4^n$ is composite.

(b) If $n > 1$, then $n^4 + 4$ is composite.

10. Show that there are infinitely many primes of the form $6k + 5$.

11. Let $f(x_1, \ldots, x_n)$ be a polynomial in n variables with integer coefficients. Show that f cannot assume prime values for all values of its variables.

12. (a) If $p \geq 5$ is a prime, then $p^2 + 2$ is composite.

(b) If p is an odd prime and not equal to 5, then either $p^2 - 1$ or $p^2 + 1$ is divisible by 5. Can you tell when which is the case?

13. If $p \geq q \geq 5$ and p and q are both primes, then $24 \mid p^2 - q^2$.

14. Show that 3, 5 and 7 are the only triple of primes of the form n, $n + 2$ and $n + 4$.

15. Suppose that $n > 1$ is an integer. If n is not of the form $6k + 3$, then $n^2 + 2^n$ is a composite.

16. (a) Use the result of Problem 15(a) of Problem Set 1.2 to show that there are infinitely many primes.

(b) Use the result of Problem 16(b) of Problem Set 1.2 to show that there are infinitely many primes.

1.4. UNIQUE FACTORIZATION

We are now ready to answer the question asked above about the factorization of integers into primes. Since this theorem is the basis for much of what follows we shall give two very distinct proofs of this result. It is somewhat amazing, given the fundamental nature of this result, that it was Gauss, in [33], who published the first proof.

The first proof requires the Lemma which follows. In this form the proof was essentially the version of the proof that was known to Euclid, except that he had no way to state the theorem in geometric language. The second proof is based on the well-ordering principle, that is, Proposition 2.

Lemma 1.19.1 (Euclid). Let p be a prime. If $p \mid ab$ then $p \mid a$ or $p \mid b$. More generally, if $p \mid a_1 \cdots a_n$, then $p \mid a_i$ for some i, $1 \le i \le n$.

Proof. Note that if p is a prime and $p \nmid a$, then we must have $(p, a) = 1$. For if $1 < d = (a, p)$, then $d \mid p$ and $d \mid a$. Thus $d = p$, which contradicts $p \nmid a$.

Now suppose $p \mid ab$. If $p \mid a$, we're done. If $p \nmid a$, then $(p, a) = 1$. Thus, by Corollary 1.9.2, we have $p \mid b$.

The second statement is an easy induction argument. ∎

Theorem 1.19 (Unique Factorization, The Fundamental Theorem of Arithmetic). Let $n > 1$ be an integer. Then the factoring of n into prime factors is unique apart from the order of the prime factors.

Proof I. Suppose there is an integer n with two factorizations. Say

$$n = P_1 \cdots P_t \text{ and } n = Q_1 \cdots Q_u,$$

where the P_i, $1 \le i \le t$ and the Q_j, $1 \le j \le u$ are prime numbers. If we divide out any common primes, we may then write

$$p_1 \cdots p_r = q_1 \cdots q_s,$$

where the p_i are the remaining P_i, the q_j are the remaining Q_j so that no p_i is a q_j, $1 \le i \le r$, $1 \le j \le s$. Thus we have $p_1 \mid q_1 \cdots q_s$, and so, by Lemma 1.19.1, this implies that $p_1 \mid q_j$ for some j, $1 \le j \le s$. Since both p_1 and q_j are primes we have $p_1 = q_j$, which is a contradiction. Thus the P_i are just the Q_j in some other order and $t = u$.

Proof II (due to Zermelo). Let us call an integer that can be factored into primes in more than one way abnormal. We wish to show that the set of abnormal integers is empty.

If the set of abnormal integers is not empty, then, by Proposition 2, it has a least element, say n. Now the same prime P cannot appear in two different factorizations of n for if it did, then n/P would be abnormal with $n/P < n$.

We suppose

$$n = p_1 \cdots p_r \text{ and } n = q_1 \cdots q_s$$

where no p is a q and no q is a p. We may suppose that p_1 is the least p and q_1 is the least q, since order is unimportant. Since n is composite we have $p_1^2 \leq n$ and $q_1^2 \leq n$, by Theorem 1.12. Thus we have $p_1 q_1 \leq n$ and since $p_1 \neq q_1$ we see we really have $p_1 q_1 < n$.

Let

$$N = n - p_1 q_1.$$

Then $0 < N < n$, and so is not abnormal. Since $p_1 \mid n$ we see that $p_1 \mid N$, by (e) of Theorem 1.2, and similarly $q_1 \mid N$. Thus $p_1 q_1 \mid N$ since both p_1 and q_1 appear in the unique factorization of N. Thus, by (e) of Theorem 1.2, we have $p_1 q_1 \mid n$, and so $q_1 \mid (n/p_1)$, that is, $q_1 \mid p_2 \cdots p_r$. Since $n/p_1 < n$ this latter factorization is unique. Thus q_1 must be a p, which is a contradiction.

Thus the set of abnormal integers must be empty.   ∎

Since the primes that divide a given number are unique, up to order, we may reorder them so as to group all equal prime factors together and arrange them in numerical order. Let $p_1, \ldots, p_r$ be the distinct primes dividing n, with $p_1 < p_2 < \cdots < p_r$. Then we may write

$$n = p_1^{e_1} \cdots p_r^{e_r},$$

where $e_i \geq 1$, $1 \leq i \leq r$, and is the number of times the prime p_i appears in the prime factorization of n. This representation of n is called the **canonical factorization** of n. Theorem 1.19 then says that the canonical factorization of a positive integer is unique.

Example 1.7. (a) Find the canonical factorization of 117.

(b) Find the canonical factorization of 1950.

(c) Find the canonical factorization of 10!.

Solution. (a) We have

$$117 = 3^2 \cdot 13.$$

(b) We have

$$1950 = 2 \cdot 3 \cdot 5^2 \cdot 13.$$

(c) We have

$$10! = 2^8 \cdot 3^4 \cdot 5^2 \cdot 7.$$ □

Before continuing along we should give an example to show that while unique factorization may seem obvious one does not have to change the situation too much before it becomes false. To show this we use an example created by David Hilbert.

Let $H = \{4k + 1: k$ a nonnegative integer$\} = \{1, 5, 9, 13, \ldots\}$. We can then define divisibility as in Definition 1.1 and we call an element h of H a **Hilbert prime** if $h > 1$ and if $h = ab$, for some a and b in H, then $a = 1$ or $b = 1$. This is basically the same definition as Definition 1.4. The first few Hilbert primes are:

$$5, 9, 13, 17, 21, 29, 33, 37, 41, 49.$$

Note that $25 = 5 \cdot 5$ and $45 = 5 \cdot 9$ are composite numbers in H.

We have

$$693 = 9 \cdot 77 = 21 \cdot 33.$$

As we saw in the list above, 9, 21 and 33 are Hilbert primes. It is also clear that 77 is a Hilbert prime since it is not divisible by any numbers of the form $4k + 1$. Thus the element 693 in H has two distinct prime factorizations.

This example shows the need of not taking the obvious for granted. (See Problem 4 below for a little more on the set of Hilbert primes.)

Sometimes, in stating results or theorems, it is convenient to allow zero exponents on primes. This, of course, takes the uniqueness away, since multiplying by 1 does not change anything, but if one wants to list all primes that divide any one of a finite collection of positive integers, then it is quite useful. For example, we have

$$20 = 2^2 \cdot 5, \ 30 = 2 \cdot 3 \cdot 5, \text{ and } 36 = 2^2 \cdot 3^2,$$

as canonical factorizations. To emphasize the fact that at least one of the primes 2, 3, or 5 divides at least one of the numbers 20, 30, or 36 we could rewrite them as

$$20 = 2^2 \cdot 3^0 \cdot 5, \ 30 = 2 \cdot 3 \cdot 5, \text{ and } 36 = 2^2 \cdot 3^2 \cdot 5^0.$$

Note that we haven't added any new prime factors to any of our given integers.

Corollary 1.19.1. Let

$$a = p_1^{\alpha_1} \cdots p_r^{\alpha_r} \text{ and } b = p_1^{\beta_1} \cdots p_r^{\beta_r}$$

where $\alpha_i, \beta_i \geq 0$, $1 \leq i \leq r$. Then

(a) $a \mid b$ if and only if $\alpha_i \leq \beta_i$, $1 \leq i \leq r$;

(b) $(a, b) = 1$ if and only if $\alpha_i > 0$ implies $\beta_i, = 0$ and $\beta_i, > 0$ implies

$$\alpha_i = 0, \quad 1 \leq i \leq r.$$

Proof. (a) If $a \mid b$, then there is an integer c such that

$$b = ac.$$

By Theorem 1.19 we must have

$$c = p_1^{\sigma_1} \cdots p_r^{\sigma_r}$$

where $\sigma_i \geq 0$, for $1 \leq i \leq r$. Thus, if we equate exponents, we have, for $1 \leq i \leq r$,

$$\beta_i = \alpha_i + \sigma_i$$

The result follows.

The second part follows in a similar way from the definition of greatest common divisor. ∎

Corollary 1.19.2. Let a and b be positive integers and let

$$a = p_1^{\alpha_1} \cdots p_r^{\alpha_r} \text{ and } b = p_1^{\beta_1} \cdots p_r^{\beta_r},$$

where $\alpha_i, \beta_i \geq 0$, $1 \leq i \leq r$. Then

$$(a,b) = p_1^{\min(\alpha_1,\beta_1)} \cdots p_r^{\min(\alpha_r,\beta_r)}$$

The proof is an easy consequence of the definition of greatest common divisor and (a) of Corollary 1.19.1.

Example 1.8. Find $(1500, 4455)$.

Solution. We have

$$1500 = 2^2 \cdot 3 \cdot 5^3 \text{ and } 4455 = 3^4 \cdot 5 \cdot 11.$$

Thus, by Corollary 1.19.2,

$$(1500, 4455) = 2^0 \cdot 3^1 \cdot 5^1 \cdot 11^0 = 15.$$ □

We give some further illustrations of Theorem 1.19 in the following examples.

Example 1.9. Find all integers m whose logarithm base 10 is a rational number.

Solution. Suppose

$$\log_{10} m = p/q$$

where p and q are integers such that $(p, q) = 1$. This is equivalent to

$$10^p = m^q.$$

Now the only primes dividing 10 are 2 and 5. Thus, by Theorem 1.19, we must have

$$m = 2^a \cdot 5^b,$$

for some positive integers a and b. Thus

$$2^p \cdot 5^p = 10^p = m^q = 2^{aq} \cdot 5^{bq}.$$

Thus, by Theorem 1.19, we have

$$aq = p = bq.$$

Thus $a = b$ and $q \mid p$. Since $(p,q) = 1$, we see $q = 1$. Thus we have

$$m = 10^a,$$

and so the only positive integers whose logarithms base 10 are rational numbers are the powers of ten.   □

Example 1.10. Show that $x = \sqrt[3]{5}$ is irrational.

Solution. Since $1^3 = 1$ and $2^3 = 8$ we see that x is not an integer. Suppose there exist positive integers m and n such that

$$(1.1) \qquad\qquad x = m/n.$$

Let the "canonical" factorizations of m and n be

$$m = p_1^{\alpha_1} \cdots p_r^{\alpha_r} \text{ and } n = p_1^{\beta_1} \cdots p_r^{\beta_r},$$

where $a_i, b_i \geq 0$, for $1 \leq i \leq r$. If we cube both sides of (1.1) and multiply by n^3, we get

$$(1.2) \qquad\qquad 5 p_1^{3\alpha_1} \cdots p_r^{3\alpha_r} = p_1^{3\beta_1} \cdots p_r^{3\beta_r}.$$

If 5 appears as one of the p_i, then on the left hand side of (1.2) it appears with exponent $3b_i + 1$ and on the right hand side of (1.2) with exponent $3a_i$. This is impossible. If 5 is not one of the p_i, then again we have an impossibility since 5 divides the left hand side of (1.2), but not the right hand side of (1.2).

Thus $\sqrt[3]{5}$ must be irrational.   □

We close this section with a concept that will be useful in our later work. It is, as we shall see below, closely related to the greatest common divisor.

Definition 1.5. Let $a_1, \ldots, a_n$ be nonzero integers. If m is an integer such that $a_i \mid m$, for $1 \leq i \leq n$, then m is called a **common multiple** of the $\{a_i\}$. The smallest positive common multiple of the numbers $a_1, \ldots, a_n$ is called the **least common multiple** and is denoted by

$$[a_1, \ldots, a_n].$$

The least common multiple is well-defined since the set of positive common multiples is nonempty, for example, either $a_1 \cdots a_n$ or $-a_1 \cdots a_n$ is a member. The following result is an immediate corollary of Corollary 1.19.1 and Definition 1.4.

Theorem 1.20. Let $a_1, \ldots, a_n$ be positive integers and suppose that, for $1 \leq i \leq n$,

$$a_i = p_1^{\alpha_{1,i}} \cdots p_r^{\alpha_{r,i}},$$

where $\alpha_{i,j} \geq 0$ for $1 \leq i \leq r$, $1 \leq j \leq n$. Then

$$[a_1, \ldots, a_n] = p_1^{\max(\alpha_{1,1}, \ldots, \alpha_{1,n})} \cdots p_r^{\max(\alpha_{r,1}, \ldots, \alpha_{r,n})}$$

Using this result we obtain the following corollaries immediately.

Corollary 1.20.1. If b is a common multiple of $a_1, \ldots, a_n$ and $h = [a_1, \ldots, a_n]$, then $h \mid b$. Equivalently, the set of common multiples of $a_1, \ldots, a_n$ coincides with the set $\{nh: n \text{ an integer}\}$.

Corollary 1.20.2. If m is a positive integer, then

$$[ma_1, \ldots, ma_n] = m[a_1, \ldots, a_n].$$

Corollary 1.20.3. If a and b are nonzero integers, then

$$[a, b](a, b) = |ab|.$$

Note the result of Corollary 1.20.3 is not necessarily true for more than two terms. For some related results see Problem 9 below.

Example 1.11. Find $[963, 657]$.

Solution. In Example 1, we found that $(657, 963) = 9$. Thus, by Corollary 1.20.3, we have

$$[963, 657] = \frac{963 \cdot 657}{(963, 657)}$$
$$= \frac{963 \cdot 657}{9}$$
$$= 70299. \qquad \square$$

Problem Set 1.4

1. Find the canonical factorizations of 7917, 9973, 99991, 82798848, 8105722663500, 20!, 50!, $10^{10} - 1$ and $2^{24} - 1$.

2. Find all positive integers x such that $x(x + 30)$ is a perfect square; a perfect cube.

3. (a) A natural number is said to be square-free if and only if it is divisible by the square of no integer greater than 1. If n is a natural number show that n can be uniquely written in the form

$$n = m^2 k$$

where k is square-free.

(b) Suppose that there are h primes $p \leq x$, so that $\pi(x) = h$. How many squares can be made out of these primes that are $\leq x$? How many square-free numbers can be made out of these primes? Show that we have the following lower bound for $\pi(x)$:

$$\pi(x) \geq \frac{\log x}{2 \log 2}.$$

4. Let $H = \{4k + 1: k \text{ a nonnegative integer}\}$. Above we defined the notion of a Hilbert prime.

   (a) Is every prime, in $\mathbf{Z}$, of the form $4k + 1$ a Hilbert prime?

   (b) Do there exist infinitely many Hilbert primes which are not primes in $\mathbf{Z}$?

   (c) Show that we can extend H to a larger set H^+ in which factorization is unique. Factorize 693 in H^+.

5. If a and b are natural numbers and $(a, b) = [a, b]$, then $a = b$.

6. Prove that $[a, b](a, b) = |ab|$ without the use of Theorems 1.19 and 1.20.

7. Prove that $(a + b, [a, b]) = (a, b)$.

8. Show that if a, b and c are integers, then

   (a) $(a, [b, c]) = [(a, b), (a, c)]$ and

   (b) $[a, (b, c)] = ([a, b], [a, c])$.

9. Let a, b and c be integers. Suppose that $(a,b,c)[a,b,c] = |abc|$. Then $(a,b) = (b, c) = (c, a) = 1$.

10. Let a, b and c be integers.

    (a) Show that $[a, b, c](ab, bc, ca) = |abc|$.

    (b) Show that $(a, b, c)[ab, bc, ca] = |abc|$.

    (c) Generalize the results of (a) and (b) to the case of n integers, $n > 2$.

11. Let $p_1 = 2, p_2 = 3, \ldots$ be the primes in increasing order so that p_k is the kth prime. Show that $p_1 p_2 \cdots p_n + 1$ cannot be a square or a cube for any n. What can be said about higher powers?

12. (a) Let g and m be given integers. Prove that there exist integers x and y such

that

$$[x, y] = m \text{ and } (x, y) = g$$

if and only if $g \mid m$.

(b) Find all natural numbers a and b such that $[a, b] = 72$ and $(a, b) = 12$; $[a, b] = 540$ and $(a, b) = 18$.

(c) Suppose $ab = 22122100$ and $[a,b] = 550550$. Find all possible values of the natural numbers a and b.

13. Give the details of the proof of Corollary 1.19.2.

14. Suppose that for $i = 1, \dots , n$ we have

$$a_i = p_1^{\alpha_{1,i}} \cdots p_r^{\alpha_{r,i}}$$

Show that

$$(a_1,\dots,a_n) = p_1^{\min(\alpha_{1,1},\dots,\alpha_{1,n})} \cdots p_r^{\min(\alpha_{r,1},\dots,\alpha_{r,n})}$$

15. Show that if $a^2 \mid b^2$, then $a \mid b$. Give a counterexample to the statement that $a^2 \mid b^3$ implies $a \mid b$.

16. If $a_1 \cdots a_n = b^k$ and for $1 \le i < j \le n$ implies that $(a_i, a_j) = 1$, then for $j = 1,$ $\dots , n$ there is an integer c_j such that $a_j = c_j^k$.

17. If p is a prime and $(p, r) = 1$, then for any natural number e we have $(r, p^e m) = (r, m)$.

18. Find the smallest natural number n so that $945n$ is a perfect square. Find the smallest natural number n so that $1148n$ is a perfect cube.

19. (a) Let $a_0, a_1, \dots , a_n$ be integers, with a_n nonzero. Show that if c/d is a root of the polynomial equation

$$f(x) = a_n x^n + \cdots + a_1 x + a_0 = 0,$$

with $(c, d) = 1$, then $c \mid a_n$ and $d \mid a_0$.

(b) Use this result to show that if k is a natural number, then $\sqrt[k]{n}$ is rational if and only if it is an integer.

(c) Prove the following result. Let $f(x)$ be as in part (a). If a_0, a_n and $f(1)$ are all odd, then $f(x)$ has no rational roots.

(d) Let $f(x)$ be as in part (a) and let a be an integer such that $f(a) = 0$. If $(p,q) = 1$ and $f(p/q) = 0$, then $(p - aq) \mid f(a)$.

20. Prove that if $a > 0$, then $[a, a + 2]$ equals $a(a + 2)/2$ if a is even and equals $a(a + 2)$ if a is odd.

21. Give the details for the proof of Theorem 1.20.

22. Let m and n be positive integers. Show that there are divisors r of m and s of n such that $(r, s) = 1$ and $rs = [m, n]$.

1.5. SOME FACTORIZATION METHODS, I

In this section we wish to apply some of the results of the previous sections to discuss ways of factorizing integers. Most of these methods provide an algorithm for obtaining proper factors of a given integer, but do not necessarily provide a prime factor. In some sense we can consider this an inductive procedure, that is, we assume we have the prime factors of everything less than n and now we want to deal with n.

This sort of idea is what is behind least prime factor tables. Knowing that 3 is the least prime factor of $n = 2378511$ does not immediately tell us what the other prime factors of n are, but all we have to do is look into the table for the number n /3= 792837. We find that 3 is the least prime factor of 792837 and that 792837/3 = 264279, which is again divisible by 3. We have 264279/3 = 88093 and that 3 is no longer a divisor. Thus we have discovered that in the canonical factorization of 2378511, 3 occurs to the third power. Now we look for the least prime factor of 88093 and find that it is 88093, that is, 88093 is a prime. Thus the canonical factorization of 2378511 is $3^3 \cdot 88093$.

This procedure requires that we have a table handy. If we don't, then we must proceed in some other way. The remainder of this section is devoted to two of these methods, the first of which is due to Euler and the second of which is due to Fermat. Further methods will be discussed later in Section 4.6.

The first method requires that we be able to write the number we want to factorize in a special way, namely as a sum of two squares. In fact we want to write the number as a sum of two squares in at least two ways. Later, in Section 6.3, we shall see when a number is representable as a sum of two squares, but for the moment we will proceed without this knowledge.

Lemma 1.21.1. Let n, n_1 and n_2 be natural numbers such that

$$n \mid n_1 n_2, \ n \nmid n_1 \text{ and } n \nmid n_2.$$

Then the number

$$d = n_1/(n_1, n_1 n_2/n)$$

is a divisor of n such that $1 < d < n$.

Proof. We have $n_1/d = (n_1 \ n_1 n_2/n)$, which is a natural number. Thus there exist integers k and m such that

$$n_1 = (n_1/d)k \text{ and } n_1 n_2/n = (n_1/d)m.$$

By Corollary 1.10.1, we see that $(k, m) = 1$. Thus we have

$$k = d \text{ and } n_2 d = nm.$$

Since $(d, m) = 1$, we have, by Corollary 1.10.2, that $d \mid n$. If $d = 1$, then $n_2 = nm$. Since $n \nmid n_2$, this can't happen. If $d = n$, then $n \mid n_1$ which also can't happen. Thus $1 < d < n$ and $d \mid n$.   ■

Theorem 1.21 (Euler). Suppose $n = a^2 + b^2 = c^2 + d^2$, where a, b, c and d are natural numbers with $a \geq b$, $c \geq d$, $a > c$ and $(a, b) = (c, d) = 1$. Then the number

$$\theta = (ac + bd)/(ac + bd, ab + cd)$$

is a divisor of n such that $1 < \theta < n$.

Proof. If $n = a^2 + b^2 = c^2 + d^2$, then

$$n^2 = (ac + bd)^2 + (ad - bc)^2 = (ad + bc)^2 + (ac - bd)^2.$$

Also, since

$$(ac + bd)(ad + bc) = n(ab + cd),$$

we have $n \mid (ac + bd)(ad + bc)$. If $n \mid (ac + bd)$, then $ad - bc = 0$ and so $a/b = c/d$. Since $(a, b) = (c, d) = 1$, this implies, by Corollary 1.10.2, that $a = c$, which is a contradiction. If $n \mid (ad + bc)$, then $ac - bd = 0$. This implies, again by Corollary 1.10.2, that $a = d$, which is also a contradiction, since $a > c \geq d$.

If we let

$$n_1 = ac + bd \text{ and } n_2 = ad + bc,$$

then

$$n \mid n_1 n_2, n \nmid n_1 \text{ and } n \nmid n_2.$$

The result then follows from Lemma 1.21.1.   ■

Example 1.12. Let $n = 493$. Then

$$493 = 22^2 + 3^2 = 18^2 + 13^2$$

Here, then, $a = 22$, $b = 3$, $c = 18$ and $d = 13$. Let

$$\theta = (22 \cdot 18 + 3 \cdot 13) / (22 \cdot 18 + 3 \cdot 13, 22 \cdot 3 + 18 \cdot 13)$$
$$= 435 / (435, 300) = 435 / 15 = 29.$$

Thus $29 \mid 493$. Indeed, $493 = 29 \cdot 17$.   □

Rather than trying all possible sums of two squares one can use the following result, which is also useful in the second method due to Fermat.

Theorem 1.22. We have

$$1 + 3 + 5 + \cdots + (2m - 1) = m^2.$$

Proof. We use a method of proof that goes back at least to Gauss. Let

$$S = 1 + 3 + 5 + \cdots + (2m - 1).$$

Then we can also write

$$S = (2m - 1) + \cdots + 5 + 3 + 1.$$

Thus

$$2S = 1 + (2m - 1) + \cdots + 5 + (2m - 5) + 3 + (2m - 3) + 1 + (2m - 1) = (2m)(m)$$

and the result follows. ∎

Thus, one need only subtract off successive odd numbers up to $2\sqrt{n} - 1$ to see if n is expressible as a sum of two squares in two different ways. While only two different expressions as sums of two squares are needed, the more such expressions that are obtained mean the more possible different factors that can be obtained. We illustrate this in the next example.

Example 1.13. Factor $N = 629$.

Solution. We first wish to write N as a sum of two squares, say

$$N = a^2 + b^2,$$

with $a \geq b$. Then the largest a can be is $\sqrt{N}$. In this case we find that $\sqrt{N}$ lies between 25 and 26 and we have

$$629 - 25^2 = 629 - 625 = 4 = 2^2.$$

We got lucky and hit one representation as a sum of two squares right away. Now we need a second to apply Euler's method. We again use Theorem 1.22, but in reverse since we are subtracting. If we set this up as a table, we would get the following.

$$
\begin{array}{lll}
a = 25 & N - a^2 = 4 = & 2^2 \\
       & 2a - 1 = \underline{49} & \\
24   & 53 & \\
       & \underline{47} & \\
23   & 100 = & 10^2
\end{array}
$$

Thus we find that

$$629 = 23^2 + 10^2$$

Here we have $a = 25$, $b = 2$, $c = 23$ and $d = 10$. Thus $\theta = 595/(595,280) = 595/35 = 17$. Thus 17 is a factor of 629 and indeed we find that

$$629 = 17 \cdot 37. \qquad \square$$

Problem 3 below is a generalization of this method and applies to representations in the form $x^2 + Ay^2$. One obtains a similar result for how to find factors of such integers. This leads to another question in regard to such representations. Suppose we can show that a given number N has only one representation in such a form. Could we conclude a converse of our results, that is, that N is prime? Unfortunately the answer is not always yes. For some values of A this is true, but for others it is not. Euler called such values of A idoneal (or convenient) numbers and it is still not known what all of them are. For an overview on this subject see [31 or 105].

The above method still has a defect with which we must deal—it does not always work for every n. Also the calculation involved can be quite large before we get to the sums of squares. The following method uses Theorem 1.22 above and always works to give factors when applied to an odd number. This method requires us to write a number as a difference of two squares, but unlike the above method we start from a specific square and proceed in a definite manner. The restriction to odd numbers is, of course, no real loss since even numbers have the obvious factor of some power of 2.

The basic idea is the following. Suppose N is an odd number and that $N = mn$, where $m \geq n$. Since N is odd we see that both m and n are odd and therefore if we define

$$a = (m + n)/2 \text{ and } b = (m - n)/2,$$

we see that $m = a + b$ and $n = a - b$. Thus we have

$$N = a^2 - b^2$$

Thus to every factorization of N there corresponds a representation of N as a difference of squares, and so the question becomes one of finding an a and a b. Thus we need to write

$$a^2 = N + b^2.$$

This implies that $a \geq \sqrt{N}$. The procedure is then to evaluate

$$D(a) = a^2 - N$$

for $a = [\sqrt{N}] + 1$, $[\sqrt{N}] + 2$, ... until we find a value of a for which $D(a)$ is a perfect square b^2. We illustrate with an example.

Example 1.14. Factor $N = 25273$.

Solution. We see that $\sqrt{N}$ lies between 158 and 159. We have

$$D(159) = 159^2 - 25273 = 8$$

and since this is not a square we must move on to the next number. If we use Theorem 1.22, then to get to the next case we need only add $2 \cdot 159 + 1 = 319$, and so we can produce the following table.

$a =$ 159	$D(a) = 8$
	$2a + 1 =\underline{319}$
160	327
	$\underline{321}$
161	648
	$\underline{323}$
162	971
	$\underline{325}$
163	1296

Since $1296 = 36^2$ we find that

$$25273 = 163^2 - 36^2 = (163 + 36)(163 - 36) = 199 \cdot 127. \qquad \square$$

From the way this is set up it is clear that this method works best if the two factors to be found are relatively close to each other and therefore to $\sqrt{N}$. If this is the case, then the two factors can be found relatively fast, as was shown in the previous example. However, if the two factors are not so close, or perhaps N is composed of several factors, then the process can go on for quite awhile. If we suspect this is the case, then we may want to modify Fermat's method. Instead of factorizing N we try to factorize mN, where m is a small known factor. The hope is that m will be chosen so that mN now has two factors that are relatively close to each other. We illustrate this in the following example.

Example 1.15. Factorize $N = 574793$.

Solution. We see that $\sqrt{N}$ lies between 758 and 759. If we begin the process as in the previous example we get the following string of a's: 759, 760 ..., 786 and no square in sight. Thus we start trying values of m.

We try $m = 3$. Then $3N = 1724379$ and we see that N is between 1313 and 1314. We repeat the process and get this sequence of a's: 1314, 1315, ..., 1325 and still no square.

We try $m = 5$. Then $5N = 2873965$ and we see that N lies between 1695 and 1696. This time our sequence goes: 1696, 1697, ..., 1703 and bingo! we find that

$$D(1703) = 1703^2 - 2873965 = 26244 = 162^2.$$

Thus

$$2873965 = 1703^2 - 162^2 = (1703 + 162)(1703 - 162) = 1865 \cdot 1541.$$

With this small of a case it is not hard to find the factor after dividing out the 5, that is, we find that

$$574793 = 373 \cdot 1541,$$

but when the numbers are larger and division is not so easy to do we can resort to computing $(a + b, N)$, where we have found that $mN = a^2 - b^2$. This will always produce a nontrivial factor of N. In our case we find that $(1865, 574793) = 373$. □

Now that we have found our factors we can see that the process with just N would have had to wait until $a = 957$, a rather long way off, but with $3N$ we were almost there since $a = 1330$ would have worked. Sometimes the moral of the story, when factorizing numbers is to be patient.

The Fermat method is capable of further generalization. One of these generalizations is contained in the idea of a *factor base*. Part of the idea is to look for difference of squares that N divides rather than try to write N, itself, as a difference of squares. See [54, pp. 132-143] for more details.

Problem Set 1.5

1. Use Euler's method to factor the following numbers: 221, 1769, 2501, 6641, 8633, 10001, 19109 and 1000009.

2. If N is odd and

$$N = a^2 + b^2 = c^2 + d^2$$

where a, b, c and d are integers with $a > c$ and d - b, a and c even, then

$$N = ((D/2)^2 + (u/2)^2)(s^2 + t^2),$$

where $D = (a - c, d - b)$, $s = (a - c)/D$, $t = (d - b)/D$ and $u = (d - b)/s$. Use this result on the numbers in Problem 1 and compare the results.

3. Suppose that A is a positive integer. If $n = a^2 + Ab^2 = c^2 + Ad^2$, where a, b, c and d are positive integers with $a \geq c$, $(a, Ab) = (c, Ad) = 1$, then

$$\theta = (ac + Abd)/(ac + Abd, ab + cd)$$

is a proper divisor of n. Can you produce a corresponding generalization of Problem 2?

4. Use the generalization of Theorem 1.21 given in Problem 3 to factor the following numbers.

   (a) 451, 697, 817, 1763 and 569449. (Take $A = 2$.)

   (b) 589, 1273, 2257, 2881 and 2873161. (Take $A = 3$.)

5. Use Fermat's method to factor the following numbers: 8927, 13837, 57479, 14327581, 2027651281.

6. Use the modified Fermat's method to factor the following numbers: 68987, 141467, 19578079, 29895581, 123456789, 170118759.

7. Show that if we used $m = 2$ in the modified Fermat's method, then we would never be able to factor an odd number N. Show that the same result is obtained if we took m to be any even number that is not divisible by 4. (Hint: show that a number that is twice an odd number cannot be written as a difference of two squares.)

8. Show that if N has a factor within $\sqrt[4]{N}$ of $\sqrt{N}$, then the Fermat method will work on the first try.

9. Let N be a positive integer and let x be the smallest positive integer such that $x^2 - N$ is the square of an integer. Let y be a positive integer such that $y^2 = x^2 - N$. Show that $x - y$ is the largest divisor of N that is less than or equal to $\sqrt{N}$.

10. Let p be a given prime. Find an upper bound on the prime q, $q > p$, such that if $n = pq$, then it takes Fermat's method exactly one step to factorize n. (Hint: if the process starts at $t = \left[\sqrt{p}\right] + 1$, show that it stops when $t = (p + q)/2$.) What are the bounds if it takes exactly two steps? three steps?

ADDITIONAL PROBLEMS FOR CHAPTER ONE

General Problems

1. Let a be a positive integer. If p is the least positive integer that divides a and $p > 1$, then p is a prime.

2. If $a > 1$ is an integer, calculate the following greatest common divisors.
   - (a) $(a^m - 1, a^n + 1)$.
   - (b) $(a^m + 1, a^n + 1)$.

3. Let n be a positive integer.
   - (a) If n is odd and not a multiple of 3, then n^2 is of the form $24k + 1$
   - (b) Show that $30 \mid (n^5 - n)$ and $42 \mid (n^7 - n)$. Can you generalize these results?
   - (c) Show that $504 \mid (n^3 - 1)n^3(n^3 + 1)$.

4. Let a and b be integers.
   - (a) If a is even and b is odd, then $(a, b) = (a/2, b)$.
   - (b) If a and b are even, then $(a, b) = 2(a/2, b/2)$.
   - (c) If a and b are odd, then $|a - b|$ is even.

5. Let m be a positive integer and a and b be integers, $a, b > 1$. Then

$$((a^m - b^m)/(a - b), a - b) = (a - b, m(a, b)^{m - 1}).$$

6. If n is an odd positive integer, then $n \mid (2^{n!} - 1)$.

7. Find $(2k - 1, 9k + 4)$ as a function of k.

8. Calculate $(5n + 1, 12n + 1)$ for all integers n.

9. If $n > 1$ is an integer, then

$$(2^{4n + 2} + 1)/5$$

   is composite.

10. If $(a,b) = 1$, then

$$(a^3 + b^3, a^2 + b^2) \mid (a - b).$$

11. Let a, b and c be integers. Then
   - (a) $(a, b)(b, c)(c, a)[a, b, c] = |abc|(a, b, c)$;
   - (b) $[a, b][b, c][c, a](a, b, c) = |abc|[a, b, c]$.
   - (c) Generalize both of these results to $n \geq 4$ integers.

12. If $(a, b) = 1$, then, for all positive integers s and t, we have $(a^s, b^t) = 1$.

13. If $(m, n) = 1$, then, for any integers p and q, $(pm + qn; mn) = (p, n)(q, m)$.

14. If $L_n = [1, 2, \dots , n]$, then

$$\sum_{n=1}^{+\infty} L_n^{-1}$$

converges.

15. Let a be a given integer. Find all integers x such that $x(x + a)$ is a perfect square.

16. If $a + b \geq c + 2$ and $2ab = c^2$, then $a^2 + b^2$ is composite.

17. Find all integers n such that if $1 \leq d \leq \sqrt{N}$, then $d \mid n$.

18. Prove that every composite odd number can be represented as a sum of consecutive odd numbers, but that no prime can. What can be said about even numbers?

19. Prove that every power of a natural number n, $n > 1$, can be represented as a sum of n positive odd integers.

20. If p is a prime, then, for any positive integer n, we have $p \mid \left(\binom{n}{p} - \left[\frac{n}{p} \right] \right)$.

21. Let r and s be positive integers and let

$$r/s = 0.k_1 k_2 \ \dots$$

be the decimal expansion of r/s. Let

$$\sigma_j = 10^j(r/s) - (10^{j-1}k_1 + 10^{j-2}k_2 + \cdots + k_j).$$

Prove that there exist at least two values of j, say m and n, such that $\sigma_m = \sigma_n$.

22. If a and b are given positive integers, develop a procedure to solve the system

$$x + y = a \text{ and } [x, y] = b(x, y).$$

Apply this to the case $a = 667$ and $b = 120$.

23. Denote by $\omega(n)$ the number of distinct prime divisors of n. Show that

$$\omega(n) \leq \log n / \log 2.$$

24. If a and b are positive integers and $b > 2$, then $(2^b - 1) \nmid (2^a + 1)$.

25. Determine all values of the positive integer n such that $3 \mid (2^n + 1)$.

26. Let a and b be odd positive integers. Prove that

$$2^n \mid (a^3 - b^3) \text{ if and only if } 2^n \mid (a - b).$$

27. Let $n > 1$ be an integer. Show that $2^n \nmid (3^n + 1)$.

28. Calculate $\left(\binom{2n}{1}, \binom{2n}{3}, \cdots, \binom{2n}{2n-1} \right)$.

29. Suppose that d and n are relatively prime positive integers. Show that, for all integers a and b, we have

$$n! \mid a(a + d)(a + 2d) \cdots (a + (n - 1)d).$$

30. Suppose $(a, b) = 1$. Then, in the arithmetic progression $\{ak + b : k \text{ a nonnegative integer}\}$, there exist arbitrarily long strings of elements in the progression that are composite.

31. (a) Show that if a, u and v are integers, then $(a, uv) = (a, (a, u)v)$.

    (b) Show that if a, u_1, ... , u_n are integers, then

$$(a, u_1 \cdots u_n) = (a, (a, u_1) \cdots (a, u_n)).$$

32. Show that no three successive odd numbers, except 3, 5 and 7, can all be primes.

33. If we try to modify the proof of Theorem 1.14 to prove that there are an infinite number of primes of the form $8k + 7$, it doesn't work. Why not?

34. From Euclid's proof of Theorem 1.11 show that if $n \geq 2$, then

$$p_{n + 1} \leq p_1 p_2 \cdots p_n - 1.$$

Show that this implies that, for $n \geq 1$,

$$p_n \leq 2^{2^{n-1}}$$

What kind of estimate does this give for $\pi(x)$?

35. Suppose that $(m, n) = 1$ and $d \mid mn$. Show that we can write $d = pq$, where $p \mid m$ and $q \mid n$.

36. What are the last two digits of a perfect square?

37. Suppose a, b, c, m and n are positive integers such that $(a, b) = 1$ and

$$a \mid c^m - 1 \text{ and } b \mid c^n - 1.$$

Show that

$$ab \mid c^{[m, n]} - 1.$$

38. Suppose that n is a positive integer with the property that if p is a prime and $p \mid n$, then $p^2 \mid n$. Show that there are integers a and b such that $n = a^3 b^2$.

39. Show that if $p \mid (ra - b)$ and $p \mid (rc - d)$, then $p \mid (ad - bc)$.

40. Let 2, 3, ... , p_n be the first n primes and let $N = 2 \cdot 3 \cdots p_n$. If $N = ab$, then $a + b$ has a prime divisor greater than p_n.

41. If n is an odd integer, $n > 5$, and there exist relatively prime integers a and b such that

$$a - b = n \text{ and } a + b = p_1 p_2 \cdots p_k,$$

where the p's are the primes $\leq \sqrt{n}$, then n is a prime.

42. Assume that g is a positive integer and that n is an integer. Prove that there exist integers x and y satisfying $(x, y) = g$ and $xy = n$ if and only if $g^2 \mid n$. Find integers x and y such that $(x, y) = 3$ and $xy = 36$.

43. Show that if $a \mid bc$ and $(a, b) = d$, then $a \mid dc$.

44. Show that if a, b and c are positive integers, then

$$([a, b], [b, c], [c, a]) = [(a, b), (b, c), (c, a)].$$

45. Show that every positive integer n can be uniquely written in the form

$$n = a_k 3^k + a_{k-1} 3^{k-1} + \cdots + a_1 3 + a_0,$$

where the a_j are either -1, 0 or 1, for $j = 0, 1, \ldots , k$.

46. Show that, for every positive integer n, there is an integer a_n such that

$$0 < a_n - (\sqrt{3} + 1)^{2n} < 1 \text{ and } 2^{n+1} \mid a_n.$$

47. Show that $\sqrt[q]{n}$ is irrational for every positive integer $n, n > 1$.

48. Show that the sum of the squares of three consecutive odd numbers increased by 1 is a multiple of 12.

49. Show that if we write $x = yu + v$, with $0 < v < y$, $x = va + r$, with $0 \leq r < v$ and $uy = vb + s$, with $0 \leq s < v$, then $r = s$.

50. Let a, b and c be positive integers. Let $g_1 = (a, b)$, $g_2 = (b, c)$, $g_3 = (c, a)$, $G = (a, b, c)$, $h_1 = [a, b]$, $h_2 = [b, c]$, $h_3 = [c, a]$ and $L = [a, b, c]$. Show that

$$\frac{L}{G} = \sqrt{\frac{h_1 h_2 h_3}{g_1 g_2 g_3}}.$$

Generalize to $n \geq 4$ integers.

51. Let a and b be natural numbers.

    (a) If n is a natural number and $n \mid a^n - b^n$, then

    $$n \mid (a^n - b^n)/(a - b).$$

(b) If p is an odd prime, then

$$(a + b, (a^p + b^p)/(a + b)) = 1 \text{ or } p.$$

Determine when each cases arises.

52. Prove that

$$(ah, bk) = (a, b)(h, k)(a/(a, b), k/(h, k))(b/(a, b), h/(h,k))$$

Derive the result of Problem 20c of Section 1.2 from this result.

53. Prove the following result of Lamé. Suppose that $b \geq a > 0$ and that when we apply the Euclidean algorithm to find (a, b) we get the sequence of equations

$$r_k = q_{k + 2} r_{k + 1} + r_{k + 2}, \text{ with } 0 < r_{k + 2} < r_{k + 1},$$

for $k = -1, 0, 1, \ldots, n - 2$, where $r_{-1} = b$ and $r_0 = a$, and we have

$$r_{n - 1} = q_{n + 1} r_n.$$

(a) Show that $b \geq 2r_1$ and $a \geq 2r_2$.

(b) If $k \geq 1$, then $r_k \geq 2r_{k + 2}$.

(c) Show that $b \geq 2^{n/2}$ so that $n \leq 2(\log_{10} b/ \log_{10} 2)$.

(d) Show that if $b \leq 10^6$, then the Euclidean algorithm requires at most 40 steps.

54. Let N be a positive integer with a total of n distinct divisors and let P be the product of all these divisors. Show that

$$N^n = P^2.$$

55. If $(m - 1, n + 1) = 1$, then $n + 1$ divides $\binom{m}{n}$.

56. Let a and b be natural numbers and let $m = [a, b]$.

(a) Show that there exist positive integers r and s such that $ra = m = sb$.

(b) Show that the smallest positive integer n such that $b \mid na$ is $n = r = m/a = [a, b]/a$.

57. Let a, b, c, q and n be positive integers such that $a = qb + c$, with $0 \leq c < b$.

(a) Show that $b \mid na$ if and only if $b \mid nc$.

(b) Prove that the least natural number n such that $b \mid na$ is also the least natural number n such that $b \mid nc$.

(c) Show that $[a, b]/a = [b, c]/c$.

58. (a) Let n be a positive integer and write $n = 1000q + r$, where $0 \leq r < q$. Show that if $c = 7, 11$ or 13, then $c \mid n$ if and only if $c \mid q - r$.

(b) Test $n = 514824017659$ for divisibility by 7, 11 and 13.

(c) Can you produce other divisibility criteria in a similar manner?

59. Nicomachus in his book [72] wrote extensively on polygonal (or figurate) numbers. The kth r-gonal number is $p(k, r) = (k/2)(2 + (k - 1)(r - 2))$. Their name derives from the ability to picture them as polygons. For example, for the triangular numbers, $p(k, 3)$, we have the following picture.

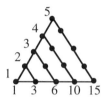

Prove the following properties of polygonal numbers:

(a) $p(n, 6) = n + 4p(n - 1, 3)$;

(b) $9p(n, 3) + 1 = p(3n + 1, 3)$;

(c) $p(n, r) + p(n - 1, 3) = p(n, r + 1)$;

(d) $p(n, 3) + (r - 3)p(n - 1, 3) = p(n, r)$;

(e) $8(r - 2)p(m, r) + (r - 4)^2 = \{(r - 2)(2m - 1) + 2\}^2$;

(f) $p(m + n, r) = p(m, r) + p(n, r) + mn(r - 2)$;

(g) $p(m, r) + p(m, s) = p(m, r + s - 2) + m$.

Computer Problems

1. Determine the range of integers that your computer can handle.

2. (a) Write a program that computes the quotient and remainder in the division algorithm.

   (b) If we require the remainder in
   $$a = bq + r$$
   to satisfy $-|b|/2 < r \le |b|/2$, instead of $0 \le r < |b|$, modify your program from (a) to do this.

3. Write a program to implement the Sieve of Eratosthenes and find all the primes less than 5000.

4. (a) Write a program to determine whether a number is a prime or composite. If the number is composite, have the program give the least prime factor. Create a least prime factor table for the numbers 2 to 1000.

   (b) Modify the program of part (a) to check for twin primes, that is primes p such that $p + 2$ is also prime. Do the same for the primes triples p,

$p + 2$, $p + 6$ and p, $p + 4$, $p + 6$. Finally find all the prime decades, that is, prime quadruples of the form p, $p + 2$, $p + 6$ and $p + 8$ that are less than 1000. (A little thought first will help in the last program. What arithmetic progression must p lie in?)

(c) Can you modify the program to compute H-primes? Find all the H-primes less than 1000.

5. Modify the program in 59(a) to find primes in an arbitrary arithmetic progression. Include the conditions of Dirichlet's theorem (Theorem 1.14) in your program.

6. Modify the program in 59(a) to find primes of the form $an^2 + bn + c$, where a, b and c are given integers. Check the quadratics $n^2 + 1$, $n^2 \pm 2$, $n^2 + n + 41$, say for $n \le 500$. Can you make any conjectures from these results about the prime divisors of these polynomials?

7. (a) Write a program to implement the Euclidean algorithm.

(b) Compute (1831, 1632), (31567, 1805), (52674, 32768) and (1234567, 9876541).

(c) Modify the program to use the result of Theorem 1.7 so that you can find the values of s and t such that

$$as + bt = (a, b)$$

Calculate the values of s and t for the pairs in (b).

(d) Modify the program in (a) to calculate the gcd of any number of integers. Calculate (234, 468, 788), (2139, 6669, 7863), (1234, 2346, 3468, 12480).

(e) Modify the program in (a) to calculate the least common multiple of two integers. Adapt it to calculate the least common multiple of any number of integers. Calculate the lcms for the pairs in (b) and for the tuples in (d).

(f) Use the results in Problem 4 above to write a program that calculates the greatest common divisor of two integers using only subtraction and division by two. Apply it to the pairs in part (b) and compare the running times.

8. Write a program that checks a given integer n to see if it can be written as a sum of two squares and if the number can so be represented, find all representations. Test the numbers less than 501. Test the integers in the interval [1500, 2000].

9. Write a program to implement Euler's method of factorization. Test any numbers from the last part of Problem 66 that have more than one representation.

10. (a) Write a program to implement Fermat's method of factorization. Test the numbers 8496167, 60096559, 24498451, 123456789, 110450171 and 593505099.

    (b) Modify the program to do trial division first up to some specified limit, say 1000. Factorize the numbers in part (a) above with this program and compare the running times.

11. In using Fermat's method of factorization one looks at $x^2 - N$ and keeps increasing x until we get a perfect square. Instead of this we might consider looking at $r = x^2 - y^2 - N$ and check to see when it is equal to zero. If r is positive, then we increase y, while if r is negative, we increase x. Of course, to get to the next square we only have to add (subtract) $2x + 1$ $(2y + 1)$. Write a program to implement this version of Fermat's factorization method. Test it on the numbers in 68(a) and compare the results.

12. Write a program that computes the canonical factorization of an integer. Use it on the integers $2 \leq n \leq 500$. Use it on 3644 and 396900.

13. (a) Let b be a positive integer, say $2 \leq b \leq 10$. Write a program to find the b-ary expansion of a given integer n.

    (b) For $b = 2, 3, 7, 9$ find the b-ary expansions of 22, 1813, 357611 and 12345678.

    (c) Modify the program of part (a) to deal with the base $b \geq 11$. The problem here is that you will have to deal with the new "digits."

14. (a) Write a program to find the Cantor expansion of a given positive integer n.

    (b) Find the Cantor expansions for $n = 25, 1803$ and 156577.

15. Write a program to calculate Fibonacci and Lucas numbers. Calculate as many as your computer will let you calculate.

16. Write a program to calculate binomial coefficients. Three possible ways of doing this use the following properties of binomial coefficients:

    (a) $\binom{n}{m} = \dfrac{n!}{m!(n-m)!}$

    (b) $\binom{n}{m} = \left(\dfrac{n}{m}\right)\binom{n-1}{m-1}$ and

(c) $\begin{pmatrix} n+1 \\ m \end{pmatrix} = \begin{pmatrix} n \\ m \end{pmatrix} + \begin{pmatrix} n \\ m-1 \end{pmatrix}$

17. Write a program to compute the kth r-gonal number (see Problem 59 above.) Calculate the first 20 triangular, pentagonal and hexagonal numbers. Can you see any patterns? If so, prove them.

18. Goldbach conjectured in 1742 that every even integer n greater than 4 can be written as a sum of two odd primes. This conjecture has never been proved. Write a program to verify this conjecture for all even numbers up to 200. List all representations as a sum of two odd primes. Do you see any patterns? How does the number of representations behave? Does it seem to increase, remain a constant or decrease?

CHAPTER TWO
Congruences

2.1. CONGRUENCES

In Definition 1.3 we defined the notion of relatively prime. In the following definition we wish to extend this notion.

Definition 2.1: If $\{a_1, \ldots, a_n\}$ is a set of integers and if $(a_i, a_j) = 1$, $1 \leq i < j \leq n$, then $a_1, \ldots, a_n$ are said to be **pairwise relatively prime** or **relatively prime in pairs**.

Note that $a_1, \ldots, a_n$ pairwise relatively prime implies that $a_1, \ldots, a_n$ are relatively prime, but that the converse does not hold. For example, we have $(2, 6, 7) = 1$, but $(2, 6) = 2$.

Definition 2.2: Let a and b be integers and let m be a positive integer. We say a is **congruent** to b modulo m, written

$$a \equiv b \pmod{m},$$

if and only if $m \mid (a - b)$. If $m \nmid (a - b)$, then a and b are said to be **incongruent** modulo m and we denote this by

$$a \not\equiv b \pmod{m}.$$

The concept of congruence will be of both a conceptual and notational convenience in our study of divisibility theory.

Example 2.1: We have

$$3 \equiv 7 \pmod{2}, \quad 12 \equiv 0 \pmod{3}, \quad 29 \cdot 71 \equiv 1 \pmod{7^3}. \qquad \square$$

The following theorem gives the basic properties of congruences.

Theorem 2.1: Let m be a positive integer.

(a)   For all integers a we have

$$a \equiv a \pmod{m}.$$

(b)   For all integers a and b we have

$$a \equiv b \ (\text{mod } m) \text{ if and only if } b \equiv a \ (\text{mod } m).$$

(c)   For all integers a, b and c we have that
if $a \equiv b \ (\text{mod } m)$, $b \equiv c \ (\text{mod } m)$, then $a \equiv c \ (\text{mod } m)$.

(d)   If a is any integer, then

$$m \mid a \text{ if and only if } a \equiv 0 \ (\text{mod } m).$$

(e)   If $\{a_1, \ldots, a_n\}$, $\{b_1, \ldots, b_n\}$ and $\{k_1, \ldots, k_n\}$ are any sets of integers such that $a_i \equiv b_i \ (\text{mod } m)$, $1 \leq i \leq n$, then

$$\sum_{i=1}^{n} k_i a_i \equiv \sum_{i=1}^{n} k_i b_i \ (\text{mod } m).$$

(f)   If a, b, c and d are any integers, then

$$a \equiv b \ (\text{mod } m) \text{ and } c \equiv d \ (\text{mod } m) \text{ implies } ac \equiv bd \ (\text{mod } m).$$

(g)   If $a \equiv b \ (\text{mod } m)$ and n is a natural number, then

$$a^n \equiv b^n \ (\text{mod } m)$$

(h)   If $f(x_1, \ldots, x_n)$ is a polynomial with integer coefficients and $\{a_1, \ldots, a_n\}$ and $\{b_1, \ldots, b_n\}$ are sets of integers such that $a_i \equiv b_i \ (\text{mod } m)$ for $1 \leq i \leq n$, then $f(a_1, \ldots, a_n) \equiv f(b_1, \ldots, b_n) \ (\text{mod } m)$.

Proof By (a) of Theorem 1.1, we have $m \mid 0$. Since $a - a = 0$, the result, (a), follows.

(b)   If $a \equiv b \ (\text{mod } m)$, then there is an integer d such that $a - b = md$. Thus $b - a = m(-d)$ and the result follows.

(c)   If $a \equiv b \ (\text{mod } m)$ and $b \equiv c \ (\text{mod } m)$, then there exist integers j and k such that $a - b = mj$ and $b - c = mk$. Since $a - c = a - b + b - c = m(j + k)$, the result follows.

(d)   This follows immediately from Definition 2.2.

(e)   If $a_i \equiv b_i (\text{mod } m)$, $1 \leq i \leq n$, then there exist integers d_i, $1 \leq i \leq n$, such that

$$a_i - b_i = md_i, \ 1 \leq i \leq n.$$

Then $(k_i a_i) - (k_i b_i) = m(k_i d_i)$, for $1 \leq i \leq n$. Thus

$$\sum_{i=1}^{n} k_i a_i - \sum_{i=1}^{n} k_i b_i = m \left(\sum_{i=1}^{n} k_i d_i \right) \pmod{m}$$

and the result follows from Definition 2.2.

(f)   If $a \equiv b \pmod{m}$ and $c \equiv d \pmod{m}$, then there exist integers j and k such that

$$a - b = mj \text{ and } c - d = mk$$

or

$$a = b + mj \text{ and } c = d + mk.$$

Then

$$ac = (b + mj)(d + mk) = bd + m(bk + dj + jk)$$

and the result follows.

(g)   This follows easily by induction from 6).

(h)   This follows easily from 5), 6), and 7).   ■

Note that (a), (b), and (c) of Theorem 2.1 show that $\equiv$ is an equivalence relation. Thus we may partition the integers into *congruence classes* modulo m, for any positive integer m.

Example 2.2: Find the last two digits of $N = 123^{456}$.

Solution: One approach to this problem is pure brute force, that is, we actually calculate N and then pick off the last two digits. If we were careful, we might be able to do this without actually calculating all of the digits of N. However, this approach requires an immense amount of calculation for anyone, or anything, doing the problem this way.

What we want to do is write N using the division algorithm (Theorem 1.3), that is, we want to write

$$N = 100k + r,$$

where $0 \leq r < 100$. The integer r is then the last two digits of N. Thus, by Definition 2.2, we are looking for a nonnegative integer r such that

$$N \equiv r \pmod{100}.$$

Then, if we do our raising of powers in terms of congruences, we will have no more than two digit numbers to raise to powers and we will also be able to apply part 7) of

Theorem 2.1.

We have

$$123 \equiv 23 (\text{mod} 100),$$
$$123^2 \equiv 23^2 = 529 \equiv 29 (\text{mod} 100),$$
$$123^3 \equiv 2329 = 667 \equiv 67 (\text{mod} 100),$$
$$123^4 \equiv 29^2 = 841 \equiv 41 (\text{mod} 100),$$
$$123^5 \equiv 2341 = 943 \equiv 43 (\text{mod} 100),$$
$$123^6 \equiv 67^2 = 4489 \equiv 89 (\text{mod} 100),$$
$$\vdots$$
$$123^{10} \equiv 43^2 = 1849 \equiv 49 (\text{mod} 100),$$

and so,

$$123^{20} \equiv 49^2 = 2401 \equiv 1 \ (\text{mod} \ 100).$$

This last result is very fortunate. It means that if 20 divides the exponent to which we are raising 123, then we'll always get a result that is congruent to 1 modulo 100 since

$$123^{20k} = (123^{20})^k \equiv 1^k = 1 \ (\text{mod} \ 100).$$

If we note that

$$456 = 20 \cdot 22 + 16,$$

we see that

$$123^{456} = 123^{20 \cdot 22 + 16} \equiv 123^{16} = 123^{10+6} \equiv 49 \cdot 89 = 4361 \equiv 61 \ (\text{mod} \ 100)$$

Thus the last two digits of N are 61.

There is another way to do the exponentiation that is very easy, especially on a computer, and doesn't require that we get lucky like we did in the above computation. This is based on the simple idea that

$$a^{2^{n+1}} = (a^{2^n})^2,$$

for any integer n. The procedure is then to find the binary expansion of the exponent and then do a sufficient amount of squaring to get the powers of two that we need.

We have

$$456 = 256 + 128 + 64 + 8.$$

We also have

$$123 \equiv 23 (\mathrm{mod}\,100),$$
$$123^2 \equiv 23^2 \equiv 29 (\mathrm{mod}\,100),$$
$$123^4 \equiv 29^2 \equiv 41 (\mathrm{mod}\,100),$$
$$123^8 \equiv 41^2 \equiv 81 (\mathrm{mod}\,100),$$
$$123^{16} \equiv 81^2 \equiv 61 (\mathrm{mod}\,100)$$
$$123^{32} \equiv 61^2 \equiv 21 (\mathrm{mod}\,100)$$
$$123^{64} \equiv 21^2 \equiv 41 (\mathrm{mod}\,100)$$
$$123^{128} \equiv 41^2 \equiv 81 (\mathrm{mod}\,100)$$
$$123^{256} \equiv 81^2 \equiv 61 (\mathrm{mod}\,100)$$

and therefore

$$123^{456} = 123^{256+\,128+\,64+\,8} \equiv 61 \cdot 81 \cdot 41 \cdot 81 = 16409061 \equiv 61 \ (\mathrm{mod}\ 100),$$

This gives us, as we would expect, the same answer as above, but with less effort and/or luck. Of course, if we were doing this on a computer, we would be calculating the remainders as we went along and not waiting until the end as we did in this example. $\square$

Definition 2.3: Let m be a positive integer. If $x \equiv a$ (mod m), then a is called a **residue** of x modulo m. If $0 \le a \le m - 1$, then a is called the **least residue** of x modulo m.

It is easy to see that the least residue is unique as implied by the definition and that the set of all possible least residues, $\{0, 1, \dots, m - 1\}$, exhaust all residues modulo m. Thus when we partition the integers into congruence classes modulo m we get a representative of the partition by using the least residues modulo m. Any set of m numbers that is congruent to exactly one of these least residues is called a **complete residue system** modulo m. Thus, if A is a complete residue system modulo m, then

1) $|A| = m$ and
2)   if a and b are in A, then $a \not\equiv b$ (mod m).

Example 2.3: If $m = 8$, then $\{0,1,2,3,4,5,6,7\}$ is the set of least residues modulo 8 and $\{64,9,18,3,36,45,-2,-1\}$ is a complete residue system modulo 8. $\square$

Definition 2.4: We say a **exactly divides** b, written

$$a \parallel b,$$

if and only if $a \mid b$ and $(a, b/a) = 1$.

Theorem 2.2: If $a \equiv b$ (mod m_i), $1 \leq i \leq n$, then

$$a \equiv b \bmod [m_1, \ldots, m_n]).$$

In particular, if the $\{m_i\}$ are relatively prime in pairs, then

$$a \equiv b \pmod{m_1 \cdots m_n}.$$

Proof Let p be a prime and suppose that $p^c \| [m_1, \ldots, m_n]$, with $c \geq 1$. Then, by Theorem 1.20, we have $p^c \mid m_i$ for some i, $1 \leq i \leq n$. Since $m_i \mid (a - b)$ and $p^c \mid m_i$ we have, by 3) of Theorem 1.1, that $p^c \mid (a - b)$. Thus, by unique factorization, we have that $[m_1, \ldots, m_n]$ divides $a - b$ and the first part follows. If the $\{m_i\}$ are pairwise relatively prime, and $p^c \| [m_1, \ldots, m_n]$ with $c \geq 1$, then there exist a unique m_k such that $p^c \| m_k$, where $1 \leq k \leq n$. Thus $p^c \| m_1 \cdots m_n$, and so $[m_1, \ldots, m_n] = m_1 \cdots m_n$. The result follows.   ∎

In contrast to the integers themselves we do not have a straight cancellation law for congruences. We do have the following modified cancellation law.

Theorem 2.3: If $(m, n) = d$, then $ma \equiv mb$ (mod n) if and only if $a \equiv b$ (mod (n/d)).

Proof: By definition, we have that $ma \equiv mb$ (mod n) if and only if there exists an integer t such that

$$ma - mb = nt,$$

which happens if and only if $(m/d)(a - b) = (n/d)t$. Now $(m/d, n/d) = 1$, by Corollary 1.10.1, and so, by Corollary 1.10.2, $(m/d) \mid t$. Thus, we have

$$a - b = (n/d)(t/(m/d)) = (n/d)T,$$

where T is an integer. Thus $a \equiv b$ (mod (n/d)).

If $a \equiv b$ (mod (n/d)), then $a - b = (n/d)t$, for some integer t. Thus $m(a - b)$ $= n(tm/d)$. Since tm/d is an integer we have $ma \equiv mb$ (mod n).   ∎

Corollary 2.3.1: If $(m, n) = 1$, then $ma \equiv mb$ (mod n) if and only if $a \equiv b$ (mod n).

Corollary 2.3.2: If $\{a_1, \ldots, a_n\}$ is a complete residue system modulo m, (k, m) $= 1$ and q is an integer, then $\{ka_1 + q, \ldots, ka_n + q\}$ is a complete residue system modulo m.

Proof: By Theorem 2.1 part 5) and Corollary 2.3.1, we see that

$$ka_i + q \equiv ka_j + q \ (\text{mod } m)$$

implies $a_i \equiv a_j \ (\text{mod } m)$. Since $\{a_1, \ldots, a_n\}$ is a complete residue system modulo m this implies that $a_i = a_j$. Thus $ka_j + q = ka_j + q$, and so $\{ka_1 + q, \ldots, ka_n + q\}$ is a complete residue system modulo m. ∎

We finish this section with a result relating congruences and greatest common divisors.

Theorem 2.4: If $x \equiv y \ (\text{mod } m)$, then $(x, m) = (y, m)$.

Proof: If $x \equiv y \ (\text{mod } m)$, then there is an integer z such that

$$x - y = mz.$$

Since $(x, m) \,|\, x$ and $(x, m) \,|\, m$ we see, by 5) of Theorem 1.1, that $(x, m) \,|\, y$. Thus, by Theorem 1.8, we have $(x, m) \,|\, (y, m)$. Similarly $(y, m) \,|\, (x, m)$. Since greatest common divisors are positive, we have, by 7) of Theorem 1.1, that $(x, m) = (y, m)$. ∎

Problem Set 2.1

1. If a and b are any integers and m is a positive integer, show that $a \equiv b \ (\text{mod } m)$ if and only if a and b leave the same non-negative remainder upon division by m.

2. Show that if $a \equiv b \ (\text{mod } m)$ and $d \,|\, m$, then $a \equiv b \ (\text{mod } d)$.

3. Find the last three digits of 2^{10000}. Find the last four digits of 9^{9^9}.

4. If $a \equiv b \ (\text{mod } p^r)$, $r \geq 1$, then for any non-negative s we have

$$a^{p^s} \equiv b^{p^s} \ (\text{mod } p^{r+s}).$$

(Hint: use induction on s.)

5. Let

$$N = \sum_{j=0}^{m} a_j 10^j$$

be the decimal expansion of the integer N. Prove the following divisibility criteria.

(a) $3 \mid N$ if and only if $3 \mid a_0 + \cdots + a_m$.

(b) $4 \mid N$ if and only if $4 \mid a_0 + 2a_1$.

(c) $11 \mid N$ if and only if $11 \mid a_0 - a_1 + \cdots + (-1)^m a_m$.

(d) Find similar divisibility criteria for 7, 37 and 101.

6. Let

$$f(x) = a_n x^n + \cdots + a_1 x + a_0$$

be a polynomial with integer coefficients. Show that if for d consecutive integers the corresponding values of f are all divisible by d, then $d \mid f(x)$ for all integers x. Give an example to show that even if $d > 1$ and $(a_0, \ldots, a_n) = 1$, this can still happen.

7. Show that the square of every integer is congruent to either 0 or 1 modulo 8. Use this result to show that no number of the form $8k + 7$ can be the sum of three squares.

8. Show that if $(a, 6) = 1$, then $a^2 \equiv 1 \pmod{24}$. What happens when $(a, 6) > 1$?

9. Show that $a^5 \equiv a \pmod{30}$ for all positive integers a. Show that for all positive integers a we have that $a^7 \equiv a \pmod{42}$. Can you prove analogous results for other exponents or other moduli?

10. Suppose that $ar \equiv b \pmod{m}$ and that $br \equiv a \pmod{m}$ for some integer r. Show that $a^2 \equiv b^2 \pmod{m}$. Find a nontrivial example of integers a, b, r and m that satisfy this result.

11. Recall the definition of triangular numbers from Problem 59 of the Additional Problems for Chapter One, namely, any number t of the form $n(n + 1)/2$ for some nonnegative integer n. Show that no triangular number has last digit 2, 4, 7 or 9. Can you generalize this result to other classes of polygonal numbers?

12. Show that the following results hold for all natural numbers n.

(a) $7 \mid 5^{2n} + 3 \cdot 2^{5n-2}$.

(b) $43 \mid 6^{n+2} + 7^{2n+1}$.

13. Show that if a is an odd integer, then, for all natural numbers n,

$$a^{2^n} \equiv 1 \pmod{2^{n+2}}$$

14. Find all values of $n > 1$ for which

$$1! + 2! + \cdots + n!$$

is a perfect square. (Hint: check the allowable units digits for perfect squares.)

15. A *repunit*, R_n for n, a positive integer, is a number of the form $(10^n - 1)/9$, that is all the digits of R_n are 1's. Show the following divisibility results.

(a) $9 | R_n$ if and only if $9 | n$.

(b) $11 | R_n$ if and only if n is even.

(c) $41 | R_n$ if and only if $5 | n$.

(d) $9901 | R_n$ if and only if $12 | n$.

(e) Derive similar results for other primes. Factor R_{14}.

16. If k is a positive integer, let

$$S_k(n) = 1^k + 2^k + \cdots + (n - 1)^k$$

(a) Show that if n is an odd positive integer, then

$$S_1(n) \equiv 0 \pmod{n}.$$

What can be said if n is even?

(b) Show that if n is odd or a multiple of 4, then

$$S_3(n) \equiv 0 \pmod{n}.$$

What happens if n is even, but not a multiple of 4?

(c) Find all values of n such that

$$S_2(n) \equiv 0 \pmod{n}.$$

2.2. THE EULER PHI FUNCTION

There is a subset of the complete residue system that will be of use to us in our further work.

Definition 2.5: A **reduced residue system** modulo m is a set of integers $\{r_1, \ldots, r_n\}$ such that

(a) $(r_j, m) = 1$, for $1 < i < n$.

(b) $r_i \equiv r_j \pmod{m}$, for $1 < i < j < n$, and

(c) if $(x, m) = 1$, then $x \equiv r_i \pmod{m}$ for some r_i of the set.

The easiest way to produce a reduced residue system is to take a complete residue system and delete those elements not relatively prime to the modulus.

Example 2.4: If $m = 8$, then the reduced residue systems corresponding to the complete residue systems of Example 2.3 are $\{1, 3, 5, 7\}$ and $\{9, 3, 45, -1\}$.

Corresponding to Corollary 2.3.2 we have the following result, whose proof is analogous to that of the corollary.

Theorem 2.5: Let $(a, m) = 1$. If $\{r_1, \ldots, r_n\}$ is a reduced residue system modulo m, then $\{ar_1, \ldots, ar_n\}$ is a reduced residue system modulo m.

There is no quick rule to tell us how many elements there are in a reduced residue system modulo m. The function that counts the number of elements in a reduced residue system has many useful properties for our work below. Therefore, after giving its definition we shall prove a few of its basic properties. For more of its properties see Section 7.6.

Definition 2.6: (Euler's phi-function) Let m be a positive integer. Then the **Euler phi function**, denoted by $\varphi(m)$, is the cardinality of the set $\{1 < n < m : (n, m) = 1\}$.

Example 2.5: We have $\varphi(8) = 4$, $\varphi(7) = 6$ and $\varphi(100) = 40$.

The following theorem allows us to calculate $\varphi(n)$, providing we know its prime factorization.

Theorem 2.6: $\varphi(1) = 1$. If $n > 1$, then

$$\varphi(n) = n \prod_{p|n} (1 - \frac{1}{p}),$$

where the notation means that the product is over all primes that divide n.

Proof: Clearly $\varphi(1) = 1$.

Now suppose $n > 1$ and $n = p_1^{a_1} \cdots p_r^{a_r}$ is its canonical factorization. Then $(m, n) > 1$ if and only if at least one p_i, $1 \le i \le r$, divides m. If we let $P = p_1 \cdots p_r$, then $(m, n) > 1$ if and only if $(m, P) > 1$. Suppose $t \mid n$, then the number of multiples of t that are $< n$ is equal to (n/t), since these multiples are $t \cdot 1, t \cdot 2, \ldots, t \cdot (n/t)$. Thus, by the combinatorial principle of inclusion-exclusion, we have

$$\varphi(n) = n - \sum_i (n / p_i) + \sum_{i,j} (n / p_i p_j) - \sum_{i,j,k} (n / p_i p_j p_k) + \cdots + (-1)^r (n / p_1 p_2 \cdots p_r)$$

$$= n \left\{ 1 - \sum_i (1 / p_i) + \sum_{i,j} (1 / p_i p_j) - \sum_{i,j,k} (1 / p_i p_j p_k) + \cdots + (-1)^r (1 / p_1 p_2 \cdots p_r) \right\}$$

$$= n(1 - 1 / p_1)(1 - 1 / p_2) \cdots (1 - 1 / p_r)$$

$$= n \prod_{p|n} (1 - 1 / p).$$

This proves the result. ∎

Example 2.6: Calculate $\varphi(m)$ when

(a) $m = 360$ and

(b) $m = p^e$, where p is a prime and e is a positive integer.

Solution: (a) We have $m = 2^3 \cdot 3^2 \cdot 5$. Thus, by Theorem 2.6,

$$\varphi(360) = 360 \prod_{p|360}(1 - \tfrac{1}{p}) = 360(1 - \tfrac{1}{2})(1 - \tfrac{1}{3})(1 - \tfrac{1}{5}) = 360 \cdot \tfrac{1}{2} \cdot \tfrac{2}{3} \cdot \tfrac{4}{5} = 96.$$

(b) Since the only prime that divides m is p, we have

$$\varphi(p^e) = p^e(1 - 1/p) = p^{e-1}(p - 1).$$

In particular, $\varphi(p) = p - 1$.

As a corollary of Theorem 2.6 we have the following useful result.

Corollary 2.6.1: Let $(m, n) = 1$. Then $\varphi(mn) = \varphi(m)\varphi(n)$.

Proof: Let the canonical factorizations of m and n be

$$m = p_1^{a_1} \cdots p_r^{a_r} \text{ and } n = q_1^{b_1} \cdots q_1^{bs}.$$

Since $(m, n) = 1$ no p_i is a q_j and conversely. By Theorem 2.6, we have

$$\varphi(mn) = mn \prod_{p|mn}(1 - 1/p)$$

$$= mn \prod_{p|mn}(1 - 1/p)(1 - 1/q)$$

$$= \left\{m\prod_{p|n}(1 - 1/p)\right\}\left\{n\prod_{p|n}(1 - 1/q)\right\}$$

$$= \varphi(m)\varphi(n).$$

This proves the result.   ■

Corollary 2.6.2: If $(m, n) = 1$, a runs through a reduced residue system modulo m and b runs through a reduced residue system mod n, then $bm + an$ runs through a reduced residue system mod mn.

Proof: To prove the result we must show three things:

(a) the numbers of the form $bm + an$ are relatively prime to mn,

(b) these numbers are incongruent mod mn, and

(c) there are $\varphi(mn)$ numbers that are incongruent to each other.

Proof of (a): Let $d = (an + bm, mn)$, where a is any element of a reduced residue system mod m and b is any element of a reduced residue system mod n. If $d > 1$, let p be a prime divisor of d, which exists by Theorem 1.2. Then $p \mid mn$ and $p \mid (an + bm)$. By Lemma 1.19.1 $p \mid mn$ implies $p \mid m$ or $p \mid n$, but not both since $(m, n) = 1$. We may assume $p \mid m$ as proof goes the same if $p \mid n$. Since $p \mid m$ and $p \mid (an + bm)$ we see, by (e) of Theorem 1.1, that $p \mid an$. Thus, by Lemma 1.19.1, $p \mid a$ or $p \mid n$. However, $p \mid n$, since $(m, n) = 1$, and $p \mid a$, since $(a, m) = 1$. This contradiction shows that we must have $d = 1$ and this proves the first part.

Proof of (b): Let a_1 and a_2 be distinct elements in a reduced residue system mod m. Let b_1 and b_2 be distinct elements in a reduced residue system mod n. If

$$b_1 m + a_1 n \equiv b_2 m + a_2 n \ (\text{mod } mn),$$

then $a_1 n \equiv a_2 n \ (\text{mod } m)$. Since $(m, n) = 1$, we have, by Corollary 2.3.1, that $a_1 \equiv a_2 \ (\text{mod } m)$, which is a contradiction. Thus $b_1 m + a_1 n$ and $b_2 m + a_2 n$ are incongruent modulo mn, which proves the second part.

Proof of (c): There are $\varphi(m)$ elements in any reduced residue system mod m and $\varphi(n)$ elements in any reduced residue system mod n. Thus, if a runs through a reduced residue system mod m and b runs through a reduced residue system mod n, then there are $\varphi(m)\varphi(n)$ numbers of the form $an + bm$. Since $(m, n) = 1$, we have $\varphi(m)\varphi(n) = \varphi(mn)$, by Corollary 2.6.1, which proves the third part. ∎

The following theorem gives us an inductive way to calculate $\varphi(n)$, providing we know $\varphi(d)$ for any positive d such that $d \mid n$.

Theorem 2.7: If $n \geq 1$, then

$$\sum_{d \mid n} \varphi(d) = n,$$

where the sum is over all positive integers d that divide n.

Proof: We will prove this theorem by induction on the number of distinct primes that divide n. Note that the result is trivially true for $n = 1$ since $d = 1$ is the only divisor and $\varphi(1) = 1$.

Suppose that n is a prime power, say $n = p^e$, where p is a prime and e is a positive integer. Then the only positive integers that divide p^e are $1, p, p^2, \ldots, p^e$. Thus

$$\sum_{d|n} \varphi(d) = \varphi(1) + \varphi(p) + \varphi(p^2) + \cdots + \varphi(p^e)$$

$$= 1 + (p-1) + p(p-1) + \cdots + p^{e-1}(p-1)$$
$$= 1 + p - 1 + p^2 - p + \cdots + p^e - p^{e-1} = p^e = n,$$

where we used (b) of Example 2.6. Thus the result holds for prime powers.

Suppose that the result holds for all positive integers n that have k or fewer distinct primes dividing them, where $k \geq 1$. Let N be a positive integer that has $k + 1$ distinct primes dividing it and let p be one of them. If $p^e \| N$, then we may write

$$N = p^e n$$

where $(p, n) = 1$ and n has k distinct primes dividing it. If d ranges over all positive divisors of n, then $d, pd, \ldots, p^e d$ range over all divisors of N. By Corollary 2.6.1, we have $\varphi(p^j d) = \varphi(p^j)\varphi(d)$, where $0 \leq j \leq e$ and $d \mid n$. Thus

$$\sum_{d|N} \varphi(d) = \sum_{d|n} \varphi(d) + \sum_{d|n} \varphi(pd) + \cdots + \sum_{d|n} \varphi(p^e d)$$

$$= \sum_{d|n} \varphi(d)\{1 + \varphi(p) + \cdots + \varphi(p^e)\} = np^e = N,$$

by the induction hypothesis. Thus the result holds for N and the theorem follows by mathematical induction. ∎

Example 2.7: Without using the canonical factorizations compute $\varphi(12)$.

Solution: The positive divisors of 12 are $d = 1, 2, 3, 4, 6$ and 12. Now $\varphi(1) = 1$, $\varphi(2) = 1$, $\varphi(3) = 2$ (since both 2 and 3 are primes), and $\varphi(4) = 2$ (since the only numbers less than 4 and relatively prime to 4 are the odd numbers 1 and 3). For $\varphi(6)$ we have

$$6 = \varphi(6) + \varphi(3) + \varphi(2) + \varphi(1) = \varphi(6) + 2 + 1 + 1,$$

and so

$$\varphi(6) = 2.$$

Finally

$$12 = \varphi(12) + \varphi(6) + \varphi(3) + \varphi(2) + \varphi(1) = \varphi(12) + 2 + 2 + 1 + 1,$$

and so

$$\varphi(12) = 6. \qquad \square$$

This is obviously not the best way to compute $\varphi(n)$, but it requires the least amount of previous knowledge.

Problem Set 2.2

1. Produce a table of values of $\varphi(n)$ for $1 < n < 20$.
2. Show that if $n > 2$, then $\varphi(n)$ is even.
3. (a). Show that

$$\sum_{\substack{n=1 \\ (n,N)=1}}^{N} n = N\varphi(N)/2.$$

   (b) Let $\{r_1, \ldots, r_n\}$ be any reduced residue system modulo m, where $m > 2$. Show that

   $$r_1 + \cdots + r_n \equiv 0 \pmod{m}.$$

4. (a) Show that if a and b are positive integers such that $(a, b) = d$, then

   $$\varphi(ab) = d\,\varphi(a)\,\varphi(b)/\varphi(d).$$

   (b) Show that

   $$\varphi(m)\,\varphi(n) = \varphi((m, n))\,\varphi([m, n])$$

5. If m and k are positive integers, then

   $$\varphi(m^k) = m^{k-1}\,\varphi(m)$$

6. Let n be a positive integer. If n is odd, show that $\varphi(2n) = \varphi(n)$ and if n is even, then $\varphi(2n) = 2\varphi(n)$

7. What is the largest prime divisor of $\varphi(100!)$?

8. (a) If $S_k(n)$ is defined as in Problem 16 of Problem Set 2.1, show that

   $$S_k(ab) \equiv bS_k(a) + aS_k(b) \pmod{a}.$$

   (b) If $(a, b) = 1$, then

   $$S_k(ab)/ab - S_k(a)/a - S_k(b)/b$$

   is an integer. Generalize this result to any number of factors that are relatively prime in pairs.

9. Show that if $\varphi(n) = n-1$, then n is a prime.

10. Suppose $d \mid n$. Then the number of integers m $1 \le m \le n$, such that $(m, n) = d$ is $\varphi(n/d)$.

2.3. CONGRUENCES OF THE FIRST DEGREE

As in the theory of equations, where the principle aim is to solve polynomial equations, so too in the theory of congruences the principle aim, beyond developing divisibility theory, is to solve polynomial congruences. That is, we wish to find all integers x such that

$$f(x) \equiv 0 \ (\text{mod } m),$$

where f is a given polynomial with integer coefficients and m is a given positive integer. More so than in the theory of equations much of the work can be reduced to solving linear congruences, singly or in systems, as we shall see below. For this reason we begin with the linear case. We begin this with a famous theorem due to Euler which is a generalization of a famous theorem of Fermat.

Theorem 2.8: (Euler) If $(a, m) = 1$, then

$$a^{\varphi(m)} \equiv 1 \ (\text{mod } m).$$

Proof: Let $\{r_1, \dots, r_{\varphi(m)}\}$ be a reduced residue system modulo m. Then, by Theorem 2.5, $\{ar_1, \dots, ar_{\varphi(m)}\}$ is also a reduced residue system modulo m, since $(a, m) = 1$. Thus

$$(ar_1)\cdots(ar_{\varphi(m)}) \equiv r_1 \cdots r_{\varphi(m)} \ (\text{mod } m)$$

or

$$a^{\varphi(m)} r_1 \cdots r_{\varphi(m)} \equiv r_1 \cdots r_{\varphi(m)} \ (\text{mod } m).$$

Since $\left(m, r_1 \cdots r_{\varphi(m)}\right) = 1$, the result follows from Theorem 2.3.   ■

Example 2.8: In Example 2.5 we saw that $\varphi(360) = 96$. Thus if $(a, 360) = 1$, we have

$$a^{96} \equiv 1 \ (\text{mod } 360).$$

For example, if we take $a = 7$ and $a = 359$, we have

$$7^{96} \equiv 1 \ (\text{mod } 360) \text{ and } 359^{96} \equiv 1 \ (\text{mod } 360).$$

Example 2.9: Find the last three digits of 127^{2003}.

Solution: We can proceed somewhat as in Example 2.2. However, with the aid of Euler's theorem we can simplify things somewhat. Again the problem is to find a positive integer r such that

$$127^{2003} \equiv r \ (\text{mod } 1000).$$

We have $\varphi(1000) = 1000(1/2)(4/5) = 400$. Since $(127, 1000) = 1$ we see that for any non-negative integer k we have

$$127^{400k} \equiv 1 \ (\mathrm{mod} \ 1000).$$

Since $2003 = 400 \cdot 5 + 3$ we see that

$$127^{2003} \equiv 127^3 \ (\mathrm{mod} \ 1000).$$

Since an easy calculation shows that $127^3 \equiv 383 \ (\mathrm{mod} \ 1000)$ we see that the last three digits are 383. Since $\varphi(p) = p - 1$ we have the following corollary.

Corollary 2.8.1: (Fermat) If p is a prime and $p \nmid a$ then $a^{p-1} \equiv 1 \ (\mathrm{mod} \ p)$. For any integer a we have $a^p \equiv a \ (\mathrm{mod} \ p)$.

Proof: The first statement follows immediately from the theorem. The second also follows readily by multiplying the congruence $a^{p-1} \equiv 1 \ (\mathrm{mod} \ p)$ by a if $p \nmid a$. If $p \mid a$ then both a^p and a are congruent to 0 modulo p, and so congruent to each other, by (c) of Theorem 2.1.

We would like to give an independent proof of this latter fact. The proof is by induction on a.

For $a = 1$ it is clear that $a^p \equiv a \ (\mathrm{mod} \ p)$. Suppose that $a^p \equiv a \ (\mathrm{mod} \ p)$ is true for $a = m$, where $m \geq 1$. Then for $a = m + 1$, we have

$$(m+1)^p = m^p + 1 + \sum_{k=1}^{p-1} \binom{p}{k} m^{p-k} \equiv m^p + 1 \ (\mathrm{mod} \ p)$$
$$\equiv m + 1 \ (\mathrm{mod} \ p),$$

by the induction hypothesis. We have used the fact that $p \mid \binom{p}{k}$ if $1 < k < p - 1$, which follows since there are no factors in the denominator of $\binom{p}{k}$ to cancel the p in the numerator. The result then follows by mathematical induction.

We give now an application of Fermat's Theorem. The following result is the companion to Theorem 1.15.

Corollary 2.8.2: There are infinitely many primes of the form $4k + 1$.

Proof: Let $n > 1$ be an integer and let $N = (n!)^2 + 1$. If p is the least prime factor of N, then $p > n$ (elsewise, $p \mid n!$, and so not N). Since $(n!)^2 \equiv -1 \ (\mathrm{mod} \ p)$, we have

$$(n!)^{p-1} \equiv (-1)^{(p-1)/2} \ (\mathrm{mod} \ p).$$

Since $p \nmid n!$, we have, by Fermat's Theorem,

$$(n!)^{p-1} \equiv 1 \pmod{p}.$$

Thus, by (c) of Theorem 2.1, we have

$$(-1)^{(p-1)/2} \equiv 1 \pmod{p}.$$

Since p is an odd prime, because N is an odd number, this implies that

$$(-1)^{(p-1)/2} = 1.$$

Now an odd prime p is either $\equiv 1 \pmod 4$ or $\equiv 3 \pmod 4$. If $p \equiv 3 \pmod 4$, then $(-1)^{(p-1)/2} = -1$. Thus we must have $p \equiv 1 \pmod 4$. Since $p > n$ and n is arbitrary the result follows. ∎

We now apply Theorem 2.8 to the subject of this section: solving linear congruences.

Corollary 2.8.3: If $(a, m) = 1$, then $ax \equiv b \pmod m$ has a unique solution modulo m for any integer b, that is, any two solutions of this congruence must be congruent to each other modulo m.

Proof: By Theorem 2.8, we have $a(a^{\varphi(m)-1}b) = a^{\varphi(m)}b \equiv b \pmod m$. Thus the congruence has a solution $x_0 = a^{\varphi(m)}b$.

Now let x and y be any two solutions of this congruence. Then

$$x \equiv a^{\varphi(m)}x = a^{\varphi(m)-1}ax \equiv a^{\varphi(m)-1}b \equiv a^{\varphi(m)-1}ay \equiv a^{\varphi(m)}y \equiv y \pmod m,$$

which proves the uniqueness modulo m and finishes the proof. ∎

Example 2.10: Solve the congruence $7x \equiv 3 \pmod{15}$.

Solution: Since $(7, 15) = 1$ we know that the congruence is solvable and has one incongruent solution modulo 15. We have $\varphi(15) = 15(2/3)(4/5) = 8$ and

$$7 \equiv 7, \ 7^2 = 49 \equiv 4 \text{ and } 7^4 \equiv 4^2 \equiv 1 \pmod{15}.$$

Thus

$$7^7 \equiv 7 \cdot 4 \cdot 1 = 28 \equiv 13 \pmod{15},$$

and so, as in the proof of Corollary 2.8.3, the solution to our congruence is

$$x \equiv 13 \cdot 3 = 39 \equiv 9 \pmod{15}. \qquad \square$$

If $(a, m) > 1$, there may be no solution to the congruence. The following theorem gives the necessary and sufficient conditions for the solvability of the general linear congruence and the number of incongruent solutions when solutions exist.

Theorem 2.9: Let $(a, m) = d$. The congruence $ax \equiv b$ (mod m) is solvable if and only if $d \mid b$. If it is solvable, then it has exactly d incongruent solutions, that is, these d solutions are incongruent to each other modulo m and any solution of the congruence must be congruent to one of the d numbers.

Proof: If $d = 1$, the result is that of Corollary 2.8.3, and so we may suppose that $d > 1$. If $ax \equiv b$ (mod m), then there exists an integer y such that

$$ax - my = b.$$

Thus, by (e) of Theorem 1.1, we have $d \mid b$.

Suppose $d \mid b$ and let $a = da_1$, $m = dm_1$ and $b = db_1$. By Theorem 2.3 this congruence is equivalent to

(2.1) $$a_1 x \equiv b_1 \text{ (mod } m_1).$$

Since $(a_1, m_1) = 1$, by Corollary 1.10.1, this congruence has, by Corollary 2.8.3, exactly one solution modulo m_1. Let t_1 be a representative of its congruence class modulo m_1. Then there exists an integer y such that if x is any solution to the congruence (2.1),

$$x = t_1 + ym_1.$$

Now

$$t_1 + ym_1 \equiv t_1 + zm_1 \text{ (mod } m)$$

if and only if

$$m \mid m_1(y - z)$$

if and only if

$$d \mid (y - z).$$

Thus there are exactly d solutions and they are given by

$$t_1, \ t_1 + m_1, t_1 + 2m_1, \ ..., \ t_1 + (d - 1)m_1.$$

This completes the proof of the theorem.   ■

Example. 2.11: Solve $20x \equiv 4$ (mod 32).

Solution: Since $(20, 32) = 4$ and $4 \mid 4$, there will be 4 incongruent solutions modulo 32. Dividing the given congruence by 4 produces the equivalent congruence

$$5x \equiv 1 \text{ (mod } 8).$$

From the proof of Corollary 2.8.3 we see that all solutions of this congruence satisfy

$$x \equiv 5^{\varphi(8)-1} \cdot 1 = 5^3 \cdot 1 \equiv 5 \ (\mathrm{mod}\ 8).$$

Thus, by the proof of Theorem 2.9, we see that all solutions of the original congruence satisfy

$$x \equiv 5,\ 13,\ 21,\ \text{or } 29 \ (\mathrm{mod}\ 32). \qquad \square$$

The following result gives a necessary and sufficient condition for a number to be a prime. As will be seen from the statement of the theorem the condition is not useful for actual calculations.

Corollary 2.9.1: (Wilson) Let n be a positive integer. Then n is a prime if and only if

$$(n - 1)! \equiv -1 \ (\mathrm{mod}\ n).$$

Proof: Suppose n is composite. Then n has a prime factor $q < n$. Thus $q \leq n - 1$, and so $q \mid (n - 1)!$. Thus, by (e) of Theorem 1.1, we have $q \nmid \{(n - 1)! + 1\}$ and consequently $n \nmid \{(n - 1)! + 1\}$.

Suppose n is a prime p. If $p = 2$ or 3, then the result is clearly true, and so we may suppose $p \geq 5$. Let $i \leq j \leq p - 1$. Then $(j, p) = 1$ and, by Theorem 2.9, there exists a unique i, $0 \leq i \leq p - 1$, such that

$$ij \equiv 1 \ (\mathrm{mod}\ p)$$

Clearly $i \not\equiv 0 \ (\mathrm{mod}\ p)$, since $p \nmid 1$ Thus $1 \leq i \leq p - 1$. If we associate to each j its i, we see that i gets associated to $j\sim$ since

$$ji \equiv ij \equiv 1 \ (\mathrm{mod}\ p).$$

Now 1 is associated to itself as is $p - 1$, and so consider j with $2 \leq j \leq p - 2$. Then $(j - 1, p) = (j + 1, p) = 1$ and so $j^2 - 1 = (j + 1)(j - 1) \not\equiv 0 \ (\mathrm{mod}\ p)$. Thus for each of these j's we see that it is not associated with itself, and so the numbers $2, \ldots, p - 2$ can be paired off, each with its associate, which is different from itself. Thus

$$2 \cdot 3 \cdots (p - 2) \equiv 1 \ (\mathrm{mod}\ p).$$

Thus

$$(p - 1)! = (p - 1)(p - 2) \cdots 2 \equiv (p - 1) \cdot 1 \equiv -1 \ (\mathrm{mod}\ p),$$

which is our result.   ■

We can apply this result to proving a result due to Fermat and Euler.

Corollary 2.9.2: Let p be a prime. Then the congruence

$$x^2 \equiv -1 \ (\text{mod } p)$$

is solvable if and only if $p = 2$ or $p \equiv 1 \ (\text{mod } 4)$.

Proof: Suppose $p = 2$. Then $x = 1$ is clearly a solution to

(2.2) $x^2 \equiv -1 \ (\text{mod } p)$.

Suppose that $p \equiv 1 \ (\text{mod } 4)$, say $p = 4k + 1$, for some positive integer k. If we apply Wilson's theorem with this p, we get

(2.3) $1 \cdot 2 \cdot 3 \cdots (2k)(2k+1) \cdots (4k-2)(4k-1)(4k) \equiv -1 \ (\text{mod } p)$.

Since $p = 4k + 1$ we see that modulo p we have

$$2k \equiv -(2k+1), (2k-1) \equiv -(2k+2), \ldots, 2 \equiv -(4k-1) \text{ and } 1 \equiv -4k \ (\text{mod } p).$$

Thus we can rewrite (2.3) as

$$1 \cdot 2 \cdot 3 \cdots (2k) \cdot (2k)(2k-1) \cdots 3 \cdot 2 \cdot 1 \cdot (-1)^{2k} \equiv -1 \ (\text{mod } p),$$

that is,

$$\left\{ (2k)! \right\}^2 \equiv -1 \ (\text{mod } p),$$

and so (2.2) is solvable for p.

Now suppose that (2.2) is solvable for p. Suppose $p > 2$ and $p \equiv 3 \ (\text{mod } 4)$. If a is a solution of (2.2), then

$$a^2 \equiv -1 \ (\text{mod } p),$$

and so, by (g) of Theorem 2.1, we have

$$a^{4k+2} \equiv (-1)^{2k+1} = -1 \ (\text{mod } p).$$

On the other hand, Fermat's theorem (Corollary 2.8.1) says that

$$a^{4k+2} \equiv 1 \ (\text{mod } p).$$

Since this is a contradiction we see that $p = 2$ or $p \equiv 1 \ (\text{mod } 4)$ and finish the proof of the theorem.   ■

This gives the immediate corollary which gives another proof of Corollary 2.8.2.

Corollary 2.9.3: Let n be an integer and let p be a prime that divides $n^2 + 1$. Then

$p = 2$ or $p \equiv 1 \pmod 4$.

To prove Corollary 2.8.2 let $\{p_1, \ldots, p_n\}$ be a collection of primes of the form $4k + 1$ and let

$$N = (2p_1 \cdots p_n)^2 + 1.$$

By Theorem 1.11, there is a prime p such that $p \mid N$. Since N is odd we see that $p > 2$. Thus, by Corollary 2.9.3, we see that p is of the form $4k + 1$. Since p is a prime we have that $p \nmid 1$, and so, by (e) of Theorem 1.2, we see that $p \neq p_i$, $1 \leq i \leq n$. Thus our list of primes is not complete. Thus there are infinitely many primes of the form $4k + 1$.

This proof is the paradigm of many similar proofs as we shall see in later sections of this book.

We close this section with a related result about solutions of congruences. This result is about approximating integers in congruences and will be of use later in this text. We give one of these applications to close the section.

Theorem 2.10: (Thue) Let n be an integer which is not a square and let a be an integer relatively prime to n. Then there exist integers x and y such that

$$ax \equiv y \pmod n,$$

where $0 < x < \sqrt{n}$ and $0 < |y| < \sqrt{n}$.

Proof: Consider all the numbers

$$ax + y$$

with $0 \leq x < \sqrt{n}$ and $0 \leq |y| < \sqrt{n}$. This set has $\left(\left[\sqrt{n}\right] + 1\right)\left(\left[\sqrt{n}\right] + 1\right) > n$ numbers in it. Thus, by the Pigeon-Hole Principle, there are at least two numbers in this set that are congruent to each other modulo n.

Suppose $(x_1, y_1) = (x_2, y_2)$, as ordered pairs, and

$$ax_1 + y_1 \equiv ax_2 + y_2 \pmod n.$$

Then

$$a(x_1 - x_2) \equiv y_2 - y_1 \pmod n.$$

Since $0 \leq x_1, x_2, y_1, y_2 < \sqrt{n}$ we have

$$0 \leq |x_1 - x_2| < \sqrt{n} \text{ and } 0 \leq |y_1 - y_2| < \sqrt{n}.$$

We must have strict inequality at the lower end for if one of these differences is zero the other must be, since $(a, n) = 1$. Thus, to satisfy the theorem, we can take $x = |x_1 - x_2|$ and $y = \pm|y_1 - y_2|$.   ∎

We give an application of this theorem the problem of which primes can be written as a sum of two squares. The result is due to Fermat.

Corollary 2.10.1: (Fermat) Let p be a prime of the form $4k + 1$, then p can be uniquely written in the form

$$p = m^2 + n^2,$$

where m and n are positive integers and $m > n$.

Proof: By Corollary 2.9.2 we know that the congruence $x^2 \equiv -1 \pmod{p}$ has a solution, say a. By Thue's theorem we know there exist integers x and y such that $ax \equiv \pm y \pmod{p}$, with $0 < x, y < \sqrt{p}$. Thus, if we square both sides of the congruence we get

$$a^2x^2 - y^2 \equiv 0 \pmod{p}.$$

Since $a^2 \equiv -1 \pmod{p}$ we see, by (d) of Theorem 2.1, that there is an integer t such that

$$x^2 + y^2 = pt.$$

Since $0 < x, y < \sqrt{p}$ we see that $0 < x^2, y^2 < p$ or $0 < x^2 + y^2 < 2p$. Thus we must have $t = 1$. Thus there are integers x and y such that

$$p = x^2 + y^2.$$

Assume that there are two representations for the prime p, say

$$p = a^2 + b^2 = c^2 + d^2$$

where $a > b$, $c > d$ and $a > c$. Then, by Theorem 1.21, we see that

$$t = (ac + bd)/(ac + bd, ab + cd)$$

is a proper divisor of p, that is, $1 < t < p$. Since p is a prime it has no proper divisors. Thus p has only one representation as a sum of two squares and the theorem is proved.   ∎

Problem Set 2.3

1. Prove the following results
   (a) An integer n is a prime if and only if $(n - 2)! \equiv 1 \pmod{n}$.
   (b) If $n > 4$ is composite, then $n \mid (n - 1)!$.
   (c) An integer $p > 3$ is a prime if and only if $2(p - 3)! + 1 \equiv 0 \pmod{p}$.
   (d) Let p be a prime and $h + k = p - 1$, with h and k non-negative integers. Then

$$h!k! + (- 1)^h \equiv 0 \pmod{p}.$$

2. (a) Show that

$$1^2 \cdot 3^2 \cdots (p-2)^2 \equiv (-1)^{(p+1)/2} \pmod{p}.$$

   (b) Show that

$$2^2 \cdot 4^2 \cdots (p-1)^2 \equiv (-1)^{(p+1)/2} \pmod{p}.$$

3. Show that, for $n > 2$,

$$\prod_{\substack{1 \le x \le n \\ (x,n)=(x+1,n)=1}} x \equiv 1 \pmod{n}.$$

4. Let a, m and n be positive integers, with $m < n$. If

$$\left(a^{2^m} + 1, a^{2^n} + 1\right) = d,$$

   then $d = 1$ if a is even and $d = 2$ if a is odd. (Hint: if p is a common prime factor, then

$$a^{2^m} \equiv -1 \pmod{p}.$$

   Raise this congruence to the 2^{n-m} power. Compare with Problem 16 of Problem Set 1.2.)

5. If p is an odd prime and $0 < a < p - 1$, then

$$\binom{p-1}{a} \equiv (-1)^a \pmod{p}.$$

6. Let $n > 1$. Then n and $n + 2$ are both primes if and only if

$$4((n - 1)! + 1) \equiv 0 \pmod{n(n + 2)}.$$

7. If a is an integer, then

$$a^m \equiv a^{m-\varphi(m)} \pmod{m}.$$

8. If p is a prime and a is an integer, then

$$a^p + (p-1)!a \equiv 0 \pmod{p}.$$

9. Solve the following linear congruences.

(a) $5x \equiv 1 \pmod 7$ (d) $187x \equiv 2 \pmod{503}$

(b) $14x \equiv 5 \pmod{45}$ (e) $79x \equiv 2 \pmod{153}$

(c) $9x \equiv 21 \pmod{12}$ (f) $182x \equiv 7 \pmod{203}$

10. Prove the following theorem due to A. Scholz. Let n be a positive integer greater than 1 and let e and f be two positive integers such that

$$ef > n, e > 1 \text{ and } n \geq f.$$

Then for any integer a, with $(a, m) = 1$, then there exists integers x and y such that

$$ax \equiv y \pmod{n},$$

where $0 < x < f$ and $0 < |y| < e$.

11. (a) If $(a, n) = (b, n) = 1$, then the congruence

$$ax + by \equiv c \pmod{n}$$

has n incongruent solutions modulo n.

(b) Solve

(i) $6x + 15y \equiv 9 \pmod{17}$.   (iii) $6x + 3y \equiv 0 \pmod 9$.

(ii) $2x + 3y \equiv 1 \pmod 7$.   (iv) $10x + 5y \equiv 9 \pmod{15}$.

12. (a) Show that the system of simultaneous linear congruences

$$ax + by \equiv c \pmod{m}$$
$$cx + dy \equiv f \pmod{m}$$

has a solution if $(ad - bc, m) = 1$. (Hint: modify the approach for solving systems of equations.)

(b) Find the general solution of the systems:

(i) $x + 3y \equiv 5 \pmod{13}$ (ii) $2x + 3y \equiv 4 \pmod 8$

$\quad$ $7x + 9y \equiv 11 \pmod{13}$ $\quad$ $5x + 6y \equiv 7 \pmod 8$

(iii) $3x + 7y \equiv 10 \pmod{14}$ (iv) $2x + 7y \equiv 2 \pmod 5$

$\quad$ $11x - 8y \equiv -8 \pmod{14}$ $\quad$ $3x - y \equiv 11 \pmod 5$

13 Let n be an odd integer which is not a multiple of 5. If R_m denotes the m digit repunit (see Problem 15 of Problem Set 2.1), then there is an m such that $n \mid R_m$.

14. Let

$$S_n = 1^n + 2^n + \ldots + (p - 1)^n,$$

where p is a prime. For $n = p - 2, p - 1, p, p + 1$ and $p + 2$ determine the values of S_n modulo p.

15. Let a be a positive integer. Then there are no integers m and n so that

$$4mn - m - n^a$$

is a perfect square. (Hint: split into two cases: a even and a odd.)

2.4. THE CHINESE REMAINDER THEOREM

In the previous section we solved congruences of degree one in one unknown, but only one at a time. In this section we shall study systems of simultaneous linear congruences in a single unknown. Unlike equations where there would usually be no solution, under the right conditions we always have a solution.

Theorem 2.11. (Chinese Remainder Theorem). Let $m_1, \ldots, m_r$ be positive integers that are pairwise relative prime and let $a_1, \ldots, a_1$ be any r integers. Then the system of congruences

$$x \equiv a_1 (\bmod\ m_1), \ldots, x \equiv a_1 (\bmod\ m_r)$$

has a solution which is unique modulo $m_1 \cdots m_r$.

Proof. Let $M = m_1 \cdots m_r$. Then $m_j \mid M$ and $(M/m_j, m_j) = 1$, for $1 \le j \le r$. Thus, by Theorem 2.9, there exist integers b_j such that, for $1 \le j \le r$,

$$(M/m_j)b_j \equiv 1 \ (\bmod\ m_j).$$

Also, if $i \ne j$, then

$$(M/m_j)b_j \equiv 0 \ (\bmod\ m_i).$$

Let

$$w = \sum_{j=1}^{n} (M / m_j)b_j a_j.$$

Then, for $1 \le i \le r$,

$$w = \sum_{j=1}^{n}(M\,/\,m_j)b_j a_j \equiv (M\,/\,m_i)b_i a_i \equiv a_i \pmod{m_i},$$

and so w is a common solution.

If x and y are common solutions, then $x \equiv a_i \equiv y \pmod{m_i}$, for $1 \le i \le r$. Thus, by Theorem 2.2, we have $x \equiv y \pmod{M}$, which is the second part of the theorem. ∎

Example 2.9. Solve the system

$$x \equiv 2 \pmod 3,\ x \equiv 3 \pmod 4 \text{ and } x \equiv 4 \pmod 5.$$

Solution. Here $m_1 = 3$, $m_2 = 4$ and $m_3 = 5$ and $a_1 = 2$, $a_2 = 3$ and $a_3 = 4$. Then $M = 3 \cdot 4 \cdot 5 = 60$. Let $M_1 = M/m_1 = 20$, $M_2 = M/m_2 = 15$ and $M_3 = M/m_3 = 12$. Then we need to find b_1, b_2 and b_3 so that

$$20b_1 \equiv 1 \pmod 3,\ 15b_2 \equiv 1 \pmod 4 \text{ and } 12b_3 \equiv 1 \pmod 5.$$

As we only need one solution of each of these congruences we can take $b_1 = 2$, $b_2 = 3$ and $b_3 = 3$. From the proof of Theorem 2.10 we then see that our solution is given by

$$x \equiv 20 \cdot 2 \cdot 2 + 15 \cdot 3 \cdot 3 + 12 \cdot 3 \cdot 4 = 359 \equiv -1 \pmod{60}. \qquad \square$$

There are many possible generalizations of Theorem 2.10 and some of these will be met in the problem set below. We will need in our work below only the version given here.

Problem Set 2.4

1. Solve the following systems of congruences.
   (a) $x \equiv 3 \pmod 5$ and $x \equiv 4 \pmod 7$.
   (b) $x \equiv 3 \pmod 6$, $x \equiv -1 \pmod{25}$ and $x \equiv 1 \pmod 7$.
   (c) $x \equiv -2 \pmod{11}$, $x \equiv 13 \pmod{28}$ and $x \equiv 7 \pmod{45}$.
   (d) $x \equiv 2 \pmod{13}$, $x \equiv 2 \pmod{35}$, $x \equiv 17 \pmod{87}$ and $x \equiv 7 \pmod{23}$.

2. (a) Modify the result of Theorem 2.10 to solve systems of the form:

   $$a_1 x \equiv b_1 \pmod{m_1}, \ldots, a_r x \equiv b_r \pmod{m_r},$$

   where, for $1 \le i \le r$, $(a_i, m_i) = 1$ and, for $1 \le i < j \le r$, $(m_i, m_j) = 1$.
   (b) Modify Theorem 2.10 to cover the case when we allow $(a_i, m_i) > 1$.

(c) Solve the following systems of congruences.

(i) $2x \equiv 3 \pmod 5$ and $4x \equiv 3 \pmod 7$.

(ii) $5x \equiv \pmod{13}$, $x \equiv 2 \pmod{35}$, $3x \equiv 13 \pmod{87}$ and $x \equiv 7 \pmod{23}$.

(iii) $4x \equiv 6 \pmod 8$ and $3x \equiv 9 \pmod{27}$.

3. (a) Find all positive integers less than 1000 which leave a remainder of 1 when divided by 2, a remainder of 2 when divided by 3, a remainder of 4 when divided by 5 and a remainder of 6 when divided by 7.

(b) Find the least positive integer which leaves a remainder of 5 when divided by 13, a remainder of 3 when divided by 12 and a reminder of 2 when divided by 35.

(c) Find four consecutive integers which are divisible by 2, 3, 5 and 7, respectively.

4. (a) Let m and n be positive integers and let $d = (m, n)$. Then the system of congruences

$$x \equiv a \pmod m \text{ and } x \equiv b \pmod n$$

is solvable if and only if $d \mid (a - b)$. If the congruence is solvable, then all solutions are congruent modulo $[m, n]$.

(b) Let $\{m_1, \ldots, m_r\}$ be a set of positive integers and let, for $1 \le i < j \le r$, $d_{ij} = (m_i, m_j)$. Then the system of congruences

$$x \equiv a_1 \pmod{m_1}, \ldots, x \equiv a_r \pmod{m_r}$$

is solvable if and only if $d_{ij} \mid (a_i - a_j)$, for all i and j. If the congruence is solvable, then the solution is unique modulo $[m_1, \ldots, m_r]$.

(c) Modify the above results to apply to the system of congruences

$$a_1x \equiv b_1 \pmod{m_1}, \ldots, a_rx \equiv b_r \pmod{m_r},$$

where, for $1 \le i \le r$, $(a_i, m_i) = 1$.

5. Solve, if possible, the following systems of congruences.

(a) $x \equiv 47 \pmod{84}$ and $x \equiv 11 \pmod{20}$.

(b) $x \equiv 47 \pmod{84}$ and $x \equiv -11 \pmod{20}$.

(c) $3x \equiv 7 \pmod{10}$ and $5x \equiv 9 \pmod{12}$.

(d) $x \equiv 2 \pmod{15}$, $x \equiv 7 \pmod{10}$ and $x \equiv 5 \pmod 6$.

(e) $8x \equiv 3 \pmod{15}$, $5x \equiv 7 \pmod{21}$ and $7x \equiv 4 \pmod{13}$.

6. When eggs in a basket are removed 2, 3, 4, 5 and 6 at a time there remain 1, 2, 3, 4 and 5 eggs, respectively. What is the minimum possible number of eggs that were originally in the basket? (Brahmagupta, 7th century India)

7. Let a and b be positive integers such that $(a, b) = 1$. Use Theorem 2.10 to prove that if c is a given integer, then there are infinitely many integers m such that $(a + bm, c) = 1$. (Compare with Problem 27(a) of Problem Set 1.2.)

8. (a) Find an integer having remainders 1, 2, 5 and 5 when divided by 2, 3, 6 and 12, respectively. (Yin-hing, 8th century China)

   (b) Find an integer having remainders 2, 3, 4 and 5 when divided by 3, 4, 5 and 6, respectively. (Bhaskara, 12th century)

9. Obtain the two incongruent solutions modulo 210 of the system
$$2x \equiv 3 \ (\text{mod } 5), \ 4x \equiv 2 \ (\text{mod } 6) \text{ and } 3x \equiv 2 \ (\text{mod } 7).$$

10. If $x \equiv r \ (\text{mod } m)$ and $x \equiv s \ (\text{mod } m + 1)$, then $x \equiv r(m + 1) - sm \ (\text{mod } m(m + 1))$.

11. Show that if n is an integer, then n must satisfy at least one of the following congruences.
$$x \equiv 0 \ (\text{mod } 2), \ x \equiv 0 \ (\text{mod } 3), \ x \equiv 1 \ (\text{mod } 4),$$
$$x \equiv 1 \ (\text{mod } 6) \text{ and } x \equiv 11 \ (\text{mod } 12).$$

12. Solve the system of congruences
$$3x^2 + x \equiv 0 \ (\text{mod } 5) \text{ and } 2x + 3 \equiv 0 \ (\text{mod } 7).$$
to find all solutions incongruent modulo 35. (Hint: solve the first congruence by trial and error (there are two incongruent solutions) and solve the second by the methods of Section 2.3. Combine the solutions via Theorem 2.11.)

2.5. POLYNOMIAL CONGRUENCES

In this section we shall begin our study of the congruence

(2.4) $f(x) \equiv 0 \ (\text{mod } m),$

where
$$f(x) = a_n x^n + a_{n-1} x^{n-1} + \cdots a_1 x + a_0$$

with $n \geq 1$ and, for $0 \leq i \leq n$, the a_i are integers. The problem will basically be handled by eventually reducing the problem of solving the congruence (mod m) to solving the congruence (mod p), where $p \mid m$ and p is a prime. We begin the reduction in this section and reduce the problem to prime power congruences.

We begin with some general definitions.

Definition 2.6: If u is an integer such that $f(u) \equiv 0$ (mod m), then u is said to be a **solution** of (2.4).

By (g) of Theorem 2.1, we see that if $u \equiv v$ (mod m), then v is also a solution of (2.4) if u is. This leads to the following definition.

Definition 2.7: Let r_1, ..., r_m be a complete residue system modulo m. The **number of solutions** of (2.4) is the number of r_i such that $f(r_i) \equiv 0$ (mod m).

Thus he number of solutions of (2.4) is at most m.

The first step in the reduction process mentioned above is the following obvious result.

Theorem 2.12: If $d \mid m$, $d > 0$ and u is a solution of (2.4), then u is a solution of

$$f(x) \equiv 0 \text{ (mod } d).$$

Write $m = p^{e_1} \cdots p^{e_r}$ in the canonical factorization of m. Then Theorem 2.12 says that solutions of (2.4) give solutions of

(2.5) $$f(x) \equiv 0 \text{ (mod } p_i^{e_i})$$

for $1 \le i \le r$. We wish to show the converse, namely that if we know the solutions to the congruences (2.5) we can obtain the solutions to the congruence (2.4).

It is clear that if one of the congruences in (2.5) has no solutions, then (2.4) can have no solutions. Suppose that all the congruences in (2.5) are solvable, say the ith congruence has k_i solutions: $r_{i,1}, \ldots, r_{i,k_i}$. Then these are incongruent modulo $p_i^{e_i}$ and every solution is congruent to one of them. If u is a solution of (2.4), then for each i there must exist a j_i. such that

$$u \equiv r_{i,j_i} \text{ (mod } p_i^{e_i}),$$

$1 \le i \le r$. The converse of this last statement holds by Theorem 2.11, since the prime power divisors of m are relative prime in pairs. If we determine integers b_i, $1 \le i \le r$, by solving

$$m p_i^{-e_i} b_i \equiv 1 \text{ (mod } p_i^{e_i}),$$

then, as in the proof of the Chinese Remainder Theorem, we see that the solutions of (2.4) are of the form

$$u \equiv \sum_{i=1}^{r} (m / p_i^{e_i}) b_i r_{i,j_i} \text{ (mod } m).$$

Since the $m/p_i^{e_i}$ are independent of j_i we can calculate these first and then insert the various r_{i,j_i}. For each choice of $j_1, \ldots, j_r$ we get a different solution u of (2.4). Since the j_i can take k_i different values we have the following result.

Theorem 2.13: Let $N(m)$ denote the number of solutions of $f(x) \equiv 0 \pmod{m}$. Then, with m as above, we have that the number of solutions of (2.4) satisfies

$$N(m) = \prod_{i=1}^{r} N(p_i^{e_i}).$$

Example 2.10: Solve $x^3 + 2x - 3 \equiv 0 \pmod{45}$.

Solution: To solve the given congruence it suffices to solve each of the following two congruences and then combine their solutions by the Chinese Remainder Theorem. We have to solve

$$x^3 + 2x - 3 \equiv 0 \pmod{5} \text{ and } x^3 + 2x - 3 \equiv 0 \pmod{9}.$$

By trial we see that the solutions are $r_{5,1} = 1$, $r_{5,2} = 3$, $r_{9,1} = 1$, $r_{9,2} = 2$, and $r_{9,3} = 6$. By Theorem 2.13 the original congruence has $2 \cdot 3 = 6$ solutions.

We need to find b_1 and b_2 such that

$$5b_1 \equiv 1 \pmod{9} \text{ and } 9b_2 \equiv 1 \pmod{5}.$$

This gives $b_1 = 2$ and $b_2 = 4$. Thus all solutions u, of the original congruence, must satisfy

$$u \equiv 5 \cdot 2r_{9,i} + 9 \cdot 4r_{5,j} \equiv 10r_{9,i} + 36r_{5,j} \pmod{45}.$$

Putting in the various values of $r_{9,i}$ and $r_{5,j}$ we find that

$$u \equiv 1, 6, 11, 28, 33, 38 \pmod{45}. \qquad \square$$

Problem Set 2.5

1. Solve the following congruences.
   (a) $x^5 + x^3 + x^2 + 2 \equiv 0 \pmod{15}$.
   (b) $x^4 + 6x^2 + 5 \equiv 0 \pmod{60}$.
   (c) $2x^3 + x^2 + 2 \equiv 0 \pmod{35}$.
   (d) $2x^3 + x^2 + 1 \equiv 0 \pmod{35}$.
   (e) $2x^3 + x^2 + 3 \equiv 0 \pmod{35}$.

2. (a) Prove the following analogue of the factor theorem from algebra. Let $f(x)$ be a polynomial with integer coefficients and let m be a positive integer. If

a is a root of the congruence

$$f(x) \equiv 0 \ (\mathrm{mod} \ m),$$

then there is a polynomial $g(x)$, with integer coefficients, such that

$$f(x) \equiv (x - a)g(x) \ (\mathrm{mod} \ m)$$

and conversely.

(b) Find a polynomial for $g(x)$ for the congruence

$$x^2 + 1 \equiv 0 \ (\mathrm{mod} \ 65)$$

using the root $a = 8$.

3. Prove the following analogue of the Lagrange interpolation formula. Let m and n be positive integers and let $\{a_1, \ldots, a_n\}$ be n integers that are incongruent modulo m and let $\{b_1, \ldots, b_n\}$ be any set of integers. Then there exists a polynomial $f(x)$ whose integral coefficients are unique modulo m such that, $1 \leq i \leq n$,

$$f(a_i) \equiv b_i \ (\mathrm{mod} \ m).$$

4. (a) If $f(x)$ is a polynomial with integer coefficients and satisfying

$$f(a) \equiv f'(a) \equiv \cdots \equiv f^{(r)}(a) \equiv 0 \ (\mathrm{mod} \ m),$$

where $(r!, m) = 1$, then there exists a polynomial $g(x)$ with integer coefficients and $\deg(g) = \deg(f) - (r + 1)$ such that

$$f(x) \equiv (x - a)^{r+1} g(x) \ (\mathrm{mod} \ m).$$

(b) Apply the result of (a) with $a = 3$ to the polynomial congruence

$$x^4 + 6x^3 + 2x^2 + 4x + 2 \equiv 0 \ (\mathrm{mod} \ 11).$$

(c) Suppose $(r!, m) = 1$ and define, for $0 \leq j \leq r$, a_j by

$$(j!)a_j \equiv 1 \ (\mathrm{mod} \ m).$$

Let $f(x)$ be a polynomial with integer coefficients and degree r. If a is an integer, then

$$f(x) \equiv \sum_{j=0}^{r} a_j f^{(j)}(a)(x - a)^j \ (\mathrm{mod} \ m).$$

(d) Illustrate the result the result of (c) with $a = 1$, $n = 11$ and $f(x)$ the polynomial of (b).

5. Let $F(k)$ denote the number of solutions to the congruence

$$f(x) \equiv k \ (\text{mod } m).$$

Show that

$$\sum_{k=0}^{m-1} F(k) = m.$$

6. Find the number of solutions of the congruence

$$24x^3 + 218x^2 + 121x + 17 \equiv 0 \ (\text{mod } 2 \cdot 3 \cdot 5 \cdot 7 \cdot 11 \cdot 13).$$

7. Let t_n denote the nth triangular number. (See Problem 59 of the Additional Problems for Chapter One.)

(a) Show that

$$t_1^2 + \cdots + t_n^2 = t_n(3n^3 + 12n^2 + 13n - 2)/30.$$

(b) For what values of n does n divide the sum of the squares of the first n triangular numbers?

8. Solve the following congruences.

(a) $x^3 + 1 \equiv 0 \ (\text{mod } 5)$.

(b) $x^3 - 1 \equiv 0 \ (\text{mod } 5)$.

(c) $x^3 - 1 \equiv 0 \ (\text{mod } 15)$.

9. Suppose $p = n^2 + a^2$ is a prime. Solve the congruence $x^4 \equiv a^2 \ (\text{mod } p)$.

2.6. PRIME POWER MODULI

From the previous section we saw that to solve

$$f(x) \equiv 0 \ (\text{mod } m)$$

it suffices to solve, for each prime p dividing m,

$$f(x) \equiv 0 \ (\text{mod } p^a)$$

where $p^a \| m$. In this section we shall further reduce the problem down to prime moduli.

Let r be a solution of

(2.6) $$f(x) \equiv 0 \ (\text{mod } p^s).$$

Then $f(r) \equiv 0 \ (\text{mod } p^t)$ for any $1 \le t \le s$. Let $x_{s,1}, \ldots, x_{s,n_s}$ be the solutions of (2.6). There may or may not be any solutions at all. Suppose $s \ge 2$. If there is a solution $x_{s,i}$ of (2.6) then there exists a solution x_{s-1,j_i} of

(2.7) $$f(x) \equiv 0 \pmod{p^{s-1}}$$

such that

$$x_{s,i} \equiv x_{s-1,j_i} \pmod{p^{s-1}}.$$

Thus there exists an integer v_{s-1} such that

$$x_{s,i} = x_{s-1,j_i} + v_{s-1}p^{s-1}.$$

Since $f(x)$ is a polynomial with integer coefficients we have

$$f(x) = a_n x^n + a_{n-1}x^{n-1} + \cdots a_1 x + a_0,$$

where, for $0 \le k \le n$, a_k is an integer and $n = \deg(f) \ge 1$. Thus, for $1 \le j \le n$, we have

$$f(x) = \sum_{k=0}^{n} a_k k(k-1)\cdots(k-j+1)x^{k-j} = \sum_{k=j}^{r} a_k k(k-1)\cdots(k-j+1)x^{k-j}.$$

Thus

$$\frac{f^{(j)}(x)}{j!} = \sum_{k=j}^{r} \frac{a_k}{j!} k(k-1)\cdots(k-j+1)x^{k-j}$$

$$= \sum_{k=j}^{r} a_k \frac{k!}{j!(k-j)!}x^{k-j} = \sum_{k=j}^{r} a_k \binom{k}{j}x^{k-j}$$

which is a polynomial of degree $n - j$ with integer coefficients. Note also that $f^{(t)}(x)$ is identically zero if $t > n$. Thus, by Taylor's Theorem,

$$f(x+h) = f(x) + f'(x)h + \tfrac{1}{2!}f''(x)h^2 + \cdots + \tfrac{1}{n!}f^{(n)}(x)h^n.$$

Thus, taking $h = v_{s-1}p^{s-1}$, we get

$$0 \equiv f(x_{s,j}) = f(x_{s-1,j_i} + v_{s-1}p^{s-1})$$

$$\equiv f(x_{s-1,j_i}) + f'(x_{s-1,j_i})v_{s-1}p^{s-1} \pmod{p^s}.$$

Since $f(x_{s-1,j_i}) \equiv 0 \pmod{p^{s-1}}$ we divide by p^{s-1} and use Theorem 2.3. This gives the congruence

$$f'(x_{s-1,j_i})v_{s-1} \equiv -p^{1-s}f(x_{s-1,j_i}) \pmod{p}.$$

Conversely, if

(2.8)
$$f'(x_{s-1,j})v \equiv -p^{1-s}f(x_{s-1,j}) \pmod{p},$$

then, by Taylor's Theorem,

$$f(x_{s-1,j} + vp^{s-1}) \equiv 0 \pmod{p^s}.$$

The above procedure gives a way of finding solutions of (2.6) with $s \geq 2$, if we know solutions of (2.7). For each root, $x_{s-1,j}$ of (2.7) we find all possible v given by (2.8). Then each $x_{s-1,j} + vp^{s-1}$ will be a solution of (2.6). Of course, for some $x_{s-1,j}$ there may be no v and so there will be no corresponding x_s

If we combine the above remarks we obtain the following theorem.

Theorem 2.14: Let x_0 be a solution of (2.7), with $s \geq 2$. Then the number of solutions of (2.6) corresponding to x_0 is

(a) none, if $f'(x_0) \equiv 0 \pmod{p}$ and x_0 is not a solution of (2.6),

(b) one, if $f'(x_0) \not\equiv 0 \pmod{p}$, and

(c) p, if $f'(x_0) \equiv 0 \pmod{p}$ and x_0 is a solution of (2.6)

Thus to solve (2.6), with $s \geq 2$, we start with the solutions, if any, $x_{1,j}$, of

$$f(x) \equiv 0 \pmod{p}.$$

Fix one of them, say x_{1,j_i}, and solve the congruence (2.8) with $s = 2$. This gives us solutions of (2.6) with $s = 2$, namely,

$$x_{2,k} = x_{1,j_i} + v_1 p,$$

one for each root v_1 of (2.8). We then take these solutions, $x_{2,k}$, solve for v_2 in (2.8) and get the solutions of (2.6) with $s = 3$. We continue in this way until we get to the desired value of s.

Note that we have $x_{s,k} \equiv x_{s-1,j_i} \pmod{p}$, for $s \geq 2$. Thus, instead of (2.8), we can solve for v_{s-1} from the equivalent congruence

(2.9)
$$f(x_{1,j_i})v_{s-1} \equiv -p^{1-s}f(x_{s-1,k}) \pmod{p}.$$

This allows us to build up to the solutions of (2.6).

Example 2.11: Solve the following congruences.

(a) $x^2 + x + 7 \equiv 0 \pmod{27}$.

(b) $x^2 + x + 7 \equiv 0 \pmod{81}$.

(c) $x^2 + x + 7 \equiv 0 \pmod{343}$.

Solution of (a) By trial we find that $x \equiv 1 \pmod 3$ is the only solution to $x^2 + x + 7$ $\equiv 0 \pmod 3$. Let $f(x) = x^2 + x + 7$. Then $f'(x) = 2x + 1$ and $f'(1) \equiv 0 \pmod 3$. There is only one x_1 and (2.9) reduces to

$$0 \equiv -3^{1-s} f(x_{s-1,k}) \pmod 3.$$

Thus there are no solutions, v_{s-1} if $f(x_{s-1,k}) \not\equiv 0 \pmod{3^s}$ and that $v_{s-1} \equiv 0,\ 1,\ -1$ $\pmod 3$ if $f(x_{s-1,k}) \equiv 0 \pmod{3^s}$. We find

$$x_{1,1} \equiv 1 \pmod 3, f(x_{1,1}) = 9, \text{ and so } v_1 \equiv -0,1,-1 \pmod 3;$$
$$x_{2,1} \equiv 1 \pmod 9, f(x_{2,1}) = 9, \text{ and so no } v_2;$$
$$x_{2,2} \equiv 4 \pmod 9, f(x_{2,2}) = 27, \text{ and so } v_2 \equiv 0,1,-1 \pmod 3;$$
$$x_{2,3} \equiv -2 \pmod 9, f(x_{2,3}) = 9, \text{ and so no } v_2;$$
$$x_{3,1} \equiv 4 \pmod{27};$$
$$x_{3,2} \equiv 13 \pmod{27};$$
$$x_{3,3} \equiv -5 \pmod{27};$$

Solution of (b) Continuing from (a) we see that $f(x_{3,1}) = 27$, $f(x_{3,2}) = 189$ and $f(x_{3,3}) = 27$. Thus the congruence has no solutions, since

$$27,189 \not\equiv 0 \pmod{81}.$$

Solution of (c) We find that the solution to $f(x) \equiv 0 \pmod 7$ are $x \equiv 0, -1 \pmod 7$. Also $f'(0) = 1$ and $f'(-1) = -1$. Thus there will be just one $x_{s,1}$ corresponding to $x_{1,1} = 0$ and one $x_{s,2}$ corresponding to $x_{1,2} = -1$. The congruences (2.9) become

$$v_{s-1} \equiv -7^{1-s} f(x_{s-1,1}) \pmod 7$$

and

$$v_{s-1} \equiv -7^{1-s} f(x_{s-1,2}) \pmod 7$$

We find

$$x_{1,1} = 0, f(x_{1,1}) = 7, v_1 = -1, x_{2,1} = -7, f(x_{2,1}) = 49, v_2 = -1, x_{3,1} = -56$$

and

$$x_{1,2} = -1, f(x_{1,2}) = 7, v_1 = 1, x_{2,2} = 6, f(x_{2,2}) = 49, v_2 = 1, x_{3,2} = 55.$$

Thus the solutions to $x^2 + x + 7 \equiv 0 \pmod{73}$ are

$$x \equiv -56 \text{ or } 55 \pmod{73}. \qquad \square$$

Problem Set 2.6

1. Solve the following congruences.
   (a) $x^5 - 6x^4 - 3x^3 - 2x^2 + 3x \equiv 0 \pmod 9$.
   (b) $x^4 + 6x^2 + 5 \equiv 0 \pmod{16}$.
   (c) $x^3 + 2x^2 + 4x + 8 \equiv 0 \pmod{125}$.
   (d) $x^3 + 3x^2 + x + 5 \equiv 0 \pmod{1125}$.

2. Use the results of this section to solve the congruence

   $$ax \equiv 1 \pmod{p^s},$$

   where a is an integer relatively prime to p and s is a positive integer.

3. Assume that w is a solution of the polynomial congruence

   $$f(x) \equiv 0 \pmod p.$$

   Prove that a complete list of candidates for solutions of the congruence

   $$f(x) \equiv 0 \pmod{p^2}$$

   can be taken to be $\{w, w + p, w + 2p, \ldots, w + (p - 1)p\}$.

4. Generalize the method of this section and show that if m is a positive integer, then for all positive integers n we have

   $$(u + mv)^n \equiv u^n + nu^{n-1}mv \pmod{m^2}.$$

   Can you generalize this result to higher powers?

5. (a) Let $f(x)$ be a polynomial with integer coefficients and let p be a prime. Show that if

   $$f(w) \equiv 0 \pmod{p^n}$$

   for all positive integers n, then $f(w) = 0$.

   (b) Generalize (a) by showing that if there is an finite set A such that for each positive integer n there is an x_n in A such that

   $$f(x_n) \equiv 0 \pmod{p^n},$$

   then there is a w in A such that $f(w) = 0$.

6. (a) Let $h > 0$ be the number of elements in a complete solution to

   $$f(x) \equiv 0 \pmod{p^2}$$

   and suppose that $p \mid h$. Show that the congruence

$$f(x) \equiv 0 \ (\text{mod } p^n)$$

has a solution for all positive integers n.

(b) For what positive integers h does there exist a polynomial f such that a complete solution to the congruence

$$f(x) \equiv 0 \ (\text{mod } 25)$$

has exactly h solutions.

7. Suppose that $f(x) \equiv 0 \ (\text{mod } p^k)$ has $x = a$ as a solution and that $p \nmid f'(a)$. Let A be an integer so that

$$Af'(a) \equiv 1 \ (\text{mod } p^{2k}).$$

If $y = a - Af(a)$, then $f(y) \equiv 0 \ (\text{mod } p^{2k})$.

8. Show that the congruence

$$x^{p-1} - 1 \equiv 0 \ (\text{mod } p^e)$$

has $p - 1$ solutions.

2.7. PRIME MODULI

We now must solve the congruence

(2.10) $$f(x) \equiv 0 \ (\text{mod } p)$$

in order to finish the general problem. Unfortunately, there are no general methods available at this stage and so we shall have to content ourselves with making a few general statements regarding the congruence (2.10).

We take

$$f(x) = a_n x^n + a_{n-1} x^{n-1} + \cdots a_1 x + a_0,$$

where, for $0 \leq i \leq n$, the a_i are integers and $a_n \not\equiv 0 \ (\text{mod } p)$.

Theorem 2.15: If $n \geq p$, then every integer is a solution of (2.10) or else there exists a polynomial $g(x)$, with integer coefficients and leading coefficient 1, such that the degree of $g(x)$ modulo p is less than p and the solutions of

$$g(x) \equiv 0 \ (\text{mod } p)$$

are precisely the same as those of (2.10).

Proof: By Fermat's Theorem we know that the congruence

$$x^p - x \equiv 0 \pmod{p}$$

has every integer as a solution. Since $n \geq p$, we may write

$$f(x) = (x^p - x)q(x) + r(x),$$

where $q(x)$ and $r(x)$ are polynomials with integer coefficients and either $r(x)$ is identically zero or $\deg(r) < p$. Thus

$$f(u) \equiv r(u) \pmod{p}$$

for every integer u. If $r(x)$ is identically zero or every coefficient of $r(x)$ is divisible by p, then every integer is a solution of (2.10).

The only other possibility is that

$$r(x) = b_m x^m + \cdots + b_0,$$

where $m \geq 1$, $b_m \neq 0$ and at least one coefficient is not divisible by p. Let b_k be the coefficient of largest subscript k such that $(b_k, p) = 1$. Then there is an integer b such that $bb_k \equiv 1 \pmod{p}$. Clearly $r(x) \equiv 0 \pmod{p}$ and $br(x) \equiv 0 \pmod{p}$ have the same solutions. If we take $g(x) = x^k + b(b_{k-1}x^{k-1} + \cdots + b_0)$ the result follows. ∎

Example 2.12: Take $p = 7$ and $f(x) = 3x^{11} + 2x^8 + 5$.

Solution: If we do the division long hand we find that

$$3x^{11} + 2x^8 + 5 = (x^7 - x)(3x^4 + 2x) + 3x^5 + 2x^2 + 5.$$

Since $3 \cdot 5 \equiv 1 \pmod{7}$ we can take, for the $g(x)$ of Theorem 2.15,

$$g(x) = x^5 + 3x^2 + 4.$$

The importance of this theorem is that it allows us to work with polynomials of lower degree than we originally started with if we so desire. Our next theorem gives us an upper bound on the number of possible solutions to the congruence (2.10).

Theorem 2.16 (Lagrange): If $f(x)$ is a polynomial of degree n, then (2.10) has at most n solutions.

Proof. The proof will be by induction on n.

If $n = 0$, then $f(x) = a_0$ and since $a_0 \not\equiv 0 \pmod{p}$ there are no solutions of (2.10) in this case.

If $n = 1$, then the congruence is $a_1 x + a_0 \equiv 0 \pmod{p}$, which, by Theorem

2.9, has exactly one solution modulo p.

Assume that the result holds for polynomial congruences (2.10) with the $\deg(f) < n$. Suppose now that $\deg(f) = n$ and that (2.10) has more than n solutions, say $u_1, u_2, ..., u_n, u_{n+1}$ are $n + 1$ of them with $u_j \not\equiv u_i \pmod p$, $1 \leq i < j \leq n + 1$.

Let $g(x) = f(x) - a_n(x - u_1)\cdots(x - u_n)$. Since the $a_n x^n$ coefficient cancels we see that $\deg(g) < n$ or $g(x)$ is identically zero. The degree of the congruence

$$(2.11) \qquad\qquad\qquad g(x) \equiv 0 \pmod p$$

may be less than $\deg(g)$ since the first terms of $g(x)$ may have coefficients divisible by p. Thus, if h is the degree of the congruence (2.11) and $k = \deg(g)$, we have $h \leq k < n$. But (2.11) has the solutions $u_1, ..., u_n$ which is impossible by the induction hypothesis unless either $g(x)$ is identically zero or all of its coefficients are divisible by p. In either case, we have

$$g(u) \equiv 0 \pmod p$$

for all integers u. Thus, for all integers u,

$$f(u) \equiv a_n(u - u_1)\cdots(u - u_n) \pmod p$$

and in particular,

$$0 \equiv f(u_{n+1}) - a_n(u_{n+1} - u_1)\cdots(u_{n+1} - u_n) \pmod p.$$

This contradicts Lemma 1.16.1 since $a_n \not\equiv 0 \pmod p$ and all the u_i, $1 \leq i \leq n + 1$, are incongruent modulo p.

Thus it must be that (2.10) has no more than n solutions.   ∎

Corollary 2.16.1: If

$$b_n x^n + \cdots + b_0 \equiv 0 \pmod p$$

has more than n solutions, then all the coefficients, b_j, $0 \leq j \leq n$, must be divisible by p.

This follows from the proof of the theorem.

The following corollary of Lagrange's theorem relates to a type of polynomial congruence that we will deal with in Section 9 below.

Corollary 2.16.2. Let p be a prime and let d be an integer such that $d \mid p - 1$. Then the congruence

$$x^d - 1 \equiv 0 \pmod p$$

has exactly d distinct incongruent solutions modulo p.

Proof. Since $d \mid p - 1$ there is a polynomial $g(x)$, with integer coefficients and degree $p - 1 - d$ such that

(2.12) $$x^{p-1} - 1 = (x^d - 1)g(x).$$

Since p is a prime the congruence

$$x^{p-1} - 1 \equiv 0 \ (\mathrm{mod}\ p)$$

has $p - 1$ incongruent solutions, by Fermat's theorem. Thus, by (2.12), we see that all of these solutions must satisfy either the congruence

$$x^d - 1 \equiv 0 \ (\mathrm{mod}\ p)$$

or the congruence

$$g(x) \equiv 0 \ (\mathrm{mod}\ p)$$

The first congruence has at most d solutions by Lagrange's theorem and the second congruence has at most $p - 1 - d$ solutions. If the first solution has less than d incongruent solutions, the congruence $x^{p-1} - 1 \equiv 0 \ (\mathrm{mod}\ p)$ would have less than $d + (p - 1 - d) = p - 1$ incongruent solutions in contradiction to Fermat's theorem. Thus the congruence $x^d - 1 \equiv 0 \ (\mathrm{mod}\ p)$ must have exactly d incongruent solutions. ∎

Theorem 2.17: The congruence (2.10), where $a_n = 1$, has exactly n solutions if and only if $f(x)$ is a factor of $x^p - x$ modulo p, that is,

(2.13) $$x^p - x = f(x)q(x) + ps(x),$$

where $q(x)$ and $s(x)$ are polynomials with integer coefficients and either $s(x)$ is identically zero or it is a polynomial of degree less than n.

Proof: If (2.10) has n solutions, then we must have $n \leq p$ since any residue system $(\mathrm{mod}\ p)$ has p elements in it.

If we divide $x^p - x$ by $f(x)$ we get

$$x^p - x = f(x)q(x) + r(x),$$

where $r(x)$ is identically zero or degree of $r(x)$ is less than n. If u is a solution of (2.10), then, by Fermat's Theorem, $u^p - u \equiv 0 \ (\mathrm{mod}\ p)$. Thus $r(u) \equiv 0 \ (\mathrm{mod}\ p)$ for each solution u. If $r(x)$ is not identically zero, then $r(x)$ is a polynomial of degree less than n such that the congruence $r(x) \equiv 0 \ (\mathrm{mod}\ p)$ has n solutions. Thus, by Corollary 2.16.1, all the coefficients of $r(x)$ must be divisible by p, that is,

$$r(x) = ps(x),$$

where $s(x)$ is a polynomial with integer coefficients.

Conversely, if (2.13) holds, then $f(u)q(u) = u^p - u - ps(u) \equiv 0$ (mod p), for all integers u. Thus $f(x)q(x) \equiv 0$ (mod p) has p solutions. We have that $q(x)$ is a polynomial of degree $p - n$ with leading coefficient not divisible by p. Thus, by Theorem 2.16, the congruence

$$q(x) \equiv 0 \ (\text{mod } p)$$

has $p - n$ solutions, say $v_1, ..., v_k$ with $k \le p - n$. If u is any one of the other $p - k$ residues modulo p, then $(q(u), p) = 1$ and $f(u)q(u) \equiv 0$ (mod p). Thus for these u we have $f(u) \equiv 0$ (mod p), by Lemma 1.16.1. Thus (2.10) has at least $p - k > p - (p - n) = n$ solutions. Combined with Theorem 2.16 we see that (2.10) has exactly n solutions. ∎

Example 2.13. Illustrate Theorem 2.17 when $f(x) = x^2 + 3x + 2$ and $p = 5$.

Solution. By trial and error we see that the solutions to $f(x) \equiv 0$ (mod 5) are $x \equiv 3$ or 4 (mod 5). Thus the congruence has two incongruent solutions and we find that

$$x^5 - x = (x^2 + 3x + 2)(x^3 - 3x^2 + 7x - 15) + 5(3x + 1). \qquad \square$$

We need $a_n = 1$ only to be able to divide $x^p - x$. Since $a_n \not\equiv 0$ (mod p) we have, by Theorem 2.9, that there exists an integer a such that

$$aa_n \equiv 1 \ (\text{mod } p).$$

If we let $g(x) = af(x) - (aa_n - 1)x^n$, then the congruence

$$g(x) \equiv 0 \ (\text{mod } p)$$

has the same solutions as the congruence (2.10) and $g(x)$ has leading coefficient 1. Thus $f(x)$ has exactly n solutions if and only if $g(x)$ divides $x^p - x$ modulo p.

As an application of these results we shall prove the following theorem.

Theorem 2.18 (Wolstenhome): Let $p > 3$ be a prime. Then the numerator of the fraction

$$1 + \frac{1}{2} + \frac{1}{3} + \cdots + \frac{1}{p-1}$$

is divisible by p^2.

Proof: Let A_n be the nth elementary symmetric polynomial on the numbers 1, 2,..., $p - 1$. Then we have

(2.14) $(x - 1)(x - 2)\cdots(x - (p - 1)) = x^{p-1} - A_1x^{p-2} + \cdots + A_{p-1}.$

If $p{\not|}x$, then, by Fermat's Theorem, $x^{p-1} \equiv 1 \pmod{p}$. Thus, as in the proof of Theorem 2.16, we have

$$x^{p-1} - 1 \equiv (x - 1)(x - 2) \cdots (x - (p - 1)) \pmod{p}.$$

Thus

$$(x^{p-1} - 1) - (x^{p-1} - A_1x^{p-2} + \cdots + A_{p-1}) \equiv 0 \pmod{p}.$$

Wilson's Theorem, in the present notation, states that

$$A_{p-1} \equiv -1 \pmod{p}.$$

Thus, we have

$$A_1x^{p-2} + \cdots + A_{p-2}x \equiv 0 \pmod{p}$$

for all x. Thus, by Corollary 2.16.1, we must have $p \,|\, A_i,\ 1 \le i \le p - 2$.

Let $x = p$ in (2.14). Then we have

$$(p - 1)! = p^{p-1} - A_1p^{p-2} + \cdots - A_{p-2}p + A_{p-1}.$$

Since $A_{p-1} = (p - 1)!$ we see that

$$p^{p-2} - A_1p^{p-3} + \cdots + A_{p-2} = 0.$$

Since $p > 3$ and $p \,|\, A_i,\ 1 \le i \le p - 2$, we have $p^2 \,|\, A_{p-2}$. Since

$$A_{p-2} = (p-1)!\left(1 + \frac{1}{2} + \cdots + \frac{1}{p-1}\right)$$

and $p{\not|}p - 1$ we see that p^2 must divide the numerator of

$$1 + \frac{1}{2} + \frac{1}{3} + \cdots + \frac{1}{p-1}.$$

This completes the proof.   ■

Problem Set 2.7

1. Deduce Wilson's Theorem (Corollary 2.9.1) from the result of equation (2.15).

2. Let

$$S = \sum_{1 \le i < j \le p-1} ij = 1 \cdot 2 + 1 \cdot 3 + \cdots + 1 \cdot (p-1) + 2 \cdot 3 + 2 \cdot 4 + \cdots + (p-2)(p-1).$$

(a) Deduce from the result that $p \,|\, A_i$ that

$$S \equiv 0 \pmod{p}.$$

(b) Derive as closed form expression for S and then prove the result of (a).

3. Illustrate the remarks after Theorem 2.17 with $f(x) = 2x + 7$ and $p = 7$.

4. Use Corollary 2.16.1 and the methods of the previous section to show that if p is a prime and d is a positive integer such that $d \mid p - 1$, then the congruence

$$x^d - 1 \equiv 0 \pmod{p^n}$$

has exactly d incongruent roots modulo p^n for all positive integer values of n.

5. (a) Show that if a polynomial of degree n with integer coefficients has values at $n + 1$ consecutive integers that are divisible by a prime p, then all the coefficients of the polynomial are divisible by p.

   (b) Give an example to show that one cannot delete the requirement that the values of the function be at consecutive integers.

6. Let $f(x) = a_n x^n + a_{n-1} x^{n-1} + \cdots a_1 x + a_0$ be a polynomial of degree n with integer coefficients. Suppose that the congruence

$$f(x) \equiv 0 \pmod{p}$$

has the n incongruent solutions $x \equiv x_1, \ldots, x_n \pmod{p}$. Prove that the analogues of Viete's formulas hold, namely

$$a_{n-1} \equiv -a_n S_1 \pmod{p},$$
$$a_{n-2} \equiv a_n S_2 \pmod{p},$$
$$\vdots$$
$$a_0 \equiv (-1)^n a_n S_n \pmod{p},$$

where S_i is the ith symmetric polynomial on the $x_1, \ldots, x_n$

7. Show that, in the notation of the proof of Theorem 2.18,

$$A_{p-2} \equiv p A_{p-3} \pmod{p^3}.$$

2.8. PRIMITIVE ROOTS

We now turn to the solution of a special type of congruence, namely,

(2.16) $$ax^n \equiv b \pmod{m},$$

where n, a, b, and m are given integers, with n and m positive. As we have seen to solve (2.16) we need only solve

(2.17) $$ax^n \equiv b \pmod{p}$$

for all primes p which divide m. Unlike the previous state of affairs for general polynomials, for the congruence (2.17) we can give a necessary and sufficient condition for its solvability, determine the number of solutions and give a procedure for solving the congruence when it is solvable. We begin with some preliminary results.

Definition 2.8: If $x^n \equiv a$ (mod m) has a solution, then a is called an n<u>th</u> **power residue modulo** m.

Definition 2.9: Let m be a positive integer and a be any integer relatively prime to m. If h is the least positive integer such that

$$a^h \equiv 1 \pmod{m},$$

then h is called the **order of** a **modulo** m or one says that a **belongs to the exponent** h **modulo** m.

Theorem 2.19: If a belongs to the exponent h modulo m, then $h \mid \varphi(m)$. Moreover, $a^j \equiv a^k$ (mod m) if and only if $h \mid (j - k)$.

Proof: By Euler's Theorem we know that if $(a, m) = 1$, then

$$a^{\varphi(m)} \equiv 1 \pmod{m}.$$

Thus by definition 2.9, $h \le \varphi(m)$. Write

$$\varphi(m) = hq + r, \ \ 0 \le r < h.$$

Then

$$a^r \equiv a^r (a^h)^q = a^{r+hq} = a^{\varphi(m)} \equiv 1 \pmod{m}.$$

Thus $r = 0$, since h is the least positive integer with this property and so $h \mid \varphi(m)$.

To prove the second part we may assume $j > k$. Since $(a, m) = 1$, we have $(a^k, m) = 1$. Thus

$$a^j \equiv a^k \pmod{m} \text{ if and only if } a^{j-k} \equiv 1 \pmod{m}.$$

The result then follows as in the proof of the first part.   ∎

Thus to find the order of an integer modulo m we need only check through the divisors of $\varphi(m)$. Note also that if $a \equiv b$ (mod m), then a and b have the same order modulo m. So to find the order of any integer a modulo m, when $(a, m) = 1$, we can find the least residue of a modulo m, which by Theorem 2.4 is also relatively prime to m, and calculate its order modulo m.

Example 2.13: Let $m = 12$. We will determine the order of the numbers is the least reduced residue system modulo 12, which is $\{1, 5, 7, 11\}$. We see $\varphi(12) = 4$, and so the order of any of these numbers must be a divisor of 4, that is, either 1, 2 or 4. We find that 1 has order 1, 5 has order 2, 7 has order 2 and 11 has order 2.   □

As an application of Theorem 2.19 we give another primality test before returning to our main topic.

Theorem 2.20: Let p be an odd prime, $h < p$, $n = hp + 1$ or $hp^2 + 1$, where $2^h \not\equiv 1 \pmod{n}$ and $2^{n-1} \equiv 1 \pmod{n}$. Then n is a prime.

Proof: We write $n = hp^b + 1$, where $b = 1$ or 2.
Let d be the order of 2 modulo n. Then, by Theorem 2.19, we have that $d \mid h$ and $d \mid n - 1$. Thus $d \mid hp^b$ and so, by Theorem 1.16, since $b \geq 1$, we have $p \mid d$. Also, by Theorem 2.19, $d \mid \varphi(n)$, and so $p \mid \varphi(n)$. Suppose that $n = p_1^{a_1} \cdots p_k^{a_k}$. Then, by Theorem 2.6,

$$\varphi(n) = p_1^{a_1-1} \cdots p_k^{a_k-1}(p_1 - 1) \cdots (p_k - 1).$$

Since $p \mid n - 1$ we have $p \nmid n$. Thus p is not one of the p_i, $1 \leq i \leq k$, and so, by Lemma 1.16.1, we must have that $p \mid (p_i - 1)$ for some i, $1 \leq i \leq k$. Thus n has a prime factor P such that $P \equiv 1 \pmod{p}$. Let $n = Pm$. Since $n \equiv 1 \equiv P \pmod{p}$ we must have $m \equiv 1 \pmod{p}$. If $m > 1$, then

$$n = (up + 1)(vp + 1),$$

where $1 \leq u \leq v$, and so

(2.18) $hp^{b-1} = uvp + u + v.$

If $b = 1$, then, from (2.18), we see that

$$h = uvp + u + v.$$

Thus $p \leq uvp < h < p$, which is a contradiction. Thus, in the case $b = 1$, we must have $m = 1$, that is, n is a prime.

If $b = 2$, then, from (2.18), we have

$$hp = uvp + u + v.$$

Thus $p \mid u + v$, and so, by (g) of Theorem 1.1, we have $u + v \geq p$. Thus $2v \geq u + v \geq p$ or $v > p/2$. Also $uv < h < p$. Thus

$$uv \leq p - 2 \text{ or } u \leq (p-2)/v < 2(p-2)/p < 2.$$

Thus $u = 1$ and so $v \geq p - 1$ and $uv \geq p - 1$, which is a contradiction. Thus, in the case $b = 2$, we must have $m = 1$, that is, n is a prime.   ∎

Example 2.14. Take $p = 5$, $b = 2$. Then the possible list of h to be used is $h = 1$, 2, 3, 4 and the corresponding n are 26, 51, 76, and 101. The only possible candidate for primality is 101, since 26 and 76 are even and 3 divides 51. Here $h = 4$. We have

$2^4 = 16 \equiv 1 \pmod{101}$. To show that 101 is a prime we must show that $2^{100} \equiv 1$ (mod 101). By using the various parts of Theorem 2.1 we have the following table of computations.

$$2^7 = 128 \equiv 27 \pmod{101}$$
$$2^{14} \equiv 27^2 \equiv 22 \pmod{101}$$
$$2^{28} \equiv 22^2 \equiv 80 \pmod{101}$$
$$2^{56} \equiv 80^2 \equiv 37 \pmod{101}$$
$$2^{70} \equiv 22 \cdot 37 \equiv 6 \pmod{101}$$
$$2^{100} \equiv 2^2 \cdot 2^{28} \cdot 2^{70} \equiv 4 \cdot 6 \cdot 80 \equiv 1 \pmod{101}.$$

Thus 101 is a prime.   □

Theorem 2.21: If a belongs to the exponent h mod m, then a^k belongs to the exponent $h/(h, k)$ modulo m.

Proof. By Theorem 2.19 $(a^k)^j \equiv 1 \pmod{m}$ if and only if $h \mid kj$. But $h \mid kj$ if and only if $\{h/(h,k)\} \mid \{k/(h,k)\}j$, which happens if and only if $\{h/(h,k)\} \mid j$, by Corollary 1.14.2.

By definition, if j is the order of a^k, then it is the least positive integer such that $(a^k)^j \equiv 1 \pmod{m}$. Thus $j = h/(h, k)$.   ∎

Definition 2.10: If a belongs to the exponent $\varphi(m)$ modulo m, then a is called a **primitive root** modulo m.

Example 2.15. Find the primitive roots of 7 in the least reduced residue system.

Solution. Since 7 is a prime the least reduced residue system is $\{1, 2, 3, 4, 5, 6\}$. We compute the following table:

n	exponent to which n belongs
2	
3	6
4	3
5	6
6	2

Thus 3 and 5 are the primitive roots of 7.   □

    The following theorem tells us which numbers can have primitive roots.

Theorem 2.22: If p is an odd prime, then there are $\varphi(p-1)$ primitive roots mod p. The only integers having primitive roots are 1, 2, 4, p^e and $2p^e$, where e is a positive integer and p is an odd prime.

Proof. Each integer a, $1 \leq a \leq p - 1$, belongs to some exponent h modulo p, where $h \mid (p - 1)$. If a belongs to h, then $(a^k)^h \equiv 1 \pmod{p}$ for all positive integers k. By Theorem 2.20 the numbers $1, a, a^2, ..., a^{h-1}$ are all incongruent modulo p. Thus, by Theorem 2.16, these h numbers are all the solutions to

$$x^h \equiv 1 \pmod{p}.$$

By Theorem 2.21 just $\varphi(h)$ of these numbers belong to the exponent h modulo p and the others belong to smaller exponents modulo p. Also any integer a that belongs to h modulo p is a solution of $x^h \equiv 1 \pmod{p}$. Thus for every integer h such that $h \mid p - 1$ there will be either $\varphi(h)$ or no a, $1 \leq a \leq p - 1$, such that a belongs to the exponent h modulo p. Let $\psi(h)$ denote the number of integers in $1 \leq a \leq p - 1$ that belong to the exponent h modulo p. Then, for all $h \mid p - 1$, we have

(2.19) $\psi(h) \leq \varphi(h).$

Also

$$\sum_{h \mid p-1} \psi(h) = p - 1 = \sum_{h \mid p-1} \varphi(h),$$

by Theorem 2.7 and the fact that every integer belongs to some exponent modulo p (there can only be at most p incongruent values of a^k as k runs through the positive integers). Thus

$$\sum_{h \mid p-1} \left(\psi(h) - \varphi(h) \right) = 0.$$

By (2.19), $\psi(h) - \varphi(h) \leq 0$ for all $h \mid p - 1$. Thus we must have

$$\psi(h) = \varphi(h)$$

for all $h \mid p - 1$. In particular $\psi(p-1) = \varphi(p-1) > 0$; that is, there are $\varphi(p-1)$ primitive roots modulo p, which proves the first part of the theorem.

From Theorem 2.6, we see that if $n > 2$, then $\varphi(n)$ is even. Let

$$m = 2^f \prod_{i=1}^{k} p_i^{e_i},$$

where f is a nonnegative integer $k \geq 1$ and the e_i, $1 \leq i \leq k$, are positive integers. If $(a, m) = 1$, then we have

$$a^{\varphi(p_i^{e_i})} \equiv 1 \pmod{p_i^{e_i}}$$

and

$$a^{\varphi(m/p_i^{e_i})} \equiv 1 \pmod{m / p_i^{e_i}}.$$

Suppose that either $k \geq 2$ or $k = 1$ and $f \geq 2$. Then both $\varphi(p_i^{e_i})$ and $\varphi(m / p_i^{e_i})$ are even, and so $\frac{1}{2}\varphi(p_i^{e_i})\varphi(m / p_i^{e_i})$ is an integer for which we have

$$a^{\frac{1}{2}\varphi(p_i^{e_i})\varphi(m/p_i^{e_i})} \equiv 1 \pmod{p_i^{e_i}}$$

and

$$a^{\frac{1}{2}\varphi(p_i^{e_i})\varphi(m/p_i^{e_i})} \equiv 1 \pmod{m / p_i^{e_i}}.$$

Thus, by Theorem 2.2 and Corollary 2.6.1, we have

$$a^{\frac{1}{2}\varphi(m)} \equiv 1 \pmod{m}$$

for any integer a such that $(a, m) = 1$. Thus an integer m cannot have primitive roots if $k \geq 2$ or if $k = 1$ and $f \geq 2$.

We now show that 2^n has no primitive root if $n \geq 3$. Let a be an odd integer, say $a = 2b + 1$, where b is a nonnegative integer. Then

$$a^2 = 4b(b + 1) + 1 = 8c + 1,$$

$$a^4 = 64c^2 + 16c + 1 = 16d + 1,$$

etc., where, in general, an easy induction argument will show that

$$a^{2^{n-2}} = 1 + 2^n g$$

or $n \geq 3$. If $n \geq 3$, then $\varphi(2^n) = 2^{n-1} > 2^{n-2}$. Since

$$a^{2^{n-2}} \equiv 1 \pmod{2^n}$$

for $n \geq 3$ we see that we can have no primitive roots in this case either.

Thus the only possible moduli possessing primitive roots are $m = 1, 2, 4, p^e$ or $2p^e$, where $e \geq 1$ and p is an odd prime.

Note that 3 is a primitive root of 4 and that 1 is a primitive root of both 1 and 2.

Let $m = p^e$, $e \geq 2$, and let a be a primitive root modulo p, which exists by the first part of the theorem. Let $b = a + pt$. Then, by the Binomial Theorem, we have

$$b^{p-1} \equiv a^{p-1} + (p - 1)a^{p-2}pt \pmod{p^2}.$$

Choose t so that $b^{p-1} = 1 + n_1 p$ with $n_1 \not\equiv 0 \pmod p$. Since $a^{p-1} \equiv 1 \pmod p$ we have that $a^{p-1} = 1 + np$ for some integer n. Thus $b^{p-1} \equiv 1 + np + p(p - 1)a^{p-2}t \pmod{p^2}$ and so there exists an integer N such that

$$b^{p-1} = 1 + np + p(p - 1)a^{p-2}t + p^2N = 1 + p(n + (p - 1)\,a^{p-2}t + pN).$$

We have $n + (p - 1)a^{p-2}t + pN \equiv n - a^{p-2}t \pmod{p}$. Thus we may take $t = 1$ if $p \mid n$, since $p \nmid a$, and we may take $t = 0$ if $p \nmid n$. Thus we see that we can take $b = a$ if $a^{p-1} \not\equiv 1 \pmod{p^2}$ and $b = a + p$ if $a^{p-1} \equiv 1 \pmod{p^2}$.

Note that $(1 + np^{j-1})^p \equiv 1 + np^j \pmod{p^{2j-1}}$ so that, by induction, we see that

$$b^{p^{j-1}} = 1 + n_j p^j,$$

where $n_j \equiv n_{j-1} \pmod{p^{j-1}}$. Then $n_j \equiv n_1 \not\equiv 0 \pmod{p}$. For $e \geq 2$ let h be the order of $b \bmod p^e$. Then $h \mid p^{e-1}(p - 1)$, and so, by Theorem 1.16, we have that $h = p^s d$, where $0 \leq s \leq e - 1$ and $d \mid p - 1$. Thus, since $d \mid p - 1$,

$$b^{p^s(p-1)} \equiv 1 \pmod{p^e}$$

and so, since $b^{p^s(p-1)} = 1 + n_{s+1}p^{s+1}$, we have

$$1 + n_{s+1}p^{s+1} \equiv 1 \pmod{p^e}.$$

Thus $s \geq e - 1$, and so $s = e - 1$. Also, since, by Fermat's Theorem, $b^{p^s} \equiv b \pmod{p}$, we have

$$b^d \equiv b^{p^s d} \equiv 1 \pmod{p}.$$

Since $b \equiv a \pmod{p}$ and a is a primitive root modulo p, we have, by Theorem 2.19, that $(p - 1) \mid d$. Thus $d = p - 1$ and so $h = \varphi(p^e)$. Thus b is a primitive root modulo p^e.

Suppose $m = 2p^e$ and let a be a primitive root modulo p^e, which exists by the above. Let $b = a$ or $a + p^e$, whichever is odd. Then, for any positive integer h we have $b^h \equiv 1 \pmod{2}$. By Theorem 2 19, we have $b^h \equiv a^h \equiv 1 \pmod{p^e}$ if and only if $p^{e-1}(p - 1) \mid h$. Thus, by the Chinese Remainder Theorem, we see that b is a primitive root mod $2p^e$. ∎

Note that, by the proof, if g is a primitive root of p^2, then g is a primitive root of p^e for any $e \geq 2$.

Although Theorem 2.22 guarantees the existence of primitive roots, indeed $\varphi(p-1)$ of them, for any prime p it does not tell how to find them. The following procedure is due to Gauss and can be used to determine a primitive root g modulo p. The rest of them will then be found, by Theorem 2.21, among the numbers g^k, $1 \leq k \leq p - 1$, where $(k, p - 1) = 1$.

Let $1 < a < p$ (in practice one usually takes $a = 2$). Let a belong to the exponent e modulo p, where $e \mid p - 1$. If $e = p - 1$, we're done having found a primitive root. If $e < p - 1$, then there exists a b, $1 < b < p - 1$, such that b is not congruent to a positive power of a. Let b belong to the exponent f. If $f = p - 1$, we're done. If $f < p - 1$, then there are two cases:

$$\text{(I)} \quad f \not\equiv 0 \pmod{e}$$

and

$$\text{(II)} \quad f \equiv 0 \pmod{e}.$$

Case I. Let $m = [f, e]$. Then, by Corollary 1.19.3, we may write $m = e_1 f_1$ where $e_1 \mid e, f_1 \mid f$ and $(e_1, f_1) = 1$. Write $e = e_1 e_2$ and $f = f_1 f_2$. Then a^{e_2} belongs to the exponent e_1 and b^{f_2} belongs to the exponent f_1, by Theorem 2.21. Thus $a^{e_2} b^{f_2}$ belongs to the exponent $e_1 f_1 = m$ modulo p. Since e has a factor not dividing f we have $m > f$. Also $m \geq e$. If $m = e$, then $f \mid e$. If $f = e$, then we have a contradiction to $f \not\equiv 0 \pmod{e}$. If $f \mid e$ and $f < e$, then all incongruent solutions, $b, b^2, \ldots, b^f$, of $x^f \equiv 1 \pmod{p}$ are also solutions of $x^e \equiv 1 \pmod{p}$. However, all solutions to $x^e \equiv 1 \pmod{p}$ are congruent to one of $a, a^2, \ldots, a^e$. Since b is not congruent to a power of a this can't happen. Thus $m > e$. Thus we have a number, $a^{e_2} b^{f_2}$ belonging to the exponent m modulo p with $m > \max(e, f)$.

Case II. Here we have $f \geq e > 1$. Now $x^f \equiv 1 \pmod{p}$ has the f incongruent solutions $b, b^2, \ldots, b^f$. Also each of the e solutions of $x^e \equiv 1 \pmod{p}$, namely $a, a^2, \ldots, a^e$, is a solution of $x^f \equiv 1 \pmod{p}$, since $e \mid f$. If $f = e$, then b is a solution of $x^e \equiv 1 \pmod{p}$, and so congruent to a power of a. Thus we must have $f > e$. Thus we have a number, b, belonging to the exponent f mod p with $f > e$.

In either case we obtain a sequence of integers which belong to a strictly increasing sequence of exponents modulo p all of which, by Theorem 2.19, must divide $p - 1$. Thus we eventually reach an integer that has exponent $p - 1$, that is, we obtain a primitive root modulo p.

By the remark after the proof of Theorem 2.22, this suffices to find primitive roots of all numbers that possess them.

Example 2.16: Find a primitive root of the following numbers.

    (a)   71.

    (b) 71^2.

    (c) $2 \cdot 71^3$.

Solution: of (a) We take $a = 2$ and find that 2 belongs to the exponent 35 mod 71. The residues of the successive powers of 2 are

$$2, 4, 8, 16, 32, 64, 57, 43, 15, 30, 60, 49, 27, 54, 37, 3, 6, 12, 24,$$
$$48, 25, 50, 29, 58, 45, 19, 38, 5, 10, 20, 40, 9, 18, 36, 1.$$

The smallest number not in the sequence is 7, which we find belongs to the exponent 70 and so is a primitive root of 71.

We could shorten this by noting that $70 \equiv -1 \pmod{71}$ and so 70 belongs to the exponent 2. Since 2 belongs to the exponent 35 and $(2, 35) = 1$ we see that $(-1) \cdot 2 \equiv 69 \pmod{71}$ belongs to the exponent $2 \cdot 35 = 70$, and so is a primitive root of 71. We could also select a number near 71 among the residues of the powers of 2. For example, $64 \equiv 26 \pmod{71}$ and since $(6, 35) = 1$ we see that 64 belongs to the exponent 5. Thus $(-1) \cdot 64 \equiv 7 \pmod{71}$ belongs to the exponent $2 \cdot 35 = 70$, and so is a primitive root of 71.

Solution of (b) Note that in finding a primitive root of 71^2 we are also finding a primitive root of 71^k for any $k \geq 2$. By the remark in the proof of Theorem 2.22 a primitive root of 71 will be 7 or $7 + 71 = 78$ depending on whether $7^{70} \not\equiv 1 \pmod{71^2}$ or not. We find $7^{70} \equiv 1563 \not\equiv 1 \pmod{71^2}$. Thus 7 is a primitive root modulo 71^2.

Solution of (c) By (b) and proof of Theorem 2.22 we know that 7 is a primitive root of 71^3. Thus, by the proof of Theorem 2.22, we see that either 7 or $7 + 71^3$ is a primitive root of $2 \cdot 71^3$, whichever is odd. Thus 7 is a primitive root of $2 \cdot 71$. $\square$

In Table IV at the end of the book we give a list of the least positive primitive roots for all primes less than 1000.

By the proof of Theorem 2.22 we know that if p is an odd prime and $h \mid p - 1$, then there are $\varphi(h)$ integers a, $1 \leq a \leq p - 1$, belonging to the exponent h modulo p. We can use Theorem 2.21 and the existence of primitive roots to find all the integers of a given exponent. We illustrate this with the following example.

Example 2.18. Find all integers of order 5 modulo 71.

Solution. By Example 2.17, we know that 7 is a primitive root modulo 71. Now, by the remarks after Theorem 2.22, we know that $1 \leq a \leq 70$ implies there is a k, $0 \leq k \leq 70$, such that

$$7^k \equiv a \pmod{71}.$$

By Theorem 2.21, we know that the order of 7^k is $70/(k, 70)$ modulo 71. If we want this order to be 5, then we need $(k, 70) = 14$. Thus $k = 14, 28, 42$ and 56. Since $\varphi(5) = 4$ we see that we have all of the numbers belonging to the exponent 5 modulo 71. We have

$$7^{14} \equiv 7^{2+4+8} \equiv 49 \cdot 58 \cdot 27 \equiv 54 \pmod{71}$$
$$7^{28} \equiv (7^{14})^2 \equiv 54^2 \equiv 5 \pmod{71}$$
$$7^{42} \equiv (7^{14})^3 \equiv 54^3 \equiv 57 \pmod{71}$$
$$7^{56} \equiv (7^{28})^2 \equiv 5^2 \equiv 25 \pmod{71}.$$

Thus the four numbers of order 5 modulo 71 are 5, 25, 54 and 57. ☐

Problem Set 2.8

1. Find the quadratic, cubic and quartic residues modulo 7; modulo 9.

2. (a) Find the exponents to which 1, 2, 3, 4, 5 and 6 belong modulo 7; modulo 11.

   (b) Find the exponents to which 11 and 19 belong modulo 35.

   (c) Find the exponents to which 10 and 26 belong modulo 37.

3. Let a belong to the exponent e_1 modulo m_1 and to the exponent e_2 modulo m_2. Show that a belongs to the exponent $[e_1, e_2]$ modulo $[m_1, m_2]$.

4. Let p and q be distinct odd primes and m an integer with $m \equiv a \pmod{p}$ and $m \equiv b \pmod{q}$. If a belongs to the exponent e modulo p and b belongs to the exponent f modulo q, then m belongs to the exponent $[e, f]$ modulo pq.

5. Let p be an odd prime. Then a belongs to the exponent 3 modulo p if and only if $a + 1$ belongs to the exponent 6 modulo p.

6. If a belongs to the exponent $2k$ modulo p, then $a^k \equiv -1 \pmod{p}$.

7. Let $n \mid p - 1$. Then the number of integers of order n modulo p is $\varphi((p-1)/n)$.

8. (a) Find all primitive roots of 41, 82 and 1681.

   (b) Find a primitive root of the prime 10007.

9. (a) If $ab \equiv 1 \pmod{m}$ and a belongs to the exponent h modulo m, then b belongs to the exponent h modulo m.

   (b) If $p > 3$, then the product of all primitive roots is congruent to 1 modulo p.

10. Let g be a primitive root modulo p. Then $p - g$ is a primitive root modulo p if $p \equiv 1 \pmod{4}$ and belongs to the exponent $(p - 1)/2$ if $p \equiv 3 \pmod{4}$.

11. Use primitive roots to prove Wilson's Theorem.

12. If m possesses a primitive root, then it possesses $\varphi(\varphi(m))$ primitive roots

13. Show that

$$1^k + 2^k + \cdots + (p-1)^k \equiv \begin{cases} 0 \pmod{p}, & \text{if } p - 1 \nmid k \\ -1 \pmod{p}, & \text{if } p - 1 \mid k \end{cases}.$$

14. (a) 2 is a primitive root modulo p if $p = 4q + 1$ and q is an odd prime.

    (b) 3 is a primitive of all primes p of the form $2^n + 1$, where $n > 1$.

(c) 2 is a primitive root of p if $p = 2q + 1$, q is a prime and $q \equiv 1 \pmod{4}$.

(d) $p - 2$ is a primitive root of p if $p = 2q + 1$, q is a prime and $q \equiv 3 \pmod{4}$.

15. If g is a primitive root modulo p^2, then g is a primitive root of p.

16. (a) Show that the odd prime divisors of $n^4 + 1$ are of the form $8k + 1$. (Hint: use Theorem 2.19.)

   (b) Show that there are infinitely many primes of the form $8k + 1$. (Hint: consider the number $(2p_1 \cdots p_r)^4 + 1$, where $p_1, \ldots, p_r$ is any collection of primes of the form $8k + 1$. Now apply (a).)

17. Show that if g_1 and g_2 are both primitive roots modulo p, then $g_1 g_2$ is not a primitive root modulo p.

18. If m does not possess primitive roots, then there is an integer a such that $(a, m) = 1$ and $\text{ord}_m a = 2$.

19. If $a > 1$ and $n > 0$, show that $n \mid \varphi(a^n - 1)$.

2.9. INDICES AND BINOMIAL AND EXPONENTIAL CONGRUENCES

The great usefulness of primitive roots is illustrated by the following theorem, which is a corollary of Theorem 2.19 .

Theorem 2.23: Suppose m has a primitive root g. Then $g^j \equiv g^k \pmod{m}$ if and only if $j \equiv k \pmod{\varphi(m)}$. In particular, we have $g^j \equiv 1 \pmod{m}$ if and only if $\varphi(m) \mid j$. The set $\{g, g^2, \ldots, g^{\varphi(m)}\}$ forms a reduced residue system modulo m so that if $(a, m) = 1$, then there exists a unique j, $1 \leq j \leq \varphi(m)$, such that $a \equiv g^j \pmod{m}$.

The integer j defined in the last sentence of Theorem 2.23 is called the **index of a, relative to g, modulo m** and is denoted by $j = \text{ind}_g a$. If the primitive root remains fixed in a particular discussion, then we will sometimes write $j = \text{ind} a$.

Example 2.17: Let $m = 9$. Then a primitive root of 9 is $g = 2$ and we have the following table of indices.

a	$\text{ind}_2 a$
1	0
2	1
4	2
5	5
7	4
8	3

As the following theorem will show there is a great similarity between indices and logarithms.

Theorem 2.24: If m has a primitive root g and if $(MN, m) = 1$,

(a) $\text{ind}_g M \equiv \text{ind}_g N \pmod{\varphi(m)}$ if and only if $M \equiv N \pmod m$,

(b) $\text{ind}_g (MN) \equiv \text{ind}_g M + \text{ind}_g N \pmod{\varphi(m)}$,

(c) $\text{ind}_g 1 \equiv 0 \pmod{\varphi(m)}$,

(d) $\text{ind}_g g \equiv 1 \pmod{\varphi(m)}$,

(e) $\text{ind}_g (M^k) \equiv k \text{ind}_g M \pmod{\varphi(m)}$, when k is an integer, and

(f) $\text{ind}_g (-1) \equiv \varphi(m) / 2 \pmod{\varphi(m)}$, if $m > 2$.

Proof: By definition we have

$$M \equiv g^{\text{ind}_g M} \pmod m \text{ and } N \equiv g^{\text{ind}_g N} \pmod m.$$

Thus, by Theorem 2.23, $M \equiv N \pmod m$ if and only if

$$g^{\text{ind}_g M} \equiv g^{\text{ind}_g N} \pmod m \text{ if and only if } \text{ind}_g M \equiv \text{ind}_g N \pmod{\varphi(m)}.$$

This proves (a).

By definition $MN \equiv g^{\text{ind}_g (MN)} \pmod m$. Also

$$MN \equiv g^{\text{ind}_g M} g^{\text{ind}_g N} = g^{\text{ind}_g M + \text{ind}_g N} \pmod m.$$

Thus, by (a), we have $\text{ind}_g (MN) \equiv \text{ind}_g M + \text{ind}_g N \pmod{\varphi(m)}$, which proves (b).

Since $g^{\varphi(m)} \equiv 1 = g^0 \pmod m$ and $g^1 \equiv g \pmod m$ we see that (c) and (d) follow from (a).

Since $(M, m) = 1$ we know that M^{-1} can be defined by $MM^{-1} \equiv 1 \pmod m$, and so if k is a positive integer, we have $M^{-k} \equiv (M^{-1})^k \pmod m$. Thus we may restrict ourselves to positive integers k, since $M^0 \equiv 1 \pmod m$. In this case we have

$$g^{\text{ind}_g M^k} \equiv M^k \equiv (g^{\text{ind}_g M})^k = g^{k \text{ind}_g M} \pmod m$$

and the result follows by (a).

If $m > 2$, then $\varphi(m)$ is even. If $m = 4$, then $g = 3$ and the result is true. If $m = p^e$, where $p > 3$, we have

(2.20) $(g^{\varphi(m)/2} + 1)(g^{\varphi(m)/2} - 1) = g^{\varphi(m)} - 1 \equiv 0 \pmod m.$

Since g is a primitive root we must have $g^{\varphi(m)/2} + 1 \equiv 0 \pmod m$ since $p \geq 3$ implies that p cannot divide both $g^{\varphi(m)/2} + 1$ and $g^{\varphi(m)/2} - 1$. If $m = 2p^e$, where $p \geq 3$, then we again have the congruence (2.20). Since g must be odd, both $g^{\varphi(m)/2} + 1$ and $g^{\varphi(m)/2} - 1$ are even, that is, divisible by 2. The same argument as before shows that

p^e can only divide one of the two terms $g^{\varphi(m)/2}+1$ and $g^{\varphi(m)/2}-1$. Since g is a primitive root it must be that $g^{\varphi(m)/2}+1\equiv 0 \pmod{m}$. Thus, in every case where $m>2$ has a primitive root, we have

$$g^{\varphi(m)/2}\equiv -1 \pmod{m}.$$

The result then follows by (a).   ∎

Since $\varphi(p)=p-1$ we have the following corollary for an odd prime p.

Corollary 2.24.1: Let p be an odd prime, g a primitive root of p and $(MN, p)=1$. Then we have the following results:

(a) $\text{ind}_g M \equiv \text{ind}_g N \pmod{p-1}$ if and only if $M \equiv N \pmod{p}$;

(b) $\text{ind}_g(MN) \equiv \text{ind}_g M + \text{ind}_g N \pmod{p-1}$;

(c) $\text{ind}_g(M^k) \equiv k\,\text{ind}_g M \pmod{p-1}$ for k an integer.

Sometimes it is useful to use a different primitive root than the one started with. This may be accomplished through the following "change of base" formula.

Theorem 2.25: If g_1 and g_2 are distinct primitive roots of m, then

(a) $\text{ind}_{g_1} N \equiv (\text{ind}_{g_2} M)(\text{ind}_{g_1} g_2) \pmod{\varphi(m)}$

and

(b) $(\text{ind}_{g_1} g_2)(\text{ind}_{g_2} g_1) \equiv 1 \pmod{\varphi(m)}$.

Proof: We have

$$N \equiv g_1^{\text{ind}_{g_1} N} \pmod{m} \text{ and } N \equiv g_2^{\text{ind}_{g_2} N} \pmod{m}.$$

Also

$$g_2 \equiv g_1^{\text{ind}_{g_1} g_2} \pmod{m}.$$

Thus

$$g_1^{\text{ind}_{g_1} N} \equiv N \equiv (g_1^{\text{ind}_{g_1} g_2})^{\text{ind}_{g_2} N} \equiv g_1^{(\text{ind}_{g_2} N)(\text{ind}_{g_1} g_2)} \pmod{m}$$

and (a) follows from (a) of Theorem 2.24.

Part (b) follows from (d) of Theorem 2.24 upon taking $N=g_1$ in (a).   ∎

Logarithms are of great help in solving certain types of exponential equations and in a similar way one can use indices to help solve certain types of congruences. To illustrate the general problems we will make use of the table of indices for the modulus 9, which was given above.

In the following we will assume the modulus m possesses a primitive root g and we will write $\text{ind}_g N$ as $\text{ind}\,N$.

Problem 1: Solve $ax^n \equiv b$ (mod m).

Solution: In terms of indices the given congruence is equivalent, by Theorem 2.24, to the congruence

$$\text{ind}a + n(\text{ind}x) \equiv \text{ind}b \ (\text{mod} \ \varphi(m))$$

or

$$n(\text{ind}x) \equiv \text{ind}b - \text{ind}a \ (\text{mod} \ \varphi(m)).$$

Thus, by Theorem 2.9, a necessary and sufficient condition for the solvability of the original congruence is that

$$(n, \ \varphi(m)) \,|\, (\text{ind}b - \text{ind}a)$$

and if this condition is fulfilled, then there will be $(n, \ \varphi(m))$ incongruent solutions.

Example 2.18: Solve the following congruences.

  (a) $4x^4 \equiv 5$ (mod 9).

  (b) $4x^4 \equiv 7$ (mod 9).

Solution of (a) Taking indices we have $4\text{ind}x + \text{ind}4 \equiv \text{ind}5$ (mod 6). From the table $\text{ind}4 = 2$ and $\text{ind}5 = 5$. Thus we must solve

$$4\text{ind}x \equiv 5 - 2 = 3 \ (\text{mod} \ 6).$$

Since $(4, 6) = 2$ and $2 \nmid 3$ we see that this congruence has no solution.

Solution of (b) Taking indices and using the table gives the congruence

$$4\text{ind}x \equiv 4 - 2 = 2 \ (\text{mod} \ 6)$$

to solve. Since $(4, 6) = 2$ and $2 \,|\, 2$, there will be 2 solutions. These are given by $\text{ind}x \equiv 2$ or 5 (mod 6), which, using the table of indices again, says that

$$x \equiv 4 \text{ or } 5 \ (\text{mod} \ 9). \qquad \square$$

Problem 2: Solve $a^x \equiv b$ (mod m).

Solution: In terms of indices the given congruence is equivalent to, by Theorem 2.24, the congruence

$$x(\text{ind}a) \equiv \text{ind}b \ (\text{mod} \ \varphi(m)).$$

Thus a necessary and sufficient condition for the solvability of the original congruence is that $(\text{ind}a, \ \varphi(m)) \,|\, \text{ind}b$ and if this condition is fulfilled, then there will be $(\text{ind}a, \varphi(m))$ solutions.

Example 2.19: Solve the following congruences.

(a) $5^x \equiv 4 \pmod 9$.

(b) $4^x \equiv 7 \pmod 9$.

Solution of (a) Taking indices gives the congruence

$$x\,\text{ind}5 \equiv \text{ind}4 \pmod 6.$$

Using the table of indices gives the congruence

$$5x \equiv 2 \pmod 6.$$

Since $(5, 6) = 1$, there will be a unique solution mod 6, namely

$$x \equiv 4 \pmod 6.$$

Solution of (b) Taking indices and using the table we obtain the congruence

$$2x \equiv 4 \pmod 6.$$

Since $(2,6) = 2$ and $2|4$ there will be two solutions, namely

$$x \equiv 2 \text{ or } 5 \pmod 6. \qquad \square$$

Problem 3: Determine the exponent to which a belongs modulo m, when $(a, m) = 1$.

Solution: We must find the least positive integer h such that

$$a^h \equiv 1 \pmod m.$$

Taking indices gives the congruence

$$h(\text{ind}a) \equiv \text{ind}1 \equiv 0 \pmod{\varphi(m)}.$$

Let $d = (\text{ind}a,\ \varphi(m))$. Then we have

$$h \equiv 0 \pmod{\varphi(m)/d},$$

and so $h = \varphi(m)/d$. Note that by Theorem 2.25 it is immaterial which primitive root of m we use, since, from (b) of Theorem 2.25, we see that $(\text{ind}_{g_1}g_2, \varphi(m)) = (\text{ind}_{g_2}g_1, \varphi(m)) = 1$, and so, by Theorem 2.4, we have

$$(\text{ind}_{g_1} N, \varphi(m)) = ((\text{ind}_{g_2}g_1)(\text{ind}_{g_1}g_2), \varphi(m)) = (\text{ind}_{g_2} N, \varphi(m)).$$

Example 2.20: Find the order of 7 modulo 9.

Solution; We have $\varphi(9) = 6$ and $\text{ind}7 = 4$. Since $(6, 4) = 2$ we see that the order of 7 modulo 9 is $6/2 = 3$. $\qquad \square$

As we have mentioned several times, to find the primitive roots of those numbers possessing them it suffices to have the primitive roots of the odd primes.

Also given one primitive root, g, we obtain all the rest from g^k, where $(k, \varphi(m))$ = 1, by Theorem 2.21. With this in mind we give a table of the least positive primitive root for all odd primes less than 50.

p	g	p	g	p	g	p	g	p	g
3	2	11	2	19	2	31	3	43	3
5	2	13	2	23	5	37	2	47	5
7	3	17	3	29	2	41	6		

By use of the Chinese Remainder Theorem we can now solve the congruences $ax^n \equiv b$ (mod m) and $a^x \equiv b$ (mod m) for any odd number m, or any number, m, such that the power of 2 that exactly divides m is not greater than the number of distinct odd prime divisors of m. The question remains: what do we do if we have too high a power of 2. It turns out that we can get an almost-primitive root modulo 2^a, $a \geq 3$, and that we will be able to use this to solve the problems considered in this section for the modulus 2^a, $a \geq 3$.

Theorem 2.26: The exponent of 5 modulo 2^a, $a \geq 3$, is 2^{a-2}.

Proof: In the proof of Theorem 2.22, we showed that if m is odd, then

$$m^{2^{a-2}} \equiv 1 \text{ (mod } 2^a).$$

Let h be the exponent to which 5 belongs modulo 2^a. If $h < 2^{a-2}$, then $h \mid 2^{a-3}$, that is $h = 2^b$ and $b \leq a - 3$. Thus

$$5^{2^{a-3}} \equiv 1 \text{ (mod } 2^a).$$

By the binomial theorem we have

$$5^{2^{a-3}} = (1 + 2^2)^{2^{a-3}} = 1 + 2^{a-3} \cdot 2^2 + 2^a t,$$

since all terms after the second have a factor of at least $2^4 \cdot 2^{a-4} = 2^a$. Thus

$$5^{2^{a-3}} \equiv 1 + 2^{a-1} \text{ (mod } 2^a),$$

which is a contradiction. Thus $h = 2^{a-2}$. ∎

By Theorem 2.19 and Theorem 2.26, we see that $5, 5^2, ..., 5^{2^{a-2}}$ are incongruent modulo 2^a and similarly $-5, -5^2, ..., -5^{2^{a-2}}$ are incongruent modulo 2^a. Since $5 \equiv 1$ (mod 4) implies $5^b \not\equiv -1$ (mod 2^a) if $a \geq 2$, we see that the union of these two sets are also pairwise incongruent. Thus we have $2^{a-1} = \varphi(2^a)$ numbers which are pairwise incongruent, and so form a reduced residue system modulo 2^a. Thus, if m is odd then there exist nonnegative integers u and k such that

$$m \equiv (-1)^u 5^k \pmod{2^a}$$

which $u = 0$ or 1 and $0 \le k \le 2^{a-2}$, for $a \ge 3$. One can then use this reduced residue system, as we used primitive roots above, to solve the same problems considered above. We illustrate this with the following theorem.

Theorem 2.27: Let $a \ge 3$ and m an odd number. Then the congruence

(2.21) $$x^n \equiv m \pmod{2^a}$$

has

    (a)   one solution if n is odd and

    (b)   if $n = 2N$, where N is odd, then, there are four solutions, if $a \equiv 1 \pmod 8$, and no solutions, if $a \not\equiv 1 \pmod 8$.

Proof: By the remarks after Theorem 2.26 we know that there exist nonnegative integers p, v, k and y such that

$$x \equiv (-1)^v 5^y \pmod{2^a} \text{ and } m \equiv (-1)^p 5^k \pmod{2^a}.$$

If $n \equiv 1 \pmod 2$, then (2.21) is equivalent to

$$(-1)^v 5^{ny} \equiv (-1)^p 5^k \pmod{2^a}.$$

Thus $v = p$ and we have

(2.22) $$ny \equiv k \pmod{2^{a-2}}.$$

Since $(n, 2^{a-2}) = 1$, (2.22) has a unique solution, in y, modulo 2^{a-2} and so (2.21) has a unique solution modulo 2^a. This proves (a).

    If $n = 2N$, where N is odd, then (2.21) is equivalent to

$$5^{2Ny} \equiv (-1)^p 5^k \pmod{2^a}.$$

Thus $p = 0$ and we have

(2.23) $$2my \equiv k \pmod{2^{a-2}}.$$

Because $5^{2Ny} \equiv 1 \pmod 8$, since $5^2 = 25 \equiv 1 \pmod 8$, we see that we must have $m \equiv 1 \pmod 8$ if there is to be a solution at all. Since N is odd we have $(2N, 2^{a-2}) = 2$. Since $m \equiv 1 \pmod 8$ implies $2 \mid k$ we see that (2.23) has two solutions. Thus (2.21) has 4 solutions since we may take $v = 0$ or 1. ∎

    For an extension of this result see Problem 14 below.

Example 2.21: Solve $11x^3 \equiv 17 \pmod{56}$.

Solution: We have $56 = 2^3 \cdot 7$ so that our given congruence is equivalent to the system of congruences

$$11x^3 \equiv 17 \pmod 7 \text{ and } 11x^3 \equiv 17 \pmod 8.$$

Now $11x^3 \equiv 17 \pmod 7$ is equivalent to $4x^3 \equiv 3 \pmod 7$. Since 3 is a primitive root of 7, by Example 2.15, and ind4 $= 4$ and ind3 $= 1$, the latter congruence is equivalent to

$$3\,\text{ind}x \equiv -3 \pmod 6$$

or

$$\text{ind}x \equiv -1 \pmod 2,$$

whose solutions are given by

$$\text{ind}x \equiv 1, 3 \text{ or } 5 \pmod 6.$$

Converting back to the modulus 7 we find

$$x \equiv 3, 5 \text{ or } 6 \pmod 7.$$

Now $11x^3 \equiv 17 \pmod 8$ is equivalent to $3x^3 \equiv 1 \pmod 8$. Since $3 \cdot 3 \equiv 1$ $\pmod 8$ and $x^2 \equiv 1 \pmod 8$, because x must be odd since we must have $(11x^3, 8)$ $= (17, 8) = 1$, the latter congruence is equivalent to

$$x \equiv 3 \pmod 8.$$

Thus, by the Chinese Remainder Theorem, the solutions to the original congruence are

$$x \equiv 3, 19 \text{ or } 27 \pmod{56}. \qquad \qquad \square$$

Problem Set 2.9

1. Solve the following congruences
   (a) $5x^6 \equiv 3 \pmod{17}$ (d) $5x^6 \equiv 6 \pmod{34}$
   (b) $5x^6 \equiv 3 \pmod{289}$ (e) $5x^6 \equiv 6 \pmod{289}$
   (c) $5x^6 \equiv 3 \pmod{578}$ (f) $5x^6 \equiv 6 \pmod{578}$.

2. For what values of b, $1 \le b \le 28$, is $8x^7 \equiv b \pmod{29}$ solvable?

3. For what values of n and a is the congruence $3^x \equiv a \pmod{2^n}$ solvable?

4. Solve the following congruences.
   (a) $7^x \equiv 5 \pmod{17}$ (d) $6^x \equiv 7 \pmod{13}$
   (b) $5 \cdot 7^x \equiv 2 \pmod{23}$ (e) $3 \cdot 5^x \equiv 1 \pmod 9$
   (c) $7 \cdot 11^x \equiv 15 \pmod{23}$ (f) $8^x \equiv 3 \pmod{43}$.

5. For what values of a, $1 \le a \le 40$, is $10^x \equiv a \pmod{41}$ solvable?

6. (a) Give necessary and sufficient conditions for the solvability of

$$a^x \equiv b \pmod{2^m},$$

where a and b are odd and $m \geq 3$.

   (b) Solve $7^x \equiv 23 \pmod{32}$.

7. If p is a prime of the form $8k + 1$, solve the congruence $x^4 \equiv -1 \pmod{p}$.

8. Prove Theorem 2.23.

9. Let $(a, m) = 1$, where m is a number with primitive roots. Show that a necessary and sufficient condition that the congruence

$$x^n \equiv a \pmod{m}$$

be solvable is that

$$a^{\varphi(m)/d} \equiv 1 \pmod{m},$$

where $d = (n, \varphi(m))$.

10. Construct a table of indices for $m = 7, 11$ and 14.

11. Solve the congruences of Problem 9 of Problem Set 2.3.

12. Let p be an odd prime and let a be an integer. Suppose that p does not divide a and that n is an integer, $n \geq 2$, which is also not divisible by p. If the congruence

$$x^n \equiv a \pmod{p^m}$$

is solvable for $m = 1$, then it is also solvable for all integers $m > 1$.

13. Solve the congruence $x^x \equiv x \pmod{11}$.

14. Let $a \geq 2$ be an integer. Show that if n is even, then an integer m is an nth power residue modulo 2^a if and only if $m \equiv 1 \pmod{(4n, 2^a)}$. (This is an extension of Theorem 2.27.)

15. Prove that $\mathrm{ind}(p - a) \equiv \mathrm{ind}\, a + (p - 1)/2 \pmod{p - 1}$.

16. Show that the index of a number a belonging to the exponent h is a multiple of $(p - 1)/h$; in fact one can choose a primitive root g so that

$$\mathrm{ind}_g a = (p - 1)/h.$$

17. (a) If p is a prime of the form $6k + 5$, the least positive residues of the sequence

$$1^3, 2^3, \ldots, (p - 1)^3$$

coincides with $1, 2, \ldots, p - 1$ in some order. If p is of the form $6k + 1$, then each of these residues repeats itself three times.

   (b) Show that the sum of the cubic residues in the interval $1 \leq n \leq p - 1$ is equal to $p(p - 1)/2$ if p is of the form $6k + 5$, but is equal to $p(p - 1)/6$ if p is of the form $6k + 1$.

2.10. AN APPLICATION TO CRYPTOGRAPHY

In this section we discuss two applications of congruences to the problem of cryptography.

The making of coded messages has probably been done since people first had languages and had to get messages to friends without enemies intercepting the message. One of the first recorded uses of coded messages was due to Julius Caesar and is therefore called the **Caesar Cipher**. Before we discuss this code we give some definitions.

Definition 2.11. Given a fixed alphabet a **code** is a mapping between the words of this alphabet. The domain of this map is called the set of **plain texts** and the range is called the set of **cipher texts**.

The Caesar cipher is obtained by shifting the alphabet by three letters and then substituting for the letters in the plain text for the corresponding letters in the shifted alphabet. If we lay out the alphabet we get the following table:

A	B	C	D	E	F	G	H	I	J	K	L	M
D	E	F	G	H	I	J	K	L	M	N	O	P

N	O	P	Q	R	S	T	U	V	W	X	Y	Z
Q	R	S	T	U	V	W	X	Y	Z	A	B	C

Thus, if our plain text message is

(2.24)   I LOVE NUMBER THEORY,

then the corresponding cipher text is

L ORYH QXPEHU WKHRUB.

If we were trying to break this code, that is, find the plain text without having to know beforehand the particular code, then knowing this is in the English language tells us that G is either A or I, since these are the only one letter words in English. Thus we usually regroup the plain text in blocks of equal size to avoid this problem. Thus we might write the plain text as

(2.25)   ILOV ENUM BERT HEOR YABC,

where the ABC at the end is just to fill out the block. Since ABC is just as obvious as a one letter word we might just fill the block with random letters. Since the person we are sending this to knows how to decipher this message it really doesn't matter that much. Thus, with the plain text, as in (2.25), the cipher text becomes

(2.26)   LORY HQXP EHUW KHRU BDEF.

Another way of avoiding some of these problems and allowing for the expansion of the plain text to include other items, like punctuation, for example, is to decide to go to numerical equivalences. Thus A → 00, B → 01, ..., Y → 24 and Z → 25. Thus the plain text message of (2.24) becomes

08 11142104 132012010417 190704141724

and the (2.25) version would now be blocks of four digits, that is, two letters, and would look like

(2.27)   0811 1421 0413 2012 01204 1719 0704 1417 2400,

where the 00 at the end is just added to fill out the block.

In number theoretic terms what the Caesar cipher does is to take a letter and add three to its numerical equivalent. However, since our alphabet wraps around, that is X → A, Y → B and Z → C, when we add three and get a number greater than 25 we must subtract off 26. Thus 26 → 00, 27 → 01 and 28 → 02, just as above. This is just what happens with congruence. Thus, if P stands for a plain text letter and C for the corresponding cipher text letter, then we must have

$$C \equiv P + 3 \ (\mathrm{mod}\ 26),$$

with $0 \le C \le 25$. Thus (2.27) becomes

1215 1825 0817 2416 0508 2123 1107 1720 0103,

which then translates back to (2.26).

The Caesar cipher is an example of a type of cipher called a **shift transformation**, that is, a transformation of the form

$$C \equiv P + k \ (\mathrm{mod}\ 26),$$

where k is any integer. Since complete residue system modulo 26 has exactly 26 distinct values we see that there are 26 different shift transformations.

Though we are mainly interested in ways to encipher and not decipher it should be mentioned that shift transformations are not very secure. If we know the encipherment was done by a shift transformation, then, since there are only 26 different shifts possible, we just run through all of them until we break the code. Thus we should try to make the transformation more complex.

This leads to the class of **affine transformations**. These are of the type

(2.28) $$C \equiv aP + b \ (\text{mod } 26),$$

where $0 \le C \le 25$ and a and b are integers with $(a, 26) = 1$. We require that $(a, 26) = 1$ so that as P runs through a complete residue system modulo 26 so does C, by Corollary 2.3.2. There are $\varphi(26) = 12$ values for a and 26 values of b. Thus there are $12 \cdot 26 = 312$ distinct affine transformations.

Note that $(a, 26) = 1$ implies that there is an integer a' so that

$$aa' \equiv 1 \ (\text{mod } 26).$$

Thus, from (2.28), we get

$$P \equiv a'(C - b) \ (\text{mod } 26),$$

with $0 \le P \le 25$.

Example 2.24. Given the affine transformation

(2.28) $$C \equiv 5P + 3 \ (\text{mod } 26)$$

find the inverse transformation.

Solution. We have $55 = 25 \equiv -1 \ (\text{mod } 26)$. Thus $5(-5) \equiv -25 \equiv 1 \ (\text{mod } 26)$. Since $-5 \equiv 21 \ (\text{mod } 26)$ we see that the inverse transformation is

(2.29) $$P \equiv 21(C - 3) \ (\text{mod } 26). \qquad \square$$

Example 2.25.

    (a) Use the affine transformation (2.28) to encipher the plain text message (2.27).

    (b) If it is known that the affine transformation (2.28) was used to encipher

QVJR PUMX URLX,

decipher this message.

Solution of (a). Using the congruence (2.28) we see that (2.27) is transformed into

1706 2104 2316 2511 0823 1020 1223 2110 1903,

which is equivalent to

RG VE XQ ZL IX KU MX VK TD

or

RGVE XQZL IXKU MXVK TD.

We may or may not want to stick some filler on at the end.

Solution to (b). In Example 2.24 we found the inverse transformation of the transformation (2.28), namely (2.29), which is

$$P \equiv 21(C - 3) \text{ (mod 26)}.$$

First we give the numerical equivalence of our message, namely

1621 0917 1520 1223 2017 1123.

Applying (2.29) gives

1314 2208 1819 0704 1908 1204

and undoing the numerical equivalence gives the message

NOWI STHE TIME

or

NOW IS THE TIME.

Something that might be noticed in the first part of Example 2.25 is all the X's that show up in the final coded message. These correspond to the original letter E. When one is trying to decipher a message one of the first things one does is make up frequency tables. This gives a count of how many times all the letters appear in the message. In the English language E, T and N are the three most frequent letters that occur in a typical text. Thus one would, to start with, assume that the most frequently occurring letter is an E and see what this does to the code. For more on this and various generalizations see the book by Sinkov [116]. (See also Problem 7 below.)

If one looks closely at the affine transformations, say the one used in Example 2.25, one sees that it really leaves the frequencies alone. For

$$C \equiv 5P + 3 \text{ (mod 26)}$$

just shifts by 3 and then skips every fifth letter. Thus E, T and N, say, have all moved exactly the same way, so that their relative frequency will still be the same. One can try to mix this up with what are called **polygraphic codes**. (What is talked about above are **monographic codes**.) In a **digraphic code**, for example, we would take our plaintext blocks, $P_1 P_2$, and instead of just finding $C_1 C_2$ by

$$C_1 \equiv aP_1 + b \ (\text{mod } 26) \text{ and } C_2 \equiv aP_2 + b \ (\text{mod } 26)$$

we obtain $C_1 C_2$ by

$$C_1 \equiv aP_1 + bP_2 \ (\text{mod } 26)$$

$$C_2 \equiv cP_1 + dP_2 \ (\text{mod } 26),$$

where, as usual, $0 \le C_1, C_2 \le 25$. In order to ensure a unique decipherment we require that $(ad - bc, 26) = 1$. (See Problem 12 of Problem Set 2.3 as well as Problems 10 and 11 below.)

Polygraphic codes overcome the problems mentioned above, but are vulnerable when one does frequency analysis on blocks of letters. For example, in the English language TH and HE are the two most frequent two letter blocks. More on this can be found in the book of Sinkov mentioned above.

One way of getting past the cryptanalysis is to use what are called **exponential codes**. The prime ingredient here is Euler's Theorem (Theorem 2.8.) In a sense what exponentiation codes do is to randomize the alphabet enough to make frequency analysis useless. We will discuss one of these codes, namely the RSA system (named after its inventors Rivest, Shamir and Adleman.) In this system one almost gives away the whole system (hence the name **public key codes**), except for a key ingredient. Even the process for finding this key ingredient is known, but the computer time needed to process this is enormous.

We begin with two large primes, say p and q, and let $n = pq$. Let e be a number such that $(e, \varphi(n)) = 1$. We then take our plaintext message and write it in numerical blocks with an even number of digits and the largest possible size. We then form the corresponding cipher block C by

$$C \equiv P^e \ (\text{mod } n),$$

where $0 < C < n$. To decipher this block we need an integer d that is the multiplicative inverse of e modulo $\varphi(n)$. Since we assume that $(e, \varphi(n)) = 1$ we know that such a d exists. Since $ed \equiv 1 \ (\text{mod } \varphi(n))$ we can write $ed = \varphi(n)t + 1$, for some integer t. Then, by Euler's Theorem,

$$C^d \equiv (P^e)^d = P^{ed} = P^{t\varphi(n)+1} = P(P^{\varphi(n)})^t \equiv P \ (\text{mod } n),$$

whenever $(P, n) = 1$. This gets us back to where we started. The pair (e, n) is called the **enciphering key** and the pair (d, n) is called the **deciphering key**.

Suppose $(P, n) > 1$. What can we say in this case? In this case we cannot appeal to Euler's Theorem, but we can appeal to Fermat's Little Theorem (Corollary 2.8.1.)

Since $n = pq$, where p and q are primes, if $(P, n) > 1$, then $p \mid P, q \mid P$ or possibly both. First we take the case $p \nmid P$. Then, by Fermat's Theorem, we have

$$P^{p-1} \equiv 1 \ (\text{mod } p).$$

Since $\varphi(n) = (p-1)(q-1)$ and $ed = t\varphi(n) + 1 = t(p-1)(q-1) + 1$ we have

$$C^d \equiv (P^e)^d = P^{ed} = P^{t(p-1)(q-1)+1} = P(P^{t(q-1)})^{p-1} \equiv P \ (\text{mod } p).$$

If $p \mid P$, then we use the other part of Fermat's Theorem which says that

$$P^p \equiv P \ (\text{mod } p).$$

Then

$$C^d \equiv (P^e)^d = P^{t(p-1)(q-1)+1} = P(P^{t(q-1)})^p \, P^{-t(q-1)} \equiv P^{t(q-1)} P^{-t(q-1)} P \equiv P \ (\text{mod } p).$$

Thus we have

$$C^d \equiv P \ (\text{mod } p) \text{ and } C^d \equiv P \ (\text{mod } q),$$

even if $(P, n) > 1$, and so, by Theorem 2.2, since $n = pq$,

$$C^d \equiv P \ (\text{mod } n).$$

Note that we can tell the world what e and n are because what is needed is d. If p and q are large enough, then the calculation of $\varphi(n)$, either by factoring or brute force, will use an enormous amount of computer time. At the present time factorizing numbers of more than 200 digits is way out of the question, and so we choose p and q to be primes of about 100 digits each.

It should be noted that brute force is no easier than factorization. Since $n = pq$ we have $\varphi(n) = (p-1)(q-1)$. Thus

$$p + q = n - \varphi(n) + 1 \text{ and } p - q = \sqrt{(p+q)^2 - 4n}.$$

Thus, if we know $\varphi(n)$ we easily find p and q and conversely, if we have p and q, then $\varphi(n)$ is easily found.

There are also a few precautions that should be taken so that n can't be factored too easily. For details see the books of Koblitz [61] or Bressoud [12].

Example 2.26. Use the enciphering key (13, 2773) to encode the message

GAUSS IS NUMBER ONE.

Solution. If we write this in blocks of 4 digits, we see that the numerical equivalent of the message is

$$0600 \quad 2018 \quad 1808 \quad 1813 \quad 2012 \quad 0104 \quad 1714 \quad 1304.$$

The enciphering of the first block would give

$$C \equiv (0600)^{13} \equiv 2049 \ (\text{mod } 2773)$$

where we have used the binary expansion of 13: $13 = 8 + 4 + 1$, to simplify the exponentiation. If we continue in the same way we get the following enciphered message:

$$2049 \quad 0904 \quad 0313 \quad 2053 \quad 2135 \quad 0503 \quad 1441 \quad 1545.$$

This is as far as enciphering goes: a sequence of numbers. To decipher this the receiver must have the deciphering key. See the next example. □

It might be remarked that we take blocks of four letters here since that is as large as we can do with our modulus 2773. If n were a 200 digit integer we could take considerably larger blocks of digits, but with a four digit n it makes no sense to take larger blocks because n will only give us four digit blocks.

Example 2.27. Given that the following message was enciphered with the same enciphering key that was used in Example 2.26 decipher it.

$$0300 \quad 0439 \quad 2405 \quad 2234 \quad 1743 \quad 0031 \quad 2683.$$

Solution. The enciphering key used in the previous example was (13, 2773) and we have $2773 = 47 \cdot 59$. Thus $\varphi(2773) = 2668$. The deciphering key is $(d, 2773)$, where $13d \equiv 1 \ (\text{mod } 2668)$. To solve this congruence we could use Euler's Theorem again, since the modulus is even and small. Since

$$13^{\varphi(2668)} \equiv 1 \ (\text{mod } 2668)$$

we see that we can take $d \equiv 13^{\varphi(2668)-1} \equiv 821 \ (\text{mod } 2668)$.

Then the first block of the deciphered code would be

$$(0300)^{821} \equiv 1700 \ (\text{mod } 2773).$$

If we continue in this way, we obtain the rest of the deciphered blocks, namely

$$1700 \quad 1200 \quad 1320 \quad 0900 \quad 1311 \quad 0821 \quad 0418.$$

Alphabetically this is

$$\text{RA MA NU JA NL IV ES}$$

or

$$\text{RAMANUJAN LIVES.} \qquad \square$$

Problem Set 2.10

1. Use the Caesar cipher to

   (a) encipher the message

   MATHEMATICS;

   (b) decipher the message

   WKHPH VVDJH.

2. Consider the affine transformation $C \equiv 15P + 7 \pmod{26}$, $0 \le C \le 25$.

   (a) Encipher the message

   MADAM IM ADAM.

   (b) Decipher the message

   0920 1504 0902 0608 0920 1503.

3. Consider the affine transformation $C \equiv 3P + 11 \pmod{26}$, $0 \le C \le 25$.

   (a) Encipher the message

   LIVE LONG AND PROSPER.

   (b) Decipher the message

   1625 0100 0110 1606 2313 0601 2525.

4. Suppose $(a, 26) = 1$. Does the affine transformation $C \equiv aP + b \pmod{26}$ ever have fixed letters, that is $C \equiv P \pmod{26}$? If so, how many for a given value of a?

5. Given two affine transformations

   $$C \equiv aP + b \pmod{26} \text{ and } C \equiv cP + d \pmod{26},$$

   where $(a, 26) = (c, 26) = 1$, show that they will give distinct messages, in general, unless $a \equiv c \pmod{26}$ and $b \equiv d \pmod{26}$.

6. Use the affine transformation $C \equiv 2P + 1 \pmod{26}$ to encipher LONG HAUL. Now try to decipher. What happens?

7. Suppose we have a sufficiently large enciphered text and we see that T and Z are the two most frequently appearing letters. If this text was enciphered with an affine transformation: $C \equiv aP + b \pmod{26}$, what are the most likely values of a and b?

8. To try and make cryptanalysis harder we might try **product ciphers**, that is, we encipher a text and then encipher it again. Use the two affine transformations $C \equiv 5P + 3 \pmod{26}$ and $C \equiv 7P + 25 \pmod{26}$ to produce the product cipher of OVER AND OVER.

9. Suppose $(a, 26) = (c, 26) = 1$. What single affine transformation is equivalent to the product of the two affine transformations $C \equiv aP + b$ (mod 26) and $C \equiv cP + d$ (mod 26)? What affine transformation is equivalent to the product cipher of Problem 7?

10. Consider the digraphic transformation

$$C_1 \equiv 3P_1 + 5P_2 \text{ (mod 26)}$$

$$C_2 \equiv 7P_1 + 13P_2 \text{ (mod 26)}.$$

(a) Encipher the message

MADAM IM ADAM.

(b) Decipher the message

1416 1915 0315 2020 0315 1915 1513 1012.

11. Consider the digraphic transformation

$$C_1 \equiv 11P_1 + 4P_2 \text{ (mod 26)}$$

$$C_2 \equiv P_1 + 13P_2 \text{ (mod 26)}.$$

(a) Encipher the message

HEARTBREAK HOTEL.

(b) Decipher the message

0205 0217 0519 1806.

12. How many pairs of letters remain unchanged when encipherment is done with the digraphic cipher

(a) $C_1 \equiv P_1 + 3P_2$ (mod 26)
$C_2 \equiv 7P_1 + 19P_2$ (mod 26).

(a) $C_1 \equiv 11P_1 + 3P_2$ (mod 26)
$C_2 \equiv 5P_1 + 17P_2$ (mod 26).

(c) What can be said about the general digraphic cipher?

13. (a) Use the enciphering key $(13, 2813)$ to encipher

MADAM IM ADAM.

(b) Given the enciphering key from (a) decipher

1743 2697 0111 1529 2254 2103 0269 0111 0444 0430.

14. (a) Use the enciphering key $(17, 2747)$ to encipher

LOST IN SPACE.

(b) Given the enciphering key from (a) decipher

$$1970 \ 0667 \ 1674 \ 1165 \ 1004 \ 2082 \ 1059 \ 1239$$

15. Given that $n = pq$, p and q distinct primes, find p and q given that

(a) $n = 1947$ and $\varphi(n) = 1840$;

(b) $n = 5123$ and $\varphi(n) = 4968$;

(c) $n = 64722701$ and $\varphi(n) = 64706400$.

16. Suppose we get a P such that $(P, n) > 1$. Show that we can factor n and hence decipher the message.

2.11. PSEUDOPRIMES

We begin with a converse of Fermat's Theorem (Corollary 2.8.1.)

Theorem 2.28. Let $n > 1$ and suppose that there exists an integer a such that

$$a^{n-1} \equiv 1 \ (\mathrm{mod} \ n),$$

but if d is any proper divisor of $n - 1$, then

$$a^d \not\equiv 1 \ (\mathrm{mod} \ n).$$

Then n is a prime.

Proof. Let M be a set of positive integers m such that

$$a^m \equiv 1 \ (\mathrm{mod} \ n).$$

Since $n - 1 \in M$, by hypothesis, we see that M is not empty, and so it has a least element. Let d be the least positive element of M. If m is in M, we write $m = dk + r$, $0 \le r < d$. Then

$$1 \equiv a^m = (a^d)^k a^r \equiv a^r \ (\mathrm{mod} \ n).$$

Since d is the least positive element of M we see that $r = 0$. Thus m in M implies that $d \mid m$. Since $n - 1$ is in M we see that $d \mid (n - 1)$. The hypothesis also states that

$$a^d \not\equiv 1 \ (\mathrm{mod} \ n),$$

for any proper divisor of $n - 1$. If $d = 1$, then $a \equiv 1 \ (\mathrm{mod} \ n)$, and so $a^k \equiv 1 \ (\mathrm{mod} \ n)$ for all integers k, including the proper divisors of $n - 1$. Since this contradicts the hypothesis we must have $d = n - 1$.

By Theorem 2.8 we have

$$a^{\varphi(n)} \equiv 1 \ (\mathrm{mod} \ n).$$

Thus $n - 1 \mid \varphi(n)$. Thus $n - 1 \le \varphi(n)$. By definition, we have $\varphi(n) \le n - 1$. Thus $\varphi(n) = n - 1$. If n is composite, say $n = rs$, $1 < r, s < n$, then $\varphi(n) \le n - 2 < n - 1$. Thus n must be a prime. ∎

The immediate question that should come to mind is "Why not the straight converse?" That is, why not

$$a^{n-1} \equiv 1 \pmod{n} \text{ implies that } n \text{ is a prime?}$$

Unfortunately this result is false as is shown by the following example.

Example 2.28. Let $n = 341 = 11 \cdot 31$. Now, by Fermat's Theorem, we have that

$$2^{10} \equiv 1 \pmod{11},$$

and so

$$2^{340} = (2^{10})^{34} \equiv 1 \pmod{11}.$$

Now

$$2^5 \equiv 1 \pmod{31},$$

and so

$$2^{340} = (2^5)^{68} \equiv 1 \pmod{31}.$$

Thus, by Theorem 2.2, we have that

$$2^{340} \equiv 1 \pmod{341}. \qquad \square$$

Still we can use the contrapositive of Fermat's Theorem to prove that a number is composite. That is, if $a^{n-1} \not\equiv 1 \pmod{n}$, for some integer a, $(a, n) = 1$, then n is not a prime.

Example 2.29. Show that 341 is composite using Fermat's Theorem.

Solution. We can't take $a = 2$ as the above example shows. However,

$$3^{340} = 3^{256+64+16+4} \equiv 245 \cdot 81 \cdot 245 \cdot 81 \equiv 56 \pmod{341}.$$

Thus, by the contrapositive of Fermat's theorem, we see that 341 is not a prime. $\square$

This leads to the following definition.

Definition 2.12. Let a be a positive integer. We say that the composite integer n is a **pseudoprime to the base** a or a **base** a **pseudoprime** if $a^{n-1} \equiv 1 \pmod{n}$.

Thus 341 is a base 2 pseudoprime.

Pseudoprimes to any given base are fairly rare. For example, if we go up to 10^{10} we find that there are 455052512 primes, whereas there are only 14884 pseudoprimes to the base 2. However, for any given base there are infinitely many pseudoprimes.

Theorem 2.29 (Cipolla). For each integer a, $a > 1$, there exist infinitely many base a pseudoprimes.

Proof. Let p be an odd prime such that $p \nmid a(a^2 - 1)$. There are infinitely many of these since $a(a^2 - 1)$, which is even, has only a finite number of prime factors. Let

(2.30)
$$n = \frac{a^{2p} - 1}{a^2 - 1} = \frac{a^p - 1}{a - 1} \cdot \frac{a^p + 1}{a + 1}.$$

Thus n is a composite integer. We wish to show that n is a pseudoprime to the base a.

From (2.30) we see that

(2.31)
$$(a^2 - 1)(n - 1) = a^{2p} - a^2 = a(a^{p-1} - 1)(a^p + a).$$

Now $a^p + a$ is an even integer and by Fermat's Theorem we have that $p \mid (a^{p-1} - 1)$, since $p \nmid a$ by hypothesis. Also, since $p - 1$ is even we see that $(a^2 - 1) \mid (a^{p-1} - 1)$. By hypothesis $p \nmid (a^2 - 1)$, and so $p(a^2 - 1) \mid (a^2 - 1)(n - 1)$. Thus

$$2p(a^2 - 1) \mid (a^2 - 1)(n - 1),$$

and so $2p \mid (n - 1)$.

If we let $n = 2pk + 1$, we see that

$$a^{n-1} = a^{2pk} = \{1 + n(a^2 - 1)\}^k \equiv 1 \pmod{n},$$

by (2.31). Thus n is a pseudoprime to the base a. Since different values of p yield different n the result follows. ∎

Even though this result says that there are infinitely many pseudoprimes for any given base they are still quite rare, as mentioned above. Thus, if we were trying to test an odd number n for primality we could check the congruence $2^{n-1} \equiv 1 \pmod{n}$. If this doesn't hold, then we know that n is composite. If it does hold, then we have a prime or a pseudoprime to the base 2. We could then check the congruence $3^{n-1} \equiv 1 \pmod{n}$, as we did in Example 2.29. If this congruence also holds we could try $b^{n-1} \equiv 1 \pmod{n}$ for various values of b for which $(b, n) = 1$. If we can find one value of b for which the congruence does not hold, then we know that n is composite.

The question arises: how many values of b must we try before we achieve success or failure, that is, before we know that n is prime or composite? The following definition gives the worse case scenario.

Definition 2.13. Let n be a composite number and assume that

$$a^{n-1} \equiv 1 \pmod{n}$$

for all integers a for which $(a, n) = 1$. Then n is called a **Carmichael number**.

Such numbers were first investigated in the early 1900's by R. D. Carmichael. Unfortunately he found some and it has just recently been proved that there are infinitely many of them.

Example 2.30. Show that $n = 561$ is a Carmichael number.

Solution. Note that $n = 3 \cdot 11 \cdot 17$. Thus $(a, 561) = 1$ implies that $(a, 3) = (a, 11) = (a, 17) = 1$. If we apply Fermat's Theorem we see that

$$a^2 \equiv 1 \pmod{3}$$
$$a^{10} \equiv 1 \pmod{11}$$

and

$$a^{16} \equiv 1 \pmod{17}.$$

Thus

$$a^{560} = (a^2)^{280} \equiv 1 \pmod{3}$$
$$a^{560} = (a^{10})^{56} \equiv 1 \pmod{11}$$

and

$$a^{560} = (a^{16})^{35} \equiv 1 \pmod{17}.$$

Thus, by Theorem 2.2,

$$a^{560} \equiv 1 \pmod{561}$$

and therefore 561 is a Carmichael number.   ■

Unlike the general pseudoprime we can exactly characterize the Carmichael numbers. The above example gives an idea of how this will work. To do this we introduce a function, analogous to the Euler phi function, that was first introduced by Carmichael.

Definition 2.14. Define the function $\lambda : \mathbf{N} \to \mathbf{N}$ by

(a) $\lambda(1) = 1$;

(b) $\lambda(p^a) = \varphi(p^a)$, if p is an odd prime and $a \geq 1$;

(c)   (i) $\lambda(2^a) = 1$, if $a = 1$ or 2;

 (ii) $\lambda(2^a) = \frac{1}{2}\varphi(2^a)$, if $a \geq 3$.

If m is a natural number and $m = 2^a p_1^{a_1} \cdots p_r^{a_r}$, then we define

$$\lambda(m) = [\lambda(2^a), \lambda(p_1^{a_1}), \ldots, \lambda(p_r^{a_r})],$$

where [,] denotes the least common multiple.

The function $\lambda(m)$ is sometimes called the **universal exponent modulo** m because of the following result.

Theorem 2.30. Let m be an integer, $m \geq 2$, and suppose that $(a, m) = 1$. Then

$$a^{\lambda(m)} \equiv 1 \pmod{m}.$$

Moreover, for each m there is there is an integer a such that $\lambda(m) = \operatorname{ord}_m a$.

Proof. Write m in its canonical factorization, that is

(2.32) $$m = 2^a p_1^{a_1} \cdots p_r^{a_r}.$$

Suppose $(a, m) = 1$. Note that if $p^c \| m$, p odd, then $\lambda(p^c) = \varphi(p^c)$, and so

(2.33) $$a^{\lambda(p^c)} \equiv 1 \pmod{p^c},$$

by Theorem 2.2, since $\lambda(p^c) \mid \lambda(m)$ for each $p^c \| m$.

If m is even, so that $a \geq 1$, then a must be odd, if $(a, m) = 1$. thus

$$a \equiv 1 \pmod{2} \text{ or } a^{\lambda(2)} \equiv 1 \pmod{2}$$

$$a^2 \equiv 1 \pmod{4} \text{ or } a^{\lambda(4)} \equiv 1 \pmod{4},$$

and from the proof of Theorem 2.22 we see that

(2.34) $$a^{\lambda(2^a)} \equiv 1 \pmod{2^a},$$

if $a \geq 3$. Thus, for all possibilities we have $a^{\lambda(2^a)} \equiv 1 \pmod{2^a}$, and so, by Theorem 2.2, (2.33) and (2.34), we have

$$a^{\lambda(m)} \equiv 1 \pmod{m}$$

for all a with $(a, m) = 1$.

For the second part of we choose an integer a such that

$$a \equiv 5 \pmod{2^a}, a \equiv g_1 \pmod{p_1^{a_1}}, \ldots, a \equiv g_r \pmod{p_r^{a_r}},$$

where, for $1 \leq k \leq r$, g^k is a primitive root modulo $p_k^{a_k}$. Let $L = \operatorname{ord}_m a$. Then

$$a^L \equiv 1 \pmod{p_k^{a_k}}, 1 \leq k \leq r.$$

Thus, by the choice of a, we see that we must have $\lambda(p_k^{a_k}) \mid L$, since $\lambda(p_k^{a_k}) = \varphi(p_k^{a_k})$, for $1 \leq k \leq r$. By Theorem 2.26, we see that $\lambda(2^a) \mid L$. Thus $\lambda(m) \mid L$. Since $L = \operatorname{ord}_m a$ and $a^{\lambda(m)} \equiv 1 \pmod{m}$, by the first part of the theorem, we have, by Theorem 2.19, that $L \mid \lambda(m)$. Thus $\lambda(m) = L$ and the result follows. ∎

Example 2.31. Let $m = 120$. Construct the number a that is guaranteed by the second part of the proof of Theorem 2.30.

Solution. We have $m = 2^3 \cdot 3 \cdot 5$. Thus $\lambda(m) = [\frac{1}{2}\varphi(8), \varphi(3), \varphi(5)] = [2,2,4] = 4$. The number a is to satisfy

$$a \equiv 5 \ (\text{mod } 8), \ a \equiv 2 \ (\text{mod } 3) \text{ and } a \equiv 2 \ (\text{mod } 5),$$

since 2 is a primitive root of both 3 and 5. If we use the Chinese Remainder Theorem to solve this we find that $a \equiv 77 \ (\text{mod } 120)$. Thus

$$77^4 \equiv 1 \ (\text{mod } 120)$$

and (as is easily checked) $77^a \not\equiv 1 \ (\text{mod } 120)$, for $a = 1, 2$ or 3. $\qquad\square$

We are now ready for our characterization of Carmichael numbers.

Theorem 2.31. Let $C > 2$. Then C is a Carmichael number if and only if $C = p_1 \cdots p_r$, where $r \geq 3$ and the p_k, $1 \leq k \leq r$, are distinct primes such that $(p_k - 1) \,|\, (C - 1)$, $1 \leq k \leq r$.

Proof. Let C be a Carmichael number. Then $(b, C) = 1$ implies that

$$b^{C-1} \equiv 1 \ (\text{mod } C).$$

By Theorem 2.30 there is an integer a, $(a, C) = 1$, such that $\text{ord}_C a = \lambda(C)$. Thus, by Theorem 2.19, we must have $\lambda(C) \,|\, (C - 1)$. Since $C > 2$ we see that $\lambda(C)$ is even, and so C must be odd.

Suppose $C = p_1^{a_1} \cdots p_r^{a_r}$, and $a_k \geq 2$ for some k, $1 \leq k \leq r$. Then we have

$$\lambda(p_k^{a_k}) = \varphi(p_k^{a_k}) = p_k^{a_k-1}(p_k - 1) \,|\, \lambda(C) \,|\, (C - 1).$$

Thus, since $a_k \geq 2$, we have $p_k \,|\, (C - 1)$. This contradicts the fact that $p_k \,|\, C$. Thus we must have $a_k = 1$, $1 \leq k \leq r$. Also, as above,

(2.35) $(p_k - 1) \,|\, (C - 1),$

for $1 \leq k \leq r$.

Since Carmichael numbers are composites we see that $r \geq 2$. Suppose $r = 2$, that is, $C = pq$, where p and q are distinct primes, with $p > q$, say. Then, by (2.35), we have

$$(p - 1) \,|\, (C - 1) = pq - 1 = (p - 1)q + q - 1.$$

Thus $(p - 1) \,|\, (q - 1)$. This is impossible. Thus $r \geq 3$.

Now suppose $C = p_1 \cdots p_r$, $r \geq 3$, with $(p_k - 1) \mid (C - 1)$, $1 \leq k \leq r$. Let $(b, C) = 1$. Then $(b, p_k) = 1$, $1 \leq k \leq r$. Thus, by Fermat's Theorem,

$$b^{p_k - 1} \equiv 1 \pmod{p_k}.$$

Since $(p_k - 1) \mid (C - 1)$ we see that

$$b^{C-1} \equiv 1 \pmod{p_k},$$

for $1 \leq k \leq r$. Thus, by Theorem 2.2, we have

$$b^{C-1} \equiv 1 \pmod{C}.$$

Thus C is a Carmichael number.   ∎

We can expand on the argument in the above proof where we showed that a Carmichael number must have at least three prime divisors to discuss particular forms of the number.

Example 2.32. Find all Carmichael numbers of the forms $3pq$ and $5pq$.

Solution. Let $A \geq 3$ be a fixed prime. We wish to investigate the Carmichael numbers of the form Apq, where $A < p < q$. By Theorem 2.31 we must have

(2.36) $(A - 1) \mid (Apq - 1)$, $(p - 1) \mid (Apq - 1)$ and $(q - 1) \mid (Apq - 1)$.

As in the proof of Theorem 2.31 this means that

$$(p - 1) \mid (Aq(p - 1) + Aq - 1) \text{ and } (q - 1) \mid (Ap(q - 1) + Ap - 1).$$

Thus we have that $(p - 1) \mid (Aq - 1)$ and $(q - 1) \mid (Ap - 1)$, and so we let $(p - 1)u = Aq - 1$ and $(q - 1)v = Ap - 1$, where u and v are integers. Thus

$$(p-1)uv = (Aq-1)v = A(q-1)v + (A-1)v = A(Ap-1) + (A-1)v$$
$$= A^2 p - A + (A-1)v = A^2(p-1) + (A-1)(A+v).$$

Thus we have $(p - 1) \mid (A - 1)(A + v)$, and so

$$(p-1)(q-1) \mid (A^2 - A)(q-1) + (A-1)(q-1)v$$
$$= (A^2 - A)(q-1) + (A-1)(Ap-1)$$
$$= (A^2 - A)q + (A^2 - A)p - A^2 + A - A + 1$$
$$= (A^2 - A)(p+q) + 1 - A^2.$$

Note that if x and y are any real numbers, then

(2.37)
$$(x-1)(y-1)-(A^2-A)(x+y)+A^2-1$$
$$= (x-A^2+A-1)(y-A^2+A-1)-(A-1)^2(A^2+1).$$

Now, if x and y are sufficiently large, then the right hand side of (2.37) is positive, that is

(2.38) $(x-1)(y-1) > (A^2-A)(x+y)+1-A^2.$

Thus, if p and q are primes, this says that

$$(p-1)(q-1)I\{(A^2-A)(p+q)+1-A^2\}.$$

If $A = 3$, we see that $p \geq 13$ and $q \geq 17$ imply that the inequality (2.38) holds. Thus we need only try pairs (p, q) with $3 < p \leq 11$ and $p < q$. For example, if we take $p = 5$, we need, by (2.36), $(q-1)|(3 \cdot 5 - 1) = 14$. Thus $q - 1 = 1, 2, 7$ or 14. Since none of these values yield a prime value greater than $p = 5$ we see that $p \neq 5$. Similarly, for $p = 7$ we find no values for q. Finally, for $p = 11$ we see that $q = 17$ will work. This gives the smallest Carmichael number $C = 3 \cdot 11 \cdot 17 = 561$.

If $A = 5$, then we see that $p \geq 41$ and $q \geq 43$ imply that the inequality (2.38) holds. Thus we want pairs (p, q) such that $5 < p \leq 37$ and $p < q$. The search is a little longer than the one for $A = 3$, but in the end we find that there are only three Carmichael numbers of the form $5pq$, namely

$$5 \cdot 13 \cdot 17 = 1105, \ 5 \cdot 17 \cdot 29 = 2465 \text{ and } 5 \cdot 29 \cdot 73 = 10585. \quad \square$$

As can be seen in the above example, the amount of numerical checking can get somewhat extensive as A gets larger. First one needs values of x and y so that the right hand side of (2.37) is positive. This then gives an upper bound on the values of p and q to check the requirements of (2.36) against. At some point a computer becomes handy. Similar remarks apply to finding Carmichael numbers with 4 or more prime factors.

So where do Carmichael numbers leave us in our search for a primality test based on the "converse" to Fermat's Little Theorem? They certainly show that in certain rare cases we cannot find a base b for which the congruence $b^{n-1} \equiv 1 \pmod{n}$ fails. However, if we examine this congruence a little more we are led to a generalization of pseudoprimes that will serve our purposes.

Suppose n is an odd integer, say $n = 2k + 1$. If n is a prime, then we know that

$$b^{2k} = b^{n-1} \equiv 1 \pmod{n}.$$

Thus n divides $b^{2k} - 1 = (b^k + 1)(b^k - 1)$. If n is a prime, then we know, by Lemma 1.19.1, that n divides $b^k + 1$ or n divides $b^k - 1$. Since $(b^k + 1) - (b^k - 1) = 2$ and n is an odd prime we see that n divides exactly one of these factors. Thus, if n is an odd prime, we have that exactly one of the congruences

(2.39) $$b^k \equiv 1 \pmod{n} \text{ or } b^k \equiv -1 \pmod{n}$$

holds. The hope is that if n is composite, at least one of these congruences would fail since some of n's prime divisors divide one factor and some divide the other.

Example 2.33. Let $n = 561$. Show that for $b = 5$, then the congruences in (2.39) fail.

Solution. We have $561 = 2 \cdot 280 + 1$ so that $k = 280$. If we note that $280 = 256 + 16 + 8$, we find that

$$5^{280} \equiv 67 \pmod{561}.$$

Since both congruences fail we see that 561 is composite, as we already know. □

Unfortunately there are some pseudoprimes that get by the (2.39) hurdle for their base. This leads us farther down the factorization path.

Suppose that n is odd and that

$$n - 1 = 2^e k,$$

where $e \geq 1$ and k is odd. Then

(2.40) $$b^{n-1} - 1 = (b^k - 1)(b^k + 1)(b^{2k} + 1) \cdots (b^{2^{e-1}k} + 1),$$

and so if n were indeed a prime, then n must divide at least one of the factors on the right hand side of (2.40). An argument similar to the above shows that n cannot divide more than one of these factors. Thus if n is an odd prime and we write $n - 1 = 2^e k$, as above, then exactly one of the congruences

(2.41) $$b^{2^j k} \equiv -1 \pmod{n}, 0 \leq j \leq e - 1 \text{ or } b^k \equiv 1 \pmod{n}$$

must hold. This leads to the following definition.

Definition 2.14. Let n be an odd composite integer and b a positive integer that is relative prime to n. Then n is a **strong pseudoprime to the base b** if at least one of the congruences in (2.41) holds.

It should be noted that a strong pseudoprime to the base b is a pseudoprime to the base b.

One can show that the first strong pseudoprime to the base 2 is 2047, whereas the first pseudoprime is 341. The good thing about strong pseudoprimes is that they are even rarer than pseudoprimes. Indeed one should try an integer n for several bases to see if one of congruences in (2.41) holds. For example, the first strong pseudoprime for the bases 2, 3 and 5 is 25326001, whereas the first pseudoprime for all three bases is 1729.

The next logical question might be about strong Carmichael numbers: composite numbers that are strong pseudoprimes to all bases they are relatively prime to. Fortunately such things don't occur. The following result, which we will not prove here (but whose proof is outlined in Problem 15 below) shows that this cannot occur.

Theorem 2.32 (Miller). If n is an odd composite positive integer, then n is a strong pseudoprime to the base b for at most $(n - 1)/4$ bases b with $1 \leq b \leq n - 1$.

This result also shows that if n is a strong pseudoprime for more than $(n - 1)/4$ bases, then n must be a prime. However, the computations involved in showing this for any particular value of n turn out to be worse than simple trial division. One is better off using the remarks above about strong pseudoprimes to several bases. See [12. pp. 75-78] for more details.

Problem Set 2.11

1. Show that the following numbers are pseudoprimes to the given base.
   (a) 91 for the base 3,
   (b) 45 for the bases 17 and 19,
   (c) 15 for the base 11,
   (d) 1105 and 1905 for the base 2.
   (e) Find all bases for which 26 is a pseudoprime.
2. Let $F_n = 2^{2^n} + 1$ be a Fermat number. If F_n is not a prime, then F_n is a base 2 pseudoprime.
3. If p is a prime, then $M_p = 2^p - 1$ is either a prime or a base 2 pseudoprime.
4. (a) If n is a pseudoprime to the bases a and b, then it is a pseudoprime to the base ab.
   (b) Suppose $ab \equiv 1 \pmod{n}$, where $(a, n) = 1$. If n is a pseudoprime to the base a, then n is a pseudoprime to the base b.
5. (a) Let p and q be distinct odd primes and let a be a positive integer such that

$$a^{pq-1} \equiv 1 \pmod{pq}.$$

Show that

$$a^{q-1} \equiv 1 \pmod{p} \text{ and } a^{p-1} \equiv 1 \pmod{q}.$$

(b) Use this result to find the three smallest pseudoprimes for the bases 2 and 3.

6. Compute $\lambda(n)$ for $n = 25, 36, 100, 335, 1000$ and $10!$.

7. Show that if $(m, n) = 1$, then $\lambda(mn) = [\lambda(m), \lambda(n)]$.

8. Show that if $e = \text{ord}_m a$, then $e \mid \lambda(m)$.

9. Show that the following integers are Carmichael numbers.

(a) 2821   (c) 29341

(b) 15841   (d) 118901521.

10. If $6m + 1$, $12m + 1$ and $18m + 1$ are all primes, then $(6m + 1)(12m + 1)(18m + 1)$ is a Carmichael number. Show that 1729 and 294409 are Carmichael numbers.

11. Find all Carmichael numbers of the form $7 \cdot 23q$, where $q > 23$ is an odd prime.

12. Show that there are only finitely many Carmichael numbers of the form pqr, where $p < q < r$ and p, q and r are distinct odd primes with p fixed.

13. Show that the following integers are strong pseudoprimes to the given bases.

(a) 2047 for the base 2, indeed it is the smallest,

(b) 25 for the base 7,

(c) 341 is a pseudoprime to the base 2, but not a strong pseudoprime to the base 2,

(d) 25326001 for the bases 2, 3 and 5.

14. Suppose n is a pseudoprime to the base 2 and $N = 2^n - 1$. Then N is a strong pseudoprime to the base 2. Show that there are infinitely many strong pseudoprimes to the base 2.

15. The following exercise outlines the proof of Miller's Theorem (Theorem 2.32.)

(a) Show that if p is an odd prime and e and q are positive integers, then the number of incongruent solutions of the congruence

$$x^{q-1} \equiv 1 \pmod{p^e}$$

is $(q, p^{e-1}(p - 1))$.

(b) Let $N = 2^a n$, where a is a nonnegative integer and n is an odd positive integer. Suppose $p - 1 = 2^b t$, where b and t are positive integers with t odd. Show that there are $2^a(t, n)$ incongruent solutions of

$$x^N \equiv -1 \pmod{p}.$$

(c) Suppose $n = p_1^{a_1} \cdots p_r^{a_r}$ with some $a_k \geq 2$, $1 \leq k \leq r$. Use the Chinese Remainder Theorem to prove Miller's Theorem in this case. (Hint: use (a) to count the number of solutions of the congruence

$$x^{n-1} \equiv 1 \pmod{n}.$$

Use the fact that $a_k \geq 2$ to estimate this number and show that it is less than $(n - 1)/4$.)

(d) Suppose $n = p_1 \cdots p_r$. Use the Chinese Remainder Theorem to count the number of solutions of

$$x^{2^j t_i} \equiv -1 \pmod{p_i}$$

where $p_i - 1 = 2^{b_i} t_i$, with b_i a positive integer and t_i an odd positive integer.

(e) Suppose $n = p_1 \cdots p_r$, with $r \geq 3$. Prove Miller's Theorem in this case.

(f) Suppose $n = pq$. Prove Miller's Theorem in this case.

16. For what values of m is $\lambda(m) = \varphi(m)$?

17. Show that the function $\lambda(n)$ can be used to solve the linear congruence $ax \equiv b \pmod{n}$. Solve $6x \equiv 7 \pmod{145}$.

ADDITIONAL PROBLEMS FOR CHAPTER TWO

GENERAL PROBLEMS

1. Let m and n be integers and p a prime such that $p \nmid mn$. Let $r \geq 1$ be such that $p^r \| (m - n)$.

   (a) If p is odd, then $p^{r+1} \| (m^p - n^p)$.

   (b) If $p = 2$ and $r > 1$, then $2^{r+1} \| (m^2 - n^2)$.

2. (a) Let $(a, m) = 1$ and let x_1 be a solution of $ax \equiv 1 \pmod{m}$. Let

$$x_s = \frac{1}{a} - \frac{1}{a}(1 - ax_1)^s$$

   for $s = 1, 2, \ldots$. Then x_s is an integer and $ax_s \equiv 1 \pmod{m^s}$.

   (b) If $a = \pm 1$, then the solution of $ax \equiv 1 \pmod{m^s}$ is $x \equiv a \pmod{m^s}$.

   (c) If $a = \pm 2$, then m is odd and

$$x \equiv \tfrac{1}{2}(1 - m^s) \cdot \tfrac{1}{2}a \pmod{m^s}$$

is the solution of $ax \equiv 1 \pmod{m^s}$.

(d) For all other a show that the solution of $ax \equiv 1 \pmod{m^s}$ is $x \equiv k \pmod{m^s}$, where k is the nearest integer to

$$-\frac{1}{a}(1 - ax_1)^s.$$

(e) Can these results be generalized to $ax \equiv b \pmod{m^s}$?

3. (a) If p is a prime and a, b and c are integers such that $p \nmid ab$, then the congruence

$$ax^2 + by^2 \equiv c \pmod{p}$$

is solvable.

(b) Show that $ax^m + by^n \equiv c \pmod{p}$ has the same number of solutions as $ax^\mu + by^\nu \equiv c \pmod{p}$, where

$$\mu = (m, p-1) \text{ and } \nu = (n, p-1).$$

4. (a) If $p > 5$, p a prime, and $m \geq 1$, then $(p - 1)! + 1 = p^m$ is impossible.

(b) Find all positive integers n such that $(n - 1)! + 1 = n^2$.

5. Show that if $(A, M) = 1$, $m = p_1^{a_1} \cdots p_r^{a_r}$ and y is chosen so that $y \equiv y_k \pmod{p_k}$, $1 \leq k \leq r$, where

$$y_k = \begin{cases} 0 & \text{if } p_k \mid A \\ 1 & \text{if } p_k \nmid A \end{cases}.$$

Then $(A + My) = 1$.

6. (a) Find all incongruent solutions to $x + y + z \equiv 1 \pmod{7}$.

(b) Find all incongruent solutions to $x + 2y + 3z \equiv 0 \pmod{11}$.

7. Let

$$F(a,q) = \frac{a^q - 1}{a - 1},$$

where q is an odd prime. If $p \mid F(a, q)$, then either

(a) $a \equiv 1 \pmod{p}$ and $p = q$

or

(b) $a \not\equiv 1 \pmod{p}$ and $p = 2qk + 1$ for some integer k.

8. Let g and h be primitive roots modulo m. If $\text{ind}_g h = \text{ind}_h g$, what can be said about the relationship between g and h.

9. Show that if a, b and c are integers such that $a^3 + b^3 = c^3$, then $3 \mid abc$.

10. Let n be a positive integer and let p_k denote the $k\underline{th}$ prime. Then there exists a sequence of n consecutive integers $m_1, \ldots, m_n$ such that $p_k \mid m_k$, $1 \le k \le n$.

11. Show that the equation $x^3 + 117y^3 = k$ has no solutions in integers if $k = 5$. Can one make a similar statement for other values of k? (Hint: work modulo 9.)

12. Let $\{a_k\}$ be a set of integers. What are necessary and sufficient conditions so that the system of congruences

$$x \equiv 2a_k^2 \pmod{2a_k - 1}$$

   is solvable.

13. Let $(m, n) = 1$. Suppose a has order e modulo m and has order f modulo n. Show that a has order $[e, f]$ modulo mn.

14. Let p be a prime of the form $4n + 3$. If $a^{(p-1)/2} \equiv 1 \pmod{p}$, then $x = \pm a^{n+1}$ satisfies $x^2 \equiv a \pmod{p}$.

15. If n and k are integers, then $n^{2k+4} \equiv n^{2k} \pmod{20}$. Show that this is still true if the 4 is replaced by 8.

16. Let p be an odd prime. Suppose that a is an integer less than p that belongs to the exponent m modulo p. Define the integer b by $b \equiv a^{m-1} \pmod{p}$, $b < p$. Show that either a or b belongs to the exponent mp^{r-1} modulo p^r.

17. (a) If $m \equiv 1 \pmod{4}$ and 2^s is the highest power of 2 dividing $m - 1$, show that, for $r \ge 3$, m belongs to the exponent 2^{r-s} modulo 2^r.

    (b) If $m \equiv 3 \pmod{4}$ and 2^s is the highest power of 2 dividing $m + 1$, show that if $r > s$, then m belongs to the exponent 2^{r-s} modulo 2^r and if $r = s$, then m belongs to the exponent 2 modulo 2^r.

18. If $m \ge 1$ is an integer, show that $10^{10^m} \equiv 4 \pmod{7}$. Prove an analogous result when 7 is replaced by 13.

19. Show that if p is an odd prime, then
    (a) $p \mid a^r + 1$ if and only if $\mathrm{ord}_p a$ is even
    and
    (b) $p \mid a^{2s+1} - 1$ if and only if $\mathrm{ord}_p a$ is odd.

20. Suppose $a^4 + 1 \equiv 0 \pmod{p}$ and $b^8 \equiv 1 \pmod{p}$, where p is an odd prime. Show that b, ab, a^2b and a^3b are solutions to

$$x^8 + 1 \equiv 0 \pmod{p}$$

21. Consider the system of congruences

$$x \equiv a_1 \pmod{m_1}, \ldots, x \equiv a_r \pmod{m_r},$$

where the a_i are integers and the m_i are pairwise relatively prime, $1 \leq i \leq r$.

(a) Define the integers b_i, $1 \leq i \leq r$, by

$$b_i \equiv \delta_{ij} \pmod{m_j}$$

where the δ_{ij} is the Kronecker delta. Show that

$$\sum_{i=1}^{n} a_i b_i$$

is a solution to the system of congruences. (This is an analogue of the Lagrange Interpolation Formula and is apparently due to M. Riesz.)

(b) The following is an analogue of Newton's Divided Difference Formula. Define the integers d_i, $1 \leq i \leq r$, by

$$d_1 \equiv a_1 \pmod{m_1}$$
$$d_1 + d_2 m_1 \equiv a_2 \pmod{m_2}$$
$$\vdots$$
$$d_1 + d_2 m_1 + d_3 m_1 m_2 + \cdots + d_r m_1 m_2 \cdots m_{r-1} \equiv a_r \pmod{m_r}.$$

Then

$$x = d_1 + \sum_{i=2}^{r} d_i \prod_{j=1}^{i-1} m_j$$

is a solution of the system of congruences.

22. Show that if p is an odd prime, then

$$3 \cdot 2^{p-1} - 2 \equiv \pm \binom{p-1}{[p/4]} \pmod{p^2}.$$

Determine when each sign occurs.

23. Let p be a prime. Prove the following results.

(a) If p is odd, then $(p + 1)/2$ is the multiplicative inverse of 2 modulo p.

(b) If $p \neq 3$, then either $(p + 1)/3$ or $(p + 2)(p + 1)/6$ is the multiplicative inverse of 3 modulo p, depending on the residue class of p modulo 3.

(c) Determine the multiplicative inverse of 6 modulo p if $p > 3$.

24. Let $p \geq 5$ be a prime. Show that there are at least two distinct primes P and Q such that $1 < P, Q < p -1$ and $P^{p-1}, Q^{p-1} \not\equiv 1 \pmod{p^2}$.

25. Let $\pi_2(x)$ denote the number of primes $p \leq x$ such that p and $p + 2$ are both primes. Show that

$$\pi_2(x) = 2 + \sum_{7 \leq n \leq x} \sin\left(\frac{\pi}{2} n \left[\frac{n!}{n+2}\right]\right) \sin\left(\frac{\pi}{2} n \left[\frac{(n-2)!}{n}\right]\right).$$

26. Let $P(x) = 2x^3 + 3x^2 - 1$ and define the sequence $\{a_n\}$ by

$$a_1 = 4, a_{n+1} = P(a_n) - 10^{2^n}\left[P(a_n)/10^{2^n}\right], n > 1.$$

Show that if $e = 2^{n-1}$, then an satisfies $x^3 \equiv x \pmod{10^e}$.

27. Let $p > 3$ be a prime and let m, $1 \leq m \leq p - 1$, be a fixed integer. For each k, $1 \leq k \leq p - 1$, determine x_k by $kx_k \equiv m \pmod{p}$. Determine n_k by $kx_k = m + pn_k$. Show that

$$\sum_{k=1}^{p-1} kn_k \equiv \frac{p-m}{2} \pmod{p}.$$

28. If we label the days of the week 0 = Sunday, ..., 6 = Saturday, determine what day of the week a given date falls. In particular, what day of the week was 5 February 1948?

29. Show that n and $n + k$, $n > k$, k even, are both primes if and only if

$$(k!)^2[(n-1)! + 1] + n(k! - 1)(k-1)! \equiv 0 \pmod{n(n+k)}.$$

(Hint: see Problem 6 of Problem Set 2.3.)

30. This problem is designed to prove a version of Chevalley's Theorem. We begin with some notation and definitions. In what follows f, g and h will be polynomials with integer coefficients in n variables. We let deg(f) denote the total degree of f. Let p be a fixed prime.

We say that f and g are **equivalent** modulo p, written $f \sim g$, if for all sets $\{a_1, ..., a_n\}$ of integers, with $1 \leq a_k \leq p$, for $1 \leq k \leq n$, we have

$$f(a_1, ..., a_n) \equiv g(a_1, ..., a_n) \pmod{p}.$$

We say that f is **congruent** to g modulo p, written $f \equiv g$, if all the coefficients of the corresponding monomials of f and g are congruent modulo p. Finally, we say that f is **reduced** modulo p if f has degree less than p in each of its variables.

(a) Show that if $f \equiv g$, then $f \sim g$. Find an example to show that the converse does not hold.

(b) Given a polynomial f show that we can find a reduced polynomial F such that $\deg(F) \leq \deg(f)$ and $F \sim f$ modulo p.

(c) Show that if f and g are reduced polynomials and $f \sim g$ modulo p, then $f \equiv g$ modulo p.

(d) (Chevalley) Suppose f and g are polynomials in n variables with degrees less than n. Show that

(i) if the congruence

$$f(x_1, \ldots, x_n) \equiv 0 \pmod{p}$$

is soluble, then it has at least two solutions.

(ii) if g is a homogeneous polynomial, then the congruence

$$g(x_1, \ldots, x_n) \equiv 0 \pmod{p}$$

has a solution distinct from $(0, \ldots, 0)$.

31. Show that if $(m, n) = 1$, then

$$m^{\varphi(n)} + n^{\varphi(m)} \equiv 1 \pmod{mn}.$$

32. If $p = ef + 1$ is a prime and N is the number of solutions of the congruence $n^e \equiv a \pmod{p}$ with $1 \le n \le p - 1$, then

$$N = \begin{cases} e & \text{if } a^f \equiv 1 \pmod{p} \\ 0 & \text{else} \end{cases}.$$

33. Let $u_k = 4 \cdot 14^k + 1$, for $k \ge 0$. Show that

$$u_{2k+1} \equiv 0 \pmod{3} \text{ and } u_{2k} \equiv 0 \pmod{5}.$$

Prove an analogous result for the sequence $w_k = 521 \cdot 12^k + 1$, $k \ge 0$. (Hint: you must split the even numbers.)

34. The following is a generalization of Thue's Theorem. Let m, a_k, $1 \le k \le n$, be integers and let t_k, $1 \le k \le n$, be real numbers with $t_1 \cdots t_n = m$. Show that the congruence

$$a_1 x_1 + \cdots + a_n x_n \equiv 0 \pmod{m}$$

always has a solution satisfying $|x_k| \le t_k$, $1 \le k \le n$. Prove an analogous generalization of Scholz's Theorem. (See Problem 10 of Problem Set 2.3.)

35. Show that a necessary and sufficient condition for the congruence

$$a_1 x_1 + \cdots + a_n x_n + b \equiv 0 \pmod{m}$$

to have a solution is that $(a_1, \ldots, a_n, b) \mid m$. If this is the case, then there are $m^{n-1}(a_1, \ldots, a_n, m)$ incongruent solutions modulo m.

36. Use the theory of indices to find the remainder of 31658712 when it is divided by 17.

37. Prove that if p and q are odd primes with $q > 3$ and $q \mid R_p$, then $q = 2kp + 1$ for some integer k. Find the smallest prime divisors of R_5 and R_{11}.

38. (a) Show that 3 is a primitive root modulo 17 and construct the corresponding table of indices.

    (b) Using the result of (a) show that 5 is a primitive root of 17.

    (c) Use the change of base formula to construct the corresponding table of indices for the primitive root 5.

39. If p and $q = 2p + 1$ are both primes, show that if $p \equiv 1$ (mod 4), then $p + 1$ is a primitive root modulo q and if $p \equiv 3$ (mod 4), then p is a primitive root modulo q.

40. (a) Show that if p is an odd prime and $q = 8p + 1$ is a prime, then ± 6 are primitive roots modulo q and ± 3 is a primitive root modulo q if $p \neq 5$.

    (b) Show that if p is an odd prime and $q = 16p + 1$ is prime, then ± 3 and ± 6 are primitive roots modulo q.

41. Let p be an odd prime and suppose $p \equiv 1$ (mod 4). Show that if $3\varphi(p-1) > p-1$, then there exists an integer k, $1 \leq k \leq p - 1$, $(k, p - 1) = 1$ and k is a primitive root modulo p.

42. Show that if $a \equiv b$ (mod k^n), then $a^k \equiv b^k$ (mod k^{2n}).

43. An integer n is said to be **square-free** if $p|n$ implies that $p^2 \nmid n$. If k is a positive integer, show that there exist k consecutive integers that are not square-free.

44. Let $S_k(n)$ be as in Problem 16 of Problem Set 2.1. Show that if $p_1, \ldots, p_k$ are primes such that $p_j \mid m$ and $(p_j - 1) \mid 2n$, $1 \leq j \leq k$, then

$$\frac{S_{2n}(m)}{m} + \frac{1}{p_1} + \cdots + \frac{1}{p_k}$$

   is an integer. (See Problem 8 of Problem Set 2.2. This result is a variation of the Von Staudt Theorem.)

45. Show that if p is a prime, then

$$\binom{pa}{pb} \equiv \binom{a}{b} \ (\text{mod } p).$$

46. Let a and m be relatively prime integers and let

$$x \equiv a - 12 \sum_{k=1}^{m-1} k \left\lfloor \frac{ka}{m} \right\rfloor \ (\text{mod } m).$$

   Show that $ax \equiv 1$ (mod m).

47. Show that if p is a prime, then

$$\sum_{j=0}^{p} \binom{p}{j}\binom{p+j}{j} \equiv 2^p + 1 \ (\text{mod } p^2).$$

48. Show that if $k \geq 3$, then the solutions to $x^2 \equiv 1 \pmod{2^k}$ are 1, $2^{k-1} - 1$, $2^{k-1} + 1$ and $2^k - 1$.

49. Suppose $N = p_1 \cdots p_r$ is a square-free integer. Show that

$$\sum_{k=1}^{r} \left(\frac{N}{p_k} \right)^{p_k - 1} \equiv 1 \pmod{N}.$$

50. If $k > 0$ and $m \geq 1$, then show that $x \equiv 1 \pmod{m^k}$ implies that $x^m \equiv 1 \pmod{m^{k+1}}$.

51. For which positive integers k does $kx \equiv 1 \pmod{k(k+1)/2}$ have a solution?

52. Find the smallest positive integer m such that $2^a \mid m$ and $3^a \mid (m + 1)$, where a is a given positive integer.

53. Let $p > 3$ be a prime.

   (a) If m and n are any positive integers, show that

   $$\binom{mp - 1}{p - 1} \equiv \binom{np - 1}{p - 1} \pmod{p^3}.$$

   (b) Show that if a and n are positive integers, then

   $$\binom{ap^n}{p} \equiv ap^{n-1} \pmod{p^{2n+1}}.$$

54. If n is a positive integer, let

   $$A_n = \left\{ a^n - a : a = 2, 3, \ldots \right\}.$$

   Show that the greatest common divisor of the elements in A_n is the product of all those primes p such that $(p - 1) \mid (n - 1)$.

55. Prove the following result of E. Lucas. Let p be a prime and let a and b be positive integers. Suppose that

   $$a = a_0 + a_1 p + \cdots + a_r p^r$$
   $$b = b_0 + b_1 p + \cdots + b_r p^r,$$

   where, for $0 \leq k \leq r$, we have $0 \leq a_k < p$ and $0 \leq b_k < p$. Then

   $$\binom{a}{b} \equiv \binom{a_0}{b_0}\binom{a_1}{b_1} \cdots \binom{a_r}{b_r} \pmod{p}.$$

   Can you prove analogous results for higher powers of p?

56. Let a, b and m be positive integers. Show that

   $$a^{\varphi(m)} b \equiv b \pmod{m}$$

   if and only if $(ab, m) = (b, m)$.

57. The following is an analog of the Lagrange Interpolation Formula from numerical analysis.

   (a) Let n be a nonnegative integer and let p be a prime. Suppose that $a_0, \ldots, a_n$ are pairwise incongruent modulo p and that $b_0, \ldots, b_n$ are any integers. Show that there exists a polynomial f of degree at most n such that $f(a_k) \equiv b_k \pmod{p}$, $0 \le k \le n$. (Hint: let

   $$g(x) = \prod_{k=0}^{n}(x - a_k).$$

   Show that $g'(a_k) \not\equiv 0 \pmod{p}$, $0 \le k \le n$.)

   (b) As an application of this result, show that if $p \nmid a$ and b is a primitive root modulo p, then

   $$\mathrm{ind}_b a = -1 + \sum_{j=1}^{p-2}(b^{-j} - 1)^{-1}a^j,$$

   where the inverses are taken modulo p.

   (c) If $a \not\equiv 1 \pmod{p}$, then show that the results of (b) reduces to

   $$\mathrm{ind}_b a = \sum_{j=1}^{p-2}(1 - b^j)^{-1}a^j.$$

58. If p is a prime and n is a positive integer so that $(n, p - 1) = 1$, show that the integers

   $$1^n, \ 2^n, \ \ldots, \ (p - 1)^n$$

   form a complete set of residues modulo p. Prove the converse also.

59. Let m be a positive integer and let $f(x)$ be a polynomial with integer coefficients. Let N be the number of incongruent solutions to the congruence

   $$f(x) \equiv 0 \pmod{m}.$$

   Show that

   $$mN = \sum_{k=0}^{m-1}\sum_{a=0}^{m-1} e^{2\pi i k f(a)/m}.$$

   What does this equation say about the number of solutions of the linear congruence

   $$Ax \equiv B \pmod{m}?$$

COMPUTER PROBLEMS

   1. Given three numbers a, b and m write a program to determine whether or not $a \equiv b \pmod{m}$.

2. (a) Given a positive integer m, write a program to determine the least nonnegative of entered integers.

   (b) Let $m = 10001$. Determine the least nonnegative remainders of the numbers -50034, -123679, 453722, 1834677, 10811231 and 453572891.

3. (a) Let a, m and n be three given positive integers. Write a program to implement the second method of Example 2.2 to find the value of a^n modulo m.

   (b) Compute a table of values of a^n modulo m for $1 < m \le 15$, $1 \le n \le 20$ and $1 \le a < m$.

   (c) Modify the program in (b) to only note those a's, n's and m's so that $a^n \equiv 1 \pmod{m}$.

4. (a) Write a program that solves the congruence $ax \equiv b \pmod{m}$. Make sure it checks that the equation is solvable.

   (i) Use Euler's Theorem (Corollary 2.8.8) in one version.

   (ii) Use the Euclidean Algorithm, as in Example 1.6, to produce another version.

   (b) Solve the following congruences.

   (i) $63131x \equiv 5743 \pmod{81214}$.

   (ii) $7023x \equiv 8131 \pmod{11303}$.

   (iii) $271828x \equiv 637421 \pmod{89375}$.

5. (a) Write a program to implement the Chinese Remainder Theorem to solve simultaneous congruences.

   (b) Solve the following systems of congruences.

   (i) $11x \equiv 38 \pmod{45}$ and $6x \equiv 46 \pmod{122}$.

   (ii) $7x \equiv 47 \pmod{55}$ and $13x \equiv 97 \pmod{128}$.

   (iii) $-17x \equiv 49 \pmod{73}$, $-16x \equiv 8 \pmod{237}$ and $4x \equiv 5 \pmod{37}$.

6. (a) Write a program to solve systems of congruences in two variables:

$$ax + by + f \equiv 0 \pmod{m}$$
$$cx + dy + e \equiv 0 \pmod{m}.$$

   (See Problem 12 of Problem Set 2.3.)

   (b) Solve the systems:

   (i)
$$x + 2y + 1 \equiv 0 \pmod{7}$$
$$2x + 3y - 1 \equiv 0 \pmod{7}.$$

   (ii)
$$2x + 3y + 5 \equiv 0 \pmod{13}$$
$$7x - 2y + 11 \equiv 0 \pmod{13}.$$

   (iii)
$$x + y + 1 \equiv 0 \pmod{5}$$
$$4x + 5y + 3 \equiv 0 \pmod{5}.$$

7. A prime p is called a **Wilson prime** if $(p - 1)! \equiv -1 \pmod{p^2}$. Write a program to find Wilson primes. Find the first three Wilson primes.

8. (a) Write a program to solve polynomial congruences of the form $f(x) \equiv 0 \pmod{p}$ by trial and error. Solve the following congruences:

   (i) $x^3 + 5x^2 - 6x + 7 \equiv 0 \pmod 7$.

   (ii) $x^7 + 3x + 2 \equiv 0 \pmod{11}$.

   (You may find Horner's method of evaluating polynomials, namely,

   $$a_n x^n + a_{n-1} x^{n-1} + \ldots + a_1 x + a_0 = a_0 + x(a_1 + x(a_2 + \cdots + x(a_{n-1} + a_n x) \cdots)),$$

   to be of some use.)

   (b) Write a program to solve polynomial congruences of the form $f(x) \equiv 0 \pmod{p^n}$, $n \geq 2$, as illustrated in Example 2.11. Solve the congruences:

   (i) $x^3 + 5x^2 - 6x + 7 \equiv 0 \pmod{7^3}$

   (ii) $x^7 + 3x + 2 \equiv 0 \pmod{11^4}$.

   (c) Write a program to solve polynomial congruences for a general modulus. Solve the congruence

   $$x^3 + 4x^2 - 2x + 15 \equiv 0 \pmod{3025}.$$

9. (a) Write a program to compute the least primitive root for all primes less than 100.

   (b) Modify the program in (a) so that it finds all primitive roots of a given prime. Find all the primitive roots of $p = 73$, 101 and 199.

   (c) Modify the program in (a) to determine the primitive roots of p^2 and hence for p^n, $n \geq 2$.

10. Write a program so that given a prime p and a primitive root of p, g, one obtains a table of indices. Construct a table of indices for the pairs (p, g): $(7, 3)$, $(41, 6)$ and $(103, 5)$.

11. Write a program, using indices, to solve congruences of the form $ax^n \equiv b \pmod m$. Solve the following congruences:

    (a) $3x^7 \equiv 5 \pmod 7$.

    (b) $4x^3 \equiv 1 \pmod 7$.

    (c) $x^5 \equiv 1 \pmod{41}$.

    (d) $6x^{11} \equiv 5 \pmod{103}$.

12. Write a program, using indices, to solve congruences of the form $ab^x \equiv c \pmod m$. Solve the following congruences:

   (a) $2^x \equiv 3 \pmod 7$.

   (b) $3 \cdot 5^x \equiv 7 \pmod{41}$.

   (c) $11^x \equiv 13 \pmod{103}$.

13. (a) Write a program which takes two given numbers (a, b) and uses them in an affine code $C \equiv aP + b \pmod{26}$. Encode

   (i)   There is never enough time. (Use $(11, 17)$.)

   (ii) Sometimes the night is better. (Use $(3, 24)$.)

   (iii) This makes sense. (Use $(15, 8)$.)

   (b) Write a program which can be used to decode the affine code $C \equiv aP + b$ $\pmod{26}$. Decode the messages

   (i)   BXPH GTMC TXUT. (Use $(3, 7)$.)

   (ii) ZLTR UFYL OOCU ZYPX QFPLY. (Use $(5, 11)$.)

   (iii) LCBE IPUH JQBL TFVS. (Use $(11, 3)$.)

14. (a) Use brute force, that is try all possible values of a and b to decipher the message:

                     SKHN UHNE CHEC UPCU JEWP JCUK.

   (b) Use frequency analysis to decipher the message in (a). Use the fact the e, n, d and a are the most common letters in the plaintext to determine a and b and hence decipher the message.

15. (a) Given a quadruple (a, b, c, d) such that $(ad - bc, 26) = 1$ write a program to encode by the digraphic code
$$C_1 \equiv aP_1 + bP_2 \pmod{26}$$
$$C_2 \equiv cP_1 + dP_2 \pmod{26}.$$

   (b) Encode the messages:

   (i)   Who truly knows me. (Use $(1, 2, 3, 5)$.)

   (ii) In manifold being. (Use $(5, 3, 11, 6)$.)

   (iii) Everywhere present. (Use $(11, 13, 2, 5)$.)

16. (a) Write a program to decode digraphic codes.

   (b) Decode the messages:

   (i)   YWYQ WIMY FFEY. (Use $(3, 7, 6, 11)$.)

   (ii) PYKQ UMDJ RPPM IHEC FIQE. (Use $(13, 4, 9, 5)$.)

17. (a) Write a program to encipher using the public key (e, n).

   (b) Use the key $(3, 731)$ to encode the message

              I am the beginning, the life-span and the end of all.

(c) Use the key $(11, 2701)$ to encode the message

I come as Time, the master of the peoples.

18. (a) Write a program to decipher using the public key (e, n).
    (b) Use the public key $(7, 2747)$ to decode the message

    0054 1513 2725 0528 2596 1294 1116 2466

    1811 2717 0267 0533 2592 1801 2627 1811.

    (c) Use the key $(5, 2881)$ to decode the message

    1718 0580 0507 2479 1077 0855 2551 2046 1316

    0216 1176 0935 0543 2177 1724 0399 0223 0348.

19. Write a program to determine when $6m + 1$, $12m + 1$ and $18m + 1$ are all simultaneously prime. Find the first five Carmichael numbers of this form.

20. Find all Carmichael numbers of the form $7pq$, where $7 < p < q$. Can you extend this type of search much farther on your machine in a feasible length of time?

21. (a) Write a program that computes $\lambda(n)$ given the prime factorization of n. Compute $\lambda(n)$ for $1 \le n \le 100$.
    (b) Write a program that uses $\lambda(n)$ to solve linear congruences of the form $ax \equiv b \pmod{m}$. Solve the following congruences:
    (i) $7x \equiv 9 \pmod{153}$.
    (ii) $15x \equiv 17 \pmod{17} \pmod{403}$.
    (iii) $127x \equiv 5 \pmod{1337}$.

22. This exercise outlines another method of factorization, namely the Pollard rho method. Let m be a composite number and let p be its least prime factor. The idea is to find a set of integers $x_1, \ldots, x_k$ such that their least nonnegative remainders are distinct modulo m, but not modulo p. Then we have $x_i \equiv x_j \pmod{p}$, for some $1 \le i < j \le k$, but $x_i \not\equiv x_j \pmod{m}$. Thus $(x_i - x_j, m)$ is a nontrivial divisor of m since p divides $x_i - x_j$, but m does not divide $x_i - x_j$. Finding $(x_i - x_j, m)$ is easy by the Euclidean Algorithm. We describe a way to find a set of integers without all the trial and error.

Let f be a polynomial with integer coefficients and let x_1 be any integer. We define the sequence $\{x_k\}$ by $x_{k+1} \equiv f(x_k) \pmod{m}$, for $k \ge 1$ with $0 \le x_{k+1} < m$. If d is a positive integer and for some i and j we have $x_i \equiv x_j \pmod{d}$, then, by (h) of Theorem 2.1, we have $x_{i+1} \equiv f(x_i) \equiv f(x_j) \equiv x_{j+1} \pmod{d}$.

(a) Show that if $x_i \equiv x_j \pmod{d}$ for some i and j, then the sequence $\{x_k\}$ is periodic. Show that there is an s such that $x_s \equiv x_{2s} \pmod{d}$.

Thus we look at $(x_{2k} - x_k, m)$ and when we find a k such that $1 < x_{2k} - x_k < m$, we have found a nontrivial factor of m. In practice one often takes $f(x) = x^2 + 1$ and $x_1 = 2$.

(b) Write a program to implement the rho method with $x_1 = 2$ and $f(x) = x^2 + 1$. Use this program to factor 10001, 36287, 8496167 and 593505099.

(c) Modify the program in (b) to use other polynomials and other starting values. Use this program to factorize $n = 96139$ with

   (i) $x_1 = 2$, $f(x) = x^2 + 1$.

  (ii) $x_1 = 3$, $f(x) = x^2 + 1$.

 (iii) $x_1 = 2$, $f(x) = x^2 - 1$.

 (iv) $x_1 = 2$, $f(x) = x^3 + x + 1$.

  (v) $x_1 = 3$, $f(x) = x^3 + x + 1$.

Compare the results obtained. Which is faster?

(d) Use the program in (b) to factorize $n = 9973$. What happened?

(e) What happens if we take $f(x)$ to be a linear polynomial?

CHAPTER THREE
Quadratic Residues

3.1. INTRODUCTION

In the last chapter we concerned ourselves with solving the general polynomial congruence. In this chapter we return to a special case:

(3.1) $$ax^2 + bx + c \equiv 0 \ (\text{mod } m)$$

For this case, unlike the general case, we can make good progress on the problem of when this congruence is solvable. In fact it was the success on quadratic congruences that spurred research on higher degree congruences. See, for example, [20].

From the results of the previous chapter we know that we must first solve

(3.2) $$ax^2 + bx + c \equiv 0 \ (\text{mod } p)$$

for every prime p that divides m.

Since the case $p = 2$ is easy we will assume that p is an odd prime, except in Section 5 below where we will do powers of 2. If $p \mid a$, then we have the congruence

$$bx + c \equiv 0 \ (\text{mod } p),$$

which is no longer a quadratic and may be solved by the methods of Section 2.3. If $p \nmid a$, then multiply the congruence in (3.2) by $4a$. Since $(p, 4a) = 1$ (3.2) is equivalent to

$$4a^2x^2 + 4abx + 4ac \equiv 0 \ (\text{mod } p)$$

or

$$(2ax + b)^2 \equiv b^2 - 4ac \ (\text{mod } p).$$

Thus we must solve the system

$$z^2 \equiv D \ (\text{mod } p)$$
$$2ax + b \equiv z \ (\text{mod } p),$$

where $D = b^2 - 4ac$. Since $(p, 2a) = 1$ we know the linear congruence is solvable so the problem is to determine when the congruence

(3.3) $z^2 \equiv D \pmod{p}$

is solvable.

This last problem will take up the bulk of this chapter. In Section 6 we will return to the congruence (3.1).

The above remarks are historically misleading. The idea of quadratic residues (Definition 3.1) did not arise as a special case of the general theory of congruences. If anything it was the other way around. See the books by Cox [20] and Weil [126] for the historical route. What basically happened was that Euler was trying to prove/generalize results of Fermat and was led to the study of quadratic residues.

As we saw in Chapter 2 the solution of the congruence (3.3), when $D = -1$, is intimately connected to the problem of whether or not a prime can be written as a sum of two squares. Suppose that p is a prime and given the integer A we can write

$$p = x^2 + Ay^2.$$

Since p is a prime we see that we must have $(p, y) = 1$. Thus, by Theorem 2.9, there is an integer w such that

$$yw \equiv 1 \pmod{p},$$

and so

$$(xw)^2 \equiv -A \pmod{p}.$$

Thus (3.3) is solvable modulo p when $D = -A$. Further research by Euler led him to arrive at the "Law of Quadratic Reciprocity," (Theorem 3.5), but in a somewhat different form than is presented in this theorem. Euler's form is that presented in Theorem 3.7. He did not give a proof, but from numerical evidence he was convinced of its truth. From Euler the path led to Legendre and finally ended with a complete proof by Gauss.

3.2. QUADRATIC RESIDUES

Definition 3.1: If there exists an integer x such that $x^2 \equiv a \pmod{m}$, where (a, m) = 1, then a is called a **quadratic residue** of m. If there does not exist such an x, then a is called a **quadratic nonresidue** of m.

From the remarks in the introduction and the general theory of polynomial congruences it suffices to find the quadratic residues of the odd primes. We begin with the following theorem.

Theorem 3.1: Let p be an odd prime. Then every reduced residue system mod p contains exactly $(p - 1)/2$ quadratic residues and exactly $(p - 1)/2$ quadratic nonresidues mod p. The quadratic residues belong to the residue classes containing the numbers

$$1^2, 2^2, 3^2, \ldots, ((p - 1)/2)^2.$$

Proof: Note that all of these numbers are distinct mod p. If not then we would have

$$x^2 \equiv y^2 \pmod{p},$$

and so

$$p \mid (x + y) \text{ or } p \mid (x - y),$$

by Lemma 1.19.1. Since $1 \le x \le (p - 1)/2$ and $1 \le y \le (p - 1)/2$ we have $2 \le x + y \le p - 1$ and $(3 - p)/2 \le x - y \le (p - 3)/2$. Thus $p \mid (x - y)$, and so $x = y$. Since $(p - k)^2 \equiv k^2 \pmod{p}$ we see that every quadratic residue mod p is congruent to one of these numbers. ∎

Example 3.1: Find all the quadratic residues of $p = 17$; $p = 31$.

Solution: The quadratic residues of 17 are in the residue classes represented by

$$1^2, 2^2, 3^2, 4^2, 5^2, 6^2, 7^2, 8^2$$

or taking least positive residues

$$1, 4, 9, 16, 8, 2, 15, 13.$$

Similarly the least positive residues of the quadratic residues of 31 are given by

$$1, 2, 4, 5, 7, 8, 9, 10, 14, 16, 18, 19, 20, 25, 28. \qquad \square$$

Definition 3.2: Let p be an odd prime. The **Legendre symbol**, $\left(\frac{a}{p}\right)$, is defined by

$$\left(\frac{a}{p}\right) = \begin{cases} +1 & \text{if } x^2 \equiv a \pmod{p} \text{ is solvable} \\ 0 & \text{if } p \mid a \\ -1 & \text{if } x^2 \equiv a \pmod{p} \text{ is not solvable} \end{cases}.$$

Example 3.2: Take $p = 31$. Then a reduced residue system is the set of all positive integers less than 31. This gives

$$\left(\tfrac{1}{31}\right) = \left(\tfrac{2}{31}\right) = \left(\tfrac{4}{31}\right) = \left(\tfrac{5}{31}\right) = \left(\tfrac{7}{31}\right) = \left(\tfrac{8}{31}\right) = \left(\tfrac{9}{31}\right) = \left(\tfrac{10}{31}\right)$$

$$= \left(\tfrac{14}{31}\right) = \left(\tfrac{16}{31}\right) = \left(\tfrac{18}{31}\right) = \left(\tfrac{19}{31}\right) = \left(\tfrac{20}{31}\right) = \left(\tfrac{25}{31}\right) = \left(\tfrac{28}{31}\right)$$

$$= +1$$

and

$$\left(\tfrac{3}{31}\right) = \left(\tfrac{6}{31}\right) = \left(\tfrac{11}{31}\right) = \left(\tfrac{12}{31}\right) = \left(\tfrac{13}{31}\right) = \left(\tfrac{15}{31}\right) = \left(\tfrac{17}{31}\right) = \left(\tfrac{21}{31}\right)$$

$$= \left(\tfrac{22}{31}\right) = \left(\tfrac{23}{31}\right) = \left(\tfrac{24}{31}\right) = \left(\tfrac{26}{31}\right) = \left(\tfrac{27}{31}\right) = \left(\tfrac{29}{31}\right) = \left(\tfrac{30}{31}\right)$$

$$= -1,$$

as desired. $\qquad\square$

The following theorem gives the basic properties of the Legendre Symbol.

Theorem 3.2: Let p be an odd prime and $(a, p) = (b, p) = 1$.

(a) $\left(\dfrac{a}{p}\right) \equiv a^{(p-1)/2} \pmod p$,

(b) $\left(\dfrac{a}{p}\right)\left(\dfrac{b}{p}\right) = \left(\dfrac{ab}{p}\right)$,

(c) if $a \equiv b \pmod p$, then $\left(\dfrac{a}{p}\right) = \left(\dfrac{b}{p}\right)$,

(d) $\left(\dfrac{a^2}{p}\right) = 1$,

(e) $\left(\dfrac{a^2 b}{p}\right) = \left(\dfrac{b}{p}\right)$

and

(f) $\left(\dfrac{1}{p}\right) = +1$ and $\left(\dfrac{-1}{p}\right) = (-1)^{(p-1)/2}$.

Proof of (a). If $\left(\tfrac{a}{p}\right) = +1$, then $x^2 \equiv a \pmod p$ is solvable and has a solution, say x_0. Then

$$a^{(p-1)/2} \equiv x_0^{p-1} \equiv +1 = \left(\dfrac{a}{p}\right) \pmod p.$$

If $\left(\tfrac{a}{p}\right) = -1$, then $x^2 \equiv a \pmod p$ is not solvable. Thus to each j in $1 \le j \le p - 1$ we can associate a unique i such that

$$ji \equiv a \pmod p,$$

where $0 \le i \le p - 1$. Since $p \nmid a$ we know $i \ne 0$. Since $\left(\frac{a}{p}\right) = -1$, no j is associated to itself, and so we can pair of the integers $1 \le j \le p - 1$ with their associates in $1 \le i \le p - 1$ so that $ji \equiv a \pmod{p}$. There are $(p - 1)/2$ such pairs and multiplying these pairs together gives, by Wilson's Theorem (Corollary 2.9.1),

$$a^{(p-1)/2} \equiv (p - 1)! \equiv -1 = \left(\frac{a}{p}\right) \pmod{p}.$$

(Note that if $\left(\frac{a}{p}\right) = 0$, then the result still holds since $\left(\frac{a}{p}\right) = 0$ implies $p \mid a$, and so $a^{(p-1)/2} \equiv 0 \pmod{p}$.)

Proof of (b) We have, by (a),

$$\left(\frac{a}{p}\right)\left(\frac{b}{p}\right) \equiv a^{(p-1)/2}b^{(p-1)/2} \equiv (ab)^{(p-1)/2} \equiv \left(\frac{ab}{p}\right) \pmod{p}.$$

Since the values of the Legendre symbol are +1, 0 or -1 only and p is an odd prime we see that $\left(\frac{a}{p}\right)\left(\frac{b}{p}\right) = \left(\frac{ab}{p}\right)$.

Proof of (c) If $a \equiv b \pmod{p}$, then, by (c) and (g) of Theorem 2.1,

$$\left(\frac{a}{p}\right) \equiv a^{(p-1)/2} \equiv b^{(p-1)/2} \equiv \left(\frac{b}{p}\right) \pmod{p}$$

and so $\left(\frac{a}{p}\right) = \left(\frac{b}{p}\right)$ by the argument above.

Proof of (d) We need only take $x = a$ to get a solution of $x^2 \equiv a^2 \pmod{p}$. Thus $\left(\frac{a^2}{p}\right) = +1$.

Proof of (e) This follows from (b) and (d).

Proof of (f) We have $\left(\frac{1}{p}\right) = \left(\frac{1^2}{p}\right) = +1$ by (d).

By (a) we have

$$\left(\frac{-1}{p}\right) \equiv (-1)^{(p-1)/2} \pmod{p}$$

and since both sides are either + 1 or -1 we see they must be equal. ∎

Part (a) of Theorem 3.2 is known as Euler's Criterion. At first glance it may look like an impractical way of determining the quadratic character of a number. However, if we recall the short cut to the evaluation of a^n modulo m discussed in Example 2.2 that uses the binary expansion of n, then the prospect of using Euler's Criterion shouldn't seem so daunting, especially on a computer. We illustrate with an example.

Example 3.3. Use Euler's Criterion to calculate $\left(\frac{-15}{61}\right)$.

Solution. We have

$$\left(\frac{-15}{61}\right) \equiv (-15)^{(p-1)/2} = (-15)^{30} = 15^{30} \pmod{61}.$$

Now

$$30 = 16 + 8 + 4 + 2,$$

and so

$$15^{30} = 15^{16} \cdot 15^8 \cdot 15^4 \cdot 15^2.$$

We have

$$15 \equiv 15 \pmod{61}$$
$$15^2 \equiv 15^2 \equiv 42 \pmod{61}$$
$$15^4 \equiv 42^2 \equiv 56 \pmod{61}$$
$$15^8 \equiv 56^2 \equiv 25 \pmod{61}$$
$$15^{16} \equiv 25^2 \equiv 15 \pmod{61}.$$

Thus

$$15^{30} \equiv 42 \cdot 56 \cdot 25 \cdot 15 \equiv 1 \pmod{p}.$$

Thus, by Euler's Criterion,

$$\left(\frac{-15}{61}\right) = +1.$$

□

Corollary 3.2.1: -1 is a quadratic residue of an odd prime p if and only if $p \equiv 1 \pmod 4$.

This corollary was proved in Chapter Two as Corollary 2.9.2 and was proved as a corollary to Wilson's Theorem. Later this fact was used to show that there exist positive integers x and y such that

$$p = x^2 + y^2$$

when $p \equiv 1 \pmod 4$. (See Corollary 2.10.1.) We shall see below how similar results may be obtained for other primes.

By (b) and (e) of Theorem 3.2 we see that it suffices to calculate $\left(\frac{2}{p}\right)$ and $\left(\frac{q}{p}\right)$, where p and q are odd primes. In Theorem 3.4 below we shall calculate $\left(\frac{2}{p}\right)$ and in Theorem 3.5, in the next section, we shall see how to calculate $\left(\frac{q}{p}\right)$. On the way to these results we will obtain two results that deal with calculating $\left(\frac{a}{p}\right)$ in general.

Theorem 3.3 (Gauss): Let p be an odd prime and $(a, p) = 1$. Let n denote the number of least nonnegative residues modulo p among the numbers

$$a, 2a, 3a,..., \{(p - 1)/2\}a$$

which exceed $p/2$. Then we have

$$\left(\frac{a}{p}\right) = (-1)^n.$$

Proof: Let $r_1, ..., r_n$ denote the residues that exceed $p/2$ and let $s_1, ..., s_k$ denote the remaining residues. Then the r_i and s_j are all distinct and none of them is zero since $p \nmid a$. Also $n + k = (p - 1)/2$. Now $0 < p - r_i < p2$, $1 \leq i \leq n$, and the numbers $p - r_i$ are distinct. If $p - r_i = s_j$ for some i and j, then, since $r_i \equiv ua$ and $s_j \equiv va$ for some $1 \leq u, v \leq (p - 1)/2$, we have $p - ua = va$. Since $p \nmid a$, this implies that $p \mid (u + v)$, which is impossible since $2 \leq u + v \leq p - 1$. Thus the numbers $p - r_1, ..., p - r_n, s_1, ..., s_k$ are all distinct, all greater than zero, all less than $p/2$ and are $n + k = (p - 1)/2$ in number. Thus they are the numbers $1, 2,..., (p - 1)/2$ in some order. Thus

$$(p - r_1)\cdots(p - r_n)s_1\cdots s_k = 1\cdots\left(\tfrac{p-1}{2}\right),$$

and so

$$(-1)^n r_1\cdots r_n s_1\cdots s_k \equiv \left(\tfrac{p-1}{2}\right)! \pmod{p}.$$

By definition of the r_i and s_j this is equivalent to

$$(-1)^n a \cdot 2a\cdots\left(\tfrac{p-1}{2}\right)a \equiv \left(\tfrac{p-1}{2}\right)! \pmod{p}$$

or, since $(p, ((p - 1)/2)!) = 1$,

$$(-1)^n a^{(p-1)/2} \equiv 1 \pmod{p}.$$

Thus

$$(-1)^n \equiv a^{(p-1)/2} \equiv \left(\frac{a}{p}\right) \pmod{p}$$

and since $(-1)^n$ and $\left(\frac{a}{p}\right)$ are either $+1$ or -1 we see that they are equal. ∎

Example 3.4: Compute $\left(\frac{6}{17}\right)$ using Theorem 3.3.

Solution: Here $(p - 1)/2 = 8$. We have

$6 \equiv 6, 2 \cdot 6 \equiv 12, 3 \cdot 6 \equiv 1, 4 \cdot 6 \equiv 7, 5 \cdot 6 \equiv 13, 6 \cdot 6 \equiv 2, 7 \cdot 6 \equiv 8, 8 \cdot 6 \equiv 14 \pmod{17}$.

Thus $n = 3$ and so

$$\left(\frac{6}{17}\right) = (-1)^3 = -1,$$

which says that 6 is a quadratic nonresidue of 17. □

Recall that $[x]$ denotes the greatest integer in x, that is, $[x]$ is the unique integer n such that

$$n \leq x < n + 1.$$

Theorem 3.4 (Eisenstein): If p is an odd prime and $(a, 2p) = 1$, then

$$\left(\frac{a}{p}\right) = (-1)^t,$$

where

$$t = \sum_{j=1}^{(p-1)/2} \left[\frac{ja}{p}\right].$$

Also we have

$$\left(\frac{2}{p}\right) = (-1)^{(p^2-1)/8}.$$

Proof: We continue the notation from the proof of Theorem 3.3.

The r_i and s_j are just the least positive remainders obtained upon dividing the integer ja by p, $1 \leq j \leq (p - 1)/2$. By Corollary 1.10.1, the quotient in this process is $[ja/p]$. If $(a, p) = 1$, we have

$$\sum_{j=1}^{(p-1)/2} ja = \sum_{j=1}^{(p-1)/2} p\left[\frac{ja}{p}\right] + \sum_{j=1}^{n} r_j + \sum_{j=1}^{k} s_j.$$

Also we have

$$\sum_{j=1}^{(p-1)/2} j = \sum_{j=1}^{n}(p - r_j) + \sum_{j=1}^{k} s_j = np - \sum_{j=1}^{n} r_j + \sum_{j=1}^{k} s_j.$$

Subtracting these two equations gives

$$(a-1)\sum_{j=1}^{(p-1)/2} j = p\left(\sum_{j=1}^{(p-1)/2}\left[\frac{ja}{p}\right] - n\right) + 2\sum_{j=1}^{n} r_j.$$

Now

$$\sum_{j=1}^{(p-1)/2} j = \frac{p^2 - 1}{8},$$

and so if we take congruences mod 2 we get

$$(a-1)\frac{p^2 - 1}{8} \equiv \sum_{j=1}^{(p-1)/2} \left[\frac{ja}{p}\right] - n \pmod{2}.$$

If a is odd, then $(a-1)\dfrac{p^2 - 1}{8} \equiv 0 \pmod{2}$, and so

$$n \equiv \sum_{j=1}^{(p-1)/2} \left[\frac{ja}{p}\right] \pmod{2}$$

and the first result follows from Theorem 3.3. If $a = 2$, then

$$n \equiv \frac{p^2 - 1}{8} \pmod{2},$$

since $[2j/p] = 0$ for $1 \le j \le (p - 1)/2$, and so the second result follows. ∎

Corollary 3.4.1: 2 is a quadratic residue of p if and only if $p \equiv 1$ or $7 \pmod 8$.

Example 3.5: Calculate $\left(\frac{7}{19}\right)$ using Theorem 3.4.

Solution: Here $(p - 1)/2 = 9$. Also

$$t = \sum_{j=1}^{9} \left[\frac{7j}{19}\right] = 12.$$

Thus

$$\left(\frac{7}{19}\right) = (-1)^{12} = +1,$$

and so 7 is a quadratic residue of 19.   □

As a corollary of Corollary 3.4.1 we give another primality test which is due to Euler.

Corollary 3.4.2 (Euler): Let $k > 1$ and let $p = 4k + 3$ be a prime. Then $2p + 1$ is a prime if and only if

$$2^p \equiv 1 \pmod{(2p + 1)}.$$

Proof: Suppose $2p + 1$ is a prime. Since $2p + 1 \equiv 7 \pmod 8$ we see that 2 is a quadratic residue of $2p + 1$ by Corollary 3.4.1. Then by (a) of Theorem 3.2, we have

$$2^p = 2^{(2p+1-1)/2} \equiv \left(\frac{2}{p}\right) = 1 \ (\text{mod } 2p+1).$$

Suppose

$$2^p \equiv 1 \ (\text{mod } 2p + 1).$$

In Theorem 2.20 we let $h = 2$ and $n = 2p + 1$. Then $h < p$ and $2^h = 4 \not\equiv 1 \ (\text{mod } n)$. Also

$$2^{n-1} \equiv 2^{2p} \equiv 1 \ (\text{mod } n).$$

Thus $n = 2p + 1$ is a prime. ∎

Let $M_p = 2^p - 1$. These numbers are called **Mersenne numbers** and those that are primes are called **Mersenne primes**. These are candidates for large primes and this corollary gives a condition for them to be composite. If $2p + 1$ is prime, then $(2p + 1) \mid M_p$. Since $k > 1$, we have $p > 3$. Thus

$$M_p = 2^p - 1 > 2p + 1,$$

and so M_p is composite. For example,

$$23 \mid M_{11}, 47 \mid M_{23}, 167 \mid M_{83}, 263 \mid M_{131},$$
$$359 \mid M_{179}, 383 \mid M_{191}, 479 \mid M_{239}, 503 \mid M_{251}.$$

Mersenne primes are of importance in the study of even perfect numbers, among other things. (See Theorem 7.15 as well as Theorem 7.55 for another primality test for Mersenne primes.)

Problem Set 3.2

1. Let r_1, r_2 and r be quadratic residues modulo p, an odd prime, and n_1, n_2 and n be quadratic nonresidues modulo p. Then $r_1 r_2$ and $n_1 n_2$ are both quadratic residues, while rn is a quadratic nonresidue.

2. Use Wilson's Theorem (Corollary 2.9.1) to show that if m is the number of quadratic nonresidues modulo p in the interval $[1, (p - 1)/2]$, where $p > 2$, then for $p \equiv 3 \ (\text{mod } 4)$ we have

$$\left(\frac{p-1}{2}\right)! \equiv (-1)^m \ (\text{mod } p).$$

What happens when $p \equiv 1 \ (\text{mod } 4)$? How many quadratic residues does such a prime have in $[1, (p - 1)/2]$?

3. (a) Let p be an odd prime. Then every primitive root, g, of p is a quadratic nonresidue modulo p. All quadratic nonresidues are congruent to one of g,

g^3, ..., g^{p-2} and all quadratic residues are congruent to one of g^2, g^4, ..., g^{p-1}.

(b) Use the result of (a) to show that the product of the quadratic residues is
$$\equiv \begin{cases} -1 & \text{if } p \equiv 1 \pmod 4 \\ +1 & \text{if } p \equiv 3 \pmod 4 \end{cases}.$$

(c) What can be said about the product of the quadratic nonresidues?

4. Calculate the value of the Legendre symbols $\left(\frac{133}{17}\right), \left(\frac{1009}{13}\right), \left(\frac{227}{7}\right)$ and $\left(\frac{11}{229}\right)$.

5. If $p = 4m + 1$ and $d \mid m$, then $\left(\frac{d}{p}\right) = +1$. (Hint: let q be a prime divisor of m and consider $q = 2$ and $q > 2$ separately.)

6. If p is a prime of the form $2^s + 1$, then every quadratic nonresidue is a primitive root of p.

7. Show that the number of solutions of the congruence
$$ax^2 + bx + c \equiv 0 \pmod p,$$
if $p \nmid a$, is
$$1 + \left(\frac{b^2 - 4ac}{p}\right).$$

8. (a) Prove that
$$\sum_{n=1}^{p-1} \left(\frac{n}{p}\right) = 0.$$

In general, if $p \nmid a$ show that
$$\sum_{m=0}^{p-1} \left(\frac{am + b}{p}\right) = 0.$$

What happens if $p \mid a$?

(b) If $p \equiv 1 \pmod 4$, show that
$$\sum_{n=1}^{p-1} n \left(\frac{n}{p}\right) = 0.$$

(Hint: if n goes from 1 to $p - 1$, so does $p - n$.)

(c) If $p \equiv 1 \pmod 4$, then
$$\sum_{\substack{n=1 \\ \left(\frac{n}{p}\right)=+1}}^{p-1} n = \frac{p(p-1)}{4}.$$

(Hint: evaluate
$$\sum_{\substack{n=1 \\ \left(\frac{n}{p}\right)=+1}}^{p-1} n + \sum_{\substack{n=1 \\ \left(\frac{n}{p}\right)=-1}}^{p-1} n.)$$

(d) Let m be the number of quadratic residues modulo p in the interval $(0, p/2)$.
Show that

$$m = \sum_{k=1}^{(p-1)/2} \left\{ \left[\frac{2k^2}{p} \right] - 2 \left[\frac{k^2}{p} \right] \right\}.$$

If $p \equiv 1 \pmod 4$, show that this sum reduces to $(p - 1)/4$.

9. Suppose that $u^2 \equiv a \pmod p$. Show that u and $p - u$ are the only solutions, modulo p, of the congruence $x^2 \equiv a \pmod p$.

10. (a) Show that $x^4 \equiv -1 \pmod p$ has a solution if and only if $p \equiv 1 \pmod 8$. (Hint: use (a) of Theorem 3.2.)

(b) Show that there are infinitely many primes of the form $8k + 1$. (Hint: modify the proof of Corollary 2.8.2 that is given after Corollary 2.9.3.)

11. Use Theorem 3.4 to calculate $\left(\frac{3}{13} \right), \left(\frac{7}{17} \right), \left(\frac{9}{19} \right)$ and $\left(\frac{13}{23} \right)$.

12.(a) Suppose that $p = 4k + 1$ and $q = 2p + 1$ are both primes. Show that 2 is a primitive root modulo q.

(b) If p and $q = 4p + 1$ are both primes, then 2 is a primitive root modulo q.

13. (a) Determine the value of $\left(\frac{-2}{p} \right)$.

(b) Modify the proof of Corollary 2.10.1 to show that if p is of the $8k + 1$ or $8k + 3$, then p can be uniquely expressed in the form $x^2 + 2y^2$.

14. Show that if $p \geq 7$ is a prime, then there are always two consecutive quadratic residues of p. (Hint: show that at least one of 2, 5 or 10 is a quadratic residue of p.)

15. Suppose p is an odd prime and a is an integer such that $(a, p) = 1$ and . Then $x^2 \equiv a \pmod{p^n}$ has exactly two solutions modulo p^n for every integer n. (Hint: Problem 9 above is the case $n = 1$.)

16. Show that if p and q are primes with $q = 2p + 1$ and $0 < m < \sqrt{q+1}$, then m is a primitive root modulo q if only if it is a quadratic nonresidue modulo q.

17. Show that $\left(\frac{2}{p} \right) = (-1)^{[(p+1)/4]}$.

18. Let q be the least positive quadratic nonresidue modulo p, where p is an odd prime.

(a) Show that q is a prime.

(b) Show that $q < \sqrt{p} + 1$. (Hint: show that $q, 2q, \ldots, (q - 1)q$ are all less than p.)

3.3. THE LAW OF QUADRATIC RECIPROCITY

In this section we shall give a result that is of great use in calculating $\left(\frac{q}{p}\right)$, where both p and are odd primes. The main result (Theorem 3.5) was empirically discovered by Euler, explicitly stated by Legendre and was first proved by Gauss. This result may be one of the most proved in mathematics, certainly in number theory. As of 1963 there were more than 150 proofs known, a good many of which were mere modifications of each other, no doubt.

Theorem 3. 5: Let p and q be distinct odd primes. Then

$$\left(\frac{p}{q}\right)\left(\frac{q}{p}\right) = (-1)^{\frac{p-1}{2} \cdot \frac{q-1}{2}}.$$

Proof: Let S be the set of all pairs of integers x and y such that $1 \leq x \leq (p - 1)/2$ and $1 \leq y \leq (q - 1)/2$. Then S has

$$\frac{p-1}{2} \cdot \frac{q-1}{2} = (p-1)(q-1)/4$$

elements in it. We separate S into two mutually disjoint sets S_1 and S_2 where

$$S_1 = \{(x,y) \in S: qx > py\} \text{ and } S_2 = \{(x,y) \in S: qx < py\}.$$

Note that there are no $(x, y) \in S$ such that $qx = py$ since $p \nmid q$ (p and q are distinct primes) and $p \nmid x$ ($1 \leq x \leq (p - 1)/2$). We may also write

$$S_1 = \{(x,y) \in S: 1 \leq x \leq (p-1)/2 \text{ and } 1 \leq y \leq qx/p\},$$

and so

$$|S_1| = \sum_{x=1}^{(p-1)/2} \left[\frac{qx}{p}\right].$$

Similarly we have

$$S_2 = \{(x,y) \in S: 1 \leq y \leq (q-1)/2 \text{ and } 1 \leq x \leq py/q\},$$

and so

$$|S_2| = \sum_{y=1}^{(q-1)/2} \left[\frac{py}{q}\right].$$

Since $S_1 \cap S_2 = \emptyset$ and $S_1 \cup S_2 = S$ we have

$$|S| = |S_1| + |S_2|.$$

Thus

$$\frac{p-1}{2} \cdot \frac{q-1}{2} = \sum_{x=1}^{(p-1)/2} \left[\frac{qx}{p} \right] + \sum_{y=1}^{(q-1)/2} \left[\frac{py}{q} \right].$$

Thus

$$\left(\frac{p}{q} \right) \left(\frac{q}{p} \right) = (-1)^{\frac{p-1}{2} \cdot \frac{q-1}{2}}.$$

by Theorem 3.4.   ∎

Since the value of $\left(\frac{q}{p} \right)$ is either +1 or -1, if p and q are odd primes and $p \neq q$, we have that $\left(\frac{q}{p} \right)\left(\frac{q}{p} \right) = +1$. This proves the following corollary.

Corollary 3.5.1: If p and q are distinct odd primes, then

$$\left(\frac{p}{q} \right) = \left(\frac{q}{p} \right)(-1)^{\frac{p-1}{2} \cdot \frac{q-1}{2}}.$$

We are now set to calculate the value of Legendre symbols.

Example 3.6: (a) Is $x^2 \equiv -42 \pmod{61}$ solvable?
(b) Determine the value of $\left(\frac{3}{p} \right)$, where $p > 3$.

Solution of (a). The congruence is solvable if and only if $\left(\frac{-42}{61} \right) = +1$. By (b) of Theorem 3. 2 we have

$$\left(\frac{-42}{61} \right) = \left(\frac{-1}{61} \right)\left(\frac{2}{61} \right)\left(\frac{3}{61} \right)\left(\frac{7}{61} \right).$$

Applying our results above we have

$$\left(\frac{-1}{61} \right) = (-1)^{(61-1)/2} = (-1)^{30} = +1,$$

$$\left(\frac{2}{61} \right) = (-1)^{(61^2-1)/8} = (-1)^{60 \cdot 62/8} = (-1)^{15 \cdot 31} = -1,$$

$$\left(\frac{3}{61} \right) = \left(\frac{61}{3} \right)(-1)^{(2/2)(60/2)} = \left(\frac{1}{3} \right)(-1)^{30} = +1,$$

$$\left(\frac{7}{61} \right) = \left(\frac{61}{7} \right)(-1)^{(6/2)(60/2)} = \left(\frac{5}{7} \right)(-1)^{90} = \left(\frac{5}{7} \right)$$

$$= \left(\frac{7}{5} \right)(-1)^{(4/2)(6/2)} = \left(\frac{2}{5} \right)(-1)^6 = \left(\frac{2}{5} \right) = (-1)^{(5^2-1)/8} = (-1)^3 = -1.$$

Thus

$$\left(\frac{-42}{61} \right) = (+1)(-1)(+1)(-1) = +1,$$

and so the equation is solvable, indeed the solutions, found by trial, are $x \equiv 18$ and 43 (mod 61).

Solution of (b) We have, by Theorem 3.5,

$$\left(\frac{3}{p}\right) = \left(\frac{p}{3}\right)(-1)^{(2/2)(p-1)/2} = \left(\frac{p}{3}\right)(-1)^{(p-1)/2}.$$

Now

$$\left(\frac{p}{3}\right) = \begin{cases} \left(\frac{1}{3}\right) = +1 & \text{if } p \equiv 1 \ (\text{mod } 3) \\ \left(\frac{-1}{3}\right) = -1 & \text{if } p \equiv 2 \ (\text{mod } 3) \end{cases}.$$

Also

$$(-1)^{(p-1)/2} = \begin{cases} +1 & \text{if } p \equiv 1 \ (\text{mod } 4) \\ -1 & \text{if } p \equiv 3 \ (\text{mod } 4) \end{cases}.$$

Thus, by the Chinese Remainder Theorem,

$$\left(\frac{3}{p}\right) = \begin{cases} +1 & \text{if } p \equiv 1 \text{ or } 11 \ (\text{mod } 12) \\ -1 & \text{if } p \equiv 5 \text{ or } 7 \ (\text{mod } 12) \end{cases},$$

as desired. $\qquad\square$

We use the result of (b) of Example 3.6 in the following theorem which is modeled after Corollary 2.10.1.

Theorem 3.6: Every prime of the form $6k + 1$ can be written in the form $p = x^2 + 3y^2$ for some positive integers x and y.

Proof: Suppose $p = 6k + 1$ for some positive integer k. Then

$$\left(\frac{-3}{p}\right) = \left(\frac{3}{p}\right)\left(\frac{-1}{p}\right) = \left(\frac{3}{p}\right)(-1)^{(p-1)/2}.$$

If k is even, then $p \equiv 1 \ (\text{mod } 12)$ and so $\left(\frac{3}{p}\right) = +1$ and $(p - 1)/2$ is even. If k is odd, then $p \equiv 7 \ (\text{mod } 12)$ and so $\left(\frac{3}{p}\right) = -1$ and $(p - 1)/2$ is odd. In either case we see that

$$\left(\frac{3}{p}\right) = +1$$

if $p \equiv 1 \ (\text{mod } 6)$.

Thus there is an integer a, $(a, p) = 1$, such that

(3.4) $a^2 + 3 \equiv 0 \pmod{p}$.

By Thue's Theorem, Theorem 2.10, we know there exist integers x and y such that

$$ay \equiv x \pmod{p}$$

and $0 < |x|, |y| < \sqrt{p}$. Thus

(3.5) $a^2 y^2 - x^2 \equiv 0 \pmod{p}$.

If we multiply (3.4) by $-y^2$ and add it to (3.5), then we obtain (after multiplying by -1)

$$x^2 + 3y^2 \equiv 0 \pmod{p}.$$

From the restrictions on x and y we see that

$$0 < x^2 + 3y^2 < 4p.$$

Thus $x^2 + 3y^2 = pt$, where $t = 1, 2,$ or 3.

If $t = 3$, then $3 \mid x$ Thus $x = 3z$ and $p = y^2 + 3z^2$.

If $t = 2$, then x and y must have the same parity. In either case $4 \mid x^2 + 3y^2$ and since p is an odd prime $4 \nmid 2p$. Thus this case is impossible.

If $t = 1$, then $p = x^2 + 3y^2$.   ∎

By considering congruences modulo 6 we see that if $p > 3$ can be written in the form $p = x^2 + 3y^2$, then p must be of the form $6k + 1$.

We give an application of Theorem 3.5 toward the calculation of Legendre symbols. This is Euler's form of the statement of quadratic reciprocity.

Theorem 3.7: Let p be an odd prime and for any prime $q > p$ define the positive integer r as follows:

if $p \equiv 1 \pmod 4$, then $q = kp + r$, $0 < r < p$,

and

if $p \equiv 3 \pmod 4$, then $q = 4kp \pm r$, $0 < r < 4p$, $r \equiv 1 \pmod 4$.

Then

$$\left(\frac{p}{q}\right) = \left(\frac{r}{p}\right).$$

Proof: If $p \equiv 1 \pmod 4$, then, by (c) of Theorem 3.2 and Theorem 3.5, we have

$$\left(\frac{p}{q}\right) = \left(\frac{q}{p}\right) = \left(\frac{r}{p}\right).$$

If $p \equiv 3$ (mod 4) we first show that r exists as stated. Write

$$q = 4kp + r_0, \ 0 \le r_0 \le 4p$$

If $r_0 \equiv 1$ (mod 4), then take $r = r_0$ and if $r \equiv 3$ (mod 4), then take $r = 4p - r_0$. The uniqueness of r is clear.

If $q = 4kp + r$, then $q \equiv r \equiv 1$ (mod 4), and so we again have

$$\left(\frac{p}{q}\right) = \left(\frac{q}{p}\right) = \left(\frac{r}{p}\right).$$

If $q = 4kp - r$, then $q \equiv -r \equiv 3$ (mod 4). By (c) and (g) of Theorem 3.2 and Theorem 3.5, we have

$$\left(\frac{p}{q}\right) = -\left(\frac{q}{p}\right) = -\left(\frac{-r}{p}\right) = -\left(\frac{-1}{p}\right)\left(\frac{r}{p}\right) = \left(\frac{r}{p}\right).$$

This proves the result. ■

Example 3.7: Find all primes that have 23 as a quadratic residue.

Solution: Now the list of quadratic residues, r, of 23, $0 < r < 92$ and $r \equiv 1$ (mod 4) is given by

$$R = \{1, 9, 13, 25, 29, 41, 49, 73, 77, 81, 85\}.$$

Thus, by Theorem 3.7, the odd primes q having 23 as a quadratic residue are those of the form

$$92k \pm r, r \in R. \qquad \square$$

We conclude our discussion of quadratic reciprocity by giving an application to another primality test similar to Theorem 2.20.

Theorem 3.8: Let $m \ge 2$ and $h < 2^m$. Suppose $n = h2^m + 1$ is a quadratic nonresidue mod p for some odd prime p. Then n is a prime if and only if

$$(3.6) \qquad p^{(n-1)/2} \equiv -1 \ (\text{mod } n).$$

Proof: Suppose n is a prime. Since $n \equiv 1$ (mod 4) we have, by Theorem 3.5,

$$\left(\frac{p}{n}\right) = \left(\frac{n}{p}\right) = -1.$$

Then, by Euler's criterion, (a) of Theorem 3.2, the congruence (3.6) follows.

Suppose now that (3.6) holds. Let q be any prime factor of n and let d be the order of p mod q. Then

$$p^{(n-1)/2} \equiv -1, p^{n-1} \equiv 1, p^{q-1} \equiv 1 \pmod{q}.$$

Thus, by Theorem 2.19,

$$p \nmid \tfrac{1}{2}(n-1), d \mid (n-1) \text{ and } d \mid (q-1),$$

that is,

$$d \nmid 2^{m-1}h, d \mid 2^m h \text{ and } d \mid (q-1).$$

Thus $2^m \mid d$ and so $2^m \mid (q-1)$. Thus $q = 2^m x + 1$ for some integer x. Since $n \equiv 1 \equiv q \pmod{2^m}$ we have $n/q \equiv 1 \pmod{2^m}$. Thus

$$n = (2^m x + 1)(2^m y + 1)$$

where $x \geq 1$ and $y \geq 0$. Then

$$2^m xy < 2^m xy + x + y = h < 2^m.$$

Thus $y = 0$ and $n = q$ is a prime.   ∎

Example 3.8: Let $h = 1$ and $m = 2^k$. Then $n = F_k = 2^{2^k} + 1$. The numbers F_k are called **Fermat numbers** and those values of F_k that are prime are called **Fermat primes**. It was conjectured by Fermat that all the Fermat numbers are Fermat primes, but Euler showed that $641 \mid F_5$. Indeed

$$2^{2^5} = 2^{32} = 16 \cdot 2^{28} = (641 - 5^4)2^{28} = 641m - (5 \cdot 2^7)^4$$
$$= 641m - (641 - 1)^4 = 641n - 1,$$

where m and n are positive integers (in fact $m = 2^{28}$ and $n = 6700417$).

Now $F_k \equiv 2 \pmod 3$, and so F_k is a quadratic nonresidue of 3. Thus a necessary and sufficient condition for F_k to be a prime is that

$$F_k \mid \left(3^{(F_k - 1)/2} + 1\right).$$

This is a result due to Pepin.   □

Fermat primes come into the theory of the construction of regular polygons with only ruler and compass. Gauss showed that such a polygon of n sides can be so constructed if n is of the form a power of two times a product of distinct Fermat primes. For a sketch of the proof see [86, pp. 346-358].

Problem Set 3.3

1. Determine the values of $\left(\frac{a}{p}\right)$ for all odd primes p, $p \neq a$, where $a = 5, 6, 7$ and 10.

2. (a) Find all primes that have 37 as a quadratic residue.

   (b) Find all primes that have 41 as a quadratic nonresidue.

3. (a) Prove the following result. If a and b are given natural numbers, then a prime p has at most one representation in the form

   $$p = ax^2 + by^2$$

   where x and y are positive integers.

   (b) Determine which primes can be represented in the form $x^2 + 7y^2$.

4. (a) Prove the following generalization of Example 3.8. Let p be a positive integer of the form $p = 4^k + 1$, where k is a positive integer. Show that p is a prime if and only if

   $$3^{(p-1)/2} \equiv -1 \pmod{p}.$$

   (b) Can you replace 3 by 5 or 7?

5. Show that the equation $y^2 = x^3 + 7$ has no solution in integers x and y. (Hint: first prove that x must be odd and then if we write $x = 2k + 1$ see what polynomial in k $y^2 + 1$ becomes.)

6. Show that $2^a - 1$ does not divide $3^b - 1$ when $a > 1$ and a and b have the same parity.

7. Prove the following result about prime divisors of repunits. If p is a prime of the form $20n \pm 1$ or $20n - 7$ and $2p + 1$ is a prime, then $(10^p - 1)/9$ is composite and divisible by $2p + 1$.

8. Show that if $p \equiv q \pmod{20}$ are two primes, then

   $$\left(\frac{5}{p}\right) = \left(\frac{5}{q}\right).$$

9. Find all primes p such that

   $$\left(\frac{11}{p}\right) = \left(\frac{13}{p}\right).$$

10. Show that if p and q are odd primes and $q = 2p + 1$, then

    $$\left(\frac{p}{q}\right) = \left(\frac{-1}{p}\right).$$

11. Use the result of Problem 1 above and the fact that $3161 = 1792 - 5762$ to determine the prime factors of 3161. (Hint: show that if $p \mid (x^2 + Ay^2)$, where A is an integer, then $\left(\frac{-A}{p}\right) = +1$.

12. (a) If p and q are distinct odd primes with $p \equiv 1 \pmod 4$, then $\left(\frac{p}{q}\right) = +1$ if and only if we have

$$q \equiv p + a(p - 1) \pmod{2p},$$

where $\left(\frac{a}{p}\right) = +1$.

(b) If p and q are distinct odd primes with $p \equiv 3 \pmod 4$, then $\left(\frac{p}{q}\right) = +1$ if and only if we have

$$q \equiv \pm\{3p + a(p + 1)\} \pmod{4p},$$

where $\left(\frac{a}{p}\right) = +1$.

13. (a) If $p \equiv \pm 1 \pmod{4a}$, then $\left(\frac{a}{p}\right) = +1$.

(b) If $p \equiv 2a - 1 \pmod{4a}$ and $a \equiv 1 \pmod 4$, then $\left(\frac{a}{p}\right) = +1$.

3.4. THE JACOBI SYMBOL

We give in this section the first of two generalizations of the Legendre symbol. While neither of these generalizations relate directly to quadratic residues they are both of use in calculating the Legendre symbol.

Definition 3.3: Let $(P, Q) = 1$, $Q > 0$ and Q an odd integer. The **Jacobi symbol**, denoted by $\left(\frac{P}{Q}\right)$, is defined as follows. If $Q = q_1 \cdots q_s$, as a product of primes, which are not necessarily distinct, then

$$\left(\frac{P}{Q}\right) = \prod_{j=1}^{s} \left(\frac{P}{q_j}\right),$$

where $\left(\frac{P}{q_j}\right)$ are the Legendre symbol, for $j = 1, \ldots, s$. $\left(\frac{P}{Q}\right) = 0$, if $(P, Q) > 1$.

Note that if Q is an odd prime, then the Jacobi and Legendre symbols coincide. Also the values of the Jacobi symbol are $+1$, 0, or -1. However, if $\left(\frac{P}{Q}\right) = +1$, then it does not necessarily follow that P is a quadratic residue of Q, though if $\left(\frac{P}{Q}\right) = -1$, then P is a quadratic nonresidue of Q since P must be a quadratic nonresidue of at least one prime divisor of Q, say q, and so

$$x^2 \equiv P \pmod q$$

and hence

$$x^2 \equiv P \pmod{Q}$$

is not solvable.

As we shall see directly below the Legendre and Jacobi symbols possess almost the same properties.

Theorem 3. 9: Suppose Q and Q' are positive odd numbers and that $(PP', QQ') = 1$. Then we have

(a) $\left(\dfrac{P}{Q}\right)\left(\dfrac{P}{Q'}\right) = \left(\dfrac{P}{QQ'}\right)$,

(b) $\left(\dfrac{P}{Q}\right)\left(\dfrac{P'}{Q}\right) = \left(\dfrac{PP'}{Q}\right)$,

(c) $\left(\dfrac{P^2}{Q}\right) = \left(\dfrac{P}{Q^2}\right) = +1$,

(d) $\left(\dfrac{P(P')^2}{Q(Q')^2}\right) = \left(\dfrac{P}{Q}\right)$

and

(e) if $P \equiv P' \pmod{Q}$, then $\left(\dfrac{P}{Q}\right) = \left(\dfrac{P'}{Q}\right)$.

Proof of (a). Let $Q = q_1 \cdots q_s$ and $Q' = q_1' \cdots q_r'$, as a product of primes. Then

$$\left(\frac{P}{Q}\right)\left(\frac{P}{Q'}\right) = \prod_{j=1}^{s}\left(\frac{P}{q_j}\right)\prod_{k=1}^{r}\left(\frac{P}{q_k'}\right) = \prod_{l=1}^{r+s}\left(\frac{P}{q_l''}\right),$$

where $\{q_l''\}_{l=1}^{r+s} = \{q_j\}_{j=1}^{s} \cup \{q_k'\}_{k=1}^{r}$, and by the definition of the Jacobi symbol this latter product is simply $\left(\frac{P}{QQ'}\right)$.

Proof of (b). This follows by the definition of the Jacobi symbol and (b) of Theorem 3.2.

Proof of (c). This follows from (a) and (b).

Proof of (d). This follows from (a) and (b).

Proof of (e). Let $Q = q_1 \cdots q_s$, as a product of primes. Then
$$P' \equiv P \pmod{q_j}, \ 1 \le j \le s.$$

Thus

$$\left(\frac{P}{q_j}\right) = \left(\frac{P'}{q_j}\right), \ 1 \le j \le s,$$

by (c) of Theorem 3.2, and the result follows by the definition of the Jacobi symbol. ∎

Lemma 3.10.1: Let $a_1, ..., a_n$ be odd integers. Then

(a) $\displaystyle\sum_{j=1}^{n} \frac{a_j - 1}{2} \equiv \frac{1}{2} \prod_{j=1}^{n} (a_j - 1) \pmod{2}$

and

(b) $\displaystyle\sum_{j=1}^{n} \frac{a_j^2 - 1}{8} \equiv \frac{1}{8} \prod_{j=1}^{n} (a_j^2 - 1) \pmod{2}$.

Proof: The general cases of both of these results are easy induction arguments that are based on the case $n = 2$. We will only prove the case $n = 2$ for both (a) and (b).

(a) We have, if a and b are odd integers,

$$\frac{ab - 1}{2} - \left(\frac{a - 1}{2} + \frac{b - 1}{2} \right) = \frac{(a - 1)(b - 1)}{2} \equiv 0 \pmod{2}.$$

The result follows.

(b) If $a \equiv 1 \pmod{2}$ and $b \equiv 1 \pmod{2}$, then

$$(a^2 - 1)(b^2 - 1) \equiv 0 \pmod{16}.$$

Thus

$$a^2 b^2 - 1 \equiv (a^2 - 1) + (b^2 - 1) \pmod{16},$$

and so we have

$$\frac{a^2 b^2 - 1}{8} \equiv \frac{a^2 - 1}{8} + \frac{b^2 - 1}{8} \pmod{2},$$

by Theorem 2.3. ■

Theorem 3.10: Let Q be a positive odd integer. Then

(a) $\displaystyle\left(\frac{-1}{Q} \right) = (-1)^{(Q-1)/2}$

and

(b) $\displaystyle\left(\frac{2}{Q} \right) = (-1)^{(Q^2 - 1)/8}$.

Proof. Let $Q = q_1 \cdots q_s$. Then, by definition and (f) of Theorem 3.2,

$$\left(\frac{-1}{Q} \right) = \prod_{j=1}^{s} \left(\frac{-1}{q_j} \right) = \prod_{j=1}^{s} (-1)^{(q_j - 1)/2} = (-1)^{\frac{1}{2}\sum_{j=1}^{s}(q_j - 1)}.$$

By (a) of Lemma 3.10.1, we have

$$\frac{1}{2}\sum_{j=1}^{s}\left(q_j-1\right)\equiv\frac{q_1\cdots q_s-1}{2}=\frac{Q-1}{2}\ (\mathrm{mod}\ 2).$$

Thus

$$\left(\frac{-1}{Q}\right)=(-1)^{(Q-1)/2},$$

which proves (a).

By definition and Theorem 3.4, we have

$$\left(\frac{2}{Q}\right)=\prod_{j=1}^{s}\left(\frac{2}{q_j}\right)=\prod_{j=1}^{s}(-1)^{(q_j^2-1)/8}=(-1)^{\frac{1}{8}\sum_{j=1}^{s}\left(q_j^2-1\right)}.$$

By (b) of Lemma 3.10.1, we have

$$\frac{1}{8}\sum_{j=1}^{s}\left(q_j^2-1\right)\equiv\frac{q_1^2\cdots q_s^2-1}{8}=\frac{Q^2-1}{8}\ (\mathrm{mod}\ 2),$$

and so

$$\left(\frac{2}{Q}\right)=(-1)^{(Q^2-1)/8},$$

which proves (b).   ■

Theorem 3.11: Let P and Q be odd and positive. If $(P,Q)=1$, then

$$\left(\frac{P}{Q}\right)\left(\frac{Q}{P}\right)=(-1)^{\frac{P-1}{2}\cdot\frac{Q-1}{2}}.$$

Proof: Let $P=p_1\cdots p_r$ and $Q=q_1\cdots q_s$ Then, by definition and (b) of Theorem 3.2,

$$\left(\frac{P}{Q}\right)=\prod_{j=1}^{s}\left(\frac{P}{q_j}\right)=\prod_{j=1}^{s}\prod_{i=1}^{r}\left(\frac{p_i}{q_j}\right)=\prod_{j=1}^{s}\prod_{i=1}^{r}\left(\frac{q_j}{p_i}\right)(-1)^{\frac{p_i-1}{2}\cdot\frac{q_j-1}{2}}$$

$$=\left(\frac{Q}{P}\right)(-1)^{\sum_{j=1}^{s}\sum_{i=1}^{r}\frac{p_i-1}{2}\cdot\frac{q_j-1}{2}}.$$

Now

$$\sum_{j=1}^{s}\sum_{i=1}^{r}\frac{p_i-1}{2}\cdot\frac{q_j-1}{2}=\sum_{j=1}^{s}\frac{q_j-1}{2}\cdot\sum_{i=1}^{r}\frac{p_i-1}{2}\equiv\frac{P-1}{2}\cdot\frac{Q-1}{2}\ (\mathrm{mod}\ 2),$$

by (a) of Lemma 3.10.1, and so

$$\left(\frac{P}{Q}\right)=\left(\frac{Q}{P}\right)(-1)^{\frac{P-1}{2}\cdot\frac{Q-1}{2}}.$$

Since $\left(\frac{Q}{P}\right)=\pm1$ in this case the result follows.   ■

Example 3.9: Use the Jacobi symbol to calculate $\left(\frac{105}{317}\right), \left(\frac{213}{499}\right)$ and $\left(\frac{87}{131}\right)$.

Solution: We have

$$\left(\frac{105}{317}\right) = \left(\frac{317}{105}\right) = \left(\frac{2}{105}\right) = (-1)^{(105^2-1)/8} = +1;$$

$$\left(\frac{213}{499}\right) = \left(\frac{499}{213}\right) = \left(\frac{73}{213}\right) = \left(\frac{67}{73}\right) = \left(\frac{73}{67}\right) = \left(\frac{6}{67}\right) = \left(\frac{2}{67}\right)\left(\frac{3}{67}\right)$$

$$= (-1)^{(67^2-1)/8}\left(\frac{3}{67}\right) = -\left(\frac{3}{67}\right) = \left(\frac{67}{3}\right) = \left(\frac{1}{3}\right) = +1$$

and

$$\left(\frac{87}{113}\right) = -\left(\frac{113}{87}\right) = -\left(\frac{44}{87}\right) = -\left(\frac{11}{87}\right) = \left(\frac{87}{11}\right) = \left(\frac{10}{11}\right)$$

$$= \left(\frac{2}{11}\right)\left(\frac{5}{11}\right) = (-1)^{(11^2-1)/8}\left(\frac{5}{11}\right) = -\left(\frac{5}{11}\right)$$

$$= -\left(\frac{11}{5}\right) = -\left(\frac{1}{5}\right) = -1,$$

as desired.   □

As one can see the benefit of the Jacobi symbol is that we no longer require prime numbers in the "denominator" of the symbol so that we can keep applying Theorem 3.11 óver and over. The only requirement is that the "denominator" be odd.

Problem Set 3.4

1. Evaluate the Jacobi symbols: $\left(\frac{-23}{83}\right), \left(\frac{51}{71}\right), \left(\frac{71}{73}\right)$ and $\left(\frac{-35}{97}\right)$.

2. Let $(P, Q) = 1$ with P and Q both odd positive integers greater than 1. Let μ be the number of integers in the sequence

$$1 \cdot Q, 2 \cdot Q, \ldots, \tfrac{1}{2}(P-1) \cdot Q$$

whose least positive residues modulo P are greater than $P/2$. Show that for the Jacobi symbol we have

$$\left(\frac{Q}{P}\right) = (-1)^\mu.$$

This generalizes the result of Gauss' Theorem (Theorem 3.3). (Hint: let $R(x) = x - \left[x + \tfrac{1}{2}\right]$ and let $\text{sgn} x = +1$ if $x > 0$ and -1 if $x < 0$. Start by showing that

$$(-1)^\mu = \text{sgn}\left(\prod_{h=1}^{(P-1)/2} R(hQ/P)\right).)$$

3. (a) Let a and b be positive integers, with b odd. For the Jacobi symbol show that we have

$$\left(\frac{a}{2a+b}\right) = \begin{cases} \left(\frac{a}{b}\right) & \text{if } a \equiv 0 \text{ or } 1 \ (\text{mod } 4) \\ -\left(\frac{a}{b}\right) & \text{if } a \equiv 0 \text{ or } 1 \ (\text{mod } 4) \end{cases}.$$

(b) Let a, b and c be positive integers, with $(a, b) = 1$, b odd and $b < 4ac$. For the Jacobi symbol show that we have

$$\left(\frac{a}{4ac-b}\right) = \left(\frac{a}{b}\right).$$

4. If n is an odd positive integer which is not a perfect square, show that there is at least one integer b, $1 < b < n$, $(b, n) = 1$, such that for the Jacobi symbol we have

$$\left(\frac{b}{n}\right) = -1.$$

5. Let n be an odd, square-free integer greater than 3. Let

$$A_n = \left\{a \ (\text{mod } n): \left(\frac{a}{n}\right) = +1\right\}$$

and

$$B_n = \left\{b \ (\text{mod } n): \left(\frac{b}{n}\right) = -1\right\},$$

where $\left(\frac{c}{n}\right)$ is the Jacobi symbol. Show that

$$\sum_{a \in A_n} a = \sum_{b \in B_n} b \equiv 0 \ (\text{mod } n).$$

6. This exercise discusses a method for computing the Jacobi symbol. Let a and b be relatively prime integers with $a > b > 0$.

(a) Let $U_0 = a$ and $U_1 = b$. Apply the division algorithm to obtain

$$U_0 = U_1 q_1 + R_0$$

with $0 \leq R_0 < U_1$. Write and use the division algorithm to get

$$U_1 = U_2 q_2 + R_1$$

with $0 \leq R_1 < U_2$. We proceed in this way until we get $U_n = 1$, for some n. At the $(k - 1)$st step we have

$$U_{k-2} = U_{k-1} q_{k-1} + 2^{P_{k-1}} U_k.$$

Show that we can express the Jacobi symbol $\left(\frac{U_{k-2}}{U_{k-1}}\right)$ in terms of $\left(\frac{U_{k-1}}{U_k}\right)$.

(b) Let

$$T = \sum_{k=2}^{n} P_{k-1}(U_{k-1}^2 - 1)/8 + \sum_{k=2}^{n}(U_{k-1} - 1)(U_k - 1)/4.$$

show that

$$\left(\frac{a}{b}\right) = (-1)^T.$$

This result is somewhat akin to Eisenstein's Theorem, Theorem 3.3.

3.5. THE KRONECKER SYMBOL

In the Jacobi symbol we required the "denominator" of the symbol to be odd. In this section we define the Kronecker symbol to partly overcome this problem. In many problems of advanced number theory the Kronecker symbol naturally arises. Though we won't be able, in this book, to go into any of these more advanced applications we shall, in the next section, give a simple application of the Kronecker symbol.

Throughout the rest of the section D will be an integer, positive or negative, which is not a perfect square and which is either $\equiv 0$ or 1 (mod 4).

Definition 3.4: Let M be a positive integer and let p denote an arbitrary prime, even or odd. Then the **Kronecker symbol**, denoted by $\left(\frac{D}{M}\right)$ is defined as follows:

(a) If $p \mid D$, then $\left(\frac{D}{p}\right) = 0$;

(b) If D is odd, then

$$\left(\frac{D}{2}\right) = \begin{cases} +1 & \text{if } D \equiv 1 \ (\text{mod } 8) \\ -1 & \text{if } D \equiv 5 \ (\text{mod } 8) \end{cases};$$

(c) If $p > 2$ and $p{\nmid}D$, then $\left(\frac{D}{p}\right)$ is the Legendre symbol;

(d) If $M = p_1 \cdots p_r$ as a product of primes, then

$$\left(\frac{D}{M}\right) = \left(\frac{D}{p_1}\right) \cdots \left(\frac{D}{p_r}\right).$$

Note that if D is odd then by Theorem 3.10, we have

$$\left(\frac{D}{2}\right) = \left(\frac{2}{|D|}\right).$$

Also note that $\left(\frac{D}{1}\right) = 1$ and that if $(D, M) > 1$, then $\left(\frac{D}{M}\right) = 0$.

Theorem 3.12: Let $M > 0$ and $(D, M) = 1$. Then

(a)   if D is odd, we have

$$\left(\frac{D}{M}\right) = \left(\frac{M}{|D|}\right).$$

(b)   if $D = 2^b d$, where d is odd, then

$$\left(\frac{D}{M}\right) = \left(\frac{2}{M}\right)^b (-1)^{\frac{d-1}{2} \cdot \frac{M-1}{2}} \left(\frac{M}{|d|}\right),$$

where both symbols on the right are Jacobi symbols.

Proof of (a). If D is odd, then $D \equiv 1 \pmod 4$. Let $M = 2^l m$, where m is odd and positive and l is nonnegative. By (e) of Definition 3.4, we have

$$\left(\frac{D}{M}\right) = \left(\frac{D}{2^l m}\right) = \left(\frac{D}{2}\right)^l \left(\frac{D}{m}\right) = \left(\frac{2}{|D|}\right)^l \left(\frac{D}{m}\right).$$

By Theorem 3.11, (a) of Theorem 3.9 and (a) of Theorem 3.10, we have

$$\left(\frac{D}{M}\right) = \left(\frac{2}{|D|}\right)^l \left(\frac{m}{|D|}\right) = \left(\frac{2^l m}{|D|}\right) = \left(\frac{M}{|D|}\right).$$

Proof of (b). Let D be even. Then

$$\left(\frac{D}{M}\right) = \left(\frac{2^b d}{M}\right) = \left(\frac{2}{M}\right)^b \left(\frac{d}{M}\right).$$

and so, by Theorem 3.11, since M is odd and positive, we have

$$\left(\frac{D}{M}\right) = \left(\frac{2}{M}\right)^b (-1)^{\frac{d-1}{2} \cdot \frac{M-1}{2}} \left(\frac{M}{|d|}\right),$$

by (a) Theorem 3.10.   ∎

Theorem 3.13: The Kronecker symbol satisfies

(a) $\left(\dfrac{D}{M_1 M_2}\right) = \left(\dfrac{D}{M_1}\right)\left(\dfrac{D}{M_2}\right),$

(b) $\left(\dfrac{D}{M_1}\right) = \left(\dfrac{D}{M_2}\right),$ if $M_1 \equiv M_2 \pmod{|D|}$

and

(c) $\left(\dfrac{D}{M}\right) = -1$ for suitable M.

Proof of (a). This follows from (e) of Definition 3.4.

Proof of (b). By Theorem 2.4, we have $(D, M_1) = (D, M_2)$.

If $(D, M_1) > 1$, then $(D, M_2) > 1$, and so

$$\left(\frac{D}{M_1}\right) = 0 = \left(\frac{D}{M_2}\right).$$

If $(D, M_1) = 1$, then $(D, M_2) = 1$. Suppose D is odd, then $D \equiv 1 \pmod 4$ and so by Theorem 3.12 and (e) of Theorem 3.9, we have

$$\left(\frac{D}{M_1}\right) = \left(\frac{M_1}{|D|}\right) = \left(\frac{M_2}{|D|}\right) = \left(\frac{D}{M_2}\right).$$

Suppose $D = 2^b d$ is even, where d is odd. Then, by Theorem 3.12,

$$\left(\frac{D}{M_1}\right) = \left(\frac{2}{M_1}\right)^b (-1)^{\frac{d-1}{2} \cdot \frac{M_1-1}{2}} \left(\frac{M_1}{|d|}\right)$$

and

$$\left(\frac{D}{M_2}\right) = \left(\frac{2}{M_2}\right)^b (-1)^{\frac{d-1}{2} \cdot \frac{M_2-1}{2}} \left(\frac{M_2}{|d|}\right).$$

By (e) of Theorem 3.9, we have

$$\left(\frac{M_1}{|d|}\right) = \left(\frac{M_2}{|d|}\right),$$

since $d \mid D$. Since $4 \mid D$, and so $M_1 \equiv M_2 \pmod 4$, we have

$$(-1)^{\frac{d-1}{2} \cdot \frac{M_1-1}{2}} = (-1)^{\frac{d-1}{2} \cdot \frac{M_2-1}{2}}.$$

If $b = 2$, then obviously

$$\left(\frac{2}{M_1}\right)^b = \left(\frac{2}{M_2}\right)^b.$$

If $b \geq 3$, then $8 \mid D$ and so $M_1 \equiv M_2 \pmod 8$. Thus, by (b) of Theorem 3.10, we have

$$\left(\frac{2}{M_1}\right)^b = \left(\frac{2}{M_2}\right)^b.$$

Proof of (c). Let D be odd . Then $|D|$ is not a perfect square, since either $|D| = D$,

if $D > 0$, and if $D < 0$, then $|D| \equiv 3 \pmod 4$. For suitable p we therefore have $|D|$
$= p^l g$, where $p > 2$, l is odd and $p \nmid g$, with g odd. Let s be a quadratic nonresidue
mod p (which exists by Theorem 3.1). Choose, by Theorem 2.11, a positive integer
M so that

$$M \equiv s \pmod p \text{ and } M \equiv 1 \pmod g .$$

Then $(D, M) = 1$ and, by (a) of Theorem 3.12, we have

$$\left(\frac{D}{M}\right) = \left(\frac{M}{|D|}\right) = \left(\frac{M}{p}\right)^l \left(\frac{M}{g}\right) = \left(\frac{s}{p}\right)^l \left(\frac{1}{g}\right) = (-1)^l = -1.$$

Let D be even, $D = 2^b d$, with d odd. If b is odd, choose $M > 0$ so that

$$M \equiv 5 \pmod 8 \text{ and } M \equiv 1 \pmod{|d|}.$$

This is possible, by Theorem 2.11, since $(8, d) = 1$. Then $(D, M) = 1$ and, by (b) of
Theorem 3.12 and (b) of Theorem 3.10, we have

$$\left(\frac{D}{M}\right) = \left(\frac{2}{M}\right)^b (-1)^{\frac{d-1}{2} \cdot \frac{M-1}{2}} \left(\frac{M}{|d|}\right) = \left(\frac{2}{M}\right) \cdot 1 \cdot \left(\frac{1}{|d|}\right) = -1.$$

If b is even, then d is not a perfect square and so, by (b) of Theorem 3.12, if (D, M)
$= 1$ and $M > 0$, we have

$$\left(\frac{D}{M}\right) = (-1)^{\frac{d-1}{2} \cdot \frac{M-1}{2}} \left(\frac{M}{|d|}\right).$$

If $d \equiv 3 \pmod 4$, choose $M > 0$ so that

$$M \equiv -1 \pmod 4 \text{ and } M \equiv 1 \pmod{|d|},$$

which is possible by Theorem 2.11, since $(4, d) = 1$. Then $(D, M) = 1$ and

$$\left(\frac{D}{M}\right) = (-1)^{\frac{d-1}{2}} \left(\frac{1}{|d|}\right) = -1.$$

If $d \equiv 1 \pmod 4$, then for $(D, M) = 1$ and $M > 0$, we have

$$\left(\frac{D}{M}\right) = \left(\frac{M}{|d|}\right).$$

Now $|d|$ is not a perfect square, since $|d| = d$ or $|d| = -d \equiv 3 \pmod 4$. Then for
suitable p we have $|d| = p^l g$, where l is odd, $p > 2$, $p \nmid g$ and g is odd. Let s be a
quadratic nonresidue of p (which exists by Theorem 3.1) and choose $M > 0$, by

Theorem 2.11, so that

$$M \equiv s \pmod{p}, \; M \equiv 1 \pmod{g} \text{ and } M \equiv 1 \pmod{2}.$$

Then $(D, M) = 1$ and

$$\left(\frac{D}{M}\right) = \left(\frac{M}{p^l g}\right) = \left(\frac{M}{p}\right)^l\left(\frac{M}{g}\right) = \left(\frac{s}{p}\right)^l\left(\frac{1}{g}\right) = (-1)^l = -1,$$

which proves (c). ∎

Theorem 3 .14: We have

$$\left(\frac{D}{|D|-1}\right) = \begin{cases} +1 & \text{if } D > 0 \\ -1 & \text{if } D < 0 \end{cases}.$$

Proof: Suppose D is odd. Then, by (a) of Theorem 3 .12, we have

$$\left(\frac{D}{|D|-1}\right) = \left(\frac{|D|-1}{|D|}\right) = \left(\frac{-1}{|D|}\right) = (-1)^{\frac{|D|-1}{2}} = \begin{cases} +1 & \text{if } D > 0 \\ -1 & \text{if } D < 0 \end{cases}.$$

Suppose D is even, $D = 2^b d$, with $b \geq 2$ and d odd. Then, by (b) of Theorem 3.12, we have

$$\left(\frac{D}{|D|-1}\right) = \left(\frac{2}{|D|-1}\right)^b (-1)^{\frac{d-1}{2}}\left(\frac{|D|-1}{d}\right).$$

Now

$$\left(\frac{2}{|D|-1}\right)^b = 1;$$

if $b = 2$, this is obvious, and if $b \geq 3$, then $|D| \equiv 7 \pmod{8}$ and the result follows by (b) of Theorem 3 .10 . We have

$$(-1)^{\frac{d-1}{2}}\left(\frac{|D|-1}{|d|}\right) = (-1)^{\frac{d-1}{2}}\left(\frac{-1}{|d|}\right) = (-1)^{\frac{d-1}{2}\cdot\frac{|d|-1}{2}} = \begin{cases} +1 & \text{if } D > 0 \\ -1 & \text{if } D < 0 \end{cases},$$

since $|d| \| D|$. ∎

Theorem 3.15: If $N > 0$, $M > 0$ and $N \equiv -M \pmod{|D|}$, then

$$\left(\frac{D}{N}\right) = \begin{cases} \left(\frac{D}{M}\right) & \text{if } D > 0 \\ -\left(\frac{D}{M}\right) & \text{if } D < 0 \end{cases}.$$

Proof: We have, by (b) of Theorem 3.13,

$$\left(\frac{D}{N}\right) = \left(\frac{D}{|D|M - M}\right) = \left(\frac{D}{M(|D| - 1)}\right) = \left(\frac{D}{M}\right)\left(\frac{D}{|D| - 1}\right)$$

and the result follows by Theorem 3.14. ∎

Problem Set 3.5

1. (a) Show that the Jacobi symbol can be expressed in terms of the Kronecker symbol as follows: if $m > 0$, m odd and $(m, n) = 1$, then

$$\left(\frac{n}{m}\right) = \begin{cases} 1 & \text{if } m \text{ is a square} \\ \left(\frac{m}{|n|}\right) & \text{if } m \equiv 1 \pmod 4 \text{ is not a square} \\ \left(\frac{-m}{|n|}\right) & \text{if } m \equiv 3 \pmod 4 \end{cases}$$

and

$$\left(\frac{n}{m}\right) = \begin{cases} 1 & \text{if } n \text{ is a square} \\ \left(\frac{4n}{m}\right) & \text{if } n \text{ is not a square} \end{cases}$$

   (b) Show that the Kronecker symbol can be expressed in terms of the Jacobi symbol as follows: if $d \equiv 0$ or $1 \pmod 4$, is not a perfect square and $m > 0$, then

$$\left(\frac{d}{m}\right) = \begin{cases} 0 & \text{if } (m,d) > 1 \\ \left(\frac{m}{|d|}\right) & \text{if } (m,d) = 1 \text{ and } d \equiv 1 \pmod 4. \\ \left(\frac{d/4}{m}\right) & \text{if } (m,d) = 1 \text{ and } d \equiv 0 \pmod 4 \end{cases}$$

   (c) Let $a \equiv 1 \pmod 4$ and $b = 2^k b_1$, $2 \nmid b_1$. Use the definition of $\left(\frac{a}{b_1}\right)$ (a Jacobi symbol) and the Kronecker symbols $\left(\frac{a}{2}\right)$ and $\left(\frac{a}{b}\right) = \left(\frac{a}{2}\right)^k \left(\frac{a}{b_1}\right)$ to find a relation between $\left(\frac{a}{b}\right)$ and $\left(\frac{b}{a}\right)$. Do the same when $a \equiv 0 \pmod 4$. (Caution: if $a = 2^k a_1$, $k > 0$, and $b \equiv 3 \pmod 4$, then $\left(\frac{a}{b}\right)$ is defined, but $\left(\frac{b}{a}\right)$ is not.)

2. Evaluate the Kronecker symbols $\left(\frac{7}{12}\right)$, $\left(\frac{11}{20}\right)$ and $\left(\frac{103}{200}\right)$.

3. Show that if d is not a square and $d \equiv 0$ or $1 \pmod 4$, then

$$\left(\frac{d}{-1}\right) = \begin{cases} +1 & \text{if } d > 0 \\ -1 & \text{if } d < 0 \end{cases}.$$

4. Let $d \equiv 0 \pmod 4$ be not a square. Show that, if $(m, d) = 1$, then

$$\left(\frac{d}{m}\right) = (\text{sgn} d)^{(m-1)/2}\left(\frac{|d|}{|m|}\right).$$

3.6. THE SOLUTION OF $x^2 \equiv D$ (MOD M)

Now that we know how to calculate $\left(\frac{P}{Q}\right)$ for various combinations of P and Q we can return to the problem discussed in the introduction to this chapter: solve

$$ax^2 + bx + c \equiv 0 \ (\text{mod } m).$$

Since we allow m to be even and don't necessarily assume $(a, m) = 1$, an equivalent congruence is, by Theorem 2.3,

$$4ax^2 + 4abx + 4ac \equiv 0 \ (\text{mod } 4am),$$

or

$$(2ax + b)^2 \equiv b^2 - 4ac \ (\text{mod } 4am).$$

This leads to the system of equations

$$z \equiv b^2 - 4ac \ (\text{mod } 4am)$$

$$2ax + b \equiv z \ (\text{mod } 4am).$$

Since the linear congruence in this system is solvable if and only if $2a \mid (z - b)$, and we can find all solutions if this is the case, we turn our attention to the quadratic congruence.

The remainder of this section is devoted to giving necessary and sufficient conditions for the solvability of

(3.7) $x^2 \equiv D \ (\text{mod } M),$

where, for simplicity, we assume $(D, M) = 1$, and of determining the number of solutions when it is solvable.

By Theorem 2.13, we know that the number of solutions of (3.7) is the product, over all prime powers exactly dividing M, of the number of solutions of

(3.8) $x^2 \equiv D \ (\text{mod } p^e)$

where $p^e \| M$. Also, it is clear that if (3.7) is solvable, then

(3.9) $x^2 \equiv D \ (\text{mod } p)$

is solvable for all primes p dividing M. We wish to show the converse: if (3.9) is solvable for all $p \mid M$, then (3.7) is solvable. We could do this by the theory of

Chapter Two, but we shall carry out the procedure here in a constructive way.

We begin with the odd prime powers p^e. Let D be a quadratic residue of p. Since $x^2 \equiv (p - x)^2$ (mod p) we see that (3.9) has exactly two solutions modulo p if $p \nmid D$. We now show that (3.8) has exactly two solutions modulo p^e. Let α be a solution of (3.9) and define the integers P_e and Q_e by

$$P_e + Q_e \sqrt{D} = \left(\alpha + \sqrt{D}\right)^e$$
$$P_e - Q_e \sqrt{D} = \left(\alpha - \sqrt{D}\right)^e,$$

for $e = 1, 2, \ldots$. Then

$$P_e \pm Q_e \sqrt{D} = \left(P_{e-1} \pm Q_{e-1} \sqrt{D}\right)\left(\alpha \pm \sqrt{D}\right),$$

and so

$$P_e = \alpha P_{e-1} + D Q_{e-1} \text{ and } Q_e = P_{e-1} + \alpha Q_{e-1}.$$

We have

$$P_e + \alpha Q_e = 2\alpha P_{e-1} + Q_{e-1}\left(D + \alpha^2\right)$$
$$\equiv 2\alpha\left(P_{e-1} + \alpha Q_{e-1}\right) \text{ (mod } p\text{)}.$$

Thus, by induction, we have

$$P_e + \alpha Q_e \equiv (2\alpha)^e \text{ (mod } p\text{)},$$

and so $p \nmid (P_e + Q_e)$. Also $P_e - \alpha Q_e = Q_{e-1}(D - \alpha^2)$ is divisible by p. Thus (Q_e, p) $= (2\alpha Q_e, p) = 1$. Now

$$P_e^2 - D Q_e^2 = (\alpha^2 - D)^e \equiv 0 \text{ (mod } p^e\text{)}.$$

Since $(Q_e, p) = 1$, there exists an integer $\overline{Q_e}$ such that

$$Q_e \overline{Q_e} \equiv 1 \text{ (mod } p^e\text{)}$$

and if we let $\beta = P_e \overline{Q_e}$, then

$$\beta^2 \equiv D \text{ (mod } p^e\text{)}.$$

Thus every quadratic residue mod p leads to a quadratic residue modulo p^e, $e \geq 1$. If x is any solution of (3.8), then

$$(x - \beta)(x + \beta) \equiv 0 \text{ (mod } p^e\text{)} .$$

If $p \mid (x + \beta)$ and $p \mid (x - \beta)$, then $p \mid 2\beta$ or $p \mid \beta$, which can't happen since (β, p)

= 1. Thus either $x \equiv \beta$ (mod p^e) or $x \equiv -\beta$ (mod p^e). Thus there are exactly two solutions to (3.8).

Example 3.10: Solve $x^2 \equiv 2$ (mod 7^3) .

Solution: A solution to $x^2 \equiv 2$ (mod 7) is $\alpha = 3$. Thus

$$P_1 = 3 \text{ and } Q_1 = 1,$$
$$P_2 = 3 \cdot 3 + 2 \cdot 1 = 11 \text{ and } Q_2 = 3 + 1 \cdot 3 = 6,$$
$$P_3 = 3 \cdot 11 + 2 \cdot 6 = 45 \text{ and } Q_3 = 11 + 6 \cdot 3 = 29.$$

Now $\overline{29}$ (mod 7^3) is 71 (mod 7^3) . Thus

$$\beta = 45 \cdot 71 \equiv 108 \text{ (mod } 7^3) .$$

Thus the two roots are 108 and 7^3 - 108 = 235. □

We now take up the case $p = 2$. Here D must be odd, since we require $(D, M) = 1$. Here we give a constructive solution to

(3.10)
$$x^2 \equiv D \text{ (mod } 2^e),$$

$e \geq 1$. If $e = 1$, then $D \equiv 1$ (mod 2) and there is only one solution. If $e = 2$, then (3.10) is solvable if and only if $D \equiv 1$ (mod 4), in which case there are two solutions $x \equiv 1$ or - 1 (mod 4) . For $e \geq 3$ (3.10) is solvable only if $D \equiv 1$ (mod 8). Suppose there is an a such that $\alpha^2 \equiv D$ (mod 2^e), $e \geq 3$. Since x and D are both odd $x^2 \equiv \alpha^2$ (mod 2^e) is equivalent to

$$\frac{x + \alpha}{2} \cdot \frac{x - \alpha}{2} \equiv 0 \text{ (mod } 2^{e-2}).$$

But of the two numbers $(x + \alpha)/2$ and $(x - \alpha)/2$ only one can be even and the other is odd. Thus with the proper choice of sign

$$(x \pm \alpha)/2 \equiv 0 \text{ (mod } 2^{e-2})$$

or

$$x \equiv \pm\alpha \text{ (mod } 2^{e-1})$$

or

$$\dot{x} \equiv \pm\alpha + 2^{e-1} \text{ (mod } 2^e).$$

Since these numbers are distinct modulo 2^e we see that if we have one solution to (3.10) we have four solutions to (3.10). To finish we must therefore show that if $D \equiv 1$ (mod 8), then (3.10) is solvable when $e \geq 3$. Again we shall give a constructive procedure for finding the solution.

We begin with a different problem: find a polynomial

$$P_n(x) = 1 + \alpha_1 x + \cdots + \alpha_n x^n,$$

where $\alpha_n \neq 0$, so that there exists a polynomial $Q(x)$ such that

(3.11) $$P_n^2(x) - 1 - x = x^{n+1} Q(x).$$

If we take derivatives in (3.11) we get

$$2P_n(x)P_n'(x) - 1 = x^n R(x),$$

where $R(x)$ is some polynomial. If we multiply by $P_n(x)$ and use (3.11) we get

(3.12) $$2(1 + x)P_n'(x) - P_n(x) = x^n S(x),$$

for some polynomial $S(x)$. Now the degree of the left hand side of (3.12) is n. Thus $S(x)$ is a constant polynomial, say $S(x) = c$. Thus, for some real number c, we have

(3.13) $$2(1 + x)P_n'(x) - P_n(x) = cx^n.$$

Suppose $P_n(x)$ is a polynomial that satisfies (3.13) and $P_n(0) = 1$. If we let $P_n'^2(x) = 1 + x + g(x)$, then we have, by (3.13),

(3.14) $$(1 + x)g'(x) - g(x) = cx^n P_n(x).$$

Now $g(0) = 0$. Thus

$$g(x) = x^m f(x),$$

where $f(x)$ is a polynomial and $x^{-1}f(x)$ is not. Put this representation of g into (3.14) and divide by x^{m-1}. This gives

$$[m(1 + x) - x]f(x) + x(1 + x)f'(x) = cx^{n+1-m}P_n(x).$$

Thus we must have $m \geq n + 1$ or else $mf(0) = 0$, which it is not. If $m > n + 1$, then $c = 0$, in which case we again see that $f(0) = 0$, which is not the case. Thus

$$g(x) = x^{n+1}f(x),$$

where $f(x)$ is a polynomial, and so any solution of (3.13) is a solution of (3.11). We put $P_n(x) = 1 + \alpha_1 x + \cdots + \alpha_n x^n$, into (3.13) and equate coefficients of like powers of x. This gives, for $k = 0, 1, \ldots, n - 1$,

$$\alpha_{k+1} = -\frac{2k - 1}{2k + 2}\alpha_k, \quad \alpha_0 = 1.$$

Thus, by a simple induction argument,

$$\alpha_k = (-1)^{k-1} \frac{1 \cdot 3 \cdot 5 \cdots (2k-3)}{2 \cdot 4 \cdot 6 \cdots (2k)}.$$

Now

$$\frac{1 \cdot 3 \cdot 5 \cdots (2k-3)}{2 \cdot 4 \cdot 6 \cdots (2k)} = \frac{(2k)!}{2^{2k} k! k!} = \frac{1}{2^{2k}} \binom{2k}{k} = \frac{G}{2^{2k-1}}$$

where, since, as is easy to show (see Problem 1(c) of Problem Set 1.2) $2\binom{2k}{k}$, G is

an integer. Thus

$$(2k-1)2^{2k-1}\alpha_k \in \mathbf{Z} \text{ and } (2k)2^{2k-1}\alpha_k \in \mathbf{Z}.$$

Thus there is an integer g_k such that

(3.15)
$$\alpha_k = \frac{g_k}{2^{2k-1}}.$$

If r is an integer we see, by (3.15), that $P_n(8r)$ is a polynomial with integer coefficients since $3k \geq 2k - 1$, for $k \geq 0$. Also the coefficient of r^k is divisible by 2^{k+1}. Thus, by (3.11), we see that all the coefficients of $P_n^2(8r) - 1 - 8r$ are divisible by 2^{n+3}. Thus, if $D \equiv 1 \pmod{8}$, then $D = 1 + 8r$ for some integer r, and so $P_{e-3}(8r)$ is a solution of (3.10).

Example 3.11: Solve $x^2 \equiv 17 \pmod{2^7}$.

Solution: We have

$$P_4(x) = 1 + \tfrac{1}{2}x - \tfrac{1}{8}x^2 + \tfrac{1}{16}x^3 - \tfrac{5}{128}x^4,$$

and so

$$P_4(8r) = 1 + 4r - 8r^2 + 32r^3 - 160r^4.$$

Since $17 = 1 + 8 \cdot 2$ we take $r = 2$ and get

$$P_4(16) = 1 + 8 - 32 + 256 - 2560 = -2327 \equiv -23 \pmod{2^7}.$$

Thus the four solutions of our congruence are -23, 23, 41 and -41. □

If we combine the above with Theorem 2.13 we arrive at the following theorem.

Theorem 3.16: A necessary and sufficient condition for

(3.16)
$$x^2 \equiv D \pmod{M}$$

to be solvable, where $(D, M) = 1$, is that D be a quadratic residue of all odd prime divisors of M and that if $2\|M$, then D is odd, if $4\|M$, then $D \equiv 1$ (mod 4) and if $8\,|\,M$, then $D \equiv 1$ (mod 8). If (3.16) is solvable and

$$M = 2^\alpha\, p_1^{\alpha_1} \cdots p_r^{\alpha_r},$$

where, for $1 \le i \le r$, the p_i are distinct odd primes, then the number of solutions of (3.16) is $2^{r+\sigma}$, where

$$\sigma = \begin{cases} 0 & \text{if } \alpha = 0 \text{ or } 1 \\ 1 & \text{if } \alpha = 2 \\ 4 & \text{if } \alpha \ge 3 \end{cases}.$$

This also holds when $r = 0$, that is, M is a power of 2.

We can use the Kronecker symbol of the previous section to give another expression for the number of solutions in the case that $M \equiv 0$ (mod 4).

Theorem 3.17: Let $k > 0$ and let $(D, k) = 1$. The number of solutions of

(3.17) $x^2 \equiv D$ (mod $4k$),

where $D \equiv 1$ or 0 (mod 4)is,

$$2 \sum_{d|k} \left(\frac{D}{d} \right),$$

where d runs through the square-free positive divisors of k

Proof: If D is odd, then $D \equiv 1$ (mod 4) and so $(D, 4k) = 1$. If $p^l\|4k$, then, by Theorem 3.16 and Definitions 3.2 and 3.4, we have that the number of solutions of

$$x^2 \equiv D \pmod{p^e}$$

is (since $4\,|\,4k$, $l = 1$ does not occur when $p = 2$)

$$\begin{cases} 2 & \text{if } p = 2,\ l = 2 \\ 2\left(1 + \left(\frac{D}{2}\right)\right) & \text{if } p = 2,\ l > 2. \\ 1 + \left(\frac{D}{p}\right) & \text{if } p > 2 \end{cases}$$

By Theorem 2.13, the number of solutions of (3.17) is then

$$2\prod_{p|k}\left(1 + \left(\frac{D}{p}\right)\right) = 2\sum_{d|k}\left(\frac{D}{d}\right),$$

by (a) of Theorem 3.13, since for $p = 2$ we have $l = 2$ if $2\nmid k$ and $l > 2$ if $2\,|\,k$.

If D is even, then $D \equiv 0$ (mod 4) and so k must be odd. The congruence

$$x^2 \equiv D \equiv 0 \pmod 4$$

has two solutions ($x \equiv 0$ or 2 (mod 4)) and

$$x^2 \equiv D \pmod{p^l}$$

has $1+\left(\frac{D}{p}\right)$ solutions if $p^l \| k$ with $l > 0$. Thus, by Theorem 2.13, the number of solutions of (3.17) is

$$2\prod_{p|k}\left(1+\left(\frac{D}{p}\right)\right) = 2\sum_{d|k}\left(\frac{D}{k}\right),$$

where, as above, the sum is over the square free divisors of k. ■

Example 3.12: Find the number of solutions of

$$x^2 \equiv 17 \pmod{4 \cdot 3 \cdot 7 \cdot 13}.$$

Solution: Here $D = 17 \equiv 1 \pmod 4$ and so we may apply Theorem 3.17. The number of solutions is

$$2\sum_{\substack{d|3\cdot7\cdot13 \\ d \text{ square-free}}}\left(\frac{17}{d}\right) = 2\left\{\left(\frac{17}{1}\right)+\left(\frac{17}{3}\right)+\left(\frac{17}{7}\right)+\left(\frac{17}{13}\right)\right.$$

$$\left.+\left(\frac{17}{3\cdot7}\right)+\left(\frac{17}{3\cdot13}\right)+\left(\frac{17}{7\cdot13}\right)+\left(\frac{17}{3\cdot7\cdot13}\right)\right\}$$

where the symbols on the right hand side are Legendre-Jacobi symbols. Now

$$\left(\frac{17}{1}\right) = +1, \quad \left(\frac{17}{3}\right) = \left(\frac{2}{3}\right) = -1,$$

$$\left(\frac{17}{7}\right) = \left(\frac{3}{7}\right) = -1 \text{ and } \left(\frac{17}{13}\right) = \left(\frac{4}{13}\right) = +1.$$

Thus, by the definition of the Jacobi symbol, the number of solutions to this congruence is

$$2\{(+1)+(-1)+(-1)+(+1)+(-1)(-1)+(-1)(+1)+(-1)(+1)+(-1)(-1)(+1)\}=0,$$

in other words the equation is not solvable. This is the result obtained from Theorem 3.16 since 17 is a quadratic nonresidue of 3 and 7. □

Problem Set 3.6

1. Solve, if possible,

(a) $x^2 + 5x + 1 \equiv 0 \pmod{41}$

(b) $5x^2 + 7x + 1 \equiv 0 \pmod{31}$

(d) $x^2 \equiv 2 \pmod{457^3}$

(e) $x^2 \equiv 3 \pmod{23^2 \cdot 109^3}$

   (c) $x^2 + 6x - 154 \equiv 0 \pmod{399}$ (f) $x^2 \equiv 5 \pmod{29^3 \cdot 53^2}$.

2. Show that if $k \geq 3$, then the solutions to

$$x^2 \equiv 1 \pmod{2^k}$$

are 1, $2^{k-1} - 1$, $2^{k-1} + 1$, $2^k - 1$.

3. Suppose n is a quadratic residue modulo the odd prime p. Consider the equation

(3.18) $x^2 \equiv n \pmod{p}$.

   (a)   If $p = 4k + 3$, show that $x \equiv n^{k+1} \pmod{p}$ is a solution to (3.18).

   (b)   If $p = 8k + 5$ and

      (i) $n^{2k+1} \equiv 1 \pmod{p}$, show that $x \equiv n^{k+1} \pmod{p}$ is a solution to (3.18).

      (ii) $n^{2k+1} \equiv -1 \pmod{p}$, show that $x \equiv (4n)^{k+1}((p + 1)/2) \pmod{p}$ is a solution to (3.18).

   (c)   Choose an integer h so that $\left(\frac{h^2 - 4n}{p}\right) = -1$ and define the sequence $\{v_n\}$ by the recursion

$$v_1 = h, v_2 = h^2 - 4n \text{ and, for } k \geq 3, \ v_k = hv_{k-1} - nv_{k-2}.$$

      (i)   Show that, for $k \geq 1$,

$$v_{2k} = v_k^2 - 2n^k$$

      and

$$v_{2k+1} = v_k v_{k+1} - hn^k.$$

      (ii)   Show that a solution of (3.18) is given by

$$x \equiv v_{(p+1)/2}\left(\frac{p+1}{2}\right) \pmod{p}.$$

4. Use the results of Problem 3 to solve

   (a) $x^2 \equiv 17 \pmod{19}$,

   (b) $x^2 \equiv 7 \pmod{29}$,

   (c) $x^2 \equiv 65 \pmod{101}$,

   (d) $x^2 \equiv 61 \pmod{113}$.

   (e)   Use 3(c) to solve $x^2 \equiv 73 \pmod{127}$.

5. Show that the number of quadratic residues in a reduced system for the composite modulo m is $\varphi(m)2^{-r-\sigma}$, where r and σ have the same meaning as

in Theorem 3.16.

6. Use the methods of Chapter Two to show that if $x^2 \equiv a \pmod{p}$ is solvable, then $x^2 \equiv a \pmod{p^n}$ is solvable for all $n \geq 1$. Here p is an odd prime. How are the two sets of solutions related?

7. How many solutions do the following congruences have?

(a) $x^2 \equiv -1 \pmod{61}$.

(b) $x^2 \equiv -1 \pmod{244}$.

(c) $3x^2 + 5x + 1 \equiv 0 \pmod{1564}$.

Additional Problems for Chapter Three

General Problems

1. Show that if p is an odd prime, then

(a) $\left(\dfrac{-1}{p}\right) = \sin\left(\dfrac{\pi p}{2}\right)$,

(b) $\left(\dfrac{-2}{p}\right) = \sqrt{2}\sin\left(\dfrac{\pi p}{2}\right)$

and

(c) $\left(\dfrac{-3}{p}\right) = \dfrac{2}{\sqrt{3}}\sin\left(\dfrac{\pi p}{2}\right)$.

Are there other similar results?

2. Let p be an odd prime and let $S_1 \cup S_2 = \{1,\ldots,p-1\}$, with $S_1, S_2 \neq \varnothing$, be two sets of integers such that the product of two elements in the same set is in S_1 and the product of an element in S_1 and an element in S_2 is in S_2. Prove that S_1 is the set of quadratic residues modulo p and S_2 is the set of quadratic nonresidues modulo p. (Hint: use primitive roots modulo p.)

3. If $\left(\dfrac{n}{p}\right) = -1$, then show that

$$\sum_{d \mid n} d^{(p-1)/2} \equiv 0 \pmod{p}.$$

4. Suppose that $n = kp^3 + 1$, where k is even, not a cube and $k < 2\sqrt{2p+2} - 2$. Suppose $2^k \not\equiv 1 \pmod{n}$ and $2^{n-1} \equiv 1 \pmod{n}$. Prove that n is a prime.

5. Prove the following result. If $a = 2^\alpha p_1^{\alpha_1} \cdots p_r^{\alpha_r}$, where $p_k \equiv -1 \pmod 6$, $1 \leq k \leq r$, then $a^k - 1 = n^3$ implies that $k = 0$, $n = 0$ or $a = 2$, $k = 1$.

6. Let p be a prime, $p > 5$. Show that there exist primes q and r, both less than p such that r is a quadratic residue modulo p and q is a quadratic nonresidue

modulo p.

7. Let p be an odd prime and let a and b be integers that are not both divisible by p.

   (a) If n is an integer no divisible by p, define m by $nm \equiv 1 \pmod{p}$. Show that

   $$\left(\frac{2ab - (a^2n + b^2m)}{p} \right) = (-1)^{(p-1)/2} \left(\frac{n}{p} \right).$$

   (b) Suppose $p \mid (a^2 + b^2)$ and $p \nmid ab$. Show that $\left(\frac{ab}{p} \right) = \left(\frac{2}{p} \right)$.

8. Suppose $p \equiv 1 \pmod 4$ and write $p = a^2 + b^2$, where a is odd. Prove that $\left(\frac{a}{p} \right) = +1$ and $\left(\frac{b}{p} \right) = (-1)^{(p-1)/4}$.

9. Show that $n^2 + (n + 1)^2 + \cdots + (n + k)^2 = m^2$ is impossible whenever $1^2 + 2^2 + \cdots + k^2$ is a quadratic nonresidue modulo $k + 1$.

10. The following is a generalization of Wilson's Theorem that is due to Gauss. Let n be a positive integer, $n > 2$, and let $r = \varphi(n)$. If $\{a_1, \ldots, a_r\}$ is a reduced residue system modulo n, show that

   $$a_1 \cdots a_r = \begin{cases} -1 \pmod n & \text{if } n = 4, p^e \text{ or } 2p^e \\ +1 \pmod n & \text{else} \end{cases}.$$

   Note that the product gives -1 precisely when n possesses primitive roots. (Hint: split the a_k into solutions of $x^2 \equiv 1 \pmod n$ and into nonsolutions. Then calculate the subproducts separately.)

11. Suppose that p is an odd prime and a and b are integers such that $p \nmid ab$. Show that the number of solutions to the congruence

   $$ax^2 + by^2 \equiv 1 \pmod p$$

   is $p - \left(\frac{-ab}{p} \right)$. What happens if we replace 1 by another integer?

12. (a) Prove that an integer is a square if and only if it is a quadratic residue of every prime.

   (b) Show that the number $\sqrt{21^{36} + 36^{21}}$ is irrational.

13. (a) Prove that p can be written in the form $p = 2x^2 + 3y^2$ if and only if $p \equiv 5$ or $11 \pmod{24}$.

   (b) Prove that p can be written in the form $p = x^2 + 6y^2$ if and only if $p \equiv 1$ or $7 \pmod{24}$. 1 or 7 (mod 24).

   (c) Prove that p can be written in the form $p = 2x^2 + 5y^2$ if and only if $p \equiv 7$, 13, 23 or 37 (mod 24).

   (Hint: for (a) and (c) consider the congruence $2z^2 + d \equiv 0 \pmod p$, for an appropriate value of d.)

14. Show that unless p is a Fermat prime, i.e. $p = 2^{2^k} + 1$ for some nonnegative integer k, then p has a quadratic nonresidue which is not a primitive root. (Hint: show that $(p-1)/2 - \varphi(p-1) > 0$ unless p is a Fermat prime.)

15. Show that a prime factor p of a Fermat number $F_n = 2^{2^n} + 1$ must be of the form $2^{n+2}k + 1$. (Hint: show that $\mathrm{ord}_p 2 = 2^{n+1}$ and then use Euler's criterion in connection with the quadratic character of 2.)

16. Show that the number of solutions N of the congruence

$$ax^2 + bx + c \equiv 0 \pmod{m}$$

is given by

$$mN = \sum_{p=0}^{m-1}\sum_{q=0}^{m-1} e^{2\pi i p(aq^2 + bq + c)/m}.$$

17. (a) Show that if P is a square-free odd integer greater than 1, then

$$\sum_m \left(\frac{m}{P}\right) = 0,$$

where the symbol denotes the Jacobi symbol and the sum is over a reduced residue system modulo P. (Hint: use Problem Set 3.4.)

(b) Let A be a reduced residue system modulo P, where P is as in (a). Let μ be the number of incongruent elements a in A such that $\left(\frac{a}{P}\right) = +1$ and let ν be the number of incongruent elements b of A such that $\left(\frac{b}{P}\right) = -1$. Show that $\mu = \nu = \varphi(P)/2$.

(c) In this problem we will determine necessary and sufficient conditions on primes p so that there exists an integer n with $p \mid (n^2 - P)$. Since P is odd we know that we can make $n^2 - P$ even and also $q \mid P$ implies that we can find n so that $q \mid (n^2 - P)$. Thus we are looking for p such that $p \nmid 2P$. We say that p divides $x^2 - P$ if there is an integer n such that $p \mid (n^2 - P)$.

   (i) Suppose $\pm P \equiv 1 \pmod 4$ with $P > 0$. Let $a_1, \ldots, a_m$ be the odd integers in the interval $(0, 2P)$ such that $\left(\frac{a_k}{P}\right) = +1$ and let $b_1, \ldots, b_m$ be odd integers in the interval $(0, 2P)$ such that $\left(\frac{b_k}{P}\right) = -1$. If, show that p divides $x^2 - P$ if and only if $p \equiv a_k \pmod{2P}$, $1 \le k \le m$. Similarly show that q is not a divisor of $x^2 - P$ if and only if $q \equiv b_k \pmod{2P}$, $1 \le k \le m$.

   (ii) Suppose $\pm P \equiv 3 \pmod 4$ with $P > 0$. Let $a_1, \ldots, a_m$ be the odd integers in the interval $(0, 4P)$ such that $\left(\frac{a_k}{P}\right) = +1$ and let $b_1, \ldots, b_m$ those for which $\left(\frac{b_k}{P}\right) = -1$. Show that a prime p, $p \nmid 2P$, is a divisor of $x^2 - P$ if and only if either

$$p \equiv a_k \pmod{4P}, \ 1 \le k \le m$$

or

$$p \equiv b_k \pmod{4P}, \ 1 \le k \le m.$$

(iii) Determine analogous results for the case when P is an even square-free integers.

(d) Determine the prime divisors of the following binomial expressions.

(i) $x^2 - 33$.

(ii) $x^2 + 17$.

(iii) $x^2 - 18$.

(iv) $x^2 + 26$.

18. Suppose p and q are primes and $q = p + 2$. Show that there is an integer a such that $p \mid (a^2 - q)$ if and only if there is an integer b such that $q \mid (b^2 - p)$.

19. (a) Show that the equation $y^2 + 4a^2 = x^3 + (4b - 1)^3$ has no solutions in integers x and y if a has no prime factors of the form $4k + 3$.

(b) Show that the equation $y^2 - 3b^2 = x^3 - a^3$ has no solutions in integers x and y if $a \equiv 1 \pmod 4$, $b \equiv \pm 2 \pmod 6$ and b has no prime factors of the form $12k \pm 5$.

(c) Show that the equation $y^2 - 2b^2 = x^3 - a^3$ has no solutions in integers x and y if $a \equiv 2$ or $4 \pmod 8$ and b is odd and has no prime factors of the form $8k \pm 3$.

(d) Show that the equation $y^2 - kb^2 = x^3 - k^3a^3$ has no integer solutions when $a \equiv -1 \pmod 4$, $b \equiv 0 \pmod 2$, k is square-free, $k \equiv 3 \pmod 4$, $(k, b) = 1$, if $k \equiv 2 \pmod 3$ and a and b have no common prime factor p if $\left(\frac{k}{p}\right) = -1$.

(Hints: first eliminate residue classes modulo 4 that x can lie in and then factor the sum/difference of cubes that result. Now use the quadratic character of the quadratic factor.) For similar results see [68, ch. 26].

20. Show that

$$\left(\frac{2}{p}\right) = \prod_{j=1}^{(p-1)/2} 2\cos\frac{2\pi j}{p},$$

where p is an odd prime.

21. Let p be a prime of the form $4k + 3$. Show that

$$\prod_{\substack{r=1 \\ \left(\frac{r}{p}\right)=+1}}^{p-1} 2\sin\frac{\pi r}{p} = \prod_{\substack{s=1 \\ \left(\frac{s}{p}\right)=-1}}^{p-1} 2\sin\frac{\pi s}{p} = \sqrt{p}.$$

22. Calculate the value of the product

$$\prod_{j=1}^{(p-1)/2} 2\sin\frac{k^2 j\pi}{p}.$$

23. Let p be an odd prime and partition the set $\{1, 2, \ldots, p - 1\}$ into subsets of consecutive quadratic residues and consecutive quadratic nonresidues. Let $r(p)$ be the number of consecutive triples of quadratic residues and let $n(p)$ be the number of consecutive quadratic nonresidues. Let s be the number of sets that are singletons, e the number of endpoints of these sets and i the number of interior points.

   (a) Show that $r(p) + n(p) = i$.

   (b) If $p \equiv 3 \pmod 4$, show that

$$r(p) + n(p) = \begin{cases} (p-3)/8 & \text{if } p \equiv 3 \pmod 8 \\ (p-7)/8 & \text{if } p \equiv 7 \pmod 8 \end{cases}.$$

   (c) What can be said if $p \equiv 1 \pmod 4$?

   (These results are due to Monzingo.)

24. Under the notation of Problem 23, if $(a, p) = 1$, define $\bar{a}$ by $a\bar{a} \equiv 1 \pmod p$.

   (a) Show that if $p \equiv 1 \pmod 4$, then a is a quadratic nonresidue singleton if and only if $\bar{a}$ is a quadratic nonresidue interior point.

   (b) Show that if $p \equiv 3 \pmod 4$, then if $a \not\equiv \bar{a} \pmod p$, a is a quadratic residue singleton if and only if $\bar{a}$ is a quadratic residue left endpoint.

25. Show that

$$\sum_{\substack{r=1 \\ \left(\frac{r}{p}\right)=+1}}^{p-1} r = \frac{1}{24}p(p^2 - 1) - p\sum_{k=1}^{(p-1)/2}\left[\frac{k^2}{p}\right].$$

   If $p \equiv 1 \pmod 4$, show that this sum reduces to $p(p - 1)/4$.

26. Suppose $p \equiv 3 \pmod 4$ is a prime.

   (a) Show that

$$-\left\{2-\left(\frac{2}{p}\right)\right\}\frac{1}{p}\sum_{a=1}^{p-1}a\left(\frac{a}{p}\right)=\sum_{a=1}^{(p-1)/2}\left(\frac{a}{p}\right).$$

(Hint: break up the sum on the right hand side into even and odd values of a and recall that as a goes from 1 to p - 1 so does p - a.)

(b) Show that

$$-\left\{1-2\left(\frac{2}{p}\right)\right\}\sum_{a=1}^{(p-1)/2}a\left(\frac{a}{p}\right)=p\frac{1-\left(\frac{2}{p}\right)}{2}\sum_{a=1}^{(p-1)/2}\left(\frac{a}{p}\right).$$

What can be said if $p \equiv -1$ (mod 8)?

27. Let p be a prime.

(a) If $p \equiv 1$ (mod 4), show that

$$\sum_{0<n<p/2}\left(\frac{n}{p}\right)=0.$$

(b) If $p \equiv 3$ (mod 8), show that

$$\sum_{0<n<p/4}\left(\frac{n}{p}\right)=0.$$

(c) If $p \equiv 7$ (mod 8), show that

$$\sum_{p/4<n<p/2}\left(\frac{n}{p}\right)=0.$$

(d) If $p \equiv 5$ (mod 8), show that

$$\sum_{0<n<p/6}\left(\frac{n}{p}\right)=0.$$

(e) If $p \equiv 7$ (mod 8), show that

$$\sum_{p/6<n<p/2}\left(\frac{n}{p}\right)=0.$$

(Hints: start with the result

$$\sum_{n=1}^{p-1}\left(\frac{n}{p}\right)=0$$

(Problem 8(a) of Problem Set 3.2) and split the range of summation up as appropriate. Then use the quadratic character of -1 and ±2.)

28. Let, for an odd prime p,

$$S(a) = \sum_{m=1}^{p-1} \left(\frac{m^2 + a}{p} \right).$$

(a) Show that $S(a)$ depends only on $\left(\frac{a}{p}\right)$ and, in particular, if $p \mid a$, then $S(a)$
   $= p - 1$. (Hint: if $p \nmid k$, then $S(a) = \left(\frac{k^2}{p}\right) S(a) = S(ak^2)$.)

(b) Show that

$$\sum_{a=0}^{p-1} S(a) = 0.$$

(Hint: invert the order of summation.)

(c) Show that

   (i)   if $\left(\frac{a}{p}\right) = -1$, then $S(a) = 0$

   and

   (ii) if $\left(\frac{a}{p}\right) = +1$, then $S(a) = -2$.

   More compactly we can write this as follows. If $p \nmid a$, then
   $S(a) = -\left(1 + \left(\frac{a}{p}\right)\right)$.

   (Hint: for (i) note that for $p \nmid n$,

$$S(a) = S(an^2) = \sum_{m=0}^{p-1} \left(\frac{m^2 + an^2}{p} \right),$$

   and so $\left(\frac{a}{p}\right) = -1$ implies that

$$S(a) = \left(\frac{a}{p} \right) \sum_{m=1}^{p-1} \left(\frac{m^2 a + a^2 n^2}{p} \right).$$

   If one sums on n, $1 \le n \le p - 1$, show that $S(a) = -S(a)$. For (ii) use
   (b) and split the sum by the values of $\left(\frac{a}{p}\right)$. Then use (a) and (i).)

29. Let p be an odd prime, $f(x) = ax^2 + bx + c$ and $d = b^2 - 4ac$. This exercise
   generalizes the result of Problem 28. Let

$$T(f) = \sum_{m=0}^{p-1} \left(\frac{f(m)}{p} \right).$$

   Show that

   (a) if $p \nmid ad$, then $T(f) = -\left(\frac{a}{p}\right)$;

   (b) if $p \mid a$, $p \nmid d$ or $p \nmid a$, $p \mid d$, then $T(f) = (p-1)\left(\frac{a}{p}\right)$;

   and

   (c) if $p \mid a$ and $p \mid d$, then $T(f) = (p-1)\left(\frac{c}{p}\right)$.

(Hint: the case $p \mid a$ and $p \nmid d$ follows from Problem 8 of Problem Set 3.2. If $p \nmid a$, then show that

$$\left(\frac{a}{p}\right)T(f) = \left(\frac{4a}{p}\right)T(f) = \sum_{m=0}^{p-1}\left(\frac{(2am+b)^2 - d}{p}\right) = \sum_{k=0}^{p-1}\left(\frac{k^2 - d}{p}\right).)$$

30. This exercise proves a result of E. Jacobsthal that gives a construction for a and b such that $p = a^2 + b^2$, if $p \equiv 1 \pmod 4$. Suppose $(k, p) = 1$ and let

$$U(k) = \sum_{n=0}^{p-1}\left(\frac{n(n^2 + k)}{p}\right).$$

Show that

(a) $U(k)$ is even,

(b) $U(kt) = \left(\frac{t}{p}\right)U(k)$

and

(c) if $\left(\frac{r}{p}\right) = +1$ and $\left(\frac{n}{p}\right) = -1$, then

$$p = \left(\tfrac{1}{2}U(r)\right)^2 + \left(\tfrac{1}{2}U(n)\right)^2.$$

31. Let $p \equiv 5 \pmod 8$ be a prime. Show that

$$\sum_{n=1}^{p-1}\left(\frac{n^5 + 1}{p}\right) = 0.$$

32. Let $a \geq 2$ be even and let p be a prime. Show that

(a) if $p \equiv 3 \pmod 4$, then

$$\sum_{n=1}^{p-1}\left(\frac{n}{p}\right)\left(\frac{n^a + 1}{p}\right) = 0;$$

and

(b) If $p = a(2k + 1) + 1$ and $a \equiv 0 \pmod 4$, then

$$\sum_{n=1}^{p-1}\left(\frac{n}{p}\right)\left(\frac{n^a + 1}{p}\right) = 0.$$

33. Let p be an odd prime and let

$$N_1 = \#\left\{1 \leq n \leq p - 2: \left(\tfrac{n}{p}\right) = \left(\tfrac{n+1}{p}\right) = +1\right\},$$

$$N_2 = \#\left\{1 \leq n \leq p - 2: \left(\tfrac{n}{p}\right) = \left(\tfrac{n+1}{p}\right) = -1\right\},$$

$$N_3 = \#\left\{1 \leq n \leq p - 2: \left(\tfrac{n}{p}\right) = +1, \left(\tfrac{n+1}{p}\right) = -1\right\}, \text{ and}$$

$$N_4 = \#\left\{1 \leq n \leq p - 2: \left(\tfrac{n}{p}\right) = -1, \left(\tfrac{n+1}{p}\right) = +1\right\}.$$

Show that

$$N_1 = \left\{p - 4 - \left(\tfrac{-1}{p}\right)\right\} / 4, N_2 = \left\{p - 2 + \left(\tfrac{-1}{p}\right)\right\} / 4,$$

$$N_3 = \left\{p - \left(\tfrac{-1}{p}\right)\right\} / 4, \text{ and } N_4 = \left\{p - 2 + \left(\tfrac{-1}{p}\right)\right\} / 4.$$

(Hint: for example,

$$4N_1 = \sum_{n=1}^{p-2}\left\{1 + \left(\frac{n}{p}\right)\right\}\left\{1 + \left(\frac{n+1}{p}\right)\right\}$$

and use Problem 29.)

Can you extend these results to

$$N(\varepsilon, \eta, h) = \#\left\{n: \left(\frac{n}{p}\right) = \varepsilon, \left(\frac{n+h}{p}\right) = \eta\right\},$$

where $\varepsilon, \eta \in \{1, -1\}$?

34. If $p \equiv 1$ (mod 4) is a prime, then there is a unique integer a such that $p = a^2 + b^2$ with $a \equiv -1$ (mod 4) and b even. Show that

$$\sum_{k=0}^{p-1}\left(\frac{k^3 + 6k^2 + k}{p}\right) = \begin{cases} 2\left(\tfrac{2}{p}\right)a & \text{if } p \equiv 1 \text{ (mod 4)} \\ 0 & \text{if } p \equiv 3 \text{ (mod 4)} \end{cases}.$$

(Hint: find a transformation that will take $k^3 + 6k^2 + k$ into $(k^2 - 1)(k^2 - 2)$ modulo p and use Problem 30.)

35. (a) If $p = 8q + 1$, $q \equiv 2$ (mod 3), $q > 5$ and p and q are both primes, show that 3 is a primitive root modulo p.

   (b) If $p = 8q + 1$, $q \equiv 2$ or 4 (mod 5) with p and q both primes, show that 5 is a primitive root modulo p.

   (c) If $p = 4q + 1$, $q \equiv 1$ (mod 3) and p and q are both primes. show that 13 is a primitive root modulo p.

   (d) Can you make a general theorem?

   (e) Let p and $q = 2p + 1$ be primes with $p \equiv 3$ (mod 4). If g is a quadratic nonresidue modulo q and $g \not\equiv -1$ (mod q), then g is a primitive root modulo q. (Hint: see Problem 14(d) of Problem Set 2.8.)

   The last problems deal with higher degree residues. We say that a is an **nth degree residue** modulo p if the congruence

$$x^n \equiv a \pmod{p}$$

(See Section 2.9 for some more general remarks on higher degree residues.)

36. Show that if a prime p is of the form $3k + 2$, then all integers in a reduced residue system modulo p are cubic residues. Show that if p is a prime of the form $3k + 1$, then one-third of the members of a reduced residue system are cubic residues.

37. Show that if $p \equiv 3 \pmod 4$ is a prime, then the biquadratic residues and the quadratic residues coincide.

38. Show that -1 is a biquadratic residue modulo the prime p if and only if $p \equiv 1 \pmod 8$.

39. Let a be an integer and $p > 3$ be a prime such that $p \nmid a$. Let a be a cubic residue modulo p and suppose that $r^3 \equiv a \pmod p$. Show that

$$x^2 + xr + r^2 \equiv 0 \pmod p$$

has two solutions different from r if and only if $p \equiv 1$ or $7 \pmod{12}$.

40. (a) Use the identity $x^4 + 4 = ((x + 1)^2 + 1)((x + 1)^2 - 1)$ to show that -4 is a biquadratic residue modulo p if and only if $p \equiv 1 \pmod 4$.

 (b) Show that if $p \equiv 1 \pmod 4$, then $(p - 1)/4$ is a biquadratic residue modulo p.

 (c) Show that the fourth roots of $(p - 1)/4$ consist of two pairs of consecutive integers.

41. This exercise outlines Dirichlet's determination of the biquadratic character of p, where $p \equiv 1 \pmod 4$ is a prime. Write $p = a^2 + b^2$, where a is odd. Show that

 (a) $\left(\dfrac{a+b}{p}\right) = (-1)^{((a+b)^2-1)/8}$;

 (Hint: $2p = (a + b)^2 + (a - b)^2$.);

 (b) $(a + b)^2 \equiv 2ab \pmod p$;

 (c) $(a + b)^{(p-1)/2} \equiv (2ab)^{(p-1)/4} \pmod p$;

 (d) if we define f by $af \equiv b \pmod p$, then $f^2 \equiv -1 \pmod p$ and $2^{(p-1)/4} \equiv f^{ab/2} \pmod p$

and

 (e) 2 is a biquadratic residue modulo p if and only if p can be written in the form $p = A^2 + 64B^2$ for some integers A and B.

Computer Problems

1. Write a program to compute and count the number of quadratic residues modulo n for $2 \le n \le 100$.

2. Write programs to compute the value of the Legendre symbol $\left(\dfrac{a}{p}\right)$, $1 \le a \le p - 1$, using

(a) Gauss' Lemma (Theorem 3.3);

(b) Eisenstein's Lemma (Theorem 3.4);

and

(c) Euler's Criterion ((a) of Theorem 3.2.).

(d) Compute $\left(\frac{a}{p}\right)$ for $p \leq 101$ and compare the speed of each method (a) - (c) and the method of Problem 1.

3. Write a program to compute the values of Jacobi symbols. (Hint: see Problem 6 of Problem Set 3.4.) Modify the program to compute the values of Legendre symbols. Make a table of values of $\left(\frac{a}{p}\right)$ for $1 \leq a \leq b - 1$ and $1 \leq b \leq 50$.

4. Write a program to compute Kronecker symbols. Compute the values of the Kronecker symbols $\left(\frac{a}{n}\right)$, where $a \equiv 0$ or $1 \pmod 4$. Make a table of values for $2 \leq a \leq 100$.

5. (a) Write a program to solve quadratic congruences modulo a prime.

   (b) Solve the following congruences.

   (i) $x^2 + 122x - 4697 \equiv 0 \pmod{9787}$.

   (ii) $x^2 + 34x + 2789 \equiv 0 \pmod{9001}$.

   (iii) $x^2 \equiv 31 \pmod{4987}$.

   (iv) $2x^2 + 5x + 3 \equiv 0 \pmod{29}$.

   (c) Modify the program using the Chinese Remainder Theorem to handle quadratic congruences for composite moduli.

   (d) Solve the following congruences.

   (i) $9x^2 + 995x - 47 \equiv 0 \pmod{1001}$.

   (ii) $x^2 \equiv -1 \pmod{3599}$.

   (iii) $x^2 \equiv 2 \pmod{457^2}$

   (iv) $x^2 \equiv 5 \pmod{29^3 \cdot 53^2}$.

6. Write a program to calculate sums of Legendre symbols of the form

$$S(h,l,k) = \sum_{ph/k<n<pl/k} \left(\frac{n}{p}\right).$$

Compute several of these sums for various choices of h, k and l. Can you make any conjectures analogous to Problem 27 above? Can you prove any of them?

CHAPTER FOUR
Approximation of Real Numbers

4.1. INTRODUCTION

In this chapter we are concerned with the subject of Diophantine approximation. This subject is vast and we can barely scratch the surface in these notes. For a more detailed survey of this subject, see, for example, [14].

Diophantine approximation is mainly concerned with the approximation of real numbers by rational numbers (hence, the term "Diophantine"). We will discuss two different ways of doing this: by way of Farey fractions and by way of continued fractions. Farey fractions are a sequence of rationals that can be used in approximation problems and exist irrespective of the real numbers we are trying to approximate. Continued fractions can be looked on as an algorithm for producing a series of rationals that can be used as a sequence of progressively better approximations for a given real number.

One can consider the problem in the following way. If α is a given real number, can we find a rational number p/q, $q > 0$, (or a sequence of rational numbers $\{p_n / q_n\}_{n=1}^{+\infty}$) such that

$$|\alpha - p/q| < \varepsilon,$$

where ε is "sufficiently small." We shall see below (Theorem 4.1) that if we do not demand that ε be too small we can always solve this problem. We shall see also (Theorem 4.6) that in general we cannot get too good of an approximation that works for all real numbers.

We begin with a result that illustrates one of the standard methods in this subject: the famous pigeon-hole principle, which states that if we have N boxes and more than N objects to be distributed among these boxes, then some box has at least two objects.

Lemma 4.1.1 (Dirichlet). Let α be a real number. Then for any positive integer t there exist relatively prime integers x and y, $1 \leq y \leq t$, such that

(4.1) $|\alpha - x / y| < 1 / yt.$

Proof. Let t be an arbitrary positive integer and divide the internal $[0, 1)$ (or $(0, 1]$ if you prefer) into t subintervals of equal length. $[(h - 1)/t, h/t)$, where $1 \le h \le t$, h an integer.

Now let y be a nonnegative integer and let x be the least integer $\ge \alpha y$. Then

$$0 \le x - \alpha y < 1.$$

(If we chose the interval $(0, 1]$ we would take x to be $[\alpha y + 1]$ so that $0 < x - \alpha y \le 1$.) If we now let y run from 0 to t we get $t + 1$ numbers $x - \alpha y$, all of which belong to the interval $[0, 1)$.

Since there are only t subintervals we see that there exists a positive integer $h \le t$ and two distinct pairs of integers x_1, y_1 and x_2, y_2 such that

$$(h - 1) / t \le x_1 - \alpha y_1, x_2 - \alpha y_2 < h / t.$$

If we let

$$x = x_2 - x_1 \text{ and } y = y_2 - y_1,$$

then

$$|x - \alpha y| < 1/t.$$

If we suppose $y_2 > y_1$, then y is one of the numbers 1, ..., t, and so we have

$$|\alpha - x/y| < 1/yt.$$

Since $1 \le y \le t$ we obtain the result stated as it is clear that we may suppose x and y relatively prime.   ∎

Note that since $y \le t$ this yields the following weaker result: There exist relatively prime integers x and y, with $y > 1$, such that

$$|\alpha - x / y| < 1 / y^2.$$

We illustrate the technique of the lemma with the following example.

Example 4.1. Find rational approximations to $\sqrt[3]{2}$ and $\sqrt{2}$ satisfying (4.1) with $t = 5$.

Solution. The subintervals are $(0, .2]$, $(.2, .4]$, $(.4, .6]$, $(.6, .8]$, and $(.8, 1]$. We allow k to be 0, 1, 2, 3, 4 or 5 and take

$$x_k = 1 + [k\alpha] \text{ and } y_k = k.$$

When $\alpha = \sqrt[3]{2}$ we get the series

$$x_0 = 1,\ x_1 = 2,\ x_2 = 3,\ x_3 = 4,\ x_4 = 6 \text{ and } x_5 = 7.$$

We find that $.6 < x_1 - ay_1,\ x_5 - ay_5 \leq .8$. Thus

$$x = 7 - 2 = 5 \text{ and } y = 5 - 1 = 4.$$

This gives

$$\left|\sqrt[3]{2} - \tfrac{5}{4}\right| \approx .009 < \tfrac{1}{16} = .0625.$$

Now take $\alpha = \sqrt{2}$. This gives the sequence

$$x_0 = 1,\ x_1 = 2,\ x_2 = 3,\ x_3 = 4,\ x_4 = 6 \text{ and } x_5 = 8.$$

We find that $.8 < x_0 - \alpha y_0,\ x_5 - \alpha y_5 \leq 1$. Thus

$$x = 8 - 1 = 7 \text{ and } y = 5 - 0 = 5.$$

This gives

$$\left|\sqrt{2} - \tfrac{7}{5}\right| \approx .014 < \tfrac{1}{25} = .04. \qquad \square$$

If α is a rational number, say $\alpha = a/b$, $(a, b) = 1$ and $b > 0$, then

$$1/y^2 > |x/y - a/b| = |bx - ay|/by \geq 1/by,$$

since a, b, x and y are integers. Thus we must have $y < b$ and so we have only a finite number of number of solutions to the approximation problem in this case. On the other hand, if α is irrational, the following theorem applies.

Theorem 4.1. Let α be a real irrational number. Then the inequality

$$|\alpha - x/y| < 1/y^2$$

has infinitely many solutions in relatively prime integers x and y.

Proof. Let t_1 be a positive integer. Then, by (4.1), we can find a pair of relatively prime integers x_1 and y_1, with $1 \leq y_1 \leq t_1$, such that

$$\alpha_1 = |\alpha - x_1/y_1| < 1/y_1^2.$$

Since α is irrational we know that α_1 is not zero, and so we can find an integer $t_2 > 1/\alpha_1$. Now choose relatively prime integers x_2 and y_2, with $1 \leq y_2 \leq t_2$, such that

$$\alpha_2 = |\alpha - x_2/y_2| < 1/y_2 t_2 \leq 1/t_2 < \alpha_1.$$

Repeating this process we obtain an infinite sequence of successively decreasing positive numbers

$$\alpha_1 > \alpha_2 > \cdots > \alpha_k > \cdots,$$

where

$$\alpha_k = |\alpha - x_k / y_k| < 1 / y_k^2,$$

which proves our result.   ∎

Thus the problem is set before us. Given a real irrational number α, what functions $f(\alpha, q)$ can be found so that

1) $f(\alpha, q) \to 0$ as $q \to +\infty$

and

2)   there are infinitely many positive integers q and corresponding integers p such that

$$|\alpha - p / q| < f(\alpha, q).$$

Theorem 4.1 states that we can take $f(\alpha, q) = q^{-2}$. Our problem is therefore to find the smallest possible f that will work for all real irrationals α or for a given infinite class of real irrational α.

For an extension of many of the results of this chapter to the approximation of complex numbers see [83, ch. 4].

Problem Set 4.1

1. Find rationals a/b, $0 < b \le 10$, that satisfy the inequality

$$|\theta - a / b| < 1 / 11b$$

for $\theta = \sqrt{2}, \sqrt[3]{10}, \sqrt{33}$ and $\sqrt[4]{101}$.

2. Let $p \equiv 1 \pmod 4$ be a prime. Use Lemma 4.1.1 to show that we can write

$$p = a^2 + b^2$$

for some integers a and b. (Hint: let n be an integer such that $n^2 + 1 \equiv 0 \pmod p$) and then approximate n/p with $t = \left[\sqrt{p}\right]$.

3. Use Lemma 4.1.1 to prove that if p is a odd prime, then among the integers -1, 1, 2, ..., $\left[\sqrt{p}\right]$ there is an $e\underline{th}$ power nonresidue modulo p for any $e \mid (p - 1)$, $e \ge 2$. (Hint: let n be any $e\underline{th}$ power nonresidue modulo p and proceed much as in Problem 2.)

4.2. FAREY FRACTIONS

Let n be a given positive integer. We define the **Farey sequence** of order n, written F_n, to be the set of rational numbers written in increasing order and defined inductively as follows. Let $F_n = \{0/1, 1/1\}$ and if $n \geq 2$, then given F_{n-1} we form F_n by adding to F_{n-1} all fractions of the form $(a + a_1)/(b + b_1)$, where a/b and a_1/b_1 are consecutive terms in F_{n-1} and $b + b_1 < n$. Thus $F_2 = \{0/1, (0 + 1)/(1 + 1), 1/1\}$. Continuing in this way, we get the following table of Farey sequences of order $n \leq 5$.

$n = 1$: $\{0/1, 1/1\}$

$n = 2$: $\{0/1, 1/2, 1/1\}$

$n = 3$: $\{0/1, 1/3, 1/2, 2/3, 1/1\}$

$n = 4$: $\{0/1, 1/4, 1/3, 1/2, 2/3, 3/4, 1/1\}$

$n = 5$: $\{0/1, 1/5, 1/4, 1/3, 2/5, 1/2, 3/5, 2/3, 3/4, 4/5, 1/1\}$

If we look at this table, several interesting properties become apparent. Some of them are the subject of the following theorems.

Theorem 4.2. If a/b and c/d are consecutive fractions in F_n, with $a/b < c/d$ then $cb - ad = +1$.

Proof. This is clearly true when $n = 1$. Suppose the result holds for F_{n-1}. Then any consecutive fractions in F_n will be either (with a/b and c/d consecutive in F_{n-1})

a) a/b and c/d, in which case the result holds by assumption, or

b) a/b and $(a + c)/(b + d)$, in which case we have

$$(a + c)b - a(b + d) = cb - ad = +1,$$

   or

c) $(a + c)/(b + d)$ and c/d, in which case we have

$$c(b + d) - (a + c)b = cb - ad = +1.$$

The result is proved by mathematical induction. ∎

Corollary 4.2.1. Every fraction a/b in F_n, for $n \geq 1$, is reduced, that is, $(a, b) = 1$.

Theorem 4.3. If a/b and c/d are consecutive fractions in any row, then among all rationals with values between the two $(a + c)/(b + d)$ is the unique fraction with the smallest denominator.

Proof. First note that $(a + c)/(b + d)$ will be the first fraction to be inserted between a/b and c/d in the appropriate F_n as n increases and it will appear in F_{b+d}. Since the rationals are in increasing order in F_n, for any n, we have

$$a/b < (a + c)/(b + d) < c/d.$$

Now let x/y be a rational between a/b and c/d, so that $a/b < x/y < c/d$. Then

(4.2)
$$\begin{aligned}
c/d - a/b &= (c/d - x/y) + (x/y - a/b) \\
&= (cy - dx)/dy + (bx - ay)/by \\
&\geq 1/dy + 1/by = (b + d)/bdy,
\end{aligned}$$

since $cy - dx$ and $bx - ay$ are positive integers, and so we have

$$(b + d)/bdy \leq (cb - ad)/bd = 1/bd,$$

by Theorem 4.2. Thus $y \geq b + d$. If $y > b + d$, then x/y does not have the least denominator and so we are done. If $y = b + d$, then the inequality in (4.2) must be an equality. Thus $cy - dx = 1 = bx - ay$. If we solve this system, we see that $x = a + c$ and $y = b + d$ and the theorem is proved.   ∎

Theorem 4.4. If $0 \leq x \leq y$ and $(x, y) = 1$, then x/y appears in F_n for all $n \geq y$.

Proof. This is clear for $y = 1$. Suppose the result holds for $y = y_0 - 1$, with $y_0 > 1$. If $y = y_0$, then x/y cannot be in F_{y-1} by definition, and so must lie between two consecutive fractions, say $a/b < c/d$, in F_{y-1}. Thus $a/b < x/y < c/d$. Since

$$a/b < (a + c)/(b + d) < c/d$$

and a/b and c/d are consecutive, we see that $(a + c)/(b + d) \notin F_{y-1}$. Thus $b + d > y - 1$, by assumption. By Theorem 4.3, we know that $y \geq b + d$ and so we must have $y = b + d$. The uniqueness part of Theorem 4.3 implies that $x = a + c$. Thus $x/y = (a + c)/(b + d)$ enters in F_y, and so in all later sequences.   ∎

Corollary 4.4.1. $F_n = \{a/b : 0 \leq a/b \leq 1, (a, b) = 1 \text{ and } 0 < b \leq n\}$.

We define the extended Farey sequence of order n, written $\mathbf{F}_n$, to be the set of all reduced fractions with denominators not exceeding n, written in increasing order. Thus

$$\mathbf{F}_n = \bigcup_{a \in Z}(F_n + a) = \{a/b: (a, b) = 1 \text{ and } 1 \leq b \leq n\}.$$

For example,

$$\mathbf{F}_3 = \{\ldots, -1/1, -2/3, -1/2, -1/3, 0/1, 1/3, 1/2, 2/3, 1/1, 4/3, 3/2, 5/3, 2/1, \ldots\}.$$

The elements of F_n are called the **Farey fractions of order** n. Note that any two consecutive elements of F_n, a/b and c/d satisfy the equality of Theorem 4.2 as well as the inequality $b + d > n$.

Problem Set 4.2

1. List the Farey sequences of orders 11 and 12.

2. Prove that the number of terms in the Farey sequence of order n is

$$\phi(1) + \phi(2) + \cdots + \phi(n).$$

3. Let a/b, c/d and e/f be any three consecutive fractions in the Farey sequence of order n. Show that

$$c/d = (a + e)/(b + f).$$

4. (a) Let a/b and c/d be the fractions immediately to the right and left of $1/2$ in the Farey sequence of order n. Show that

$$b = d = 1 + 2[(n - 1)/2].$$

   (b) Show that $a + c = b$, in the notation of part (a).

   (c) Show that if a/b and c/d are symmetric about $1/2$, then $a/b + c/d = 1$.

5. (a) Use Farey fractions to solve the Diophantine equation $ax + by = c$, when it is solvable, where a, b and c are given integers. (Hint: consider first the case that $(a, b) = 1$ and apply Theorem 4.2.)

   (b) Solve the equations:

   (i) $5x - 6y = 87$;

   (ii) $11x + 71y = 32$;

   (iii) $243x - 256y = 101$.

6. Show that if a/b and c/d are consecutive fractions in a Farey sequence of order n, then $b + d \geq n + 1$.

7. Show that the following algorithm produces the Farey sequence of order n. Let $a_0 = 0$, $b_0 = a_1 = 1$ and $b_1 = n$. Then, for $k \geq 0$, we have

$$a_{k+2} = \left[(b_k + n) / b_{k+1}\right]a_{k+1} - a_k$$

and

$$b_{k+2} = \left[(b_k + n) / b_{k+1}\right]b_{k+1} - b_k.$$

8. Show that if a/b is the immediate predecessor of c/d, in the Farey sequence of order n, then $(b - a)/b$ is the immediate successor of $(d - c)/d$ and vice-versa.

9. (a) Show that if a/b is a term of F_n, then so is $(b - a)/b$.

   (b) Show that the sum of the terms of the numerators of the fractions in F_n is one half the sum of the denominators.

   (c) Show that the sum of the fractions in the Farey sequence is one half the number of terms. (Hint: see Problem 1 above.)

10. Let a/b and c/d be consecutive fractions in F_n. Show that

    (a) $(a, c) = (b, d) = 1$;

    (b) if $n \geq 2$, then $c \neq d$;

    (c) between two consecutive elements of F_n there is at most one element of F_{n+1};

    (d) $(a + c, b + d) = 1$;

    (e) F_{b+d} is the first Farey sequence to include an element between a/b and c/d; it includes exactly one such element, namely $(a + b)/(c + d)$.

11. Let a/b be a fraction in F_n. Show that the element of F_n that follows a/b is c/d, where $(c, d) = 1$ is the unique pair of integers such that $bc - ad = 1$ and $0 \leq n - b < d \leq n$.

12. Let a/b and c/d run through all pairs of adjacent fractions in the Farey series of order n, $n > 1$. Show that

$$\min\left(\frac{c}{d} - \frac{a}{b}\right) = \frac{1}{n(n - 1)} \quad \text{and} \quad \max\left(\frac{c}{d} - \frac{a}{b}\right) = \frac{1}{n}.$$

13. Let a/b and c/d be fractions such that $ad - bc = 1$, with $b, d > 0$. Show that a/b and c/d are consecutive fractions in $F_{\max(b, d)}$.

14. Let $b_1, \ldots, b_k$ run through the denominators of all the terms of F_n when they are written in increasing order from $0/1$ to $1/1$ so that $b_1 = 1$ and $b_k = 1$. Show that

$$\sum_{j=1}^{k-1} \frac{1}{b_j b_{j+1}} = 1.$$

4.3. APPROXIMATION BY RATIONALS, I

We begin with the application of Farey fractions to the approximation of irrationals. This result gives a slight improvement to the inequality (4.1).

Theorem 4.5. Let α be a real irrational and let N be a positive integer. Then there exists a fraction h/k, with $k < N$, such that

$$|\alpha - h/k| < 1/(N+1)k.$$

Proof. In F_N we know we can find two consecutive terms, say a/b and c/d, such that

$$a/b < \alpha < c/d.$$

Since α is irrational we know that either

$$a/b < \alpha < (a+c)/(b+d) \text{ or } (a+c)/(b+d) < \alpha < c/d.$$

Also, since a/b and c/d are consecutive in F_N, we know that $(a + c)/(b + d)$ is not in F_N. Thus $b + d \geq N + 1$. Then we either have

$$0 < \alpha - a/b < (a+c)/(b+d) - a/b = 1/b(b+d) \leq 1/(N+1)b$$

or

$$0 < c/d - \alpha < c/d - (a+c)/(b+d) = 1/d(b+d) \leq 1/(N+1)d.$$

Either case proves the result.   ∎

Example 4.2. Find a rational p/q such that

$$\left| \sqrt[3]{2} - p/q \right| < 1/11q.$$

Solution. We take $N = 10$ and look at F_{10}. We have $\left[\sqrt[3]{2} \right] = 1$, and so we find $\sqrt[3]{2} - 1$ in F_{10}. We find

$$\tfrac{1}{4} < \sqrt[3]{2} - 1 < \tfrac{2}{7}.$$

We have

$$\left| \sqrt[3]{2} - \tfrac{5}{4} \right| \approx .009^+ < \tfrac{1}{44} \text{ and } \left| \sqrt[3]{2} - \tfrac{9}{7} \right| \approx .025^+ > \tfrac{1}{77}.$$

Thus we take $p/q = 5/4$.   □

Note that this implies Theorem 4.1. It also implies the following slight improvement on Theorem 4.1.

Corollary 4.5.1. Let α be a real irrational. Then there exist infinitely many pairs of relatively prime integers p and q, $q > 0$, such that

$$|\alpha - p/q| < 1/q(q+1).$$

Thus we may take the approximating function $f(a, q) = 1/q(q + 1)$.

Lemma 4.6.1. If x and y are positive integers, then not both of the inequalities

$$\frac{1}{xy} \geq \frac{1}{\sqrt{5}}\left(\frac{1}{x^2} + \frac{1}{y^2} \right) \text{ and } \frac{1}{x(x+y)} \geq \frac{1}{\sqrt{5}}\left(\frac{1}{x^2} + \frac{1}{(x+y)^2} \right)$$

can hold.

Proof. We rewrite the inequalities as

$$\sqrt{5}xy \geq y^2 + x^2 \text{ and } \sqrt{5}x(x+y) \geq (x+y)^2 + x^2.$$

If we add them, we get

$$\sqrt{5}(x^2 + 2xy) \geq 3x^2 + 2xy + 2y^2,$$

and so

$$2y^2 - 2(\sqrt{5} - 1)xy + (3 - \sqrt{5})x^2 \leq 0.$$

Multiplying by 2 and completing the square gives

$$(2y - (\sqrt{5} - 1)x)^2 \leq 0,$$

which is impossible since x and y are integers and $\sqrt{5}$ is irrational. ∎

Lemma 4.6.2. If a/b and c/d are consecutive fractions in F_N, then

$$|a/b - (a+c)/(b+d)| \leq 1/b(N+1)$$

and

$$|c/d - (a+c)/(b+d)| \leq 1/d(N+1).$$

Proof. By Theorem 4.2 we have

$$|a/b - (a+c)/(b+d)| = |ad - bc|/b(b+d) \leq 1/b(N+1),$$

since $b + d > N + 1$. The second inequality is proved similarly. ∎

We use these lemmas to prove the following theorem of Hurwitz which improves Theorem 4.1 and at the same time shows that we cannot improve it much more.

Theorem 4.6 (Hurwitz). Let α be any real irrational number. Then there exist infinitely many relatively prime integers p and q, with $q > 0$, such that

(4.3) $$|\alpha - p/q| < 1/\sqrt{5}q^2.$$

The constant $\sqrt{5}$ is best possible in the sense that if $\sqrt{5}$ is replaced by any larger number, then there exist irrational numbers for which (4.3) fails to hold with this replacement.

Proof. Let n be a positive integer. Then there are two consecutive rationals in F_n, say a/b and c/d, such that

$$a/b < \alpha < c/d.$$

We shall show that one of a/b, c/d or $(a + c)/(b + d)$ will serve for p/q in (4.3).

Suppose $\alpha < (a + c)/(b + d)$. If the result does not hold, then we have

$$\alpha - a/b \geq 1/\sqrt{5}b^2, (a+c)/(b+d) - \alpha \geq 1/\sqrt{5}(b+d)^2$$

and

$$c/d - \alpha \geq 1/\sqrt{5}d^2.$$

If we add these inequalities we obtain

$$\frac{c}{d} - \frac{a}{b} \geq \frac{1}{\sqrt{5}}\left(\frac{1}{d^2} + \frac{1}{b^2}\right)$$

and

$$\frac{a+c}{b+d} - \frac{a}{b} \geq \frac{1}{\sqrt{5}}\left(\frac{1}{(b+d)^2} + \frac{1}{b^2}\right),$$

and so

$$1/bd = (cb - ad)/bd = c/d - a/b \geq (b^{-2} + d^{-2})/\sqrt{5},$$

by Theorem 4.2, and

$$1/b(b+d) = \{(a+c)b - (b+d)a\}/b(b+d) \geq (b^{-2} + (b+d)^{-2})/\sqrt{5},$$

which contradicts Lemma 4.6.1. Thus at least one of a/b, c/d or $(a + c)/(b + d)$ will serve as p/q in this case.

If $\alpha > (a + c)/(b + d)$, then a similar contradiction to Lemma 4.6.1 is arrived at if we assume none of a/b, c/d or $(a + c)/(b + d)$ will serve as p/q in (4.3).

By Lemma 4.6.2 and the fact that p/q can be taken to be one of a/b, c/d or $(a + c)/(b + d)$, we have

$$\begin{aligned}
|\alpha - p/q| &< |c/d - a/b| \\
(4.4) \qquad &= |c/d - (a+c)/(b+d)| + |(a+c)/(b+d) - a/b| \\
&\leq d^{-1}(n+1)^{-1} + b^{-1}(n+1)^{-1} \leq 2/(n+1),
\end{aligned}$$

since b and d are positive integers. Let p_1/q_1 be the fraction that satisfies (4.3) that we have found. Since α is irrational we have $|\alpha - p/q| > 0$. Choose $n > 2|\alpha - p_1/q_1|^{-1}$. Then from F_n we can produce a p/q that, by (4.4), satisfies

$$|\alpha - p/q| \leq 2/(n+1) < |\alpha - p_1/q_1|.$$

This gives us the infinite sequence of rationals and proves the first part of the theorem.

To prove the second part of the theorem we need only exhibit one number for which the result of (4.3) would fail. Take $\alpha = (1 + \sqrt{5})/2$, so that $\overline{\alpha} = (1 - \sqrt{5})/2$. Then

$$(x - \alpha)(x - \overline{\alpha}) = x^2 - x - 1.$$

Thus for any integers p and q, with $q > 0$, we have

(4.5)
$$\begin{aligned}
|p/q - \alpha||p/q - \overline{\alpha}| &= |(p/q - \alpha)(p/q - \overline{\alpha})| \\
&= |p^2/q^2 - p/q - 1| \\
&= q^{-2}|p^2 - pq - q^2|.
\end{aligned}$$

The expression of the left hand side of (4.5) cannot be zero since both α and $\overline{\alpha}$ are irrational. Thus $|p^2 - pq - q^2|$ is a positive integer, and so

(4.6)
$$|p/q - \alpha||p/q - \overline{\alpha}| \geq 1/q^2.$$

Now suppose we have an infinite sequence of rationals p_k/q_k, with $q_k > 0$ and some positive real number c such that

(4.7)
$$|p_k/q_k - \alpha| < 1/cq_k^2.$$

Then

$$q_k\alpha - (cq_k)^{-1} < p_k < q_k\alpha + (cq_k)^{-1}.$$

Thus for any given value of q_k there are only a finite number of possible values for p_k. Thus we must have $q_k \to +\infty$ as $k \to +\infty$. By (4.6), (4.7) and the triangle inequality, we have

$$q_k^{-2} \leq |p_k/q_k - \alpha||p_k/q_k - \overline{\alpha}| < (cq_k^2)^{-1}((cq_k^2)^{-1} + \sqrt{5}),$$

and so

$$c < (cq_k^2)^{-1} + \sqrt{5}.$$

Letting $k \to +\infty$ we have

$$c \leq \lim_{k \to +\infty}((cq_k^2)^{-1} + \sqrt{5}) = \sqrt{5},$$

which finishes the proof of the theorem.   ■

The question might be asked if we could replace $\sqrt{5}$ by some larger number if we are willing to ignore some irrationals. The answer to this is yes as we shall see below in Section 8. On the other hand, Robinson has proved the following partial improvement.

For every positive ε and every irrational ξ there exist infinitely many rationals A/B such that

$$-1/(\sqrt{5} - \varepsilon)B^2 < A/B - \xi < 1/(\sqrt{5} + 1)B^2.$$

The coefficient of B in the denominator on the right cannot be replaced by any larger constant.

The proof of this result would take us too far afield. See [104]. However, we can prove a theorem that includes both Theorem 4.1 and Theorem 4.6.

Theorem 4.7. Given any irrational ξ and any nonnegative real number t, there exist infinitely many rational numbers h/k such that

(4.8) $$\frac{-1}{(1+4t)^{1/2}k^2} < \xi - \frac{h}{k} < \frac{t}{(1+4t)^{1/2}k^2}.$$

Moreover, the statement holds if $\xi - h/k$ in (4.8) is replaced by $h/k - \xi$.

Note that if we take $t = 0$ we have Theorem 4.1 and if we take $t = 1$ we have Theorem 4.6. First we prove two lemmas.

Lemma 4.7.1. Let ξ be any irrational and t any nonnegative real number. Let a/b and c/d be rational numbers with positive denominators such that $bc - ad = 1$ and

(4.9) $$a/b < (a + c)/(b + d) < \xi < c/d.$$

Then (4.8) holds with h/k replaced by at least one of a/b, $(a + c)/(b + d)$, or c/d.

Proof. Define λ and μ by

$$\lambda = (1 + 4t)^{-1/2} \text{ and } \mu = t(1 + 4t)^{-1/2}.$$

Then $\mu = (1 - \lambda^2)/4\lambda$ and $0 < \lambda < 1$. Suppose that the lemma is false so that

(4.10) $$\begin{cases} \xi - a/b \geq \mu/b^2 \\ \xi - (a+c)/(b+d) \geq \mu/(b+d)^2. \\ c/d - \xi \geq \lambda/d^2 \end{cases}$$

Adding the first and third of these inequalities and also the second and the third we see that

$$c/d - a/b = 1/bd \geq \mu/b^2 + \lambda/d^2$$

and

$$c/d - (a+c)/(b+d) = 1/d(b+d) \geq \mu/(b+d)^2 + \lambda/d^2.$$

Thus

(4.11) $\qquad \lambda b^2 - bd + \mu d^2 \le 0$ and $\lambda(b+d)^2 - d(b+d) + \mu d^2 \le 0.$

Adding these we obtain

(4.12) $\qquad 2\lambda b^2 + (2\lambda - 2)bd + (\lambda + 2\mu - 1)d^2 \le 0.$

Since $\lambda^2 = 4\lambda\mu = 1$ we see that the quadratic form in b and d has discriminant zero, and so must be a perfect square. Indeed, if we multiply (4.12) by $2\lambda > 0$ we can rewrite (4.12) as

$$(2\lambda b + (\lambda - 1)d)^2 \le 0.$$

Since equality must hold, if this is to be true, we see that equality holds throughout (4.10), (4.11), and (4.12) and also that

$$\lambda = d / (d + 2b),$$

that is, λ is rational. Now the third relation in (4.10) is

$$c / d - \xi = \lambda / d^2,$$

which implies that ξ is rational. This is a contradiction and so the lemma holds. ∎

Lemma 4.7.2. Let ξ be a given irrational number and let r be a given positive integer. Then for all n sufficiently large the two fractions a/b and c/d adjacent to ξ in F_n have denominators larger than r, that is, $b > r$ and $d > r$.

Proof. Let $m_1, ..., m_r$ be the integers nearest to ξ, 2ξ, ..., $r\xi$, that is

$$m_k = \begin{cases} [k\xi] & \text{if } 0 < k\xi - [k\xi] < 1/2 \\ [k\xi + 1] & \text{if } 1/2 < k\xi - [k\xi] < 1 \end{cases}.$$

Thus, if q is any integer and $j = 1, 2, ..., r$, we have

$$|j\xi - m_j| \le |j\xi - q|$$

or

(4.13) $\qquad |\xi - m_j / j| \le |\xi - q / j|.$

Choose n to be an integer such that

$$n > \max\{|\xi - m_j / j|^{-1} : j = 1, 2, ..., r\}.$$

Then

(4.14) $\qquad n^{-1} < \min\{|\xi - m_j / j| : j = 1, 2, ..., r\}.$

Thus, by (4.13) and (4.14), we have

$$(4.15) \qquad\qquad n^{-1} < \left| \xi - q/j \right|$$

for any integer q. Now the difference between adjacent fractions in $\boldsymbol{F}_n$ does not exceed $1/n$ since $\boldsymbol{F}_n$ contains all fractions with denominator n. Thus

$$\left| \xi - a/b \right| < \left| c/d - a/b \right| \le 1/n$$

and

$$\left| \xi - c/d \right| < \left| c/d - a/b \right| \le 1/n.$$

Comparing these inequalities with (4.15) establishes the result. ∎

We now prove Theorem 4.7.

Let a_1/b_1 and c_1/d_1 be two consecutive fractions in $\boldsymbol{F}_1$ between which x lies. Then $b_1 c_1 - a_1 d_1 = 1$, by Theorem 4.2. In the case that

$$a_1/b_1 < (a_1 + c_1)/(b_1 + d_1) < \xi < c_1/d_1$$

we apply Lemma 4.7.1 with a/b and c/d replaced by a_1/b_1 and c_1/d_1, respectively. This would give us one solution of (4.8). On the other hand, if

$$a_1/b_1 < \xi < (a_1 + c_1)/(b_1 + d_1) < c_1/d_1,$$

let the positive integer j be defined by

$$(4.16) \qquad\qquad \frac{a_1}{b_1} < \frac{(j+1)a_1 + c_1}{(j+1)b_1 + d_1} < \xi < \frac{ja_1 + c_1}{jb_1 + d_1}.$$

This can be done since $(ja_1 + c_1)/(jb_1 + d_1)$ tends to a_1/b_1 as j tends to infinity. Now apply Lemma 4.7.1 with a/b and c/d replaced by a_1/b_1 and $(ja_1 + c_1)/(jb_1 + d_1)$, respectively. The condition (4.9) is replaced by (4.16) and $bc - ad = 1$ holds since

$$b_1(ja_1 + c_1) - a_1(jb_1 + d_1) = b_1 c_1 - a_1 d_1 = 1.$$

Thus, in this case also, we have a solution of (4.8). Let h_1/k_1 denote the solution of (4.8) obtained by use of $\boldsymbol{F}_1$.

Now apply Lemma 4.7.2 to choose n sufficiently large so that the fractions adjacent to in $\boldsymbol{F}_n$ have denominators greater than k_1. Thus we get a_2/b_2 and c_2/d_2 in $\boldsymbol{F}_n$ with

$$a_1/b_1 < \xi < c_1/d_1.$$

Then we repeat the preceding argument. That is, if

$$a_2/b_2 < (a_2 + c_2)/(b_2 + d_2) < \xi < c_2/d_2,$$

we apply Lemma 4.7.1 directly, and if not choose j so that

$$a_2 / b_2 < [(j+1)a_2 + c_2] / [(j+1)b_2 + c_2] < \xi < (ja_2 + c_2) / (jb_2 + d_2).$$

In either case we get a solution h_2/k_2 of (4.8) with k_2 one of the numbers

$$b_2, d_2, b_2 + d_2, jb_2 + d_2 \text{ or } (j+1)b_2 + d_2.$$

It is clear that all the fractions in the preceding inequalities are in their lowest terms (just as above). Thus h_2/k_2 is distinct from h_1/k_1. The process proceeds by induction and the result follows. ∎

Problem Set 4.3

1. Show that if a/b and c/d are consecutive terms in the Farey sequence of order n, $n > 1$, then at least one of them satisfies the inequality

$$|\theta - p/q| < 1/2q^2,$$

where $p/q = a/b$ or c/d and θ is an irrational number between a/b and c/d.

2. (a) Find rational numbers a_1/b_1 and a_2/b_2 so that $\left|\sqrt{2} - a/b\right| < 1/\sqrt{5}b^2$.

   (b) Find rational numbers a_1/b_1 and a_2/b_2 so that $\left|\sqrt{3} - a/b\right| < 1/2\sqrt{2}b^2$.

3. Find rational approximations of the following irrational numbers to the desired approximation.

   (a) $\sqrt{10}$ to five decimal places.

   (b) $(1+\sqrt{2})/2$ to six decimal places.

   (c) e to three decimal places.

4.4. CONTINUED FRACTIONS

In the previous section we saw that we could approximate every irrational number by infinitely many distinct rationals to better and better degree. Unfortunately, though we know the approximating rationals are in each Farey sequence of every order, the proof did not tell us which ones they are, unless we construct the entire Farey sequence of that order. To alleviate this problem we will discuss continued fractions, which, among other things, will provide us with a sequence of rationals that satisfy the inequality of Theorem 4.6 that are tied solely to the number we wish to approximate. Though our concern is with approximating irrationals we shall begin with the rational case.

Let u_0/u_1 be a rational number with $u_1 > 0$ and $(u_0, u_1) = 1$. If we apply the Euclidean algorithm (Theorem 1.6) we get a sequence of the following sort

$$u_0 = u_1 a_0 + u_2 \qquad 0 < u_2 < u_1$$

$$u_1 = u_2 a_1 + u_3 \qquad 0 < u_3 < u_2$$

(4.17)
$$\vdots$$

$$u_{j-1} = u_j a_{j-1} + u_{j+1} \qquad 0 < u_{j+1} < u_j$$

$$u_j = u_{j+1} a_j.$$

If we write $\xi_i = u_i / u_{i+1}$, $0 \le i \le j$, then (4.17) states that

(4.18)
$$\begin{cases} \xi_i = a_i + 1 / \xi_{i+1}, 0 \le i \le j-1 \\ \xi_j = a_j \end{cases}.$$

Thus

$$\xi_0 = a_0 + \cfrac{1}{a_1 + \cfrac{1}{\xi_2}} = a_0 + \cfrac{1}{a_1 + \cfrac{1}{a_2 + \cfrac{1}{\xi_3}}}$$

and so on. Thus

(4.19)
$$\frac{u_0}{u_1} = \xi_0 = a_0 + \cfrac{1}{a_1 + \cfrac{1}{a_2 + \cfrac{}{\ddots \; + \cfrac{1}{a_{j-1} + \cfrac{1}{a_j}}}}}$$

This, (4.19), is called the **continued fraction expansion** of the rational number u_0/u_1. The integers a_i are called the **partial quotients** and the numbers ξ_i are called the **complete quotients** of u_0/u_1. We usually write (4.19) in the more condensed form $[a_0; a_1, \ldots, a_j]$

(4.20)
$$u_0 \big/ u_1 = [a_0; a_1, \ldots, a_j].$$

Note that a_0 can be positive, negative or zero, but all further partial quotients must be strictly positive. Note also that if $j > 1$, then $a_j = [u_j/u_{j+1}]$ and $0 < u_{j+1} < u_j$ imply that $a_j > 1$.

One may generalize the notation (4.20). If $x_0, x_1, \ldots, x_j$ are any real numbers, with $x_1, \ldots, x_j$ all positive, then we define

$$[x_0; x_1, \ldots, x_j] = x_0 + \cfrac{1}{x_1 + \cfrac{1}{x_2 + \cfrac{\ddots}{\ddots + \cfrac{1}{x_{j-1} + \cfrac{1}{x_j}}}}}$$

If the x_i are all integers, then the continued fraction is said to be **simple**. The following obvious formulas will be useful in what follows:

(4.21)
$$[x_0; x_1, \ldots, x_j] = x_0 + \frac{1}{[x_1; x_2, \ldots, x_j]}$$

$$= \left[x_0; x_1, \ldots, x_{j-2}, x_{j-1} + \frac{1}{x_j} \right].$$

Example 4.3. Find the finite simple continued fraction expansions of the rational numbers 6/7, 15/11 and -31/17.

Solution. We have

$$6 = 7 \cdot 0 + 6$$
$$7 = 6 \cdot 1 + 1$$
$$6 = 1 \cdot 6,$$

so that 6/7 = [0; 1, 6]. We have

$$15 = 1 \cdot 11 + 4$$
$$11 = 4 \cdot 1 + 3$$
$$4 = 3 \cdot 1 + 1$$
$$3 = 1 \cdot 3$$

so that 15/11 = [1; 1, 1, 3]. We have

$$-31 = 17(-2) + 3$$
$$17 = 3 \cdot 5 + 2$$

so that -31/17 = [-2; 5, 1, 2]. □

In Example 4.3 we showed that 15/11 = [1; 1, 1, 3]. It is clear that we also have 15/11 = [1; 1, 1, 2, 1]. As is the case with decimal expansions of rationals (see Section 4.9, below) one can give a criterion for uniqueness of the expansion as a finite simple continued fraction.

Theorem 4.8. If $\left[a_0;a_1,...,a_j\right]=\left[b_0;b_1,...,b_n\right]$ as finite simple continued fractions and if $a_j > 1$ and $b_n > 1$, then $j = n$ and $a_i = b_i$ for $i = 0, 1, ..., n$.

Proof. For $i > 0$, let

$$y_i = \left[b_i;b_{i+1},...,b_n\right].$$

Then, by (4.20), we have

(4.22) $\qquad y_i = \left[b_i;b_{i+1},...,b_n\right] = b_i + \dfrac{1}{\left[b_{i+1},b_{i+2},...,b_n\right]} = b_i + \dfrac{1}{y_{i+1}}.$

Thus we have $y_i > b_i$ and $y_i > 1$ for $i = 1, 2, ..., n - 1$ and $y_n = b_n > 1$. Thus we have

$$b_i = \left[y_i\right],$$

for all $0 < i < n$.

Let ξ_i be as defined above, $0 \le i \le j$. Then the hypothesis that the two continued fractions are equal can be expressed as $y_0 = \xi_0$. Since $\xi_i = {u_i}/{u_{i+1}}$ we see that $\xi_{i+1} > 1$ for all $i \ge 0$. Thus $a_i = [\xi_i]$, for $0 \le i \le j$. Thus

$$b_0 = \left[y\right]_0 = \left[\xi_0\right] = a_0$$

and so

$$\frac{1}{\xi_1} = \xi_0 - a_0 = y_0 - b_0 = \frac{1}{y_1}$$

that is, $\xi_1 = y_1$ and $a_1 = [\xi_1] = [y_1] = b_1$.

We now proceed by mathematical induction. Assume that

$$\xi_i = y_i \text{ and } a_i = b_i.$$

Then

$$1/\xi_{i+1} = \xi_i - a_i = y_i - a_i = 1/y_{i+1}.$$

Thus

$$\xi_{i+1} = y_{i+1} \text{ and } a_{i+1} = [\xi_{i+1}] = [y_{i+1}] = b_{i+1}.$$

Suppose $j < n$. The above argument shows that $\xi_j = y_j$ and $a_j = b_j$. By (4.18), we have $\xi_j = a_j$ and, by (4.22), we have $y_j > b_j$, which is impossible. Thus $j \ge n$ and a similar argument shows that this only works when $j = n$. ∎

We now finish the topic of finite simple continued fractions with the following theorem.

Theorem 4.9. Any finite simple continued fraction represents a rational number. Conversely, any rational number can be expressed as a finite simple continued fraction and in exactly two ways.

Proof. The first assertion can be proved by mathematical induction on the number of terms in the continued fraction by use of the formula

$$[a_0; a_1, \ldots, a_j] = a_0 + \frac{1}{[a_1; a_2, \ldots, a_j]}.$$

The second assertion follows from the development of u_0/u_1 into a simple continued fraction at the beginning of this section together with the identity

$$[a_0; a_1, \ldots, a_{j-1}, a_j] = [a_0; a_1, \ldots, a_{j-1}, a_j - 1, 1]$$

and Theorem 4.8. ∎

If we wish to deal with irrationals we must therefore look beyond finite simple continued fractions. One way would be to remove the simple part of the expansion, but a simpler method is to go to infinity.

Let a_0, a_1, a_2, ... be a sequence of integers, with $a_i > 0$ for $i \geq 1$. We define two associated sequences of integers $\{h_n\}$ and $\{k_n\}$ by

(4.23)
$$\begin{cases} h_{-2} = 0, h_{-1} = 1, h_i = a_i h_{i-1} + h_{i-2}, \text{ for } i \geq 0, \\ k_{-2} = 1, k_{-1} = 0, k_i = a_i k_{i-1} + k_{i-2}, \text{ for } i \geq 0, \end{cases}$$

Note that $k_0 = 1$, $k_1 = a_1 k_0 \geq k_0$, $k_2 > k_1$, $k_3 > k_2$, and so on. Thus we have

$$1 \leq k_0 \leq k_1 < k_2 < \cdots < k_n < \cdots.$$

We begin with some properties of the sequences $\{h_n\}$ and $\{k_n\}$.

Theorem 4.10. Let x be any positive real number. Then

$$[a_0; a_1, \ldots, a_{n-1}, x] = \frac{x h_{n-1} + h_{n-2}}{x k_{n-1} + k_{n-2}},$$

where for $n = 0$, the result is to be interpreted as

$$x = \frac{x h_{-1} + h_{-2}}{x k_{-1} + k_{-2}}.$$

Proof. The case $n = 0$ follows directly from (4.23). For $n = 1$, the result is

$$[a_0;x] = \frac{xh_0 + h_{-1}}{xk_0 + k_{-1}},$$

which is also true, by (4.23) and the fact that $[a_0; x] = a_0 + 1/x$.

We proceed by induction and assume the result is true for $n = N$, that is,

$$[a_0;a_1,\ldots,a_{N-1},x] = \frac{xh_{N-1} + h_{N-2}}{xk_{N-1} + k_{N-2}}.$$

Then, for $n = N + 1$, we have, by (4.21) and (4.23),

$$[a_0;a_1,\ldots,a_{N-1},a_N,x] = [a_0;a_1,\ldots,a_{N-1},a_N + 1/x] = \frac{(a_N + 1/x)h_{N-1} + h_{N-2}}{(a_N + 1/x)k_{N-1} + k_{N-2}}$$

$$= \frac{x(a_N h_{N-1} + h_{N-2}) + h_{N-1}}{x(a_N k_{N-1} + k_{N-2}) + k_{N-1}} = \frac{xh_N + h_{N-1}}{xk_N + k_{N-1}},$$

which is the result for $n = N + 1$.   ∎

Theorem 4.11. Let $r_n = [a_0; a_1, \ldots, a_n]$ for n a nonnegative integer. Then $r_n = h_n/k_n$.

Proof. We apply Theorem 4.10 with $x = a_n$. Then, by (4.23),

$$r_n = [a_0;a_1,\ldots,a_n] = \frac{a_n h_{n-1} + h_{n-2}}{a_n k_{n-1} + k_{n-2}} = \frac{h_n}{k_n},$$

which is the result.   ∎

Theorem 4.12. We have, for $i > 1$,
   (a) $h_i k_{i-1} - h_{i-1} k_i = (-1)^{i-1}$ and $r_i - r_{i-1} = (-1)^{i-1}/k_i k_{i-1}$;
   (b) $h_i k_{i-2} - h_{i-2} k_i = (-1)^i a_i$ and $r_i - r_{i-2} = (-1)^i a_i / k_i k_{i-2}$;
   (c) $(h_i, k_i) = 1$.

Proof. By (4.23), we have $h_{-1}k_{-2} - h_{-2}k_{-1} = 1$. We continue by induction and assume that $h_{i-1}k_{i-2} - h_{i-2}k_{i-1} = (-1)^{i-2}$. Then, by (4.23),

$$h_i k_{i-1} - h_{i-1} k_i = (a_i h_{i-1} + h_{i-2})k_{i-1} - h_{i-1}(a_i k_{i-1} + k_{i-2})$$
$$= -(h_{i-1}k_{i-2} - h_{i-2}k_{i-1}) = (-1)^{i-1}.$$

If we divide the first equation in (a) by $k_i k_{i-1}$ we get, by Theorem 4.11, the second equation, which completes the proof of (a).

The first equation of (a) immediately implies (c).

To prove (b) note that, by (4.23), $h_0 k_{-2} - h_{-2}k_0 = a_0$. In general, we have, by (4.23),

$$h_i k_{i-2} - h_{i-2} k_i = (a_i h_{i-1} + h_{i-2}) k_{i-2} - h_{i-2} (a_i k_{i-1} + k_{i-2})$$
$$= a_i (h_{i-1} k_{i-2} - h_{i-2} k_{i-1}) = (-1)^i a_i,$$

by (a). The second part of (c) follows from the first upon division by $k_i k_{i-2}$ and completes the proof of the theorem. ∎

Theorem 4.13. Let the sequence $\{r_n\}$ be defined as in Theorem 4.11. Then we have the following chain of inequalities

(4.24) $$r_0 < r_2 < r_4 < \cdots < r_5 < r_3 < r_1.$$

Moreover, $\lim\limits_{n \to +\infty} r_n$ exists and for every nonnegative j we have

(4.25) $$r_{2j} < \lim_{n \to +\infty} r_n < r_{2j+1}$$

Proof. The identities of Theorem 4.12 for $r_i - r_{i-1}$ and $r_i - r_{i-2}$ imply that $r_{2j} < r_{2j+2}, r_{2j-1} > r_{2j+1}$ and $r_{2j} < r_{2j-1}$, since the k are positive for $k_i > 0$ and the a_i are positive for $i \geq 1$. Note that

$$r_{2n} < r_{2n+2j} < r_{2n+2j-1} < r_{2j-1},$$

which proves (4.24). Since the even subscripted terms are monotonically increasing and bounded above by r_1 and the odd subscripted terms are monotonically decreasing and bounded below by r_0 we see that both sequences have a limit. These limits must be the same since $r_i - r_{i-1}$ tends to 0 as i tends to $+\infty$ because the integers k_i are increasing to $+\infty$ with i. The last inequality, (4.25), follows from (4.24) and the existence of the limit. ∎

We are led to make the following definition.

Definition 4.1. Let $a_0, a_1, a_2, \ldots$ be an infinite sequence of integers, with $a_i > 0$ for $i > 1$. We define the **infinite simple continued fraction** $[a_0; a_1, a_2, \ldots]$ to be the value of the limit

$$\lim_{n \to +\infty} [a_0; a_1, \ldots, a_n].$$

Thus by Theorem 4.13, the infinite simple continued fractions are defined. The rational number $[a_0; a_1, a_2, \ldots, a_n] = h_n / k_n$ is called the **nth convergent** to the infinite simple continued fraction. In the case of a finite simple continued fraction $[a_0; a_1, a_2, \ldots, a_m]$ is called the **mth convergent** to $[a_0; a_1, a_2, \ldots, a_n]$ if $m \leq n$.

Since finite simple continued fractions are rational, and conversely, the following result should be no surprise.

Theorem 4.14. The value of any infinite simple continued fraction $[a_0; a_1, a_2, ...]$ is irrational.

Proof: Let $\theta = [a_0; a_1, a_2, ...]$. By Theorem 4.13, we have

$$r_n < \theta < r_{n+1} \text{ or } r_{n+1} < \theta < r_n$$

for any n, depending on whether n is even or odd. Thus

$$0 < |\theta - r_n| < |r_{n+1} - r_n|.$$

If we multiply by k_n and use (a) of Theorem 4.12 we have

(4.26) $$0 < |k_n\theta - h_n| < 1/k_{n+1}$$

If θ were rational, then there would be integers a and b, $b > 0$, such that $\theta = a/b$. Thus, by (4.26), we would have

(4.27) $$0 < |k_n a - h_n b| < b/k_{n+1}.$$

Since the integers k_{n+1} increase to $+\infty$ there exists a value of n so that $b < k_{n+1}$. Thus for this value of n (and all larger values of n) we have, from (4.27),

$$0 < |k_n a - h_n b| < 1,$$

which is impossible for the integer $k_n a - h_n b$. Thus θ is irrational.   ■

To finish showing that Definition 4.1 gives us a well-defined object we must show that if we take two different infinite sequences of integers we get different irrational numbers. We accomplish this in the following two results.

Theorem 4.15. Let $\theta = [a_0; a_1, a_2, ...]$ be a simple continued fraction. Then $a_0 = [\theta]$ and if $\theta_1 = [a_1; a_2, ...]$, then $\theta = a_0 + 1/\theta_1$.

Proof. By Theorem 4.13 we have $r_0 < \theta < r_1$, that is $a_0 < \theta < a_0 + 1/a_1$. Since $a_1 \geq 1$ we see that $a_0 < \theta < a_0 + 1$, and so $a_0 = [\theta]$. Also, by (4.21), we have

$$\theta = \lim_{n \to +\infty} [a_0; a_1, ..., a_n] = \lim_{n \to +\infty} \left(a_0 + \frac{1}{[a_1; a_2, ..., a_n]} \right)$$

$$= a_0 + \lim_{n \to +\infty} \frac{1}{[a_1; a_2, ..., a_n]} = a_0 + 1/\theta_1,$$

which is the second result.   ■

Theorem 4.16. Two distinct infinite simple continued fractions converge to different values.

Proof. Suppose $[a_0; a_1, a_2, \ldots] = \theta = [b_0; b_1, b_2, \ldots]$. By Theorem 4.15 we have

$$a_0 = [\theta] = b_0$$

and

$$a_0 + \frac{1}{[a_1; a_2, \ldots]} = b_0 + \frac{1}{[b_1; b_2, \ldots]}.$$

Thus $[a_1; a_2, \ldots] = [b_1; b_2, \ldots]$ and the result follows by a simple induction argument. ∎

Example 4.4. Evaluate the infinite continued fractions $[1; 1, 1, \ldots]$, $[3; 2, 5, 1, 1, \ldots]$ and $[3; 1, 1, 6, 1, 1, 6, \ldots]$.

Solution. Let $\phi = [1; 1, 1, \ldots]$. Then, by Theorem 4.15, we have

$$\phi = 1 + 1/\phi$$

or

$$\phi^2 - \phi - 1 = 0.$$

Thus, solving the quadratic equation, we have

$$\phi = (1 \pm \sqrt{5})/2,$$

but since all the partial quotients are positive we see (since $\phi > r_0 = a_0$) that we must have

$$\phi = (1 + \sqrt{5})/2.$$

Let $\eta = [3; 2, 5, 1, 1, \ldots]$. Then, by the first part,

$$\eta = 3 + \cfrac{1}{2 + \cfrac{1}{5 + \cfrac{1}{\phi}}} = \frac{38\phi + 7}{11\phi + 2} = \frac{655 + \sqrt{5}}{190}.$$

Finally, let $\rho = [3; 1, 1, 6, 1, 1, 6 \ldots]$. Applying Theorem 4.15 again we have

$$\rho = 3 + \cfrac{1}{1 + \cfrac{1}{1 + \cfrac{1}{6 + \rho - 3}}}.$$

After the algebraic simplification we arrive at the equation satisfied by ρ:

$$2\rho^2 - 25 = 0.$$

Thus, since $\rho > 0$ (all its partial quotients are positive), we see that

$$\rho = \sqrt{25/2}. \qquad \square$$

With these three examples in mind we make the following definition.

Definition 4.2. Let $\theta = [a_0; a_1, \ldots]$ be an infinite simple continued fraction. Then is said to be **periodic of period** r, if there exists a positive integer N such that if $n \geq N$, then $a_{r+n} = a_n$ and r is the least positive integer for which this is true.

If θ is periodic of period r, then one can write θ in the form

$$\theta = [a_0; a_1, \ldots, a_{N-1}, a_N, \ldots, a_{N+r-1}, a_N, \ldots, a_{N+r-1}, \ldots],$$

which we write shortly as $[a_0; a_1, \ldots, a_{N-1}, \overline{a_N, \ldots, a_{N+r-1}}]$

$$\theta = [a_0; a_1, \ldots, a_{N-1}, \overline{a_N, \ldots, a_{N+r-1}}].$$

As might be guessed from the examples above is periodic if and only if is a quadratic irrational. This is indeed the case as we shall see in the next section.

We can compute (at least in the periodic case) the irrational number corresponding to the infinite simple continued fraction. We now describe the process by which we can produce the infinite simple continued fraction given the irrational.

Let ξ be a given irrational real number. We define a sequence of integers and irrationals as follows:

$$(4.28) \qquad \begin{cases} \xi_0 = \xi & a = [\xi_0] \\ a = [\xi_i] & \xi_{i+1} = (\xi_i - a_i)^{-1}, \end{cases}$$

for $i \geq 0$. Clearly a_i, $i \geq 0$, are integers and the ξ_i, $i \geq 0$, are irrational since $\xi_0 = \xi$ is irrational. Since ξ_{i-1} is irrational we have

$$a_{i-1} < \xi_{i-1} < a_{i-1} + 1$$

and, for $i \geq 1$,

$$0 < \xi_i^{-1} = \xi_{i-1} - a_{i-1} < 1$$

Thus $\xi_i > 1$ and so $a_i \geq 1$, for $i \geq 1$. We obtain, successively, by use of (4.28)

$$\xi = \xi_0 = a_0 + \xi_1^{-1} = [a_0; \xi_1]$$
$$= [a_0; a_1 + \xi_2^{-1}] = [a_0; a_1, \xi_2]$$
$$\vdots$$
$$= [a_0; a_1, \ldots, a_{m-2}, a_{m-1}, \xi_m]$$

for any integer m.

By Theorem 4.10 we have

$$\xi = [a_0; a_1, \ldots, a_{m-1}, \xi_m] = \frac{\xi_m h_{m-1} + h_{m-2}}{\xi_m k_{m-1} + k_{m-2}},$$

and so, by Theorem 4.12, we have

$$\xi - r_{m-1} = \xi - h_{m-1} / k_{m-1} = \frac{\xi_m h_{m-1} + h_{m-2}}{\xi_m k_{m-1} + k_{m-2}} - \frac{h_{m-1}}{k_{m-1}}$$

(4.29)
$$= \frac{-(h_{m-1} k_{m-2} - h_{m-2} k_{m-1})}{k_{m-1}(\xi_m k_{m-1} + k_{m-2})} = \frac{(-1)^{m-1}}{k_{m-1}(\xi_m k_{m-1} + k_{m-2})}$$

Since $k_m \to +\infty$ as $m \to +\infty$ and $x_m > 0$ we see that as $m \to +\infty$ we have $x - r_m \to 0$. Thus

$$\xi = \lim_{m \to +\infty} r_m = \lim_{m \to +\infty} [a_0; a_1, a_2, \ldots, a_m] = [a_0; a_1, a_2, \ldots].$$

Note that if we apply this algorithm to we get

$$\xi_n = [a_n; a_{n+1}, a_{n+2}, \ldots].$$

We summarize our results about continued fractions in the following theorem.

Theorem 4.17. Any real irrational ξ is uniquely expressible, by the algorithm that gives equation (4.28), as an infinite simple continued fraction $[a_0; a_1, a_2, \ldots]$. Conversely, any such continued fraction determined by integers a_i, which are positive for $i \geq 1$, represents a real irrational number ξ.

Example 4.5. Find the simple continued fraction expansions of $\sqrt{2}, \sqrt{7}, \sqrt{3/2}$ and the first five partial quotients for $\sqrt[3]{2}$.

Solution. Let $\xi_0 = \sqrt{2}$. Then we have the following table:

$a_0 = [x_0] = 1$ and $\xi_1 = (\sqrt{2} - 1)^{-1} = \sqrt{2} + 1$

$a_1 = [x_1] = 2$ and $\xi_2 = (\sqrt{2} + 1 - 2)^{-1} = (\sqrt{2} - 1)^{-1} = \xi_1$.

Thus we have periodicity (the period is 1) and

$$\sqrt{2} = [1; \overline{2}].$$

Let $\xi_0 = \sqrt{7}$. Then we have the following table:

$a_0 = [x_0] = 2$ and $\xi_1 = (\sqrt{7} - 2)^{-1} = (\sqrt{7} + 2)/3$

$a_1 = [x_1] = 1$ and $\xi_2 = ((\sqrt{7} + 2)/3 - 1)^{-1} = (\sqrt{7} + 1)/2$

$a_2 = [x_2] = 1$ and $\xi_3 = ((\sqrt{7} + 1)/2 - 1)^{-1} = (\sqrt{7} + 1)/3$

$a_3 = [x_3] = 1$ and $\xi_4 = ((\sqrt{7} + 1)/3 - 1)^{-1} = \sqrt{7} + 2$

$a_4 = [x_4] = 4$ and $\xi_5 = (\sqrt{7} + 2 - 4)^{-1} = (\sqrt{7} + 2)/3 = \xi_1$.

We have periodicity (the period is 4) and

$$\sqrt{7} = [2;\overline{1,1,1,4}].$$

Let $\xi_0 = \sqrt{3/2}$. Then we have the following table:

$a_0 = [x_0] = 1$ and $\xi_1 = (\sqrt{3/2} - 1)^{-1} = \sqrt{6} + 2$

$a_1 = [x_1] = 4$ and $\xi_2 = (\sqrt{6} + 2 - 4)^{-1} = (\sqrt{6} + 2)/2$

$a_2 = [x_2] = 2$ and $\xi_3 = ((\sqrt{6} + 2)/2 - 2)^{-1} = \sqrt{6} + 2 = \xi_1.$

Thus we have periodicity (the period is 2) and

$$\sqrt{3/2} = [1;\overline{4,2}].$$

Since $\sqrt[3]{2}$ is not a quadratic irrational we know by the remarks above we cannot have periodicity. Let $\xi_0 = \sqrt[3]{2}$. Then we have the following table:

$a_0 = [x_0] = 1$ and $\xi_1 = (\sqrt[3]{2} - 1)^{-1} = \sqrt[3]{4} + \sqrt[3]{2} + 1$

$a_1 = [x_1] = 3$ and $\xi_2 = (\sqrt[3]{4} + \sqrt[3]{2} + 1 - 3)^{-1}$

$a_2 = [x_2] = 1$ and $x_3 = (x_2 - 1)^{-1}$

$a_3 = [x_3] = 5$ and $x_4 = (x_3 - 5)^{-1}$

$a_4 = [x_4] = 1.$

We did not explicitly state ξ_3 and ξ_4 since they are somewhat complicated. We have

$$\sqrt[3]{2} = [1;3,1,5,1,\ldots]. \qquad \square$$

This completes our general discussion of continued fractions. We turn now to a discussion of a special class of continued fractions, namely periodic continued fractions. Afterward we will see how continued fractions are of use in the approximation of real numbers. In the process we will give another proof of Hurwitz' Theorem (Theorem 4.6.)

This section and the next three barely scratch the surface of the theory of continued fractions. See, for example, [18] or [60] and for more depth and several generalizations see [89].

Problem Set 4.4

1. Find the simple continued fraction expansions of the rational numbers 7/6, 13/2, -176/85, -16/81 and 1001/999.

2. Find the rational numbers whose finite simple continued fraction expansions are [1; 2, 3], [-1; 2, 1, 3, 1, 4], [1; 1, 1, 1, 1, 1] and [-7; 9, 11, 13, 100].

3. If h_s / k_s is the sth convergent to $[a_0;a_1,\ldots,a_n], a_0 > 0$, then

$$h_n / h_{n-1} = [a_n;a_{n-1},\ldots,a_0] \text{ and } k_n / k_{n-1} = [a_n;a_{n-1},\ldots,a_1].$$

4. (a) Let $a/b = [a_0; a_1, \ldots, a_n], (a, b) = 1$ and let h_i / k_i be its convergents. Use Theorem 4.12 to find the solution to

$$ax + by = 1.$$

Solve the equation $Ax + By = C$, where A, B and C are integers such that $(A, B) \mid C$.

   (b) Solve the equations:
   (i) $11x - 13y = 57$;
   (ii) $127x + 257y = 331$;
   (iii) $57x + 101y = 1009$.

5. Write the simple continued fraction expansions of the quadratic irrationals

$$(1 + \sqrt{6})/2, (\sqrt{6} - 1)/3 \text{ and } \sqrt{99}.$$

Find the first five partial quotients for the irrational numbers

$$\sqrt[3]{3} \text{ and } \sqrt[12]{2}.$$

6. (a) Show that

$$[a_0; a_1, \ldots, a_n] > [a_0; a_1, \ldots, a_n + c]$$

for all positive values of c if n is odd, but that this is false if n is even. What is the truth when n is even?

   (b) Find necessary and sufficient conditions so that

$$[a_0; a_1, \ldots, a_n] < [b_0; b_1, \ldots, b_n].$$

7. Let ξ be an irrational number with continued fraction expansion $[a_0; a_1, a_2, \ldots]$. Show that
   (a) if $a_1 > 1$, then $-\xi = [-a_0 - 1; 1, a_1 - 1, a_2, a_3, \ldots]$
   (b) if $a_1 = 1$, then $-\xi = [-a_0 - 1; a_2 + 1, a_3, a_4, \ldots]$.

8. Show that the nth convergent of $1/\xi$ is the reciprocal of the $(n - 1)$st convergent of ξ for any irrational $\xi > 1$. (Hint: Theorem 4.12.)

9. Find the values of the following continued fractions:
   (a) $[k; 2k, 2k, 2k, \ldots]$, where k is a positive integer;
   (b) $[1; 2, 3, 1, 2, 3, \ldots]$;
   (c) $[1; 2, 3, 1, 2, 1, 2, \ldots]$;
   (d) $[1; 2, 3, 1, 3, 1, 3, \ldots]$;
   (e) $[1; 2, 3, 1, 1, 1, 1, \ldots]$.

10. Let $\alpha = [a_0; a_1, \ldots, a_m]$ be a simple continued fraction with $a_0 > 0$. Let h_n / k_n

be the nth convergent of α. Show that

(a) if $h_{n-1} = k_n$, then $k_n^2 + (-1)^n \equiv 0 \pmod{h_n}$;

(b) if $k_n^2 + (-1)^n \equiv 0 \pmod{h_n}$ and $h_n > k_n$, then $h_{n-1} = k_n$;

(c) $a_n = a_{m-n}, 0 \le n \le m$, if and only if $k_m^2 + (-1)^m \equiv 0 \pmod{h_m}$.

11. (a) Recall that the Fibonacci numbers are defined by $F_0 = 0$, $F_1 = 1$ and, for $n \ge 0$,

$$F_{n+2} = F_{n+1} + F_n.$$

Let $h_n / k_n = [a_0; a_1, \dots, a_n]$. Show that, for $n \ge 0$, we have $k_n \ge F_{n+1}$.

(b) Show that, for $n \ge 2$, we have $k_n > k_{n-1}$.

12. Suppose $h_n / k_n = [a_0; a_1, \dots, a_n]$. Show that, as matrices,

$$\begin{pmatrix} h_n & h_{n-1} \\ k_n & k_{n-1} \end{pmatrix} = \begin{pmatrix} a_0 & 1 \\ 1 & 0 \end{pmatrix} \begin{pmatrix} a_1 & 1 \\ 1 & 0 \end{pmatrix} \cdots \begin{pmatrix} a_n & 1 \\ 1 & 0 \end{pmatrix}.$$

13. Let ξ be the value of the periodic simple continued fraction $[\overline{b;a}]$ and h_n/k_n be the nth convergent. Show that, for $n \ge 2$, we have

$$h_{n+2} - (ab+2)h_n + h_{n-2} = 0.$$

Is there an analogous result for the denominators k_n?

14. Let x be the value of the continued fraction $[a_0; a_1, a_2, \dots]$ and write, for k a positive integer, $x = [a_0; a_1, a_2, \dots, a_k, x_k]$. Show that

(a) $x - \dfrac{h_n}{k_n} = \dfrac{(-1)^{n-1}}{k_n(k_n x_n + k_{n-1})}$;

(b) $x_k = -\dfrac{k_{n-1}x - h_{n-1}}{k_n x - h_n}$;

(c) $x_1 \cdots x_n = \dfrac{(-1)^n}{h_n - k_n x}$.

15. Let h_n/k_n be the nth convergent to $[a_0; a_1, \dots, a_m]$ and let p_n^s / q_n^s be the nth convergent to $[a_s; a_{s+1}, \dots, a_m]$, for $s \ge 1$. Show that

$$h_n k_{n-r} - h_{n-r} k_n = (-1)^{n-r+1} q_n^{n-r+1}.$$

4.5. PERIODIC CONTINUED FRACTIONS

In this section we discuss the properties of periodic continued fractions. Besides being of interest in itself some of this material will find applications in the theory of Pell's equation (Section 6.7.) The big theorem is the following.

Theorem 4.18: Any periodic simple continued fraction is a quadratic irrational and conversely.

Proof. Let

$$\xi = [b_0; b_1, \ldots, b_j, \overline{a_0, a_1, \ldots, a_{r-1}}]$$

and let

$$\theta = [\overline{a_0; a_1, \ldots, a_{r-1}}] = [a_0; a_1, \ldots, a_{r-1}, \theta].$$

Then, by Theorem 4.10, we have

$$\theta = \frac{\theta h_{r-1} + h_{r-2}}{\theta k_{r-1} + k_{r-2}}.$$

This gives rise to a quadratic equation in θ, and so θ is either rational or a quadratic irrational. By Theorem 4.14 we see that θ cannot be rational. By Theorem 4.10, we can write ξ in terms of θ and we obtain

$$\xi = [b_0; b_1, \ldots, b_j, \theta] = \frac{\theta m + p}{\theta n + q},$$

where p/q and m/n are the last two convergents of $[b_0; b_1, \ldots, b_j]$. Since $\theta = (a + \sqrt{b})/c$, for some integers a, b and c, we see that ξ is again of this form, since, by Theorem 4.14, cannot be rational.

To prove the converse let $\xi_0 = (a + \sqrt{b})/c$, where a, b and c are integers with $b > 0$ and $c \neq 0$. Since ξ_0 is irrational we see that b is not a perfect square. Then multiplying the numerator and denominator by $|c|$ gives

$$\xi_0 = \frac{ac + \sqrt{bc^2}}{c^2} \text{ or } \xi_0 = \frac{-ac + \sqrt{bc^2}}{-c^2},$$

depending on whether $c > 0$ or $c < 0$. In either case we may write

$$\xi_0 = (m + \sqrt{d})/q_0,$$

where $q_0 | (d - m_0^2)$, with d, m_0 and q_0 integers, $q_0 \neq 0$, and $d > 0$ not a perfect square. This allows us to get a simple formulation of the continued fraction expansion of ξ_0. We shall show that, for $i \geq 0$,

(4.30)
$$\begin{cases} a_i = [\xi_i], & \xi_i = (m_i + \sqrt{d})/q_i \\ m_{i+1} = a_i q_i - m_i, & q_{i+1} = (d - m_{i+1}^2)/q_i \end{cases}.$$

We take $a_0 = [\xi_0]$ with ξ_0, m_0 and q_0 as above. If we have ξ_i, m_i, q_i and a_i we define m_{i+1} and q_{i+1} as in (4.30) and let

$$\xi_{i+1} = (m_{i+1} + \sqrt{d})/q_{i+1} \text{ and } a_{i+1} = [\xi_{i+1}].$$

We wish to show m_i and q_i are integers such that $q_i \neq 0$ and $q_i|(d - m_i^2)$ for all $i \geq 0$. This holds for $i = 0$ so suppose it holds for $i = k - 1$. Then $m_k = a_{k-1}q_{k-1} - m_{k-1}$

Then

$$q_k = (d - m_k^2)/q_{k-1} = (d - m_{k-1}^2)/q_{k-1} + 2a_{k-1}m_{k-1} - a_{k-1}^2 q_{k-1},$$

and so, by the induction hypothesis, q_k is an integer. Since d is not a square we see that q_k is not zero. Since q_{k-1} is an integer and

$$q_{k-1} = (d - m_k^2)/q_k$$

we see that $q_k|(d - m_k^2)$ This finishes the induction.

Now we must prove the periodicity. We have

$$\xi_i - a_i = (-a_iq_i + m_i + \sqrt{d})/q_i = (\sqrt{d} - m_{i+1})/q_i$$
$$= (d - m_{i+1}^2)/(q_i(\sqrt{d} + m_{i+1})) = q_{i+1}/(\sqrt{d} + m_{i+1}) = \xi_{i+1}^{-1}.$$

This shows that

$$\xi_0 = [a_0; a_1, \ldots].$$

Let $\overline{\xi_i}$ be the conjugate of ξ_i, that is

$$\overline{\xi_i} = (m_i - \sqrt{d})/q_i.$$

Since the conjugate of a quotient equals the quotient of the conjugates we have

$$\overline{\xi_0} = \frac{\overline{\xi_n}h_{n-1} + h_{n-2}}{\overline{\xi_n}k_{n-1} + k_{n-2}}$$

or, solving for $\overline{\xi_n}$

(4.31)
$$\overline{\xi_n} = -\frac{k_{n-2}}{k_{n-1}}\left(\frac{\overline{\xi_0} - h_{n-2}/k_{n-2}}{\overline{\xi_0} - h_{n-1}/k_{n-1}}\right).$$

Now as $n \to +\infty$ both h_{n-2}/k_{n-2} and h_{n-1}/k_{n-1} approach $\xi_0 \neq \overline{\xi_0}$. Thus the expression is parentheses on the right hand side of (4.31) approaches 1 as $n \to +\infty$. Thus there exists a positive integer N such that if $n > N$, then the expression in parentheses on the right hand side of (4.31) is positive. But ξ_n is positive if $n \geq 1$ and $\xi_n - \overline{\xi_n} > 0$ for $n > N$. By the definition of ξ_n we see that $2\sqrt{d}/a_n > 0$ and so $q_n > 0$ if $n > N$. By (4.30) we see that

$$a_nq_{n+1} = d - m_{n+1}^2 \leq d, \quad q_n \leq q_nq_{n+1} \leq d$$

and

$$m_{n+1}^2 < m_{n+1}^2 + q_n q_{n+1} = d$$

so that

$$|m_{n+1}| < \sqrt{d}$$

for $n > N$. Since d is a fixed positive integer we conclude that q_n and m_{n+1} can assume only a fixed number of possible values of $n > N$. Hence the ordered pairs (m_n, q_n) can assume only a fixed number of possible pair values for $n > N$. Thus there exist distinct positive integers, j and k, such that

$$m_j = m_k \text{ and } q_j = q_k$$

with $j < k$. By (4.30) we see that $\xi_j = \xi_k$. Thus

$$\xi_0 = [a_0; a_1, \ldots, a_{j-1}, \overline{a_j, \ldots, a_{k-1}}],$$

that is, ξ_0 is periodic. ∎

We now determine that subclass of real quadratic irrationals that have purely periodic continued fraction expansions, that is,

$$\xi = [\overline{a_0; a_1, \ldots, a_j}].$$

First we make a definition.

Definition 4.3: We say that a quadratic irrational ξ is **reduced** if $\xi > 1$ and its conjugate $\overline{\xi}$ satisfies $-1 < \overline{\xi} < 0$.

Theorem 4.19: The continued fraction expansion of the real quadratic irrational number ξ is purely periodic if and only if it is reduced.

Proof: Suppose $\xi > 1$ and $-1 < \overline{\xi} < 0$. If we let $\xi_0 = \xi$ and take the conjugate in the definition of ξ_k, (4.28), we have

$$\overline{\xi_{i+1}^{-1}} = \overline{\xi_i} - a_i.$$

Now $a_i \geq 1$ for all $i \geq 0$ since $x_0 > 1$. Thus, if $\overline{\xi_i} < 0$, then $\overline{\xi_{i+1}^{-1}} < -1$, and so $-1 < \overline{\xi_{i+1}} < 0$. Since $\overline{\xi_i} = a_i + \overline{\xi_{i+1}^{-1}}$ we see, by mathematical induction, that $-1 < \overline{\xi_i} < 0$ for all $i \geq 0$. Since $\overline{\xi_i} = a_i + \overline{\xi_{i+1}^{-1}}$ we have

$$0 < -\overline{\xi_{i+1}^{-1}} - a_i < 1,$$

that is,

$$a = [-1 / \overline{\xi_{i+1}}].$$

Now ξ is a quadratic irrational so we must have, by Theorem 4.18, that $\xi_j = \xi_k$ for some integers j and k, with $0 < j < k$. Then $\overline{\xi_j} = \overline{\xi_k}$ and

$$a_{j-1} = [-1/\overline{\xi_j}] = [-1/\overline{\xi_k}] = a_{k-1}.$$

Also

$$\xi_{j-1} = a_{j-1} + 1/\xi_j = a_{k-1} + 1/\xi_k = \xi_{k-1}.$$

Thus $\xi_j = \xi_k$ implies $\xi_{j-1} = \xi_{k-1}$. Doing this j times yields

$$\xi_0 = \xi_{k-j}.$$

Thus

$$\xi = \xi_0 = [\overline{a_0; a_1, \ldots, a_{k-j-1}}].$$

To prove the converse we assume $\xi = [\overline{a_0; a_1, \ldots, a_{n-1}}]$ with $a_0, a_1, \ldots, a_{n-1}$ positive integers. Then

$$\xi > a_0 \geq 1.$$

Also

$$\xi = [a_0; a_1, \ldots, a_{n-1}, \xi] = \frac{\xi h_{n-1} + h_{n-2}}{\xi k_{n-1} + k_{n-2}},$$

which implies that ξ is a zero of the quadratic polynomial

$$f(x) = x^2 k_{n-1} + x(k_{n-2} + h_{n-1}) - h_{n-2}.$$

Since ξ and $\overline{\xi}$ are the zeros of $f(x)$ we need only show that the other zero of $f(x)$ is in the interval $(-1, 0)$. To do this it suffices to show that $f(0)f(-1) < 0$. We have

$$f(0) = -h_{n-2} < 0,$$

since $a_i > 0$ for $i \geq 0$. If $n > 1$, then

$$\begin{aligned}
f(-1) &= k_{n-1} - k_{n-2} + h_{n-1} - h_{n-2} \\
&= (k_{n-2} + h_{n-2})(a_{n-1} - 1) + k_{n-3} + h_{n-3} \\
&\geq k_{n-3} + h_{n-3} > 0.
\end{aligned}$$

If $n = 1$, then

$$f(-1) = k_0 - k_{-1} + h_0 - h_{-1} = a_0 > 0$$

and the proof is complete.   ∎

Example 4.16: Find the continued fraction expansion of $\xi_0 = (1 + \sqrt{17})/4$.

Solution: We have the following table:

$a_0 = [\xi_0] = [(1 + \sqrt{17})/4] = 1$ and $\xi_1 = ((1 + \sqrt{17})/4 -)^{-1} = (3 + \sqrt{17})/2$;

$a_1 = [\xi_1] = 3$ and $\xi_2 = ((3 + \sqrt{17})/2 - 3)^{-1} = (3 + \sqrt{17})/4$;

$a_2 = [\xi_2] = 1$ and $\xi_3 = ((3 + \sqrt{17})/4 - 1)^{-1} = (1 + \sqrt{17})/4 = \xi_0$.

Thus the cycle starts over, and so

$$(1 + \sqrt{17})/4 = [\overline{1;3,1}].$$

For a certain class of quadratic irrationals we can identify the continued fraction expansions even closer.

Theorem 4.20: Let D be a positive nonsquare integer. Then the simple continued fraction expansion of $\sqrt{D}$ has the form

$$\sqrt{D} = [a_0; \overline{a_1, a_2, \ldots, a_{r-1}, 2a_0}],$$

where $a_0 = [\sqrt{D}]$. Moreover, with $\xi_0 = \sqrt{D}$, $q_0 = 1$ and $m_0 = 0$ in (4.30) we have $q_i = 1$ if and only if $r \mid i$ and $q_i = -1$ holds for no i, where r is the length of the least period in the continued fraction expansion of $\sqrt{D}$.

Proof: Let $\xi = [\sqrt{D}] + \sqrt{D}$. Then ξ satisfies the hypothesis of Theorem 4.19. Thus

$$[\sqrt{D}] + \sqrt{D} = [\overline{a_0; a_1, \ldots, a_{r-1}}] = [a_0; \overline{a_1, \ldots, a_{r-1}, a_0}],$$

where r is the smallest integer for which such an expression holds. Note that $\xi_i = [a_i; a_{i+1}, \ldots]$ is purely periodic for all values of i and that $\xi_0 = \xi_r = \xi_{2r} = \cdots$. Also $\xi_0, \xi_1, \ldots, \xi_{r-1}$ are all distinct since r is minimal, and so $\xi_i = \xi_0$ if and only if $r \mid i$.

In (4.30), with $\xi_0 = [\sqrt{D}] + \sqrt{D}$, we may take $q_0 = 1$ and $m_0 = [\sqrt{D}]$ since $1 \mid (D - [\sqrt{D}]^2)$. Then, for all $j \geq 0$, we have

$$(m_{jr} + \sqrt{D})/q_{jr} = \xi_{jr} = \xi_0 = (m_0 + \sqrt{D})/q_0 = [\sqrt{D}] + \sqrt{D}.$$

Thus

(4.33) $$m_{jr} - q_{jr}[\sqrt{D}] = (q_{jr} - 1)\sqrt{D}$$

and so $q_{jr} = 1$ since the left hand side of (4.33) is rational while the right hand side is not if $q_{jr} \neq 1$. If $q_i = 1$, then $\xi_i = m_i + \sqrt{D}$ and, since ξ_i is purely periodic, we have that

$$-1 < m_i - \sqrt{D} < 0$$

or

$$\sqrt{D} - 1 < m_i < \sqrt{D},$$

that is,

$$m_i = [\sqrt{D}].$$

Thus $\xi_i = \xi_0$, and so $r \mid i$.

If $q_i = -1$, then $\xi_i = -m_i - \sqrt{D}$ and so, by Theorem 4.19, we have

$$-m_i - \sqrt{D} > 1 \quad \text{and} \quad -1 < -m_i + \sqrt{D} < 0.$$

Thus

$$\sqrt{D} < m < -\sqrt{D} - 1,$$

which is impossible. Thus q_i is never -1;

Now $a_0 = [\sqrt{D} + [\sqrt{D}]] = 2[\sqrt{D}]$, and so

$$\sqrt{D} = -[\sqrt{D}] + (\sqrt{D} + [\sqrt{D}])$$
$$= -[\sqrt{D}] + [2[\sqrt{D}]; \overline{a_1, \ldots, a_{r-1}, a_0}]$$
$$= [[\sqrt{D}]; \overline{a_1, \ldots, a_{r-1}, 2[\sqrt{D}]}].$$

When we apply (4.30) to $\sqrt{D} + [\sqrt{D}]$ with $q_0 = 1$ and $m_0 = [\sqrt{D}]$, we have $a_0 = 2[\sqrt{D}], m_1 = [\sqrt{D}]$ and $q_1 = D - [\sqrt{D}]^2$. Now apply (4.30) to $\sqrt{D}$ with $q_0 = 1$ and $m_0 = 0$. Then $a_0 = [\sqrt{D}], m_1 = [\sqrt{D}]$ and $q_1 = D - [\sqrt{D}]^2$. Thus only the value of a_0 is different, but the values of m_1 and q_1 are the same in both cases. Since $\xi_i = (m_i + \sqrt{D}) / q_i$ we see that further application of these equations yield same values of a_i, of m_i and of q_i in both cases. Thus the expansions of $[\sqrt{D}] + \sqrt{D}$ and $\sqrt{D}$ differ only by the values of a_0 and m_0. This gives our result. ∎

See Problem 10 below for an extension of this result.

For quadratic irrationals we can make the calculations a little more systematic. By (4.30), we have

$$q_{i+1} = (D - m_{i+1}^2) / q_i = (D - (a_i q_i - m_i)^2) / q_i$$
$$= (D - m_i^2) / q_i - a_i^2 q_i + 2a_i m_i = q_{i-1} - a_i(a_i q_i - m_i) + a_i m_i$$
$$= q_{i-1} + a_i(m_i - m_{i+1}).$$

Starting with $\xi_0 = (m_0 + \sqrt{D}) / q_0$, with $q_0 \mid (D - m_1^2)$ we obtain

$$a_0 = [(m_0 + \sqrt{D})/q_0], m_1 = a_0 q_0 - m_0, q_1 = (D - m_1^2)/q_0$$
$$a_1 = [(m_1 + \sqrt{D})/q_1], m_2 = a_1 q_1 - m_1, q_2 = q_0 + a_1(m_1 - m_2)$$
$$\vdots$$
$$a_{i-1} = [(m_{i-1} + \sqrt{D})/q_{i-1}], m_i = a_{i-1}q_{i-1} - m_{i-1}, q_i = q_{i-2} + a_{i-1}(m_{i-1} - m_i).$$

The formula $a_i q_{i+1} = D - m_{i+1}^2$ can be used as a check. We illustrate this with two examples.

Example 4.17. Calculate the continued fraction expansion of

(a) $\sqrt{29}$

(b) $\dfrac{1+\sqrt{2}}{2}$.

Solution of (a): We have $\sqrt{29} = (m_0 + \sqrt{29})/q_0$, so that $m_0 = 0$ and $q_0 = 1$. Then
$$a_0 = [\sqrt{29}] = 5, m_1 = 5 \cdot 1 - 0 = 5, q_1 = (29 - 25)/1 = 4$$
$$a_1 = [(5+\sqrt{29})/4] = 2, m_2 = 2 \cdot 4 - 5 = 3, q_2 = 1 + 2(5-3) = 5;$$
$$a_2 = [(3+\sqrt{29})/5] = 1, m_3 = 1 \cdot 5 - 3 = 2, q_3 = 4 + 1(3-2) = 5;$$
$$a_3 = [(2+\sqrt{29})/5] = 1, m_4 = 1 \cdot 5 - 2 = 3, q_4 = 5 + 1(2-3) = 4;$$
$$a_4 = [(3+\sqrt{29})/4] = 2, m_5 = 2 \cdot 4 - 3 = 5, q_5 = 5 + 2(3-5) = 1;$$
$$a_5 = [5+\sqrt{29}] = 10, m_6 = 10 \cdot 1 - 5 = 5, q_6 = 4 + 10(5-5) = 4.$$

By Theorem 4.20, we see that we are done. Thus

$$\sqrt{29} = [5; \overline{2,1,1,2,10}].$$

In general we stop when the values of m and q simultaneously repeat. For then the next a will repeat, and so we have started the next period.

Solution of (b). If we let $\xi_0 = \dfrac{1+\sqrt{2}}{2}$, then $m_0 = 1, q_0 = 2$, but 2 does not divide $2 - 1^2 = 1$. Thus we modify ξ_0 by multiplying the numerator and denominator by 2. Thus we start with

$$\xi_0 = \frac{2+\sqrt{8}}{4}, m_0 = 2, q_0 = 4 \text{ with } 4 \mid (8 - 2^2) = 4.$$

Then

$$a_0 = \left[\frac{2+\sqrt{8}}{4}\right] = 1, m_1 = 1 \cdot 4 - 2 = 2, q_1 = \frac{8 - 2^2}{4} = 1$$

and

$$a_1 = \left[\frac{2+\sqrt{8}}{1}\right] = 4, m_2 = 4 \cdot 1 - 2 = 2, q_2 = 4 + 4(2-2) = 4.$$

Since $m_2 = m_0$ and $q_2 = q_0$ we see that $a_2 = a_0$, etc. Thus

$$\frac{1+\sqrt{2}}{2} = [\overline{1;4}]. \qquad\qquad \square$$

We can say more about the expansion of $\sqrt{D}$ given by Theorem 4.20. As may have become apparent from the examples and exercises the continued fraction expansion of $\sqrt{D}$ is symmetric, that is, if we write

$$\sqrt{D} = [a_0; \overline{a_1, \ldots, a_{r-1}, 2a_0}],$$

then $a_k = a_{r-k}, 1 \le k \le r-1$. Before proving this result we need a further result about the numbers discussed in Theorem 4.19.

Lemma 4.21.1: Suppose $\xi = [a_0; a_1, a_2, \ldots]$, with $a_0 > 0$, has convergents h_n / k_n. Then we have

(a) $h_n / h_{n-1} = [a_n; a_{n-1}, \ldots, a_1, a_0]$

and

(b) $k_n / k_{n-1} = [a_n; a_{n-1}, \ldots, a_2, a_1].$

Proof: By (4.23), we have

$$[a_1; a_0] = a_1 + \frac{1}{a_0} = \frac{a_0 a_1 + 1}{a_0} = \frac{h_1}{h_0}$$

and

$$h_{n+1} = a_{n+1} h_n + h_{n-1}.$$

Assume that, for $n = m \ge 1$, we have

$$\frac{h_m}{h_{m-1}} = [a_m; a_{m-1}, \ldots, a_1, a_0].$$

Then, for $n = m + 1$, we have

$$\frac{h_{m+1}}{h_m} = a_{m+1} + \frac{h_{m-1}}{h_m} = a_{m+1} + \frac{1}{[a_m; a_{m-1}, \ldots, a_1, a_0]} = [a_{m+1}; a_m, a_{m-1}, \ldots, a_1, a_0]$$

and (a) follows by mathematical induction.

Part (b) is proved similarly. ∎

Lemma   4.21.2: If ξ is a reduced quadratic irrational and we write $\xi = [\overline{a_0; a_1, \ldots, a_n}]$, then we have

$$\frac{-1}{\bar{\xi}} = [\overline{a_n; a_{n-1}, \ldots, a_1, a_0}].$$

Proof: Let

$$\beta = [\overline{a_n; a_{n-1}, \ldots, a_1, a_0}].$$

Then, by Theorem 4.10,

$$\beta = [a_n; a_{n-1}, \ldots, a_1, a_0, \beta] = \frac{\beta p_n + p_{n-1}}{\beta q_n + q_{n-1}},$$

(4.34)

where p_n / q_n and p_{n-1} / q_{n-1} are the nth and $(n-1)$st convergents of β. Now, by Lemma 4.21.1, we have

(4.35)
$$\begin{cases} \dfrac{h_n}{h_{n-1}} = [a_n; a_{n-1}, \ldots, a_1, a_0] = \dfrac{p_n}{q_n} \\ \dfrac{k_n}{k_{n-1}} = [a_n; a_{n-1}, \ldots, a_2, a_1] = \dfrac{p_{n-1}}{q_{n-1}} \end{cases}.$$

Since p_n / q_n and p_{n-1} / q_{n-1} are convergents to β we must have that $(p_{n-1}, q_{n-1}) = (p_n, q_n) = 1$, by Theorem 4.12. Also, by Theorem 4.12, we have $(h_n, h_{n-1}) = (k_n, k_{n-1}) = 1$. Thus, by (4.35), we have $p_n = h_n, q_n = h_{n-1}, p_{n-1} = k_n$ and $q_{n-1} = k_{n-1}$.

Thus, by (4.34), we have

$$\beta = \frac{\beta h_n + k_n}{\beta h_{n-1} + k_{n-1}}$$

or

$$h_{n-1} \beta^2 + (k_{n-1} - h_n) \beta - k_n = 0$$

or

$$k_n (-1 / \beta)^2 + (k_{n-1} - h_n)(-1 / \beta) - h_{n-1} = 0.$$

From the proof of Theorem 4.19 we know that ξ and $\bar{\xi}$ satisfy the equation (4.32). Thus we must have $\bar{\xi} = -1 / \beta$. Thus

$$-1 / \bar{\xi} = \beta = [\overline{a_n; a_{n-1}, \ldots, a_1, a_0}]. \qquad \blacksquare$$

Theorem 4.21. Let D be a positive nonsquare integer. If we write

$$\sqrt{D} = [a_0; \overline{a_1, \ldots, a_{r-1}, 2a_0}],$$

then we have $a_k = a_{r-k}, 1 \le k \le r-1$.

Proof: We have, as in the proof of Theorem 4.20,

$$[\sqrt{D}] + \sqrt{D} = [2a_0; \overline{a_1, \ldots, a_{r-1}, 2a_0}] = [\overline{2a_0; a_1, \ldots, a_{r-1}}].$$

Thus, by Lemma 4.21.2, we have

$$\frac{1}{\sqrt{D} - [\sqrt{D}]} = [\overline{a_{r-1}; a_{r-2}, \ldots, a_1, 2a_0}].$$

By Theorem 4.20, we also have

$$\sqrt{D} - [\sqrt{D}] = [0; \overline{a_1, \ldots, a_{r-1}, 2a_0}],$$

and so

$$\frac{1}{\sqrt{D} - [\sqrt{D}]} = [\overline{a_1; a_2, \ldots, a_{r-1}, 2a_0}].$$

Since, by Theorem 4.17, the continued fraction expansion of an irrational is unique the partial quotients of the two expansions must agree. Thus

$$a_1 = a_{r-1}, a_2 = a_{r-2}, \ldots, a_{r-1} = a_1,$$

that is, $a_k = a_{r-k}, 1 \le k \le r-1$. ∎

Problem Set 4.5

1. Find the simple continued fraction expansions of

   (a) $\sqrt{7}$ (d) $\sqrt{21}$

   (b) $\sqrt{11}$ (e) $\sqrt{22}$

   (c) $\sqrt{12}$ (f) $\sqrt{94}$

2. Find the simple continued fraction expansions of

   (a) $(1 - \sqrt{6})/2$ (d) $(1 + \sqrt{21})/3$

   (b) $\frac{1}{2}\sqrt{3}$ (e) $(1 + \sqrt{21})/7$

   (c) $(2 - \sqrt{5})/3$ (f) $(3 + \sqrt{5})/8$

3. Determine the quadratic irrationals whose simple continued fraction expansions are given.

   (a) $[1; \overline{2, 3}]$ (d) $[1; \overline{2}]$

   (b) $[1; 2, \overline{3}]$ (e) $[1; \overline{1, 2}]$

   (c) $[\overline{1; 2, 3}]$ (f) $[1; \overline{1, 1, 2}]$

4. Determine the quadratic irrationals whose simple continued fraction expansions are given.

(a) $[\overline{2;8}]$

(d) $[9;\overline{3,18}]$

(b) $[\overline{1;12}]$

(e) $[\overline{1;2,1,1,4}]$

(c) $[7;\overline{7,14}]$

(f) $[\overline{2;2,14,2}]$

5. Let $n \geq 2$ be an integer.

(a) Show that $\sqrt{n^2+1} = [n;\overline{2n}]$.

(b) Show that $\sqrt{n^2-1} = [n-1;\overline{1,2n-2}]$.

(c) Find the simple continued fraction expansions of

$$\sqrt{101}, \sqrt{257}, \sqrt{10001}, \sqrt{99}, \sqrt{255} \text{ and } \sqrt{9999}.$$

6. Let n be an integer.

(a) If $n \geq 1$, then $\sqrt{n^2+n} = [n;\overline{2,2n}]$.

(b) If $n \geq 2$, then $\sqrt{n^2-n} = [n-1;\overline{2,2n-2}]$.

(c) Find the simple continued fraction expansions of

$$\sqrt{110}, \sqrt{272}, \sqrt{10100}, \sqrt{90}, \sqrt{240} \text{ and } \sqrt{9900}.$$

7. (a) Show that if n is a positive integer, then $2\sqrt{n^2+1} = [2n;\overline{n,4n}]$.

(b) Find the simple continued fraction expansions of

$$2\sqrt{82}, \sqrt{680} \text{ and } \sqrt{60520}.$$

8. Determine the simple continued fraction expansions of $\sqrt{n^2+2}$ and $\sqrt{n^2-2}$, if $n \geq 2$ is an integer.

9. Let D be a positive nonsquare integer and suppose that

$$\sqrt{D} = [a_0;\overline{a_1,a_2}].$$

Show that, in the notation of Theorem 4.18,

(a) $m_j = a_0, q_{2j} = 1$, for $j \geq 1$.

and

(b) $q_{2j-1} = D - a_0^2 \neq 0$, for $j \geq 1$.

(Hint: use Theorem 4.20 and expand the continued fraction to show that the relation between D and a_0, a_1 and a_2.)

10. (a) Let N be a positive nonsquare integer and let M be a positive integer. Show that we can write $\sqrt{N}/M$ in the form

$$\sqrt{N}/M = [a_0;\overline{a_1,a_2,\ldots,a_{r-1},2a_0}],$$

where $a_0 = [\sqrt{N}/M]$.

(b) Show that in the expansion in (a) we have $a_k = a_{r-k}, 1 \le k \le r$.

(c) Show that if any real number x has an expansion of the form

$$x = [a_0; \overline{a_1, \ldots, a_{r-1}, 2a_0}],$$

then x must be the square root of a rational number greater than 1.

11. (a) Show that if r is the period of the simple continued fraction expansion of $(P + \sqrt{R})/Q$, then $r < 2R$. (Hint: recall the proof of Theorem 4.18.)

(b) Show that the partial quotients satisfy $a_n < 2R$, for $n \ge 1$.

12. Let $\xi = (a + \sqrt{b})/c$ be a quadratic irrational with a, b and c integers with $b > 0$. Show that ξ is reduced if and only if $0 < a < \sqrt{b}$ and $\sqrt{b} - a < c < \sqrt{b} + a$.

13. Let D be a positive nonsquare integer. For what values of the positive integer a is the quadratic irrational $([\sqrt{D}] + \sqrt{D})/a$ reduced?

4.6. APPROXIMATION BY RATIONALS, II

We take up the problem of approximation again, however, this time we will be able to be more concrete about the infinite sequence of rationals "produced" by Hurwitz' theorem.

Let ξ be an irrational and let $\{h_n/k_n\}$ be the sequence of convergents for ξ.

Theorem 4.22. Let n be a nonnegative integer. Then

$$|\xi - h_n/k_n| < 1/k_n k_{n+1} \text{ and } |\xi k_n - h_n| < 1/k_{n+1}.$$

Proof. The second inequality follows from the first upon multiplication of both sides by k_n. By (4.29), we have

$$|\xi - h_n/k_n| = k_n^{-1}(\xi_{n+1}k_n + k_{n-1})^{-1} < k_n^{-1}(a_{n+1}k_n + k_{n-1})^{-1}$$

and the result follows from (4.23). ∎

Corollary 4.22.1. Let n be a nonnegative integer. Then

$$|\xi - h_n/k_n| < k_n^{-2}.$$

Proof. This follows immediately from Theorem 4.22 since the k_n are increasing, that is, $k_n + 1 > k_n$. ∎

This corollary makes concrete Theorem 4.1 by actually producing an infinite sequence of rationals that satisfy the inequality of the theorem.

Theorem 4.23. We have, for any consecutive convergents h_n/k_n and h_{n+1}/k_{n+1},

$$\left| \xi - h_{n+1} / k_{n+1} \right| < \left| \xi - h_n / k_n \right|.$$

In fact, we have the stronger inequality

$$\left| \xi k_{n+1} - h_{n+1} \right| < \left| \xi k_n - h_n \right|.$$

Proof. We show that the second inequality implies the first. Since $k_n \le k_{n+1}$ we have

$$\left| \xi - h_{n+1} / k_{n+1} \right| = \left| \xi k_{n+1} - h_{n+1} \right| / k_{n+1} < \left| \xi k_n - h_n \right| / k_{n+1}$$
$$\le \left| \xi k_n - h_n \right| / k_n = \left| \xi - h_n / k_n \right|.$$

By (4.28), we have $a_n + 1 > x_n$. Thus, by (4.23), we have

$$\xi_n k_{n-1} + h_{n-2} < (a_n + 1) k_{n-1} + k_{n-2}$$
$$= k_n + k_{n-1} \le a_{n+1} k_n + k_{n-1} = k_{n+1}.$$

This inequality and (4.29) imply

$$\left| \xi - h_n / k_n \right| = k_n^{-1} (\xi_{n+1} k_n - h_{n-1})^{-1} > k_n^{-1} k_{n+2}^{-1}.$$

If we multiply by k_n and use Theorem 4.21 we get

$$\left| \xi k_n - h_n \right| > k_{n+2}^{-1} > \left| \xi k_{n+1} - h_{n+1} \right|,$$

which is our result.   ∎

These two theorems show that the convergents give successively better approximations as well as giving an estimate for the degree of approximation. The next theorem shows that the convergents h_n/k_n give the best rational approximation among all rationals with denominator of k_n or less.

Theorem 4.24. If a/b is a rational number with $b > 0$ and such that $\left| \xi - a/b \right| < \left| \xi - h_n / k_n \right|$, for some $n \ge 1$, then $b > k_n$. In fact, if $\left| \xi b - a \right| < \left| \xi k_n - h_n \right|$, for some $n \ge 0$, then $b > k_{n+1}$.

Proof. Again we shall show that the second part of the theorem implies the first. Suppose that the first part is false so that there exists some rational a/b, $b \le k_n$, such that

$$\left| \xi - a/b \right| < \left| \xi - h_n / k_n \right|,$$

Multiplying this inequality by $b \le k_n$ gives

(4.36) $$\left| \xi b - a \right| < \left| \xi k_n - h_n \right|,$$

which contradicts the second part of the theorem which asserts that we must have $b \ge k_{n+1}$ if (4.36) holds, since $k_n < k_{n+1}$, for $n \ge 1$.

We prove the second part of the theorem by an indirect argument also. Suppose $|\xi b - a| < |\xi k_n - h_n|$, and $b < k_{n+1}$. Let x and y be solutions to the system of equations

$$xk_n + yk_{n+1} = b \text{ and } xh_n + yh_{n+1} = a.$$

By (a) of Theorem 4.12 we know that the determinant of the system is ± 1. Thus x and y will be integers.

Note that neither x nor y can be zero. If $x = 0$, then $b = yk_{n+1}$, and so $y \neq 0$, indeed $y > 0$. Thus $b > k_{n+1}$, since y a positive integer implies that $y \geq 1$. This contradicts the hypothesis $b < k_{n+1}$. If $y = 0$, then $a = xh_n$, $b = xk_n$, and so

$$|\xi b - a| = |\xi x k_n - x h_n| = |x||\xi k_n - h_n| > |\xi k_n - h_n|,$$

since $x \neq 0$ implies that $|x| > 1$. Again we obtain a contradiction.

If $y < 0$, then $xk_n = b - yk_{n+1}$ shows that $x > 0$. If $y > 0$, then $b < k_{n+1}$ implies that $b < yk_{n+1}$, and so xk_n, and hence x, must be negative. Thus x and y have opposite signs.

By Theorems 4.12 and 4.13 it follows that $\xi k_n - h_n$ and $\xi k_{n+1} - h_{n+1}$ have opposite signs, and so $x(\xi k_n - h_n)$ and $y(\xi k_{n+1} - h_{n+1})$ have the same sign. We have

$$\xi b - a = x(\xi k_n - h_n) + y(\xi k_{n+1} - h_{n+1}),$$

and so

$$|\xi b - a| = |x(\xi k_n - h_n)| + |y(\xi k_{n+1} - h_{n+1})|$$
$$= |x||\xi k_n - h_n| + |y||\xi k_{n+1} - h_{n+1}|$$
$$> |x||\xi k_n - h_n| \geq |\xi k_n - h|,$$

which is a contradiction to (4.36).   ■

The next theorem shows that if the rational approximation is sufficiently good, then the rational number must be a convergent.

Theorem 4.25. Let ξ denote any irrational number. If there is a rational number a/b, with $b \geq 1$, such that

$$|\xi b - a| < 1/2b^2,$$

then a/b equals one of the convergents of the simple continued fraction expansion of ξ.

Proof. We need only consider the case $(a, b) = 1$. Suppose a/b is not a convergent. Then the inequalities $k_n \leq b < k_{n+1}$ determine an integer n. For this n we cannot have

$|\xi b - a| < |\xi k_n - h_n|$, since, by Theorem 4.22, this would imply $b \geq k_{n+1}$. Thus we must have

$$|\xi k_n - h_n| \leq |\xi b - a| < 1/2b$$

or

$$|\xi - h_n / k_n| < 1/2bk_n.$$

Since $a/b \neq h_n / k_n$ and $bh_n - ak_n$ is an integer we find that

$$1/bk_n \leq |bh_n - ak_n| / bk_n = |h_n / k_n - a/b|$$
$$\leq |\xi - h_n / k_n| + |\xi - a/b| < 1/2bk_n + 1/2b^2.$$

Thus $b < k_n$, which is a contradiction. ∎

Example 4.18. Show that the rational number 377/233 is a convergent of the simple continued fraction expansion of $(1 + \sqrt{5})/2$.

Solution: Recall that, by Example 4.4, $(1 + \sqrt{5})/2 = [1;\bar{1}]$. One could just calculate convergents, but there is no guarantee when, if ever, we'd run into 377/233.

However, with the aid of Theroem 4.25 it is relatively painless.

We have

$$\left| \frac{1 + \sqrt{5}}{2} - \frac{377}{233} \right| < 8.2378 \cdot 10^{-6} < \frac{1}{2 \cdot 233^2}.$$

Thus, by Theorem 4.25, 377/233 must be a convergent of $(1 + \sqrt{5})/2$.

We close this section with another proof of Hurwitz' theorem (Theorem 4.6), which uses the theory of continued fractions.

We begin with a simple fact. Note that if $x \geq 1$, then $x + x^{-1}$ is increasing with x. If we let $x = (1 + \sqrt{5})/2$, then $x + x^{-1} = \sqrt{5}$. Thus if $x + x^{-1} < \sqrt{5}$, then $x < (1 + \sqrt{5})/2$ and $x^{-1} > (1 + \sqrt{5})/2$.

To prove the first part of Hurwitz's theorem, that is, that given any real irrational ξ there exists infinitely many rationals h/k such that

(4.37) $|\xi - h/k| < 1/\sqrt{5}k^2,$

we shall show that at least one of any three consecutive convergents for ξ satisfy the inequality (4.37).

Let $q_n = k_n / k_{n-1}$. We shall show that if (4.37) is false for both $h/k = h_{j-1} / k_{j-1}$ and $h/k = h_j / k_j$, then we must have

(4.38)
$$q_j + q_j^{-1} < \sqrt{5}.$$

If (4.37) fails for these two values of h/k, then

(4.39)
$$\left| \xi - h_{j-1} / k_{j-1} \right| + \left| \xi - h_j / k_j \right| \geq 1 / \sqrt{5} k_{j-1}^2 + 1 / \sqrt{5} k_j^2.$$

By Theorem 4.12, we know ξ lies between h_{j-1} / k_{j-1} and h_j / k_j. Thus, by (a) of Theorem 4.12, we have

(4.40)
$$\left| \xi - h_{j-1} / k_{j-1} \right| + \left| \xi - h_j / k_j \right| = \left| h_{j-1} / k_{j-1} - h_j / k_j \right| = 1 / k_j k_{j-1}.$$

Combining (4.39) and (4.40) gives

$$k_j / k_{j-1} + k_{j-1} / k_j \leq \sqrt{5},$$

which proves (4.38) since the right hand side is irrational.

Now suppose (4.37) fails to hold for $h / k = h_i / k_i, i = n-1, n$ and $n+1$. By (4.38), with $j = n$ and $j = n + 1$, we have

$$q_n^{-1} > (\sqrt{5} - 1) / 2 \text{ and } q_{n+1} < (\sqrt{5} + 1) / 2.$$

By (4.23) we have $q_{n+1} = a_{n+1} + q_n^{-1}$, and so

$$(\sqrt{5} + 1) / 2 > q_{n+1} = a_{n+1} + q_n^{-1} > a_{n+1} + (\sqrt{5} - 1) / 2 > 1 + (\sqrt{5} - 1) / 2 = (\sqrt{5} + 1) / 2,$$

which is a contradiction. This proves the first part of Hurwitz' theorem.

The second part of Hurwitz' theorem states that the $\sqrt{5}$ cannot be replaced by any larger value if (4.37) is to hold for all real irrationals ξ. Again the counter example is $(\sqrt{5} + 1) / 2$, though this time expressed in terms of continued fractions.

Let $\xi = [1; \bar{1}]$, which, by Example 4.4, is equal to $(\sqrt{5} + 1) / 2$. By (4.29) it is easy to show, by induction, that $\xi_i = (\sqrt{5} + 1) / 2$, for all $i \geq 0$. A simple calculation shows that $h_0 = k_0 = k_1 = 1$ and $h_1 = k_2 = 2$. By (4.23), we have $h_i = h_{i-1} + h_{i-2}$ and $k_i = k_{i-1} + k_{i-2}$. Thus, for $n \geq 1$, we have $k_n = h_{n-1}$. Thus

$$\lim_{n \to +\infty} \frac{k_{n-1}}{k_n} = \lim_{n \to +\infty} \frac{k_{n-1}}{h_{n-1}} = \frac{1}{x} = \frac{\sqrt{5} - 1}{2}.$$

and

$$\lim_{n \to +\infty} \left(\xi_{n+1} + \frac{k_{n-1}}{k_n} \right) = \tfrac{1}{2}(\sqrt{5} + 1) + \tfrac{1}{2}(\sqrt{5} - 1) = \sqrt{5}.$$

Thus, if c is any constant greater than $\sqrt{5}$, then

$$\xi_{n+1} + k_{n-1} / k_n > c$$

can hold for at most a finite number of values of n. By (4.29), we see that

$$\left|\xi - h_n / k_n\right| = k_n^{-2}\left|\xi_{n+1} + k_{n-1} / k_n\right|^{-1} < 1 / c k_n^2$$

holds for at most a finite number of values of n. Thus there can only be a finite number of rationals h/k satisfying $\left|\xi - h/k\right| < 1/ck^2$, since, by Theorem 4.25, any such rational must be a convergent to ξ.

Problem Set 4.6

1. Let $\theta = [1; 2, 1, 3, 1, 4, 1, 5, ...]$. Find the best rational approximation h/k with
   (a) $k < 10$;
   (b) $k < 20$; and
   (c) $k < 500$.

2. Let $\alpha = [a_0; a_1,...]$ be an irrational number. Show that

   $$\frac{1}{(a_{n+1} + 2)k_n^2} < \left|\alpha - \frac{h_n}{k_n}\right| < \frac{1}{a_{n+1}k_n^2},$$

   where h_n/k_n are the convergents to α.

3. Show that if h_i/k_i are convergents to ξ, then

   $$k_n\left|k_{n-1}\xi - h_{n-1}\right| + k_{n-1}\left|k_n\xi - h_n\right| = 1.$$

4. (a) Find rationals a_1/b_1 and a_2/b_2 so that $\left|\sqrt{2} - a/b\right| < 1/\sqrt{5}b^2$.
   (b) Find rationals a_1/b_1 and a_2/b_2 so that $\left|\sqrt{3} - a/b\right| < 1/\sqrt{5}b^2$.

5. If h_n/k_n is the nth convergent of the infinite simple continued fraction expansion of the real number ξ, show that

   $$\left|\xi - h_n / k_n\right| > 1 / 2k_n k_{n+1}.$$

6. (a) Show that 49171/18089 is a convergent of the simple continued fraction expansion of e.
   (b) Show that 1146408/361913 is a convergent of the simple continued fraction expansion of π.

7. Find rational approximations of the following irrational numbers to the desired approximation.
   (a) $\sqrt{10}$ to five decimal places.
   (b) $(1 + \sqrt{2})/2$ to six decimal places.
   (c) e to three decimal places.

8. Let ξ be a positive irrational number whose simple continued fraction has convergents h_n/k_n. Show that at least one of the following inequalities must hold for all $n \geq 1$:

$$\left|\xi - h_n / k_n\right| < 1 / 2k_n^2 \text{ or } \left|\xi - h_{n+1} / k_{n+1}\right| < 1 / 2k_{n+1}^2.$$

(Hint: use Theorem 4.13 to show that

$$\left|\xi - h_n / k_n\right| + \left|\xi - h_{n+1} / k_{n+1}\right| = \left|h_n / k_n - h_{n+1} / k_{n+1}\right| = 1 / k_n k_{n+1}.)$$

4.7. SOME FACTORIZATION METHODS, II

In this section we wish to show how we can use continued fractions to factorize numbers. Like many of today's factorization methods this one starts with the ideas of Fermat's factorization method. (See Section 1.5.) Recall that in Fermat's method the idea was to write the number to be factorized, N, in the form

$$N = x^2 - y^2,$$

where $x - y \neq 1$. However, we don't really need to do this well. Suppose instead that we just had

$$x^2 \equiv y^2 \pmod{N},$$

with $0 < y < x < N$. Then $x^2 - y^2 = Nt$ and if we assume that $x + y < N$ then we see that $N \!\!\not|\,(x + y)$ (since $0 < x + y < N$) and $N \!\!\not|\,(x + y)$ (since $0 < x - y < N$.) Thus we see that $(N, x + y)$ and $(N, x - y)$ are both nontrivial divisors of N (that is, not equal to 1 or N.) Both of these greatest common divisors can be quickly found using the Euclidean Algorithm (Theorem 1.6.)

Example 4.19. Find a nontrivial factor of 141.

Solution. Note that

$$34^2 - 13^2 = 987 \equiv 0 \pmod{141}.$$

Since $0 < 13 < 34 < 141$ and $34 + 13 = 47 < 141$ we see that $(141, 47) = 47$ and $(141, 21) = 3$ give nontrivial divisors of 141. Indeed $141 = 3 \cdot 47$.

The problem is then to find such x and y. This is where the continued fractions come into the picture. We use the following result, which will also be of use in our work of Pell's equation. First, we recall some of the notation from the proof of Theorem 4.18, namely (4.30). We begin with

$$\xi_0 = \sqrt{N}, a_0 = [\sqrt{N}], m_0 = 0 \text{ and } q_0 = 1$$

and for $n \geq 0$ we define

(4.41)
$$\xi_n = (m_n + \sqrt{N}) / q_n, a_n = [\xi_n], m_{n+1} = a_n q_n - m_n$$
and
$$q_{n+1} = (N - m_{n+1}^2) / q_n.$$

Theorem 4.26. Let N be a positive nonsquare integer and let h_n/k_n denote the nth convergent to the simple continued fraction expansion of $\sqrt{N}$. We have

$$h_n^2 - Nk_n^2 = (-1)^{n-1} q_{n+1}.$$

Proof. Now $\sqrt{N} = [a_0; a_1, a_2, \ldots]$ implies that $\sqrt{N} = [a_0; a_1, \ldots, a_n, \xi_{n+1}]$, and so, by Theorem 4.10,

$$\sqrt{N} = \frac{\xi_{n+1} h_n + h_{n-1}}{\xi_{n+1} k_n + k_{n-1}}.$$

If we put into this relation the value of ξ_{n+1}, which, by (4.41), is $(m_{n+1} + \sqrt{N}) / q_{n+1}$, we obtain

$$\sqrt{N} = \frac{(m_{n+1} + \sqrt{N})h_n + q_{n+1}h_{n-1}}{(m_{n+1} + \sqrt{N})k_n + q_{n+1}k_{n-1}}.$$

If we multiply this out and collect terms we have

(4.42) $\quad (m_{n+1}k_n + q_{n+1}k_{n-1} - h_n)\sqrt{N} + (m_{n+1}h_n + q_{n+1}h_{n-1} - Nk_n) = 0.$

Since N is not a square we know, by unique factorization (Theorem 1.19), that $\sqrt{N}$ is irrational. Since the right hand side of (4.42) has a zero coeeficient of $\sqrt{N}$ so must the left hand side. Thus

$$m_{n+1}k_n + q_{n+1}k_{n-1} = h_n$$

and

$$m_{n+1}h_n + q_{n+1}h_{n-1} = Nk_n,$$

and so, by Theorem 4.12,

$$h_n^2 - Nk_n^2 = h_n(m_{n+1}k_n + q_{n+1}k_{n-1}) - k_n(m_{n+1}h_n + q_{n+1}h_{n-1})$$
$$= (h_n k_{n-1} - k_n h_{n-1})q_{n+1} = (-1)^{n-1}q_{n+1}.$$

This proves the result. ∎

What this theorem shows is that for every $n \geq 0$ we have

$$h_n^2 \equiv (-1)^{n-1} q_{n+1} \pmod{N}.$$

If we can find an odd value of n so that q_{n+1} is a square, say $q_{n+1} = m^2$, then we have that

$$h_n^2 \equiv m^2 \pmod{N},$$

which we may be able to use to factorize N as in Example 4.19. Thus the procedure is to hunt through the values of q_n, with even index, and see if we can't find a square. Of course, there is no guarantee that this process will give us the result we want right off. We illustrate with two examples.

Example 4.20. Use the continued fraction algorithm to factorize
  (a) 1007
  (b) 9991.

Solution of (a). We produce the following table of values in accordance with (4.41) and (4.23).

k	1	1	2	3	4	5	6
a_k	31	1	2	1	2	1	62
m_k	0	31	15	19	19	15	31
q_k	1	46	17	38	17	46	1
h_k	31	32	95	127	349	476	29881

Now $q_6 = 1$ is a square. Thus, by Theorem 4.26, we have

$$4762^2 \equiv 1 = 1^2 \pmod{1007}.$$

Therefore, by the remarks before Example 4.19, both $(1007, 477)$ and $(1007, 475)$ give nontrivial divisors of 1007. If we use the Euclidean algorithm to calculate these greatest common divisors, we find that

$$(1007, 477) = 53 \text{ and } (1007, 475) = 19.$$

Indeed, $1007 = 19 \cdot 53$.

Solution of (b). Proceeding as in (a) we form the following table.

k	0	1	2	3	4	5	6
a_k	99	1	21	4	1	1	1
m_k	0	99	91	98	74	31	55
q_k	1	190	9	43	105	86	81
h_k	99	100	2199	8896	11095	19991	31086

Now $q_2 = 9$ is a square. Thus

$$100^2 \equiv 9 = 3^2 \ (\text{mod } 9991).$$

If we use the Euclidean algorithm we find that

$$(103, 9991) = 103 \text{ and } (97, 9991) = 97.$$

Indeed, $9991 = 97 \cdot 103$. $\qquad\qquad\qquad\qquad\qquad\qquad\qquad\qquad\qquad\qquad$ $\square$

To illustrate what might go wrong let us move a little farther along in our table. Note that $q_6 = 81$ is a square. Thus

$$19991^2 \equiv 81 = 9^2 \ (\text{mod } 9991).$$

Thus $(20000, 9991)$ and $(19982, 9991)$ are factors of 9991. If we use the Euclidean algorithm we find that

$$(20000, 9991) = 1 \text{ and } (19982, 9991) = 9991.$$

Of course, th reason for this is that $19991 \equiv 9 \ (\text{mod } 9991)$, and so $19982 \equiv 0 \ (\text{mod } 9991)$. Thus, in a program to implement the continued fraction factorization method, we should check, after finding a square value for q_k, say m^2, that $h_{k-1} \not\equiv m \ (\text{mod } N)$.

In closing we might mention that even nonsquare values of q_k are useful. Suppose $q_k = q_l = m$, where k and l have the same parity. Then we have

$$(h_k h_l)^2 \equiv (-1)^{k+l} q_k^2 = m^2 \ (\text{mod } N).$$

Thus the method is still applicable. Recall that if r, the period of the continued fraction, is even, then $q_{r-k} = q_k$, and so we can apply this idea with the middle pair of q's and then the next symmetric pair and so on.

Problem 4.7

1. Show that if N is an odd composite integer other than a prime power, then the equation $x^2 \equiv y^2 \ (\text{mod } N)$ has a solution $x = a$, $y = b$ such that $a \not\equiv \pm b \ (\text{mod } N)$. (Hint: use the Chinese Remainder Theorem (Theorem 2.11) and Problem 15 of Problem Set 3.3.)

2. Use the following congruences to find factors of the moduli.
   (a) $26^2 \equiv 9^2 \ (\text{mod } 119)$ $\qquad\qquad$ (d) $123^2 \equiv 64^2 \ (\text{mod } 1003)$
   (b) $34^2 \equiv 5^2 \ (\text{mod } 377)$ $\qquad\qquad$ (e) $354^2 \equiv 335^2 \ (\text{mod } 1007)$
   (c) $85^2 \equiv 48^2 \ (\text{mod } 703)$ $\qquad\qquad$ (f) $98^2 \equiv 57^2 \ (\text{mod } 1271)$

3. Use the continued fraction algorithm to factor the following numbers:

(a) 1357   (c) 7009

(b) 2911   (d) 8023

4. Use the continued fraction algorithm to factor 65021.

5. The following result illustrates another way to solve the congruence of Problem 1. Let N be a positive integer and let $p_1,\ldots,p_r$ be primes. Suppose that there exist integers $n_1,\ldots,n_m$ such that, for $1 \le j \le m$,

$$n_j^2 \equiv (-1)^{e_{0j}} p^{e_{1j}} \cdots p^{e_{rj}} \pmod{N},$$

where, for $0 \le j \le r$,

$$e_{j1} + \cdots + e_{jm} = 2e_j.$$

Show that if $a = n_1 \cdots n_m$ and $b = (-1)^{e_0} p_1^{e_1} \cdots p_r^{e_r}$, then

$$a^2 \equiv b^2 \pmod{N}.$$

The set of primes $p_1,\ldots,p_r$ and -1 are called the **factor base**.

6. Let N be a positive integer and let $p_1,\ldots,p_m$ be primes. If the integers q_k are defined in (4.41) suppose for some set of indices $k_1,\ldots,k_t$ we have

$$q_{k_j} = p_1^{k_{1j}} \cdots p_m^{k_{mj}}, \quad 1 \le j \le t.$$

Show that n can be factored if $k_1 + \cdots + k_t$ is even and, for $1 \le j \le m$, $k_{1j} + \cdots + k_{tj}$ is even.

7. Use the factor base $\{-1, 2, 3\}$ as well as the continued fraction expansion of $\sqrt{9073}$ to factor 9073.

8. Use the remark at the end of the section to factor the numbers in Problems 3 and 4 above.

4.8. EQUIVALENT NUMBERS

Definition 4.4. If ξ and η are any two real numbers, we say ξ that is **equivalent** to η if there exists integers a, b, c, and d with $ad - bc = \pm 1$ and if

(4.43) $$\xi = \frac{a\eta + b}{c\eta + d}.$$

Note that by taking $a = d = 1$ and $b = c = 0$ we see that every number is equivalent to itself. Also, if. (4.43) holds, then

$$\eta = \frac{-d\xi + b}{c\xi - a}$$

and $(-d)(-a) - bc = ad - bc = \pm 1$. Thus η is equivalent to ξ.

Theorem 4.27. The notion of equivalence in Definition 4.4 is an equivalence relation.

Proof. After the remarks above all that is left to show is that equivalence is transitive. Suppose ξ is equivalent to η and η is equivalent to ζ Then there exist integers a, b, c, d and a_1, b_1, c_1, d_1 such that $ad - bc = +1$, $a_1d_1 - b_1c_1 = \pm 1$,

$$\xi = \frac{a\eta + b}{c\eta + d}, \text{ and } \eta = \frac{a_1\zeta + b_1}{c_1\zeta + d_1}.$$

If we let $A = aa_1 + bc_1, B = ab_1 + bd_1, C = ca_1 + dc_1$ and $D = cb_1 + dd_1$, then we have

$$AD - BC = (ad - bc)(a_1d_1 - b_1c_1) = \pm 1$$

and

$$\xi = \frac{A\zeta + B}{C\zeta + D}.$$

Thus ξ is equivalent to ζ and the theorem is proved. ∎

Corollary 4.27.1. Any two rational numbers are equivalent.

Proof. Let h/k be a reduced rational number, that is, $(h, k) = 1$. Then, by Theorem 1.5 there exist integers h_1 and k_1 such that

$$hk_1 - kh_1 = 1.$$

Thus

$$\frac{h}{k} = \frac{h_1 \cdot 0 + h}{k_1 \cdot 0 + k} = \frac{a0 + b}{c0 + d}.$$

where $ad - bc = -1$. Thus h/k is equivalent to 0, and so, by Theorem 4.27, to all other rationals. ∎

For irrationals there is an easy criterion for equivalence in terms of their continued fraction expansion.

Lemma 4.28.1. If

$$x = \frac{P\zeta + R}{Q\zeta + S},$$

where $\zeta > 1$ and P, Q, R and S are integers such that $Q > S > 0$ and $PS - QR = \pm 1$, then R/S and P/Q are two consecutive convergents to the simple continued fraction whose value is x. If R/S is the $(n - 1)$st convergent and P/Q is the nth, then ζ is the $(n + 1)$st complete quotient.

Proof. We write P/Q as a continued fraction, say

(4.44)
$$P / Q = [a_0; a_1, \ldots, a_n] = h_n / k_n.$$

By Theorem 4.9 we may take an expansion with n even or odd. Choose n so that

(4.45)
$$PS - QR = \pm 1 = (-1)^{n-1}.$$

Since $(P, Q) = 1$ and $Q > 1$ we see that h_n and k_n satisfy the same conditions, and so (4.44) and (4.45) imply that $P = h_n$ and $Q = k_n$. Thus

$$h_n S - k_n R = PS - QR = (-1)^{n-1} = h_n k_{n-1} - h_{n-1} k_n$$

or

(4.46)
$$h_n(S - k_{n-1}) = k_n(R - h_{n-1}).$$

Since $(h_n, k_n) = 1$, (4.46) implies that

(4.47)
$$k_n | (S - k_{n-1}).$$

Since $k_n = Q > S > 0$ and $k_n \geq k_{n-1} > 0$ we have $|S - k_{n-1}| < k_n$. which is inconsistent with (4.47) unless $S - k_{n-1} = 0$. Thus

$$S = k_{n-1} \text{ and } R = h_{n-1},$$

that is,

$$x = \frac{h_n \zeta + h_{n-1}}{k_n \zeta + k_{n-1}},$$

which, by Theorem 4.10, implies that

$$x = [a_0; a_1, \ldots, a_n, \zeta].$$

Let the continued fraction expansion of ζ be $[a_{n+1}; a_{n+2}, \ldots]$, where $a_{n+1} = [\zeta] > 1$. Thus

$$x = [a_0; a_1, \ldots, a_n, a_{n+1}, a_{n+2}, \ldots],$$

which is a simple continued fraction with consecutive convergents $h_{n-1}/k_{n-1} = R/S$ and $h_n/k_n = P/Q$ and ζ as the $(n + 1)$st complete quotient. ∎

Theorem 4.28. Two irrational number ξ and η are equivalent if and only if

$$\xi = [a_0; a_1, \ldots, a_m, c_0, c_1, c_2, \ldots]$$

(4.48)   and

$$\eta = [b_0; b_1, \ldots, b_n, c_0, c_1, c_2, \ldots],$$

that is, the sequence of partial quotients in ξ after the mth is the same as the sequence in η after the nth.

Proof. Suppose ξ and η are given by (4.48) and let $\omega = [c_0; c_1, c_2, \ldots]$. Then, by Theorem 4.10,

$$\xi = [a_0; a_1, \ldots, a_m, \omega] = \frac{h_m \omega + h_{m-1}}{k_m \omega + k_{m-1}},$$

where, by (a) of Theorem 4.12, $h_m k_{m-1} - k_m h_{m-1} = \pm 1$. Thus ξ and ω are equivalent and, since a similar argument shows that η and ω are equivalent, we have, by Theorem 4.27, that ξ and η are equivalent.

Suppose now that η and ξ are equivalent so that there exist integers a, b, c, and d such that

$$\eta = \frac{a\xi + b}{c\xi + d} \text{ and } ad - bc = \pm 1,$$

where we may suppose $c\xi + d > 0$ (otherwise replace a by $-a$, b by $-b$, c by $-c$, and d by $-d$). If we develop ξ into a continued fraction, suppose

$$\xi = [a_0; a_1, \ldots, a_{r-1}, a_r, \ldots] = [a_0; a_1, \ldots, a_{r-1}, \xi_r] = \frac{h_{r-1} \xi_r + h_{r-2}}{k_{r-1} \xi_r + k_{r-2}}.$$

Thus, by Theorem 4.27,

$$\eta = \frac{P\xi_r + R}{Q\xi_r + S},$$

where

$$P = ah_{r-1} + bk_{r-1}, R = ah_{r-2} + bk_{r-2}, Q = ch_{r-1} + dk_{r-1} \text{ and } S = ch_{r-2} + dk_{r-2}.$$

Then P, Q, R and S are integers and, by (a) of Theorem 4.12, we have

$$PS - QR = (ad - bc)(h_{r-1} k_{r-2} - k_{r-1} h_{r-2}) = \pm 1.$$

By Corollary 4.22.1, we have

$$h_{r-1} = \xi k_{r-1} + \delta / k_{r-1} \text{ and } h_{r-2} = \xi k_{r-2} + \delta_1 / k_{r-2},$$

where $|d|$, $|d_1| < 1$. Thus

$$Q = (c\xi + d)k_{r-1} + c\delta / k_{r-1} \text{ and } S = (c\xi + d)k_{r-2} + c\delta_1 / k_{r-2}.$$

Since $c\xi + d > 0, k_{r-1} > k_{r-2} > 0$ and k_{r-1} and k_{r-2} tend to infinity, we have $Q > S > 0$ for r large enough. For such r we have

$$\eta = \frac{P\xi_r + R}{Q\xi_r + S},$$

where $PS - QR = \pm 1$, $Q > S > 0$, and $\xi_r > 1$, and so, by Lemma 4.28.1, we have

$$\eta = [b_0; b_1, \ldots, b_m, \xi_r] = [b_0; b_1, \ldots, b_m, a_r, a_{r+1}, \ldots],$$

for some $b_0, b_1, \ldots, b_m$. This finishes the proof of the theorem. ∎

With this result in hand we can now shed a little more light on Hurwitz' theorem. We define the Markov constant for a number ξ to be the least upper bound of the numbers λ such that

$$|\xi - h/k| < 1/\lambda k^2$$

has infinitely many solutions in integers h and k and denote it by $M(\xi)$. Hurwitz' theorem says that $M(\xi) \geq \sqrt{5}$ for all irrational numbers ξ.

Theorem 4.29. If ξ and η are equivalent, then $M(\xi) = M(\eta)$. If ξ is equivalent to $(\sqrt{5} + 1)/2$, then $M(\xi) = \sqrt{5}$. If ξ is irrational and not equivalent to $(\sqrt{5} + 1)/2$, then $M(\xi) \geq \sqrt{8}$. If ξ is equivalent to $\sqrt{2}$, then $M(\xi) = \sqrt{8}$. If ξ is irrational and not equivalent to either $(\sqrt{5}+1)/2$ or $\sqrt{2}$, then $M(\xi) \geq 17/6$.

Proof. Let $\xi = [a_0; a_1, \ldots]$ be an irrational number. By (4.29), we have

$$M(\xi) = \limsup_{n \to +\infty} \left(\xi_{n-1} + \frac{k_{n-1}}{k_n} \right).$$

Now

$$\xi_{n+1} = [a_{n+1}; a_{n+2}, \ldots]$$

and

$$\frac{k_{n-1}}{k_n} = \frac{k_{n-1}}{k_{n-1}a_n + k_{n-2}} = \frac{1}{a_n + k_{n-2}/k_{n-1}}$$

$$= \frac{1}{a_n + \dfrac{1}{a_{n-1} + k_{n-3}/k_{n-2}}}$$

$$= \ldots = [a_n; a_{n-1}, \ldots, a_2 + k_0/k_1] = [a_n; a_{n-1}, \ldots, a_0].$$

Thus

$$M(\xi) = \limsup_{n \to +\infty} \{ [a_n; a_{n-1}, \ldots, a_1] + [a_{n+1}; a_{n+2}, \ldots] \}.$$

If $\eta = [a_0'; a_1', \ldots]$ is equivalent to ξ, then, by Theorem 4.28, we have $\xi_j = \eta_m$ and $a_j = a_m'$ for all sufficiently large j and m for which $j - m$ has a suitable fixed value t. If the convergents of η are h_n'/k_n', then for such j and m the continued fraction expansions of k_{j-1}/k_j and k'_{m-1}/k'_m have the same partial quotients at the beginning and the interval of agreement can be made arbitrarily long by choosing j and m sufficiently large. Suppose they agree on the first $r+1$ partial quotients and that $u_s/v_s, s = 0,1,\ldots,j$ are the common convergents and that

$$\frac{k_{j-1}}{k_j} = \frac{u_{j-1}\alpha_j + u_{j-2}}{v_{j-1}\alpha_j + v_{j-2}} \text{ and } \frac{k'_{m-1}}{k'_m} = \frac{u_{j-1}\alpha'_j + u_{j-2}}{v_{j-1}\alpha'_j + v_{j-2}}$$

Since $[\alpha_j] = [\alpha'_j] \geq 1$ we have

$$\left| \frac{k_{j-1}}{k_j} - \frac{k'_{j-1}}{k'_j} \right| = \frac{|\alpha_j - \alpha'_j|}{(v_{j-1}\alpha_j + v_{j-2})(v_{j-1}\alpha'_j + v_{j-2})} \leq \frac{1}{v_{j-1}^2},$$

and so

$$\lim_{\substack{j\to+\infty \\ j-m=l}} \{(\xi_j + k_{j-1}/k_j) - (\eta_m + k'_{m-1}/k'_m)\} = \lim_{\substack{j\to+\infty \\ j-m=l}} (k_{j-1}/k_j - k'_{m-1}/k'_m) = 0.$$

Thus $M(\xi) = M(\eta)$.

To prove the second part we need only note that

$$M((\sqrt{5}+1)/2) = \lim_{k\to+\infty} \{[1;\bar{1}] + [1;1,\ldots,1]_k\} = (1+\sqrt{5})/2 + ((1+\sqrt{5})/2)^{-1} = \sqrt{5},$$

where $[1;1,\ldots,1]_k$ denotes a continued fraction with k 1's in it.

To prove the third part we may suppose $a_n \geq 2$ for infinitely many n. If $a_n \geq 3$ for infinitely many n, then it is clear that $M(\xi) \geq 3$. Since $\sqrt{8} < 3$ we need only consider those ξ's for which a_n is either 1 or 2 for all large n. If there are infinitely many 1's and 2's, then there are infinitely many n such that $a_n = 1$ and $a_{n+1} = 2$. Since the value of a continued fraction is always at least equal to its convergent with index 2, we have

$$[a_{n+1}; a_{n+2}, \ldots] \geq 2 + 1/(a_{n+2} + 1/a_{n+3}) \geq 2 + 1/(2 + 1/1) = 7/3$$

and

$$[a_n; a_{n-1}, \ldots, a_1] \geq 1/(1 + 1/a_{n-1}) \geq 1/(1 + 1/1) = 1/2.$$

Thus

$$M(\xi) \geq 7/3 + 1/2 = 17/6 > \sqrt{8}.$$

If $a_k = 2$ for all large k, then is equivalent to $\sqrt{2} = [1;\bar{2}]$ and

$$M(\xi) = \lim_{k\to+\infty} \{[\bar{2}] + [2;2,\ldots,2]_k\} = (\sqrt{2}+1) + (\sqrt{2}-1) = \sqrt{8},$$

■

which proves the theorem.

There is more that can be said on the subject of $M(\xi)$. For such a discussion see [14, ch. II].

Problem Set 4.8

1. Prove that if a, b, c and d are integers with $ad - bc = \pm 1$ and $c > d > 0$ and if ξ and $\eta, \eta > 1$, are such that

$$\xi = \frac{a\eta + b}{c\eta + d},$$

then b/d and a/c are successive convergents of the simple continued fraction expansions of ξ. Also η is the corresponding complete quotient. Thus, for an appropiate value of n,

$$a = h_{n-1}, b = h_{n-2}, c = k_{n-1}, d = k_{n-2} \text{ and } \eta = \xi_n.$$

2. (a) Are the numbers $\sqrt{5}$ and $(1+\sqrt{5})/2$ equivalent?
   (b) Are the numbers $\sqrt{3}$ and $(1+\sqrt{3})/2$ equivalent?

3. Prove that if an irrational number ξ is not equivalent to $(1+\sqrt{5})/2$ or to $\sqrt{2}$, then $M(\xi) \geq 5/\sqrt{221}$.

4. Show that if α is irrational either there are infinitely many irreducible fractions a/b satisfying

$$|\alpha - a/b| < \frac{1}{b^2\sqrt{m^2 + 4}}$$

or there is an n_0 such that $a_n < m$ for all $n > n_0$, where the a_n are the partial quotients of α.

5. Show that if

$$|\alpha - h/k| \geq \frac{1}{k_n^2\sqrt{m^2 + 4}}$$

for each $n = t - 1$, $n = t$ and $n = t + 1$, then $a_{t+1} < m$.

4.9. DECIMAL REPRESENTATION

Let $b > 1$ be an integer. In this section we shall discuss some results regarding the b-ary expansion of real numbers. We are interested in representing a real number x in the form

$$x = N + \sum_{k=1}^{+\infty} \frac{a_k}{b^k},$$

where N is an integer and $0 \leq a_k \leq b - 1$ and a_k is an integer, for $k \geq 1$. The decimal expansion would be the case $b = 10$. Since the integer N does not influence the properties of the fractional part we shall assume $0 \leq x < 1$ so that $N = 0$.

In Theorem 1.4 we showed that every integer has a unique base b expansion. We now extend this result to real numbers in the interval $[0, 1)$.

Theorem 4.30. Let x be a real number in the interval $[0, 1)$ and let $b > 1$ be an integer. Then x can be uniquely written in the form

(4.49) $$x = \sum_{k=1}^{+\infty} \frac{a_k}{b^k},$$

where the a_k are integers such that $0 \leq a_k \leq b - 1$, for $k \geq 1$, and we require that for every positive integer m there is a $k \geq m$ such that $a_k \neq b - 1$.

Proof. The proof of existence proceeds musch as in the proof of Theorem 1.4 except that we multiply by b instead of divide by b. Let

$$a_1 = [bx].$$

Since $0 \leq x < 1$ we see that $0 \leq bx < b$, and so $0 \leq a_1 \leq b - 1$. If we let $x_1 = bx - a_1 = bx - [bx]$, then $0 \leq x_1 < 1$. Also we have

$$x = \frac{a_1}{b} + \frac{x_1}{b}.$$

For $k \geq 2$, we define

$$a_k = [bx_{k-1}] \text{ and } x_k = bx_{k-1} - a_k.$$

Then $0 \leq x_{k-1} < 1$ implies $0 \leq bx_{k-1} < b$, and so $0 \leq a_k \leq b - 1$ and $0 \leq x_k < 1$. Thus, if n is any positive integer, we have

(4.50) $$x = \frac{a_1}{b} + \frac{a_2}{b^2} + \cdots + \frac{a_n}{b^n} + \frac{x_n}{b^n}.$$

Since $0 \leq x_n < 1$ we have $0 < x_n/b^n < 1/b^n$. Thus

$$\lim_{n \to +\infty} \left\{ x - \left(\frac{a_1}{b} + \cdots + \frac{a_n}{b^n} \right) \right\} = \lim_{n \to +\infty} \frac{x_n}{b^n} = 0$$

and therefore

$$x = \sum_{k=1}^{+\infty} \frac{a_k}{b^k}.$$

Now we show uniqueness. Recall the formula for the sum of an infinite geometric progression: if a and r are real numbers with $|r| < 1$, then

$$\sum_{k=0}^{+\infty} ar^k = \frac{a}{1-r}.$$

(See Problem 1 below.)

Suppose we have two representations for x, say

(4.51)
$$\sum_{j=1}^{+\infty} \frac{a_j}{b^j} = x = \sum_{j=1}^{+\infty} \frac{c_j}{b^j},$$

where, for all $j \geq 1$, $0 \leq a_j, c_j \leq b - 1$ and for every positive integer m there are integers r and s such that $r, s \geq m$ with $a_r \neq b - 1$ and $c_s \neq b - 1$. Let k be the least subscript so that $a_k \neq c_k$. We may assume that $a_k < c_k$ since the case $a_k > c_k$ is handled by simply switching the roles of the two expressions in (4.51). By (4.51), we have

$$0 = \sum_{j=1}^{+\infty} (c_j - a_j)b^{-j} = \frac{c_k - a_k}{b^k} + \sum_{j=k+1}^{+\infty} (c_j - a_j)b^{-j}$$

or

(4.52)
$$(c_k - a_k)b^{-k} = \sum_{j=k+1}^{+\infty} (a_j - c_j)b^{-j}.$$

Since we assumed that $c_k > a_k$ and the coefficients are integers we have that $c_k - a_k \geq 1$, and so $(c_k - a_k)b^{-k} \geq b^{-k}$. Also

(4.53)
$$\sum_{j=k+1}^{+\infty} (a_j - c_j)b^{-j} \leq \sum_{j=k+1}^{+\infty} \frac{b-1}{b^j}$$

$$= \frac{b-1}{b^k} \sum_{j=0}^{+\infty} \frac{1}{b^j} = \frac{b-1}{b^k} \cdot \frac{1}{1-1/b} = b^{-k}.$$

Note that we have equality in (4.53) if and only if $a_j - c_j = b - 1$, for all $j \geq k + 1$. Since $a_j \leq b - 1$ and $c_j \geq 0$ we see that $a_j - c_j = b - 1$ if and only if $a_j = b - 1$ and $c_j = 0$. By hypothesis there is a $j \geq k + 1$ so that $a_j < b - 1$. Thus $a_j - c_j \neq b - 1$ for all $j \geq k + 1$, and so we have strict inequality in (4.53). Thus

$$\sum_{j=k+1}^{+\infty} \frac{a_j - c_j}{b^j} < \frac{1}{b^k} \leq \frac{c_k - a_k}{b^k}.$$

This contradicts (4.52). Thus the expression of x must be unique.   ∎

We denote this unique base b expansion of the number x, (4.49), by

$$(.a_1 a_2 \cdots)_b.$$

Example 4.21.

   (a) Find the base 3 expansion of 1/5.

   (b) Find the base 6 expansion of 5/18.

Solution of (a). We have the following table with $x = 1/5$:

$$a_1 = [3/5] = 0, \; x_1 = 3/5 - 0 = 3/5$$
$$a_2 = [9/5] = 1, \; x_2 = 9/5 - 1 = 4/5$$
$$a_3 = [12/5] = 2, \; x_3 = 12/5 - 2 = 2/5$$
$$a_4 = [6/5] = 1, \; x_4 = 6/5 - 1 = 1/5.$$

Since $x_4 = x$ we see that the values of a_k will repeat. Thus

$$1/5 = (.01210121\ldots)_3.$$

Solution of (b). We have the following table with $x, = 5/18$:

$$a_1 = [30/18] = 1, \; x_1 = 30/18 - 1 = 12/18$$
$$a_2 = [72/18] = 4, \; x_2 = 72/18 - 4 = 0.$$

Thus all further values of a_k will be equal to 0. Thus

$$5/18 = (.14)_6.$$ □

This example motivates the following definitions.

Definition 4.5. Suppose that the real number x, $0 \leq x < 1$, has the base b expansion $(.a_1 a_2 \cdots)_b$.

   (a) We say that the expansion is **terminating** if there is a positive integer N such that if $k \geq N$, then $a_k = 0$.

   (b) We say that the expansion is **periodic** if there are positive integers p and N such that if $n \geq N$, then

(4.54) $a_{n+p} = a_n.$

The least positive integer p such that (4.54) holds for some N is called the **period**

length of the base b expansion of x.

Note that a periodic expansion can always be written in the form

$$(.a_1 \cdots a_{N-1} c_1 \cdots c_p c_1 \cdots c_p \cdots)_b,$$

which we abbreviate as

$$(.a_1 \cdots a_{N-1} \overline{c_1 \cdots c_p})_b.$$

The part of the expansion at the beginning, $a_1 \cdots a_{N-1}$, is called the **pre-period** and the periodic part, $c_1 \cdots c_p$, is called the **period**.

The next result allows us to classify those reals with terminating base b expansions.

Theorem 4.31. A real number x, $0 \le x < 1$, has a terminating base b expansion if and only if x is a rational number which can be written as $x = h/k$, $0 \le h < k$, $(h, k) = 1$, where if p is a prime and $p \mid k$, then $p \mid b$.

Proof. If we suppose that x has a terminating expansion, say

$$x = (.a_1 \cdots a_m)_b,$$

then

(4.55) $$x = \frac{a_1}{b} + \cdots + \frac{a_m}{b^m} = \frac{a_1 b^{m-1} + \cdots + a_m}{b^m},$$

and so we see that x is a rational, say h/k, with $(h, k) = 1$. Since $0 \le x < 1$ we see that $0 \le h < k$. By (4.55), we have

$$hb^m = k(a_1 b^{m-1} + \ldots + a_m).$$

If p is a prime and $p \mid k$, then $p \mid hb^m$. Since $(h, k) = 1$ we see that $p \mid b^m$, and so $p \mid b$.

To prove the converse suppose that $x = h/k$, $0 \le h < k$, $(h, k) = 1$ and if $p \mid k$, then $p \mid b$, where p is a prime. Then, by unique factorization (Theorem 1.19), we see that there is an integer m so that $k \mid b^m$. (For example, we could take m to be the largest exponent in the canonical factorization of k.) If we define j by $b^m = kj$, then j is a positive integer and we have

$$b^m x = b^m h/k = hj.$$

Since hj is a positive integer we can write, by Theorem 1.4,

$$hj = a_r b^r + \ldots + a_0,$$

where r is a nonnegative integer and $0 \le a_s \le b - 1$, $0 \le s \le r$. Thus

$$x = \frac{hj}{b^m} = \frac{a_r b^r + \cdots + a_0}{b^m} = a_r b^{r-m} + \cdots + a_0 b^{-m}.$$

Since $hj < b^m$ we see that $r < m$, and so we have

$$x = (.0 \cdots 0 a_r a_{r-1} \cdots a_1 a_0)_b,$$

which is a terminating base b expansion.   ■

We specialize this result to two well-known bases.

Corollary 4.31.1.

  (a) A rational number has a terminating decimal (base 10) expansion if and only if its denominator is divisible only by powers of 2 or 5.

  (b) A rational number has a terminating sexigesmal (base 60) expansion if and only if its denominator is divisible only by powers of 2, 3 or 5.

The second result was apparently known to the Babylonians who used a base of 60. At least it appears so from the tables of reciprocals that they produced (for example, 1/7 is often omitted) as well as problems in their problem texts.

We can, of course, turn a terminating expansion into a nonterminating one since

$$(b-1)b^{-(k+1)} + (b-1)b^{-(k+2)} + \cdots = b^{-k},$$

as we saw in the proof of Theorem 4.30. This gives us

$$(.a_1 \cdots a_m)_b = (.a_1 \cdots (a_m - 1)(b-1)(b-1) \cdots)_b.$$

For example $(.14)_6 = (.13555\cdots)_6 = (.13\bar{5})_6$. In this case we see that the expansion has become periodic. The next result classifies all real numbers with periodic base b expansions.

Theorem 4.32. Let b be a positive integer, $b > 1$.

  (a) A periodic base b expansion represents a rational number.

  (b) If $x = h/k$ is a rational number with $0 < h < k$ and $(h, k) = 1$, write $k = NM$, where $p \mid N$ implies $p \mid b$ and $(b, M) = 1$. Let μ be the smallest positive integer such that $N \mid b^\mu$ and let $m = \text{ord}_M b$. Then the period length of the base b expansion of x is m and the length of the pre-period is μ.

Proof of (a). Suppose x has the periodic expansion

$$x = (.c_1 \cdots c_k \overline{a_1 \cdots a_r}).$$

Then we can write

$$x = \frac{c_1}{b} + \cdots + \frac{c_k}{b^k} + \frac{a_1}{b^{k+1}} + \cdots + \frac{a_r}{b^{k+r}} + \frac{a_1}{b^{k+r+1}} + \cdots + \frac{a_r}{b^{k+2r}} + \cdots$$

$$= \frac{c_1}{b} + \cdots + \frac{c_k}{b^k} + \left(\frac{a_1}{b^{k+1}} + \cdots + \frac{a_r}{b^{k+r}} \right)\left(1 + \frac{1}{b^r} + \frac{1}{b^{2r}} + \cdots \right)$$

$$= \frac{c_1}{b} + \cdots + \frac{c_k}{b^k} + \left(\frac{a_1}{b^{k+1}} + \cdots + \frac{a_r}{b^{k+1}} \right)\frac{b^r}{b^r - 1}.$$

Thus x is a rational number since it is a sum of rational numbers.

Proof of (b). Let $x = h/k$ with $0 < h < k$ and $(h, k) = 1$. Write $k = NM$, where $p \mid N$ implies $p \mid b$ and $(M, b) = 1$. Let $m = \mathrm{ord}_M b$ and let μ be the smallest positive integer such that $N \mid b^\mu$.

Since $N \mid b^\mu$ there is a positive integer n so that $Nn = b^\mu$. Thus

$$(4.56) \qquad\qquad b^\mu x = b^\mu \frac{h}{NM} = \frac{hn}{M}.$$

By Theorem 1.3, we can write

$$(4.57) \qquad\qquad \frac{hn}{M} = c + \frac{d}{M},$$

where c and d are positive integers and $0 < d < M$. Note that since $0 < h/k < 1$ we have that $0 < b^\mu x = hn/M < b^\mu$, and so $0 < c < b^\mu$. Since $(h, k) = 1$ we see that we must also have $(d, M) = 1$.

If $M = 1$, then the base b expansion of x terminates, by Theorem 4.31, since $p \mid k$ implies $p \mid b$.

Suppose $M > 1$. Since $m = \mathrm{ord}_M b$, we have $b^m \equiv 1 \pmod{M}$, and so

$$(4.58) \qquad\qquad b^m \frac{d}{M} = \frac{(rM + 1)d}{M} = rd + \frac{d}{M},$$

where r is an integer. Let $y_0 = d/M$ and suppose it has the base b expansion $(.a_1a_2a_3\ldots)_b$, where, for $k \geq 1$,

$$a_k = [by_{k-1}] \text{ and } y_k = by_{k-1} - [by_{k-1}].$$

Then we can write, by (4.50),

$$(4.59) \qquad b^m \frac{d}{M} = b^m \left(\frac{a_1}{b} + \cdots + \frac{a_m}{b^m} + \frac{y_m}{b^m} \right) = a_1 b^{m-1} + \ldots + a_m + y_m.$$

If we equate the fractional parts in (4.58) and (4.59), we have, since, $0 \leq y_k < 1$,

$$y_m = d/M = y_0.$$

Thus, by the recursion relation for the a_k's, we see that $a_{k+m} = a_k$, for $k \geq 1$. Thus d/M has a periodic expansion of the form

$$\frac{d}{M} = (.\overline{a_1 \cdots a_m})_b.$$

By Theorem 1.4, we can write $c = c_r b^r + \ldots + c_1 b + c_0$, where r is a nonnegative integer and $0 \leq c_j \leq b - 1$, $0 \leq j \leq r$. As in the proof of Theorem 4.31 we see that $r \leq \mu$. If we combine these results with (4.56) and (4.57) we see that

$$b^\mu x = (c_r \cdots c_0 . \overline{a_1 \cdots a_m})_b$$

or

$$x = (.0 \cdots 0 c_r \cdots c_0 \overline{a_1 \cdots a_m})_b.$$

To finish the proof we must show that in any other arrangement the pre-period length is not shorter than μ and the period length is not shorter than m.

Suppose we also have

$$x = (.d_1 \cdots d_s \overline{f_1 \cdots f_t})_b.$$

Then, as above,

$$x = \frac{d_1}{b} + \cdots + \frac{d_s}{b^s} + \frac{f_1}{b^{s+1}} + \cdots + \frac{f_t}{b^{s+t}} + \frac{f_1}{b^{s+t+1}} + \cdots + \frac{f_t}{b^{s+2t}} + \cdots$$

$$= \frac{d_1}{b} + \cdots + \frac{d_s}{b^s} + \left(\frac{f_1}{b^{s+1}} + \cdots + \frac{f_t}{b^{s+t}} \right) \frac{b^t}{b^t - 1}$$

$$= \frac{(d_1 b^{s-1} + \cdots + d_s)(b^t - 1) + f_1 b^{t-1} + \cdots + f_t}{b^s (b^t - 1)}.$$

Since $x = h/k$, with $(h, k) = 1$, we see that $k \mid b^s(b^t - 1)$. Thus we have $N \mid b^s$ and $M \mid (b^t - 1)$. Thus $s \geq \mu$, by the minimality of μ, and, by Theorem 2.19, $m \mid t$. Thus the pre-period length can't be shorter than μ and the period length can't be shorter than m.   ∎

The next two corollaries are restatements of (b) of Theorem 4.32 and follow easily. (See Problem 18 below.)

Corollary 4.32.1. Suppose $b = p_1^{\beta_1} \cdots p_r^{\beta_r}$, with $\beta_j > 0$, $1 \leq j \leq r$, and $k = p_1^{\alpha_1} \cdots p_r^{\alpha_r} M$, with $(M, b) = 1$ and $\alpha_j \geq 0$, $1 \leq j \leq r$. Then the length of the pre-period of any rational number h/k, $(h, k) = 1$, is the smallest integer greater than or equal to $\max(\alpha_1/\beta_1, \ldots, \alpha_r/\beta_r)$.

Corollary 4.32.2. Suppose $b = p_1 \cdots p_r$ is square-free and write $k = p_1^{\alpha_1} \cdots p_r^{\alpha_r} M$, with $(M, b) = 1$ and $\alpha_j \geq 0$, $1 \leq j \leq r$. Then the length of the pre-period of any rational number h/k, $(h, k) = 1$, is $\max(\alpha_1, \ldots, \alpha_r)$.

The following corollary specializes these results to two concrete cases.

Corollary 4.32.3.

(a) Write $k = 2^\alpha 5^\beta M$, where $(M, 10) = 1$ and $\alpha, \beta \geq 0$. Let $\mu = \max(\alpha, \beta)$ and $m = \operatorname{ord}_M 10$. Then any rational number h/k, $0 < h < k$, $(h, k) = 1$, can be written in the form

$$\frac{h}{k} = .a_1 \cdots a_\mu \overline{c_1 \cdots c_m}.$$

(b) Write $k = 2^\alpha 3^\beta 5^\gamma M$, where $(M, 60) = 1$ and $\alpha, \beta, \gamma \geq 0$. Let μ be the smallest integer greater than or equal to $\max(\alpha/2, \beta, \gamma)$ and let $m = \operatorname{ord}_M 60$. Then any rational number h/k, $0 < h < k$, $(h, k) = 1$, can be written in the form

$$\frac{h}{k} = (.a_1 \cdots a_\mu \overline{c_1 \cdots c_m})_{60}.$$

Besides giving us the form of the rational numbers Theorem 4.32 can also be used to identify irrational numbers. The following class of irrationals were discovered by J. Liouville.

Example 4.22. Let $b > 1$ be an integer. Show that the number with base b expansion

$$x = \sum_{j=1}^{+\infty} b^{-j!} = (.110001000000000000000001\cdots)_b$$

is irrational.

Solution. By Theorem 4.32 x is rational if and only if it can be written in a periodic base b expansion. Since the factorials are not periodic we see that the 1's will not occur periodically. Thus x must be irrational.   □

Unfortunately "simpler" numbers, like e and π, are not so easy to deal with since we do not know the pattern of their decimal expansion. Their irrationality must be proved in other ways. (See Problems 35 and 36 of Additional Problems.)

Problem Set 4.9

1. Prove that if a and r are real numbers and $|r| < 1$, then

$$\sum_{k=0}^{+\infty} ar^k = \frac{a}{1-r}.$$

(Hint: let $S_n = \sum_{k=0}^{n} ar^k$ and consider the form of rS_n.)

2. Find the decimal expansions of the following numbers.
   (a) 4/5   (d) 11/400
   (b) 10/11   (e) 1/101
   (c) 5/8   (f) 3/1111

3. Find the base 6 expansion of the following numbers.
   (a) 2/3   (d) 1/30
   (b) 1/5   (e) 13/24
   (c) 1/4   (f) 1/25

4. Find the fraction, in lowest terms, that is represented by each of the following decimal expansions.
   (a) $.123$ (d) $.\overline{123}$
   (b) $.12\overline{3}$ (e) $.\overline{321}$
   (c) $.1\overline{23}$ (f) $.2\overline{13}$

5. Find the fraction, in lowest terms, that is represented by each of the following expansions.
   (a) $(.11)_6$ (d) $(.0021)_3$
   (b) $(.\overline{11})_7$ (e) $(.002\overline{1})_3$
   (c) $(.01\overline{3})_8$ (f) $(.00\overline{21})_3$

6. Find the length of the pre-period and the period length of the decimal expansions of the following numbers.
   (a) 10/103   (d) 100/109
   (b) 23/49   (e) 16/17
   (c) 1/14   (f) 55/67

7. Find the length of the pre-period and the period length of the base 60 expansions of each of the following rational numbers.
   (a) 6/7   (d) 10/121
   (b) 7/30   (e) 3/14
   (c) 11/120   (f) 9/14

8. Suppose $pq = 10^k - 1$, where k is a positive integer. What is the relation between the periodic parts of the decimal expansions of $1/p$ and $1/q$? Can this be extended to general base b expansions?

9. Suppose that the real number x has a purely periodic base b expansion, that is, $x = (.\overline{a_1 \cdots a_m})$. Show that

$$x = \frac{(a_1 \cdots a_m)_b}{b^m - 1},$$

where $(a_1 \cdots a_m)_b = a_1 b^{m-1} + \cdots + a_{m-1}b + a_m$.

10. (a) Suppose x has a periodic base b expansion of period length m and y has a periodic base b expansion of period length n. Show that the period length of the base b expansion of $x + y$ is $[m, n]$. (Here $[m, n]$ denotes the least common multiple.)

    (b) Can you find a corresponding result for multiplication?

11. Let b and m be positive integers with $m \geq 2$. Show that the period length of the base b expansion of $1/m$ is $m - 1$ if and only if b is a primitive root of m.

12. Find all positive numbers m, $(m, 10) = 1$, such that the decimal expansion of $1/m$ has period length

    (a) 2   (c) 4

    (b) 3   (d) 5

13. Find the base b expansions of the following numbers.

    (a) $1/(b - 1)$ (c) $1/(b - 1)^2$

    (b) $1/(b + 1)$ (d) $1/(b + 1)^2$

14. Let $b_1, b_2, b_3, \ldots$ be an infinite sequence of positive integers greater than 1. Show that every real number can be represented as

$$a_0 + \frac{a_1}{b_1} + \frac{a_2}{b_1 b_2} + \frac{a_3}{b_1 b_2 b_3} + \cdots,$$

where $a_0, a_1, a_2, \ldots$ are integers such that $0 \leq a_k < b_k$, for $k \geq 1$. (See Problem 17(a) of Problem Set 1.2.)

15. If b is a positive integer, $b > 1$, show that

$$\sum_{k=1}^{+\infty} b^{-n^k}$$

is irrational for any positive integer $k, k \geq 2$.

16. Let n be a positive integer. Show that there exist at most a finite number of primes p such that $1/p$ has period length n.

17. Suppose $1/d$ has period length m and $1/f$ has period length n. Show that if $d \mid f$, then $m \mid n$.

18. Prove Corollary 4.32.1.

19. Show that

$$\frac{1}{n} + \frac{1}{n+1} + \frac{1}{n+2}$$

has a non-terminating decimal expansion that has a non-periodic part.

Additional Problems for Chapter Four

General Problems

1. (a) Show that if $n \mid (a^2 + 3)$, where n is odd and $(3, n) = 1$, then there exist integers s and t with $n = t^2 + 3s^2$. (Hint: use Lemma 4.1.1 and approximate a/n.)

   (b) Show that if $n \mid (a^2 + 2)$, where a is odd, then there exist integers s and t such that $n = t^2 + 2s^2$.

   (c) Show that if $n \mid (a^2 + 5)$ with $(a, 5) = 1$, then there exist integers s and t such that either $n = t^2 + 5s^2$ or $2n = t^2 + 5s^2$.

2. Let $A_n/B_n = [a_0; a_1, \ldots, a_n]$, show that

$$\frac{A_n + A_{n-1}}{B_n + B_{n-1}} = [a_0; a_1, \ldots, a_{n-1}, a_n + 1]$$

(which equals A_{n+1}/B_{n+1} if $a_{n+1} = 1$) and

$$\frac{A_n - A_{n-1}}{B_n - B_{n-1}} = [a_0; a_1, \ldots, a_{n-1}, a_n - 1]$$

(which equals A_{n-2}/B_{n-2} if $a_n = 1$.)

3. Let $\alpha = [a_0; 1, a_2, a_3, \ldots]$ and $\beta = [a_0; 1 + a_2, a_3, \ldots]$. Find a relationship between α and β.

4. This problem gives a construction for the integers x and y so that if $p \equiv 1 \pmod 4$ is a prime, then $p = x^2 + y^2$. The first method is due to Hermite and the second part is a refinement due to Brillhart. Let x_0 be a solution to $x^2 \equiv -1 \pmod p$ with $0 < x_0 < p/2$.

   (a) Hermite's method. If we expand x_0/p into a simple continued fraction to the point where the convergents h_n/k_n satisfy $h_{n+1} < \sqrt{p} < h_{n+2}$, show that

$$p = (x_0 k_{n+1} - p h_{n+1})^2 + (k_{n+1})^2.$$

(b) Brillhart's modification. Apply the Euclidean algorithm to the fraction p/x_0 producing a sequence of remainders $R_1, R_2, ...$ If R_k is the first remainder less than $\sqrt{p}$, that is, $R_k < \sqrt{p} < R_{k-1}$, show that

$$p = \begin{cases} R_k^2 + R_{k+1}^2 & \text{if } R_1 > 1 \\ x_0^2 + 1 & \text{if } R_1 = 1 \end{cases}.$$

(Hint: to derive (b) from (a) relate the remainders R_n to the convergents h_n/k_n.)

(c) Use both methods on the prime $p = 3989$. Which is faster? To find x_0 find a quadratic nonresidue c and compute $x_0 \equiv c^{(p-1)/4} \pmod{p}$. Recall that 2 is a nonresidue of primes $p \equiv 5 \pmod 8$.

5. If h_n/k_n is the nth convergent to $[a_0; a_1, a_2, ...]$, show that

$$\frac{h_n}{k_n} = a_0 + \frac{1}{k_0 k_1} - \frac{1}{k_1 k_2} + \frac{1}{k_2 k_3} - \cdots + \frac{(-1)^{n-1}}{k_{n-1} k_n}.$$

6. Let a/b, c/d and f/g be consecutive fractions in a Farey sequence of order n. Show that $bf - ag = (a + f, b + g)$.

7. Let θ and $\bar\theta$ be the roots of the equation $x^2 - ax - 1 = 0$, where a is a positive integer and $\theta > 0$. Show that if h_n/k_n is the nth convergent for θ, then

$$k_{n-1} = \frac{\theta^n - \bar\theta^n}{\theta - \bar\theta}.$$

8. Let h_n/k_n be the nth convergent to the real number θ and let a_n denote its partial quotients. Show that if $a_n \le A, A > 0$ is a constant, then

$$k_n \le (\tfrac{1}{2}(A + \sqrt{A^2 + 4}))^n.$$

9. Let h_n/k_n denote the nth convergent to the real number x. Show that
(a) $(h_n^2 - k_n^2)(h_{n-1}^2 - k_{n-1}^2) = (h_n h_{n-1} - k_n k_{n-1})^2 - 1.$
(b) $(h_n^2 + k_n^2)\{(h_{n-1} h_{n-2} + k_{n-1} k_{n-2})^2 + 1\} = (h_{n-2}^2 + k_{n-2}^2)\{(h_n h_{n-1} + k_n k_{n-1})^2 + 1\}.$

10. Let $[a_0; a_1, a_2, ...]$ be a convergent continued fraction with nth convergent h_n/k_n. Show that, as determinants,

$$h_n = \begin{vmatrix} a_1 & -1 & 0 & & \\ 1 & a_2 & -1 & & \text{\Large 0} \\ & & & \ddots & \\ \text{\Large 0} & & & 1 & a_n \end{vmatrix} \quad \text{and} \quad k_n = \begin{vmatrix} a_2 & -1 & & & \\ 1 & a_3 & -1 & & \text{\Large 0} \\ & & & \ddots & \\ \text{\Large 0} & & & a_{n-1} & -1 \\ & & & 1 & a_n \end{vmatrix}.$$

11. Show that the simple continued fraction of

$$\sqrt{(3^n + 1)^2 + 3}$$

has period length $6n$, for any positive integer n.

12. Show that

(a) $\sqrt{m^2 + 4} = \begin{cases} [2n;\overline{n,4n}] & \text{if } m = 2n, n \geq 1 \\ [2n+1;\overline{n,1,1,n,4n+2}] & \text{if } m = 2n+1, n \geq 1 \end{cases}$.

(b) $\sqrt{m^2 - 4} = \begin{cases} [2n-1;\overline{n-2,1,4n-2}] & \text{if } m = 2n, n \geq 3 \\ [2n;\overline{1,n-1,2,n-1,4n}] & \text{if } m = 2n+1, n \geq 2 \end{cases}$.

13. (a) Determine the value of the continued fraction $[a;\overline{b}]$ and determine conditions on the integers a and b so that this continued fraction represents $\sqrt{N}$, for some integer N.

   (b) Determine the value of the continued fraction $[a;\overline{b,c}]$ and determine conditions on the integers a, b and c so that this continued fraction represents $\sqrt{N}$, for some integer N.

   (c) Determine the value of the continued fraction $[a;\overline{b,c,d}]$ and determine conditions on the integers a, b, c and d so that this continued fraction represents $\sqrt{N}$, for some integer N.

14. Let c be the period length of the simple continued fraction expansion of $\sqrt{N}/M$, where N and M are integers with N a positive nonsquare integer and let h_n/k_n denote its nth convergent.

   (a) If $c = 2t$, show that

   $$\frac{h_c}{k_c} = \frac{h_{t+1}k_t + h_t k_{t+1}}{k_t(k_{t+1} + k_{t-1})}.$$

   (b) If $c = 2t + 1$, show that

   $$\frac{h_c}{k_c} = \frac{h_{t+1}k_{t+1} + h_t k_t}{k_{t+1}^2 + k_t^2}.$$

15. (a) Let

   $$\alpha = \frac{1 + \sqrt{4k^2 + 1}}{2},$$

   where k is a positive integer. Show that $\alpha = [k;\overline{1,1,2k-1}]$.

   (b) Find the continued fraction expansion of

$$\frac{1+\sqrt{(2k+1)^2+1}}{2},$$

where k is a positive integer.

(c) Show that

$$\sqrt{\frac{3n^2+4n+1}{3}} = [n;\overline{1,1,1,2n}],$$

for any positive integer n.

16. We say that a rational number a/b is a **good approximation** to an irrational number ξ if

$$|\xi b - a| = \min\{|\xi y - x| : x, y \in Z, 0 < y \le b\}.$$

Show that every convergent to ξ is a good approximation. Show that the converse is also true.

17. (a) If two irrational numbers have the same convergents up to h_n/k_n, show that their continued fraction expansions have the same partial quotients up to a_n.

(b) If $\alpha < \beta < \gamma$ and α and γ have identical convergents up to h_n/k_n, show that β does also.

18. Let x, y and d be positive integers such that $x^3 - dy^3 = n$. If $y > 8|n|/3d^{2/3}$, show that x/y is a convergent for the continued fraction expansion of $\sqrt[3]{d}$. (Hint:

$$\frac{x}{y} - d^{\frac{1}{3}} = \frac{x^3 - dy^3}{y((x/y + \frac{1}{2}\sqrt[3]{d})^2 + 3d^{2/3}/4)}.\Big)$$

19. (a) Prove the following result due to Chebychev. If α is irrational and β is real, then

$$|\alpha x - y - \beta| < 3/x$$

has infinitely many solutions in integers x and y with $y > 0$.

(b) Prove that there exists a positive number C with the property that for arbitrary reals $n \ge 1$ and β there exist integers x and y, with $y > 0$, such that $x \le Cn$ and

$$|\alpha x - y - \beta| < 1/n$$

if and only if the irrational number α has bounded partial quotients in its continued fraction expansion.

20. Prove the following result due to Kronecker. Given real number α, any irrational number θ and any $\varepsilon > 0$, there exist integers h and k, $k > 0$, such that

$$|k\theta - h - \alpha| < \varepsilon.$$

(Hint: first prove the result for such that $0 \le \alpha \le 1$.)

21. (a) Find the first ten convergents to π.
    (b) Give an upper bound for $n|\sin n|$. (Hint: recall that if n is any integer, then $|\sin x| = |\sin(x + n\pi)|$ and $|\sin x| < |x|$.)

22. Show that if $x_1, x_2, \ldots, x_n$ are any n real numbers, then the system of inequalities

$$|h_i / k - x_i| < k^{-1-1/k}, \quad 1 \le i \le n,$$

has at least one solution in integers. If at least one of the $x_1, \ldots, x_n$ is irrational, then it has infinitely many such solutions.

23. Show that there exists a positive constant c with the following property: to every positive irrational x there can be found infinitely many rationals h/k, with $(h, k) = 1$, satisfying

$$|x - h / k| < c / k^2$$

with $k \equiv 1 \pmod 4$. Can this result be generalized to other arithmetic progressions?

24. (a) Show that every real number has an expansion of the form

$$a_0 + \frac{a_1}{1!} + \frac{a_2}{2!} + \frac{a_3}{3!} + \cdots,$$

where $a_0, a_1, a_2, \ldots$ are integers satisfying $0 \le a_k < k$, for $k = 1, 2, \ldots$.
    (b) Show that every rational number has a terminating expansion of the form given in (a).

25. Let p be a prime and let $b > 1$ be an integer. Suppose that

$$\frac{1}{p} = (.\overline{a_1 \cdots a_{p-1}})_b$$

so that the period length is $p - 1$. Show that if m is a positive integer with $1 \le m < p$, then

$$\frac{m}{p} = (.\overline{a_{k+1} \cdots a_{p-1} a_1 a_2 \cdots a_{k-1} a_k})_b$$

where $k = \text{ind}_b m$ modulo p. (Hint: recall Problem 11 of Problem Set 4.9.)

26. Show that if p is a prime and

$$\frac{1}{p} = (.\overline{a_1\cdots a_k})_b$$

where $k = 2t$ is an even, then $a_j + a_{j+t} = b - 1$, for $j = 1, 2, \ldots, t$.

27. Let N be a positive nonsquare integer and let $b > 1$ be an integer. Form the sequence of integer $n_0, n_1, n_2, \ldots$ by

$$n_k^2 < Nb^{2k} < (n_k + 1)^2.$$

and define the sequence of integers $q_0 = n_0$ and, for $k \geq 1$,

$$n_{k+1} = bn_k + q_k.$$

(a) Show that $bn_k \leq n_{k+1} < n_{k+1} + 1 \leq bn_k + b$, for $k \geq 0$.

(b) Show that, for $k \geq 1$, $0 \leq q_k < b - 1$.

(c) Show that the base b expansion for $\sqrt{N}$ is

$$q_0 + (.q_1q_2\ldots)_b.$$

(d) Can you generalize this to higher order roots?

Problems 28 to 34 discuss general continued fractions. Let $a_0, a_1, a_2, \ldots$ and b_1, b_2, be two sequences of positive real numbers. Consider the continued fraction

$$a_0 + \cfrac{b_1}{a_1 + \cfrac{b_2}{a_2 + \cfrac{b_3}{a_3 + \cdots}}}$$

which we denote in a more condensed manner as

$$a_0 + \frac{b_1}{a_1} + \frac{b_2}{a_2} + \frac{b_3}{a_3} + \cdots.$$

While the theory of these continued fractions is not as complete as simple continued fractions we shall indicate some results.

28. If $x = a_0 + \dfrac{b_1}{a_1} + \dfrac{b_2}{a_2} + \dfrac{b_3}{a_3} + \cdots$. we define the convergents to x to be the fractions

$$\frac{h_n}{k_n} = a_0 + \frac{b_1}{a_1} + \frac{b_2}{a_2} + \frac{b_3}{a_3} + \cdots \frac{b_n}{a_n}.$$

Show that if $h_0 = 1$, $h_1 = a_0$, $k_0 = 0$ and $k_1 = 1$, then, for $n \geq 2$,

$$h_n = a_n h_{n-1} + b_n h_{n-2}$$
$$k_n = a_n k_{n-1} + b_n k_{n-2}.$$

29. (a) Show that

$$\frac{h_n}{h_{n-1}} = a_n + \frac{b_n}{a_{n-1}} + \cdots \frac{b_1}{a_0}.$$

(b) Show that

$$\frac{k_n}{k_{n-1}} = a_n + \frac{b_n}{a_{n-1}} + \cdots \frac{b_2}{a_1}.$$

30. (a) Show that

$$h_n k_{n-1} - h_{n-1} k_n = (-1)^n b_1 \cdots b_n.$$

(b) Show that

$$\frac{h_n}{k_n} = a_0 + \frac{b_1}{k_1 k_2} - \frac{b_1 b_2}{k_2 k_3} + \cdots + (-1)^n \frac{b_1 b_2 \cdots b_n}{k_{n-1} k_n}.$$

31. Show that the odd convergents are increasing, whereas the even convergents are decreasing. Show that every odd convergent is less than, and every even convergent is greater than, the following convergents.

32. Show that the continued fraction

$$a_0 + \frac{b_1}{a_1} + \frac{b_2}{a_2} + \frac{b_3}{a_3} + \cdots.$$

converges to a real number if the series

$$\sum_{n=1}^{+\infty} \frac{a_{n-1} a_n}{b_n}$$

is divergent.

33. Given the continued fraction

$$a_0 + \frac{b_1}{a_1} + \frac{b_2}{a_2} + \frac{b_3}{a_3} + \cdots.$$

define the sequence of numbers d_k, for $k = 1, \ldots, n$, by $d_0 = a_0$, $d_1 = b_1/a_1$, $d_2 = a_2 b_1/b_2$, $\ldots$, $d_n = a_n b_{n-1} b_{n-3} \ldots / b_n b_{n-2} \ldots$. Show that

$$a_0 + \frac{b_1}{a_1} + \frac{b_2}{a_2} + \frac{b_3}{a_3} + \cdots = [d_0; d_1, d_2, \ldots].$$

34. If $a_1, a_2, \ldots$ and $b_1, b_2, \ldots$ are all positive integers, then the infinite continued fraction

$$\frac{b_1}{a_1 +} \frac{b_2}{a_2 +} \cdots$$

converges to an irrational number if there exists an integer N such that $n \geq N$ implies $a_n \geq b_n$.

Problems 35 to 39 deal with the subject of irrational numbers. We have discussed some problems regarding irrational in many of the sections of this chapter, but in this set of problems we wish to look at the problem of irrational numbers as a subject in its own right.

35. (a) Show that a real number x is rational if and only if there is an integer n such that $[nx] = nx$.

   (b) Show that a real number x is rational if and only if there is an integer n such that $[n!x] = n!x$.

   (c) Show that e is irrational. (Hint: recall that

   $$e = \sum_{k=0}^{+\infty} \frac{1}{k!}$$

   and use (b).)

   (d) Show that $\cos 1$ is irrational, where radian measure is being used.

36. (a) Let $\alpha > 1$ be rational and suppose that for a given real number β there exist infinitely many rationals h/k such that

   $$|h/k - \beta| < k^{-\alpha}.$$

   Show that β is irrational.

   (b) Show that e is irrational.

   (c) Show that $\sin 1$ is irrational.

   Let x be a real number. We say that x is an **algebraic number of degree** n if x satisfies an equation of the form

   $$a_n x^n + a_{n-1} x^{n-1} + \cdots + a_1 x + a_0 = 0,$$

   where the a_i, $0 \leq i \leq n$, are integers and x satisfies no polynomial equation of lower degree. In Chapter Ten we will discuss algebraic numbers in greater detail.

37. Show that if x is an algebraic number of degree n, then there does not exist any value of $\varepsilon > 0$ such that

   $$|x - h/k| < 1/k^{n+\varepsilon}$$

for infinitely many relatively prime integers h and k. (Hint: let $f(x) = a_n x^n + a_{n-1} x^{n-1} + \cdots + a_1 x + a_0$ be an $n\underline{th}$ degree polynomial that x is a zero of and use the mean value theorem along with a lower bound for f evaluated as a polynomial.)

We say that a real number x is **transcendental** if it is not an algebraic number for any degree $n \geq 1$.

38. (a) Show that the number

$$\sum_{k=0}^{+\infty} a_k b^{-k!}$$

is transcendental, where b is an integer greater than 1 and the a_k are bounded, that is, there is a number $B > 0$ such that $|a_k| \leq B$, for all k.

(b) Show that the number

$$[0; b^{1!}, b^{2!}, b^{3!}, \ldots]$$

is transcendental for any integer b, $b > 1$.

(c) Let $\{a_k\}$ be a sequence such that $a_{k+1}/a_k \to +\infty$, as $k \to +\infty$. Let $b > 1$ be an integer. Show that the number

$$\sum_{k=1}^{+\infty} (-1)^k b^{a_k}$$

is transcendental.

39. (a) Suppose $\alpha = [a_0; a_1, \ldots]$ is an irrational number with convergents h_n/k_n. Show that if

$$a_{n+1} > k_n^{n-1},$$

for all $n \geq 1$, then α is transcendental.

(b) Suppose $\alpha = [a_0; a_1, \ldots]$ is an irrational number. Show that if

$$a_{n+1} \geq (2^n a_1 \cdots a_n)^{n-1},$$

for all $n \geq 1$, then α is transcendental.

Computer Problems

1. (a) Write a computer program so that given a real number x and a positive integer t one can find relatively prime integers h and k such that $1 \leq k \leq t$ and

$$|x - h/k| < 1/kt.$$

(b) Find rational approximations to $\sqrt[12]{12}$ and π with $t = 25$.

2. (a) Use Problem 7 of Problem Set 4.2 to write a program that computes the Farey series of order n.

(b) Calculate F_{10}, F_{15} and F_{25}.

3. (a) Write a program that converts a rational number h/k, with $k > 0$, into a simple continued fraction.

(b) Find finite simple continued fraction representations of:

    (i)   -1193/1725   (iv) 355/113

    (ii) 30031/16579   (v) -943/1001

    (iii) 1103/87   (vi) -831/8110

4. (a) Write a program that converts a finite simple continued fraction into a rational number.

(b) Determine the rational numbers represented by:

    (i)   [0; 1, 2, 3, 4, 5]   (iv) [2; 7, 1, 8, 2, 8, 1, 8, 2, 8]

    (ii) [1; 2, 1, 2, 1, 2, 1, 2]   (v) [1; 1, 1, 1, 1, 1, 1, 1, 1, 1, 1]

    (iii) [3; 1, 4, 1, 5, 9, 2]   (vi) [5; 4, 3, 2, 1]

5. (a) Write a program that computes the partial quotients and the convergents of the simple continued fraction representation of any given irrational number x and to a given number of terms n, that is, it will output $a_0, \ldots, a_n$ and $h_1/k_1, \ldots, h_n/k_n$.

(b) Determine the first 15 partial quotients and convergents of the following irrational numbers:

    (i) $\sqrt{12}$ (iv) $\sqrt[3]{5}$

    (ii) $\sqrt{69}$ (v) $\sqrt{1+\sqrt{2}}$

    (iii) $\sqrt{1000009}$ (vi) e

(c) Determine how accurate your calculations are in (b), that is, has natural machine error (roundoff, etc.) gotten in the way of your calculations?

6. (a) Write a program to use continued fractions to solve the linear Diophantine equation $ax - by = 1$.

(b) Modify the program in (a) to solve the general equation $ax + by = c$, where $(a, b) \,|\, c$.

(c) Solve the following equations.

    (i) $101x - 103y = 1$ (iii) $3101x + 4751y = 7$

    (ii) $257x + 813y = 1000$ (iv) $4951x - 9901y = 1$.

7. (a) Write a program that computes the periodic simple continued fraction expansion of a given quadratic irrational.

   (b) Find the periodic simple continued fraction expansions of the following quadratic irrationals.

   (i) $\sqrt{46}$ (iv) $\sqrt{9991}$

   (ii) $2-\sqrt{94}$ (v) $\sqrt{1537}$

   (iii) $(1-\sqrt{76})/3$ (vi) $(-6+\sqrt{86})/5$

8. (a) Write a program that converts a periodic simple continued fraction into a quadratic irrational.

   (b) Determine the quadratic irrationals represented by the following periodic simple continued fractions.

   (i) $[\overline{1;2,3,4,5,6,7,8,9}]$ (iii) $[-3;4,5,6,7,8,\overline{9,9,21}]$

   (ii) $[17;\overline{3,5,2,1,6,34}]$ (iv) $[4;2,2,1,\overline{1,3,5,6}]$

9. (a) Write a program that finds a rational approximation to a given real number x and whose denominator does not exceed a given positive integer n.

   (b) Find rational approximations to the following real numbers whose denominators do not exceed the given natural number.

   (i) $x = \pi, n = 100$ (iii) $x = 1/e, n = 100$

   (ii) $x = e, n = 500$ (iv) $x = \pi + e, n = 50$

10. (a) Write a program that finds a rational approximation to a given real number x to with a specified accuracy ε.

    (b) Find rational approximations to the following real numbers to six decimal places.

    (i) π^2 (iii) $\sqrt[3]{3/2}$

    (ii) e^2 (iv) $(e + 1)/(e - 1)$

11. (a) Use the result of Theorem 4.26 to write a program that calculates "small" quadratic residues of a number N. How large can these quadratic residues be?

    (b) Determine quadratic residues for the following positive integers.

    (i) 10001   (iii) 9991

    (ii) 8104   (iv) 85851

12. (a) Write a program that uses the simple continued fraction expansion of $\sqrt{N}$ to factorize a positive nonsquare integer N.

    (b) Find factors of the following numbers.

        (i)   9757   (iii) 716539

        (ii) 1009091   (iv) 38807

13. (a) Modify the program in 12(a) to include the idea of factor bases as given in Problem 6 of Problem Set 4.7.

    (b) Factor the following numbers.

        (i) $2^{2^7}+1$ (iii) 30031

        (ii) 197209   (iv) 510509

14. (a) Write a program that finds the base b expansion of a given rational number, where $b > 1$ is an integer.

    (b) Find the base b expansions of the following rational numbers.

        (i)   71/83, $b = 9$ (iv) 22/7, $b = 8$

        (ii) 211/112, $b = 3$ (v)   123/457, $b = 10$

        (iii) 355/113, $b = 7$ (vi) 1/997, $b = 10$

15. (a) Write a program that converts a periodic base b expansion into a rational number in lowest terms, where $b > 1$ is an integer.

    (b) Find the rational numbers represented by the following expansions.

        (i) $(.1\overline{3112113})_5$ (iv) $(.\overline{62616})_9$

        (ii) $(.\overline{67911})_{10}$ (v) $(.3\overline{12311})_6$

        (iii) $(.\overline{43210})_7$ (vi) $(.311\overline{5611})_8$

16. (a) Write a program that determines the lengths of the pre-period and period in the base b expansion of a rational number, where $b > 1$ is an integer.

    (b) Find the lengths of the pre-periods and periods in the base b expansions of the following rational numbers.

        (i)   71/73, $b = 8$ (iv) 22/7, $b = 9$

        (ii) 211/112, $b = 7$ (v)   123/457, $b = 8$

        (iii) 355/113, $b = 6$ (vi) 1/1001, $b = 5$

17. Write a program to calculate the periods in the decimal expansions of $1/p$ for all primes p, $3 \leq p \leq 500$. How many primes have period length $p - 1$?

18. Calculate the first 100 decimals of π. (There are many faster methods than the method of Theorem 4.30, but they require extra knowledge.) Do you notice any patterns? Is there a block of digits of the form 0123456789?

19. Write a program to calculate the decimal expansion of $\sqrt{N}$ using Problem 27 above. Find the decimal expansions to 50 places of $\sqrt{2}$, $\sqrt{5}$, $\sqrt{11}$ and $\sqrt{101}$.

CHAPTER FIVE
Diophantine Equations, I

5.1. INTRODUCTION

To solve a Diophantine equation means to solve an equation integers or occasionally in rational numbers. Such equations have been around since the dawn of time, though one of the first systematic expositions of Diophantine equations, as simply equations to solve, was due to Diophantus of Alexandria, hence the name. For general references, see [77], [25, v. II], [13] and for the original work [48].

There are two primary ways of considering Diophantine equations. Let $f(x_1, ..., x_m)$ be a function of m variables $x_1, ..., x_m$. Then

1. one can ask for the solution, in integers $x_1, ..., x_m$, of the equation

$$f(x_1, ..., x_m) = 0$$

or

2. given an integer N one can ask for the solution, in integers $x_1, ..., x_{m\cdot}$, of the equation

$$f(x_1, ..., x_m) = N.$$

Technically the first problem is a special case of the second, but we prefer to separate the two cases, since one really uses different techniques for the two cases. In this chapter we shall concern ourselves with problems of the first kind and in the next chapter we shall discuss representation problems of the second kind.

5.2. PYTHAGOREAN TRIANGLES

Surely one of the first Diophantine problems to be solved was that of finding right triangles with integer sides. This would be a great help in surveying, giving easily constructed right angles. By the theorem of Pythagoras, if x and y are the lengths of the legs and z is the length of the hypotenuse, then,

(5.1) $$x^2 + y^2 = z^2.$$

If x, y and z are positive integers, then x, y and z are said to form a **Pythagorean Triangle**. The solution is said to be **primitive** if $(x, y, z) = 1$.

Theorem 5.1: All primitive solutions of (5.1), with x, y and z positive integers, are given by

$$x = u^2 - v^2, \ y = 2uv \text{ and } z = u^2 + v^2,$$

where $u > v > 0$, $(u, v) = 1$ and u and v are of opposite parity.

Proof: Note that if any two of x, y or z have a common factor, then it must also divide the third. Thus, if $(x, y, z) = 1$, then $(x, y) = (x, z) = (y, z) = 1$. Also if x and y are both odd, then $x^2 + y^2 \equiv 2 \pmod 4$, and so cannot be the square of an integer. Thus z is odd and one of x and y, say y, is even while the other, x, is odd. Say $y = 2y_1$.

We have

$$y^2 = 4y_1^2 = z^2 - x^2$$

or

$$y_1^2 = \frac{z-x}{2} \cdot \frac{z+x}{2}.$$

Since $\left(\frac{z-x}{2}, \frac{z+x}{2}\right) = 1$, we see, by unique factorization, that there exist integers u and v such that $u > v > 0$, $(u, v) = 1$ and

$$\frac{z+x}{2} = u^2 \text{ and } \frac{z-x}{2} = v^2.$$

Since z and x are both odd we see that u and v must be of opposite parity. Finally, solving for x, y and z, we have

$$z = u^2 + v^2, \ x = u^2 - v^2 \text{ and } y = 2y_1 = 2uv. \qquad \blacksquare$$

Corollary 5.1.1: All Pythagorean triangles are given by

$$x = k(u^2 - v^2), \ y = 2kuv \text{ and } z = k(u^2 + v^2),$$

where k is a positive integer and u and v are as in Theorem 5.1.

This follows from the remark in the first paragraph of the proof of Theorem 5.1. We give a short table of primitive Pythagorean triangles.

u	v	x	y	z
2	1	3	4	5
3	2	5	12	13
4	1	15	8	17

u	v	x	y	z
4	3	7	24	25
5	2	21	20	29
5	4	9	40	41
6	1	35	12	37
6	5	11	60	61
7	2	45	28	53
7	4	33	56	65
7	6	13	84	85

There is a vast literature on Pythagorean triangles, see, for example, [114].

We present a few problems related to Pythagorean triangles that are suggested by looking at the table above.

Problem 1: Find all primitive Pythagorean triangles that have one of the legs and the hypotenuse differing by a fixed positive integer k.

Solution: There are two cases depending on whether k is even or odd.

Case I: If k is odd, then it must be the hypotenuse and the even leg that differ by k since the hypotenuse is odd. Thus, by Theorem 5 .1, we must have

$$u^2 + v^2 = 2uv + k$$

or

$$(u - v)^2 = k.$$

Thus, if this problem is to have a solution we must have k a square, say $k = q^2$, for some positive integer q. Since $u > v$ we have $u = v + q$. Thus

$$x = q(2v + q), \ y = 2v(v + q) \text{ and } z = 2v^2 + 2qv + q^2,$$

where $v, q > 0$ and have opposite parity. In particular, if $q = 1$, $k = 1$ and

$$x = 2v + 1, \ y = 2v(v + 1) \text{ and } z = 2v^2 + 2v + 1 .$$

This gives the sequence of triangles $(3, 4, 5)$, $(5, 12, 13)$, $(7, 24, 25)$ $(9, 40, 41)$ etc.

Case II: Suppose k even. Since the hypotenuse is odd, it must be the hypotenuse and the odd leg that differ by k. Thus, by Theorem 5 .1, we have

$$u^2 + v^2 = u^2 - v^2 + k$$

or

$$2v^2 = k.$$

Thus, if there is to be a solution, then k must be twice a square, say $k = 2q^2$. Then $v = q$ and u is arbitrary, though it must be $> q$ and of opposite parity than q. Thus

$$x = u^2 - q^2, \ y = 2uq \text{ and } z = u^2 + q^2.$$

In particular, for $k = 2$ we have

$$x = u^2 - 1, \ y = 2u \text{ and } z = u^2 + 1,$$

where u is even and $u > 2$. This can also be written as

$$x = 4w^2 - 1, \ y = 4w \text{ and } z = 4w^2 + 1,$$

where w is a positive integer. This yields the sequence of triangles (3, 4, 5), (15, 8, 17), (35, 12, 37), (63, 16, 65), etc. $\square$

The next problem would be to determine those primitive Pythagorean triangles for which the two legs differ by some fixed constant k. Since this would lead to a type of problem we are going to consider at length in the next chapter, we will postpone the discussion of this problem until Section 6 of Chapter Six.

Problem 2: Find all Pythagorean triangles whose sides are in an arithmetic progression.

Solution: We want integers x and k such that

$$(x - k)^2 + x^2 = (x + k)^2,$$

where k is the common difference. Multiplying out gives $x^2 = 4kx$. Thus $x = 0$ or $x = 4k$. Since $x = 0$ does not yield a real triangle we must have $x = 4k$ and our triangle is $(3k, 4k, 5k)$. The only primitive triangle is (3, 4, 5). $\square$

Problem 3: Find all primitive Pythagorean triangles in which one of the sides is a square.

Solution: : There are three cases depending on which side is to be a square.

Case I: The hypotenuse is to be a square. Then we want

$$x^2 + y^2 = (z^2)^2 = z^4.$$

By Theorem 5.1 we have

$$z^2 = u^2 + v^2,$$

where $u > v$. Since u and v are of opposite parity one of them must be even. Then

$$u = 2mn, \ v = m^2 - n^2 \text{ and } z = m^2 + n^2$$

where $(m, n) = 1$, $m > n$ and m and n are of different parity, or similar expressions with u and v interchanged. This yields the solution

$$x = \left| m^4 - 6m^2n^2 + n^4 \right|, \; y = 4mn(m^2 - n^2) \text{ and } z = m^2 + n^2.$$

This gives the triangles (7, 24, 25), (119, 120, 169), etc.

Case II: The odd leg is to be a square. Then we want

$$(x^2)^2 + y^2 = x^4 + y^2 = z^2.$$

By Theorem 5.1, we then have

$$x^2 = u^2 - v^2,$$

where $u > v$, or

$$x^2 + v^2 = u^2.$$

Thus u must be odd and v must be even, since x is odd. Thus, by Theorem 4.1,

$$x = m^2 - n^2, \; v = 2mn \text{ and } u = m^2 + n^2,$$

where $(m, n) = 1$, $m > n > 0$ and m and n are of opposite parity. Thus

$$x = m^2 - n^2, \; y = 4mn(m^2 + n^2) \text{ and } z = m^4 + 6m^2n^2 + n^4.$$

This gives the triangles (9, 40, 41), (25, 312, 313), etc.

Case III: The even leg to be a square. By Theorem 5.1, this means

$$y^2 = 2uv.$$

Now $(u, v) = 1$ and one of u and v is even. Since $y = 2w$ is even, this implies

$$uv = 2w^2.$$

Thus, by unique factorization,

$$u = m^2 \text{ and } v = 2n^2 \text{ or } u = 2m^2 \text{ and } v = n^2,$$

where $(m, 2n) = 1$ or $(2m, n) = 1$. Thus

$$x = \left| m^4 - 4n^4 \right|, \; y = 2mn \text{ and } z = m^4 + 4n^4,$$

where $(m, 2n) = 1$. This gives the triangles (3, 4, 5), (77, 36, 85), etc. $\square$

One might next inquire about the possibility of two sides being squares. Fermat answered this problem in the negative. The proof we give below uses Fermat's famous method of descent: assume you have the smallest counterexample and then construct a yet smaller one.

Theorem 5.2: There does not exist a primitive Pythagorean triangles with at least two sides being squares.

Proof: There are two cases to consider: case I: $x^4 + y^4 = z^2$ and case II: $x^2 + y^4 = z^4$.

Case I: Without loss of generality we may assume $(x, y, z) = 1$. Assume such a triangle exists and let (x, y, z) be that triangle with the least positive z. By Theorem 5.1 we have

$$x^2 = u^2 - v^2, \ y^2 = 2uv \text{ and } z = u^2 + v^2,$$

where $(u, v) = 1$, $u > v > 0$ and u and v are of opposite parity. Note that we cannot have u even and v odd, since this would imply $x^2 \equiv -1 \equiv 3 \pmod 4$, which is impossible. Thus we have u odd and v even. Now $(u, 2v) = 1$ so in the equation

$$y^2 = 2uv$$

we must have 2 occur to an even power on the right hand side. Thus

$$u = u_0^2 \text{ and } v = 2v_0^2,$$

where u_0 and v_0 are positive integers. Thus

$$x^2 = u_0^2 - 4v_0^2,$$

that is,

$$u_0^2 = (2v_0^2)^2 + x^2.$$

Now $(u_0, 2v_0) = 1$. Thus, by Theorem 5.1, we have

$$u_0^2 = t^2 + w^2 \text{ and } v_0^2 = tw,$$

where $(t, w) = 1$. Thus

$$t = \alpha^2 \text{ and } w = \beta^2,$$

where $(\alpha, \beta) = 1$. Thus

$$u_0^2 = t^2 + w^2 = \alpha^4 + \beta^4.$$

We have

$$u_0^2 = u \le u^2 < z < z^2,$$

and so $0 < u_0 < z$. This contradicts the minimality of z, and so proves the first part of the theorem.

Case II: Again we let (x, y, z) be the primitive solution with the least positive z. If x is even, then by Theorem 5.1, there exist $(m, n) = 1$, $m > n$, such that

$$x = 2mn, \ y^2 = m^2 - n^2 \ \text{and} \ z = m^2 + n^2.$$

Then $z^2 > m^2$. Also $(yz)^2 = m^4 - n^4$, and so

$$n^4 + (yz)^2 = m^4$$

and we contradict the minimality of z. Thus we must have x odd, y even and z odd. We have

$$x^2 = z^4 - y^4 = (z^2 - y^2)(z^2 + y^2).$$

Since $(z^2 - y^2, z^2 + y^2) = 1$ (z and y are of opposite parity) we see there exist integers r and s with $(r, s) = 1$, $s > r$, such that

$$z^2 - y^2 = r^2 \ \text{and} \ z^2 + y^2 = s^2$$

Since $2z^2 = r^2 + s^2$, we have

$$\left(\frac{s+r}{2}\right)^2 + \left(\frac{s-r}{2}\right)^2 = z^2.$$

Since r and s are both odd we see that $\left(\frac{s+r}{2}, \frac{s-r}{2}\right) = 1$. Thus, by Theorem 5.1, there exist relatively prime integers m and n of opposite parity such that

$$\frac{s+r}{2} = m^2 - n^2, \frac{s-r}{2} = 2mn \ \text{and} \ z = m^2 + n^2$$

or

$$\frac{s-r}{2} = m^2 - n^2, \frac{s+r}{2} = 2mn \ \text{and} \ z = m^2 + n^2.$$

In either case we have

$$2x^2 = s^2 - r^2 = 8mn(m^2 - n^2).$$

Since x is even we have $x = 2x_1$ and $x_1^2 = mn(m^2 - n^2)$. Since $(m, n) = 1$ and are of opposite parity we have $(m + n, m - n) = 1$. Also

$$(m, m + n) = (m, m - n) = (n, m + n) = (n, m - n) = 1.$$

Thus there exist positive integers k, w, p, q such that

$$m = k^2, \ n = w^2, \ m - n = p^2, \ \text{and} \ m + n = q^2.$$

Thus

$$k^4 - w^4 = (pq)^2$$

or

$$w^4 + (pq)^2 = k^4.$$

Since $k^4 = m^2 < m^2 + n^2 = z < z^4$ we have $k < z$ and this contradicts the minimality of z. This proves the second part of the theorem. ∎

Corollary 5.2.1: There does not exist a pair of Pythagorean triangles such that a leg and hypotenuse of one are the legs of the other.

Proof: If the contrary held, then we could solve the system

$$x^2 + y^2 = z^2 \text{ and } x^2 + z^2 = u^2.$$

Then we have

$$x^4 + (uy)^2 = z^4,$$

which contradicts Theorem 5.2. ∎

An immediate corollary of Theorem 5.2 is the following result due to Fermat.

Corollary 5.2.2: There do not exist positive integers such that

$$x^4 + y^4 = z^4.$$

On the basis of this Fermat conjectured that

$$x^n + y^n = z^n$$

is impossible in positive integers x, y and z for any integer $n \geq 3$. Fermat claimed to have a proof of this result, but no one has since found Fermat's proof and the belief today is that Fermat's "proof" was no doubt incorrect. It has been shown to be true for all exponents divisible by a prime ≤ 125000. For a proof in the case $n = 3$ see [115, pp.384f] as well as Section 4 of Chapter 10. As of this writing it is possible that Fermat's conjecture may have finally been settled by Wiles.

For further references on the general problem and proofs of some of the known results see [27], [41, ch 9-11]. For an historical overview of the Fermat conjecture and a review of the variety of techniques used on its proof see [101]. For more on the method of descent see [112].

Problem Set 5.2

1. Find all primitive solutions of $x^2 + y^2 = z^2$ with x, y and z positive integers and $z \leq 50$.

2. Find all values of θ such that $\sin\theta$ and $\cos\theta$ are both rational.

3. Prove that if $x^2 + y^2 = z^2$, $(x, y, z) = 1$, then one of x and y is divisible by 3 and one by 4. Also show that one of x, y or z is divisible by 5.

4. Find all Pythagorean triangles whose sides are in geometric progression.

5. Show that if $n \geq 3$, then there exists a Pythagorean triangle with n as one of its legs.

6. Prove that there does not exist a Pythagorean triangle with a square area. (Hint: you need not start from scratch. This result should be obtained as a corollary to the results obtained so far.)

7. (a) Solve $x^2 + (2y)^4 = z^2$ in positive integers.
   (b) Solve $(2x)^2 + y^4 = z^2$ in positive integers.

8. Solve

$$\frac{1}{x^2} + \frac{1}{y^2} = \frac{1}{z^2}$$

   in positive integers.

9. Show that if (x, y, z) is a primitive Pythagorean triple, then the hypotenuse never exceeds a leg by 4.

10. If (x, y, z) is a primitive Pythagorean triple, show that

$$x + y \equiv x - y \equiv 1 \text{ or } 7 \pmod 8.$$

11. Find all Pythagorean triangles whose perimeter equals their area.

12. Show that in any Pythagorean triangle the product of all three legs is divisible by 60.

13. Prove that the system of equations

$$x^2 + y^2 = u^2$$
$$x^2 - y^2 = v^2$$

   has no solutions in positive integers. (Hint: try a proof by descent.)

14. Find all primitive Pythagorean triangles such that the square of the odd leg is the sum of the even leg and the hypotenuse.

15. Let m and n be generators of a Pythagorean triangle (x, y, z). Show that

$$\frac{y + z}{x} = \frac{m}{n},$$

   where x is the even leg.

16. If (x, y, z) is a Pythagorean triple and $7 \nmid xyz$, then $7 \mid (x^2 - y^2)$.

17. (a) Find all primitive Pythagorean triangles with the even leg equal to 48.

(b) Find all Pythagorean triangles whose even leg equal to 48.

(c) Find all primitive Pythagorean triangles whose odd leg equal to 231.

(d) Find all Pythagorean triangles whose odd leg equal to 231.

5.3. RELATED QUADRATIC EQUATIONS

In this section we deal with three quadratic equations that are related to the Pythagorean equation and in the next section we shall deal with a quadratic equation that generalizes most of the equations considered in this chapter. In the problem set at the end of this chapter we include a few higher degree equations to solve. Since not much is known in general about such equations we will restrict our attention to the quadratic case. For higher degree equations see the books of Mordell, Carmichael and Dickson referred to above as well as [78, ch. 7].

Theorem 5.3: The solution, in positive integers, of

$$x^2 + 2y^2 = z^2,$$

where $(x, y, z) = 1$, is given by

$$x = \left| u^2 - 2v^2 \right|, \; y = 2uv \text{ and } z = u^2 + 2v^2,$$

where $u, v > 0$ and $(u, 2v) = 1$.

Proof: If $(x, y, z) = 1$ we see that we must have x (and so z) odd. Considering congruences modulo 4 we see we must have y even. Let $y = 2w$. Then we have

$$2w^2 = \frac{z+x}{2} \cdot \frac{z-x}{2}$$

and since $\left(\frac{z+x}{2}, \frac{z-x}{2} \right) = 1$ we must have, by unique factorization,

$$\frac{z+x}{2} = u^2 \text{ and } \frac{z-x}{2} = 2v^2$$

or

$$\frac{z+x}{2} = 2u^2 \text{ and } \frac{z-x}{2} = v^2,$$

where $(u, 2v) = 1$ or $(2u, v) = 1$. The result follows. ∎

Theorem 5.4: The solution, in relative prime positive integers, of

$$x^2 + y^2 = 2z^2$$

is given by

$$x = u^2 - v^2 + 2uv, \quad y = \left| u^2 - v^2 - 2uv \right| \text{ and } z = u^2 + v^2$$

where u and v are positive relative prime integers of opposite parity.

Proof: Since x, y and z are relatively prime we see that none of them can be even (z cannot be even by congruence considerations modulo 4). Then $\frac{x+y}{2}$ and $\frac{x-y}{2}$ are both integers, and so

$$\left(\frac{x+y}{2} \right)^2 + \left(\frac{x-y}{2} \right)^2 = z^2.$$

If $x = y$, then $x = y = z = 1$ is a primitive solution. Except for this case we apply Theorem 5.1 and obtain

$$\frac{x+y}{2} = u^2 - v^2, \frac{x+y}{2} = 2uv \text{ and } z = u^2 + v^2$$

or

$$\frac{x+y}{2} = 2uv, \frac{x+y}{2} = u^2 - v^2 \text{ and } z = u^2 + v^2,$$

where u and v are as in the statement of the theorem. The result follows. ■

For related results see Problem 1 below as well as well as Problem 4 of Problem Set 5.4.

The following theorem generalizes the Pythagorean equation in another way: by adding more squares.

Theorem 5.5: All solutions of

$$x^2 + y^2 + z^2 = t^2,$$

where y and z are even, are given by

$$x = \frac{p^2 + q^2 - r^2}{r}, y = 2p, z = 2q \text{ and } t = \frac{p^2 + q^2 + r^2}{r},$$

where p and q are positive integers and r runs through the divisors of $p^2 + q^2$ less than $\sqrt{p^2 + q^2}$.

Proof: Suppose all of x, y and z were odd. Then we would have $t^2 \equiv 3 \pmod 8$, which is impossible. If only one were even, then we would have $t^2 \equiv 2 \pmod 4$, which is impossible. Thus two of x, y and z must be even, say $y = 2p$ and $z = 2q$.

Let $t - x = u$. Then

$$(x + u)^2 = x^2 + 4p^2 + 4q^2,$$

and so

$$u^2 = 4p^2 + 4q^2 - 2xu.$$

Thus u is even, say $u = 2r$, where

$$r^2 = p^2 + q^2 - rx,$$

and so $r \mid (p^2 + q^2)$. Thus

$$x = \frac{p^2 + q^2 - r^2}{r} \quad \text{and} \quad t = \frac{p^2 + q^2 - r^2}{r} + r = \frac{p^2 + q^2 + r^2}{r}.$$

Since we wish x to be positive we need $r < \sqrt{p^2 + q^2}$. ∎

Note that if we multiply the solutions of the equation

$$x^2 + y^2 + z^2 = t^2$$

by r we may rewrite the primitive solution as

$$dx = p^2 + q^2 - r^2, \, dy = 2pr, \, dz = 2qr \text{ and } dt = p^2 + q^2 + r^2,$$

where $d = (p^2 + q^2 - r^2, \, 2pr, \, 2qr, \, p^2 + q^2 + r^2)$ and p, q and r are now any integers such that $(p, q, r) = 1$.

Example 5.2: Take $p = 3$, $q = 2$, $r = 1$. Then

$$dx = 3^2 + 2^2 - 1^2 = 12$$
$$dy = 2 \cdot 3 \cdot 1 = 6$$
$$dz = 2 \cdot 2 \cdot 1 = 4$$
$$dt = 3^2 + 2^2 + 1^2 = 14,$$

and so $d = 2$ and $x = 6$, $y = 3$, $z = 2$ and $t = 7$. Indeed

$$6^2 + 3^2 + 2^2 = 7^2.$$

For the extension of this problem to sums of more squares see Problem 3 below as well as [25, v. II, ch.IX].

We close this section with one more variation on the Pythagorean theme.

Theorem 5.6. The solution, in integers, of the equation

$$x^2 + y^2 = u^2 + v^2,$$

with $x > u$, is given by

(5.3)
$$x = (ms + nr)/2 \quad u = (ms - nr)/2$$
$$v = (ns + mr)/2 \quad y = (ns - mr)/2,$$

where if m and n are both odd, then r and s must be of the same parity.

Proof. Let $s = (x + u, y + v)$. Then we may write

(5.4) $x + u = ms$ and $y + v = ns$,

where $(m, n) = 1$. If we rewrite the equations as $x^2 - u^2 = v^2 - y^2$ we see that

$$(x - u)m = (v - y)n.$$

Thus $n \mid (x - u)$ and $m \mid (v - y)$, and so there is an integer r such that

(5.5) $x - u = nr$ and $v - y = mr$.

If we combine (5.4) and (5.5), then we obtain the result (5.3).

Note that if m and n are both odd, then we have $2 \mid (mr + ns)$, and so

$$r + s \equiv 0 \pmod{2}.$$

Thus r and s must have the same parity.   ∎

Problem Set 5.3

1. (a) Solve $x^2 + 3y^2 = z^2$, in positive integers.
   (b) Solve $x^2 + 2y^2 = z^4$, in positive integers.
   (c) Solve $x^2 + py^2 = z^2$, in positive integers, where p is a prime.
   (d) Find all positive integral solutions of $x^2 + y^2 = 4z^2$.

2. Let t_n denote the nth triangular, $n(n + 1)/2$. Solve the equation

$$t_x + t_y = t_z.$$

   Can you extend this to other, nonsquare, figurate numbers?

3. Solve $x_1^2 + x_2^2 + \cdots + x_n^2 = x^2$ in positive integers, when $n \geq 4$.

4. Find all positive, primitive solutions in integers of $x^2 + y^3 = z^2$.

5. Solve the equation $t_x + y^2 = t_z$ in positive integers.

6. (a) Solve the equation $x^2 + y^2 = z^2 - 1$ in positive integers.
   (b) Solve the equation $x^2 + y^2 = z^2 + 1$ in positive integers.

7. (a) Solve, in positive integers, the equation $x^2 + xy + y^2 = z^2$.
   (b) Solve, in positive integers, the equation $x^2 - xy + y^2 = z^2$.
   (Hint: show that x and y must have the same parity.)

8. Find all values of integers j, k and l so that the sum of the k consecutive odd numbers, starting with the jth odd number, is equal to the sum of the next l consecutive odd numbers. (For example, if we take $k = 4$, $l = 2$ and $j = 3$, we have $3 + 5 + 7 + 9 = 11 + 13$.)

9. Solve, in the integers x, y and z, the equation

$$x^2 + (x + z)^2 + (x + 2z)^2 + \cdots + (x + (n - 1)z)^2 = y^2 + \cdots + (y + nz)^2.$$

(Hint: divide into two cases: $n \equiv 1, 2 \pmod 4$ and $n \equiv 0, 3 \pmod 4$.)

10. Solve, in integers, the equation

$$t_x + t_y = z^2.$$

11. Solve, in integers, the equation

$$x^2 + y^2 = t_z.$$

12. Another method of solving the equation of Theorem 5.6 is discussed in this problem. We wish to solve

$$x^2 + y^2 = u^2 + v^2,$$

in integers. Introduce the parameters p, q and r by $y = x + p$, $u = x + q$ and $v = x + r$. Substitute these into the equation and solve the resulting linear equation in x. Note that this gives a rational expression for x. Use the homogeneity of the equation to find the integer solutions. Primitive solutions can be obtained ala the remarks after Theorem 5.5.

5.4. THE EQUATION $AX^2 + BY^2 + CZ^2 = 0$

The methods used in this section illustrate some of the other methods used in the study of Diophantine equations. We proceed the main theorem with two lemmas about certain types of congruence equations. These lead to the main theorem which gives the necessary and sufficient conditions for the equation of the title to be solvable. The proof of the main result also gives bounds for a solution, if there is one. We follow the main theorem with a method of obtaining all the rest of the solutions from one given one which uses analytic geometry.

Lemma 5.6.1: Let λ, μ and v be positive reals with product $\lambda\mu v = m$, an integer. Then the congruence

$$\alpha x + \beta y + \gamma z \equiv 0 \pmod m$$

has a solution x, y and z, not all zero, such that

$$|x| \le \lambda, |y| \le \mu \text{ and } |z| \le v.$$

Proof: Let x range over the integers 0, 1, ..., $[\lambda]$, y range over the integers 0, 1, ..., $[\mu]$ and z range over the integers 0, 1, ..., $[v]$. This gives

$$(1 + [\lambda])(1 + [\mu])(1 + [v]) > \lambda \mu v = m$$

different triples (x, y, z). Thus there must be at least two distinct triples (x_1, y_1, z_1) and (x_2, y_2, z_2) such that

$$\alpha x_1 + \beta y_1 + \gamma z_1 \equiv \alpha x_2 + \beta y_2 + \gamma z_2 \equiv 0 \pmod{m}$$

or

$$\alpha(x_1 - x_2) + \beta(y_1 - y_2) + \gamma(z_1 - z_2) \equiv 0 \pmod{m}.$$

Also

$$|x_1 - x_2| \le [\lambda] \le \lambda, |y_1 - y_2| \le [\mu] \le \mu \text{ and } |z_1 - z_2| \le [v] \le v.$$

Let $x = x_1 - x_2$, $y = y_1 - y_2$ and $z = z_1 - z_2$. Since the triples are distinct not all of x, y and z can be zero and the result follows. ∎

Note that this theorem is a generalization to three variables of Thue's Theorem (Theorem 2.10). It is easy to see from the proof of this lemma that the result can be generalized to linear congruences in any number of variables. See Problem 34 of Additional Problems for Chapter Two.

Lemma 5.6.2: Suppose $ax^2 + by^2 + cz^2$ factors into linear factors modulo m and also modulo n, that is

$$ax^2 + by^2 + cz^2 \equiv (\alpha_1 x + \beta_1 y + \gamma_1 z)(\alpha_2 x + \beta_2 y + \gamma_2 z) \pmod{m}$$

and

$$ax^2 + by^2 + cz^2 \equiv (\alpha_3 x + \beta_3 y + \gamma_3 z)(\alpha_4 x + \beta_4 y + \gamma_4 z) \pmod{n}.$$

If $(m, n) = 1$, then $ax^2 + by^2 + cz^2$ factors into linear factors modulo mn.

Proof: By the Chinese Remainder Theorem (Theorem 2.11) we may choose $\alpha, \beta, \gamma, \alpha', \beta'$ and γ' to satisfy

$$\alpha \equiv \alpha_1, \beta \equiv \beta_1, \gamma \equiv \gamma_1, \alpha' \equiv \alpha_2, \beta' \equiv \beta_2, \gamma' \equiv \gamma_2 \pmod{m}$$
$$\alpha \equiv \alpha_3, \beta \equiv \beta_3, \gamma \equiv \gamma_3, \alpha' \equiv \alpha_4, \beta' \equiv \beta_4, \gamma' \equiv \gamma_4 \pmod{n}.$$

Then the congruence

$$ax^2 + by^2 + cz^2 \equiv (\alpha x + \beta y + \gamma z)(\alpha' x + \beta' y + \gamma' z)$$

holds both modulo m and modulo n, and so by Theorem 2.2 the congruence holds modulo mn. ∎

Theorem 5.6: Let a, b and c be nonzero integers such that the product abc is square-free. Then the equation

(5.6) $ax^2 + by^2 + cz^2 = 0$

is solvable in integers x, y and z, not all zero, if and only if
    (a) the integers a, b and c do not have the same sign
    (b) $-bc$, $-ac$ and $-ab$ are quadratic residues modulo a, b and c, respectively.

Proof: If (5.6) has a solution x_0, y_0 and z_0, not all zero, then it is clear that a, b and c do not have the same sign. Dividing x_0, y_0 and z_0 by (x_0, y_0, z_0), we have a solution x_1, y_1 and z_1 such that $(x_1, y_1, z_1) = 1$. We show that $(c, x_1) = 1$. If not, then there exists a prime, p, such that $p \mid c$ and $p \mid x_1$. Since abc is square-free we see that $p \nmid b$ and, since $p \mid ax_1^2$ and $p \mid cz_1^2$, and so $p \mid by_1^2$, we see that $p \mid y_1$ Thus $p^2 \mid ax_1^2 + by_1^2$, and so $p^2 \mid cz_1^2$. Since c is square-free we see $p \mid z_1$, and so $p \mid (x_1, y_1, z_1)$, which is a contradiction. Thus $(c, x_1) = 1$ and we can choose u so that $ux_1 \equiv 1(\mathrm{mod}\ c)$. Since $ax_1^2 + by_1^2 + cz_1^2 = 0$ implies $ax_1^2 + by_1^2 \equiv 0\ (\mathrm{mod}\ c)$, if we multiply by u^2b, we have

$$u^2 b^2 y_1^2 \equiv -ab\ (\mathrm{mod}\ c),$$

that is $-ab$ is a quadratic residue modulo c. Similarly one can show that $-bc$ is a quadratic residue modulo a and $-ac$ is a quadratic residue modulo b.

We now prove the converse. Note that if a, b and c are replaced by their negatives, then $-bc$, $-ac$ and $-ab$ are still quadratic residues modulo a, b and c, respectively. Since a, b and c are not all of the same sign we can always change signs, if necessary, so that one is positive and two are negative, say $a > 0$ and b, c < 0. Let r be an integer such that

$$r^2 \equiv -ab\ (\mathrm{mod}\ c)$$

and let a_1 be defined by

$$aa_1 \equiv 1\ (\mathrm{mod}\ c),$$

The Equation $ax^2 + by^2 + cz^2 = 0$

this exists since abc square-free implies that $(a, c) = 1$. Then

$$ax^2 + by^2 \equiv aa_1(ax^2 + by^2) \equiv a_1(a^2x^2 + aby^2) \ (\text{mod } c)$$
$$\equiv a_1(a^2x^2 - r^2y^2) \equiv a_1(ax - ry)(ax + ry) \ (\text{mod } c)$$
$$\equiv (x - a_1ry)(ax + ry) \ (\text{mod } c).$$

Thus $ax^2 + by^2 + cz^2$ is congruent to a product of two linear factors modulo c. In a similar way we may show that $ax^2 + by^2 + cz^2$ is congruent to a product of two linear factors modulo a and modulo b. By Lemma 5.7.2, we see that there exist integers $\alpha, \beta, \gamma, \alpha', \beta'$ and γ' such that

$$ax^2 + by^2 + cz^2 \equiv (\alpha x + \beta y + \gamma z)(\alpha'x + \beta'y + \gamma'z) \ (\text{mod } abc),$$

since abc square-free implies $(a, c) = (b, c) = (a, b) = 1$. Take $m = abc$, $\lambda = \sqrt{bc}, \mu = \sqrt{|ac|}$ and $v = \sqrt{|ab|}$ is Lemma 5.7.1. Then there exist integers x_1, y_1 and z_1 not all zero, such that

$$|x_1| \le \sqrt{bc}, |y_1| \le \sqrt{|ac|} \text{ and } |z_1| \le \sqrt{|ab|}$$

and

$$\alpha x_1 + \beta y_1 + \gamma z_1 \equiv 0 \ (\text{mod } abc).$$

Since abc is square-free we see that λ is an integer if and only if it is 1 and similarly for μ and v. Thus we have

$$x_1^2 \le bc \text{ with equality if and only if } b = c = -1$$
$$y_1^2 \le -ac \text{ with equality if and only if } a = 1, c = -1 \text{ and}$$
$$z_1^2 \le -ab \text{ with equality if and only if } a = 1, b = -1.$$

Thus, unless $b = c = -1$, we have

$$ax_1^2 + by_1^2 + cz_1^2 \le ax_1^2 < abc$$

and also we have

$$ax_1^2 + by_1^2 + cz_1^2 \ge by_1^2 + cz_1^2 > b(-ac) + c(-ab) = -2abc.$$

Thus, unless $b = c = -1$, we have

$$-2abc < ax_1^2 + by_1^2 + cz_1^2 < abc$$

and

$$ax_1^2 + by_1^2 + cz_1^2 \equiv 0 \ (\text{mod } abc),$$

and so we must have either

$$ax_1^2 + by_1^2 + cz_1^2 = 0 \text{ or } ax_1^2 + by_1^2 + cz_1^2 = -abc.$$

In the first case we have a solution of (5.6): x_1, y_1 and z_1. If the second case obtains, let $x_2 = -by_1 + x_1z_1$, $y_2 = ax_1 + y_1z_1$ and $z_2 = z_1^2 + ab$. It is easy to see that

$$ax_2^2 + by_2^2 + cz_2^2 = 0$$

and this gives a solution of (5.6). If $x_2 = y_2 = z_2 = 0$, then $z_1^2 + ab = 0$ and so $z_1^2 = -ab$ or $z_1 = \pm 1$, since ab is square-free. Thus $a = 1$ and $b = -1$ and $x_3 = 1$, $y_3 = -1$ and $z_3 = 0$ is a solution of (5.6), in which not all of x, y and z are zero. Finally we suppose $b = c = -1$. Then -1 is a quadratic residue of a and we may proceed as in Theorem 3.6 to show that there exist positive integers (since a is square-free) y_1 and z_1 such that

$$a = y_1^2 + z_1^2.$$

This gives the solution 1, y_1 and z_1 to the equation (5.6). Thus in all cases the equation (5.6) is solvable in integers x, y and z, not all of which are zero. ■

The question arises: if (5.6) is solvable how do we find all of its solutions? If $z \neq 0$ and a and b have the same sign, divide (5.6) by z^2 to get

(5.7) $aX^2 + bY^2 + c = 0,$

where X and Y are rational numbers and it is easy to see that to a solution of (5.6) we get a solution of (5.7) and to a solution of (5.7) we get a solution of (5.6). Let (p, q) be a rational solution of (5.7). Then the line

$$y - q = t(x - p)$$

intersects the curve determined by the equation (5.7) in two points if $t \neq -ap/bq$, where the other point is

$$x = \frac{-ap - 2bqt + bqt^2}{a + bt^2}$$

and

$$y = \frac{-aq - 2apt - bqt^2}{a + bt^2}.$$

If t is rational, then x and y are rational. Thus as t runs through all rational numbers, then x and y run through all rational solutions of (5.7) (when $t = -ap/bq$ we get x

$= p$, $y = q$ again and since a and b have the same sign $a + bt^2 \neq 0$ for any value of t). Let $t = u/v$, where $(u, v) = 1$, then

$$x = \frac{-apv^2 - 2bquv + bqu^2}{av^2 + bu^2}$$

and

$$y = \frac{-aqv^2 - 2apuv - bqu^2}{av^2 + bu^2}.$$

Thus if x_1, y_1 and z_1 is an integral primitive solution of (5.6), then all other integral solutions are give by

(5.8)
$$\Delta x = -ax_1u^2 - 2by_1uv + by_1v^2$$
$$\Delta y = ay_1u^2 - 2ax_1uv - by_1v^2 \quad .$$
$$\pm\Delta z = ax_1u^2 + bz_1v^2$$

where $(u, v) = 1$ and Δ is the greatest common divisor of the three members on the right hand side of (5.8).

We give two examples.

Example 5.3: (a) Solve the equation $x^2 + y^2 = z^2$.

(b) Solve the equation $2x^2 + 3y^2 = 5z^2$.

Solution of (a). An obvious solution to this equation is $x_1 = 1$, $y_1 = 0$ and $z_1 = 1$. Then (5.8) gives

$$\Delta x = u^2 - v^2, \Delta y = 2uv \text{ and } \pm\Delta z = u^2 + v^2,$$

where $\Delta = 1$ or 2 depending on whether the product uv is even or odd. If the product is odd, let $u_1 = (u + v)/2$ and $v_1 = (u - v)/2$, and get the solution

$$x = 2uv, y = u_1^2 - v_1^2 \text{ and } \pm z = u_1^2 + v_1^2,$$

which is simply a rearrangement of the case $\Delta = 1$. Thus all solutions of the equation $x^2 + y^2 = z^2$, with x odd and y even, are given by

$$x = u^2 - v^2, \ y = 2uv \text{ and } \pm z = u^2 + v^2,$$

where u and v are relatively prime integers of opposite parity.

Solution of (b) Here a solution is $x_1 = 1$, $y_1 = 1$ and $z_1 = 1$. Thus all solutions are given by (5.8), which in this case are

$$\Delta x = -2u^2 - 6uv + 3v^2$$
$$\Delta y = 2u^2 - 4uv - 3v^2 \; ,$$
$$\pm\Delta z = 2u^2 + 3v^2$$

where $(u, v) = 1$. In this case it is more complicated to actually compute Δ in general, but it is no problem to do so for each specific choice of u and v.

Problem Set 5.4

1. Show that $x^2 + 2y^2 = Dz^2$ is not solvable in integers if D is divisible by a prime p with $p \equiv 5 \pmod 8$.

2. If $p \equiv 5 \pmod 8$, show that $x^2 + 3y^2 - Dz^2 = 0$ is not solvable in positive integers.

3. If $p \equiv 3 \pmod 4$, $q = 4pN + 1$ are both primes, prove that $x^2 + qy^2 = pz^2$ has no solution in positive integers.

4. If $p \equiv 3 \pmod 4$, then $x^2 + y^2 = pz^2$ is not solvable in positive integers.

5. Show that if a, b and c are nonzero integers, not all of the same sign, and abc is squarefree, then the following three conditions are equivalent.
   (a) $ax^2 + by^2 + cz^2 = 0$ has an integer solution x, y and z with not all of them zero.
   (b) $ax^2 + by^2 + cz^2$ factors into linear factors modulo abc.
   (c) $-bc$, $-ac$ and $-ab$ are quadratic residues modulo a, b and c, respectively.

6. Let a, b and c be three integers such that abc is squarefree. Prove that $ax^2 + by^2 + cz^2 = 0$ is solvable in integers x, y and z, not all zero, if and only if the following two conditions both hold.
   (a) $-bc$, $-ac$ and $-ab$ are quadratic residues modulo a, b and c, respectively.
   (b) The congruence $ax^2 + by^2 + cz^2 \equiv 0 \pmod 8$ is solvable in integers x, y and z which are not all even.

7. Let a, b and c be three integers such that abc is squarefree. Prove that the equation $ax^2 + by^2 + cz^2 = 0$ is solvable in integers x, y and z, not all zero, if and only if the congruence

$$ax^2 + by^2 + cz^2 \equiv 0 \pmod N$$

is solvable for all integral moduli N in integers x, y and z such that $(x, y, z, N) = 1$.

8. Solve the following equations by the methods of this section.
   (a) $x^2 + 3y^2 = z^2$ (b) $x^2 + 6y^2 = z^2$
   (c) $x^2 + 2y^2 = z^2$ (d) $x^2 + y^2 = 4z^2$.

9. Suppose a and b are positive nonsquare integers such that $(a, 2b) = 1$. Such that the equation

$$ax^2 + by^2 = z^2$$

has a nontrivial solution with $(z, b) = 1$ and z odd if and only if a is a quadratic residue of b and b is a quadratic residue of a. (Hint: modify the proof of Theorem 5.8.)

Additional Problems for Chapter Five

General Problems

1. Find all triangles having integer sides and an interior angle of $60°$.
2. Solve the following problem of Diophantus. Find all Pythagorean triangles such that the hypotenuse minus each of its legs is a cube. (Diophantus gives 40, 96 and 104 as his answer.)
3. Find all integers x, y and z so that the three triangular numbers t_x, t_y and t_z are in arithmetic progression.
4. If $(x, y) = 1$ and $x^2 + y^2 = z^4$, show that $7 \mid xy$. Show also that the condition $(x, y) = 1$ is necessary.
5. Solve the system $x + y = z^2$ and $xy = t^2$.
6. If $z + (z + 1) = y^2$, show that $z + 1$ is the hypotenuse and z is a leg of a Pythagorean triangle.
7. Find a formula for three squares in arithmetic progression.
8. Find all solutions in positive integers x, y and z, with $(x, y, z) = 1$, of the equation $x^2 + y^3 = z^2$.
9. Show that if a and b are integers, not both zero, then the quadratic equation

$$x^2 + 2(a + b)x = a^2 + b^2$$

has no rational solutions. (Fermat had trouble solving this problem.)

10. Show that the only solution of $x^4 + y^4 = 2z^2$, with $(x, y) = 1$, is $x = y = z = 1$.
11. (a) Show that for neither choice of sign does the equation

$$x^4 \pm 4y^4 = z^2$$

have a solution in integers. (Hint: take $(x, y) = 1$ and if necessary divide by 4 to make x odd.)

(b) Show that $x^4 - y^4 = 3z^2$ has no solution in positive integers.

12. Show that the system of equations

$$x^2 + y^2 = u^2 \text{ and } x^2 + 2y^2 = v^2$$

has no solution in positive integers.

13. Find the integer solutions to the following system of equations

$$x + 1 = 2y^2 \text{ and } x^2 + 1 = 2z^2.$$

14. Determine all solutions in positive integers of $3x^4 + y^4 = z^2$.

15. Solve the following problems of Diophantus.

    (a) Find two numbers such that their sum is equal to the sum of their cubes.

    (b) Find a number which when added to each of two given numbers a and b makes each sum a square.

    (Since Diophantus was willing to accept rational numbers for solutions so shall we, however, it would make a good further exercise to find conditions for integer solutions.)

16. Solve, in integers, $x^3 + y^3 = z^3 + w^3$.

17. Solve the equation

$$\frac{1}{x} + \frac{1}{y} = \frac{1}{z}$$

in integers and show that for such a solution we have $(x, y) = (x, z) = (y, z) > 1$.

18. Show that the equation

$$\frac{1}{x} + \frac{1}{y} = \frac{1}{x+y}$$

is impossible in integers. What can be said about the equation

$$\frac{1}{x} + \frac{1}{y} = \frac{k}{x+y},$$

where k is a given integer?

19. Solve, in integers, $x^2 + 2y^2 = z^2 + w^2$.

20. Let a and b be given integers of equal parity and let $f(x) = (ax^2 + bx)/2$. Find all integers x, y and z so that

$$f(x) + f(y) = f(z).$$

(This generalizes the Pythagorean problem which is the case $a = 2$ and $b = 0$. It does include polygonal numbers as well. (See Problem 59 of Additional Problems for Chapter One.))

21. Find all Pythagorean triangles whose area is twice its perimeter. What happens if the multiple is neither two or one? (One is Problem 11 of Problem Set 5.2.)

22. Here are two more problems due to Diophantus.

   (a) Find two integers such that their product plus the square of either of them yields a square.

   (b) Find a square number such that when a given number h is added to it or subtracted from it one obtains squares in both cases.

23. Let a and b be given integers of equal parity and let $P_n = (an^2 + bn)/2$. Suppose that $c^2 + d^2 = e^2$. Show that
$$(4a(e+c)P_n + eb^2)^2 = (2abdn + db^2)^2 + (4a(e+c)P_n + cb^2)^2.$$

24. Solve the equation $(x^2 - 2)^2 = 2(y^2 + 1)$. (Hint: first show that y must be odd and that x must be even. Then reduce the problem to that of finding Pythagorean triples.)

25. Show that if m is composite and $m \neq 4$, then the congruence
$$x^n + y^n \equiv z^n \pmod{m}$$
is solvable. Show that the converse is also true.

26. Solve the equation $xy + xz + yz = 0$.

27. Solve, in integers, the equation $x_1^2 + \cdots + x_n^2 = y_1^2 + \cdots + y_n^2$, where $n \geq 3$.

28. Solve, in integers, the equation $x_1^2 + \cdots + x_n^2 = x_1 y_1 + \cdots + x_n y_n$, where $n \geq 2$.

29. In this problem we discuss the Diophantine equation

(5.9)
$$x^4 + ax^3 + bx^2 + cx + d = y^2,$$

where $x^4 + ax^3 + bx^2 + cx + d$ is not a perfect square for the given integers a, b, c and d.

   (a) Define $A = 4b - a^2$, $B = 64c - 8aA$ and $C = 64d - A^2$. Show that if we let $p(x) = 8x^2 + 4ax + A$, then we have
$$64(x^4 + ax^3 + bx^2 + cx + d) = p^2(x) + Bx + C.$$

   (b) Suppose that $\max(|a|, |b|, |c|, |d|) \leq M$. Show that if x is a solution to (5.9), then $|x| \leq 30M^3$. (Hint: consider the cases $Bx + C \neq 0$ and $Bx + C = 0$ separately.)

   (c) Solve the following Diophantine equations.

   (i) $y^2 = 1 + x + x^2 + x^3 + x^4$

   (ii) $y^2 = 8 + 8x + 2x^3 + x^4$

   (iii) $y^2 = x(x+1)(x+2)(x+3) + 1$

   (iv) $y^2 = 1 + x + 2x^2 + 2x^3 + x^4$

   (v) $y^2 = 1 + x + 3x^2 + 3x^3 + x^4$

30. Solve the equation $y^2 + y = x + x^2 + x^3 + x^4$. (Hint: complete the square on the left hand side and introduce new variables. Since the bounds from Problem 29 are so large perhaps partial completing the square, as in (a) of Problem 29, may be useful.)

31. Solve the equation $x^2 + y^2 = (x + y)z$ in integers.

32. Solve the following Babylonian systems of quadratics in integers. You will need to find conditions on the given integers a and b so that there is the possibility of integer solutions.

    (a) $x^2 + y^2 = a$ and $xy = b$.

    (This problem could be thought of as asking for the legs of a right triangle given the hypotenuse and area.)

    (b) $x + y = a$ and $x^2 + y^2 = b$.

33. Find two natural numbers such that their sum will divide their product.

34. In this problem we discuss a generalization of Legendre's equation (5.6). Let

$$H(x,y,z) = Ax^2 + Bxy + Cy^2 + Dxz + Eyz + Fz^2,$$

where A, B, C, D, E and F are integers.

    (a) Find conditions on the coefficients analogous to those of Theorem 5.7 to guarantee a solution to $H(x, y, z) = 0$. (Hint: complete the square and write $H(x, y, z)$ in the form

$$K(Lx + My + Nz)^2 + P(Qy + Rz)^2 + Sz^2,$$

    for appropriate integers K, L, M, N, P, Q, R and S.)

    (b) Discuss the solvability of the following equations.
    (i) $x^2 + 7xy + 3y^2 + 11xz + 9yz + 5z^2 = 0$
    (ii) $3x^2 + 9xy + 5y^2 + 13xz + 11yz + 7z^2 = 0$
    (iii) $x^2 + 2xy + 3y^2 + 6xz + 4yz + 5z^2 = 0$

    (c) Suppose $H(x, y, z)$ is irreducible over the real numbers and there exists a nontrivial solution (x_0, y_0, z_0), with $z_0 \neq 0$, say, of $H(x, y, z) = 0$. Show that $H(x, y, z) = 0$ has infinitely many solutions which are given by

$$x = -(Ax_0 + By_0 + Cz_0)r^2 - (2Cy_0 + Ez_0)rs + Cx_0s^2$$
$$y = Ay_0r^2 - (2Ax_0 + Dz_0)rs - (Bx_0 + Cy_0 + Ez_0)s^2 \ ,$$
$$z = z(Ar^2 + Brs + Cs^2)$$

    where r and s are relatively prime integers and r is nonnegative. (If $r = 0$, we take $s = 1$.)

35. This problem deals with Diophantine equations involving binomial coefficients.

(a) Find all triples of binomial in a given row that are in an arithmetic progression.

(b) Let k be a positive integer. Find integers x and y so that

$$\binom{x}{y+1} = k\binom{x}{y}.$$

(c) Let k be a positive integer. Find integers x and y so that

$$\binom{x+1}{y} = k\binom{x}{y}.$$

36. Solve the equation $x^y = y^x$ in integers; in rationals. (Hint: for the solution in integers write x and y in their canonical factorizations and notice the inequalities forced on the powers. Work similarly for solutions in the rationals except that here we take the exponents of the primes to be any integers.)

37. Find all solutions of $x^y = y^{x-y}$ in positive integers.

38. Find all integer values of x, y and z so that $4^x + 4^y + 4^z$ is a perfect square.

39. Solve the equation $4^x - 9^y = 55$ in positive integers.

40. Consider the equation

$$x^n + y^n = z^n,$$

where x, y and z are positive integers and $n \geq 3$.

(a) Prove that the equation has no solutions in positive integers.

(b) Make your proof in part (a) concise enough to fit in the margins of this book.

Computer Problems

1. (a) Write a program that finds all primitive Pythagorean triangles (x, y, z) with x even and $x^2 + y^2 = z^2$ with z less than some preassigned limit.

(b) Find all primitive Pythagorean triangles with $z \leq 200$. Do you find any interesting patterns? Can you prove them?

(c) Modify your program in (a) so that it finds all Pythagorean triangles with z less than some preassigned limit.

(d) Find all Pythagorean triangles with $z \leq 200$.

2. (a) Given an integer n find all Pythagorean triangles to which it is a side.

(b) Find all Pythagorean triangles to which 15 belongs; 102 belongs; 501 belongs.

3. Consider the equation (5.9). (See Problem 29 above.) Write a program that will check for solutions under the bound given in (b) of that problem. Solve the equation

$$y^2 = x^4 - 3x^3 - 2x^2 + 11x + 4.$$

CHAPTER SIX
Diophantine Equations, II

6.1. INTRODUCTION

In this chapter we deal with the problem of integrally representing a given integer, N, by a given function, $f(x_1, \ldots, x_m)$. Since general results are hard to come by we will restrict our attention, mainly, to the cases where f is a linear polynomial or a quadratic polynomial. For these cases one can more or less completely solve the representation problem, though we shall only deal with a few special cases.

The problems to be encountered when dealing with polynomials of degree three or higher are immense. In the exercises there are given some of the special results known. To adequately discuss these problems one needs to know more algebraic number theory than we will cover in Chapter 10, however in Section 10.3 we will indicate some of the simpler results that can be obtained. We state a famous theorem of Thue and then a related result of Baker.

Theorem 6.1 (Thue): The equation

$$(6.1) \qquad f(x, y) = a_0 x^n + a_1 x^{n-1} y + \cdots + a_{n-1} x y^{n-1} + a_n y^n = m \neq 0,$$

where $n \geq 3$ and $f(x, y)$ is irreducible over the rationals, has only a finite number of solutions.

Thus, in principle, given any such equation, we could find all of its solutions, if we knew it was solvable. Unfortunately, this is not known in general, that is, there are no general procedures for determining when (6.1) is solvable. However, there is a quantitative result due to Baker which gives bounds on the size of the solution. Thus one can, in truth, determine the solvability of (6.1). Unfortunately, the bounds are immense.

Theorem 6.2 (Baker): Let $m > 0$ and $\alpha > n + 1$. Then the solutions x and y of (6.1) satisfy the inequality

$$\max(|x|, |y|) < c \exp\left((\log m)^\alpha\right),$$

where c is an effectively computable constant depending only on n, α and the coefficients of $f(x, y)$.

For more details see [77, ch. 22].

6.2. LINEAR EQUATIONS

In the case where $f(x_1, ..., x_m)$ is a linear polynomial we can solve the representation problem completely. We begin with the case $m = 2$.

Theorem 6.3: The equation

(6.2) $ax + by = c,$

where a, b and c are given integers, has integral solutions x and y if and only if

$$(a, b) \mid c.$$

If (6.2) is solvable and x_0, y_0 is an integral solution, then all integral solutions are given by

$$x = x_0 - (b/(a, b))t \text{ and } y = y_0 + (a/(a, b))t,$$

where t runs through all integers.

Proof: If $(a, b) \nmid c$, then clearly there can be no integer solution since $(a, b) \mid (ax + by)$, by (e) of Theorem 1.1, for any integers x and y.

Suppose $(a, b) \mid c$ and let $a = (a, b)A$, $b = (a, b)B$ and $c = (a, b)C$. Then $ax + by = c$ if and only if $Ax + By = C$. By Corollary 1.14.1 we know $(A, B) = 1$ and by Theorem 1.11 we know there exist integers x_0 and y_0 such that

$$Ax_0 + By_0 = 1.$$

Thus

$$A(Cx_0) + B(Cy_0) = C,$$

and so

$$a(Cx_0) + b(Cy_0) = c$$

is solvable.

Let $g = (a, b)$ and assume $g \mid c$. Then (6.2) is solvable and let x, y be any other solution besides x_0, y_0. Then

$$ax_0 + by_0 = c = ax + by.$$

Thus

$$a(x_0 - x) = b(y - y_0).$$

Thus

(6.3) $(a/g)(x_0 - x) = (b/g)(y - y_0).$

Since $(a/g, b/g) = 1$, by Corollary 1.14.1, we see, by Corollary 1.14.2, that

(6.4) $(b/g) | (x_0 - x)$ and $(a/g) | (y - y_0).$

Thus, by (6.3) and (6.4), we see there exists an integer t such that

(6.5) $y = y_0 + (a/g)t$ and $x = x_0 - (b/g)t.$

It is clear that for any value of t the integers x and y, defined in (6.5), give a solution of (6.2).   ■

Example 6.1: Solve the equation $7x + 13y = -5$.

Solution: Since 7 and 13 are relatively prime, we see from the proof of Theorem 5.3, that to find a solution of the given equation it suffices to find a solution of the equation

$$7x + 13y = 1.$$

A solution to this is easily seen to be $x = 2$ and $y = -1$. Thus a solution to the given equation is $x_0 = -10$ and $y_0 = 5$. Thus all solutions to the given equation are given by

$$x = -10 - 13t \text{ and } y = 5 + 7t,$$

where t runs through the integers.   □

In the above example it was relatively easy to solve for a particular solution. In general, it is not such an easy matter to spot a particular solution of the equation

(6.6) $ax + by = (a, b).$

We give an algorithm for finding particular solutions of (6.6) that uses the Euclidean Algorithm. This is basically a restatement of Theorem 1.7. The main difference is that this revision fits closer to the mold of finite continued fractions ((4.17) - (4.19)).

Let a and b be given positive integers. If we apply the Euclidean Algorithm to a and b we get the sequence of equations

$$b = aq_1 + r_1$$
$$a = r_1 q_2 + r_2$$
$$\vdots$$
$$r_{n-2} = r_{n-1} q_n + r_n,$$

where $r_n = (a, b)$ is the last nonzero remainder. From the first equation we get

$$r_1 = a - bq_1 = ak_1 - bh_1$$

where $k_1 = 1$ and $h_1 = q_1$

$$r_2 = (h_1 q_2 + 1)b - k_1 q_2 a = -(ak_2 - bh_2),$$

where $h_2 = q_2 h_1 + 1$ and $k_2 = q_2 k_1$. Continuing in this way we get

$$r_m = (-1)^{m-1}(ak_m - bh_m),$$

where

$$h_{m+1} = q_{m+1} h_m + h_{m-1}$$
$$k_{m+1} = q_{m+1} k_m + k_{m-1}.$$

If we set $h_0 = 1$ and $k_0 = 0$ we can say the above recurrence relation holds for $m = 1, ..., n - 1$. Thus we get

$$r_n = (a, b) = (-1)^{n-1}(ak_n - bh_n).$$

Thus we may take $x = (-1)^{n-1}k_n$ and $y = (-1)^n h_n$ in (6.6). Note that h_n/k_n is simply the nth convergent to the simple continued fraction expansion of b/a.

Example 6.2: Find the solution to $91x + 221y = 1066$. Find all solutions in positive integers.

Solution: We have $(91, 221) = 13$ and $1066 = 13 \cdot 82$, so that the equation is solvable. We have

$$91 = 221 \cdot 0 + 91$$
$$221 = 91 \cdot 2 + 39$$
$$91 = 39 \cdot 2 + 13$$
$$39 = 13 \cdot 3.$$

Thus $q_1 = 0$, $q_2 = 2$, $q_3 = 2$ and $n = 3$. Thus we have the following table:

$$h_0 = 1 \quad h_1 = 0 \quad h_2 = 1 \quad h_3 = 2$$
$$k_0 = 0 \quad k_1 = 1 \quad k_2 = 2 \quad k_3 = 5.$$

Thus we take $u = (-1)^{3-1} \cdot 5 = 5$ and $v = (-1)^3 \cdot 2 = -2$ as solutions to $91u + 221v$

= 3. Thus we may take $x_0 = 82u = 410$ and $y_0 = 82v = 164$ as a solution to $91x + 221y = 1066$. All solutions are therefore given by

(6.7) $x = 410 + 17t$ and $y = -164 - 7t,$

where t runs through all integers.

To find all positive solutions we need to find those values of t which simultaneously make x and y, as given by (6.7), positive, that is, t must satisfy the simultaneous inequalities

$$410 + 17t > 0 \text{ and } -164 - 7t > 0$$

This is equivalent to

$$-410/17 < t < -164/7.$$

The only integer in this interval is -24. Thus there is only one positive solution, namely

$$x = 2 \text{ and } y = 4. \qquad \square$$

For more on positive solutions see problem 12 in the exercises below.

We now tackle the case of $n \geq 3$ variables, that is, the equation

(6.8) $a_1 x_1 + \cdots + a_n x_n = c,$

where $n \geq 3$ and $a_1, \ldots, a_n$ and c are given integers. By Theorem 1.13 we see that if $g = (a_1, \ldots, a_n)$, then a necessary and sufficient condition for the solvability of (6.8) is that $g \mid c$. To solve (6.8) we shall reduce it to a similar equation in $n - 1$ variables and hence down to 2 variables.

Without loss of generality we may assume not all the a_i, $1 \leq i \leq n$, are zero and that $g \mid c$. Let

(6.9) $x_{n-1} = \alpha u + \beta v$ and $x_n = \gamma u + \delta v,$

where α, β, γ and δ are integers such that $\alpha\delta - \beta\gamma = 1$. Then

$$u = \delta x_{n-1} - \beta x_n \text{ and } v = -\gamma x_{n-1} + \alpha x_n.$$

Thus u and v are integers if and only if x_{n-1} and x_n are integers. Now take $\beta = a_n / (a_{n-1}, a_n)$ and $\delta = -a_{n-1} / (a_{n-1}, a_n)$. By Corollary 1.10.1, we see that $(\beta, \delta) = 1$, and so, by Theorem 1.5, there exist integers α and γ such that $\alpha\delta - \beta\gamma = 1$. Since $a_{n-1}\beta + a_n\delta = 0$ we see that (6.8) becomes

(6.10) $a_1 x_1 + \cdots + a_{n-2} x_{n-2} + (a_{n-1}\alpha + a_n\gamma)u = c,$

which has one less variable. We have

$$a_{n-1}\alpha + a_n\gamma = -(a_{n-1}, a_n)a\delta + (a_{n-1}, a_n)\beta\gamma = -(a_{n-1}, a_n)$$

and, by Corollary 1.9.1, we have

$$(a_1, \ldots, a_{n-2}, (a_{n-1}\alpha + a_n\gamma)) = (a_1, \ldots, a_{n-2}, -(a_{n-1}, a_n)) = g.$$

Thus (6.10) is solvable if and only if (6.8) is solvable. Thus we may reduce the original equation (6.8) when $n \geq 3$, to an equation of the type covered by Theorem 6.3 and use a series of equations of the form (6.9) to recover the solutions to (6.8).

Example 6.3: Solve $x + 2y + 3z = 1$.

Solution: Since $(1, 2, 3) = 1$ we see the equation is solvable. Here $a_1 = 1$, $a_2 = 2$ and $a_3 = 3$. We have $(a_2, a_3) = (2, 3) = 1$. Thus, for this equation,

$$\beta = 3/1 = 3 \text{ and } \delta = -2/1 = -2.$$

We can choose $\alpha = 1$ and $\gamma = -1$. Then the original equation is transformed into the equation

$$x - u = 1.$$

A solution here is $x_0 = 2$ and $u_0 = 1$, and so, by Theorem 6.3, the general solution is

$$x = 2 - t \text{ and } u = 1 - t.$$

Thus

$$y = 1 \cdot u + 3 \cdot v = 1 - t + 3v$$

and

$$z = -1 \cdot u - 2v = -1 + t - 2v.$$

If we let t and v run through all integers independently of one another we obtain all solutions to the given equation. □

We close this section with two examples that illustrate other ways of solving linear Diophantine equations. The benefit of these methods is that they do not require one to find an initial solution to the equation. The two methods are somewhat similar in the way they work. One of them uses simple divisibility ideas and the other uses congruences.

Example 6.4.(a) Solve the equation

$$3x + 5y = 7.$$

   (b) Solve the equation

$$6x + 15y = 11.$$

   (c) Solve the equation

$$8x - 5y + 7z = 21.$$

Solution. The method of solution used in this example might be called the **method of reduction**. We don't reduce the number of variables, as above, but we reduce the coefficients. This method works for the same reason the Euclidean algorithm works, namely when we apply the division algorithm over and over again the nonnegative remainders must eventually become zero. In general, when we solve an equation we solve for one of the variables (usually the one with the smallest coefficient) and this then leads us to a string of conditions for integral solutions.

   (a) We have

$$3x = 7 - 5y$$

or

$$x = 2 - y + \frac{1 - 2y}{3}.$$

Thus, for integral solutions, there must be an integer t such that

$$\frac{1 - 2y}{3} = t$$

or

$$1 - 2y = t \text{ or } y = -t + \frac{1 - t}{2}.$$

Again we see that there is an integer u such that

$$\frac{1 - t}{2} = u \text{ or } t = 1 - 2u.$$

If we work backward through this string of equations we find that the general solution to the equation is

$$x = 4 - 5u \text{ and } y = 3u - 1.$$

   (b) Proceeding as above, we have

$$6x = 11 - 15y \text{ or } x = 1 - 2y + \frac{5 - 3y}{6}.$$

Thus, for integral solutions, there is an integer t such that

$$\frac{5-3y}{6} = t \quad \text{or} \quad y = \frac{5}{3} - 2t.$$

Since y and t are to be integers we see we are in an untenable position. Thus the equation has no integer solutions.

We know, by Theorem 6.3, that this equation has no solution, but this illustrates how the method handles a case like this.

(c)   We proceed as above and solve for y. We get

$$5y = 8x + 7z - 21 \quad \text{or} \quad y = x + z - 4 + \frac{3x + 2z - 1}{5}.$$

Thus we need an integer t such that

$$\frac{3x + 2z - 1}{5} = t \quad \text{or} \quad 3x + 2z - 1 = 5t \quad \text{or} \quad z = 2t - 2x + \frac{1 + x + t}{2}.$$

Thus we need an integer u such that

$$\frac{1 + x + t}{2} = u \quad \text{or} \quad x = 2u - 1 - t.$$

If we work backward we find that the general solution is

$$x = 2u - 1 - t, \, y = -u + 4t - 3 \text{ and } z = 2 - 3u + 4t. \qquad \square$$

In the next example we will solve the same three equations by the **method of congruences**. This will help illustrate the similarities and differences of the two methods.

Example 6.5.(a) Solve the equation

$$3x + 5y = 7.$$

(b)   Solve the equation

$$6x + 15y = 11.$$

(c)   Solve the equation

$$8x - 5y + 7z = 21.$$

Solution. Here we pick a variable (usually the one with the least coefficient) and reduce everything to a congruence with that coefficient as the modulus.

(a)   We have the congruence

$$5y \equiv 7 \pmod 3.$$

Since $(5, 3) = 1$ and $1 \mid 2$ we see, by Theorem 2.9, that this congruence is solvable.

Since $2 \cdot 5 \equiv 1 \pmod 3$ this congruence is equivalent to

$$y \equiv 14 \equiv 2 \pmod 3.$$

Thus there is an integer t such that $y = 2 + 3t$. If we substitute this into the equation we find that

$$7 = 3x + 5(2 + 3t) \text{ or } 3x = -3 - 15t \text{ or } x = -1 - 5t.$$

Thus the general solution is

$$x = -1 - 5t \text{ and } y = 2 + 3t.$$

(b) Here we obtain the congruence

$$15y \equiv 11 \pmod 6.$$

Now $(15, 6) = 3$ and $3 \nmid 11$. Thus, by Theorem 2.9, this congruence has no solutions. Thus the equation has no solutions.

(c) Here we obtain the congruence

$$5x + 7z \equiv 21 \pmod 5$$

or, if we reduce modulo 5, the congruence is

$$3x + 2z \equiv 1 \pmod 5.$$

We then pick one of the remaining variables to solve for, say x. This gives

$$3x \equiv 1 - 2z \pmod 5$$

or, since $3 \cdot 2 \equiv 1 \pmod 5$,

$$x \equiv 2 - 4z \equiv 2 + z \pmod 5.$$

If we take $z = t$, then there is an integer u such that $x = 2 + t + 5u$. Thus

$$21 = 8(2 + t + 5u) - 5y + 7t \text{ or } y = -1 + 3t + 8u.$$

Thus the general solution of this equation is

$$x = 2 + t + 5u, \ y = -1 + 3t + 8u \text{ and } z = t. \qquad \square$$

In both (a) and (c) of the two examples the forms of the answers are different, but a linear transformation will put them into agreement. For example, if we let $u = t + 1$, then the two part (a) answers agree. It is a little more complicated for the part (c) answers, but not too much (see Problem 4 below.)

Of course, either of these methods is applicable to equations in more variables.

Problem Set 6.2

1. Give the complete solution in integers to the following linear Diophantine equations.

   (a) $3x + 5y = 1$

   (b) $10x - 7y = 17$

   (c) $5x + 3y = 52$

   (d) $6x + 35y = 1$

   (e) $14x - 33y = 7$

   (f) $2x + 11y = 5$

   (g) Solve each of the equations (a) - (f) by the other two methods that you did not use.

2. Find all positive integer solutions to the following linear Diophantine equations.

   (a) $5x + 3y = 52$

   (b) $2x + 13y = 50$

   (c) $5x + 13y = 72$

   (d) $x + 6y = 18$

   (e) $8x + 12y = 144$

   (f) $18x - 21y = 15$

3. Give complete solutions in integers to the following linear Diophantine equations.

   (a) $5x - 2y - 4z = 1$

   (b) $2x + 3y + 7z = 15$

   (c) $x - y + 3z = 5$

   (d) $2x + 3y - z + 4w = 15$

   (e) $36x + 24y + 18z + 30w = 6$

   (f) $x - y + 2z + 5w = 17$

   (g) Solve each of the equations (a) - (f) by the other two methods that you did not use.

4. Find a linear transformation that will transform the solutions of Example 6.4(c) into the solutions of Example 6.5(c)

5. Find the positive integer solutions to the following equations.

   (a) $97x + 56y + 3z = 16047$

   (b) $17x + 15 = 13y + 11 = 10z + 3$

   (c) $23x + 12 = 17y + 7 = 10z + 3$

   These three problems are due to Regiomontanus (Johann Müller.)

   (d) $30x + 49y + 23z = 1000$

   (e) $101x + 311y + 203z = 10000$

6. If bananas sell for $.85 per pound and plums are $.23 each, how much of each type of fruit was bought if a total of $7.17 was paid for the fruit?

7. A woman goes to a post office to mail a package and the postal worker weighs the package and determines that it will cost $2.89. If the post office only has 23 cent stamps and 15 cent stamps to use how many of each must be used?

8. The following problem is due to Euler (see [30].) Divide 100 into two parts so that one part is divisible by 11 and the other part is divisible by 7.

9. Solve the following problem due to the Indian mathematician Mahaviracarya. In the forest 37 equal piles of wood apples were seen by some travelers. After 17 rotted fruits were removed the remainder of the wood apples were divided evenly among the 79 travelers. What is the least number of apples it was possible for each traveler to get?

10. Solve the following linear equations in integers x and y.

   (a) If n and m are integers, solve
   $$nx + (n + 1)y = m.$$

   (b) If n is an odd integer and m is any integer, solve
   $$nx + (n + 2)y = m.$$

   (c) If n is an integer not divisible by 3 and m is any integer, solve
   $$nx + (n + 3)y = m.$$

   (d) Generalize these results.

11. Solve the following systems of equations in integers.

   (a) $\begin{aligned} 5x + 6y + 7z &= 173 \\ 17x + 4y + 3z &= 510 \end{aligned}$

   (b) $\begin{aligned} 2x + 5y + 3z &= 324 \\ 6x - 4y + 14z &= 190 \end{aligned}$

   (c) $\begin{aligned} x + y + z + w &= 26 \\ 3x + 2y + 4z + w &= 63 \\ 2x + 2y + 2z + 4w &= 74 \end{aligned}$

   (d) $\begin{aligned} x + y + z + w &= 4 \\ 5y + 6z + 9w &= 18 \end{aligned}$

12. Here is an old problem due, in this version, to Alcuin. If 100 bushels of corn are distributed among 100 people as follows: each man gets three bushels, each woman gets 2 bushels and each child gets half a bushel, then how many men, women and children are there?

13. Prove that if a and b are relatively prime integers of different sign, then the equation $ax + by = c$ has infinitely many solutions in positive integers x and y for any integer value of c.

14. Let N denote the number of positive integer solutions of the equation $ax + by = c$, where a, b and c are given integers.

   (a) Show that
   $$-\left[\frac{-(a,b)c}{ab}\right] - 1 \le N \le -\left[\frac{-(a,b)c}{ab}\right].$$

(b) If $(a, b) = 1$ and $ab \nmid c$, then show that

$$N = \left[\frac{c}{ab} \right] \quad \text{or} \quad N = \left[\frac{c}{ab} \right] + 1.$$

Show that if $a \mid c$, then $N = [c/ab]$. (Hint: $a \mid c$, then a specific solution can easily be found.)

(c) If $ab \mid c$ and $(a, b) = 1$, show that $N = -1 + c/ab$.

15. Let a and b be positive integers such that $(a, b) = 1$. If c is a positive integer, let $N(c)$ be the number of solutions of $ax + by = c$ in positive integers x and y.

(a) Suppose r and s are integers such that $br - as = 1$. Show that

$$N(c) = \left[\frac{rc}{a} \right] - \left[\frac{sc}{b} \right] - E(c),$$

where $E(c) = 1$ or 0 according to whether $a \mid c$ or not.

(b) If n is a positive integer, show that all integers c so that $N(c) = n$ satisfy

$$(n - 1)ab + a + b \le c \le (n + 1)ab.$$

16. (a) Suppose that $(a, b, c) = 1$. Show that the solutions to

$$ax + by + cz = 1$$

are given by

$$x = \alpha\delta + c\beta t + ub / d, \quad y = \beta\delta + c\alpha t - ua / d \quad \text{and} \quad z = \gamma - dt,$$

where $d = (a, b)$, α and β are integers so that $a\alpha + b\beta = d$, γ and δ are integers so that $d\delta + c\gamma = 1$ and t and u are arbitrary integers.

(b) Use this result to solve

(i) $x + 3y + 5z = 1$ (ii) $11x + 13y + 17z = 1$.

(c) Can you extend this result to $n \ge 4$ variables?

(d) Can you make statements for the equation $ax + by + cz = d$ analogous to those of Problems 14 or 15?

6.3. SUMS OF TWO SQUARES

In this section we wish to determine those positive integers n for which the equation

(6.11) $x^2 + y^2 = n$

is solvable in integers x and y. There are many ways to do this. One can proceed by way of algebraic number theory and study the complex numbers of the form $a + bi$, where a and b are integers and $i = \sqrt{-1}$. We will do this in Chapter Ten. In this section we will use another approach, which is capable (as is the first method) of generalization to other representation problems. Let $r_2(n)$ be the number of solutions of the equation (6.11). Then the numbers n for which (6.11) is solvable are precisely those numbers n for which $r_2(n) \geq 1$. We shall see that it is relatively straightforward to calculate $r_2(n)$ and we shall give several equivalent formulas for it.

We say a solution of (6.11) is **primitive** if $(x, y) = 1$. Let $P(n)$ be the number of nonnegative primitive solutions of (6.11) and $Q(n)$ the number of primitive solutions of (6.11). In counting solutions of (6.11) we consider the solutions x_1, y_1 and x_2, y_2 to be different if $x_1 \neq x_2$ or $y_1 \neq y_2$. Thus order is taken into consideration.

Since $1 = (\pm 1)^2 + 0^2 = 0^2 + (\pm 1)^2$ we see that $r_2(1) = Q(1) = 4$ and $P(1) = 2$. If $n > 1$, let x and y be a nonnegative primitive solution of (6.11). Then $x \geq 1$ and $y \geq 1$ and $\pm x, \pm y$ is a primitive solution for all choices of sign. Thus $Q(n) = 4P(n)$. Note that if $x^2 + y^2 = n$ and $g = (x, y)$, then $g^2 \mid n$, $(x/g, y/g) = 1$ and $(x/g)^2 + (y/g)^2 = (n/g^2)$. Thus

$$r_2(n) = \sum_{d^2 \mid n} Q(n / d^2),$$

where the sum is over all squares that divide n.

Theorem 6.4: Suppose $n > 1$ is an integer. Then each nonnegative primitive solution of $x^2 + y^2 = n$ determines a unique s modulo n such that $sy \equiv x \pmod{n}$. Moreover, $s^2 \equiv -1 \pmod{n}$ and different nonnegative primitive solutions determine different s modulo n.

Proof: If x and y are a nonnegative primitive solution, then $(y, n) = 1$. Thus, by Theorem 2.9, the congruence $sy \equiv x \pmod{n}$. has a unique solution modulo n.

Let $\bar{y}$ be a solution of $y\bar{y} \equiv 1 \pmod{n}$, which exists by Theorem 2.9. Then $s \equiv x\bar{y} \pmod{n}$. Thus

$$s^2 \equiv x^2 \bar{y}^2 \equiv -y^2 \bar{y}^2 \equiv -1 \pmod{n}.$$

Suppose x, y and u, v are two nonnegative primitive solutions and s is an integer such that

$$sy \equiv x \pmod{n} \quad \text{and} \quad sv \equiv u \pmod{n}.$$

Then

$$xv \equiv syv \equiv yu \pmod{n}.$$

Since $n > 1$ any nonnegative primitive solutions must satisfy $1 \leq x < \sqrt{n}$ and $1 \leq v < \sqrt{n}$. Thus $1 \leq xv < n$. Similarly $1 \leq yu < n$. Since $n \mid (xv - yu)$ we must have xv

$= yu$. Since $(x, y) = (u, v) = 1$ we see that we must have $x = u$ and $y = v$, by Corollary 1.10.2 and (f) of Theorem 1.1. ∎

Theorem 6.5: Suppose $n > 1$ and $s^2 \equiv -1 \pmod{n}$. Then there is a nonnegative primitive solution x and y of $x^2 + y^2 = n$ such that $sy \equiv x \pmod{n}$.

Proof: As in the proof of Corollary 2.10.1 we can find u_0 and v_0 such that

$$0 < |u_0|, |v_0| \leq \sqrt{n}, \ sv_0 \equiv u_0 \pmod{n} \quad \text{and} \quad u_0^2 + v_0^2 = n.$$

If n is a square and $|u_0| = |v_0| = \sqrt{n}$, then we have

$$s\sqrt{n} \equiv \pm n \pmod{n}$$

or, by Theorem 2.3,

$$s \equiv \pm 1 \pmod{\sqrt{n}}.$$

Thus

$$s^2 \equiv 1 \pmod{\sqrt{n}}$$

Since $s^2 \equiv -1 \pmod{n}$ implies $s^2 \equiv -1 \pmod{\sqrt{n}}$ we have

$$1 \equiv -1 \pmod{\sqrt{n}}.$$

Thus $\sqrt{n} = 2$ or $n = 4$. Since $s^2 \equiv -1 \pmod 4$ is impossible we see that we must have $|u_0|, |v_0| < \sqrt{n}$.

Let $g = (u_0, v_0)$. Then $g^2 | n$. Also

$$s(v_0 / g) \equiv u_0 / g \pmod{n / g}.$$

Thus

$$\begin{aligned}
n / g^2 &= (u_0^2 + v_0^2) / g^2 \\
&\equiv (sv_0 / g)^2 + (v_0 / g)^2 \pmod{n / g} \\
&\equiv -(v_0 / g)^2 + (v_0 / g)^2 \equiv 0 \pmod{n / g},
\end{aligned}$$

which is only possible if $g = 1$, and so we have a primitive solution.

If u_0 and v_0 have the same sign, let $x = |u_0|$ and $y = |v_0|$. Then

$$sy = s(\pm v_0) \equiv \pm u_0 = x \pmod{n}.$$

If u_0 and v_0 have opposite signs let $x = |v_0|$ and $y = |u_0|$. Then

$$sy = s(\pm u_0) \equiv \pm s(sv_0) \equiv \mp v_0 = x \pmod{n}$$

Thus we have a nonnegative solution. ∎

These two theorems give us a one-to-one correspondence between the number of nonnegative primitive solutions of $x^2 + y^2 = n$ and the number of solutions of $s^2 \equiv -1 \pmod{n}$. Let $R(n)$ be the number of incongruent solutions of $s^2 \equiv -1 \pmod{n}$. Then the content of the last two theorems can be expressed by the following result.

Corollary 6.5.1: We have $R(1) = 1$. For $n > 1$ we have $P(n) = R(n)$ and $Q(n) = 4R(n)$. For $n > 1$ we have

$$r_2(n) = 4\sum_{d^2 \mid n} R(n / d^2).$$

Theorem 6.6: Let $(m, n) = 1$. Then $R(m)R(n) = R(mn)$ and $(r_2(m)/4)(r_2(n)/4) = r_2(mn)/4$.

Proof: The result for R follows from Theorem 2.13. We have, by unique factorization and Corollary 5.6.1,

$$\frac{1}{4}r_2(mn) = \sum_{d^2 \mid mn} R\left(\frac{mn}{d^2}\right) = \sum_{d_1^2 \mid m}\sum_{d_2^2 \mid n} R\left(\frac{m}{d_1^2} \cdot \frac{n}{d_2^2}\right)$$

$$= \sum_{d_1^2 \mid m} R\left(\frac{m}{d_1^2}\right)\sum_{d_2^2 \mid n} R\left(\frac{n}{d_2^2}\right)$$

$$= \left(\frac{1}{4}r_2(m)\right)\left(\frac{1}{4}r_2(n)\right),$$

by the first part of the theorem.   ∎

Functions that satisfy the multiplication criterion of Theorem 6.6 are said to be multiplicative functions. (For more on multiplicative functions see Section 7.3.) If we know that a function is multiplicative, then to calculate its values for arbitrary positive integers it suffices to calculate its values for prime powers. We will use this fact in the next theorem.

First we define a function that will be useful in calculating $r_2(n)$. Define the function $\chi(n)$ as follows:

(a) $\chi(1) = 1$,

(b) $\chi(2) = 0$,

(c) $\chi(p) = \left(\dfrac{-1}{p}\right)$, the Legendre symbol,

(so that $\chi(p) = (-1)^{(p-1)/2}$ by (f) of Theorem 3.2), and

(d) for general $n > 1$ write

$$n = \prod_{i=1}^{r} p_i^{a_i}$$

in its canonical factorization and define

$$\chi(n) = \prod_{i=1}^{r} \chi(p_i)^{a_i}.$$

Note that if n is even, then $\chi(n) = 0$ and if n is odd, then, by Definition 3.3,

(6.12) $$\chi(n) = \left(\frac{-1}{n}\right),$$

the Jacobi symbol.

For further reference we state the following properties of the function $\chi(n)$.

Theorem 6.7: (a) If a and b are any two integers, then

$$\chi(ab) = \chi(a)\chi(b).$$

(b) $\chi(n)$ satisfies the following relation

$$\chi(n) = \begin{cases} 0 & \text{if } n \equiv 0 \ (\text{mod } 2) \\ +1 & \text{if } n \equiv 1 \ (\text{mod } 4) \\ -1 & \text{if } n \equiv -1 \ (\text{mod } 4) \end{cases}.$$

(c) If $a + b \equiv 0 \ (\text{mod } 4)$, then $\chi(a) = -\chi(b)$.

Proof of (a) follows from (d) of the definition of $\chi(n)$.

Proof of (b) If $n \equiv 0 \ (\text{mod } 2)$, then $\chi(n) = 0$, by (b) of the definition of $\chi(n)$.

If $n \equiv 1 \ (\text{mod } 4)$, then n is odd and $p^e \| n$ implies that e can be any integer if $p \equiv 1 \ (\text{mod } 4)$, but e must be even if $p \equiv 3 \ (\text{mod } 4)$. Thus, by the definition of $\chi(n)$

$$\chi(n) = \prod_{\substack{p^e \| n \\ p \equiv 1 \ (\text{mod } 4)}} \chi(p)^e \prod_{\substack{p^{2e} \| n \\ p \equiv 3 \ (\text{mod } 4)}} \chi(p)^{2e}$$

$$= \prod_{\substack{p^e \| n \\ p \equiv 1 \ (\text{mod } 4)}} (-1)^{(p-1)e/2} \prod_{\substack{p^{2e} \| n \\ p \equiv 3 \ (\text{mod } 4)}} (-1)^{(p-1)2e/2} = +1.$$

If $n \equiv 3 \ (\text{mod } 4)$, then n is odd and if $p^e \| n$, then e can be anything if $p \equiv 1 \ (\text{mod } 4)$, but e must be odd for an odd number of $p \equiv 3 \ (\text{mod } 4)$. Thus

$$\chi(n) = \prod_{\substack{p^e \| n \\ p \equiv 1 \ (\text{mod } 4)}} \chi(p)^e \prod_{\substack{p^{2e} \| n \\ p \equiv 3 \ (\text{mod } 4)}} \chi(p)^{2e} \prod_{\substack{p^{2e+1} \| n \\ p \equiv 3 \ (\text{mod } 4)}} \chi(p)^{2e+1}$$

$$= (+1)(+1)(-1) = -1,$$

which proves (b) for n positive and we can use it as a definition if n is negative.

Proof of (c) From (b) we see that if $a \equiv b \pmod 4$, then $\chi(a) = \chi(b)$. Thus, if $a + b \equiv 0 \pmod 4$, then $\chi(a) = \chi(-b) = \chi(-1)\chi(b) = -\chi(b)$, since $\chi(-1) = -1$, by (b), which proves (c). ∎

Theorem 6.8: If n is a positive integer, then

(6.13) $$r_2(n) = 4\sum_{d|n} \chi(d),$$

where the sum is over all positive divisors of n.

Proof: Since $r_2(n)/4$ is a multiplicative function, by Theorem 6.6, and the proof that $\sum_{d|n}\chi(d)$ is a multiplicative function goes analogously to the proof for $r_2(n)/4$ (see Problem 1 below), to prove the theorem we need only show that the two functions agree when n is a prime power.

Now $s^2 \equiv -1 \pmod 2$ has the unique solution $s \equiv -1 \pmod 2$ and if $e \geq 2$, then $s^2 \equiv -1 \pmod{2^e}$ has no solutions since $s \equiv 0,1 \pmod 4$. Thus, by Corollary 6.5.1, we have

$$\frac{1}{4}r_2(n) = \sum_{f=0}^{[e/2]} R(2^{e-2f}) = 1.$$

since the only nonzero term on the right hand side occurs when $e - 2f = 0$ or 1 and only one of these can occur. Also

$$\sum_{d|2^e} \chi(d) = \sum_{f=0}^{e} \chi(2^f) = 1,$$

since the only nonzero term occurring is when $f = 0$. Thus the two sides of (6.13) agree when n is a power of 2.

Now consider an odd prime p. Now $s^2 \equiv -1 \pmod p$ has to solutions if $p \equiv 1 \pmod 4$ and no solutions if $p \equiv 3 \pmod 4$. By the general theory of polynomial congruences of Chapter Two applied to the congruence $x^2 + 1 \equiv 0 \pmod{p^e}$ we see that $s^2 \equiv -1 \pmod{p^e}$ has the same number of solutions for each value of $e \geq 1$, since $(x^2 + 1)' = 2x$. (We could also use the results in Section 6 of Chapter Three to derive the same conclusion.) Thus, for $e \geq 1$, we have

$$R(p^e) = \begin{cases} 2 & \text{if } p \equiv 1 \pmod 4 \\ 0 & \text{if } p \equiv 3 \pmod 4 \end{cases}.$$

Thus, for e even, we have

$$\frac{1}{4}r_2(p^e) = \sum_{f=0}^{e/2} R(p^{e-2f}) = \frac{e}{2}R(p) + R(1)$$

(6.14)

$$= \begin{cases} e+1 & \text{if } p \equiv 1 \pmod 4 \\ 1 & \text{if } p \equiv 3 \pmod 4 \end{cases},$$

since $R(1) = 1$. For e odd, we have

$$\frac{1}{4}r_2(p^e) = \sum_{f=0}^{(e-1)/2} R(p^{e-2f}) = \frac{e+1}{2}R(p)$$

(6.15)

$$= \begin{cases} e+1 & \text{if } p \equiv 1 \pmod 4 \\ 1 & \text{if } p \equiv 3 \pmod 4 \end{cases}.$$

Finally, by (6.14) and (6.15), we have

$$\sum_{d|p^e}\chi(d) = \sum_{f=0}^{e}\chi(p^f) = \chi(1) + \sum_{f=1}^{e}(-1)^{(p-1)f/2}$$

(6.16)

$$= \begin{cases} e+1 & \text{if } p \equiv 1 \pmod 4 \\ 1 & \text{if } p \equiv 3 \pmod 4 \ e \text{ even}. \\ 0 & \text{if } p \equiv 3 \pmod 4, \ e \text{ odd} \end{cases}$$

Thus the two sides of (6.13) agree for odd prime powers.   ∎

Corollary 6.8.1: $r_2(n)$ is four times the excess of the number of divisors of n of the form $4j + 1$ over those of the form $4j + 3$.

Proof: If $d = 1$, then $\chi(d) = 1$. If d is even, then $\chi(d) = 0$. Suppose d is odd. Then, by Theorem 6.8 and (6.12),

$$r_2(n) = 4\sum_{d|n}\chi(d) = 4\sum_{d|n}\left(\frac{-1}{d}\right).$$

By Theorem 3.10, we have $\left(\dfrac{-1}{n}\right) = (-1)^{(n-1)/2}$. Thus

$$r_2(n) = 4\sum_{d|n}(-1)^{(d-1)/2}.$$

Since

$$(-1)^{(d-1)/2} = \begin{cases} +1 & \text{if } d \equiv 1 \pmod 4 \\ -1 & \text{if } d \equiv 3 \pmod 4 \end{cases},$$

the result follows.   ∎

Corollary 6.8.2: The equation

$$x^2 + y^2 = n$$

is solvable if and only if the canonical factorization of n into prime powers contains no factor p^e with e odd and $p \equiv 3 \pmod 4$.

This follows immediately from (6.16) and (6.13).

Example 6.6: (a) Since $1722 = 2 \cdot 3 \cdot 7 \cdot 41$ and both 3 and 7 are $\equiv 3 \pmod 4$ and occur to odd powers we see that 1722 cannot be written as a sum of two squares.

(b)   We have $1744 = 2^4 \cdot 109$. Since $109 \equiv 1 \pmod 4$ we see that 1744 can be written as a sum of two squares. We have, from the proof of Theorem 6.8,

$$\frac{1}{4} r_2(1744) = \frac{1}{4} r_2(16) \frac{1}{4} r_2(109) = 1 \cdot 2 = 2.$$

Thus $r_2(1744) = 8$, which accounts for rearrangements and the choices of sign so that 1744 has essentially only one representation as a sum of two squares, namely

$$1744 = 40^2 + 12^2. \qquad \square$$

It might be remarked that if all we wished to prove was Corollary 6.8.2, then we could have done so in a reasonably direct manner. In fact this is the way we have chosen to proceed for sums of four squares in the next section. We chose the manner of presentation we did to illustrate other ways of preceding to the same goal. Sometimes to show something can be represented in some way it is easier, or at least no harder, to show that the number of representations is a positive integer.

To prove Corollary 6.8.2 directly we could proceed as follows.

First, we prove the algebraic identity

(6.17) $$(a^2 + b^2)(c^2 + d^2) = (ac \pm bd)^2 + (ad \mp bc)^2.$$

(This is Problem 2 below.) This shows that we need only worry about proving representability for primes. Since we also have

$$2(a^2 + b^2) = (a + b)^2 + (a - b)^2$$

(this is Problem 5 below) we can restrict ourselves to odd primes.

Second, we show that if $p \equiv 1 \pmod 4$, then there are integers a and b so that

$$p = a^2 + b^2.$$

This was done in Corollary 2.10.1.

Finally, we show that if $p \equiv 3 \pmod 4$ and $p \mid n = a^2 + b^2$ for some integers a and b, then $p \mid a$ and $p \mid b$. This can be done much as in Corollary 2.9.3. Or to do it a little differently we suppose that

$$a^2 + b^2 \equiv 0 \pmod p.$$

Then we have

$$a^2 \equiv -b^2 \pmod p.$$

If $p \nmid b$, then there is a $c \not\equiv 0 \pmod p$ such that $bc \equiv 1 \pmod p$. Thus we have

$$(ac)^2 = a^2 c^2 \equiv -b^2 c^2 \equiv -(bc)^2 \equiv -1 \pmod p,$$

which contradicts Corollary 2.9.2 or (f) of Theorem 3.2. Thus we must have $p \mid b$, and so $p \mid a$. This shows, by (6.17), that if $p \equiv 3 \pmod 4$ and $p \mid n = a^2 + b^2$, then $p \mid a$ and $p \mid b$, and so $p^2 \mid n$.

If we combine all of these results, we have Corollary 6.8.2.

Problem Set 6.3

1. Suppose that f is a multiplicative function, that is, $(m, n) = 1$ implies that $f(mn) = f(m)f(n)$. Show that if we define

$$F(n) = \sum_{d \mid n} f(d),$$

   then F is a multiplicative function. (Hint: mimic the proof of Theorem 6.6.)

2. Show that for any numbers a, b, c and d we have

$$(a^2 + b^2)(c^2 + d^2) = (ac \pm bd)^2 + (ad \mp bc)^2.$$

3. Show that if p is a prime that can be represented as a sum of two squares, then the representation is unique up to sign and order.

4. Which of the following numbers can be represented as a sum of two squares? $97, 251, 991, 1001, 1530, 3589$. Write all possible representations as a sum of two squares in the form $n = a^2 + b^2$, where $a \geq b \geq 0$.

5. Use the identity $2(a^2 + b^2) = (a + b)^2 + (a - b)^2$ to show that $r_2(2n) = r_2(n)$.

6. (a) Show that if we can write $n = a^2 + b^2$, with $(a, b) = 1$ and $d \mid n$, then there are integers e and f such that $d = e^2 + f^2$.

   (b) Solve the Diophantine equation $x^2 + y^2 = z^n$, for $n \geq 3$.

7. Suppose $p \equiv 3 \pmod 4$ is a prime. If we write $p^2 = a^2 + b^2$, for some integers a and b, show that $a = 0$ or $b = 0$.

8. Which integers can be represented as the sum of two squares of rational numbers?

9. Find positive integers n that have at least two essentially distinct representations as a sum of two squares; three representations; four representations.

10. (a) Show that if an integer n can be written as a sum of two triangular numbers, then $4n + 1$ can be written as a sum of two squares.

    (b) Show that if n is a triangular number, then each of $8n^2$, $8n^2 + 1$ and $8n^2 + 2$ can be written as a sum of two squares.

11. Prove that the distance between consecutive integers that are representable as a sum of two squares can be arbitrarily large. (Hint: use the Chinese Remainder Theorem.)

12. Show that $5x^2 + 14xy + 10y^2 = n$ is solvable if and only if $r_2(n) > 0$ and find a formula for the number of representations.

13. Let n be a positive integer. Find a necessary and sufficient condition so that n can be represented in the form $n = x^2 + dy^2$ for each of the three values of $d = 2, 3$ and 7. In each case find the number of representations of n in this form.

14. Let $n = 2^a n_1 n_2$, where a is a nonnegative integer and if $p \mid n_1$, then $p \equiv 1 \pmod 4$ and if $p \mid n_2$, then $p \equiv 3 \pmod 4$. Denote by $d(m)$ the number of positive integral divisors of the positive integer m. Then

$$r_2(n) = \begin{cases} 4d(n_1) & \text{if } n_2 \text{ is a square} \\ 0 & \text{else} \end{cases}.$$

15. Let $d_k(n)$ be the number of positive integral divisors d of n such that $d \equiv k \pmod 4$. If $(a, b) = 1$, show that

    (a) $d_1(ab) = d_1(a)d_1(b) + d_3(a)d_3(b)$

    and

    (b) $d_3(ab) = d_1(a)d_3(b) + d_3(a)d_1(b)$.

6.4. SUMS OF FOUR SQUARES

In this section we tackle the problem of representation by sums of four squares. This time we prove the existence of representations before counting the number of representations. The proof of existence is carried out along the lines laid out at the end of the last section. First we begin with an algebraic identity (Lemma 6.9.1) and secondly we find out which primes can be represented as a sum of four squares (Lemma 6.9.3.) As we shall see every number can be represented as a sum of four nonnegative squares of integers. The proof given here is essentially Euler's simplification of Lagrange's completion of Euler's first attempt at a proof.

We begin with a lemma due to Euler.

Lemma 6.9.1 (Euler): Let $x_1, x_2, x_3, x_4, y_1, y_2, y_3$ and y_4 real numbers. Then the following identity holds:

(6.18)
$$(x_1^2 + x_2^2 + x_3^2 + x_4^2)(y_1^2 + y_2^2 + y_3^2 + y_4^2)$$
$$= (x_1 y_1 + x_2 y_2 + x_3 y_3 + x_4 y_4)^2 + (x_1 y_2 - x_2 y_1 + x_3 y_4 - x_4 y_3)^2$$
$$+ (x_1 y_3 - x_3 y_1 + x_4 y_2 - x_2 y_4)^2 + (x_1 y_4 - x_4 y_1 + x_2 y_3 - x_3 y_2)^2.$$

The proof is an easy calculation which we omit. This identity is an easy consequence of the fact that the ring of quaternions has a multiplicative norm, but, of course, Euler knew nothing of quaternions.

By Theorem 1.11, it therefore suffices to prove that every prime can be written as the sum of four squares. Since $2 = 1^2 + 1^2 + 0^2 + 0^2$ we need only consider the case of odd primes. This will be accomplished in the following two lemmas.

Lemma 6.9.2: If p is an odd prime, then there exists an integer m, $1 \leq m < p$, such that

$$mp = x_1^2 + x_2^2 + x_3^2 + x_4^2,$$

for some integers $x_1, x_2, x_3,$ and x_4.

Proof: The $(p + 1)/2$ numbers x^2, $0 \leq x \leq (p - 1)/2$ are incongruent to each other modulo p since both sum and differences are too small ($x_1^2 \equiv x_2^2 \pmod{p}$ implies $p \mid (x_1 - x_2)$ or $p \mid (x_1 + x_2)$). The same is true of the $(p + 1)/2$ numbers $-1 - y^2$, $0 \leq y \leq (p - 1)/2$. Since these two groups make up $p + 1$ numbers we see, by the pigeon hole principle, that there exist integers x and y, with $0 \leq x, y < (p - 1)/2$, such that

$$x^2 \equiv -1 - y^2 \pmod{p}.$$

Then

$$x^2 + y^2 + 1^2 + 0^2 = mp.$$

Since

$$0 < mp < (p/2)^2 + (p/2)^2 + 1 = (p^2)/2 + 1 < p^2$$

we see that $1 \leq m < p$. ∎

To complete the proof we need to show that we can take $m = 1$. That is the content of the next lemma.

Lemma 6.9.3: If p is an odd prime, then there exist nonnegative integers x_1, x_2, x_3, and x_4 such that

$$p = x_1^2 + x_2^2 + x_3^2 + x_4^2.$$

Proof: Let m be the least positive integer such that

$$mp = x_1^2 + x_2^2 + x_3^2 + x_4^2.$$

By Lemma 6.9.2 we have $m < p$. If m were even, then

$$x_1 + x_2 + x_3 + x_4 \equiv 0 \pmod 2$$

and without loss of generality we may assume

$$x_1 + x_2 \equiv 0 \pmod 2 \quad \text{and} \quad x_3 + x_4 \equiv 0 \pmod 2.$$

Then

$$\frac{m}{2} p = \left(\frac{x_1 + x_2}{2}\right)^2 + \left(\frac{x_1 - x_2}{2}\right)^2 + \left(\frac{x_3 + x_4}{2}\right)^2 + \left(\frac{x_3 - x_4}{2}\right)^2,$$

which would contradict the minimality of m. Thus m must be odd. Suppose that $m > 1$, that is, since m is odd, $m \geq 3$. We will proceed by descent.

Choose y_k, $k = 1, 2, 3, 4$, such that

$$y_k \equiv x_k \pmod m \quad \text{and} \quad |y_k| < \frac{m}{2}, \ k = 1,2,3,4,$$

which can be done since $-(m-1)/2 \leq y \leq (m-1)/2$ is a complete set of residues modulo m. Then

$$y_1^2 + y_2^2 + y_3^2 + y_4^2 \equiv x_1^2 + x_2^2 + x_3^2 + x_4^2 \pmod m.$$

Thus there is an integer n such that

$$y_1^2 + y_2^2 + y_3^2 + y_4^2 = mn.$$

If $n = 0$, then $y_1 = y_2 = y_3 = y_4 = 0$, and so $m \,|\, x_k$, $1 \leq k \leq 4$. This would imply that $m^2 \,|\, \left(x_1^2 + x_2^2 + x_3^2 + x_4^2\right)$ that is, $m^2 \,|\, mp$ or $m \,|\, p$, which contradicts the fact that $1 < m < p$ and p is a prime. Also $n < m$ since $mn < 4(m/2)^2 = m^2$. Thus, by Lemma 6.9.1, $m^2 np$ is the right hand side of (6.18). Now each term on the right hand side of (6.18) is divisible by m since

$$x_k y_l - x_l y_k \equiv x_k x_l - x_l x_k \equiv 0 \pmod m, \ 1 \leq k, l \leq 4$$

and

$$x_1y_1 + x_2y_2 + x_3y_3 + x_4y_4 \equiv x_1^2 + x_2^2 + x_3^2 + x_4^2 \equiv 0 \pmod{m}.$$

Thus there exist integers z_1, z_2, z_3 and z_4 such that

$$np = z_1^2 + z_2^2 + z_3^2 + z_4^2$$

and, since $1 \leq n < m$, this contradicts the minimality of m. Thus $m = 1$. ∎

We now state the theorem that we have proved.

Theorem 6.9 (Lagrange): If n is a nonnegative integer, then there exist nonnegative integers x_1, x_2, x_3 and x_4 such that

$$n = x_1^2 + x_2^2 + x_3^2 + x_4^2.$$

We illustrate these calculations in the following example.

Example 6.7: (a) Write 19 as a sum of four squares.

    (b) Write 23 as a sum of four squares.

    (c) Write 437 and 874 as sums of four squares.

Solution of (a). Since 19 is a prime we look, as in the proof of Lemma 6.9.2, at the sets of integers x^2 and $-1 - y^2$. We have

$$\{x^2 : 0 \leq x \leq 9\} = \{0,1,4,9,16,25,36,49,64,81\}$$
$$\equiv \{0,1,4,9,16,6,17,11,7,5\} \pmod{19}$$

and

$$\{-1 - y^2 : 0 \leq y \leq 9\} \equiv \{-1,-1,-5,-10,-7,-18,-12,-8,-6\} \pmod{19}.$$

Note that $6^2 \equiv -1 - 1^2 \pmod{19}$ and, indeed,

$$2 \cdot 19 = 6^2 + 1^2 + 1^2 + 0^2.$$

(Of course, we also have $19 = 3^2 + 3^2 + 1^2$, but that's solving the problem too fast.) Since 2 is even we have, as in the proof of Lemma 6.9.3, that

$$19 = \left(\frac{6+0}{2}\right)^2 + \left(\frac{6-0}{2}\right)^2 + \left(\frac{1+1}{2}\right)^2 + \left(\frac{1-1}{2}\right)^2 = 3^2 + 3^2 + 1^2,$$

as noted above.

Solution of (b). Since 23 is a prime we can proceed as in (a) and consider the two sets of numbers

$$\{x^2 : 0 \leq x \leq 11\} \equiv \{0,1,4,9,16,2,13,3,18,12,8,6\} \pmod{23}$$

and

$$\{-1-y^2 : 0 \le y \le 11\} \equiv \{-1,-2,-5,-10,-17,-3,-14,-19,-13,-9,-7\} \pmod{23}.$$

We find that $4^2 \equiv -1-11^2 \pmod{23}$ and, indeed,

$$6 \cdot 23 = 11^2 + 4^2 + 1^2 + 0^2.$$

Since 6 is even and $11+1 \equiv 0 \pmod 2$ and $4+0 \equiv 0 \pmod 2$ we can write, as in Lemma 6.9.3,

(6.19)
$$3 \cdot 23 = \left(\frac{11+1}{2}\right)^2 + \left(\frac{11-1}{2}\right)^2 + \left(\frac{4+0}{2}\right)^2 + \left(\frac{4-0}{2}\right)^2$$
$$= 6^2 + 5^2 + 2^2 + 2^2.$$

As we continue through the proof of Lemma 6.9.3 we now have a multiplier, m, odd, and so we look for y_k such that

$$y_k \equiv x_k \pmod m \quad \text{and} \quad |y_k| < m/2.$$

Since $m = 3$ we have $m/2 = 1.5$, and so we want $|y_k| \le 1$, since the y_k are to be integers. This gives us $y_1 = 0$, $y_2 = y_3 = y_4 = -1$. This gives

(6.20)
$$0^2 + (-1)^2 + (-1)^2 + (-1)^2 = 3 \cdot 1.$$

If we combine (6.19) and (6.20) with (6.18) we obtain

$$(3 \cdot 23)(3 \cdot 1) = 9^2 + 6^2 + 3^2 + 9^2.$$

Thus

$$23 = 3^2 + 2^2 + 1^2 + 3^2.$$

as advertised. Of course, if the n in this case had not turned out to be equal to 1 we would have had to continue this reduction process by finding a new set of y_k's and dividing the resulting product by n^2. However, this must come to an end since we always have $1 \le n < m < p$.

Solution of (c) We have $437 = 19 \cdot 23$ and $874 = 2 \cdot 19 \cdot 23$. If we use (6.18), we find that

$$437 = (3^2 + 3^2 + 1^2)(3^2 + 3^2 + 2^2 + 1^2) = 20^2 + 6^2 + 1^2 + 0^2.$$

Since $2(a^2 + b^2) = (a + b)^2 + (a - b)^2$ we have

$$874 = (20+6)^2 + (20-6)^2 + (1+0)^2 + (1-0)^2$$
$$= 26^2 + 14^2 + 1^2 + 1^2.$$
□

As we can see from this last example there look to be more ways possible to represent 874 as a sum of four squares. We have, for example,

$$874 = (20+1)^2 + (20-1)^2 + (6+0)^2 + (6-0)^2 = 21^2 + 19^2 + 6^2 + 6^2.$$

We now take up the question of how many representations as a sum of four squares a given integer has. Let $r_4(n)$ denote the number of representations of n as a sum of four squares.

To accomplish this we will need our results on the number of representations as sums of two squares from the previous section. To make the discussion a little easier we introduce the sum of divisors function, $\sigma(n)$, which is defined to be

$$\sigma(n) = \sum_{d|n} d.$$

We will discuss this function in its own right in the next chapter, but in this section it will allow us to write our results more compactly.

Example 6.8. (a) Compute $\sigma(21)$.

   (b)   Compute $\sigma(28)$.

Solution of (a) The positive divisors of 21 are 1, 3, 7 and 21. Thus

$$\sigma(21) = 1 + 3 + 7 + 21 = 32.$$

Solution of (b) The positive divisors of 28 are 1, 2, 4, 7, 14 and 28. Thus

$$\sigma(28) = 1 + 2 + 4 + 7 + 14 + 28 = 56. \qquad \square$$

We begin by studying an interim function.

Theorem 6.10: Let n be an odd natural number and let $A(n)$ be the number of representations of $4n$ in the form

(6.21) $$4n = x_1^2 + x_2^2 + x_3^2 + x_4^2,$$

where x_1, x_2, x_3 and x_4 are odd. Then $A(n) = \sigma(n)$.

Proof: We obtain all solutions of (6.21) by decomposing $4n = 2l + 2m$, l and m odd, in all possible ways and then solve

$$2l = x_1^2 + x_2^2 \text{ and } 2m = x_3^2 + x_4^2.$$

If v is odd, then the equation

(6.22) $$2v = x^2 + y^2$$

implies that x and y must be odd. Thus the number of solutions of (6.22) is four times the number of solutions of (6.22) in which x and y are odd positive integers.

Thus, by Theorem 6.8,

$$A(n) = \sum_{l+m=2n} \frac{r_2(2l)}{4} \cdot \frac{r_2(2m)}{4} = \sum_{l+m=2n} \sum_{a|2l} \chi(a) \sum_{b|2m} \chi(b)$$

$$= \sum_{l+m=2n} \sum_{all} \chi(a) \sum_{b|m} \chi(b) = \sum_{l+m=2n} \sum_{all,b|m} \chi(ab)$$

$$= \sum_{a\alpha+b\beta=2n} \chi(ab),$$

since $\chi(d) = 0$ if d is even by (a) of Theorem 6.7, where the last sum is over all odd positive quadruples (a, b, α, β) such that

$$a\alpha + b\beta = 2n.$$

We first count the number of quadruples with $a = b$. Then $a \mid n$, since a is odd. Also

$$2(n/a) = \alpha + \beta$$

has n/a solutions in α and β. Since a is odd $\chi(a) = \pm 1$ and so $\chi(a^2) = \chi(a)^2 = 1$. Thus the contribution of these terms is

$$\sum_{a|n} (n/a) = \sum_{d|n} d = \sigma(n).$$

To finish the proof we must show that

$$\sum_{\substack{a\alpha+b\beta=2n \\ a \neq b}} \chi(ab) = 0$$

or, what is the same thing by symmetry, we must show that

$$\sum_{\substack{a\alpha+b\beta=2n \\ a>b}} \chi(ab) = 0.$$

For this it suffices to pair off the solutions of $a\alpha + b\beta = 2n$, $a > b$, one to one in such a way that for every pair of quadruples (a, b, α, β) and $(a_1, b_1, \alpha_1, \beta_1)$ with $a\alpha + b\beta = 2n = a_1\alpha_1 + b_1\beta_1$ we have

$$\chi(ab) + \chi(a_1b_1) = 0.$$

To do this we need only produce a rule such that

    (a)   to every quadruple (a, b, α, β) there is a quadruple $(a_1, b_1, \alpha_1, \beta_1)$ such that

$$a_1\alpha_1 + b_1\beta_1 = 2n, \; a_1 > b_1,$$

(b)   to this latter quadruple the rule assigns (a, b, α, β), and

(c)   we have

$$\chi(ab) + \chi(a_1 b_1) = 0.$$

Let

(i) $m = [b/(a - b)] \geq 0$,

(ii) $a_1 = (m + 2)\alpha + (m + 1)\beta$,

(iii) $\alpha_1 = -ma + (m + 1)b$,

(iv) $b_1 = (m + 1)\alpha + m\beta$

and

(v) $\beta_1 = (m + 1)a - (m + 2)b$.

Now it is easy to see these new numbers are all odd by taking congruences modulo 2 since a, b, α and β are all odd. By definition we see that all of them are positive. Since $m \geq 0$ we see that $a_1 > b_1$. Finally we have

$$\begin{aligned} a_1\alpha_1 + b_1\beta_1 &= -m(m+2)a\alpha - m(m+1)a\beta + (m+1)(m+2)b\alpha + (m+1)^2 b\beta \\ &\quad + (m+1)a\alpha + m(m+1)\alpha\beta - (m+1)(m+2)b\alpha - m(m+2)b\beta \\ &= \left\{(m+1)^2 - m(m+2)\right\}(a\alpha + b\beta) = a\alpha + b\beta = 2n. \end{aligned}$$

This takes care of (a). We have

$$\left[\frac{b_1}{a_1 - b_1}\right] = \left[\frac{m(\alpha + \beta) + \alpha}{\alpha + \beta}\right] = m.$$

An easy computation shows that we get back a, α, b and β if we use the rule on a_1, α_1, b_1, and β_1. For example

$$(m+2)\alpha + (m+1)\beta = a\left\{-m(m+1) + (m+1)^2\right\} = a.$$

This takes care of (b). If v and w are odd, then

$$(v-1)(w-1) \equiv 0 \ (\text{mod } 4).$$

Thus

$$vw \equiv v + w - 1 \ (\text{mod } 4).$$

Thus

$$2 \equiv 2n = a\alpha + b\beta \equiv (a + \alpha - 1) + (b + \beta - 1) \ (\text{mod } 4)$$

or

$$a + b + \alpha + \beta \equiv 0 \ (\text{mod } 4).$$

Also

$$ab + a_1 b_1 \equiv (a + b - 1) + (a_1 + b_1 - 1) \ (\text{mod } 4)$$
$$\equiv a + b + a_1 + b_1 - 2 \ (\text{mod } 4)$$
$$\equiv a + b + (2m + 3)\alpha + (2m + 1)\beta - 2 \ (\text{mod } 4)$$
$$\equiv 2m(\alpha + \beta) + a + b + \alpha + \beta + 2\alpha - 2 \ (\text{mod } 4)$$
$$\equiv 0 \ (\text{mod } 4).$$

The result, (c), follows from (c) of Theorem 6.7. ∎

Theorem 6.11: If n is an odd integer, then

$$r_4(2n) = 3r_4(n).$$

Proof: In the equation

$$(6.23) \qquad\qquad 2n = x_1^2 + x_2^2 + x_3^2 + x_4^2,$$

two of the x_k must be even and two must be odd. The number of solutions in which x_1 and x_2 are even and x_3 and x_4 are odd is thus $r_4(2n)/6$. Let

$$(6.24) \qquad y_1 = \frac{x_1 + x_2}{2}, y_2 = \frac{x_1 - x_2}{2}, y_3 = \frac{x_3 + x_4}{2} \ \text{and} \ y_4 = \frac{x_3 - x_4}{2}.$$

Then from (6.23) we have

$$(6.25) \qquad\qquad n = y_1^2 + y_2^2 + y_3^2 + y_4^2,$$

where $y_1 + y_2 \equiv 0 \ (\text{mod } 2)$ and $y_3 + y_4 \equiv 1 \ (\text{mod } 2)$. Conversely, if we let

$$(6.26) \qquad x_1 = y_1 + y_2, \ x_2 = y_1 - y_2, \ x_3 = y_3 + y_4 \ \text{and} \ x_4 = y_3 - y_4,$$

then from (6.25) we get (6.23) with x_1 and x_2 even and x_3 and x_4 odd. Thus $r_4(2n)/6$ is the number of solutions of (6.25). In the equation

$$(6.27) \qquad\qquad n = y_1^2 + y_2^2 + y_3^2 + y_4^2$$

precisely one y_k is odd if $n \equiv 1 \ (\text{mod } 4)$ and this can only be y_3 or y_4 in (6.25) If $n \equiv 3 \ (\text{mod } 4)$, then exactly one y_k can be even and this too can only be y_3 or y_4. Thus (6.27) has twice as many solutions as (6.25). Thus

$$r_4(2n)/6 = r_4(n)/2,$$

which is our result. ∎

Theorem 6.12 (Jacobi): If n is an odd integer, then

$$r_4(n) = 8\sigma(n)$$

and if l is a positive integer, then

$$r_4(2^l n) = 24\sigma(n).$$

Proof: For $n > 0$ we have

(6.28) $$r_4(2n) = r_4(4n)$$

since

(6.29) $$4n = x_1^2 + x_2^2 + x_3^2 + x_4^2$$

implies $x_1 \equiv x_2 \equiv x_3 \equiv x_4$ (mod 2). Thus

(6.30) $$2n = y_1^2 + y_2^2 + y_3^2 + y_4^2,$$

where the y_k are determined by (6.24) Conversely, from (6.30) we get (6.29) with the x_k determined by (6.26). Since, in (6.29), either all of the x_k are even, $r_4(n)$ times, since the equation is equivalent to

$$n = z_1^2 + z_2^2 + z_3^2 + z_4^2,$$

with $z_k = x_k/2$, $k = 1, 2, 3, 4$, or else they are all odd, $16A(n) = 16\sigma(n)$ times (because of signs), by Theorem 6.10, we have

(6.31) $$r_4(4n) = 16\sigma(n) + r_4(n).$$

By Theorem 6.11, (6.28) and (6.31), we have

$$3r_4(n) = r_4(2n) = r_4(4n) = 16\sigma(n) + r_4(n),$$

which gives the first result. From Theorem 6.11 and (6.31) it follows that

(6.32) $$r_4(2n) = 24\sigma(n).$$

Finally, if $l > 0$, it follows from (6.28) and (6.32) that

$$r_4(2^l n) = r_4(2n) = 24\sigma(n),$$

which is the second result.   ■

Example 6.8: Let $n = 672 = 2^5 \cdot 21$. Then

$$r_4(n) = 24\sigma(21) = 24 \cdot 32 = 768,$$

by (a) of Example 6.7. Recall that each representation has a large number of essentially the same copies formed by changing order and the sign. In this case the number of distinct representations is 4 and they are given by

$$672 = 24^2 + 8^2 + 4^2 + 4^2$$
$$= 20^2 + 16^2 + 4^2 + 0^2$$
$$= 20^2 + 12^2 + 8^2 + 8^2$$
$$= 16^2 + 16^2 + 12^2 + 4^2.$$

□

Problem Set 6.4

1. Show that every sum of four odd squares can be written as a sum of four even squares. (Hint: use the identity $2(x^2 + y^2) = (x + y)^2 + (x - y)^2$ in two ways.)

2. Show that every positive integer can be written in the form

$$x^2 + 2y^2 + 3z^2 + 6w^2,$$

   where x, y, z and w are integers. (Hint: if n is a positive integer and

$$n = r^2 + s^2 + t^2 + u^2,$$

   show that 3 divides the sum of three of r, s, t or u and then rearrange the terms to pick off the multiple of 2, 3 and 6.)

3. Show that every positive integer that is a multiple of 8 can be written as a sum of eight squares of odd numbers. (Hint: consider representations of $n - 1$ as a sum of four squares.)

4. Show that if we write $n = r^2 + s^2 + t^2 + u^2$, with $r \geq s \geq t \geq u \geq 0$, then

$$\frac{1}{2}\sqrt{n} \leq r \leq \sqrt{n}.$$

5. Find all representations of the form $n = r^2 + s^2 + t^2 + u^2$, $r \geq s \geq t \geq u \geq 0$, for $n = 75, 109, 257, 1001$ and 9873. (Hint: one way is to use Problem 4, that is, look at $n - \left[\sqrt{n}\right]^2$ as a sum of three squares, and so on.)

6. Suppose $n = r^2 + s^2 + t^2 + u^2$.
   (a) If $n \equiv 3 \pmod 8$, show that $rstu \equiv 0 \pmod 4$.
   (b) If $n \equiv 7 \pmod 8$, show that $rstu \equiv 2 \pmod 4$.
   (c) If n is a prime and $n \equiv 2,3 \pmod 4$, show that at most 2 of r, s, t and u are equal.

   (Hint: recall what squares are congruent to modulo 8.)

7. If $n > 0$, find all representations of 2^n as a sum of four squares.

8. Show that if $(p, ab) = 1$, then $ax^2 + by^2 + c \equiv 0 \pmod{p}$ is solvable. (Hint: mimic the proof of Lemma 6.9.2.) Can you prove an analog of Lemma 6.9.3 based on this?

9. If m and n are relatively prime positive integers, show that

$$\frac{1}{8}r_4(mn) = \left(\frac{1}{8}r_4(m)\right)\left(\frac{1}{8}r_4(n)\right).$$

10. Let $R(n)$ denote the number of primitive solutions of

$$x_1^2 + x_2^2 + x_3^2 + x_4^2 = n.$$

(a) Show that

$$r_4(n) = \sum_{d^2 \mid n} R(n/d^2).$$

(b) Show that if $m, n > 1$ and $(m, n) = 1$, then

$$\frac{1}{8}R(mn) = \left(\frac{1}{8}R(m)\right)\left(\frac{1}{8}R(n)\right).$$

(c) Let u be a positive odd integer and let v be the largest square-free divisor of u. Show that

$$R(2^l n) = \begin{cases} 8(u/v)\sigma(v) & \text{if } l = 0 \\ 24(u/v)\sigma(v) & \text{if } l = 1 \\ 16(u/v)\sigma(v) & \text{if } l = 2 \\ 0 & \text{if } l \geq 3 \end{cases}.$$

(Hint: mimic the results of Section 6.3.)

6.5. SUMS OF OTHER NUMBERS OF SQUARES

While reading the previous two sections the reader may have begun to wonder why we skipped sums of three squares. The reason is that the problem for sums of an odd number of squares is much harder than for an even number of squares. In this section we will discuss the problem of sums of three squares as well as a few other problems involving sums of squares.

Theorem 6.13. If $n = x^2 + y^2 + z^2$, for some integers x, y and z, then n is not of the form $4^a(8b + 7)$, where a and b are nonnegative integers.

Proof. If we consider the numbers modulo 8 we see that all squares satisfy one of congruences

$$x^2 \equiv 0, 1 \text{ or } 4 \pmod 8$$

and therefore if x, y and z are any integers,

$$x^2 + y^2 + z^2 \equiv 0, 1, 2, 3, 4, 5 \text{ or } 6 \pmod 8.$$

If n is a positive integer such that

$$4n = x^2 + y^2 + z^2,$$

we see, upon taking congruences modulo 4, that x, y and z must be even. Thus we have

$$n = (x/2)^2 + (y/2)^2 + (z/2)^2.$$

Thus n and $4n$ are both representable as a sum of three squares or neither one is so representable.

If we combine these results, the theorem follows.   ∎

The converse is also true and is due to Legendre.

Theorem 6.14. If n is not of the form $4^a(8b + 7)$, for some nonnegative integers a and b, then n can be written as the sum of three squares.

We do not give the proof of this theorem since it requires material not covered in the text. Much of the material will be covered in the exercises of this chapter, and so in the Additional Problems for this chapter we will outline a proof of this theorem.

We can use this theorem to prove a result discovered by Gauss though it is part of a conjecture of Fermat. Fermat conjectured that every positive integer can be represented as a sum of at most three triangular numbers, four squares, five pentagonal numbers and so on. Gauss proved the triangular number case. Later Cauchy proved the general result and showed that from the pentagonal case onward all but four of the polygonal numbers could be taken to be 0 or 1.

Theorem 6.15. Every positive integer can be written as a sum of three triangular numbers.

Proof. Let n be a positive integer. Then, by Theorem 6.14, we see that $8n + 3$ can be written as a sum of three squares. Recalling, from the proof of Theorem 6.13, that squares are congruent to 0, 1 or 4 modulo 8 we see that if $8n + 3$ is a sum of three squares, each of the squares must be odd.

Thus there are nonnegative integers x, y and z such that

$$8n + 3 = (2x + 1)^2 + (2y + 1)^2 + (2z + 1)^2$$

or

$$8n = 4x(x + 1) + 4y(y + 1) + 4z(z + 1)$$

or

$$n = x(x + 1)/2 + y(y + 1)/2 + z(z + 1)/2. \qquad\blacksquare$$

Finally, we remark that there do exist formulas for the number of representations of a positive integer as a sum of three squares. However, the proof of these results are very much beyond the scope of this book. See [7] or [34].

Theorem 6.14 shows that there are numbers that do need four positive squares to represent them. We might ask if there are any others that can be so represented. The following theorem answers that question.

Theorem 6.16. All integers can be written as a sum of four positive squares except for the numbers 1, 3, 5, 9, 11, 17, 29, 41 or any number in one of the three infinite classes of numbers $4^m h$, where $h = 2$, 6 or 14 and $m = 0, 1, 2, \ldots$.

Proof. If $n \equiv 0 \pmod 8$, write $n = 4^m 2^a N$, where $m \geq 1$, $a = 0$ or 1 and N is odd. If we write

$$n = x^2 + y^2 + z^2 + w^2,$$

then we see that x, y, z and w must be even. Thus

$$4^{m-1} 2^a N = (x/2)^2 + (y/2)^2 + (z/2)^2 + (w/2)^2.$$

As long as $4^{m-1} 2^a \equiv 0 \pmod 8$ we can repeat this process, and so reduce the problem to a case where $n \not\equiv 0 \pmod 8$.

Suppose $n \equiv 2, 3, 4$ or $6 \pmod 8$. We first consider the case $n \geq 170$. Let $N = n - 169$. Then $N \equiv 1, 2, 3, 5$ or $6 \pmod 8$, and so, by Theorem 6.14, we see that N is either a positive square, a sum of two positive squares or a sum of three positive squares. Since $169 = 13^2 = 5^2 + 12^2 = 3^2 + 4^2 + 12^2$ we see that n can be written as a sum of four positive squares. If we check the numbers $n \equiv 2, 3, 4$ or $6 \pmod 8$ and $n \leq 169$, we see that the only numbers that cannot be represented as a sum of four positive squares are those in the list in the statement of the theorem.

Suppose $n \equiv 1 \pmod 4$. We first consider the case $n \geq 677$. Let $N = n - 676$. Then we see that $N \equiv 1$ or $5 \pmod 8$. Thus, by Theorem 6.14, we see that N is a positive square, a sum of two positive squares or a sum of three positive squares.

Since $676 = 26^2 = 10^2 + 24^2 = 6^2 + 8^2 + 24^2$ the result follows as above. Again, if we check the numbers n such that $n \equiv 1 \pmod 4$ and $n \le 676$ we see that the only exceptions are those in our list.

It still remains to be seen to be seen which multiples of 8 can be written as sums of four positive squares. As above we see that if N is a sum of four positive squares so is $4^m N$ for any nonnegative integer m. Thus we need only look through the exceptions found above.

We have $4^m \cdot 1 = 4^{m-1} \cdot 4 = 4^{m-1}(1^2 + 1^2 + 1^2 + 1^2)$, which is a sum of four positive squares if $m \ge 1$. If $m \ge 1$ and $4^m \cdot 2$ is a sum of four positive squares, say

$$4^m \cdot 2 = x^2 + y^2 + z^2 + w^2,$$

then, as above

$$4^{m-1} \cdot 2 = (x/2)^2 + (y/2)^2 + (z/2)^2 + (w/2)^2.$$

If $m - 1 \ge 1$, then we repeat this argument and find that 2 must be a sum of four positive squares, which is impossible. Thus no number of the form $4^m \cdot 2$ can be written as a sum of four positive squares. For $m \ge 1$, we find that

$$4^m \cdot 3 = 4^{m-1} \cdot 12 = 4^{m-1}(3^2 + 1^2 + 1^2 + 1^2)$$

$$4^m \cdot 5 = 4^{m-1} \cdot 20 = 4^{m-1}(3^2 + 3^2 + 1^2 + 1^2)$$

and so both of these sets of numbers can be written as a sum of four positive squares. If we continue this process we find that the only exceptions are those given in the statement of the theorem. ■

Since every number can be written as a sum of four or fewer squares, the following theorem may seem, at first glance, a little surprising, especially in light of Theorem 6.16.

Theorem 6.17. The only positive integers that cannot be written as a sum of five positive squares are the integers 1, 2, 3, 4, 6, 7, 9, 10, 12, 15, 18 and 33.

Proof. (I) By Theorem 6.16, we know that if $n > 41$ is odd, then it can be written as a sum of four positive squares. If we add either 1^2 or 2^2, we see that every odd integer greater than 45 and every even integer greater than 42 can be written as a sum of five positive squares. We can then check the numbers ≤ 45 by hand to obtain the result.

(II) We can also prove this result from scratch using the number $169 = 13^2$ $= 12^2 + 5^2 = 12^2 + 3^2 + 4^2 = 11^2 + 4^2 + 4^2 + 4^2 = 12^2 + 4^2 + 2^2 + 2^2 + 1^2$. Suppose $n \geq 170$ and let $N = n - 169$. By Theorem 6.9, we know that we can write

$$N = x_1^2 + x_2^2 + x_3^2 + x_4^2,$$

where $x_1 \leq x_2 \leq x_3 \leq x_4$. If $x_1 > 0$, we have $n = 13^2 + x_1^2 + x_2^2 + x_3^2 + x_4^2$. If $x_1 = 0$ and $x_2 > 0$, then $n = 12^2 + 5^2 + x_2^2 + x_3^2 + x_4^2$. If $x_2 = 0$ and $x_3 > 0$, we have $n = 12^2 + 3^2 + 4^2 + x_3^2 + x_4^2$. Finally, if $x_3 = 0$ and $x_4 > 0$, then $n = 11^2 + 4^2 + 4^2 + 4^2 + x_4^2$. Thus, in all cases, we can write n as a sum of five positive squares. If we check the numbers $n \leq 169$ we see that the only numbers that cannot be represented as a sum of five positive squares are those listed above. ∎

For more on sums of squares, as well as a proof of Theorem 6.14, see the book by Grosswald [42].

Problem Set 6.5

1. Write $n = 2^a n_1 n_2^2$, where $a \geq 0$ and $p \mid n_1$ implies $p \equiv 1 \pmod 4$ and $p \mid n_2$ implies $p \equiv 3 \pmod 4$. Show that n can be written as a sum of two positive squares unless $n_1 = 1$ and a is even.

2. Show that a positive square, n^2, can be written as a sum of two positive squares if and only if n has at least one prime divisor of the form $4k + 1$.

3. (a) Show that any odd positive integer can be represented in the form

$$x^2 + y^2 + 2z^2,$$

   where x, y and z are positive integers.

   (b) Show that any odd positive integer can be represented in the form

$$x^2 + y^2 + z^2 + (z + 1)^2,$$

   where x, y and z are positive integers.

4. Let $k \geq 6$. Show that the only integers that cannot be represented as a sum of k positive squares are $1, 2, 3, \ldots, k - 1, k + 1, k + 2, k + 4, k + 5, k + 7, k + 10$ and $k + 13$. This result is due to Pall.

5. Show that if $(13m)^2 \geq k + 13$, then $(13m)^2$ can be written as a sum of t positive squares for $t = 1, 2, \ldots, k$.

6. Prove that every positive integer can be represented either in the form $x^2 + y^2 + z^2$ or in the form $x^2 + y^2 + 2z^2$.

7. Prove the following theorem due to Hurwitz. The only positive integers n such

that n^2 cannot be represented as a sum of three positive squares are the numbers that are powers of 2 or five times a power of 2.

8. Show that the equation

$$x^2 + y^2 + z^2 + x + y + z = 1$$

has no solutions in integers. (Hint: complete the square.)

6.6. BINARY QUADRATIC FORMS

The sums of squares problems discussed in the preceding sections are special cases of representations by quadratic forms. In this section we discuss some results about quadratic forms and binary quadratic forms in particular. We discuss the binary case to illustrate some of the more general results which are easier to prove in this special case. We begin, however, with some general results.

Definition 6.1: An **r-ary quadratic form** is a function in the r variables x_1, ..., x_r of the form

$$F = F(x_1,\ldots,x_r) = \sum_{i,j=1}^{r} a_{ij} x_i x_j,$$

where $r > 2$ and a_{ij} are integers, $1 \le i, j \le r$, F is said to be **symmetric** if

$$a_{ij} = a_{ji}, \ 1 \le i, j \le r.$$

In the case $r = 2$ we have **binary quadratic forms** and for $r = 3$ we have **ternary quadratic forms**. The general form of a binary quadratic form is

$$ax^2 + bxy + cy^2$$

and if it is symmetric, then it has the form

$$ax^2 + 2bxy + cy^2.$$

A similar result obtains for ternary quadratic forms. After studying a few general properties of quadratic forms we will study binary quadratic forms in some detail.

Definition 6.2: The determinant

$$\det(a_{ij})$$

is called the **discriminant of the form** F and is denoted by $|F|$.

In the case of a symmetric binary quadratic form $f(x, y) = ax^2 + 2bxy + cy^2$ we have

$$|f| = \begin{vmatrix} a & b \\ b & c \end{vmatrix} = ac - b^2.$$

Definition 6.3: We say that two r-ary quadratic forms

$$F = \sum a_{ij} x_i x_j \quad \text{and} \quad G = \sum b_{kl} x_k x_l$$

are **equivalent** to each other, written $F \sim G$, if there exist r^2 integers c_{kl} such that $\det(c_{kl}) = 1$ for which the r equations

$$x_k = \sum_{l=1}^{r} c_{kl} y_l, \quad 1 \le k \le r,$$

transform $F(x_1, \ldots, x_r)$ into $G(y_1, \ldots, y_r)$.

Note that $F \sim G$ implies that

(6.33) $$b_{kl} = \sum_{m,n=1}^{r} c_{mk} a_{mn} c_{nl}$$

Theorem 6.18: If F is any r-ary quadratic form, then $F \sim F$.

Proof: Let

$$c_{kl} = \begin{cases} 1 & \text{if } k = l \\ 0 & \text{if } k \ne l \end{cases}.$$

Then

$$\det(c_{kl}) = \det I_r = 1$$

and

$$x_k = y_k, \quad 1 \le k \le r.$$

Thus $F \sim F$. ∎

Theorem 6.19: If $F \sim G$, then $G \sim F$.

Proof: Let (c_{kl}) transform F into G. Since $\det(c_{kl}) = 1$ we see that $(d_{kl}) = (c_{kl})^{-1}$ exists and has integer entries. Then

$$y_k = \sum_{l=1}^{r} d_{kl} x_l \text{ if and only if } x_k = \sum_{l=1}^{r} c_{kl} y_l.$$

Thus $F \sim G$ if and only if $G \sim F$. ∎

Theorem 6.20: If $F \sim G$ and $G \sim H$, then $F \sim H$.

Proof: Let (c_{kl}) take F into G and (d_{kl}) take G into H. Let

$$(e_{kl}) = (c_{kl}) \, (d_{kl}).$$

Thus (e_{kl}) takes F into H and since

$$\det(e_{kl}) = \det(c_{kl})\,\det(d_{kl}) = 1$$

the result follows. ∎

Thus we may partition the r-ary quadratic forms into equivalence classes.

Theorem 6.21: If $F \sim G$, the $|F| = |G|$.

Proof: Let (c_{kl}) take F into G. Then

$$|G| = \det(b_{kl}) = \det(c_{kl})\det(a_{kl})\det(c_{kl}) = \det(a_{kl}) = |F|,$$

by (6.33). ∎

Theorem 6.22: If $F \sim G$, then F and G represent the same numbers.

The proof is clear.

Definition 6.4: An r-ary quadratic form F is said to be **positive definite** (or simply **positive**) if

$$F(x_1, \ldots, x_r) > 0$$

for all values of $x_1, \ldots, x_r$ which are not all zero. A form F is said to be **negative definite** (or **negative**) if $-F$ is positive. A form F is said to be **indefinite** if F takes on both positive and negative values.

We now consider the case $r = 2$. We abbreviate the form $F(x, y) = ax^2 + bxy + cy^2$ by $\{a, b, c\}$. Let $d = 4ac - b^2$. Note that if F is a symmetric binary quadratic form, then $d = 4|F|$.

Theorem 6.23: The form $F = \{a, b, c\}$ is positive if and only if $a > 0$ and $d > 0$.

Proof: If $a \le 0$, then $F(1, 0) = a \le 0$, and so F is not positive. If $a > 0$ and $d \le 0$, then

$$F(-b, 2a) = ab^2 - 2b^2a + 4a^2c = -ab^2 + 4a^2c = ad \le 0,$$

and so F is not positive. If $a, d > 0$, then

$$4aF = 4a^2x^2 + 4abxy + acy^2 = (2ax + by)^2 + dy^2,$$

and so $F \le 0$ if and only if $x = y = 0$. Thus F is positive in this case. ∎

Our principal interest is the representation of integers by quadratic forms. In the case of positive binary quadratic forms the scope is definitely limited. As we shall see below in the case of indefinite binary quadratic forms the situation is completely different.

Theorem 6.24: Let $F = \{a, b, c\}$ be positive . Then the number of representations of an integer m, by F, is finite, possibly zero.

Proof: We may assume $m > 0$, since $m = 0$ has only one representation by a positive form and the negative integers have no representations.

Since

$$F(x,y) = \frac{1}{4a}\left\{(2ax + by)^2 + dy^2\right\}$$

we see that if $F \le m$, then

$$dy^2 \le 4am,$$

and so we have

$$-2\sqrt{am/d} \le y \le 2\sqrt{am/d}.$$

This restricts y to a finite number of values and for each of these values of y we have

$$\frac{-by - \sqrt{4am - dy^2}}{2a} \le x \le \frac{-by + \sqrt{4am - dy^2}}{2a},$$

which restricts the values of x. Thus for each value of $m > 0$ there are only a finite number of pairs of integers x and y such that $F(x, y)$ represents some positive integer $\le m$.   ∎

Corollary 6.24.1: Let $F = \{a, b, c\}$ be positive. Then the smallest positive integer representable by F can be found in a finite number of steps.

Proof: Since $a = F(1, 0)$ take $m = a - 1$ in Theorem 6.24. This gives a finite collection of x and y to search through and we can therefore find the smallest.   ∎

Example 6.10: Let $F(x, y) = 5x^2 + 14xy + 11y^2$.

    (a)   Find the least positive number that can be represented by F.

    (b)   Find all solutions of $5x^2 + 14xy + 11y^2 = 30$.

Solution of (a) We have $d = 4 \cdot 5 \cdot 11 - 14^2 = 24$. Thus the values of y are limited to the interval

$$-\sqrt{30}/3 \le y \le \sqrt{30}/3,$$

that is, $y = -1, 0, 1$. For $y = -1$ we find

$$(10x - 14)^2 \le 56$$

or $x = 1$ or 2 and we have $F(1, -1) = 2$ and $F(2, -1) = 3$. For $y = 0$ we find $x = 0$ and $F(0, 0) = 0$. For $y = 1$ we find $x = -2$ or $x = -1$ and have $F(-2, 1) = 3$ and $F(-1, 1) = 2$. Thus the smallest positive integer represented by F is 2.

Solution of (b) We take $m = 30$ in Theorem 6.24 . This gives the interval

$$-5 \leq y \leq 5.$$

The following table gives for y the corresponding range of values for x.

$y = -5$	$x = 7$
$y = -4$	$4 \leq x \leq 7$
$y = -3$	$5 \leq x \leq 9$
$y = -2$	$1 \leq x \leq 5$
$y = -1$	$-1 \leq x \leq 2$
$y = 0$	$-2 \leq x \leq 2$
$y = 1$	$-2 \leq x \leq 1$
$y = 2$	$-5 \leq x \leq -1$
$y = 3$	$-7 \leq x \leq -2$
$y = 4$	$-7 \leq x \leq -4$
$y = 5$	$x = -7$

Noting that $F(x, y) = F(-x, -y)$ we see that all solutions of $F(x, y) = 30$ are given by
the ordered pairs $(1, 1)$, $(7, -5)$, $(-7, 5)$ and $(-1, -1)$.   □

For certain positive definite binary quadratic forms we can be more definite
about the representation of some types of integers.

Theorem 6.25: Let c and d be given positive integers and let p be a prime. Then
there is at most one representation of p in the form $cx^2 + dy^2$, where x and y are
integers.

Proof: Suppose to the contrary that we have two representations:

$$p = cx^2 + dy^2 \text{ and } p = cu^2 + dv^2,$$

where x, y, u and v are positive integers. Then, if we eliminate d, we have

$$p(y^2 - v^2) = c(u^2y^2 - v^2x^2).$$

Since $c < p$ we have

$$uy \equiv \pm vx \ (\text{mod } p).$$

Also, if we multiply the two representations together, we get

$$p^2 = (cxu \pm dyv)^2 + cd(uy \mp vx)^2,$$

where the choice of upper or lower sign is arbitrary. If $uy = vx$, then we must have u
$= x$ and $v = y$, since $(x, y) = (u, v) = 1$ (p is a prime.) If $uy \neq vx$, then, since

$p \mid (uy \mp vx)$ and $c, d > 0$, we have

$$\mid uy \mp vx \mid = p.$$

Thus $c = d = 1$ and $cxu \pm yv = 0$. Thus $x = v$ and $y = u$ in this case.   ∎

Definition 6.5. We say a binary form $\{a, b, c\}$ is **reduced** if $\mid b \mid \le a \le c$ and if $a = c$ or $\mid b \mid = a$, then $0 < b \le a = c$.

Theorem 6.26. Every class of positive binary quadratic forms contains at least one reduced form.

Proof: Let $\{a_0, b_0, c_0\}$ be a given form in a given class. Let a be the least number representable by this form (and so, by Theorem 6.22, by the whole class). Then there are integers r and t such that

$$a = a_0 r^2 + b_0 rt + c_0 t^2.$$

We must have $(r, t) = 1$ for if $(r, t) = v$, then

$$a/v^2 = a_0(r/v)^2 + b_0(r/v)(t/v) + c_0(t/v)^2$$

and $a/v^2 \le a$. Thus, by Theorem 1.5, there exist integers s and u such that

$$ru - st = +1.$$

Then $\begin{pmatrix} r & s \\ t & u \end{pmatrix}$ takes $F = \{a_0, b_0, c_0\}$ into $G = \{a, B, C\}$ since

$$G(1, 0) = F(r \cdot 1 + s \cdot 0, t \cdot 1 + u \cdot 0) = F(r, t) = a.$$

Let j be an integer and consider the transformation

$$x = u - jv \text{ and } y = v.$$

This takes G into $H_j = \{a, B - 2aj, aj - Bj + C\}$. Choose j to be the nearest integer to $B/2a$. Then

$$-1/2 \le B/2a - j \le 1/2,$$

and so

$$-a \le B - 2aj \le a.$$

Thus

$$\mid B - 2aj \mid < a.$$

Let $b = B - 2aj$ and $c = aj^2 - Bj + C$, with this j. We have $H_j(0, 1) = C$. Thus, by Theorem 6.22, F represents c since $F \sim H_j$. Thus, by the definition of a, we have $c \le a$, that is, H_j is a reduced form.   ∎

Theorem 6.27. If two reduced forms are equivalent, then they are identical.

Proof: Let $F = \{a, b, c\}$ and $G = \{A, B, C\}$ be equivalent reduced forms and

suppose $a \geq A$. Let $\begin{pmatrix} \alpha & \beta \\ \gamma & \delta \end{pmatrix}$ take F into G. By taking $x \to -x$ or $y \to -y$ we may

suppose $b > 0$. Since $\alpha^2 + \gamma^2 \geq 2|\alpha\gamma|$ we have, since $0 \leq b \leq a \leq c$,

$$A = a\alpha^2 + b\alpha\gamma + c\gamma^2 \geq a\alpha^2 + c\gamma^2 - b|\alpha\gamma|$$

$$\geq a\alpha^2 + a\gamma^2 - b|\alpha\gamma| \geq 2a|\alpha\gamma| - b|\alpha\gamma| \geq a|\alpha\gamma|.$$

Since $a \geq A > 0$ we see that $|\alpha\gamma| \leq 1$. If $\alpha\gamma = 0$, then $A = a\alpha^2 + c\gamma^2 \geq a\alpha^2$
$+ a\gamma^2 \geq a$, since not both α and γ can be zero ($\alpha\delta - \beta\gamma = +1$). If $|\alpha\gamma| = 1$, then A
$\geq a$. Thus $A = a$ in either case. To finish we need only show that $b = B$ or $c = C$
since the other follows form $4ac - b^2 = 4AC - B^2$ (which is Theorem 6.21).

If $c = C$, we're done. Since $a = A$ we can rearrange to assume that $c > C$ if c
$\neq C$. Since $c > a$ we have $C \geq A = a$. Thus we can't have $|\alpha\gamma| = 1$, since $|\alpha\gamma| = 1$
implies $c\gamma^2 > a\alpha^2$, which implies $A > a$ by the above computation. Thus $a = A$, a
$< c$ and $\gamma = 0$. Since $\alpha\delta - \beta\gamma = +1$ we see that $\alpha\delta = +1$, and so we have

$$B = 2a\alpha\beta + b.$$

Thus $0 \leq b \leq a$ and $0 \leq B \leq A = a$ imply that

$$-a \leq B - b \leq a$$

and since $B - b = 2a\alpha\beta$ is a multiple of $2a$ we see that $B = b$. Thus we must have c
$= C$ and we're done. ■

Theorem 6.28. There are only finitely many reduced forms having a given
discriminant.

Proof: If $\{a, b, c\}$ is a reduced form of discriminant $d = 4ac - b^2$, then we have b
$\leq a \leq c$. Thus

$$4ac - b^2 \geq 4ac - ac > 3a^2.$$

Thus there are at most $\sqrt{d/3}$ possible values for $a > 0$. Thus there are at most $2a$
$+ 1$ different values of b since $-a \leq b \leq a$. Finally, given a, b, and d there is at most
one value of c since $d = 4ac - b^2$. ■

Example 6.11: Find all reduced forms with $d \leq 10$.

Solution: Using the estimate of Theorem 6.28 we see that

$$1 \leq a \leq \sqrt{10/3},$$

that is, $a = 1$. Since $|b| \le a$ we have $b = -1$, 0 or 1. If $b = 0$, then $d = 4c$. This gives the reduced forms

$$\{1, 0, 1\} \text{ and } \{1, 0, 2\}.$$

If $|b| = 1$, then $d = 4c - 1$. This gives

$$\{1, 1, 1\}, \{1, 1, 2\}, \{1, -1, 2\}. \qquad \square$$

Theorem 6.29: In every class of positive definite binary quadratic forms there is at least one form that satisfies

$$|b| \le a \le \sqrt{d/3}.$$

Proof: If $\{a, b, c\}$ is reduced, then $|b| \le a \le c$, and so

$$a^2 \le ac = (b^2 \pm d)/4 \le (a^2 + d)/4.$$

Thus $3a^2 \le d$ and the result follows by Theorem 6.26. ■

Corollary 6.29.1. Every positive definite binary quadratic form of discriminant 4 is equivalent to $x^2 + y^2$.

Proof: If $d = 4$, then all forms are equivalent to a form satisfying

$$|b| \le a \le 2/\sqrt{3}.$$

Thus $a = 1$, $b = 0$ and $c = 1$. ■

Our next result shows that if a number divides some number than can be represented by a form in a given class, then it too can be represented by some form in this class. For example, by Corollary 6.29.1, we see that if a number divides another number that can be represented as a sum of two squares, then it too can be represented as a sum of two squares.

Theorem 6.30. Let m be a divisor of a number that can be represented by the form $ax^2 + bxy + cy^2$ with $x = x_0$, $y = y_0$ and $(x_0, y_0) = 1$. Then m can be represented by a form $Ax^2 + Bxy + Cy^2$ with $x = u_0$, $y = v_0$, $(u_0, v_0) = 1$ and $4AC - B^2 = 4ac - b^2$.

Proof. Suppose

$$mn = ax^2 + bxy + cy^2.$$

Let $r = (n, y)$ and write $n = ru$ and $y = rv$ with $(u, v) = 1$. Then

(6.34) $$mru = ax^2 + bxrv + cr^2v^2.$$

Thus $r \mid ax^2$. Since $(x, y) = 1$ we see that $(x, r) = 1$. Thus $r \mid a$, say $a = rs$. If we divide both sides of (6.34) by r we obtain

(6.35)
$$mu = sx^2 + bxv + crv^2.$$

Since $(u, v) = 1$ we can write, by Theorem 6.3, x in the form

(6.36)
$$x = up + vq.$$

If we substitute the relation (6.36) into the equation (6.35), we obtain

(6.37)
$$mu = s(up + vq)^2 + b(up + vq)v + crv^2$$
$$= (sq^2 + bq + cr)v^2 + (2suq + bu)vp + su^2p^2.$$

We see that in (6.37) we must have $u \mid (sq^2 + bq + cr)$ since $(u, v) = 1$.

If we let

$$A = \frac{sq^2 + bq + cr}{u}, \; B = 2sq + b \text{ and } C = su,$$

we see that

$$m = Ax^2 + Bxy + Cy^2$$

with $x = v$ and $y = p$. Since $(x, y) = 1$ and $y = rv$ we see, from (6.36), that we must have $(v, p) = 1$. Finally,

$$4AC - B^2 = 4\left(\frac{sq^2 + bq + cr}{u}\right)(su) - (2sq + b)^2$$
$$= 4s(sq^2 + bq + cr) - (2sq + b)^2$$
$$= 4scr - b^2 = 4ac - b^2. \qquad \blacksquare$$

Example 6.12. Illustrate Theorem 6.30 with

$$5 \cdot 10 = 7^2 + 1^2.$$

Solution. Here we have

$$a = 1, \; b = 0, \; c = 1, \; x_0 = 7 \text{ and } y_0 = 1.$$

Then $r = (10, 1) = 1$, and so $u = 10$ and $v = 1$. Thus $s = 1$. Since we can write

$$7 = 10 \cdot 1 + 1 \cdot (-3)$$

we see that we can take $p = 1$ and $q = -3$. Thus we have

$$A = \frac{1 \cdot (-3)^2 + 0 \cdot (-3) + 1 \cdot 1}{10} = 1, \; B = 2 \cdot 1 \cdot (-3) + 0 = -6, \; C = 1 \cdot 10 = 10.$$

This gives us

$$5 = 1 \cdot 1^2 + (-6)(1)(1) + 10 \cdot 1^2,$$

which, if we complete the square, gives us the more familiar representation, in accordance with Corollary 6.29.1,

$$5 = (1 - 3 \cdot 1)^2 + 1^2 = 2^2 + 1^2. \qquad \square$$

Our last result in the theory of binary quadratic forms shows that many of our previous representation results which depend on the quadratic character of some number are special cases of a more general result.

Theorem 6.31. Let m be a positive integer which can be represented by the form $ax^2 + bxy + cy^2$ with $x = x_0$, $y = y_0$ and $(x_0, y_0) = 1$. Then $b^2 - 4ac$ is a quadratic residue modulo m.

Proof. Since $(x_0, y_0) = 1$ we know there are integers k and l such that

$$(6.38) \qquad kx_0 + ly_0 = 1.$$

Thus

$$
\begin{aligned}
(6.39) \quad 4m(al^2 - bkl + ck^2) &= 4(ax_0^2 + bx_0y_0 + cy_0^2)(al^2 - bkl + ck^2) \\
&= [k(2cy_0 + bx_0) - l(2ax_0 + by_0)]^2 \\
&\quad - (b^2 - 4ac)(kx_0 + ly_0).
\end{aligned}
$$

If we let $n = k(2cy_0 + bx_0) - l(2ax_0 + by_0)$, then, by (6.38) and (6.39), we see that

$$n^2 \equiv b^2 - 4ac \pmod{m}.$$

Thus $b^2 - 4ac$ is a quadratic residue modulo m. ∎

This finishes our limited tour through the theory of quadratic forms. There is much more that can be done. See, for example, [71, v. III, ch. 1], [1, ch. 11] or [64, part IV]. Some results for ternary quadratic forms can be found in the exercises below.

Problem Set 6.6

1. Show that the following forms are equivalent.
   (a) $2x^2 + 3y^2$ and $2x^2 + 4xy + 5y^2$.
   (b) $ax^2 + bxy + cy^2$ and $cx^2 - bxy + ax^2$.
   (c) $x^2 + y^2$ and $13x^2 + 36xy + 25y^2$.
   (d) $x^2 + y^2$ and $58x^2 + 82xy + 29y^2$.

2. Let $c_1, \ldots, c_n$ be integers such that $(c_1, \ldots, c_n) = 1$. Show that there is a matrix (c_{kl}) with $\det(c_{kl}) = 1$ and $c_{k1} = c_k$.

3. Show that $x^2 + ay^2$, where a is a positive integer, is always a reduced form.

4. Find reduced forms equivalent to the following forms.

   (a) $126x^2 + 74xy + 13y^2$.

   (b) $7x^2 + 24xy + 21y^2$.

   (c) $79x^2 + 50xy + 8y^2$.

   (d) $9x^2 + 32xy + 29y^2$.

5. Let d be a positive integer such that $-d \equiv 1 \pmod 4$. Show that the form $x^2 + xy + ((1 + d)/4)y^2$ has discriminant d and that it is reduced.

6. Find the least positive integer that can be represented by the following forms.

   (a) $4x^2 + 17xy + 20y^2$.

   (b) $2x^2 + xy + 2y^2$.

   (c) $3x^2 + 3xy + 4y^2$.

   (d) $x^2 + xy + 21y^2$.

7. Prove the converse to Theorem 6.31.

8. Find all solutions of the following equations.

   (a) $x^2 - xy + y^2 = 7$.

   (b) $3x^2 + xy + 3y^2 = 5$.

   (c) $x^2 + xy - 5y^2 = 1$.

   (d) $x^2 + xy - 7y^2 = 5$.

9. Find all reduced forms of discriminants 8, 31 and 56.

   In the following exercises we will consider ternary quadratic forms. We write

   $$F(x,y,z) = ax^2 + bxy + cxz + dy^2 + eyz + fz^2.$$

   Here the discriminant of F is

   $$D = \begin{vmatrix} 2a & b & c \\ b & 2d & e \\ c & e & 2f \end{vmatrix}.$$

10. Show that there is a binary quadratic form $K(y, z)$ such that K has discriminant $4aD$ and

    $$4aF(x,y,z) = (2ax + by + cz)^2 + K(y,z).$$

11. Show that F is positive definite if and only if

    $$a > 0, \quad \begin{vmatrix} 2a & b \\ b & 2d \end{vmatrix} > 0 \text{ and } D > 0.$$

12. Show that every class of definite ternary forms contains at least one form for
    which

$$a \leq 4\sqrt[3]{D/3}, \ 2|b| \leq a \text{ and } 2|c| \leq a.$$

13. Show that every positive definite ternary form with discriminant 8 is equivalent
    to the form

$$x^2 + y^2 + z^2.$$

6.7. THE EQUATION $X^2 - DY^2 = N$

In this section we shall consider representation problems for a special case of
indefinite binary quadratic forms. The equation of the section title is traditionally
known as the Pell equation because that is the name Euler, through faulty historical
scholarship, gave to the equation. In actual fact Pell seems to have had nothing to do
with this equation and Euler presumably meant to refer to this as Fermat's equation.
In modern times it was Fermat who first put forth this equation as a challenge to the
English mathematicians to solve and who first solved the problem, though it was
Lagrange, in 1768, who published the first solution. However, we could just as
easily refer to this as Bhaskara's equation or Brahmagupta's equation. Recent
historical scholarship has shown that the Indian mathematicians, ca. 600 AD,
possessed an algorithm that gave all solutions to this equation, but they gave no proof
that their procedure worked all the time (that is, for all admissible D) or that it gave all
the solutions. The Indian algorithm is in part a combination of Theorems 6.34 and
6.38. See [11, ch. 12].

Suppose first that $D = a^2$, $a > 0$. Then

(6.40) $$N = x^2 - Dy^2$$

implies that $N = (x - ay)(x + ay)$. If we suppose $N = \alpha\beta$, then the solutions of
(6.40) are given by

(6.41) $$x = (\alpha + \beta)/2 \text{ and } y = (\beta - \alpha)/2a.$$

It is clear that not all choices of α and β will necessarily give integer values to x and
y. In particular if $N \equiv 2 \pmod 4$, then x cannot be an integer since $N \equiv 2 \pmod 4$
and $N = \alpha\beta$ implies one of α or β is even and the other is odd.

Example 6.13. Write 435 and 604 as differences of two squares ($a = 1$) in all
possible ways.

Solution. We have $435 = 3 \cdot 5 \cdot 29$. Thus we must solve

$$x + y = 435, 145, 87, 29$$
$$x - y = 1, 3, 5, 15.$$

This gives

$$435 = 218^2 - 217^2 = 74^2 - 71^2 = 46^2 - 41^2 = 22^2 - 7^2.$$

Similarly we have $604 = 22 \cdot 151$, and so we solve

$$x + y = 604, 302, 151$$
$$x - y = 1, 2, 4.$$

Since the first and last are insoluble we see that the only solution is

$$604 = 152^2 - 150^2. \qquad \square$$

This example illustrates the following result about the case $a = 1$ whose proof follows immediately from (6.41).

Theorem 6.32: If $N \not\equiv 2 \pmod 4$, then $x^2 - y^2 = N$ is always solvable and the number of solutions is equal to one half of the number of divisors of N, if N is odd, and if $N = 2^l n$, $l \geq 2$ and n odd, then the number of solutions is one half of $(l - 1)$ times the number of divisors of n.

For other values of a one can produce similar results. (See Problem 2 below.)

Thus we may restrict our attention to the case D positive and not a square, a hypothesis that will be in force throughout the rest of this section. It will turn out for those values of D that equation (6.40) either has no solution or an infinite number of solutions.

Example 6.14: Is the equation $x^2 - 3y^2 = -1$ solvable?

Solution: No, if it was then there would exist an integer x such that

$$x^2 \equiv -1 \pmod 3,$$

which is impossible by (f) of Theorem 3.2. $\qquad \square$

While not answering the question of existence of a solution the following does indicate how one can produce more solutions from a given one. Suppose N is a positive integer and

$$N = r^2 - Ds^2 \text{ and } 1 = x^2 - Dy^2,$$

where r, s, x and y are integers with $x \neq \pm 1$. Then

$$N = N \cdot 1 = (r^2 - Ds^2)(x^2 - Dy^2) = (xr - Dsy)^2 - D(xs - ry)^2$$

and it is easy to see this actually produce another solution of (6.40). Thus, once we have proved existence, we can obtain many solutions given just one.

We shall return to the general problem at the end of this section. We begin by considering the equation with $N = 1$. It turns out, as we shall see, that this equation is the key to the whole discussion. Indeed, we shall show that as long as D is a positive nonsquare integer the equation

$$x^2 - Dy^2 = 1$$

always has a solution.

Lemma 6.33.1: Given D, there exist infinitely many pairs of natural numbers x and y which satisfy the inequality

$$\left| x^2 - Dy^2 \right| < 1 + 2\sqrt{D},$$

Proof: Since D is not a square we know $\sqrt{D}$ is irrational. Thus, by Theorem 4.1, there exist infinitely many pairs of relatively prime integers x and y such that

$$\left| \frac{x}{y} - \sqrt{D} \right| < \frac{1}{y^2}.$$

Also

$$\left| \frac{x}{y} + \sqrt{D} \right| = \left| \frac{x}{y} - \sqrt{D} + 2\sqrt{D} \right| < \frac{1}{y^2} + 2\sqrt{D}.$$

Thus

$$\left| x^2 - Dy^2 \right| = \left| x - y\sqrt{D} \right| \left| x + y\sqrt{D} \right|$$

$$< y^2 \frac{1}{y^2} \left(\frac{1}{y^2} + 2\sqrt{D} \right) \leq 1 + 2\sqrt{D},$$

which is our result. ∎

Theorem 6.33: Given D, then there exists at least one pair of natural numbers x and y which satisfy the equation

$$x^2 - Dy^2 = +1.$$

Proof: It follows from Lemma 6.32.1 that there exists at least one nonzero integer k such that

$$x^2 - Dy^2 = k$$

for infinitely many pairs of natural numbers x and y, since $0 < \left| x^2 - Dy^2 \right|$ (because

$\sqrt{D}$ is irrational), $1 + 2\sqrt{D} > 1$ (because $D \geq 2$) and $x^2 - Dy^2$ is an integer. Among these pairs there must exist at least two pairs x_1, y_1 and x_2, y_2 which satisfy the congruence conditions

(6.43) $$x_1 \equiv x_2 \pmod{|k|} \text{ and } y_1 \equiv y_2 \pmod{|k|},$$

since the remainder modulo $|k|$ of the four numbers x_1, x_2, y_1 and y_2 can only be combined in a finite number of ways (namely, at most k^4 ways). Thus we have (6.43) as well as

(6.44) $$x_1^2 - Dy_1^2 = k = x_2^2 - Dy_2^2.$$

We have

$$\left(x_1 - y_1\sqrt{D}\right)\left(x_2 + y_2\sqrt{D}\right) = x_1 x_2 - D y_1 y_2 + \left(x_1 y_2 - x_2 y_1\right)\sqrt{D}.$$

By (6.43) and (6.44) we have

$$x_1 x_2 - D y_1 y_2 \equiv x_1^2 - D y_1^2 \equiv 0 \pmod{|k|}$$
$$x_1 y_2 - x_2 y_1 \equiv x_1 y_2 - x_1 y_2 \equiv 0 \pmod{|k|}.$$

Thus, for some integers u and v, we have

$$x_1 x_2 - D y_1 y_2 = ku \text{ and } x_1 y_2 - x_2 y_1 = kv.$$

Thus

$$\left(x_1 - y_1\sqrt{D}\right)\left(x_2 + y_2\sqrt{D}\right) = k\left(u + v\sqrt{D}\right)$$
$$\left(x_1 + y_1\sqrt{D}\right)\left(x_2 - y_2\sqrt{D}\right) = k\left(u - v\sqrt{D}\right)$$

and

$$\left(x_1^2 - Dy_1^2\right)\left(x_2^2 - Dy_2^2\right) = k^2 = k^2\left(u^2 - Dv^2\right).$$

Thus, since $k \neq 0$, we have

$$u^2 - Dv^2 = +1.$$

If $v = 0$, then $x_1 y_2 = x_2 y_1$ and $u = \pm 1$. Also, in this case,

$$\left(x_1 - y_1\sqrt{D}\right)\left(x_2 + y_2\sqrt{D}\right)\left(x_2 - y_2\sqrt{D}\right) = \pm k\left(x_2 - y_2\sqrt{D}\right)$$

or dividing by k we get

$$x_1 - y_1\sqrt{D} = \pm\left(x_2 - y_2\sqrt{D}\right).$$

Thus $x_1 = \pm x_2$ and $y_1 = \pm y_2$, in this case. Since we can choose our x_1 and x_2, at the beginning, so that $|x_1| \neq |x_2|$, the theorem is proved. ∎

If x and y are positive integers that satisfy the equation (6.42), then we say that $x + y\sqrt{D}$ is a **solution** to (6.42). Among all of these solutions there exists a solution $x_1 + y_1\sqrt{D}$ which is the least in magnitude. We call such a solution the **fundamental solution** to (6.42). With respect to this we have the following result.

Theorem 6.34: Given D, the equation (6.42) has infinitely many solutions, in natural numbers, $x + y\sqrt{D}$. All solutions in positive x and y are given by

$$(6.45) \qquad\qquad x_n + y_n\sqrt{D} = \left(x_1 + y_1\sqrt{D}\right)^n,$$

where $x_1 + y_1\sqrt{D}$ is the fundamental solution and n is a positive integer.

Proof Clearly $x_n - y_n\sqrt{D} = \left(x_1 - y_1\sqrt{D}\right)^n$. Thus

$$x_n^2 - Dy_n^2 = \left(x_1^2 - Dy_1^2\right)^n = +1.$$

Thus $x_n + y_n\sqrt{D}$ is a solution of (6.42)

Suppose $u + v\sqrt{D}$ is a solution of (6.42), but is not obtainable from (6.45). Then there exists an integer n such that

$$\left(x_1 + y_1\sqrt{D}\right)^n < u + v\sqrt{D} < \left(x_1 + y_1\sqrt{D}\right)^{n+1}$$

or

$$(6.46) \qquad\qquad x_n + y_n\sqrt{D} < u + v\sqrt{D} < \left(x_n + y_n\sqrt{D}\right)\left(x_1 + y_1\sqrt{D}\right).$$

Since $\left(x_n - y_n\sqrt{D}\right)\left(x_n + y_n\sqrt{D}\right) = x_n^2 - Dy_n^2 = +1$ and $x_n + y_n\sqrt{D} > 0$, we see that $x_n - y_n\sqrt{D} > 0$. If we multiply the inequalities in (6.46) by $x_n - y_n\sqrt{D}$, we obtain

$$1 < \left(u + v\sqrt{D}\right)\left(x_n - y_n\sqrt{D}\right) < x_1 + y_1\sqrt{D}.$$

If we let $x + y\sqrt{D} = \left(u + v\sqrt{D}\right)\left(x_n - y_n\sqrt{D}\right) = ux_n - Dvy_n + (vx_n - uy_n)\sqrt{D}$, then we have $\left(u - v\sqrt{D}\right)\left(x_n + y_n\sqrt{D}\right) = x - y\sqrt{D}$. Thus

$$1 = (u^2 - Dv^2)(x_n^2 - Dy_n^2) = x^2 - Dy^2.$$

Thus $x + y\sqrt{D}$ is a solution of (6.42). By definition we have $x + y\sqrt{D} > 0$. Also $x - y\sqrt{D} = \left(x + y\sqrt{D}\right)^{-1}$ satisfies

$$0 < x - y\sqrt{D} < 1,$$

and so $x = \left(x + y\sqrt{D} + x - y\sqrt{D}\right)/2$ is positive and since $x \geq 1$ and $x - y\sqrt{D} < 1$ we see that y is positive.

Since $x + y\sqrt{D} < x_1 + y_1\sqrt{D}$ we have a contradiction to the minimality of the fundamental solution $x_1 + y_1\sqrt{D}$. Thus no such $u + v\sqrt{D}$ can exist. ∎

If we do not want to restrict ourselves to only positive solutions of (6.42), then, since $x - y\sqrt{D} = \left(x + y\sqrt{D}\right)^{-1}$, we can obtain all of these from the formula

$$x_n + y_n\sqrt{D} = \pm\left(x_1 + y_1\sqrt{D}\right)^n,$$

where, now, n is allowed to run through all integers.

The following corollary gives two further formulas for calculating solutions in positive x and y. They both follow immediately from (6.45) via the binomial theorem.

Corollary 6.34.1: If $x_1 + y_1\sqrt{D}$ is the fundamental solution of (6.42), then we have

(a) if $n \geq 1$, then

$$x_n = x_1^n + \sum_{k=1}^{[n/2]} \binom{n}{2k} x_1^{n-2k} y_1^{2k} D^k$$

and

$$y_n = \sum_{k=1}^{[(n+1)/2]} \binom{n}{2k-1} x_1^{n-2k+1} y_1^{2k-1} D^{k-1};$$

(b) if $n \geq 1$, then

$$x_{n+1} = x_1 x_n + D y_1 y_n$$

and

$$y_{n+1} = x_1 y_n + y_1 x_n.$$

The question becomes: how do we find the fundamental solution? This question will be answered shortly with the aid of the theory of continued fractions. This will provide us a way to compute the fundamental solution of (6.42) given D. In the meantime we present the following results regarding the fundamental solution of (6.42).

Theorem 6.35: Given D, let ξ and η be natural numbers satisfying the inequality

$$\xi > \left(\eta^2 / 2\right) - 1.$$

If $\alpha = \xi + \eta\sqrt{D}$ is a solution of (6.42), then it is the fundamental solution of (6.42).

Proof: The theorem is obvious if $\eta = 1$ since this implies

$$\xi^2 = D + 1,$$

if α is a solution of (6.42).

Suppose $\eta > 1$ and $x_1 + y_1\sqrt{D}$ is the fundamental solution of (6.42). If $1 \le y_1 < \eta$, then

$$D = \left(x_1^2 - 1\right) / y_1^2 = \left(\xi^2 - 1\right) / \eta^2$$

and

$$x_1^2\eta^2 - y_1^2\xi^2 = \eta^2 - y_1^2 = d > 0.$$

Thus $x_1\eta + y_1\xi = d_1$ and $x_1\eta - y_1\xi = d_2$, where $d_1 d_2 = d$. Thus

$$\xi = (d_1 - d_2) / 2y_1 \le (d-1) / 2y_1 = (\eta^2 - y_1^2 - 1) / 2y_1 \le \left(\eta^2 / 2\right) - 1,$$

which is a contradiction. If $y_1 > \eta$, then $x_1 > \xi$, and so we have a contradiction to the minimality of the fundamental solution. Thus $y_1 = \eta$, and so $x_1 = \xi$. ∎

Corollary 6.35.1: Let u and v be given natural numbers and let

$$D = u(uv^2 + 2).$$

Then $1 + uv^2 + v\sqrt{D}$ is the fundamental solution of (6.42).

This follows immediately from the theorem.

Example 6.14: Let $u = 1$ and $v = 1$ in Corollary 6.35.1. Then $D = 1 \cdot (1 \cdot 1^2 + 2)$ $= 3$. Thus the fundamental solutions to

$$x^2 - 3y^2 = +1$$

is $2 + \sqrt{3}$ and all positive solutions are therefore, given by

$$x_n + y_n\sqrt{3} = \left(2 + \sqrt{3}\right)^n,$$

where n is a positive integer. □

The following result is an immediate corollary to Theorem 4.26 since, by Theorem 4.20, we have $q_{nr} = 1$, for all $n \ge 0$.

Theorem 6.36: Let r be the length of the period of the continued fraction of $\sqrt{D}$. Then, for $n \ge 0$, we have

$$h_{nr-1}^2 - Dk_{nr-1}^2 = (-1)^{nr}.$$

Note that Theorem 4.26 gives us solutions to $x^2 - Dy^2 = N$ for certain values of N. In particular, we see that there are infinitely many solutions to

$$x^2 - Dy^2 = +1$$

and if r is odd, then there are infinitely many solutions to

$$x^2 - Dy^2 = -1.$$

Also, except for the solution $x = \pm 1$ and $y = 0$ to the equation $x^2 - Dy^2 = +1$, all solutions to $x^2 - Dy^2 = N$ come in fours using a combination of signs with $\pm x$ and $\pm y$. Thus we need only consider the solutions for which x and y are positive integers.

We first prove a lemma that appeared as Problem 8 of Problem Set 4.4.

Lemma 6.37.1. The $n\underline{\text{th}}$ convergent to the simple continued fraction expansion of $1/x$ is the reciprocal of the $(n-1)\underline{\text{st}}$ convergent of x for any irrational $x > 1$.

Proof. We have

$$x = [a_0; a_1, \ldots] \quad \text{and} \quad x^{-1} = [0; a_0, a_1, \ldots].$$

Let h_n/k_n denote the $n\underline{\text{th}}$ convergent of x and h_n' / k_n' denote the $n\underline{\text{th}}$ convergent of $1/x$. Then

$$h_0' = 0, h_1' = 1, h_2' = a_1, h_n' = a_{n-1}h_{n-1}' + h_{n-2}'$$
$$k_0 = 1, k_1 = a_1, k_{n-1} = a_{n-1}k_{n-2} + k_{n-3}$$
$$k_0' = 0, k_1' = a_0, k_2' = a_0 a_1 + 1, k_n' = a_{n-1}k_{n-1}' + k_{n-2}'$$
$$h_0 = a_0, h_1 = a_0 a_1 + 1, h_{n-1} = a_{n-1}h_{n-2} + h_{n-3},$$

by Theorem 4.11 and (4.23). The result follows by an easy mathematical induction argument. ∎

Theorem 6.37: Let $\{h_n/k_n\}$ be the convergents to $\sqrt{D}$. If N is an integer such that $|N| < \sqrt{D}$, then any positive solution $x = s$ and $y = t$ of $x^2 - Dy^2 = N$, with $(s, t) = 1$, satisfies

$$s = h_n \quad \text{and} \quad t = k_n$$

for some positive integer n.

Proof: Let P and Q be relatively prime positive integers such that

$$P^2 - \rho Q^2 = \sigma,$$

where $\sqrt{\rho}$ is irrational and $0 < \sigma < \sqrt{\rho}$. Here ρ and σ are reals, but not necessarily integers. Then

$$P/Q - \sqrt{\rho} = \sigma / Q(P + Q\sqrt{\rho}).$$

Thus

$$0 < P/Q - \sqrt{\rho} / Q(P + Q\sqrt{\rho}) = (Q^2(P/Q\sqrt{\rho} + 1))^{-1}$$

Also $0 < P/Q - \sqrt{\rho}$ implies that $P/Q\sqrt{\rho} > 1$. Thus

$$|P/Q - \sqrt{\rho}| < 1/2Q^2.$$

By Theorem 4.25 we see that P/Q is a convergent to the continued fraction expansion of $\sqrt{\rho}$.

If $N > 0$, we take $\sigma = N$, $\rho = D$, $P = s$ and $Q = t$ and the result follows . If $N < 0$, then $t^2 - (1/D)s^2 = -N/D$. We take $\sigma = -N/D$, $\rho = 1/D$, $P = t$ and $Q = s$. By Theorem 4.25 we see that t/s is a convergent in the continued fraction expansion of $1/\sqrt{D}$. The result then follows from Lemma 6.37.1.   ∎

If we combine the above results and apply them to the case $N = \pm 1$ we get the following theorem.

Theorem 6.38: All positive solutions of

$$x^2 - Dy^2 = \pm 1$$

are to be found among $x = h_n$ and $y = k_n$, where h_n/k_n are the convergents of the continued fraction expansion of $\sqrt{D}$. Let r be the period of $\sqrt{D}$. If r is even, then $x^2 - Dy^2 = -1$ has no solutions and all positive solutions of $x^2 - Dy^2 = +1$ are given by

(6.47) $x = h_{nr-1}$ and $y = k_{nr-1}$,

for $n = 1, 2, 3, \ldots$. If r is odd, then (6.47) gives all positive solutions of $x^2 - Dy^2 = -1$ for n odd and all positive solutions of $x^2 - Dy^2 = +1$ for n even.

Now the sequence of ordered pairs (h_0, k_0), (h_1, k_1), $\ldots$ will include all positive solutions of $x^2 - Dy^2 = +1$. Since $a_0 = [\sqrt{D}] > 0$ the sequence $h_0, h_1, h_2, \ldots$ is strictly increasing. If we let (x_1, y_1) be the first solution that occurs, then for any other solution (x, y) we have $x > x_1$ and $y > y_1$. Thus $x_1 + y_1\sqrt{D}$ is the fundamental solution of $x^2 - Dy^2 = +1$. By Theorem 6.38, this means that

$$x_1 + y_1\sqrt{D} = \begin{cases} h_{r-1} + k_{r-1}\sqrt{D} & \text{if } r \text{ is even} \\ h_{2r-1} + k_{2r-1}\sqrt{D} & \text{if } r \text{ is odd} \end{cases},$$

where r is the period length of $\sqrt{D}$.

In a similar way we may define the fundamental solution to the equation $x^2 - Dy^2 = -1$ as that solution, $\xi_1 + \eta_1\sqrt{D}$ with positive ξ_1 and η_1, with the least value. A similar argument as in the preceding paragraph shows that if r is odd, then the fundamental solution is determined by

$$\xi_1 + \eta_1\sqrt{D} = h_{r-1} + k_{r-1}\sqrt{D}.$$

Theorem 6.38 gives a necessary and sufficient condition for the solvability of

(6.48) $x^2 - Dy^2 = -1$.

Unfortunately, the criterion are not in terms of D, as such, but in terms of the continued fraction expansion of $\sqrt{D}$, the following result gives a partial answer.

Theorem 6.39: If $4 \mid D$ or D is divisible by a prime $\equiv 3 \pmod 4$, then (6.48) is not solvable.

Proof: If there exists a prime p such that $p \mid D$ and $p \equiv 3 \pmod 4$, then we have, for any solution of (6.48),

$$x^2 \equiv -1 \pmod p,$$

which is impossible by Corollary 3.2.1. If $4 \mid D$, then, for any solution of (6.48), we have

$$x^2 \equiv -1 \pmod 4,$$

which is also impossible. ∎

One can show that those criteria are not sufficient. First we present a result, which is an analog of Theorem 6.34.

Theorem 6.40: Suppose (6.48) is solvable and that $\xi_1 + \eta_1 \sqrt{D}$ is its fundamental solution (that is, the least positive solution with ξ_1 and η_1 positive). Then

(a) the fundamental solution to (6.42) is given by

$$x_1 + y_1 \sqrt{D} = \left(\xi_1 + \eta_1 \sqrt{D}\right)^2 = \xi_1^2 + D\eta_1^2 + 2\xi_1 \eta_1 \sqrt{D}.$$

(b) If we set $\xi_n + \eta_n \sqrt{D} = \left(\xi_1 + \eta_1 \sqrt{D}\right)^n$, for n a positive integer, then

(i) if n is odd, we get all positive solutions of (6.48) and

(ii) if n is even we get all positive solutions of (6.42).

Proof: As in the proof of Theorem 6.34 we see that

$$\xi_n^2 - D\eta_n^2 = (-1)^n$$

so that the sequence of integers defined in (b) do give solutions to their respective equations.

Since

$$\left(-\xi_1 + \eta_1 \sqrt{D}\right)\left(\xi_1 + \eta_1 \sqrt{D}\right) = +1$$

we see that

$$0 < -\xi_1 + \eta_1 \sqrt{D} < 1,$$

that is, $-\xi_1 + \eta_1 \sqrt{D}$ is the largest of the solutions of (6.48) with negative x and positive y.

Suppose our two fundamental solutions are not related by the formula of (a).

Since $\left(\xi_1 + \eta_1 \sqrt{D}\right)^2$ is a solution of $x^2 - Dy^2 = +1$ we must have

$$1 < x_1 + y_1 \sqrt{D} < \left(\xi_1 + \eta_1 \sqrt{D}\right)^2.$$

If we multiply both sides by $-\xi_1 + \eta_1 \sqrt{D}$ we have

$$-\xi_1 + \eta_1 \sqrt{D} < -\xi_0 + \eta_0 \sqrt{D} < -\xi_1 + \eta_1 \sqrt{D},$$

where $\xi_0 + \eta_0 \sqrt{D} = \left(-\xi_1 x_1 + D\eta_1 y_1\right) + \left(\eta_1 x_1 - \xi_1 y_1\right)\sqrt{D}$, which is a solution of (6.48). From the properties of $-\xi_1 + \eta_1 \sqrt{D}$ and $\xi_1 + \eta_1 \sqrt{D}$ we see we cannot have any of the following holding:

1) $\xi_0 > 0,\ \eta_0 > 0,$

2) $\xi_0 < 0,\ \eta_0 > 0,$

or

3) $\xi_0 < 0,\ \eta_0 < 0.$

If $\xi_0 > 0$ and $\eta_0 > 0$, then we would have

$$D\eta_1 y_1 > \xi_1 x_1 \quad \text{and} \quad \xi_1 y_1 > \eta_1 x_1.$$

Thus

$$\xi_1 \eta_1 y_1^2 D > \xi_1 \eta_1 x_1^2,$$

which is impossible since $Dy_1^2 - x_1^2 = -1$. The numbers ξ_0 and η_0 cannot be zero since they are a solution to (6.48). Since we have ruled out all the possibilities we see that $x_1 + y_1 \sqrt{D}$ and $\xi_1 + \eta_1 \sqrt{D}$ must be connected as in (a).

Part (ii) of (b) then follows from Theorem 6.34.

To prove part (i) of (b) we proceed as in the proof of Theorem 6.34. Suppose there is a solution of (6.48), say $u + v\sqrt{D}$, which is not of the required form. Then there exists a positive integer m such that

(6.49) $$\left(\xi_1 + \eta_1 \sqrt{D}\right)^{2m-1} < u + v\sqrt{D} < \left(\xi_1 + \eta_1 \sqrt{D}\right)^{2m+1}$$

If we multiply (6.49) by the positive number $\left(\xi_1 - \eta_1 \sqrt{D}\right)^{2m} = \xi_{2m} - \eta_{2m} \sqrt{D}$, we get

(6.50) $$-\xi_1 + \eta_1 \sqrt{D} < \xi_0 + \eta_0 \sqrt{D} < \xi_1 + \eta_1 \sqrt{D},$$

where $\xi_0 = u\xi_{2m} - v\eta_{2m} D$ and $\eta_0 = v\xi_{2m} - u\eta_{2m}$. Also $\xi_0 + \eta_0 \sqrt{D}$ is a solution of (6.48). Above we showed that (6.50) required that $\eta_0 = 0$. Thus $u = \xi_{2m}$ and $v = \eta_{2m}$. Since $\xi_{2m} + \eta_{2m} \sqrt{D}$ is not a solution of (6.48) we see that no such $u + v\sqrt{D}$ can exist.   ∎

We can use this theorem to show that the criteria of Theorem 6.39, while necessary, is not sufficient. We take $D = 34$. It is not too difficult (use Theorem 6.35) to show that $35 + 6\sqrt{34}$ is the fundamental solution of $x^2 - 34y^2 = +1$. If $\xi_1 + \eta_1\sqrt{34}$ is the fundamental solution of (6.48), then, by Theorem 6.40, we must have

$$35 = \xi_1^2 + 34\eta_1^2 \quad \text{and} \quad 6 = 2\xi_1\eta_1,$$

which is clearly impossible. Thus $x^2 - 34y^2 = -1$ is impossible in integers x and y.

We can give one class of integers D for which (6.48) is solvable.

Theorem 6.41: Let p be a prime, $p \equiv 1$ (mod 4). Then the equation

$$x^2 - py^2 = -1$$

is solvable in integers x and y.

Proof: Let $x_1 + y_1\sqrt{p}$ be the fundamental solution to

$$x^2 - py^2 = +1.$$

Then $x_1^2 - 1 = py_1^2$. If x_1 is even, then y_1 is odd and we have $-1 \equiv p$ (mod 4), which is impossible. Thus x_1 is odd, and so

$$(x_1 + 1, x_1 - 1) = 2 .$$

Thus for some choice of sign we have

$$x_1 \pm 1 = 2\xi^2 \quad \text{and} \quad x_1 \mp 1 = 2p\eta^2.$$

Thus $y_1 = 2\xi\eta$. If we eliminate x_1 we get

$$\pm 1 = \xi^2 - p\eta^2.$$

Since $\eta < y_1$ we see that we cannot take the upper sign. Thus we must take the lower sign and the result follows. ∎

Note that we have really proved more. Namely, by Theorem 6.40, if $\xi + \eta\sqrt{p}$ is the solution determined in the proof of Theorem 6.41, then it is the fundamental solution of that equation.

Before returning to the general equation, (6.39), we give some numerical examples.

Example 6.17: Solve $x^2 - Dy^2 = \pm 1$ for $D = 2, 3, 29$, and 98.

Solution:

    1) $D = 2$.

In Example 4.5 we saw that $\sqrt{2} = [1;\overline{2}]$. Thus $r = 1$, and so $x^2 - 2y^2 = -1$ is solvable and the fundamental solution is easily seen to be $1 + \sqrt{2}$. Thus, by Theorem 6.40, the fundamental solution of $x^2 - 2y^2 = +1$ is $\left(1 + \sqrt{2}\right)^2 = 3 + 2\sqrt{2}$. All positive solutions of $x^2 - 2y^2 = (-1)^n$ are given by $\left(1 + \sqrt{2}\right)^n$.

2) $D = 3$.

To compute the continued fraction expansion of $\sqrt{3}$ we use the set up given after the proof of Theorem 4.20. We have $\xi_0 = \sqrt{3} = \left(m_0 + \sqrt{3}\right)/q_0$ so that $m_0 = 0$ and $q_0 = 1$. Then

$$a_0 = [\sqrt{3}] = 1, m_1 = 1 \cdot 1 - 0 = 1, q_1 = (3 - 1)/1 = 2,$$
$$a_1 = [(1 + \sqrt{3})/2] = 1, m_2 = 12 - 1 = 1, q_2 = 1 + 1(1 - 1) = 1,$$
$$a_2 = [1 + \sqrt{3}] = 2, m_3 = 2 \cdot 1 - 1 = 1, q_3 = 2 + 2(1 - 1) = 2,$$
$$a_3 = [(1 + \sqrt{3})/2] = a_1.$$

Thus we have repetition, and so $\sqrt{3} = [1;\overline{1,2}]$.

Since $r = 2$ we see that $x^2 - 3y^2 = -1$ is not solvable (as we saw in Example 6.11). The fundamental solution of $x^2 - 3y^2 = +1$ is, by Theorem 6.38, $h_1 + k_1\sqrt{3}$. To calculate the convergents we proceed as follows. We have the table (using (4.23))

n	a	h	k
-2		0	1
-1		1	0
0	1	1	1
1	1	2	1

and so the fundamental solution is $2 + \sqrt{3}$. and all positive solutions are given by $(2 + \sqrt{3})^n$, where n is a positive integer.

3) $D = 29$.

In Example 4.17 we saw that $\sqrt{29} = [5;\overline{2,1,1,2,10}]$. Since $r = 5$ we see that $x^2 - 29y^2 = -1$ is solvable with fundamental solution $h_4 + k_4\sqrt{29}$. Our table of convergents is

n	a	h	k
-2		0	1
-1		1	0
0	5	5	1
1	2	11	2

n	a	h	k
2	1	16	3
3	1	27	5
4	2	70	13

Thus the fundamental solution of $x^2 - 29y^2 = -1$ is $70 + 13\sqrt{29}$, and so the fundamental solution to $x^2 - 29y^2 = +1$ is $(70 + 13\sqrt{29})^2 = 9801 + 1820\sqrt{29}$.

4) $D = 98$.

We have $\sqrt{98} = (m_0 + \sqrt{98}) / q_0$, so that $m_0 = 0$ and $q_0 = 1$. Then

$$a_0 = [\sqrt{98}] = 9, \ m_1 = 9 \cdot 1 - 0 = 9, \ q_1 = (98 - 9) / 1 = 17,$$

$$a_1 = [(9 + \sqrt{98}) / 17] = 1, \ m_2 = 1 \cdot 17 - 9 = 8, \ q_2 = 1 + 1(9 - 8) = 2,$$

$$a_2 = [(8 + \sqrt{98}) / 2] = 8, \ m_3 = 8 \cdot 2 - 8 = 8, \ q_3 = 17 + 8(8 - 8) = 17,$$

$$a_3 = [(8 + \sqrt{98}) / 17] = 1, \ m_4 = 1 \cdot 17 - 8 = 9, \ q_4 = 2 + 1(8 - 9) = 1,$$

$$a_4 = [9 + \sqrt{98}] = 18, \ m_5 = 18 \cdot 1 - 9 = 9, \ q_5 = 17 + 18(9 - 4) = 17.$$

Thus $\xi_5 = \xi_1$, and so

$$\sqrt{98} = [9; \overline{1, 8, 1, 18}].$$

Since $r = 4$ we see that $x^2 - 98y^2 = -1$ is not solvable and that the fundamental solution to $x^2 - 98y^2 = +1$ is $h_3 + k_3\sqrt{98}$. Our table of convergents is

n	a	h	k
-2		0	1
-1		1	0
0	9	9	1
1	1	10	1
2	8	89	9
3	1	99	10

Thus the fundamental solution is $99 + 10\sqrt{98}$. □

In the Appendix, in Table IV, we have a list of the solutions to the equations $x^2 - Dy^2 = \pm 1$, for nonsquare values of D from 2 to 100.

We return to the general equation

(6.51) $$x^2 - Dy^2 = N,$$

where D, as usual, is a positive nonsquare integer and N is a given nonzero integer. From Theorem 6.37 we know what the solutions of (6.51) are for small values of N.

We shall treat the problem from scratch in order to find all solutions of (6.51), when it is solvable, and to give constructive criteria to determine when it is solvable.

Let $r + s\sqrt{D}$ be any solution of (6.51) and $x + y\sqrt{D}$ and solution of

(6.52) $x^2 - Dy^2 = +1.$

Then, as we saw above, the number

$$u + v\sqrt{D} = (r + s\sqrt{D})(x + y\sqrt{D})$$

is another solution of (6.51), that is distinct from the solution $r + s\sqrt{D}$. Any two solutions that are related in this way (by the multiplication of a solution of (6.52)) are said to be **associated** to one another. The set of all solutions of (6.51) associated with one another form a class of solutions, which, by Theorem 6.34, is infinite. It is clear that if $u + v\sqrt{D}$ and $u_1 + v_1\sqrt{D}$ are two solutions of (6.51), then they are in the same class if and only if

$$(uu_1 - vv_1 D) / N \quad \text{and} \quad (vu_1 - uv_1) / N$$

are both integers (one needs $(u + v\sqrt{D}) / (u_1 + v_1\sqrt{D})$ to be a solution of (6.52).)

Let K be a class of solutions, say $K = \{u_i + v_i\sqrt{D}\}$. Then the set $\{u_i - v_i\sqrt{D}\}$ is also a class of solutions which is denoted by $\overline{K}$ and called the **conjugate class** to K. Usually we have $K \cap \overline{K} = \emptyset$, however, if $K = \overline{K}$, then K and $\overline{K}$ are called **ambiguous classes**.

Given the class of solutions K let $u^* + v^*\sqrt{D}$ be the solution of (6.51) determined as follows. Let v^* be the least nonnegative value of v occurring in K. If K is not ambiguous, then u^* is uniquely determined since $-u^* + v^*\sqrt{D} \in \overline{K}$. If K is ambiguous we determine u^* by $u^* \geq 0$. This solution, $u^* + v^*\sqrt{D}$, so determined, is called the **fundamental solution** of the class K. For the fundamental solution note that $|u^*|$ is the least value of $|u|$ which is possible for $u + v\sqrt{D} \in K$. Finally note that $u^* = 0$ or $v^* = 0$ if and only if K is ambiguous.

If $N = \pm 1$, then it is clear that there is only one class.

We now give criteria for finding the fundamental solutions of the various classes of solutions when (6.51) is solvable.

Theorem 6.42: Let N be a positive integer. If $u + v\sqrt{D}$ is a fundamental solution of (6.51) and $x_1 + y_1\sqrt{D}$ is the fundamental solution of (6.52), then we have

$$0 \leq v \leq y_1\sqrt{N / (2(x_1 + 1))}$$

and

$$0 \le |u| \le \sqrt{N(x_1 + 1)/2}.$$

Proof: If these are true for a given class K, then they are true also for $\overline{K}$. Thus we may suppose that $u > 0$.

We have

$$ux_1 - Dvy_1 = ux_1 - \sqrt{(u^2 - N)(x_1^2 - 1)} > 0.$$

Also

$$(u + v\sqrt{D})(x_1 - y_1\sqrt{D}) = (ux_1 - Dvy_1) + (x_1v - y_1u)\sqrt{D}$$

is another member of the class K. Since $u + v\sqrt{D}$ is the fundamental solution and $ux_1 - Dvy_1 > 0$ we must have

$$ux_1 - Dvy_1 \ge u.$$

Thus

$$u^2(x_1 - 1)^2 \ge D^2v^2y_1^2 = (u^2 - N)(x_1^2 - 1)$$

or

$$(x_1 - 1)/(x_1 + 1) \ge 1 - N/u^2.$$

Thus

$$u^2 \le (x_1 + 1)N/2$$

This gives the second inequality and the first follows from the relations

$$v^2 = (u^2 - N)/D \quad \text{and} \quad D = (x_1^2 - 1)/y_1^2. \qquad \blacksquare$$

Theorem 6.43: Let N be a positive integer and let $u + v\sqrt{D}$ be the fundamental solution for the equation

$$u^2 - Dv^2 = -N.$$

If $x_1 + y_1\sqrt{D}$ is the fundamental solution of (6.52), then

$$0 \le v \le y_1\sqrt{N/(2(x_1 - 1))}$$

and

$$0 \le |u| \le \sqrt{N(x_1 - 1)/2}.$$

Proof: If the inequalities are true for a given class K, then they are true also for $\overline{K}$ and so we may assume $u \geq 0$.

We have

$$(x_1 v)^2 = (y_1^2 + 1/D)(u^2 + N) > y_1^2 u^2,$$

and so

$$x_1 v - y_1 u > 0.$$

Now

$$(u + v\sqrt{D})(x_1 - y_1\sqrt{D}) = (ux_1 - Dvy_1) + (x_1 v - y_1 u)\sqrt{D}$$

belongs to the same class as $u + v\sqrt{D}$ and since $u + v\sqrt{D}$ is the fundamental solution we must have

$$x_1 v - y_1 u \geq v.$$

Thus

$$Dv^2(x_1 - 1)^2 \geq Dy_1^2 u^2,$$

and so, as above, we have

$$1 + N/u^2 \geq (x_1 + 1)/(x_1 - 1).$$

Thus

$$u^2 \leq (x_1 - 1)N/2,$$

which proves the second inequality. The first inequality follows from the second as in the proof of Theorem 6.42 ∎

If we combine these results we obtain the following result.

Theorem 6.44: Let N be a positive integer. Then the Diophantine equations

$$u^2 - Dv^2 = N \quad \text{and} \quad u^2 - Dv^2 = -N$$

have a finite number of classes of solutions. The fundamental solutions of all the classes can be found from the inequalities of Theorems 6.50 and 6.51. If $u^* + v^*\sqrt{D}$ is the fundamental solution of the class K, then all solutions, $u + v\sqrt{D}$, of the class K are given by

$$u + v\sqrt{D} = \pm(u^* + v^*\sqrt{D})(x_1 + y_1\sqrt{D})^n,$$

where $x_1 + y_1\sqrt{D}$ is the fundamental solution of (6.52) and n runs through all integers. If the inequalities of Theorems 6.42 or 6.43 have no solutions satisfying the respective equation, then that equation has no solution at all.

For a special case we can be more specific.

Theorem 6.45: If p is a prime, then the equations

$$x^2 - Dy^2 = p \quad \text{and} \quad x^2 - Dy^2 = -p$$

have at most one solution $u + v\sqrt{D}$ in which $u \geq 0$ and u and v satisfy the inequalities of Theorem 6.42 or 6.43. If either equation is solvable, then it has one or two classes of solutions depending on whether $p \mid 2D$ or not.

Proof: Suppose $u + v\sqrt{D}$ and $u_1 + v_1\sqrt{D}$ are two solutions of the equation satisfying the first part of the theorem. Then the numbers u, v, u_1 and v_1 are nonnegative. If we eliminate D between the equations

(6.53) $$u^2 - Dv^2 = \pm p \quad \text{and} \quad u^2 - Dv^2 = \pm p$$

we get

$$u^2 v_1^2 - u_1^2 v^2 = \pm p(v_1^2 - v^2).$$

Thus

(6.54) $$uv_1 \equiv \pm u_1 v \pmod{p}$$

for some choice of sign. If we multiply the two equations of (6.53) together we obtain

$$((uu_1 \mp Dvv) / p) - D(uv_1 \mp u_1 v) = p^2$$

or

$$((uu_1 \mp Dvv_1) / p)^2 - D((uv_1 \mp u_1 v) / p)^2 = 1$$

and if we choose the sign so that (6.54) is satisfied, then both

$$(uu_1 \mp Dvv_1) / p \quad \text{and} \quad (uv_1 \mp u_1 v) / p$$

are integers. If $uv_1 \mp u_1 v \neq 0$, then

$$|uv_1 \mp u_1 v| \geq y_1 p.$$

On the other hand, the appropriate inequalities from Theorem 6.42 or 6.43 gives

$$|uv_1 \mp u_1 v| < y_1 p.$$

Thus we must have

$$uv_1 \mp u_1 v = 0.$$

Since we must have $(u, v) = 1 = (u_1, v_1)$ we see we must have $u = u_1$ and $v = v_1$, which proves the first part of the theorem.

This shows that there are at most two classes of solutions, namely the class determined above and its conjugate class. Suppose $u + v\sqrt{D}$ and $u - v\sqrt{D}$ are two solutions satisfying the inequalities of Theorem 6.42 or Theorem 6.43. These two solutions are associated if and only if p divides the two numbers

$$2uv \quad \text{and} \quad u^2 + Dv^2 = 2Dv^2 \pm p.$$

Since $p \nmid v$ we see that this is true if and only if

$$p \mid 2u \text{ and } p \mid 2D.$$

Since $p \mid 2D$ implies $p \mid 2u$ we see that $u + v\sqrt{D}$ and $u - v\sqrt{D}$ are in the same class if and only if $p \mid 2D$.   ∎

We now give some examples of these general results.

Example 6.18: Solve the following equations.

  (a) $u^2 - 2v^2 = 119$.
  (b) $u^2 - 6v^2 = -29$.
  (c) $u^2 - 6v^2 = -2$.
  (d) $u^2 - 82v^2 = 23$.

Solution of (a) The fundamental solution to $x^2 - 2y^2 = +1$ is $3 + 2\sqrt{2}$. Thus, by Theorem 6.42, any fundamental solution must satisfy

$$0 \le v \le 2\sqrt{119/(2 \cdot 4)} = \sqrt{119/2} < 8$$

and

$$0 \le |u| \le \sqrt{119(3+1)/2} = \sqrt{238} < 16.$$

If we try all pairs in these intervals we see that the following are solutions

(6.55) $11 + \sqrt{2}, \ -11 + \sqrt{2}, \ 13 + 5\sqrt{2} \text{ and } -13 + 5\sqrt{2}$

and it is not difficult to show that these solutions are nonassociated. Thus all solutions are given by

$$u + v\sqrt{2} = (u^* + v^*\sqrt{2})(3 + 2\sqrt{2})^n,$$

where $u^* + v^*\sqrt{2}$ is one of the four fundamental solutions in (6.55) and n runs through all integers.

Solution of (b) The fundamental solution to $x^2 - 6y^2 = +1$ is $5 + 2\sqrt{6}$ and the inequalities of Theorem 6.43 are

$$0 < v \le 2\sqrt{29/(2 \cdot 4)} = \sqrt{29/2} < 4$$

and

$$0 \le |u| \le \sqrt{29(5-1)/2} = \sqrt{58} < 8.$$

If we try all pairs in these intervals we see that the following are solutions

$$5 + 3\sqrt{6} \quad \text{and} \quad -5 + 3\sqrt{6}$$

and it is not difficult to see that these solutions are nonassociated. Note that this is in accord with Theorem 6.45 since $29 \nmid 2 \cdot 6$. All solutions are given by

$$u + v\sqrt{6} = (\pm 5 + 3\sqrt{6})(5 + 2\sqrt{6})^n,$$

where n runs through all integers.

Solution of (c) Here the inequalities are

$$0 < v \le 2\sqrt{2/(2 \cdot 4)} = 1$$

and

$$0 \le |u| \le \sqrt{2(5-1)/2} = 2.$$

Thus v must equal 1 and u is 0, ± 1, ± 2 in any possible solution. We find two solutions trying the various pairs:

$$2 + \sqrt{6} \quad \text{and} \quad -2 + \sqrt{6}$$

however,

$$(2 + \sqrt{6})/(-2 + \sqrt{6}) = -5 - 2\sqrt{6},$$

which is a solution to $x^2 - 6y^2 = +1$. Thus the two solutions are associated and there is only one class of solutions which is ambiguous. This is again in accord with Theorem 6.45 since $2 \mid (2 \cdot 6)$. All solutions are given by

$$u + v\sqrt{6} = (\pm 5 + 3\sqrt{6})(5 + 2\sqrt{6})^n,$$

where n runs through the integers.

Solution of (d) The fundamental solution to $x^2 - 82y^2 = +1$ is $163 + 18\sqrt{82}$. The inequalities of Theorem 6.42 are

$$0 \le v \le 18\sqrt{82/(2 \cdot (163 + 1))} = 9\sqrt{23/82} < 5$$

and

$$0 < |u| \le \sqrt{82(163 + 1)/2} = 82.$$

If we check the values for v: 0, 1, 2, 3 and 4, we find that none of them produce integer values of u. Thus, by Theorem 6.44, the equation has no integer solutions. $\square$

Before we leave this topic we note that we can use Theorems 6.42 and 6.43 to derive a hybrid of the Euler and Fermat factorization methods discussed in Section 5.1.

Let N be a positive integer. Then, by Theorems 6.42 and 6.43, we know that if $x_1 + y_1\sqrt{D}$ is the fundamental solution to (6.42) and $x^2 - Dy^2 = \pm N$ is solvable, then a solution $u + v\sqrt{D}$ must exist with

$$0 < v < y_1\sqrt{\frac{N}{2(x_1 \pm 1)}} \quad \text{and} \quad 0 < u < \sqrt{\frac{(x_1 \pm 1)N}{2}}.$$

Since $x_1^2 + 1 = Dy_1^2$ note that the upper limit on v can be rewritten as

$$0 < v < \sqrt{\frac{(x_1 \pm 1)N}{2D}}.$$

Suppose that we have two such representations, say

(6.56) $\qquad u_1^2 - Dv_1^2 = N \quad \text{and} \quad u_2^2 - Dv_2^2 = N$

where

(6.57) $\qquad 0 < u_1 < \sqrt{\frac{(x_1 + 1)N}{2}} \quad \text{and} \quad 0 < v_1 < \sqrt{\frac{(x_1 - 1)N}{2D}}$

and

(6.58) $\qquad 0 < u_2 < \sqrt{\frac{(x_1 + 1)N}{2}} \quad \text{and} \quad 0 < v_2 < \sqrt{\frac{(x_1 - 1)N}{2D}}.$

Then we have, if we multiply the two representations in (6.56) together,

(6.59) $\qquad (u_1u_2 + Dv_1v_2)^2 - D(u_1v_2 + u_2v_1)^2 = N^2.$

If $N \mid (u_1v_2 + u_2v_1)$, then, by (6.59), we have

$$\left(\frac{u_1u_2 + Dv_1v_2}{N}\right)^2 - D\left(\frac{u_1v_2 + u_2v_1}{N}\right)^2 = 1,$$

and so $(u_1u_2 + Dv_1v_2)/N$ is an integer.

By the inequalities (6.57) and (6.58), we have

$$u_1u_2 + Dv_1v_2 < \frac{(x_1 + 1)N}{2} + \frac{(x_1 - 1)N}{2} = Nx_1.$$

Thus, if we let

$$T = \frac{u_1u_2 + Dv_1v_2}{N} \quad \text{and} \quad U = \frac{u_1v_2 + u_2v_1}{N},$$

we see that $T^2 - DU^2 = 1$ with $T < x_1$. This contradicts the minimality of $x_1 + y_1\sqrt{D}$ as a solution of (6.42). Thus we cannot have $N \mid (u_1v_2 + u_2v_1)$.

In the same way from

$$(u_1u_2 - Dv_1v_2)^2 - D(u_1v_2 - u_2v_1)^2 = N^2,$$

we can show that $N \nmid (u_1v_2 - u_2v_1)$. Now

$$(u_1v_2 + u_2v_1)(u_1v_2 - u_2v_1) = u_1^2v_2^2 - u_2^2v_1^2$$
$$= (N + Dv_1^2)v_2^2 - (N + Dv_2^2)v_1^2$$
$$= (v_2^2 - v_1^2)N \equiv 0 \pmod{N}.$$

Thus, by Lemma 1.21.1, we see that N is composite and both $(u_1v_2 \pm u_2v_1, N)$ are nontrivial divisors of N.

We can apply the same reasoning to the equations $x^2 - Dy^2 = -N$. This gives us the following result due to Chebychev.

Theorem 6.46. If the equation $x^2 - Dy^2 = \pm N$ has two solutions $u_1 + v_1\sqrt{D}$ and $u_2 + v_2\sqrt{D}$ both of which satisfy the inequalities

(6.60) $0 < u_k < \sqrt{\dfrac{(x_1 \pm 1)N}{2}}$ and $0 < v_k < \sqrt{\dfrac{(x_1 \mp 1)N}{2D}}$, $k = 1, 2,$

where $x_1 + y_1\sqrt{D}$ is the fundamental solution to (6.42), then N is composite and the two numbers $(u_1v_2 \pm u_2v_1, N)$ will both be nontrivial factors of N.

Example 6.19. Use Theorem 6.46 to find factors of 119.

Solution. By (a) of Example 6.14 we have

$$11^2 - 2 \cdot 1^2 = 119 \quad \text{and} \quad 13^2 - 2 \cdot 5^2 = 119,$$

with $u_1 = 11$, $v_1 = 1$, $u_2 = 13$ and $v_2 = 5$ satisfying the inequalities (6.60). Thus, by Theorem 6.46, both of the numbers $(55 \pm 13, 119)$ are nontrivial factors of 119. If we use the Euclidean algorithm we see that

$$(55 + 13, 119) = (68, 119) = 17$$

and

$$(55 - 13, 119) = (42, 119) = 7.$$

Indeed we have $119 = 7 \cdot 17$.   ☐

This finishes our general study of the equation $x^2 - Dy^2 = N$. In the next two sections we will apply these results to other problems. For further references on the general problem see, for example, [130] or [25, v. II, ch. 12].

Problem Set 6.7

1. Give the details of the proof of Corollary 6.34.1.

2. Solve the equation $x^2 - a^2 y^2 = n$, where $a > 1$ and n are given integers.

3. Prove that the representation of a prime by a difference of squares is unique.

4. Prove that every composite odd number can be represented as a sum of consecutive odd numbers, but no prime can. What can be said about even numbers?

5. If n is a positive integer show that the equation $x^2 - (n^2 - 1)y^2 = -1$ has no solution in integers. (Hint: find the fundamental solution to $x^2 - (n^2 - 1)y^2 = +1$.)

6. (a) Is $x^2 - 30y^2 = -1$ solvable? If so, give the general solution.

   (b) Give the general solution to $x^2 - 30y^2 = 1$.

7. (a) Is $x^2 - 74y^2 = -1$ solvable? If so, give the general solution.

   (b) Give the general solution to $x^2 - 74y^2 = 1$.

8. Give the general solutions to the following equations.

   (a) $x^2 - 23y^2 = 1$ (e) $x^2 - 29y^2 = -1$

   (b) $x^2 - 51y^2 = 1$ (f) $x^2 - 130y^2 = -1$

   (c) $x^2 - 146y^2 = 1$ (g) $x^2 - 314y^2 = -1$

   (d) $x^2 - 1112y^2 = 1$ (h) $x^2 - 1490y^2 = -1$.

9. Find the general solutions to the following equations.

   (a) $x^2 - 3y^2 = 6$ (d) $x^2 - 7y^2 = 2$

   (b) $x^2 - 14y^2 = 5$ (e) $x^2 - 82y^2 = 2$

   (c) $x^2 - 12y^2 = 4$ (f) $x^2 - 35y^2 = 9$.

10. (a) Suppose $a \mid 2n$ and $1 \le a < n$. Find the continued fraction expansion of $\sqrt{n^2 - a}$. What does this say about the solvability of either equation $x^2 - (n^2 - a)y^2 = \pm 1$?

    (b) Find the fundamental solution to the equation $x^2 - (9n^2 - 3)y^2 = 1$.

11. Answer the analogous questions for the equations $x^2 - (n^2 + a)y^2 = \pm 1$ and $x^2 - (9n^2 + 6)y^2 = 1$. Here we only require that $a \mid 2n$.

12. (a) Show that $n^2 + (n + 1)^2$ is a perfect square for infinitely many positive integers n.

    (b) Show that $2n + 1$ and $n + 1$ are perfect squares for infinitely many positive integers n.

13. Let D be a positive nonsquare integer and let n be a positive integer. Show that

there are infinitely many solutions in integers of $x^2 - Dy^2 = 1$, where $n \mid y$.

14. Let x and y be positive integers such that $x^2 - Dy^2 = \pm 1$. If all the prime factors of y divide D, then $x + y\sqrt{D}$ is the fundamental solution.

15. If n is a positive integer, find the general solution to the equation $x^2 - (n^2 - 2)y^2 = 3 - 2n$.

16. (a) If $p \equiv 1 \pmod{12}$, then there exist integers u and v such that $u^2 - 3v^2 = p$ with $0 < u < \sqrt{3p/2}$ and $0 < v < \sqrt{p/2}$.

    (b) If $p \equiv -1 \pmod{12}$, then there exist integers u and v such that $u^2 - 3v^2 = -p$ with $0 < u < \sqrt{p/2}$ and $0 < v < \sqrt{p/2}$.

17. Let D be a positive nonsquare integer such that $\sqrt{D} = [a; \overline{b, 2a}]$. Show that if h_j/k_j denotes the jth convergent, then $h_{2j}^2 - Dk_{2j}^2 = 1$ and $h_{2j+1}^2 - Dk_{2j+1}^2 = -q_1$.

18. Let t_n denote the nth triangular number, $t_n = n(n + 1)/2$, and let p_n denote the nth pentagonal number, $p_n = n(3n - 1)/2$

    (a) Solve $t_x = y^2$.
    (b) Solve $t_x = p_y$.

6.8. A PYTHAGOREAN TRIANGLE PROBLEM

Recall that a Pythagorean triangle is a right triangle all of whose sides are integers and a primitive Pythagorean triangle is a Pythagorean triangle whose sides are relatively prime. By Theorem 5.1 the sides are given parametrically as

$$(6.61) \qquad x = u^2 - v^2, \; y = 2uv \quad \text{and} \quad z = u^2 + v^2,$$

where $u > v > 0$, $(u, v) = 1$ and u and v are integers of opposite parity. At that time we passed over one problem which we are now in a position to solve completely.

Problem. Find all primitive Pythagorean triangles whose legs differ by a given constant $k > 0$.

By (6.61) this means we need to find integers u and v such that

$$u^2 - v^2 - 2uv = \pm k.$$

Completing the square gives

$$(u - v)^2 - 2v^2 = \pm k.$$

Thus we must solve the equation

$$(6.62) \qquad x^2 - 2y^2 = \pm k,$$

where $(x, y) = 1$ (since $(u, v) = 1$ implies $(u - v, v) = 1$) and x is odd. The latter condition on x implies that k must be odd. Note also that if k has a prime divisor $\equiv 3$ or 5 (mod 8), then the equation cannot be solvable since $x^2 - 2y^2 = pk_1$, $p \equiv 3$ or 5 (mod 8), implies

$$z^2 \equiv 2 \pmod{p}$$

is solvable, which it is not, by Corollary 3.4.1. Thus, if (6.62) is solvable we must have k of the form $p_1^{\alpha_1} \cdots p_r^{\alpha_r}$, where the p_i are primes $\equiv \pm 1$ (mod 8). To show that the problem is solvable in this case we need the following result.

Theorem 6.47: If p is a prime $\equiv 1$ or 7 (mod 8), then there exist two natural numbers u and v such that

$$u^2 - 2v^2 = p,$$

where $0 < u < \sqrt{2p}$ and $0 < v < \sqrt{p/2}$, and there exist two natural numbers r and s such that

$$r^2 - 2s^2 = -p,$$

where $0 < r < \sqrt{p}$ and $0 < s < \sqrt{p}$.

Proof: We need only show solvability of

$$x^2 - 2y^2 = \pm p,$$

for the inequalities then follow from Theorems 6.42 and 6.43, since the fundamental solution to $x^2 - 2y^2 = 1$ is $3 + 2\sqrt{2}$.

If $p \equiv \pm 1$ (mod 8), then we may solve $z^2 \equiv 2 \pmod{p}$, by Corollary 3.4.1. By Thue's theorem, Theorem 2.10, we know there exist integers x and y such that $0 < x < \sqrt{p}$ and $0 < y < \sqrt{p}$ and for either of those values of z we have

$$zy \equiv \pm x \pmod{p}.$$

Then

$$z^2 y^2 - x^2 \equiv 2y^2 - x^2 \pmod{p}.$$

Since $-p < 2y^2 - x^2 < 2p$ and $\sqrt{2}$ is irrational we see that

$$x^2 - 2y^2 = -p.$$

Then

$$(x + 2y)^2 - 2(x + y)^2 = p$$

and the theorem is proved.

To finish the problem we use the algebraic identity

$$(x^2 - 2y^2)(r^2 - 2s^2) = (xr - 2ys)^2 - 2(xs - yr)^2$$

and combine the results above.

Theorem 6.48: The equation (6.62) is solvable with x odd and $(x, y) = 1$ if and only if k is odd and $p \mid k$ implies that $p \equiv 1$ or $7 \pmod 8$.

If (6.62) is solvable, we may recover the parameters u and v from the relations

(6.63) $\qquad\qquad u = x + y$ and $v = y$.

Thus the original problem is solvable if and only if k satisfies the criteria of Theorem 6.48.

Example 6.20: Find all primitive Pythagorean triangles whose legs differ by 7.

Solution: We must find integers u and v such that $(u, v) = 1$, $u > v > 0$, u and v are of opposite parity and

$$(u + v)^2 - 2v^2 = \pm 7.$$

The fundamental solution to $x^2 - 2y^2 = -1$ is $1 + \sqrt{2}$, by (a) of example 6.20. Using the inequalities of Theorem 6.47 we see that the fundamental solutions of $x^2 - 2y^2 = 7$ are

$$3 + \sqrt{2} \text{ and } -3 + \sqrt{2}.$$

Thus, by Theorem 6.44, all solutions of $x^2 - 2y^2 = 7$ are given, for n an integer, by

(6.64) $\qquad\qquad x_n + y_n \sqrt{2} = \pm(3 + \sqrt{2})(1 + \sqrt{2})^n$

and

(6.65) $\qquad\qquad r_n + s_n \sqrt{2} = \pm(-3 + \sqrt{2})(1 + \sqrt{2})^n.$

Since we are after positive solutions to use in (6.63) we take the plus sign in (6.64) with $n \geq 0$ and the minus sign in (6.65) with $n \geq 1$. These solutions satisfy the following recursion relations

(6.66) $\qquad\qquad \begin{aligned} x_{n+1} &= x_n + 2y_n \\ y_{n+1} &= x_n + y_n \end{aligned}$,

with $n \geq 0$ and $x_0 = 3$ and $y_0 = 1$, and

(6.67) $\qquad\qquad \begin{aligned} r_{n+1} &= r_n + 2s_n \\ s_{n+1} &= r_n + s_n \end{aligned}$, with $n \geq 1$ and $r_1 = 1$ and $s_1 = 2$. Either of these leads to the recursion relation

(6.68)
$$u_{n+1} = 2u_n + v_n$$
$$v_{n+1} = u_n$$,

except that for (6.66) we take $n \geq 0$ and $u_0 = 4$ and $v_0 = 1$, while for (6.67) we take $n \geq 1$ and $u_1 = 3$ and $v_1 = 2$. These two sequences then give all primitive Pythagorean triangles whose legs differ by 7. The first few are given below. The first line is derived from the sequence beginning $u_0 = 4$ and $v_0 = 1$, while the second line comes from the other sequence:

(15, 8, 17)   (65, 72, 97)   (403, 396, 565)   (2325, 2332, 3293)

(5,12,13)   (55, 48, 73)   (297, 304, 425)   (1755, 1748, 2477)

This finishes the discussion of Pythagorean triples begun in Chapter 5. We include in the following problems some other problems relating to Pythagorean triples that use the results of this chapter.

Problem Set 6.8

1. Find the first five primitive Pythagorean triples whose legs differ by 31; 49; 119.

2. Modify the results of this section if we do not require that the triangle be primitive.

3. Let N be a given integer.
    (a) How many Pythagorean triples can N be the hypotenuse of?
    (b) How many primitive Pythagorean triples can it be the hypotenuse of?

4. Answer the questions of Problem 3 for the odd leg; the even leg.

5. Solve the analogous problem to that considered in this section if we wish that the sum of the legs is to be a specified number.

6.9.   THE   EQUATION $AX^2 + BXY + CY^2 + DX + EY + F = 0$

In this section we discuss the general second degree equation

(6.69) $ax^2 + bxy + cy^2 + dx + ey + f = 0$,

where not all of a, b and c are zero. From analytic geometry we know that such an equation represents a conic and what kind of conic is determined by the discriminant, $D = 4ac - b^2$. We shall list the various cases that can arise and show how to reduce the problem to one already studied.

First suppose that $aD \neq 0$. Then by completing the square we see that we may rewrite (6.69) as

$$(Dy + E)^2 + D(2ax + by + d)^2 = F,$$

where $E = 2ae - bd$ and $F = -4afD + d^2D + (2ae - bd)^2$. If $a = 0$, but $cD \neq 0$, then we may write (6.69) as

$$(Dx + E_1)^2 + D(bx + 2cy + e)^2 = F_1,$$

where $E_1 = 2cd - be$ and $F_1 = (2cd - be)^2 - 4cfD + De^2$. In either case we may make a linear transformation

$$X = Ax + By + C \text{ and } Y = Gx + Hy + K$$

so that (6.69) gets transformed into

$$X^2 + DY^2 = N.$$

If $D > 0$, then this equation has only finitely many solutions and we can obtain them by trial. If $N < 0$, then there are no solutions and if $N > 0$, then we must have $|X| \leq \sqrt{N}$ and $|Y| \leq \sqrt{N/D}$. If $D < 0$ and $-D = m^2$, a perfect square, then we have

$$(X + mY)(X - mY) = N,$$

and so there will be only a finite number of solutions in this case since both $X + mY$ and $X - mY$ must be factors of N. Finally, if $D < 0$ and $-D$ is not a perfect square, then we are back to the problems dealt with in Section 7.

Example 6.21: Solve $x^2 + 6xy + 7y^2 + 8x + 24y + 15 = 0$.

Solution: Here $D = 4 \cdot 1 \cdot 7 - 6^2 = -8$ and $a = 1$. Thus $E = 2 \cdot 1 \cdot 24 - 6 \cdot 8 = 0$ and $F = -4 \cdot 1 \cdot 15 \cdot (-8) + 8^2 \cdot (-8) + 0^2 = -32$. Thus, after canceling some common factors we obtain the equation

$$(x + 3y + 4)^2 - 2y^2 = 1.$$

Thus, if we let $X = x + 3y + 4$ and $Y = y$ we find we must solve

(6.70) $X^2 - 2Y^2 = 1.$

From (a) of Example 6.13 we know that all positive solutions of (6.70) are given by

$$X_n + Y_n \sqrt{2} = (3 + 2\sqrt{2})^n,$$

where n runs through all positive integers. This yields the simple recursion

$$X_{n+1} = 3X_n + 4Y_n \quad \text{and} \quad Y_{n+1} = 2X_n + 3Y_n,$$

for $n \geq 1$, where $X_1 = 3$ and $Y_1 = 2$. This in turn leads to a recursion for x_n and y_n, namely

$$x_{n+1} = -4x_n - 14y_n - 16 \quad \text{and} \quad y_{n+1} = 2x_n + 9y_n + 8,$$

for $n \geq 1$, where $x_1 = -7$ and $y_1 = 2$.   □

Suppose now that $a = c = 0$. Then $D = -b^2 \neq 0$ since not all of a, b and c are zero. Since b is not zero we may multiply (6.69) by b and factor. This gives

$$(bx + e)(by + d) = de - bf.$$

If we let $de - bf = \alpha\beta$, where α and β run through all pairs of divisors of $de - bf$, then we have

$$x = (\alpha - e)/b \text{ and } y = (\beta - d)/b,$$

providing these are integers.

Example 6.22: Solve $xy + 3x + 5y - 1 = 0$.

Solution: Since $b = 1$ in this equation we know that it will be solvable in integers. We rewrite the equation as

$$(x + 5)(y + 3) = 16.$$

This gives us the integer solutions

$$x = -4, -3, -1, 3 \text{ and } 11 \text{ and } y = 13, 5, 1, -1, \text{ and } -2.   □$$

Suppose $D = 0$. Since $ac = 0$ and $D = 0$ imply $b = 0$ we see that one of a and c must be nonzero. We suppose that $a \neq 0$ in what follows. If $a = 0$ and $c \neq 0$, then the details work similarly. Multiply (6.69) by $4a$ and complete the square to get

$$(2ax + by + d)^2 + (4ae - 2bd)y + 4af - d^2 = 0.$$

If we let $z = 2ax + by + d$, then we have

$$y = \frac{z^2 + 4af - d^2}{2bd - 4ae},$$

provided $bd - 2ae \neq 0$. If $bd - 2ae \neq 0$, then we must have, for there to be an integral solution in y

(6.71) $z^2 \equiv d^2 - 4af \pmod{4ae - 2bd}$.

If $bd - 2ae = 0$, then for there to be an integral solution we must have that $d^2 - 4af$ is a perfect square and the problem reduces to that of a linear congruence, which we solve as in Section 2.3. We suppose $bd - 2ae \neq 0$ and solve the congruence (6.71). If (6.71) is not solvable, then neither is (6.69). If (6.71) is solvable, then each value of z yields a value for y. Then we take that subclass of values of y that yield integral values of x from the equation

$$2ax = z - by - d.$$

Example 6.22: Solve $3x^2 + 12xy + 12y^2 + 4x - 2y - 85 = 0$.

Solution: Here $D = 4 \cdot 3 \cdot 2 - 12^2 = 0$ and $a = 3$. Multiplying by 12 transforms the equation into

$$(6x + 12y + 4)^2 - 120y - 1036 = 0.$$

Canceling out common factors and letting $z = 3x + 6y + 2$ we see that for this problem (6.71) becomes

$$z^2 \equiv 259 \ (\mathrm{mod}\ 30)$$

or

$$z^2 \equiv 19 \ (\mathrm{mod}\ 30).$$

Thus

$$z \equiv \pm 13 \text{ and } \pm 7 \ (\mathrm{mod}\ 30).$$

If $z = 30u \pm 13$, where u is an integer, then

$$y = 30u^2 \pm 26u - 3.$$

Then we must solve

$$30u \pm 13 = 3x + 6(30u^2 \pm 26u - 3) + 2$$
$$= -3x + 180u^2 \pm 156u - 16$$

or, taking the upper sign,

$$3x = -180u^2 - 126u + 29,$$

which has no integral solutions ($3 \nmid 29$), and taking the lower sign

$$3x = -180u^2 + 186u + 3$$

or

$$x = -60u^2 + 62u + 1.$$

If $z = 30u \pm 7$, where u is an integer, then

$$y = 30u^2 \pm 14u - 7.$$

Taking the upper sign we get

$$30u + 7 = 3x + 6(30u^2 + 14u - 7) + 2$$
$$= 3x + 180u^2 + 84u - 40$$

or

$$3x = -180u^2 - 54u + 47,$$

which has no integral solutions. If we take the lower sign, then we have

$$30u - 7 = 3x + 180u^2 - 84u - 40$$

or

$$3x = -180u^2 + 114u + 33$$

or

$$x = -60u^2 + 38u + 11.$$

Thus the solutions are

$$x = -60u^2 + 62u + 1 \text{ and } y = 30u^2 - 26u - 3$$

and

$$x = -60u^2 + 38u + 11 \text{ and } y = 30u^2 - 14u - 7. \qquad \square$$

For a treatment of the equation

$$Ax^2 + Bxy + Cy^2 = N$$

similar to our treatment of $x^2 - Dy^2 = N$ of section 7 see [121].

Problem Set 6.9

1. Show that if $a^2 > b$, then $x^2 + 2axy + by^2 = 1$ has infinitely many integer solutions.

2. Give the general solutions to the following equations.
   (a) $2x^2 - 3y - 5 = 0$.
   (b) $x^2 - 2xy + y^2 - x - 2y = 0$.
   (c) $x^2 + 6xy - 4y^2 - 4x - 12y - 19 = 0$.
   (d) $x^2 + 8xy + y^2 + 2x - 4y + 1 = 0$.

    (e) $3x^2 + 4xy - 7y^2 - 12 = 0.$

    (f) $5x^2 - 14xy + 7y^2 + 1 = 0.$

3. Find all squares with integer sides whose perimeters equal their areas.

4. (a) Find all right triangles with integer sides whose area is equal to the sum of its legs.

   (b) Find all right triangles with integer sides whose area is equal to a given rational multiple, say r/s, of the sum of its legs.

5. Suppose $a \neq 0$ and b, c and d are integers. Find necessary and sufficient conditions so that the equation

$$axy + bx + cy + d = 0$$

has infinitely many integer solutions. What are they?

6. Let N be a given integer. Find all integers x and y so that

$$\frac{1}{x} + \frac{1}{y} = \frac{1}{N}.$$

How many solutions does this equation have?

7. Find necessary and sufficient conditions for the solvability of the equation $ax^2 + by + c = 0.$

6.10. WARING'S PROBLEM

As mentioned in the introduction to this chapter we are concerned mainly with the representation of integers by linear and quadratic polynomials. There is one problem that deals with representation by higher degree polynomials that has been solved and we shall discuss it in this section. It is a generalization of the sums of squares problem.

In 1770 E. Waring stated, without proof, that every integer is the sum of four or fewer squares, nine or fewer cubes, nineteen or fewer biquadrates and so on. Though Waring never explained what he meant by "and so on" it was taken to mean the following. Let $k \geq 2$ be an integer. Then there exist a positive integer $s = g(k)$ such that for any given positive integer N there exist nonnegative integers $x_1, x_2, ..., x_s$ such that

$$N = x_1^k + x_2^k + \cdots + x_s^k.$$

The important thing to note is that $g(k)$ is independent of N so that this is the maximum number of kth powers needed for any positive integer N. In terms of the function $g(k)$ Waring's problem may be stated that $g(2) \leq 4$, $g(3) \leq 9$, $g(4) \leq 19$ and so on.

We know from Theorem 6.9 that $g(2) \leq 4$ (which verifies Waring in this case) and from Theorem 6.13 we know $g(2) \geq 4$. Thus $g(2) = 4$. After Waring's problem became generally known estimates were given for special values of k, but it was not until 1909 that D. Hilbert proved that $g(k)$ exists for every value of $k \geq 2$. Unfortunately Hilbert's proof merely established the existence of $g(k)$, but did not give any estimate of its size. It is now known that $g(3) = 9$ and $g(5) = 37$, but the best that is known for $g(4)$ is that $19 \leq g(4) \leq 22$. For $k \geq 6$ we have the following result.

Theorem 6.49: If

(6.72)
$$3^k - 2^k + 2 < (2^k - 1)[(3/2)^k],$$

then

$$g(k) = 2^k + [(3/2)^k] - 2.$$

If $3^k - 2^k + 2 \geq (2^k - 1)[(3/2)^k]$, let

$$N(k) = [(3/2)^k][(4/3)^k] + [(3/2)^k] + [(4/3)^k].$$

Then

$$g(k) = \begin{cases} [(3/2)^k] + [(4/3)^k] + 2^k - 3 & \text{if } 2^k < N(k) \\ [(3/2)^k] + [(4/3)^k] + 2^k - 2 & \text{if } 2^k = N(k) \end{cases}.$$

It is known that (6.72) holds for $6 \leq k \leq 200000$ and it is conjectured that (6.72) holds for all $k \geq 6$. The methods used to prove this result involve very analytical methods developed by I. M. Vinogradov in the 1930's.

While we will not prove Theorem 6.49 we will prove a lower bound for the $g(k)$ function.

Theorem 6.50: Let k be a positive integer. Then

$$g(k) \geq 2^k + [(3/2)^k] - 2.$$

Proof: If k is given, let $N = 2^k[(3/2)^k] - 1$. Then

$$N \leq 2^k(3/2)^k - 1 = 3^k - 1 < 3^k.$$

Thus we cannot use kth powers of 3 in the representations of N. Also, by the definition of N, we see that we cannot use more than $[(3/2)^k] - 1$ summands of 2^k, so that the remainders of the $k\underline{th}$ powers must be $1k$'s. In fact we must use

$$N - 2^k\{[(3/2)^k] - 1\} = 2^k[(3/2)^k] - 1 - 2^k[(3/2)^k] + 2^k = 2^k - 1$$

summands of 1^k. The total number of summands is then

$$[(3/2)^k] - 1 + 2^k - 1 = 2^k + [(3/2)^k] - 2$$

which gives us the desired result.   ∎

As the proof of the above theorem shows we run into problems when we can't use all the powers available because the number is too small. Thus we might hope that if the numbers are large enough, then we will need less $k\underline{th}$ powers. Let $G(k)$ be the least positive integer so that every integer $n \geq K(k)$ can be written as a sum of $G(k)$ of $k\underline{th}$ powers. Here $K(k)$ is some integer that depends solely on k. Theorem 6.13 shows that $G(2) = 4$. In general, we have $G(k) \leq g(k)$.

We prove the following lower bound for $G(k)$.

Theorem 6.51: For every positive integer k we have $G(k) \geq k + 1$.

Proof: Let $A(N)$ denote the number of positive integers less than or equal to N such that

$$N = x_1^k + \cdots + x_k^k,$$

where $x_i \geq 0$, $1 \leq i \leq k$, and where $0 \leq x_1 \leq x_2 \leq \cdots \leq x_k \leq [N^{1/k}]$. Then we have

(6.73)
$$A(N) \leq \sum_{x_k=0}^{[N]} \sum_{x_{k-1}=0}^{x_k} \cdots \sum_{x_1=0}^{x_2} 1.$$

Now the inner sum on the right hand side of (6.73) is

$$\sum_{x_1=0}^{x_2} 1 = \binom{x_2 + 1}{1},$$

and so the next sum is

$$\sum_{x_2=0}^{x_3} (x_2 + 1) = \frac{(x_3 + 1)(x_3 + 2)}{2} = \binom{x_3 + 2}{2}.$$

An easy induction shows that for any positive integer l

(6.74)
$$\sum_{n=0}^{m} \binom{n + l - 1}{l - 1} = \binom{m + l}{l}.$$

Thus, if we denote the sum on the right hand side of (6.73) by $B(N)$ we see that, by (6.74),

$$B(N) = \binom{\left[N^{1/k}\right]+k}{k} = \frac{\left(\left[N^{1/k}\right]+1\right)\cdots\left(\left[N^{1/k}\right]+k\right)}{k!}.$$

As $N \to +\infty$ we see that $B(N) \to N/k!$, and so for N large enough we must have $B(N) < 2N/3$. Thus, since $A(N) \leq B(N)$, we have

(6.75) $$A(N) < 2N/3.$$

However, if $G(k) \leq k$, we see that all but a finite number of n can be written as a sum of k kth powers. Thus there is some constant a so that

(6.76) $$A(N) > N - a.$$

Since (6.75) and (6.76) are incompatible for large N we see that we must have $G(k) > k$. ∎

In the exercises below we shall give some estimates for particular values of k.

For more on Waring's problem and related problems see [30], [47, ch. XXI] or [53, ch. 18]. For an introduction to the analytic machinery needed to prove the best known results see [5, ch. IV].

Problem Set 6.10

1. Prove the identity (6.74).
2. (a) Use congruences modulo 3 to determine the least number of cubes needed to represent a number of the form $9m \pm 4$.
   (b) Show that $G(3) \geq 5$.
3. (a) By looking at 4th powers modulo 16 show that $G(4) \geq 15$.
   (b) Generalize this result to give an estimate for $G(2^n)$.
4. (a) If $n = a_1^2 + a_2^2 + a_3^2 + a_4^2$, show that
   $$6n^2 = \sum_{1 \leq i < j \leq 4}\left\{\left(a_j + a_i\right)^4 + \left(a_j - a_i\right)^4\right\}.$$
   (b) Use Theorem 6.9 and the fact that every integer can be written in the form $6n + r$, with $0 \leq r < 6$, to show that $g(4) \leq 53$.
5. By considering numbers of the form $31 \cdot 16^m$, show that $G(4) \geq 16$. (Hint: $31 = 2^4 + 15 \cdot 1^4$.) (It is now known that $G(4) = 16$.)

Additional Problems for Chapter Six

General Problems

1. If a, b, c and d are positive integers with $(c, d) = 1$ and

$$a^c = b^d,$$

   show that there is a positive integer N such that $a = N^d$ and $b = N^c$ (Hint: see Problem 11 of Problem Set 6.2.)

2. (a) If a and b are relatively prime positive integers and $N > ab$, show that there exist positive integers x and y such that

$$ax + by = N.$$

   (b) Show that if $N = ab$, then no such representation exists.

3. (a) If a and b are relatively prime positive integers and $N \geq (a - 1)(b - 1)$, show that there exist nonnegative integers x and y such that

$$ax + by = N.$$

   (b) Show that for exactly half of the nonnegative integers up to $(a - 1)(b - 1)$ can be so represented.

   (c) Show that $N = ab - a - b$ cannot so be represented.

4. Let N denote the number of nonnegative solutions of

$$ax + by = c.$$

   (a) If $x = x_1$ and $y = y_1$ is one solution in nonnegative integers, show that

$$N = [y_1(a, b)/a] + [x_1(a, b)/b] + 1.$$

   (b) If, in addition, $c(a, b)/ab$ is an integer, show that

$$N = 1 + c(a, b)/ab.$$

   (c) If $c(a, b)/ab$ is not an integer, show that

$$N = [c(a, b)/ab] \text{ or } 1 + [c(a, b)/ab].$$

   If $a \mid c$, show that

$$N = 1 + [c(a, b)/ab].$$

   (Hint: see Problem 12 of Problem Set 6.2.)

5. A positive integer n is said to be **square-full** if $p \mid n$ implies that $p^2 \mid n$ for any

prime p. Show that n is square-full if and only if n can be written in the form

$$n = a^2b^3$$

for some positive integers a and b.

6. (a) Let a, b and c be integers with $ab \neq 0$ and $(a, b) = 1$. Define the sequences
b_k, c_k, q_k and g_k by $b_0 = a$, $b_1 = b$, $c_0 = c$ and for, $k \geq 1$,
$$b_{k+1} = b_{k-1} - q_k b_k, \text{ with } |b_{k+1}| \leq \tfrac{1}{2}|b_k|$$

and

$$c_k = c_{k-1} - g_k b_k, \text{ with } |c_k| \leq \tfrac{1}{2}|b_k|.$$

Let $k = n$ be the final index in this process where
$$0 < |c_k| < \tfrac{1}{2}|b_k|, \quad k = 1, 2, \ldots, n-1$$

and either

$$c_n = 0 \text{ and } b_n \neq 0 \quad \text{or} \quad c_n = \pm\tfrac{1}{2}b_n \neq 0.$$

If $c_n = 0$ and $b_n \neq 0$, let $x_n = x_{n+1} = 0$ or if $c_n = \pm\tfrac{1}{2}b_n \neq 0$, let $x = \pm\tfrac{1}{2}b_n$.
and $x_{n+1} = \mp[\tfrac{1}{2}b_{n+1}]$. Define $x_{n-1}, x_{n-2}, \ldots, x_1, x_0$ by

$$x_{k-1} = g_k - q_k x_k + x_{k+1}.$$

Show that, for $k = 0, 1, \ldots, n$, we have

$$b_k x_{k+1} + b_{k+1} x_k = c_k.$$

This method of finding solutions to $ax + by = c$ is due to Gauss.

(b) Use this method to solve the following equations.

   (i) $17x + 35y = 6$. (iii) $135x + 211y = 462$.

   (ii) $93x - 67y = 1$. (iv) $1231x - 5217y = -637$.

7. Let $(a, b) = 1$, a, $b \geq 1$ and $N \geq 1$ all be integers with $N < ab$. Divide N by a
and b and call the remainders r_1 and s_1. Let $N_1 = N - r_1 - s_1$ and divide N_1 by
a and b with remainders r_2 and s_2. Let $N_2 = N_1 - r_2 - s_2$ and continue this
process. Show that $ax + by = N$ is solvable in nonnegative integers if and only
if 0 occurs in the sequence $N_1, N_2, \ldots$.

8. Let a, b and c be positive integers and let $d = (a, b, c)$. Show that there exists a
positive integer N such that if $n > N$, then the equation

$$ax + by + cz = nd$$

has a solution with x, y and z positive integers. Can you find the minimum
value of N? (In the case of two variables this was ab. For more than two
variables such results are not known, though estimates have been given.)

9. Show that if n is an integer and $n \geq 5$, then the equation

$$nx + (n + 1)y + (n + 2)z = n^2$$

has a solution in positive integers x, y and z.

10. Show that if $n \geq 6$, then the number of solutions in positive integers of the equation

$$x + 2y + 3z = n$$

is given by $1 + [n(n - 6)/12]$.

11. (a) Let $d_1, \ldots, d_k$ be nonzero integers that are relatively prime in pairs. Let n be an integer. Show that there exist integers $n_1, \ldots, n_k$ such that

$$\frac{n}{d_1 \cdots d_k} = \frac{n_1}{d_1} + \cdots + \frac{n_k}{d_k}.$$

(b) Find integers a, b and c so that

$$\frac{5}{7 \cdot 11 \cdot 37} = \frac{a}{7} + \frac{b}{11} + \frac{c}{37}.$$

12. Let $a_1, \ldots, a_n$ and c be integers with $n \geq 2$ such that $(a_1, \ldots, a_n) \mid c$. Let $H = \max(|a_1|, \ldots, |a_n|)$ (called the **height** of the equation.) Show that the equation

$$a_1 x_1 + \cdots + a_n x_n = c$$

has a solution integers such that $|x_k| \leq |c| + (n - 1)H$, for $k = 1, \ldots, n$.

13. The following is a version of the "Cattle Problem of Archimedes." Find the smallest solution in positive integers of the resulting system of equations.

Apollo had a herd of cattle consisting of bulls and cows, one part of which was white, a second part black, a third spotted and a fourth brown.

Among the bulls, the number of white ones was one half plus one third the number of the black greater than the brown. The number of the black was one quarter plus one fifth the number of spotted greater than the brown. The number of spotted was one sixth and one seventh the number of the white greater than the brown.

Among the cows the number of white ones was one third plus one fourth of the total black cattle. The number of black was one quarter plus one fifth the total number of spotted cattle. The number of spotted was one fifth plus one sixth the total number of brown cattle. The number of brown was one sixth plus one seventh the total number of white cattle.

How many of each type of bull and each type of cow did Apollo have?

14. Suppose $n = a^2 + b^2$, where a and b are integers that are not both zero. Show that there is a nonnegative integer k such that $n \equiv 2^k \pmod{2^{k+2}}$.

15. Show that there do not exist positive integers x and y so that $4xy - x - y$ is the square of an integer. Show the same is true for the form $4xy - x - 1$.

16. (a) Let k be a positive integer. Show that there exists a positive integer n such that none of the integers $n, n + 1, \ldots, n + k - 1$ can be written as a sum of two squares.

    (b) Show that there are infinitely many positive integers n such that $n, n + 1$ and $n + 2$ are each a sum of two squares.

    (c) Can we extend the result of (b) to $n, n + 1, n + 2$ and $n + 3$?

17. (a) Let p be an odd prime and A and B integers such that $p \nmid AB$. Let n be a positive integer. Show that the congruence $x^2 + Ay^2 + B \equiv 0 \pmod{p^n}$ always has a solution. (Hint: proceed by induction on n.)

    (b) Show that if m is a positive integer such that $(m, AB) = 1$, then the congruence $x^2 + Ay^2 + B \equiv 0 \pmod{m}$ is solvable. (One can use this result to provide a constructive proof of Theorem 6.9. See [110, pp. 381-387].)

18. Show that every natural number can be written as a sum of at most ten squares of odd integers.

19. Show that every integer can be represented by the form
$$x^2 + y^2 - z^2.$$

20. Let a be an integer of the form $4n^2 + 1$, for some integer n. Show that every integer can be written in the form
$$x^2 + y^2 - az^2.$$

21. Show that if $r_8(n)$ denotes the number of representations of n as a sum of eight squares, then
$$r_8(n) = 16(-1)^n \sum_{d|n} (-1)^d d^3.$$

    (Hint: mimic, where possible, the derivation of $r_4(n)$ from $r_2(n)$.)

22. The following exercise outlines a proof of Theorem 6.14. For another proof see [124, pp. 465-474]. For this exercise one can assume the truth of Corollary 9.9.1.

    (a) Show that we need only consider n to be an odd integer and twice an odd integer to decide on the representation as a sum of three squares.

(b) Show that it suffices to find nine integers, a, b, c, d, e, f, x, y and z, so that

$$n = ax^2 + 2bxy + 2cxz + dy^2 + 2eyz + fz^2,$$

where $a > 0$, $ad - b^2 > 0$ and

$$\begin{vmatrix} a & b & c \\ b & d & e \\ c & e & f \end{vmatrix} = 1.$$

(c) If we take $c = 1$, $e = 0$, $f = n$, $x = 0$, $y = 0$ and $z = 1$, then we need $a > 0$, $g = ad - b^2 > 0$ and $d = bn - 1$.

(d) Suppose $n \equiv 2$ or 6 (mod 8). Show that there is a prime p of the form $(4v + 1)n - 1$, where v is a nonnegative integer. Then show that

$$\left(\frac{-g}{p}\right) = 1.$$

(e) Suppose $n \equiv 1, 3$ or 5 (mod 8). If we let $c = 1$ if $n \equiv 3$ (mod 8) and $c = 3$ if $n \equiv 1$ (mod 4), show that we can find a prime of the form $p = 4nv + (cn - 1)/2$, for some nonnegative integer v. Show that

$$\left(\frac{-2}{g}\right) = 1$$

and hence that

$$\left(\frac{-g}{p}\right) = 1.$$

(f) Show that in all cases $-g$ is a quadratic residue of $gn - 1$.

(g) Show that if n is not of the form $4^a(8b + 7)$, then n can be written as a sum of three squares.

23. Show that if

$$ax^2 + bxy + cy^2 = Au^2 + Buv + Cv^2,$$

where $x = pu + qv$ and $y = ru + sv$, then $B^2 - 4AC = (b^2 - 4ac)(ps - qr)^2$.

24. Show that if $ax^2 + bxy + cy^2$ is a positive definite reduced form, then $ac \leq d/3$, where $d = 4ac - b^2$.

25. We say that (x, y) is a **proper representation** of a number m by the form $ax^2 + bxy + cy^2$ if $(x, y) = 1$. Let $ax^2 + bxy + cy^2$ be a form of discriminant d and suppose (r, s) is a proper representation of m. Show that there are unique

integers a, b and n such that $rb - sa = 1$, $0 \le n < 2m$, $n^2 \equiv d \pmod{4m}$ and

$$\begin{pmatrix} r & a \\ s & b \end{pmatrix}$$

takes $ax^2 + bxy + cy^2$ into $mu^2 + nuv + lv^2$, where l is determined by $n^2 - 4ml = d$.

26. Show that two forms $ax^2 + bxy + cy^2$ and $Ax^2 + Bxy + Cy^2$ are equivalent if and only if their discriminants are equal and there exist two integers u and v such that

$$A = au^2 + buv + cv^2$$
$$2au + (b + B)v \equiv 0 \pmod{2A}.$$
$$(b - B)u + 2cv \equiv 0 \pmod{2A}$$

27. Let $p \equiv 1 \pmod 4$ be a prime and let w be an integer, $0 < w < p/2$, such that $w^2 \equiv -1 \pmod p$. Show that there exist integers a, b, c and d with $ad - bc = 1$ such that

$$px^2 + 2wxy + \left(\frac{w^2 + 1}{4}\right)y^2 = (ax + by)^2 + (cx + dy)^2.$$

28. Let $e(u)$ denote the difference between the number of divisors of n of the form $3k + 1$ and the number of divisors of the form $3k + 2$.

   (a) Show that the number of representations of n by the form $x^2 + xy + y^2$ is $6e(u)$.

   (b) Show that the number of representations of $4m$, where m is an odd integer by the form $x^2 + 3y^2$ is $e(m)$, where we only count representations with x and y both odd positive integers.

29. (a) Show that if m and k are positive integers with $k \ge 3$, then the equation $x^2 - y^2 = m^k$ is always solvable.

   (b) When is $x^2 - y^2 = m^2$ solvable?

30. (a) Solve, in positive integers, $x(x - 31) = y(y - 41)$.

   (b) More generally, solve, in positive integers, $x(x - a) = y(y - b)$, where a and b are given integers.

31. (a) Find necessary and sufficient conditions for the solvability of the system

$$x^2 + y^2 = m$$
$$x + y = n$$

   where m and n are given integers. Solve the system.

(b) Let m and n be given integers. Show that the system

$$x^2 + y^2 + z^2 + w^2 = n$$
$$x + y + z + w = m$$

is solvable if and only if either

(i) $4n - m^2 = 0$ and $4 \mid m$

or

(ii) $4n - m^2 > 0$, $m \equiv n$ (mod 2) and $4n - m^2$ is not of the form $4^a(8b + 7)$, with $a, b \geq 0$.

(c) What can be said about the three variable case?

32. Find conditions on the integer n so that the system

$$z + n = x^2 \text{ and } z - n = y^2$$

has integer solutions in z, y and z.

33. (a) Show that $2a - 1$, $2a$ and $2a + 1$ are sides of a triangle of integer area if and only if $3(a^2 - 1)$ is a square.

(b) Show that $2a$, $2a + 1$ and $2a + 2$ are sides of a triangle with integer area if and only if $3\{(2a + 1)^2 - 4\}$ is a square.

34. Suppose D is a positive nonsquare integer and n is a nonnegative integer. Prove the following results.

(a) If $x_1^2 - Dy_1^2 = -1$, then there is an integer m such that

$$(y_1\sqrt{D} - x_1)^n = \sqrt{m+1} - \sqrt{m}.$$

(b) If $x_1^2 - Dy_1^2 = 1$, then there is an integer m such that

$$(x_1 - y_1\sqrt{D})^n = \sqrt{m+1} - \sqrt{m}.$$

(c) If $x_1^2 - Dy_1^2 = \pm 1$, then there is an integer m such that

$$(x_1 + y_1\sqrt{D})^n = \sqrt{m+1} - \sqrt{m}.$$

35. Define the **Chebychev polynomials of the second kind**, $U_n(x)$, by $U_n(\cos\theta) = \sin(n\theta)/\sin(\theta)$, for $n = 0, 1, 2, \ldots$, and the **Chebychev polynomials of the first kind**, $T_n(x)$, by $T_n(\cos\theta) = \cos(n\theta)$, for $n = 0, 1, 2, \ldots$. Let D be a positive nonsquare integer.

(a) Show that $T_n^2(x) - (x^2 - 1)U_n^2(x) = 1$.

(b) If (x, y) is the fundamental solution to $x^2 - Dy^2 = 1$, show that all solutions are given by $(T_n(x_1), y_1 U_n(x_1))$, for $n \geq 1$.

36. Let k be a positive integer. Solve the following equations.

    (a) $x^2 - (k^2 + 4)y^2 = 1$.

    (b) $x^2 - (k^2 - 4)y^2 = 1$.

    (c) Is either equation solvable with 1 replaced by -1?

37. If K is a positive nonsquare integer and $p_r(n)$ denotes the $n\underline{th}$ r-gonal number, with $r \neq 4$, show that there are infinitely many integers m and n such that

$$p_r(m) = Kp_r(n).$$

38. If $(x_0, y_0) = 1$ and $x_0^2 - Dy_0^2 = N$, where D is a positive nonsquare integer, then there exists an integer k such that

$$x_0 \equiv ky_0 \pmod{N} \quad \text{and} \quad k^2 \equiv D \pmod{N}.$$

(In such a case we say that the solution (x_0, y_0) **belongs to the integer** k.)

39. This exercise outlines an alternative way of solving $x^2 - Dy^2 = N$.

    (a) Two primitive solutions of $x^2 - Dy^2 = N$ are in the same class if and only if they belong to the same integer. (See Problem 38 above.)

    (b) Let k be an integer such that $k^2 \equiv D \pmod{N}$. Let

$$M = (k^2 - D)/N.$$

Then $x^2 - Dy^2 = N$ is solvable in positive integers x and y belonging to the integer k if and only if $t^2 - Du^2 = M$ has a solution t_0, u_0 (not necessarily positive) belonging to k such that

$$(ku_0 - t_0)/M > 0 \text{ and } (u_0D - kt_0)/M > 0.$$

If this is the case, then

$$x_0 = (u_0D - kt_0)/M \text{ and } y_0 = (ku_0 - t_0)/M$$

is a solution of $x^2 - Dy^2 = N$ belonging to k.

    (c) Let k and M be as in (b). The equation $x^2 - Dy^2 = N$ has a positive solution belonging to $\pm k$ if and only if the equation $x^2 - Dy^2 = N$ has a positive solution t_0, u_0 belonging to $\pm k$. If this is the case, then

$$x_0 = \left| (u_0D \pm kt_0)/M \right| \text{ and } y_0 = \left| (ku_0 \pm t_0)/M \right|$$

is a positive solution of $x^2 - Dy^2 = N$ belonging to $\pm k$, where the sign is chosen so that $t_0 \pm ku_0 \equiv 0 \pmod{M}$.

    (d) If $N > \sqrt{D}$, then M (as defined in (b)) satisfies

$$|M| \leq \max(\sqrt{D}, |N|/4).$$

    (e) Solve $x^2 - 13y^2 = 51$ by the method above.

40. Suppose $(a, b, c) = 1$ and that the roots of $ax^2 + bx + c = 0$ are real irrationals. Show that the equation $x^2 - bxy + acy^2 = 1$ has infinitely many solutions. Give conditions to ensure that $x^2 - bxy + acy^2 = -1$ has infinitely many solutions.

41. If $k \geq 3$, k an integer, solve the equation $x^2 - kxy + y^2 = 1$.

42. Show that there are infinitely many integers n such that $p_r(n)$ is a square as long as r is not twice a square.

43. Solve the equation $x^2 - \{((4m^2 + 1)n + m)^2 + 4mn + 1\}y^2 = -1$.

44. Solve the following equations.

(a) $x^2 - 1056y^2 = 1$.   (d) $x^2 - 1937y^2 = -1$.

(b) $x^2 - 1067y^2 = 1$.   (e) $x^2 - 1130y^2 = -1$.

(c) $x^2 - 1995y^2 = 1$.   (f) $x^2 - 1490y^2 = -1$.

(g) Can you make any generalizations based on these examples?

45. Let u_1 and v_1 be the least positive integers such that $u^2 - Dv^2 = \pm 4$ when it is solvable. Then all positive solutions are given by

$$\frac{u_n + v_n \sqrt{D}}{2} = \left(\frac{u_1 + v_1 \sqrt{D}}{2} \right)^n .$$

46. (a) Suppose $D \equiv 0 \pmod 4$. Then any solution of $x^2 - Dy^2 = 4$ is of the form $x = 2u$ and $y = v$, where $u^2 - (D/4)v^2 = 1$. Prove an analogous result for $x^2 - Dy^2 = -4$ when it is solvable.

(b) Suppose $D \equiv 1 \pmod 4$. Then any solution of $x^2 - Dy^2 = 4$ is of the form $x = 2u$ and $y = 2v$, where $u^2 - Dv^2 = 1$ and similarly for $x^2 - Dy^2 = -4$ when it is solvable.

(c) Suppose $D \equiv 5 \pmod 4$. If $x_0 + y_0 \sqrt{D}$ is that solution of $x^2 - Dy^2 = 4$ which has y_0 with the least positive value and $x_0 > 0$ and if $x_1 + y_1 \sqrt{D}$ is the fundamental solution to $x^2 - Dy^2 = 1$, then

$$x_1 + y_1 \sqrt{D} = \frac{x_0 + y_0 \sqrt{D}}{2}$$

if x_0 and y_0 are even, but if x_0 and y_0 are odd, then

$$x_1 + y_1 \sqrt{D} = \left(\frac{x_0 + y_0 \sqrt{D}}{2} \right)^3 .$$

Prove an analogous result for $x^2 - Dy^2 = -4$ if it is solvable.

47. (a) Let $D > 0$ be square-free and $D \equiv 1 \pmod 4$. Show that

$$x^2 + xy + ((1 - D)/4)y^2 = 1$$

has infinitely many solutions.

(b) If (x_0, y_0) is a solution of the equation in (a), show that

$$x_n + \frac{1 + \sqrt{D}}{2} y_n = \left(x_0 + \frac{1 + \sqrt{D}}{2} y_0 \right)^n$$

are solutions for any integer value of n.

(c) Show that there exist positive integers x_0 and y_0 such that (x_0, y_0) is a solution of the equation in (a) and $x_0 + \frac{1}{2}(1 + \sqrt{D})y_0$ is least.

(d) Show that if (x_0, y_0) is determined from (c) and (x_n, y_n) from (b), then all solutions of the equation in (a) are given by $(x, y) = (\pm x_n, \pm y_n)$, for any integer n

(e) Find the fundamental solutions to $x^2 + xy - y^2 = 1$ and $x^2 + xy - 4y^2 = 1$.

(f) Repeat (a) to (e) for the equation $x^2 - xy + ((1 - D)/4)y^2 = 1$.

(g) Show that $x^2 \pm xy + ((1 - D)/4)y^2 = n$ is solvable if and only if $u^2 - Dv^2 = 4n$ is solvable.

48. If A and B are positive square-free integers, $A < B$, and C is an integer such that $|C| < \sqrt{AB}$. Let h_n/k_n be the $n\underline{\text{th}}$ convergent of the continued fraction expansion of $\sqrt{B/A}$. Define q_n from (4.41) with $N = AB$ and $q_0 = A$. Show that

$$Ax^2 - By^2 = C$$

has a solution if and only if $E = (-1)^m q_m$ for some integer m. Show that all solutions are given by $x_n = h_{m-1+nr}$ and $y_n = k_{m-1+nr}$, for $n = 0, 1, 2, \ldots$, where r is the period of the continued fraction expansion of $\sqrt{B/A}$.

49. (a) Show that if $x\sqrt{A} + y\sqrt{B}$ is a solution of $Ax^2 - By^2 = \pm 1$ and $r + s\sqrt{AB}$ is a solution of $x^2 - ABy^2 = \pm 1$, then $(xr + Bys)\sqrt{A} + (yr + Axs)\sqrt{B}$ is a solution to $Ax^2 - By^2 = \pm 1$.

(b) If both $x_1\sqrt{A} + y_1\sqrt{B}$ and $x_2\sqrt{A} + y_2\sqrt{B}$ are solutions to $Ax^2 - By^2 = \pm 1$, then $(Ax_1x_2 + By_1y_2) + (x_1y_2 + x_2y_1)\sqrt{AB}$ is a solution to $x^2 - ABy^2 = \pm 1$.

(c) Suppose $Ax^2 - By^2 = \pm 1$ is solvable and $x\sqrt{A} + y\sqrt{B}$ is its least positive solution. Show that $Ax^2 + By^2 + 2xy\sqrt{AB}$ is the fundamental solution to $x^2 - ABy^2 = \pm 1$ and conversely.

(d) Suppose $x^2 - ABy^2 = -1$ is solvable. Then $Ax^2 - By^2 = \pm 1$ is solvable for neither choice of sign if A and B are both greater than 1.

(e) If $Ax^2 - By^2 = \pm 1$ is solvable for both A and B greater than 1 and has $x_0\sqrt{A} + y_0\sqrt{B}$ as its smallest positive solution, show that all solutions to this equation are given by

$$x_n\sqrt{A} + y_n\sqrt{B} = \left(x_0\sqrt{A} + y_0\sqrt{B}\right)^n, \quad n = 0,1,2,3,\ldots.$$

(f) Find the smallest positive solutions, if they exist, of the following equations. Then find all solutions.

(i) $7x^2 - 13y^2 = 1$ (iv) $33x^2 - 74y^2 = 1$

(ii) $13x^2 - 29y^2 = 1$ (v) $5x^2 - 11y^2 = 1$

(iii) $57x^2 - 14y^2 = 1$ (vi) $21x^2 - 2y^2 = 1$

50. Show that if p and q are both primes of the form $4k + 3$, then $px^2 - qy^2 = \pm 1$ is solvable for at least one choice of sign.

51. Note that $10^2 + 11^2 + 12^2 = 13^2 + 14^2$. Find all solutions to

$$x^2 + (x + 1)^2 + (x + 3)^2 = y^2 + (y + 1)^2.$$

52. Let A and B be relatively prime positive integers such that AB is not a square. Prove that there are infinitely many pairs of triangular numbers, t_a and t_b, such that $t_a/t_b = A/B$.

53. Solve the following equations involving triangular numbers.

(a) $t_x + t_{2y} = t_{3y}$.

(b) $t_x + t_{x+1} = t_y$.

(c) $t_x + t_{x+1} + t_{x+2} = t_y$.

(d) Can you generalize these equations to other polygonal numbers?

54. Solve the equation

$$\binom{x}{y+1} = \binom{x+1}{y}$$

in positive integers x and y.

55. Consider the equation

$$ax^2 + bxy + cy^2 + dx + ey + f = 0.$$

Suppose that this equation is solvable and that the polynomial in x and y is not the product of two linear factors. Let $D = 4ac - b^2$ and $H = \max(|a|, |b|,$ $|c|, |d|, |e|, |f|)$. We suppose that $D \geq 0$ or $D = -n^2$, n a positive integer. Show that

$$\max(|x|,|y|) \leq \begin{cases} 9H^2 & \text{if } D > 0 \\ 6H^{7/2} & \text{if } D = 0 \\ 20H^4 & \text{if } D = -n^2 \end{cases}.$$

(Schinzel has shown that if $D < 0$, then the estimate is of the form

$$\max(|x|, |y|) \leq (5H)^{200H}.$$

See [81].)

56. Refine Problem 4 of Problem Set 6.10 to show that $g(4) \leq 50$.

57. By invoking Theorem 6.14 show that $g(4) \leq 45$.

58. Show that $G(3) \leq 13$.

59. Develop an identity between and sums of cubes to show that $g(6) \leq 108g(3)$ + 119.

60. (a) Determine all solutions to $x^2 + 4 = y^3$.

    (b) Show that neither $x^2 + 12 = y^3$ nor $x^2 + 16 = y^3$ has any solutions in integers.

    (c) Use congruences modulo 4 to show that $x^2 = y^3 + 23$ has no solutions in integers.

61. This is a variation on Problem 13 above. In addition to the conditions given above suppose that the total number of white bulls plus the total number of black bulls should be a square so that when they were placed on the field they formed a square figure. Similarly we want the total number of brown bulls and the total number of spotted bulls to be a triangular number so that they can be arranged in the field in the shape of a triangle.

    If you are able to solve this problem you will indeed be, in the words of the poet (Lessing) "a wise man" and "your fame will glow bright all through the world of the wise." Again we only want the smallest positive solution.

Computer Problems

1. (a) Write a program that finds all, if any, positive solutions to a given linear Diophantine equation.

   (b) Find all positive solutions of the following problems.

       (i) $5x - 3y = 7$.   (iii) $17x + 2y = 63$.

       (ii) $4x + y = 18$.   (iv) $7x + 8y = 75$.

2. (a) Write a program to solve linear Diophantine equations in three variables.

   (b) Find all solutions of the following equations.

   (i) $x + y - 3z = 5$.   (iii) $4x + 5y + 7z = -13$.

   (ii) $6x - y + 5z = 17$.   (iv) $x + y + z = 100$.

3. (a) Write a program that determines all representations of a given number as a sum of two nonnegative squares.

   (b) For $1 \leq n \leq 200$, list all representations of n, if any, as a sum of two nonnegative squares.

4. (a) Modify the program in 3(a) to check representations of the form $x^2 + Ay^2$ for any given positive value of A.

   (b) Find all representations of those values of n, $1 \leq n \leq 100$, which can be represented in the form $x^2 + Ay^2$ for $A = 2, 3, 5$ and 7.

5. (a) Write a program that for a given n determines all representations of n as a sum of four squares: $n = a^2 + b^2 + c^2 + d^2$, $0 \leq a \leq b \leq c \leq d$.

   (b) For $1 \leq n \leq 200$, write all representations of n as a sum of four squares in the above form.

6. (a) Write a program that determines all representations of a given number as a sum of three squares.

   (b) For those integers in the interval $[300, 400]$ that can be represented as a sum of three squares find all such representations.

7. (a) Let $k \geq 5$. Write a program that finds all representations of a given integer n as a sum of k nonzero squares.

   (b) For $k = 5$ and 6 find all representations of n, $100 \leq n \leq 200$ as sums of k nonzero squares. Modify your program, if necessary, and cover the integers in the interval $[1, 99]$ as well.

8. (a) For a given positive definite binary quadratic form $ax^2 + bxy + cy^2$ write a program to determine the least positive integer that it can represent.

   (b) Find the least positive integer represented by the following forms.

   (i) $21x^2 + 11xy + y^2$.   (iv) $7x^2 + 6xy + 11y^2$.

   (ii) $13x^2 + 36xy + 25y^2$.   (v) $2x^2 + 4xy + 5y^2$.

   (iii) $3x^2 + 2xy + 23y^2$.   (vi) $2x^2 - 5xy + 7y^2$.

   (c) Modify the program in (a) to determine whether or not a given number can be represented by the given form. If it can be represented determine all such representations.

(d) Find all solutions, if any, to the following equations.

    (i) $21x^2 + 11xy + y^2 = 3$.   (iv) $3x^2 + 2xy - 12y^2 = 81$.

    (ii) $2x^2 - 5xy + 7y^2 = 3$.   (v) $2x^2 - 5xy + 7y^2 = 118$.

    (iii) $2x^2 + xy + 2y^2 = 10$.   (vi) $x^2 + xy + 2y^2 = 28$.

9. (a) Write a program to find the reduced forms of a given discriminant.

  (b) Find all positive definite reduced forms of discriminant $d \le 100$.

10. (a) Write a program that finds all representations of a given integer as a difference of squares.

  (b) For those integers that are so representable in the interval $[1, 100]$ find all representations as a difference of squares.

  (c) Modify the program in (a) to find representations in the form $x^2 - a^2y^2$, where $a > 1$ is an integer.

  (d) For a = 2, 3 and 4, find all representations of n, $1 \le n \le 100$, in the form $n = x^2 - a^2y^2$.

11. (a) Write a program that finds, for a given positive nonsquare integer D, the fundamental solution to $x^2 - Dy^2 = 1$ and whenever possible $x^2 - Dy^2 = -1$.

  (b) Find the fundamental solutions for the equation $x^2 - Dy^2 = \pm 1$ for each positive nonsquare integer $101 \le D \le 300$.

  (c) Find the fundamental solution to $x^2 - Dy^2 = 1$ for $D = 807$, 1001 and 1499.

12. (a) Write a program that finds those integers n, $|n| < \sqrt{D}$, where D is a positive nonsquare integer, for which $x^2 - Dy^2 = n$ is solvable.

  (b) If D is a given positive nonsquare integer and N is a given integer, write a program that finds the least solutions to $x^2 - Dy^2 = N$.

  (c) Find solutions to the following equations, if they exist.

    (i) $x^2 - 27y^2 = 5$.   (iv) $x^2 - 6y^2 = 87$.

    (ii) $x^2 - 106y^2 = -3$.   (v) $x^2 - 6y^2 = 88$.

    (iii) $x^2 - 93y^2 = 72$.   (vi) $x^2 - 6y^2 = 89$.

13. (a) Write a program to solve the general quadratic equation

$$ax^2 + bxy + cy^2 + dx + ey + f = 0.$$

  (b) Find all solutions, if any exist, to the following equations. If the equations have infinitely many solutions, find the least positive solution.

    (i) $3x^2 - 8xy + 7y^2 - 4x + 2y - 109 = 0$.

     (ii) $9x^2 - 12xy + 4y^2 + 3x + 2y - 12 = 0.$

     (iii) $x^2 + 4xy - 11y^2 + 2x - 86y - 140 = 0.$

     (iv) $3xy + 2y^2 - 4x - 3y - 12 = 0.$

     (v) $x^2 - y^2 + 4x - 5y - 27 = 0.$

     (vi) $61x^2 + 28xy + 251y^2 + 264x + 526y + 260 = 0.$

14. Write a program to discover representations by cubes, biquadrates, etc. and find the representations for $n \leq 500$.

CHAPTER SEVEN
Arithmetic Functions

7.1. INTRODUCTION

Many times one can express certain useful facts about the properties of the integers by way of a function that counts the number of objects in a set of interest or that describes some other property of elements in a set. This chapter and the next are concerned with such functions.

Definition 7.1. A function whose domain is a subset of the integers and whose range is a subset of the complex numbers is called an **arithmetic function**.

More often than not the domain will be the set of positive integers and the range will be some subset of the nonnegative integers. To simplify the discussion we will assume that all arithmetic functions under discussion are defined at least on the positive integers, unless otherwise stated.

In the present chapter we shall discuss some of the general properties of arithmetic functions as well as deal with some of the more important arithmetic functions. As we shall see often, the values of the arithmetic functions are somewhat patternless in their behavior. In the next chapter we shall discuss the average behavior of arithmetic functions and see that on the average many of these functions are better behaved than they look at the individual integers.

7.2. DIRICHLET CONVOLUTION

In this section we study an operation defined on arithmetic functions that is very useful in deriving many of the properties of arithmetic functions. As we shall see this operation has many properties analogous to multiplication.

Definition 7.2. Let f and g be arithmetic functions. Then the **Dirichlet convolution**, or simply the **convolution**, of f and g, denoted by $f * g$, is defined by

$$(7.1) \qquad (f*g)(n) = \sum_{ij=n} f(i)g(j),$$

where the sum is over all positive integers i and j whose product is the positive integer n.

Another way of writing (7.1) is

$$(f*g)(n) = \sum_{d|n} f(d)g(n/d),$$

where the sum is over all positive integers d that divide n.

Theorem 7.1. Let f, g and h be arithmetic functions. Then

(a) $f * g = g * f$;

(b) $(f * g) * h = f * (g * h)$; and

(c) $f * (g + h) = (f * g) + (f * h)$.

Proof. Note that if d runs over all the positive divisors of n so does n/d.

(a) We have

$$(f*g)(n) = \sum_{d|n} f(d)g(n/d) = \sum_{d|n} f(n/d)g(d) = (g*f)(n).$$

(b) We have

$$((f*g)*h)(n) = \sum_{ij=n} (f*g)(i)h(j) = \sum_{ij=n}\sum_{rs=i} f(r)g(s)h(k)$$

$$= \sum_{rsk=n} f(r)g(s)h(k) = \sum_{rj=n} f(r)\sum_{sk=j} g(s)h(k)$$

$$= \sum_{rj=n} f(r)(g*h)(j) = (f*(g*h))(n).$$

(c) Finally, we have

$$(f*(g+h))(n) = \sum_{d|n} f(d)(g(n/d) + h(n/d))$$

$$= \sum_{d|n} f(d)g(n/d) + \sum_{d|n} f(d)h(n/d)$$

$$= (f*g)(n) + (f*h)(n),$$

which is the last result.   ∎

This shows that the set of arithmetic functions forms a commutative ring under convolution and pointwise addition. We denote this ring by A.

We now define some special functions in A and operators on A that will be useful in our further study.

Definition 7.3. Let $n \in \mathbf{Z}$ and $f \in A$. We define the following two arithmetic functions: $\delta(n)$

(a) $\delta(n) = \begin{cases} 1 & \text{if } n = 1 \\ 0 & \text{if } n > 1 \end{cases}$

and

(b) $e(n) = 1$ for all n.

We define the following two operators on A:

(c) if a is a real number, then

$$(T_a f)(n) = f(n)n^a$$

and

(d) $(Lf)(n) = f(n)\log(n)$

Finally, we define f^{*n} to mean f convolved with itself n times and $f^{*0} = \delta$.

Example 7.1. Compute $T_a\delta$, $T_a e$, $L\delta$ and Le.

Solution. We have

$$(T_a\delta)(n) = \delta(n)n^a = \delta(n),$$

since $\delta(1)1^a = 1$ and $\delta(n)n^a = 0$ for $n > 1$. We have

$$(T_a e)(n) = e(n)n^a = n^a,$$

for all positive integers n. We have

$$(L\delta)(n) = \delta(n)\log(n) = 0,$$

since $\delta(n) = 0$ if $n > 1$ and $\log(1) = 0$. Finally, we have

$$(Le)(n) = e(n)\log(n) = \log(n),$$

for all positive integers n.

Theorem 7.2. If $f, g \in A$, then
  (a) $f * \delta = f$,
  (b) $L(f * g) = g * Lf + f * Lg$,
  (c) $L(f + g) = Lf + Lg$,
  (d) $T_a(f * g) = (T_a f) * (T_a g)$,

and

  (e) if $k \in \mathbf{Z}^+$, then $L(f^{*k}) = k\, f^{*(k-1)} * Lf$.

Proof. In the following $n \in \mathbf{Z}^+$.

(a) We have

$$(f * \delta)(n) = \sum_{d \mid n} f(d)\delta(n / d) = f(n),$$

since $\delta(n/d) = 0$ unless $n/d = 1$, that is, $d = n$.

(b) We have

$$(L(f*g))(n) = \left\{\sum_{d|n} f(d)g(n/d)\right\}\log(n) = \sum_{d|n}\{f(d)g(n/d)(\log((d)(n/d)))\}$$

$$= \sum_{d|n}\{f(d)g(n/d)(\log(d)+\log(n/d))\}$$

$$= \sum_{d|n}\{f(d)\log(d)\}g(n/d) + \sum_{d|n} f(d)\{g(n/d)\log(n/d)\}$$

$$= (Lf*g)(n) + (f*Lg)(n).$$

(c) We have

$$(L(f+g))(n) = (f+g)(n)\log(n) = f(n)\log(n) + g(n)\log(n) = (Lf)(n) + (Lg)(n).$$

(d) We have

$$(T_\alpha(f*g))(n) = \left\{\sum_{d|n} f(d)g(n/d)\right\}n^\alpha = \sum_{d|n}\{f(d)g(n/d)((d)(n/d))^\alpha\}$$

$$= \sum_{d|n} f(d)d^\alpha g(n/d)(n/d)^\alpha = ((T_\alpha f)*(T_\alpha g))(n).$$

(e) This follows easily from (b) by induction and completes the proof of the theorem. ■

(a) of Theorem 7.2 shows that δ is an identity for convolution. It is easy to see, since convolution is commutative, that is the unique convolution identity. An operator that satisfies (b) of Theorem 7.2 is called, in general, a **derivation**. Thus A is a commutative ring with identity that has a derivation defined on it.

Definition 7.4. Let $f \in A$. Then $g \in A$ is said to be the **convolution inverse** of f if and only if

$$f * g = \delta.$$

If g exists we denote it by f^{*-1}.

It is easy to see, again because A is a commutative ring, that if f^{*-1} exists, then it is unique. This allows us to show

(7.2) $(f*g)^{*-1} = f^{*-1}*g^{*-1},$

for example, as well as the usual group properties of inverses. The following theorem gives a necessary and sufficient condition for the existence of the convolution inverse.

Theorem 7.3. If $f \in A$, then f^{*-1} exists if and only if $f(1) \neq 0$.

Proof. Suppose f^{*-1} exists. Then

$$(7.3) \qquad 1 = \delta(1) = (f * f^{*-1})(1) = f(1)f^{*-1}(1).$$

Thus $f(1) \neq 0$.

Suppose now $f(1) \neq 0$. We shall define f^{*-1} inductively. By (7.3) we see that $f^{*-1}(1) = 1/f(1)$. Suppose $n > 1$ and we have computed $f^{*-1}(1),\dots,\ f^{*-1}(n-1)$. Now $d \mid n$ and $d > 0$ implies that $d = n$ or $d < n$, by (g) of Theorem 1.2. Thus, for $n > 1$, we have

$$0 = \delta(n) = \sum_{d \mid n} f(d)f^{*-1}(n/d) = f(1)f^{*-1}(n) + \sum_{\substack{d \mid n \\ d > 1}} f(d)f^{*-1}(n/d)$$

or

$$(7.4) \qquad f^{*-1}(n) = \frac{-1}{f(1)} \sum_{\substack{d \mid n \\ d > 1}} f(d)f^{*-1}(n/d),$$

which is well-defined since $d \mid n$ and $d > 1$ imply that $1 \le n/d < n$ and we have assumed that $f^{*-1}(k)$ is known for $1 \le k \le n - 1$. ∎

Because of the importance of convolution inverses we let $A_1 = \{f \in A : f(1) \neq 0\}$. It is easy to check, by (7.2), that if $f \in A_1$, and $g \in A_1$, then $f * g \in A_1$.

Corollary 7.3.1. Let $f \in A_1$ and p be a prime. Then

$$f^{*-1}(p) = -f(p)/f^2(1).$$

Proof. The only divisors of p are p and 1. Thus, by (7.4), we have

$$f^{*-1}(p) = \frac{-1}{f(1)} \sum_{\substack{d \mid p \\ d > 1}} f(d)f^{*-1}(p/d)$$

$$= (-1/f(1))(f(p)f^{*-1}(1)) = -f(p)/f^2(1),$$

by (7.3). ∎

Theorem 7.4. Let $f \in A_1$. Then

$$Lf^{*-1} = -(f*f)^{*-1}*Lf.$$

Proof. By Example 7.1 we know that $L\delta = 0$. By Definition 7.4 we know

$$f * f^{*-1} = \delta$$

Thus, by (b) of Theorem 7.2, we have

$$0 = L\delta = L(f * f^{*-1}) = (Lf) * f^{*-1} + f * (Lf^{*-1})$$

or

$$f * (Lf^{*-1}) = -(Lf) * f^{*-1}.$$

If we convolve both sides with f^{*-1} and use (7.2), we have

$$Lf^{*-1} = -(Lf) * f^{*-1} * f^{*-1} = -(f * f)^{*-1} * (Lf),$$

by (a) of Theorem 7.1.   ■

Example 7.2. Show that $\delta \in A_1$ and find δ^{*-1}.

Solution. Since $\delta(1) = 1$ we see that $\delta \in A_1$. Since δ is the identity of the ring A we know that $\delta^{*-1} = \delta$. We can also use (7.4): if $n > 1$,

$$\delta^{*-1}(n) = \frac{-1}{\delta(1)} \sum_{\substack{d|n \\ d>1}} \delta(d)\delta^{*-1}(n/d) = 0 = \delta(n)$$

since $d > 1$ implies that $\delta(d) = 0$.   □

For e it is harder to find e^{*-1}. Indeed, we must define another arithmetic function.

Definition 7.5. The **Möbius function**, μ, is defined to be the convolution inverse of e, that is, $\mu = e^{*-1}$.

By Corollary 7.3.1 we see that $\mu(p) = -1$. In general, if $m = p_1 \cdots p_r$, where the p_i, $1 \le i \le r$, are distinct primes, then $\mu(m) = (-1)^r$. The proof is an easy induction argument on r, the number of prime factors of n. Suppose $\mu(p_1 \cdots p_r) = (-1)^r$ for $1 \le r \le n - 1$. Then by (7.4),

$$\mu(p_1 \cdots p_n) = - \sum_{\substack{d|p_1 \cdots p_n \\ d>1}} \mu(p_1 \cdots p_n / d)$$

$$= -\left\{ 1 + \sum_{i=1}^{n} \mu(p_i) + \sum_{\substack{i,j=1 \\ i \ne j}}^{n} \mu(p_i p_j) + \cdots + \sum_{\substack{i_1,\ldots,i_{n-1} \\ i_k \text{ distinct}}}^{n} \mu(p_{i_1} \cdots p_{i_{n-1}}) \right\}$$

$$= -\left\{ 1 - \binom{n}{1} + \binom{n}{2} - \cdots + (-1)^{n-1}\binom{n}{n-1} \right\}$$

$$= -(1-1)^n + (-1)^n \binom{n}{n} = (-1)^n.$$

This proves part of the following theorem which establishes two key properties of the Möbius function.

Theorem 7.5.

    (a)   If a and b are relatively prime integers, then $\mu(ab) = \mu(a)\mu(b)$.

    (b) $\mu(n) = \begin{cases} 1 & \text{if } n = 1 \\ (-1)^r & \text{if } n = p_1 \cdots p_r,\ p_i \text{ distinct} \\ 0 & \text{if there exists a prime } p \text{ such that } p^2 | n \end{cases}$

Proof of (a). We prove this result by induction on the number of prime factors of ab. Suppose ab has no prime factors. Then we must have $a = b = 1$. Since $\mu(1) = 1/e(1) = 1$, the result follows. Suppose ab has one prime factor, then ab must be a prime p. Thus $a = p$, $b = 1$ or $a = 1$, $b = p$. Again the result follows since $\mu(1) = 1$. Now suppose that $\mu(MN) = \mu(M)\mu(N)$, whenever $(M, N) = 1$ and MN has less than k prime factors (counting multiplicity) for some $k > 1$. If $(a, b) = 1$ and $d \mid ab$, then by unique factorization (Theorem 1.19) we can write $d = d_1 d_2$, where $(d_1, d_2) = 1$ and $d_1 \mid a$ and $d_2 \mid b$. Suppose ab has k prime factors and $a, b > 1$. Then, by (7.4),

$$(7.5) \qquad \mu(ab) = -\sum_{\substack{d \mid ab \\ d > 1}} \mu(ab / d) = -\sum_{\substack{d_1 \mid a \\ d_2 \mid b \\ d_1 d_2 > 1}} \mu(ab / d_1 d_2).$$

Since $d_1 d_2 > 1$ implies either $d_1 > 1$ or $d_2 > 1$ we see that $ab/d_1 d_2$ has less than k prime factors. Also, $(a/d_1, b/d_2) = 1$. Thus, by (7.5),

$$\mu(ab) = -\sum_{\substack{d_1 \mid a \\ d_2 \mid b \\ d_1 d_2 > 1}} \mu(a / d_1)\mu(b / d_2)$$

$$(7.6) \qquad = -\sum_{\substack{d_1 \mid a \\ d_1 > 1}} \mu(a / d_1) \sum_{\substack{d_2 \mid b \\ d_2 > 1}} \mu(b / d_2) - \sum_{\substack{d_2 \mid b \\ d_2 > 1}} \mu(ab / d_2) - \sum_{\substack{d_1 \mid a \\ d_1 > 1}} \mu(ab / d_1)$$

$$= -\sum_{\substack{d_1 \mid a \\ d_1 > 1}} \mu(a / d_1) \sum_{\substack{d_2 \mid b \\ d_2 > 1}} \mu(b / d_2) - \mu(a) \sum_{\substack{d_2 \mid b \\ d_2 > 1}} \mu(b / d_2) - \mu(b) \sum_{\substack{d_1 \mid a \\ d_1 > 1}} \mu(a / d_1)$$

since $(a, b/d_2) = (a/d_1, b) = 1$ for any $d_2 \mid b$ or $d_1 \mid a$ and $d_2 > 1$ and $d_1 > 1$ imply that ab/d_2 and ab/d_1 both have less than k prime factors. The definition of μ as e^{*-1} implies, since $a, b > 1$

$$0 = \sum_{d_1 \mid a} \mu(a/d_1) \sum_{d_2 \mid b} \mu(b/d_2)$$

$$= \left\{ \mu(a) + \sum_{\substack{d_1 \mid a \\ d_1 > 1}} \mu(a/d_1) \right\} \left\{ \mu(b) + \sum_{\substack{d_2 \mid b \\ d_2 > 1}} \mu(b/d_2) \right\}$$

(7.7)
$$= \mu(a)\mu(b) + \mu(a) \sum_{\substack{d_2 \mid b \\ d_2 > 1}} \mu(b/d_2) + \mu(b) \sum_{\substack{d_1 \mid a \\ d_1 > 1}} \mu(a/d_1)$$

$$+ \sum_{\substack{d_1 \mid a \\ d_1 > 1}} \mu(a/d_1) \sum_{\substack{d_2 \mid b \\ d_2 > 1}} \mu(b/d_2).$$

The result follows from (7.6) and (7.7). If $a = 1$ or $b = 1$, then $\mu(a) = 1$ or $\mu(b) = 1$ and the result follows in this case as well. This proves (a).

Proof of (b). We have already proved the first two parts of (b). We need only show that $\mu(n) = 0$ if $p^2 \mid n$ for some prime p. By (a), if $n = p_1^{\alpha_1} \cdots p_r^{\alpha_r}$, p_i all distinct primes, then

$$\mu(n) = \mu(p_1^{\alpha_1}) \cdots \mu(p_r^{\alpha_r}).$$

Thus, to prove (b), it suffices to show that $\mu(p^a) = 0$ if $a > 2$. We again proceed by induction, this time on a. We have

$$\mu(p^2) = - \sum_{\substack{d_2 \mid b \\ d_2 > 1}} \mu(p^2/d) = -\mu(p) - \mu(1) = 1 - 1 = 0.$$

Assume $\mu(p^a) = 0$ if $2 \le a \le k$. Then for $a = k + 1$ we have

$$\mu(p^{k+1}) = - \sum_{\substack{d \mid p^{k+1} \\ d > 1}} \mu(p^{k+1}) = - \sum_{j=1}^{k+1} \mu(p^j) = -\mu(p) - \mu(1) = 1 - 1 = 0,$$

by the induction hypothesis. This proves (b) and concludes the proof of the theorem. ∎

Example 7.3. Make a table of $\mu(n)$ for $n = 1, ..., 21$.

Solution . By (b) of Theorem 7.5, we have

$$\mu(1) = 1 \qquad \mu(8) = 0 \qquad \mu(15) = 1$$
$$\mu(2) = -1 \qquad \mu(9) = 0 \qquad \mu(16) = 0$$

$$\mu(3) = -1 \qquad \mu(10) = 1 \qquad \mu(17) = -1$$
$$\mu(4) = 0 \qquad \mu(11) = -1 \qquad \mu(18) = 0$$
$$\mu(5) = -1 \qquad \mu(12) = 0 \qquad \mu(19) = -1$$
$$\mu(6) = 1 \qquad \mu(13) = -1 \qquad \mu(20) = 0$$
$$\mu(7) = -1 \qquad \mu(14) = 1 \qquad \mu(21) = 1.$$

One of the uses of the Möbius function is in the following theorem.

Theorem 7.6. We have, for $f, F \in A$,

$$F(n) = \sum_{d|n} f(d) \text{ if and only if } f(n) = \sum_{d|n} \mu(d) F(n/d).$$

Proof. We have

$$(7.8) \qquad F(n) = \sum_{d|n} f(d) = \sum_{d|n} f(d) e(n/d) = (f * e)(n).$$

If we convolve both sides of (7.8) with μ and use the associativity of convolution ((b) of Theorem 7.1), we get

$$\sum_{d|n} \mu(d) F(n/d) = (\mu * F)(n) = (\mu * f * e)(n) = (\delta * f)(n) = f(n),$$

by (a) of Theorem 7.2. The other direction is proved in a similar manner. ∎

Thus we can do division in $(A_1, +, *)$. For a proof that this ring is actually a unique factorization domain see [117, pp. 29-34].

We define one final operator on A that will be of use in our later work. In particular, a special case will be of great help in proving the prime number theorem (see Section 3 of Chapter 9).

Definition 7.6. Let $f \in A_1$. We define the **van-Mangoldt operator**, Λ_f by

$$(7.9) \qquad \Lambda_f(n) = (f^{*-1} * Lf)(n).$$

Another way of writing (7.9) is

$$(7.10) \qquad (f * \Lambda_f)(n) = (Lf)(n) = f(n) \log(n).$$

The particular case of importance in prime number theory is when $f = e$. We write $\Lambda_e = \Lambda$ and call this arithmetic function the van-Mangoldt function.

Example 7.4. Compute Λ_δ and Λ_μ.

Solution. By Definition 7.6 we have

$$\Lambda_\mu = \mu^{*-1} * L\mu = e * L\mu.$$

By Theorem 7.4, (7.2) and Definition 7.5, we have

$$L\mu = -(Le)*(e*e)^{*-1} = -\mu*\mu*Le.$$

Thus, by Definition 7.5,

$$e*L\mu = -(Le)*\mu = -\Lambda.$$

By Definition 7.6 and Examples 7.1 and 7.2, we have

$$\Lambda_\delta = \delta^{*-1}*(L\delta) = \delta*(L\delta) = L\delta = 0. \qquad \square$$

Example 7.4 illustrates part of the following theorem.

Theorem 7.7. Let $f, g \in A_1$. Then

(a) $\Lambda_{f*g} = \Lambda_f + \Lambda_g$;

(b) $\Lambda_{f^{-1}} = -\Lambda_f$;

(c) $\Lambda_f(1) = 0$;

(d) if $\Lambda_f = \Lambda_g$, then there exists a constant k such that $g = kf$;

and

(e) $\Lambda_{T_\alpha f} = T_\alpha \Lambda_f$

Proof of (a). We have, by Definition 7.6,

(7.11)
$$\Lambda_{f*g} = (f*g)^{*-1}*L(f*g).$$

We apply (7.2), (a) and (b) of Theorem 7.3 and (b) and (c) of Theorem 7.1. This gives, by (7.11),

$$\begin{aligned}
\Lambda_{f*g} &= f^{*-1}*g^{*-1}*(f*Lg + g*Lf) \\
&= f^{*-1}*g^{*-1}*f*Lg + f^{*-1}*g^{*-1}*g*Lf \\
&= g^{*-1}*Lg + f^{*-1}*Lf = \Lambda_f + \Lambda_g.
\end{aligned}$$

Proof of (b). This follows from (a) and the fact that $\Lambda_\delta = 0$ (the second part of Example 7.4). We give an alternate proof. We have, by Theorem 7.4 and (7.2),

$$\begin{aligned}
\Lambda_{f^{-1}} &= f*Lf^{*-1} = f*(-(Lf)*(f*f)^{*-1} \\
&= -f*(Lf)*f^{*-1}*f^{*-1} = -f^{*-1}*(Lf) = -\Lambda_f
\end{aligned}$$

Proof of (c). We have

$$\Lambda_f(1) = (f^{*-1}*Lf)(1) = f^{*-1}(1)f(1)\log(1) = 0.$$

Proof of (d). If $\Lambda_f = \Lambda_g$, then, by Definition 7.6, we have

$$f^{*-1} * Lf = g^{*-1} * Lg$$

or

(7.12) $$g * Lf = f * Lg.$$

We proceed by induction on n. Since $f, g \in A_1$, we know that $f(1)$ and $g(1)$ are both nonzero. Let $k = g(1)/f(1)$. Take $n = 2$. Then

$$(g * Lf)(2) = g(2)f(1)\log(1) + g(1)f(2)\log(2) = g(1)f(2)\log(2)$$

and

$$(f * Lg)(2) = f(2)g(1)\log(1) + f(1)g(2)\log(2) = f(1)g(2)\log(2).$$

Thus, by (7.12), we have

$$f(1)g(2) = g(1)f(2)$$

or

$$g(2) = (g(1)/f(1))f(2) = kf(2).$$

Now suppose $g(m) = kf(m)$ for all $1 \le m \le n - 1$. Then

(7.13)
$$(g * Lf)(n) = \sum_{d|n} f(d)g(n/d)\log(d)$$
$$= f(n)g(1)\log(n) + \sum_{\substack{d|n \\ d<n}} f(d)g(n/d)\log(d)$$

and similarly

(7.14) $$(f * Lg)(n) = g(n)f(1)\log(n) + \sum_{\substack{d|n \\ d<n}} g(d)f(n/d)\log(d).$$

Since $\log(1) = 0$ we see that both sums are really over $d \mid n$ and $1 < d < n$. Thus, for these values of d, $1 \le d$, $n/d \le n - 1$. Thus $g(d) = kf(d)$ and $g(n/d) = kf(n/d)$, and so the two summations in (7.13) and (7.14) are equal. Thus, by (7.12), we have

$$f(n)g(1)\log(n) = g(n)f(1)\log(n)$$

or, since $n > 1$ implies $\log(n) > 0$,

$$g(n) = (g(1)/f(1))f(n) = kf(n),$$

which is the result for $m = n$.

Proof of (e). By Example 7.1 and (4) of Theorem 7.2, we have

$$\delta = T_\alpha \delta = T_\alpha(f * f^{*-1}) = T_\alpha f * T_\alpha f^{*-1}.$$

Thus, since convolution inverses are unique, $(T_\alpha f)^{*-1} = T_\alpha f^{*-1}$. Also

$$(LT_\alpha f)(n) = (T_\alpha f)(n)\log(n) = f(n)n^\alpha \log n = (f(n)\log n)n^\alpha = T_\alpha(Lf)(n).$$

Thus, by (d) of Theorem 7.2,

$$\Lambda_{T_\alpha f} = ((T_\alpha f)^{*-1} * LT_\alpha f) = (T_\alpha f^{*-1} * T_\alpha Lf) = T_\alpha(f^{*-1} * Lf) = T_\alpha \Lambda_f. \qquad \blacksquare$$

We close this section on general results with the following theorem, a special case of which will be our starting point in the proof of the prime number theorem.

Theorem 7.8 (Selberg). Let $f \in A_1$. Then

$$f^{*-1} * L^2 f = L\Lambda_f + \Lambda_f * \Lambda_f,$$

where $L^2 f = L(Lf)$. In particular, when $f = e$, we have

$$\mu * L^2 e = L\Lambda + \Lambda * \Lambda$$

or

$$\Lambda(n)\log(n) + \sum_{d|n}\Lambda(d)\Lambda(n/d) = \sum_{d|n}\mu(d)\log^2(n/d).$$

Proof. By (7.10), we have

$$f * \Lambda_f = Lf.$$

Thus, by (b) of Theorem 7.3, we have

$$L^2 f = L(f * \Lambda_f) = f * L\Lambda_f + \Lambda_f * Lf,$$

or, by (7.10) again,

$$L^2 f = f * L\Lambda_f + \Lambda_f * f * \Lambda_f.$$

Convolving both sides with f^{*-1} gives

$$f^{*-1} * L^2 f = L\Lambda_f + \Lambda_f * \Lambda_f,$$

as desired. The special case is immediate. $\qquad \blacksquare$

Problem Set 7.2

1. Let $f(n)$ and $a(n)$ be arithmetic functions with $f(n) > 0$, $a(n)$ real and $a(1) \neq 0$. Show that

$$g(n) = \prod_{d|n} f(d)^{a(n/d)} \text{ if and only if } f(n) = \prod_{d|n} g(n)^{a^{*-1}(n/d)}.$$

(Hint: consider logarithms.)

2. Prove that if f, g and h are three arithmetic functions such that $f \in A_1$, $g = e *$ f and $f = g * h$, then $h = \mu$.

3. Compute $\Lambda_f(p^a)$ in terms of f and f^{*-1} in as simple a manner as possible. Give an explicit formula for the case $f = e$.

4. Let $f \in A_1$. Find $(\mu f)^{*-1}$ and $(\mu^2 f)^{*-1}$.

5. (a) If $p \nmid a$ and $p^{k-1} | m$, show that $a^{mp} \equiv a^m \pmod{p^k}$.

   (b) If a and n are positive integers, show that

   $$n \left| \sum_{d|n} a^d \mu(n/d) \right..$$

   (c) If $n | (e * f)(n)$, then

   $$n \left| \sum_{d|n} a^d f(n/d) \right..$$

   (d) Show that if n is a prime, then (b) yields Fermat's Theorem (Corollary 2.8.1.)

   (Hint: prove the result for each $p^k | n$; if $p \nmid a$ use (a) and note that the sum in (b) can be separated into a collection of terms as in (a).)

6. Let $F_m = e * T_m f$ and $G_m = e * T_m g$, where f and g are in A. Show that, for any values of r and s,

   $$T_{r-s} F_s * G_r = T_{r-s} G_s * F_r.$$

7. Let f and g be arithmetic functions that satisfy $f * e = g$. Show that

   $$\sum_{d^2 | m} f(d) = \sum_{d^2 | m} g(d) \mu^2(m/d^2).$$

8. Let $\psi(n) = \prod_{d|n} \mu(d)$. Show that

   $$\psi(n) = \begin{cases} -1 & \text{if } n \text{ is a prime} \\ 0 & \text{if } n \text{ has a square factor} \\ 1 & \text{if } n \text{ is square - free and composite} \end{cases}$$

9. Show that

   $$\sum_{d^2 | n} \mu(d) = \mu^2(n).$$

10. Let f be a function of two variables defined on $\mathbf{Z} \times \mathbf{Z}$. Show that

$$\sum_{d|n} \sum_{c|d} f(c,d) = \sum_{c|n} \sum_{d|(n/c)} f(c,cd).$$

11. Define the **unitary convolution** of two arithmetic functions f and g by

$$(f \circ g)(n) = \sum_{\substack{d|n \\ (d,n/d)=1}} f(d)g(n/d).$$

Does Theorem 7.1 still hold?

12. If f and g are arithmetic functions, define an arithmetic function h by

$$h(n) = \sum_{\substack{k,m \\ [k,m]=n}} f(k)g(m).$$

If we define $F = e * f$, $G = e * g$ and $H = e * h$, show that $H(n) = F(n)G(n)$. (This result is due to von Sterneck.)

13. Let $K(m, n)$ be a function of the two positive integers m and n that takes only the values 0 or 1. Define a **K-convolution** of the arithmetic functions f and g, denoted by $f *_K g$, by

$$(f *_K g)(n) = \sum_{ab=n} f(a)g(b)K(a,b).$$

(a) Find conditions on K so that $*_K$ is commutative.

(b) Find conditions on K so that $*_K$ is associative.

14. Let f and g be arithmetic functions so that $g(1) \neq 0$. Show that

$$L(f * g^{*-1}) = (g * Lf - f * Lg) * g^{*-1} * g^{*-1}.$$

15. If f and g are arithmetic functions and $F = e * f$ and $G = e * g$, show that $g * F = f * G$.

16. Show that if n has k distinct prime divisors, then

$$\sum_{d|n} \mu^2(d) = 2^k.$$

17. Let $k > 0$ be an integer. Show that there exist an integer n such that

$$\mu(n + 1) = \mu(n + 2) = \cdots = \mu(n + k) = 0.$$

(Hint: consider the system of congruences

$$x \equiv -1 \pmod 4, x \equiv -2 \pmod 9, \ldots, x \equiv -k \pmod{p_k^2},$$

where p_k is the kth prime.) Find an n if $k = 4$.

18. (a) Let p be an odd prime. Show that the sum of all the numbers belonging to the exponent m, $m > 1$, is congruent to $\mu(m)$ modulo p.

   (b) Let $S_k(m)$ denote the sum of the $k\underline{th}$ powers of all the numbers belonging to the exponent m, $m > 1$. Show that

$$S_k(m) = \sum_{d|(m,k)} d\mu(m/d)$$

19. Give another proof of Theorem 7.8 that simply applies the L operator to both sides of (7.9).

7.3. MULTIPLICATIVE FUNCTIONS

There is a special class of arithmetic functions that are very useful in much of the study of number theory. We have met some of them in our earlier work (the Euler function of Chapter Two, for example.).

Definition 7.7. If $f \in A$, then f is said to be **multiplicative** if and only if, whenever $(m, n) = 1$, we have

(7.15) $f(mn) = f(n)f(m)$.

A multiplicative function is said to be **completely multiplicative** if and only if (7.15) holds for all positive integers m and n without restriction.

It follows from this definition that if $n = p_1^{\alpha_1} \cdots p_r^{\alpha_r}$, as a canonical factorization, then for f multiplicative we have

$$f(n) = \prod_{k=1}^{r} f(p_k^{\alpha_k})$$

and conversely. Also f is completely multiplicative if and only if $f(p^a) = (f(p))^a$. Thus to show that two multiplicative functions are equal we need only show that they agree on prime powers. To show that two completely multiplicative functions are equal we need only show that they agree on the primes.

Of the functions we have met so far we see that δ and e are completely multiplicative, while μ is only multiplicative.

Suppose we take $m = n = 1$ in (7.15) with a multiplicative function f. Then

$$f(1) = f(1 \cdot 1) = f(1)f(1).$$

Thus $f(1) = 1$ or $f(1) = 0$. If $f(1) = 0$, then, since $(n, 1) = 1$ for any positive integer n, we have

$$f(n) = f(n \cdot 1) = f(n)f(1) = 0.$$

Thus f multiplicative and $f(1) = 0$ implies that f is identically zero. In order to avoid this trivial case we assume, from now on, that when we speak of a multiplicative function f we mean one with $f(1) = 1$. Let M be the subset of A consisting of those multiplicative functions f such that $f(1) = 1$.

We begin with the following theorem whose results follow immediately from the above remarks and results of the preceding section.

Theorem 7.9. If $f \in M$, then

  (a) f^{*-1} exists,

  (b) if $g \in M$ and $\Lambda_f = \Lambda_g$, then $f = g$,

  (c) $\Lambda_f \in M$,

  (d) $T_a f \in M$ and if f is completely multiplicative so is $T_a f$,

and

  (e) $(T_\alpha f)^{*-1} = T_\alpha f^{*-1}$.

Proof of (a). Since $f \in M$ implies $f \in A_1$ the result follows from Theorem 7.3.

Proof of (b). This follows immediately from (c) of Theorem 7.7 since $g(1) = f(1) = 1$.

Proof of (c). This follows immediately from (d) of Theorem 7.7 since $\Lambda_f(1) = 0$.

Proof of (d). Let $(m, n) = 1$. Then

$$(T_a f)(mn) = f(mn)(nm)^a = f(m)m^a f(n)n^a = (T_a f)(m)(T_a f)(n),$$

so that $T_a f \in M$. If f is completely multiplicative, the above argument shows that $T_a f$ is as well.

Proof of (e). This follows immediately from (d) of Theorem 7.2 and (a). ■

A more important set of results is contained in the following theorem.

Theorem 7.10. Let $f, g \in A$.

  (a) If $f, g \in M$, then $f * g \in M$.

  (b) If f and $f * g \in M$, then $g \in M$.

  (c) If $f \in M$, then $f^{*-1} \in M$.

Proof of (a). Let $h = f * g$ and $(m, n) = 1$. Then

$$h(mn) = \sum_{d \mid mn} f(d)g(mn / d).$$

By the unique factorization theorem (Theorem 1.19) we see that if $d \mid mn$, then we may write $d = d_1 d_2$, where $d_1 \mid m$ and $d_2 \mid n$. Since $(d_1, d_2) = 1$ we have

$$h(mn) = \sum_{\substack{d_1|m \\ d_2|n}} f(d_1 d_2) g((m/d_1)(n/d_2))$$

$$= \sum_{d_1|m} f(d_1) g(m/d_1) \sum_{d_2|n} f(d_2) g(n/d_2) = h(m)h(n),$$

that is, $h \in M$.

Proof of (b). We shall assume g is not multiplicative and show that $f * g$ is also not multiplicative. Let $h = f * g$. Since g is not multiplicative, there exist integers m and n, $(m, n) = 1$, such that

$$g(mn) \neq g(m)g(n).$$

Choose a pair so that mn is minimal. If $mn = 1$, then we see that $g(1) \neq 1$, and so

$$h(1) = f(1)g(1) = g(1) \neq 1.$$

Thus $h \notin M$. If $mn > 1$, then $g(ab) = g(a)g(b)$ for all relatively prime positive integers a and b such that $ab < mn$. Also as in the proof of (a) we have

$$h(mn) = \sum_{\substack{a|m \\ b|n \\ ab>1}} f(ab)g(mn/ab) + g(mn)f(1) = \sum_{\substack{a|m \\ b|n \\ ab>1}} f(a)f(b)g(m/a)g(n/b) + g(mn)$$

$$= \sum_{a|m} f(a)g(m/a) \sum_{b|n} f(b)g(n/b) + g(mn) - g(m)g(n)$$

$$= h(m)h(n) + g(mn) - g(m)g(n).$$

Since $g(mn) \neq g(m)g(n)$ we see that $h(mn) \neq h(m)h(n)$, that is, $h \notin M$. The result follows.

Proof of (c). This follows immediately from (b) since $f \in M$ and $f * f^{*-1} = \delta \in M$. ∎

Theorem 7.11. Let $f \in M$. Then the following are equivalent:

    (a) f is completely multiplicative

and

    (b) $f^{*-1} = \mu f$.

Proof. If f is completely multiplicative, then

$$(f * \mu f)(n) = \sum_{d|n} f(d)(\mu f)(n/d) = \sum_{d|n} f(d)f(n/d)\mu(n/d)$$

$$= f(n) \sum_{d|n} \mu(n/d) = f(n)(e * \mu)(n) = f(n)\delta(n) = \delta(n),$$

since $f(1) = 1$. Thus (a) implies (b).

To prove that (b) implies (a) we need only show that $f(p^a) = f^a(p)$ for all primes p and positive integers a. We have

$$\delta(n) = (f * \mu f)(n) = \sum_{d|n} \mu(d) f(d) f(n / d).$$

Take $n = p^a$. Then we have, by (b) of Theorem 7.5,

$$0 = \sum_{j=0}^{a} \mu(p^j) f(p^j) f(p^{a-j}) = \mu(1) f(1) f(p^a) + \mu(p) f(p) f(p^{a-1})$$

$$= f(p^a) - f(p) f(p^{a-1}).$$

Thus $f(p^a) = f^a(p)$ by induction.   ∎

Corollary 7.11.1. If f is completely multiplicatively, then

$$\Lambda_f = f\Lambda.$$

Proof. Since f is completely multiplicative we know, by Theorem 7.11, that $f^{*-1} = \mu f$. Thus

$$\Lambda_f(n) = (f^{*-1} * Lf)(n) = (\mu f * Lf)(n)$$

$$= \sum_{d|n} \mu(d) f(d) f(n / d) \log(n / d)$$

$$= f(n) \sum_{d|n} \mu(d) \log(n / d) = f(n) \Lambda(n),$$

since $e(m) = 1$ for all positive integers m.   ∎

For further characterizations of completely multiplicative functions, see [3] and [97] as well as Problems 1 and 12 below.

Theorem 7.12. If $f \in M$, then

(a) $\displaystyle \sum_{d|n} \mu(d) f(d) = \prod_{p|n} (1 - f(p))$

and

(b) $\displaystyle \sum_{d|n} \mu^2(d) f(d) = \prod_{p|n} (1 + f(p)).$

Proof. The second follows from the first by taking $f = \mu f$.

Since μ and f are multiplicative we see that μf is multiplicative. Thus, by (a) of Theorem 7.10, we know that $e * \mu f$ is multiplicative, and so, by the remarks after Definition 7.7, we need only evaluate $e * \mu f$ on prime powers. We have

$$(e * \mu f)(p^a) = \sum_{j=0}^{a} e(p) \mu(p) f(p) = \mu(1) f(1) e(p) - \mu(p) f(p) e(p) = 1 - f(p)$$

for $a \in \mathbf{Z}^+$. Thus

$$\sum_{d|n} \mu(d)f(d) = (e*\mu f)(n) = \prod_{p^a\|n}(e*\mu f)(p^a) = \prod_{p|n}(1-f(p)),$$

which is our result. ∎

As an application of this theorem we solve Problem 19 of the previous problem set.

Example 7.5. Let n be a positive integer and let $\omega(n)$ denote the number of distinct prime divisors of n. Show that

$$\sum_{d|n}\mu^2(d) = 2^{\omega(n)}.$$

Solution. By (b) of Theorem 7.12 we have, if we take $f(n) = e(n)$,

$$\sum_{d|n}\mu^2(d) = \prod_{p|n}(1+e(p)) = \prod_{p|n}(1+1) = 2^{\omega(n)}.$$ □

We shall discuss the function $\omega(n)$ in more detail in Section 9 below.

We close this section with a final result on Λ_f when f is multiplicative, which illustrates one of the uses of Λ_f.

Theorem 7.13. Let $f \in M$. If n is divisible by more than one prime, then $\Lambda_f(n) = 0$.

Proof. Let $n = p^a m$, where $m > 1$ and $a > 0$ and $(p, m) = 1$. Then

$$\Lambda_f(p^a m) = \sum_{d|p^a m}f^{*-1}(d)f(p^a m / d)\log(p^a m / d)$$

$$= \sum_{j=0}^{a}\sum_{d|m}f^{*-1}(p^j d)f(p^{a-j}m / d)\log(p^{a-j}m / d)$$

$$= \sum_{j=0}^{a}\sum_{d|m}f^{*-1}(p^j)f^{*-1}(d)f(p^{a-j})f(m / d)\log(p^{a-j}m / d),$$

since f and f^{*-1} are in M. For the inner sum we have

$$\sum_{d|m}f^{*-1}(d)f(m / d)\log(p^{a-j}m / d)$$

$$= \sum_{d|m}f^{*-1}(d)f(m / d)\{\log(p^{a-j}m) - \log(d)\}$$

$$= \log(p^{a-j}m)\sum_{d|m}f^{*-1}(d)f(m / d) - \sum_{d|m}f^{*-1}(d)f(m / d)\log(d)$$

$$= \delta(m)\log(p^{a-j}m) - (f*Lf^{*-1})(m) = -\Lambda_{f^{*-1}}(m) = \Lambda_f(m),$$

by (b) of Theorem 7.7 and since $m > 1$ implies that $\delta(m) = 0$. Thus

$$\Lambda_f(p^a m) = \sum_{j=0}^{a} f^{*-1}(p^j) f(p^{a-j}) \Lambda_f(m) = \delta(p^a) \Lambda_f(m) = 0,$$

since $a > 0$.   ∎

One can get an expression for $\Lambda_f(p^a)$, when f is multiplicative, but it is somewhat complicated. For a derivation see [72]. In several of the problems in the problem set below we shall calculate $\Lambda_f(p^a)$ for certain special classes of $f \in \mathbf{M}$.

In the following four sections we study certain individual multiplicative functions that are quite common in number theory.

Problem Set 7.3

1. Prove that f is completely multiplicative if and only if f is multiplicative and
$$f(g * h) = (fg) * (fh),$$
   for any two arithmetic functions g and h.

2. Fix a positive integer N and let $f_N(n) = (n, N)$.
   (a) Show that f_N is multiplicative.
   (b) Evaluate $f_N(p^a)$ for any prime p and positive integer a.
   (c) Find f_N^{*-1}.
   (d) Find Λ_{f_N}.
   (e) Compute $(\mu * f_N)(p^a)$ for any prime p and positive integer a.

3. Suppose $f = f_1 f_2$, where f_1 and f_2 are arithmetic functions. If f_1 is completely multiplicative, show that $\Lambda_f = f_1 \Lambda_{f_2}$.

4. Let f and g be multiplicative functions with f completely multiplicative. We say that g is **strongly multiplicative** if $g(p^a) = g(p)$ for all primes p and positive integers a.
   (a) Prove that $f * g = e$ if and only if $g(p^a) = 1 - f(p)$ for all primes p, with $f(p) \neq 1$, and positive integers a.
   (b) Let $f(1) = 1$ and if $n > 1$, define
$$f(n) = \prod_{p|n} g(p)$$
   where $g \in A$. Compute Λ_f and f^{*-1}. Give the results in the special case when
$$f(n) = \gamma(n) = \prod_{p|n} p$$

the **core** or **square-free kernel** of n.

5. (a) Show that if f is multiplicative, then $f(mn) = f((m, n))f([m, n])$.

   (b) Show that if $f(1) = 1$ and $f(mn) = f((m, n))f([m, n])$, then f is multiplicative.

6. If f is a multiplicative function, define $\Lambda_{f,k}$ by

$$\Lambda_{f,k} = f^{*-1} * L^k f,$$

where $L^1 f = Lf$ and $L^k f = L(L^{k-1} f)$, for $k \geq 2$.

   (a) Show that

$$\Lambda_{f,k}(n) = \Lambda_{f,k-1}(n)\log n + \sum_{d|n} \Lambda_f(d)\Lambda_{f,k-1}(n/d).$$

   (b) Show that $\Lambda_{f,k}(n) = 0$ if n has more than k distinct prime factors.

(Hint: use induction.)

7. Prove directly that if f is multiplicative, then so is f^{*-1}.

8. Let f be a multiplicative function.

   (a) If n is odd, show that

$$\sum_{d|n} (-1)^{n/d} f(d) = -\sum_{d|n} f(d).$$

   (b) If n is even, $n = 2^s m$, $s \geq 1$ and m odd, then

$$\sum_{d|n} (-1)^{n/d} f(d) = \sum_{d|n} f(d) - 2f(2^s)\sum_{k|m} f(k).$$

   (c) What is the value of

$$\sum_{d|n} (1-(-1)^{n/d})f(d)?$$

9. (a) Let f be a multiplicative function and define the arithmetic function F by

$$F(n) = \sum_{\substack{d|n \\ (d,n/d)=1}} f(d).$$

   Show that F is also multiplicative.

   (b) Let f and g be multiplicative functions and define the arithmetic function, F, of two variables by

$$F(n,r) = \sum_{d|(r,n)} f(d)g(n/d).$$

   (i) Show that F is a multiplicative function of n.

   (ii) Is F a multiplicative function of r as well? If not, what further conditions are necessary?

(c) Let f and g be multiplicative functions and let k be a positive integer. Define the arithmetic function F by

$$F(n) = \sum_{d^k \mid n} f(n/d^k)g(d).$$

Show that F is multiplicative.

10. Let f be a multiplicative function and let k be a positive integer such that $f(k) \neq 0$. Show that the arithmetic function F_k defined by $F_k(n) = f(nk)/f(k)$ is multiplicative. (Hint: split the prime factors of k into those that divide only k from those that also divide n.)

11. Recall unitary convolution from Problem 14 of the previous problem set. Let f, g, h and k be arithmetic functions.

(a) Show that f satisfies

$$f(g \circ h) = (fg) \circ (fh)$$

for all g and h if and only if f is multiplicative.

(b) Suppose that f, g, h and k are multiplicative. Show that

$$(f \circ g)(h \circ k) = (fh) \circ (fk) \circ (gh) \circ (gk).$$

12. (a) Prove the converse of Corollary 7.11.1.

(b) Let g be an arithmetic function and define G by $G = g * e$. Show that if f is completely multiplicative, then

$$\sum_{d \mid n} f(d)G(d)f^{*-1}(n/d) = f(n)g(n).$$

(c) Suppose f is multiplicative and suppose the identity of (b) holds when G is a completely multiplicative function such that $G(p) \neq 1$ for all primes p. Show that f is completely multiplicative.

(For (b) and (c) see [3].)

(d) Show that if f is a multiplicative function, then f is completely multiplicative if and only if $f^{*-1}(p^a) = 0$ for all primes p and integers $a \geq 2$.

13. Let f be a multiplicative function and let $F = f * e$. Show that

$$\mu^2(n)f(n) = \sum_{\substack{d \mid n \\ (d,n/d)=1}} F(d)\mu^2(d)\mu(n/d).$$

14. Let f be an arithmetic function and let $F = f * f$. Show that if F is multiplicative, then so is f.

15. Let f be an arithmetic function such that $f(1) \neq 0$. Show that f is multiplicative if and only if for all m and n we have

$$f([m, n]/[d, e]) = f(m/d)f(n/e)$$

for all divisors d of m and e of n such that $(d, e) = (m, n)$.

16. (a) Show that if $S_k(n) = \mu^{*k}(n)$, then

$$S_k(n) = \prod_{p^e \| n} (-1)^e \binom{k}{e}.$$

   (b) Show that if $S_k^* = e * S_k$, then $S_k^*(n) = S_{k-1}(n)$.

17. Recall the definition of unitary convolution from Problem 14 of the previous problem set.

   (a) Show that δ is the identity for unitary convolution.

   (b) Find the unitary convolution inverse of $e(n)$.

7.4. SUM OF DIVISORS

Definition 7.8. Let a be a complex number. Then the **sum of divisors** function, σ_a, is defined by

$$\sigma_a(n) = \sum_{d \mid n} d^\alpha.$$

When $a = 1$ we write $\sigma_1(n) = \sigma(n)$ and when $a = 0$ we write $\sigma_0(n) = d(n)$.

Recall that we have seen both $\sigma(n)$ and $d(n)$ in the previous chapter in connection with the sum of squares problems. The function $\sigma(n)$ was used to express representations as sums of four squares and $d(n)$ appeared in the sum of two squares problem.

Note that we can write $\sigma = e * T_a e$. Since both e and $T_a e$ are multiplicative we see that σ is multiplicative. Thus to derive a formula for $\sigma_a(n)$ it suffices to compute $\sigma_a(p^a)$. We have, for $a \neq 0$,

$$\sigma_a(p^a) = \sum_{j=0}^{a} p^{j\alpha} = \sum (p^\alpha)^j = \frac{p^{\alpha(a+1)} - 1}{p^\alpha - 1}.$$

For $a = 0$ we have $d(p^a) = a + 1$. Thus

$$(7.16) \qquad \sigma_\alpha(n) = \begin{cases} \prod_{p^a \| n} \dfrac{p^{\alpha(a+1)} - 1}{p^\alpha - 1} & \text{if } \alpha \neq 0 \\[1em] \prod_{p^a \| n} (a+1) & \text{if } \alpha = 0 \end{cases}.$$

Example 7.5. Make a table of $\sigma(n)$ and $d(n)$ for $1 \le n \le 10$.

Solution. We have

$$\sigma(1) = 1, \ d(1) = 1 \qquad\qquad \sigma(6) = 12, d(6) = 4$$
$$\sigma(2) = 3, \ d(2) = 2 \qquad\qquad \sigma(7) = 8, d(7) = 2$$
$$\sigma(3) = 4, \ d(3) = 2 \qquad\qquad \sigma(8) = 15, d(8) = 4$$
$$\sigma(4) = 7, \ d(4) = 3 \qquad\qquad \sigma(9) = 13, d(9) = 3$$
$$\sigma(5) = 6, \ d(5) = 2 \qquad\qquad \sigma(10) = 18, d(10) = 4$$

Theorem 7.14. We have

(a) $\sigma_\alpha^{*-1} = \mu * T_\alpha \mu$,

(b) $\Lambda_{\sigma_\alpha} = \Lambda(e + T_\alpha e)$,

and

(c) $\sigma_{-\alpha} = T_{-\alpha} \sigma_\alpha$.

Proof of (a). We have $\sigma_\alpha = e * T_\alpha e$. Thus, by (7.2),

$$\sigma_\alpha^{*-1} = (e * T_\alpha e)^{*-1} = e^{*-1} * (T_\alpha e)^{*-1} = \mu * T_\alpha \mu.$$

Proof of (b). We have, by Theorems 7.1 and 7.2,

$$\Lambda_{\sigma_\alpha} = \sigma_\alpha^{*-1} * L\sigma_\alpha = \mu * T_\alpha \mu * L(e * T_\alpha e) = \mu * T_\alpha \mu * (e * LT_\alpha e + T_\alpha e * Le)$$
$$= T_\alpha \mu * LT_\alpha e + \mu * Le = T_\alpha(\mu * Le) + \mu * Le = \Lambda(e + T_\alpha e).$$

Proof of (c). We have

$$\sigma_{-\alpha} = e * T_{-\alpha} e = T_{-\alpha}(T_\alpha e * e) = T_{-\alpha} \sigma_\alpha,$$

which is the final result. ∎

As a special case of (b) of Theorem 7.16 take $a = 0$. Then $\sigma_\alpha = d$ and $\Lambda_d = 2\Lambda$.

We conclude this section with two old problems.

Definition 7.9. A positive integer n is said to be **perfect** if it is equal to the sum of its proper divisors, that is, those divisors not equal to n.

Example 7.6. Show that 6 and 28 are perfect numbers.

Solution. We have

$$6 = 1 + 2 + 3$$

and

$$28 = 1 + 2 + 4 + 7 + 14. \qquad \square$$

Note that n is perfect if and only if

$$n = \sum_{\substack{d|n \\ d<n}} d,$$

that is,

(7.17) $$\sigma(n) = 2n.$$

The case of even perfect numbers is solved in the following theorem.

Theorem 7.15 (Euclid-Euler). Let n be an even positive integer. Then n is a perfect number if and only if n is of the form

$$2^{p-1}(2^p - 1),$$

where $2^p - 1$ is a prime.

Proof. Suppose $2^p - 1$ is a prime and $n = 2^{p-1}(2^p - 1)$. Then, by (7.16) with $\alpha = 1$, we have

$$\sigma(n) = \frac{2^p - 1}{2 - 1} \cdot \frac{(2^p - 1)^2 - 1}{2^p - 2}$$
$$= (2^p - 1)(2^p - 2)2^p / (2^p - 2)$$
$$= 2 \cdot 2^{p-1}(2^p - 1) = 2n,$$

and so n is perfect, by (7.17).

Suppose n is perfect so that, by (7.17), $\sigma(n) = 2n$. Write $n = 2^a N$, where N is odd and a is a positive integer. Then $\sigma(n) = (2^{a+1} - 1)\sigma(N)$. Since n is perfect we have

$$2^{a+1} N = (2^{a+1} - 1)\sigma(N).$$

Since $(2^{a+1} - 1, 2^{a+1}) = 1$, we see that $\sigma(N) = 2^{a+1}q$, where q is a positive integer. Thus $N = (2^{a+1} - 1)q$, which, since $\sigma(N) = 2^{a+1}q$, implies that $\sigma(N) = N + q$, and so, since $N = (2^{a+1} - 1)q$ we have $q \mid N$. Since $a \geq 1$ we have $q < N$, and so N has at least two divisors: q and N. From $\sigma(N) = N + q$, we see that it has no other factors, and so, since $q < N$, we see that $q = 1$ and N is a prime. Since $N = (2^{a+1} - 1)q$ we see that N must be a prime of the form $2^{a+1} - 1$. Since

$$2^{AB} - 1 = (2^A - 1)(2^{A(B-1)} + \cdots + 2^A + 1).$$

if $A, B > 1$, we see that $a + 1$ must be a prime, say p, if N is to be a prime. Thus

$$n = 2^{p-1}(2^p - 1),$$

as required.   ∎

Recall that prime numbers of the form $2^p - 1$ are called **Mersenne primes**. There are some known conditions to determine when $2^p - 1$ is a prime given the prime p, but the problem is still somewhat intractable. See, for example, Corollary 3.4.2 or [47, pp. 223-225], as well as Theorem 7.55 below. It is still not known if there are infinitely many Mersenne primes, though most people believe so, as larger and larger ones keep getting found. With the criterion of the theorem the first four even perfect (indeed perfect) numbers are $6 = 2(2^2 - 1)$, $28 = 2^2(2^3 - 1)$, $496 = 2^4(2^5 - 1)$ and $8128 = 2^6(2^7 - 1)$.

While not much is known about odd perfect numbers their overall general structure was discovered by Euler. We give his result now. For more on odd perfect numbers see Problems 16 and 41 of the Additional Problems for this chapter.

Theorem 7.16. Let n be an odd perfect number. Then we can write

$$n = p^a \prod_{k=1}^{r} q_k^{2\alpha_k},$$

where p and $q_1, \ldots, q_k$ are all distinct odd primes with $p \equiv a \equiv 1 \pmod 4$.

Proof. Let $n = \prod_{k=1}^{r+1} q_k^{\beta_k}$. Since n is perfect we have

$$2n = \sigma(n) = \sigma(q_1^{\beta_1}) \cdots \sigma(q_{r+1}^{\beta_{r+1}}).$$

Since n is odd we know that $n \equiv 1$ or $3 \pmod 4$, and so $2n \equiv 2 \pmod 4$. Thus 2 divides $2n$, but 4 does not divide $2n$. This implies that only one of the $\sigma(q_k^{\beta_k})$ is even, say $\sigma(q_1^{\beta_1})$, and the rest, $\sigma(q_2^{\beta_2}), \ldots, \sigma(q_{r+1}^{\beta_{r+1}})$, are all odd.

If $1 \le k \le r + 1$, suppose $q_k \equiv 3 \pmod 4$. Then we have

$$\sigma(q_k^{\beta_k}) = 1 + q_k + \cdots + q_k^{\beta_k} \equiv 1 + 3 + \cdots + 3^{\beta_k} \pmod 4$$

$$\equiv 1 + (-1) + \cdots + (-1)^{\beta_k} \pmod 4 \equiv \begin{cases} 0 \pmod 4 \text{ if } \beta_k \text{ is odd} \\ 1 \pmod 4 \text{ if } \beta_k \text{ is even} \end{cases}.$$

Since we know that $4 \nmid 2n$ we see that if $q_k \equiv 3 \pmod 4$, then we must have that β_k is even. However, we know that $\sigma(q_1^{\beta_1})$, is even, and so we must have $q_1 \equiv 1 \pmod 4$.

If $1 \le k \le r + 1$, suppose $q_k \equiv 1 \pmod 4$. Then we have

$$\sigma(q_k^{\beta_k}) = 1 + q_k + \cdots + q_k^{\beta_k} \equiv 1 + \cdots + 1^{\beta_k} \equiv \beta_k + 1 \pmod 4.$$

Again $2 \le k \le r + 1$ implies that $\sigma(q_k^{\beta_k})$ is odd, and so β_k must be even in this case also. However, $\sigma(q_1^{\beta_1}) \equiv 2 \pmod 4$, and so we must have $\beta_1 \equiv 1 \pmod 4$.

If we relabel the primes, we obtain the result as stated. ∎

Corollary 7.16.1. If n is an odd perfect number, then we may write

$$n = p^a m^2,$$

where $p \equiv 1 \pmod 4$ is a prime, $a \equiv 1 \pmod 4$ and $p \nmid m$.

Proof. From Theorem 7.16 we have

$$n = p^a \prod_{k=1}^{r} q_k^{2\alpha_k} = p^a \left(\prod_{k=1}^{r} q_k^{\alpha_k} \right)^2 = p^a m^2,$$

where $p \nmid m$, since p is distinct from the q_k's. The rest follows from Theorem 7.16. ∎

Corollary 7.16.2. If n is an odd perfect number, then $n \equiv 1 \pmod 4$.

Proof. From Corollary 7.16.1 we have $n = p^a m^2$, where $p \equiv 1 \pmod 4$ and $p \nmid m,$. Since n is odd we see that m is odd, and so $m^2 \equiv 1 \pmod 4$. Thus

$$n = p^a m^2 \equiv 1^a \cdot 1 \pmod 4.$$ ∎

Definition 7.10. Let n be a positive integer. We say that n is **abundant** if $\sigma(n) > 2n$ and **deficient** if $\sigma(n) < 2n$.

We now show that if n is an odd perfect number, then it must have at least three distinct prime divisors.

Theorem 7.17. Suppose that n is odd and $n = p^a q^b$, where $a, b \ge 0$ and $q > p$. Then n is deficient.

Proof. We have, by the multiplicativity of s,

$$\frac{\sigma(n)}{n} = \frac{\sigma(p^\alpha q^\beta)}{p^\alpha q^\beta} = \frac{\sigma(p^\alpha)\sigma(q^\beta)}{p^\alpha q^\beta} = \frac{(1 + p + \cdots + p^\alpha)(1 + q + \cdots + q^\beta)}{p^\alpha q^\beta}$$

$$= \frac{p^{\alpha+1} - 1}{p^\alpha(p-1)} \cdot \frac{q^{\beta+1} - 1}{q^\beta(q-1)} = \frac{p - p^{-\alpha}}{p-1} \cdot \frac{q - q^{-\beta}}{q-1} < \frac{p}{p-1} \cdot \frac{q}{q-1}.$$

Since n is odd and $q > p$ we see that $p \ge 3$ and $q \ge 5$. Thus

$$\frac{\sigma(n)}{n} < \frac{3}{3-1} \cdot \frac{5}{5-1} = \frac{3}{2} \cdot \frac{5}{4} = \frac{15}{8} < 2$$

or $\sigma(n) < 2n$, that is, n is deficient.   ∎

Corollary 7.17.1. If n is an odd perfect number, then n has at least 3 distinct prime divisors.

Proof. If n has less than three prime divisors, then, by Theorem 7.17, n is deficient. Thus n must have at least three prime divisors.   ∎

The argument of Theorem 7.17 can be carried further. See Problem 16 of the Additional Problems below.

One can also generalize the notion of a perfect number to that of a multiply perfect number. Let k be a positive integer greater than 1 and let n be a positive integer. Then we say that n is a **multiply perfect number of order k** if and only if

$$\sigma(n) = kn.$$

Perfect numbers are of order 2 and 120 is of order 3. For a few of the known results see Problem 14 below. See also Problem 15 of the Additional Problems for this chapter.

Definition 7.11. Two positive integers m and n are said to form an **amicable pair** if

$$\sigma(m) = \sigma(n) = m + n.$$

Example 7.7 (Pythagoras). Show that 220 and 284 form an amicable pair.

Solution. We have $220 = 2^2 \cdot 5 \cdot 11$ and $284 = 2^2 \cdot 71$. Thus

$$\sigma(220) = 504 = \sigma(284) = 220 + 284.$$

To generate more pairs one can use the following old (ca. 800 AD) result.

Theorem 7.18 (Thabit ben Korrah). If $h = 3 \cdot 2^n - 1$, $t = 3 \cdot 2^{n-1} - 1$ and $s = 9 \cdot 2^{2n-1} - 1$ are all primes, then $M = 2^n ht$ and $N = 2^n s$ are an amicable pair.

Proof. We have

$$\sigma(M) = \sigma(2^n)\sigma(h)\sigma(t) = (2^{n+1} - 1)(3 \cdot 2^n)(3 \cdot 2^{n-1}) = 9 \cdot 2^{2n-1}(2^{n+1} - 1),$$

$$\sigma(N) = \sigma(2^n)\sigma(s) = (2^{n+1} - 1)(9 \cdot 2^{2n-1})$$

and

$$M + N = 2^n ht + 2^n s = 2^n \{(3 \cdot 2^n - 1)(3 \cdot 2^{n-1} - 1) + 9 \cdot 2^{2n-1} - 1\}$$
$$= 2^n \{9 \cdot 2^{2n} + 1 - 3 \cdot 2^n - 3 \cdot 2^{n-1} - 1\}$$
$$= 3 \cdot 2^{2n-1} \{3 \cdot 2^{n+1} - 2 - 1\} = 9 \cdot 2^{2n-1}(2^{n+1} - 1),$$

which gives the result. ∎

For more on amicable pairs see Problem 25 below as well as [25, vol. I, pp. 38-50], [33] and [67]. The first part of the last article gives a history of amicable numbers up to the time the article was written. The last two parts give a list of all known amicable numbers and their discoveries, as of the time that the article was written.

Problem Set 7.4

1. (a) Show that

$$\sum_{n=1}^{N} \frac{1}{n} \le 1 + \log N.$$

   (b) Show that

$$\sigma(n) \le n(1 + \log n).$$

2. Show that if a and b are positive integers with $b > 1$, then

$$\frac{\sigma(a)}{a} \le \frac{\sigma(ab)}{ab} \le \frac{\sigma(a)\sigma(b)}{ab}.$$

3. (a) For $k > 0$, show that for a prime p, $p \mid n$, we have

$$\sigma_k(pn) = \sigma_k(n)\sigma_k(p) - p^k \sigma_k(n/p).$$

   This gives a recursion relation for $\sigma_k(pn)$.

   (b) Show that

$$\sigma_k(m)\sigma_k(n) = \sum_{d \mid (m,n)} d^k \sigma_k(mn/d^2).$$

4. Show that $\sigma(n) = n + 1$ if and only if n is a prime.

5. Show that if $(m, n) > 1$, then $\sigma_\alpha(mn) < \sigma_\alpha(m)\sigma_\alpha(n)$.

6. Show that if n is composite, then $\sigma(n) > n + \sqrt{n}$.

7. If n is a positive integer, show that the equation $\sigma(m) = n$ has at most a finite number of solutions.

8. (a) Show that $\sigma_k(m)$ is odd if and only if n is a square or twice a square. What happens when $k = 0$?

   (b) Find all n such that $\sigma(n)$ is a square.

9. Show that

$$\sigma(n) = \sum_{m=1}^{n} \int_0^m \cos\left\{\frac{2\pi n[x+1]}{m}\right\} dx.$$

10. (a) Show that

$$\sum_{d\mid n} \mu^2(d)/d = \sum_{d^2\mid n} \mu(d)\sigma(n/d^2).$$

   (b) Show that $\Lambda * \sigma_\alpha = T_\alpha e * Le$.

11. Prove that 6 is the only square-free perfect number.

12. Show that if p is the least prime divisor of the perfect number n, then n has at least p distinct prime divisors.

13. Let n be a perfect number.

   (a) Show that

$$\sigma_2(n) = 2n \prod_{p^\alpha \| n} \frac{p^{\alpha+1}+1}{p+1}.$$

   (b) Show that $\sigma_{-1}(n) = 2$.

14. For $k \geq 2$, let $P_k = \{n \in \mathbf{Z}^+ : \sigma(n) = kn\}$.

   (a) Show that 120, 672 and 523776 are in P_3, 30240 and 32760 are in P_4 and 14182439040 is in P_5.

   (b) Show that if $n \in P_3$ and $3 \nmid n$, then $3n \in P_4$.

   (c) Show that if $n \in P_5$ and $5 \nmid n$, then $5n \in P_6$.

   (d) Can you generalize (b) and (c)?

   (e) If $p = (2^{n+3}-1)/(2^n+1)$ is a prime, show that

$$3 \cdot 2^{n+2} p \in P_3.$$

15. If m and n are an amicable pair, prove that

$$\sigma(m)\sigma(n) \geq 4mn$$

When is there equality?

16. Let $n \geq 28$ be an even perfect number. Show that there is an integer M such

that

$$n = \sum_{k=0}^{M} (2k+1)^3.$$

17. Let $\sigma^o(n)$ denote the sum of the odd divisors of n.

    (a) Show that

    $$\sigma^o(n) = -\sum_{d|n} (-1)^{n/d} d.$$

    (b) If n is even, show that $\sigma^o(n) = \sigma(n) - 2\sigma(n/2)$.

    (Hint: Problem 8 of Problem Set 7.3.)

    (c) Show that $\sigma^o(n)$ is a multiplicative function.

18. (a) Show that

    $$\sigma_\alpha^2(n) = \sum_{d|n} d^\alpha \sigma_\alpha(n^2/d^2).$$

    (b) If $n = \displaystyle\prod_{k=1}^{r} p_k^{\alpha_k}$, show that

    $$\sum_{d|n} \mu(d)\sigma(d) = (-1)^r p_1 \cdots p_r.$$

19. (a) If m is perfect or abundant and $n \geq 2$, then mn is abundant.

    (b) If n is deficient and $d \mid n$, then d is deficient.

20. (a) For what values of n is $21 \cdot 5^n$ abundant?

    (b) For what values of n is $2^n m$ abundant? (Here m is odd.)

21. (a) Show that there are infinitely many even abundant numbers.

    (b) Show that there are infinitely many odd abundant numbers.

    (Hint: consider the multiples of 945.)

22. Show that if n is an even perfect, then n ends in a 6 or a 28. (Hint: if $n = 2^{k-1}(2^k - 1)$, then divide into two cases: $k \equiv 1 \pmod 4$ and $k \equiv 3 \pmod 4$.)

23. Show that a square can never be a perfect number.

24. Find all positive integers n such that $\sigma^{*-1}(n) = 2n$.

25. Let k, m and n be positive integers with $k < m$. Let

    $$f = 2^k + 1 \quad \text{and} \quad g = 2^{m-k} f^2.$$

If the following numbers are all primes,

$$r_1 = f2^{m-k} - 1, \ r_2 = f2^m - 1, \ p = g(2^{m+1} - 1) + 1$$

$$q_1 = p^n[(2^m - 1)g + 2] - 1 \quad \text{and} \quad q_2 = 2^m p^n g[(2^m - 1)g + 2] - 1,$$

show that the two numbers

$$m_1 = 2^m p^n r_1 r_2 q_1 \quad \text{and} \quad m_2 = 2^m p^n q_2$$

are an amicable pair.

7.5. NUMBER OF DIVISORS

In the previous section we saw one function that counts the number of divisors of an integer n. In this section we will discuss a generalization of this function.

Definition 7.12. If n and k are positive integers, then the function $d_k(n)$ counts the number of ways of writing n as a product of k factors, where the order of the factors counts.

Clearly $d_1(n) = 1 = e(n)$ and $d_2(n) = d(n)$.

Example 7.8. Compute $d_k(200)$ for $k = 1, 2, 3, 4,$ and 5.

Solution. We have $200 = 2^3 \cdot 5^2$. Thus

$d_1(200) = e(200) = 1,$

$d_2(200) = (3 + 1)(2 + 1) = 12, \quad \text{by} \ (7.16),$

$d_3(200) = 60,$

$d_4(200) = 200,$

and

$d_5(200) = 525.$

As one can see from Example 7.9 $d_k(n)$ tends to increase with k. The problem is how to calculate $d_k(n)$ for any k and any n. First we show that d_k is multiplicative and then we evaluate $d_k(p^a)$.

Note that $d_{k+1}(n)$ counts the number of ways of writing n in the form $a_1 \cdots a_k a_{k+1}$. If we fix $a_{k+1} | n$ then n / a_{k+1} can be written as a product of k factors in $d_k(n / a_{k+1})$ ways. Thus

(7.18)
$$d_{k+1}(n) = \sum_{j|n} d_k(j),$$

for any $k \geq 1$. An easy induction argument shows that

$$d_k = e^{*k}$$

for any $k \geq 1$. Since e is multiplicative we see that d_k is multiplicative.

We have $d_1(p^a) = 1$ and $d_2(p^a) = a + 1$. We may rewrite these as

$$d_1(p^a) = \binom{a}{0} \quad \text{and} \quad d_2(p^a) = \binom{a+1}{1}.$$

With this observation we can compute $d_k(p^a)$ by using (7.18). First, we require a lemma on a binomial coefficient sum.

Lemma 7.19.1. Let m and n be nonnegative integers. Then

(7.19)
$$\sum_{j=0}^{m} \binom{j+n}{n} = \binom{m+n+1}{n+1}.$$

Proof. We proceed by induction on m. If $m = 0$, then we have

$$\sum_{j=0}^{m} \binom{j+n}{n} = \binom{n}{n} = 1 = \binom{0+n+1}{n+1}.$$

Assume (7.19) holds for $0 \leq m \leq k - 1$. Then

$$\sum_{j=0}^{k} \binom{j+n}{n} = \sum_{j=0}^{k-1} \binom{j+n}{n} + \binom{k+n}{n} = \binom{k-1+n+1}{n+1} + \binom{k+n}{n}$$

$$= (k+n)! \left\{ \frac{1}{(n+1)!(k-1)!} + \frac{1}{n!k!} \right\} = \frac{(k+n)!}{(k-1)!n!} \left\{ \frac{1}{n+1} + \frac{1}{k} \right\}$$

$$= \binom{k+n+1}{n+1}$$

which is (7.19) for $m = k$. ∎

Theorem 7.19. If $n = \displaystyle\prod_{i=1}^{r} p_i^{a_i}$ as a canonical factorization, then, for $k \geq 1$,

$$d_k(n) = \prod_{i=1}^{r} \binom{a_i + k - 1}{k - 1}.$$

Proof. It suffices, since d_k is multiplicative, to show that

(7.20)
$$d_k(p^a) = \binom{a+k-1}{k-1}.$$

We know this is true for $k = 1$ and $k = 2$ and we proceed by induction on k.

Assume (7.20) holds. Then, by (7.18) and (7.20),

$$d_{k+1}(p^a) = \sum_{j=0}^{a} d_k(p^j) = \sum_{j=0}^{a} \binom{j+k-1}{k-1} = \binom{a+k}{k},$$

which is (7.20) for $k + 1$, by Lemma 7.19.1.   ∎

Theorem 7.20. We have

    (a) $d_k^{*-1} = \mu^{*k}$

and

    (b) $\Lambda_{d_k} = k\Lambda.$

Proof of (a). Since $d_k = e^{*k}$, the result follows from (7.2) and Definition 7.5.

Proof of (b). This follows from (a) of Theorem 7.7 since $d_k = e^{*k}$ or form (a) since, by Definition 7.6,

$$\Lambda_{d_k} = d_k^{*-1} * L d_k = \mu^{*k} * L e^{*k} = \mu^{*k} * k e^{*(k-1)} * Le = k(\mu * Le) = k\Lambda,$$

by (e) of Theorem 7.2.   ∎

We now discuss a class of numbers called highly composite numbers. These were first introduced by Ramanujan in [96]. We say that a positive number $n, n > 1$, is **highly composite** if $d(m) < d(n)$ for all $m < n$. Since that time much research has been done on these numbers and there are still things that are not completely understood.

Example 7.10. Find the first five highly composite numbers.

Solution. We have, from Example 7.6, $d(2) = 2 = d(3) < d(4) = 3$, and so the first two are 2 and 4. Since $d(5) = 2 < 3$ we must go on. From the table we see that the next highly composite number is $n = 6$ since $d(6) = 4$, but there are no more ≤ 10. We have $d(11) = 2$ and $d(12) = 6$, so that 12 is highly composite. Finally, if we search through the larger table in the appendix we see that the fifth highly composite number is 24.   □

If we examine a more extensive table of highly composite numbers, such as appears in Ramanujan's paper, the following result becomes apparent.

Theorem 7.21. If $n = 2^{a_2} 3^{a_3} \cdots p_k^{a_k}$, with $a_k \geq 1$, is a highly composite number, where $2, 3, \ldots, p_k$ are the first primes, then $a_2 \geq a_3 \geq \cdots \geq a_k \geq 1$.

Proof. Suppose that some prime p is missing in the factorization of n, that is, we have

$$n = 2^{a_2} 3^{a_3} \cdots p^0 \cdots p_k^{a_k},$$

where $p < p_k$. Let $m = 2^{a_2} 3^{a_3} \cdots p^{a_k} \cdots p_{k-1}^{a_{k-1}}$. Then, by (7.16),

$$d(m) = (a_2 + 1)(a_3 + 1) \cdots (a_k + 1) \cdots (a_{k-1} + 1)$$
$$= (a_2 + 1)(a_3 + 1) \cdots (0 + 1) \cdots (a_{k-1} + 1)(a_k + 1) = d(n).$$

Since $m < n$ we see that this contradicts the fact that n is a highly composite number. Thus all of the primes in the sequence $2, 3, \ldots, p_k$ occur in the prime decomposition of n.

If the exponents do not form a decreasing sequence, then there is some pair of them, say a_r and a_s, $r < s$, with $a_r < a_s$. Let $a_r = t$ and $a_s = t + u$, $u \geq 1$. Then

$$n = 2^{a_2} 3^{a_3} \cdots p_r^t \cdots p_s^{t+u} \cdots p_k^{a_k}$$

and if we let

$$m = 2^{a_2} 3^{a_3} \cdots p_r^{t+u} \cdots p_s^t \cdots p_k^{a_k}.$$

we see that $d(n) = d(m)$, whereas $m < n$. This contradicts the fact that n is a highly composite number. Thus $a_2 \geq a_3 \geq \cdots \geq a_k$. ∎

There are many other properties of highly composite numbers that Ramanujan proved based on the table that he calculated. Some are fairly straight-forward (see Problem 20 below as well as Problem 45 of the Additional Problems below), while others require more knowledge about the distribution of primes (see Problem 13 of Problem Set 9.2.)

Problem Set 7.5

1. Let $D(n)$ denote the number of square-free divisors of n that are quadratic residues of the prime p. If n has r distinct prime divisors show that

$$D(n) = \begin{cases} 2^r & \text{if } d|n \text{ implies } \left(\dfrac{d}{p}\right) = +1 \\ 2^{r-1} & \text{else} \end{cases}.$$

2. (a) Show that if $n = 33$, 85 or 93, then

$$d(n) = d(n + 1) = d(n + 2) = 4.$$

   (b) Show that there do not exist any positive integers n such that

$$d(n) = d(n + 1) = d(n + 2) = d(n + 3) = 4.$$

3. (a) Show that

$$d(n) = 1 + \sum_{d=1}^{n-1} \left\{ \left[\frac{n}{d} \right] - \left[\frac{n-1}{d} \right] \right\}.$$

(b) Show that

$$d(n) = [\sqrt{n}] - [\sqrt{n-1}] + 2 \sum_{d=1}^{[\sqrt{n-1}]} \left\{ \left[\frac{n}{d} \right] - \left[\frac{n-1}{d} \right] \right\}.$$

4. (a) Show that $\sigma(n) / d(n) \le (n+1) / 2$ with equality if and only if n is a prime.
   (Hint: consider $f(x) = (x + n / x) / 2$ and show that $f''(x) > 0$.)

(b) Show that

$$\frac{\sigma(n)}{d(n)} \ge \frac{\left(\sqrt{n} - 1 \right)^2}{d(n)} + \sqrt{n}$$

with equality if and only if n is a prime or the square of a prime.

(c) Let p be the least prime dividing a perfect number n. Show that

$$p \le 1 + (\log n / \log 4).$$

(Hint: see Problem 12 of Problem Set 7.4.)

(d) Generalize (a) and (b) to the quotient $\sigma_k(n) / d(n)$.

5. (a) A function f is said to be **submultiplicative** if and only if $f(mn) \le f(m)f(n)$. Show that d is submultiplicative.

(b) Suppose that f is nonnegative and submultiplicative and that $F = f*e$. Show that

$$F(n) / d(n) \ge \sqrt{f(n)}.$$

(Hint: consider the arithmetic-geometric mean inequality.)

6. (a) Suppose n is an even perfect number. Show that

$$\prod_{d|n} d = n^k$$

where $n = 2^{k-1}(2^k - 1)$, with $2^k - 1$ is a prime.

(b) Show that

(i) $\prod_{d|n} d = n^{d(n)/2}$

(ii) $\prod_{d|n} d^{d(d)\mu(n/d)} = n^2.$

7. (a) Let f be a nonnegative completely multiplicative function. Show that

$$\prod_{d|n} f(d) = f(n)^{d(n)/2}.$$

(b) Find all n such that

$$\prod_{d|n} d = n^k$$

for $k = 1, 2$ and 3.

(c) Show that

$$\prod_{d|n} d = \prod_{k|m} k$$

implies that $m = n$.

8. Let $d^o(n)$ denote the number of odd divisors of n. If we write $n = 2^s m$, m odd and $s \geq 0$, show that $d^o(n) = d(m)$.

9. (a) Show that $d(2^n - 1) \geq d(n)$.

(b) Show that $d(2^n + 1) > d^o(n)$.

10. Prove that if $d(n)$ is divisible by an odd prime, then $\mu(n) = 0$.

11. (a) Show that $d(n) \leq 2\sqrt{n}$.

(b) Show that $n^k \leq \sigma_k(n) \leq n^k d(n) \leq 2n^{k+1/2}$.

12. Show that there are infinitely many n such that if $m > 1$, then $d(n) = m$.

13. (a) Find all n such that $n = d(n)$.

(b) Find all n such that $n = d_3(n)$.

14. If f is an arithmetic function and $F = f*e$, show that

$$\prod_{d|n} f(d) \leq (F(n)/d(n))^{d(n)}.$$

(Hint: consider the arithmetic-geometric mean inequality.)

15. Prove the following identities.

(a) $\displaystyle\sum_{k|n} d^2(k)\mu(n/k) = \sum_{k|n} \mu^2(k)d(n/k).$

(b) $\sigma*e = T_1 e*d.$

(c) $\displaystyle\sum_{k|n} d^3(k) = \left(\sum_{k|n} d(k)\right)^2.$

(d) $d^2(n) = (d_3*\mu^2)(n).$

(e) $d(n^2) = (d*\mu^2)(n).$

(f) $\Lambda * d_k = d_{k-1} * Le$.

(g) $\sigma_\mu * \sigma_\mu = T_\mu d * d$.

16. (a) Show that if $m \mid n$, then $d_k(m) \le d_k(n)$ for $k \ge 2$.

(b) For $k \ge 2$ show that $d_k(n) \le d_{k-1}(n)d(n)$.

(c) For $k \ge 2$ show that $d_k(n) \le d^{k-1}(n)$.

17. Show that an arithmetic function f is completely multiplicative if and only if $f * f = fd$.

18. Suppose n has r distinct prime divisors. Show that

$$d(n) \le \left(\frac{1}{r} \log \left(n \prod_{p \mid n} p \right) \right)^r \bigg/ \prod_{p \mid n} \log p.$$

19. (a) Find all integers n so that $d(n) = 21$.

(b) Find the least integer n so that $d(n) = 21$.

(c) Repeat (a) and (b) with $d(n) = 10$ and $d(n) = 105$.

20. This exercise has to do with highly composite numbers.

(a) Show that if n is highly composite and $d(m) > d(n)$, then there is at least one highly composite number k such that $n < k \le m$.

(b) Show that if m and n, $m < n$, are highly composite numbers, then $d(k) \le d(m)$ for all k with $m \le k \le n$.

(c) Show that if n is a highly composite integer, then there is a highly composite integer m such that $n < m \le 2n$.

(d) Find all of the highly composite number of the forms $2^a 3^b$ and $2^a 3^b 5^c$.

(e) Let n be a highly composite number of the form $2^a 3^b 5^c m$, where $m > 1$. Noting that $5n/6 < n$ show that

$$1 + \frac{2}{c} < \left(1 + \frac{1}{a} \right)\left(1 + \frac{1}{b} \right).$$

21. Let m be a positive integer. Show that

$$\left((k+j-1)! \right)^{\omega(m)} \sum_{r \mid m} d_k(r) d_j(m/r) = \prod_{i=0}^{k+j-2} d(\gamma^i(m)m).$$

where $\gamma(m)$ denotes the core of m and $\omega(m)$ denotes the number of distinct prime divisors of m.

22. Show that if m, n and k are positive integers, then

$$d_k(m)d_k(n) \le d_k(mn).$$

7.6. EULER'S FUNCTION

In this section we discuss the Euler function, $\varphi(n)$, which counts the number of numbers relatively prime to n and less than n. Thus

(7.21)
$$\varphi(n) = \sum_{\substack{1 \le d \le n \\ (d,n)=1}} 1.$$

In Chapter Two we proved several properties of $\varphi(n)$ and in this section we shall reprove some of them again, though from our present viewpoint of arithmetic functions.

Theorem 7.22.

(a) $\varphi = \mu * T_1 e.$

(b) φ is a multiplicative function.

(c) If n is a positive integer, then

$$\varphi(n) = n \prod_{p|n} \left(1 - \frac{1}{p}\right).$$

Proof of (a). By (7.21) and the definition of μ we have

(7.22)
$$\varphi(n) = \sum_{d=1}^{n} \sum_{k|(n,d)} \mu(k) = \sum_{d=1}^{n} \sum_{\substack{k|n \\ k|d}} \mu(k).$$

For a fixed divisor k of n we must sum over those d in the range $1 \le d \le n$ which are multiples of k. If we let $d = qk$, then $1 \le d \le n$ if and only if $1 \le q \le n/k$. Thus we can write (7.22) as

$$\varphi(n) = \sum_{k|n} \sum_{q=1}^{n/k} \mu(k) = \sum_{k|n} \mu(k) \sum_{q=1}^{n/k} 1 = \sum_{k|n} \mu(k)(n/k) = (\mu * T_1 e)(n).$$

Proof of (b) This follows immediately from (a) since both μ and $T_1 e$ are multiplicative.

Proof of (c) Since φ is multiplicative we need only evaluate $\varphi(p^a)$. From (a) we have, by (b) of Theorem 7.5,

$$\varphi(p^a) = \sum_{j=0}^{a} \mu(p^j) p^{a-j} = \mu(p^0) p^a + \mu(p) p^{a-1} = p^a - p^{a-1} = p^a \left(1 - \frac{1}{p}\right).$$

The result follows. ∎

Example 7.11. Compute $\varphi(n)$ for $n = 1, \ldots, 20$.

Solution.

$$
\begin{array}{llll}
\varphi(1) = 1 & \varphi(2) = 1 & \varphi(3) = 2 & \varphi(4) = 2 \\
\varphi(5) = 4 & \varphi(6) = 2 & \varphi(7) = 6 & \varphi(8) = 4 \\
\varphi(9) = 6 & \varphi(10) = 4 & \varphi(11) = 10 & \varphi(12) = 4 \\
\varphi(13) = 12 & \varphi(14) = 6 & \varphi(15) = 8 & \varphi(16) = 8 \\
\varphi(17) = 16 & \varphi(18) = 6 & \varphi(19) = 18 & \varphi(20) = 8
\end{array}
$$

Example 7.12. Determine the value of the sum

$$\sum_{\substack{1 \le d \le n \\ (d,n)=1}} d.$$

Solution. This is a generalization of the definition of $\varphi(n)$. We can proceed as in the proof of Theorem 7.22 or we can just use the definition of $\varphi(n)$. We will give both proofs.

As in the proof of (a) of Theorem 7.22 we have

$$\sum_{\substack{1 \le d \le n \\ (d,n)=1}} d = \sum_{d=1}^{n} d \sum_{k \mid n} \mu(k) = \sum_{d=1}^{n} d \sum_{\substack{k \mid n \\ k \mid d}} \mu(k) = \sum_{k \mid n} k\mu(k) \sum_{q=1}^{n/k} q$$

$$= \frac{1}{2} \sum_{k \mid n} k\mu(k) \left\{ \left(\frac{n}{k}\right)^2 + \left(\frac{n}{k}\right) \right\} = \frac{n}{2} \sum_{k \mid n} \mu(k)(n/k) + \frac{n}{2} \sum_{k \mid n} \mu(k)$$

$$= \begin{cases} n\varphi(n)/2 & \text{if } n > 1 \\ 1 & \text{if } n = 1 \end{cases},$$

where we used both (a) of Theorem 7.22 and Definition 7.5.

If we just want to use (7.21), then note that, for $n > 1$, we have $(0, n) = n > 1$ and $(n, n) = n > 1$. Also note that $(d, n) = (n - d, n)$. Thus

$$2 \sum_{\substack{1 \le d \le n \\ (d,n)=1}} d = \sum_{\substack{1 \le d \le n \\ (d,n)=1}} d + \sum_{\substack{1 \le d \le n \\ (d,n)=1}} d = \sum_{\substack{0 \le d \le n \\ (d,n)=1}} d + \sum_{\substack{0 \le d \le n \\ (d,n)=1}} d$$

$$= \sum_{\substack{0 \le d \le n \\ (d,n)=1}} d + \sum_{\substack{0 \le d \le n \\ (d,n)=1}} (n-d) = \sum_{\substack{0 \le d \le n \\ (d,n)=1}} n = \sum_{\substack{1 \le d \le n \\ (d,n)=1}} n = n\varphi(n)$$

and the result follows since as d runs through 0 to n so does $n - d$ (though in backward order.)   □

Theorem 7.23.

    (a) $\varphi(mn) = \varphi(m)\varphi(n)(d/\varphi(d))$, where $d = (m, n)$.

    (b) If $a \mid b$, then $\varphi(a) \mid \varphi(b)$.

    (c) $\varphi(n)$ is even for all $n \geq 3$.

    (d) If n has r distinct odd prime factors, then $2^r \mid \varphi(n)$.

*Proof of (a).*We have, by (c) of Theorem 7.22,

$$\frac{\varphi(mn)}{mn} = \prod_{p \mid mn}\left(1 - \frac{1}{p}\right) = \prod_{p \mid m}\left(1 - \frac{1}{p}\right)\prod_{p \mid n}\left(1 - \frac{1}{p}\right) / \prod_{p \mid (m,n)}\left(1 - \frac{1}{p}\right)$$

$$= (\varphi(m)/m)(\varphi(n)/n)/(\varphi(d)/d),$$

which gives the result.

Proof of (b). If $a \mid b$, then there is an integer c such that $b = ac$, with $1 \leq c \leq b$. If $c = b$, then $a = 1$, and so (b) is satisfied since $\varphi(1) = 1$. Assume $c < b$. Then

$$\varphi(b) = \varphi(ac) = \varphi(a)\varphi(c)d / \varphi(d) = d\varphi(a)(\varphi(c)/\varphi(d)),$$

where $d = (a, c)$. The result now follows by induction on b. If $b = 1$, then the result holds trivially. Assume (b) holds for all integers less than b. Then it holds for c and since $d \mid c$ we see that $\varphi(c)/\varphi(d)$ is an integer. Thus $\varphi(a) \mid \varphi(b)$, which proves (b).

Proof of (c). If $n = 2^a, a \geq 2$, then, by (c) of Theorem 7.19, we have

$$\varphi(n) = 2^a - 2^{a-1} = 2^{a-1},$$

which is even. If n has at least one odd prime factor we write

(7.23) $$\varphi(n) = n\prod_{p \mid n}\frac{p-1}{p} = \frac{n}{\prod_{p \mid n}p}\prod_{p \mid n}(p-1) = c(n)\prod_{p \mid n}(p-1),$$

where $c(n)$ is an integer. Since at least one of the $p \mid n$ is odd, we see that $\prod_{p \mid n}(p-1)$ is even and so is $\varphi(n)$. This proves (c).

Proof of (d). From (7.23) we see that each odd prime factor contributes to $\varphi(n)$ a factor of 2. Thus $2^r \mid \varphi(n)$, if n has r distinct odd prime factors. This proves (d). ■

Theorem 7.23. Let n be a positive integer. Then

    (a) $\sum_{d \mid n}\varphi(d) = n$

and

(b) $\displaystyle\sum_{d\mid n}\varphi(d)d(n/d)=\sigma(n).$

Proof. We prove both of these results by rewriting them in terms of convolutions and using the fact that $\varphi=\mu*T_1 e$ ((a) of Theorem 7.22). We have

$$\sum_{d\mid n}\varphi(d)=(e*\varphi)(n)=(e*\mu*T_1 e)(n)=(T_1 e)(n)=n,$$

by Definition 7.5 and (a) of Theorem 7.2, and

$$\sum_{d\mid n}\varphi(n)d(n/d)=(\varphi*d)(n)=(\mu*T_1 e*e*e)(n)=(e*T_1 e)(n)=\sigma(n),$$

by Definition 7.5, (a) of Theorem 7.2 and Definition 7.8.   ∎

Theorem 7.24. We have

(a) $\varphi^{*-1}=e*T_1\mu$

and

(b) $\Lambda_\varphi=-\Lambda+T_1\Lambda.$

Proof of (a). We have, by (7.2),

$$\varphi=\mu*T_1 e \text{ if and only if } \varphi^{*-1}=e*T_1\mu,$$

which gives the result.

Proof of (b) By (a), (b) and (e) of Theorem 7.7, we have

$$\Lambda_\varphi=\Lambda_{\mu*T_1 e}=\Lambda_\mu+\Lambda_{T_1 e}=-\Lambda+T_1\Lambda,$$

which is the second result.   ∎

Problem Set 7.6

1. If f is a multiplicative function, show that f is completely multiplicative if and only if $T_1 f*f^{*-1}=\varphi f$.

2. Show that if $n_1,\ldots,n_k$ are distinct even perfect numbers, then

$$\varphi(n_1\cdots n_k)=2^{k-1}\varphi(n_1)\cdots\varphi(n_k).$$

   Can anything similar be said about odd perfect numbers?

3. Show that if $(m,n)>1$, then $\varphi(mn)>\varphi(m)\varphi(n)$.

4. Show that n is a prime if and only if $\varphi(n)\mid n-1$ and $n+1\mid\sigma(n)$. Find counterexamples to show that both conditions are necessary.

5. (a) Show that if m and n are positive integers, then

$$\varphi(mn) + \varphi((m+1)(n+1)) < 2mn.$$

(b) Show that if $n \neq 4$, then

$$\varphi(n)d^2(n) \leq n^2,$$

with equality if and only if $n = 1, 2, 8$ or 12.

(c) Show that $\varphi(n)d(n) \geq n$. When is there equality?

6. Suppose a and n are integers such that $a > 1$ and $n > 0$.
   (a) Show that $n \mid \varphi(a^n - 1)$. (Hint: Euler's Theorem.)
   (b) Show that $2n \mid \varphi(a^n + 1)$.

7. Show that if $d \mid n$, $0 < d < n$, then $n - \varphi(n) > d - \varphi(d)$.

8. (a) If m and k are positive integers, prove that the number of positive integers $\leq mk$ that are relatively prime to m is $k \varphi(m)$.

   (b) If r, d and k are integers with $d \mid k$, $d > 0$, $(r, d) = 1$ and $k \geq 1$, show that the number of integers in the progression $r + td$, $1 \leq t \leq k/d$, such that $(r + td, k) = 1$ is $\varphi(k)/\varphi(d)$.

9. Find all positive integers n such that $\varphi(n) \mid n$.

10. (a) Find all n such that $\varphi(n) = 24$ and $\varphi(n) = 72$.
    (b) Develop a method to solve $\varphi(n) = N$, for general N.
    (c) Show that $\varphi(n) = 14$ is insoluble.

11. (a) Show that if n is composite, then $\varphi(n) \leq n - \sqrt{n}$.
    (b) Show that if $n > 2$, then $\sqrt{n/2} < \varphi(n) < n$.

12. Show that if

$$\varphi_k(n) = \sum_{\substack{1 \leq d \leq n \\ (d,n)=1}} d^k,$$

then

$$\sum_{d \mid m} (m/d)^3 \varphi_3(d) = \left\{ \sum_{d \mid m} (m/d)\varphi(d) \right\}^2.$$

13. Evaluate the following sums.

(a) $\displaystyle\sum_{d \mid n} d^\alpha \varphi^\beta(d)$

(c) $\displaystyle\sum_{d \mid n} (-1)^{n/d} \varphi(d)$

(b) $\displaystyle\sum_{d \mid n} d^\alpha \varphi(d^\beta)$

(d) $\displaystyle\sum_{d \mid n} \mu^2(d)\varphi^2(d)$

14. Let

$$F(n) = \sum_{k=1}^{n} (k,n).$$

Evaluate F in terms of the canonical factorization of n. (Hint: show that

$$F(n) = \sum_{d|n} d\varphi(n/d). \Big)$$

15. Prove the following identities.

   (a) $\varphi(m)\varphi(n) = \sum_{d|(m,n)} \varphi(mn/d)\mu(d)$

   (b) $\varphi * \sigma_k = T_1 \sigma_{k-1}$

   (c) $\sigma_\mu = T_{\mu-1}\varphi * \sigma_{\mu-1}$

   (d) $n/\varphi(n) = \sum_{d|n} \mu^2(d)/\varphi(d)$

   (e) $\Lambda * T_1 e = Le * \varphi$

16. Show that

$$\sigma(n) + \varphi(n) = nd(n)$$

   if and only if n is a prime.

17. Show that

$$\sum_{d|n} \varphi(d)\mu(d) = 0$$

   if and only if n is even.

18. Show that if n is a positive integer, then

$$\varphi(n)\sigma(n) > n^2 \prod_{p|n} \left(1 - p^{-2}\right)$$

19. Show that if $(a, b) = 1$ and a, b and c are positive integers, then

$$\varphi(abc)\varphi(c) = \varphi(ac)\varphi(bc).$$

20. Show that if n is a positive integer, then

$$\prod_{d|n} \varphi(d) \le (n/d(n))^{d(n)}.$$

21. Show that

$$\prod_{\substack{1 \le d \le n \\ (d,n)=1}} d = n^{\varphi(n)} \prod_{d|n} \left(d!/d^d\right)^{\mu(n/d)}.$$

22. Prove that the following congruence holds. If $n \geq 3$ is an odd integer, then, for any integer a such that $(a, n) = 1$,

$$a^{\varphi(n)/2} \equiv \pm 1 \pmod{n}.$$

23. This exercise introduces a generalization of the Euler function called the **Jordan totient function**, J_k. If n and k are positive integers, then we define

$$J_k(n) = \#\{(n_1, \ldots, n_k) \in \mathbf{Z}^{+k} : n_t \leq n, (n_t, n) = 1, t = 1, \ldots, k\}.$$

Show that the following are true.

(a) $\displaystyle\sum_{d|n} J_k(n) = n^k$

(b) J_k is multiplicative.

(c) $\displaystyle J_k(n) = n^k \prod_{p|n}\left(1 - \frac{1}{p^k}\right)$

(d) $\displaystyle J_k(n) = n^k \sum_{d|n} \mu(d) / d^k$

(e) If $d = (m, n)$, then $J_k(mn) = J_k(m)J_k(n)d^k / J_k(d)$.

What other properties of φ can be generalized to J_k?

24. Show that

$$n \left| \sum_{d|n} a^{n/d} \varphi(d) \right.$$

for any integer a.

25. Use the formula for $\varphi(n)$ to prove that there are infinitely many primes. (Hint: if there are only finitely many primes, let P be their product and calculate $\varphi(P)$.)

7.7. CHARACTERS

In this section we discuss a whole class of multiplicative functions.

Definition 7.13. Let k be a positive integer. Then χ is said to be a **character** modulo k if and only if

(1) χ is completely multiplicative,

(2) $\chi(a) = \chi(b)$, whenever $a \equiv b \pmod{k}$,

and

(3) $\chi(n) = 0$, whenever $(n, k) > 1$.

If $\chi(n) = 1$, whenever $(n, k) = 1$, we call χ the **principal character** modulo k and denote it by χ_0. A character is said to be **real** if it only takes on real values and **complex** otherwise.

We might mention that we have seen characters before in this text. The first time was in Chapter Three with the Legendre symbol. Theorem 3.1 gives the properties of the Legendre symbol and shows that $\left(\frac{n}{p}\right)$ is a character modulo p. Also the Jacobi and Kronecker symbols are characters. Finally, in the section on sums of two squares, Section 6.3, we introduced a character in order to calculate $r_2(n)$. As pointed out it was really a Jacobi symbol. All of these characters are examples of real characters.

Note that (2) of Definition 7.13 allows us to extend the domain of χ from $\mathbf{Z}^+$ to $\mathbf{Z}$.

Theorem 7.25. If $(n, k) = 1$ and χ is a character modulo k, then $\chi(n)$ is a $\varphi(k)\underline{\text{th}}$ root of unity.

Proof. By Euler's theorem (Theorem 2.8) we know that

$$n^{\varphi(k)} \equiv 1 \ (\text{mod } k),$$

if $(n, k) = 1$. Thus, by (1) and (2) of Definition 7.13, we have

$$\chi^{\varphi(k)}(n) = \chi(n^{\varphi(k)}) = \chi(1) = 1,$$

which gives the result.   ∎

Thus if χ is a real character it can only assume the values -1, 0, or 1.

We now can prove a fundamental result in the theory of characters.

Lemma 7.26.1. If $d > 0$, $(d, k) = 1$ and $d \not\equiv 1 \ (\text{mod } k)$, then there exists a character χ modulo k such that $\chi(d) \neq 1$.

Proof. First suppose that the canonical factorization of k contains an odd prime power, p^e, say. Then, by Theorem 2.22, there is a primitive root, g, modulo p^e. If $(a, k) = 1$, then $a \equiv g^t \ (\text{mod } p^e)$, $0 \le t < \varphi(p^e) = s$, where t is uniquely determined by a. Let

$$\rho = \exp(2\pi i / s) = \cos(2\pi / s) + i \sin(2\pi / s),$$

be a primitive $s\underline{\text{th}}$ root of unity. We define the character χ modulo k by $\chi(a) = \rho^t$, for $(a, k) = 1$, and $\chi(a) = 0$, otherwise. To show this defines a character we must show that conditions (1), (2), and (3) of Definition 7.13 hold.

By the definition of χ we see that (3) is fulfilled.

We have $\chi(1) = \rho^0 = 1$. Suppose $(a', k) = 1$ and $a' \equiv g^{t'} \pmod{p^e}$. Then $aa' \equiv g^{t+t'} \pmod{p^e}$, and so

$$\chi(aa') = \rho^{t+t'} = \rho^t \rho^{t'} = \chi(a)\chi(a').$$

Thus (1) holds.

If $a \equiv a' \pmod{k}$, then $a \equiv a' \pmod{p^e}$, and so $t = t'$, that is, $\chi(a) = \chi(a')$, which is condition (2). If $d \not\equiv 1 \pmod{k}$ and there exists some odd prime power p^e such that $d \not\equiv 1 \pmod{p^e}$ with $p^e \mid k$, then for the character defined above we have $\chi(d) \neq 1$, since ρ is a primitive root of unity.

Now suppose $p^e = 2^e$, where $e > 2$, since $k = 2$ and $k = 4$ possess primitive roots and we can proceed as above. By Theorem 2.26 we know that if $a = 2n + 1$, then there exists a uniquely determined integer c, $0 \leq c < 2^{e-2}$, such that

$$a \equiv (-1)^n 5^c \pmod{2^e}.$$

With ρ as above, with $s = 2^e$, we define $\chi(2n + 1) = \rho^c$ and $\chi(2n) = 0$. We see that $\chi(1) = 1$ and $\chi(a) = 0$ if $(a, k) > 1$. If $a \equiv a' \pmod{k}$, then $c = c'$ and so $\chi(a) = \chi(a')$, Thus to show χ is a character modulo k we must show that $\chi(aa') = \chi(a)\chi(a')$ for any integers a and a'. If either a or a' is even, then the result holds, and so we may assume both to be odd. Thus $(a-1)(a'-1) \equiv 0 \pmod{4}$, and so, by Lemma 3.10.1,

$$\tfrac{1}{2}(a-1) + \tfrac{1}{2}(a'-1) \equiv \tfrac{1}{2}(aa'-1) \pmod{2}.$$

Thus

$$aa' \equiv (-1)^{(aa'-1)/2} 5^{c+c'} \pmod{2^e},$$

and so

$$\chi(a)\chi(a') = \rho^c \rho^{c'} = \rho^{c+c'} = \chi(aa').$$

By construction, we see that $\chi(a) = 1$ if and only if $c = 0$, and so when $a \equiv (-1)^{(a-1)/2} \pmod{2^e}$. If $d \equiv 1 \pmod{4}$, but $d \not\equiv 1 \pmod{2^e}$, then we have $\chi(d) \neq 1$. If $d \equiv -1 \pmod{4}$, then we define our character by $\chi_1(2n+1) = (-1)^n$ and $\chi_1(2n) = 0$ so that $\chi_1(d) = -1$. The proof that χ_1 is a character is clear. Thus in either case we have constructed a character modulo k so that $\chi(d) \neq 1$. ∎

Note that if χ_1 and χ_2 are characters modulo k so is $\chi_1 \chi_2$ Also if $\overline{\chi}$ denotes the complex conjugate, then $\chi\overline{\chi}(a) = 1$.

Any character χ modulo k is completely determined by its values $\chi(n)$, $1 \le n \le k$, $(n, k) = 1$, and each of these is a $\varphi(k)$th root of unity. Since there are $\varphi(k)$ roots of unity of degree $\varphi(k)$ we see that Z, the number of characters modulo k, satisfies the inequalities

$$0 \le Z \le \varphi(k)^{\varphi(k)}.$$

Theorem 7.26. Let k be a positive integer. Then

(a) $\displaystyle\sum_{a(k)} \chi(a) = \begin{cases} 0 & \text{if } \chi \ne \chi_0 \\ \varphi(k) & \text{if } \chi = \chi_0 \end{cases},$

where the sum is over a complete residue system modulo k;

(b) $\displaystyle\sum_{\chi(k)} \chi(a) = \begin{cases} Z & \text{if } a \equiv 1 \ (\text{mod } k) \\ 0 & \text{if } a \not\equiv 1 \ (\text{mod } k) \end{cases},$

where the sum is over all Z characters modulo k;

(c) $Z = \varphi(k)$;

and

(d)   if a and b are integers, $(b, k) = 1$, then

$$\sum_{\chi(k)} \chi(a)\overline{\chi}(b) = \begin{cases} \varphi(k) & \text{if } a \equiv b \ (\text{mod } k) \\ 0 & \text{if } a \not\equiv b \ (\text{mod } k) \end{cases}.$$

Proof of (a). If $\chi = \chi_0$, then

$$\sum_{a(k)} \chi(a) = \sum_{a(k)} \chi_0(a) = \sum_{\substack{a(k) \\ (a,k)=1}} 1 = \varphi(k).$$

If $\chi \ne \chi_0$ then there exists some integer b, $(b, k) = 1$, such that $\chi(b) = 1$. Now as a runs through a complete residue system modulo k so does ab (Corollary 2.3.2). Thus

$$\sum_{a(k)} \chi(a) = \sum_{a(k)} \chi(ab) = \chi(b) \sum_{a(k)} \chi(a)$$

or

$$\{1 - \chi(b)\} \sum_{a(k)} \chi(a) = 0.$$

Since $\chi(b) \ne 1$, the result follows.

Proof of (b). If $a \equiv 1 \ (\text{mod } k)$, then $\chi(a) = 1$ for all χ, and so we have

$$\sum_{\chi(k)} \chi(a) = \sum_{\chi(k)} 1 = Z,$$

the number of distinct characters modulo k. If $(a, k) > 1$, then the second part is true since $\chi(a) = 0$ for all χ modulo k. So suppose $a \not\equiv 1 \pmod{k}$ and $(a, k) = 1$. Then, by Lemma 7.26.1, there exists a character χ_1 modulo k such that $\chi_1(a) \neq 1$. Now as χ runs through all characters modulo k so does $\chi\chi_1$, since $\chi\chi_1(b) = \chi'\chi_1(b)$ and $\chi_1(b) \neq 0$ implies that $\chi(b) = \chi'(b)$. Thus

$$\sum_{\chi(k)} \chi(a) = \sum_{\chi(k)} (\chi\chi_1)(a) = \chi_1(a) \sum_{\chi(k)} \chi(a)$$

or

$$\{1 - \chi_1(a)\} \sum_{\chi(k)} \chi(a) = 0.$$

Since $\chi_1(a) \neq 1$ the result follows.

Proof of (c). We have

$$Z = \sum_{a(k)} \sum_{\chi(k)} \chi(a) = \sum_{\chi(k)} \sum_{a(k)} \chi(a) = \varphi(k).$$

Proof of (d). Let c be defined by $bc \equiv 1 \pmod{k}$, which exists since $(b, k) =. 1$. Then $\chi(b)\chi(c) = \chi(bc) = 1$, and so $\chi(c) = \overline{\chi}(b)$. Thus

$$\sum_{\chi(k)} \chi(a)\overline{\chi}(b) = \sum_{\chi(k)} \chi(a)\chi(c) = \sum_{\chi(k)} \chi(ac).$$

Now, by (b), the last sum is

$$\begin{cases} \varphi(k) & \text{if } ac \equiv 1 \pmod{k} \\ 0 & \text{if } ac \not\equiv 1 \pmod{k} \end{cases}.$$

Since $ac \equiv 1 \pmod{k}$ if and only if $a \equiv b \pmod{k}$ the result follows and proves (d). ∎

Example 7.13. Find all characters modulo 3, 4 and 5.

Solution. Since the characters are determined by their values on a complete residue system we shall only give the values for $0 \leq n \leq k - 1$.

Modulo 3. Since $\varphi(3) = 2$ we know there will be two characters modulo 3. One of them is the principal character χ_0:

$$\chi_0(0) = 0, \chi_0(1) = 1 \text{ and } \chi_0(2) = 1.$$

The other character is easily seen to be

$$\chi_1(0) = 0, \chi_1(1) = 1 \text{ and } \chi_1(2) = -1.$$

Modulo 4. Again, since $\varphi(4) = 2$, there will be two characters modulo 4. They are given by

$$\chi_0(0) = 0, \chi_0(1) = 1, \chi_0(2) = 0 \text{ and } \chi_0(3) = 1$$

and

$$\chi_1(0) = 0, \chi_1(1) = 1, \chi_1(2) = 0 \text{ and } \chi_1(3) = -1.$$

Modulo 5. Here $\varphi(5) = 4$ and so there will be four distinct characters modulo 5. One of them is the principal character χ_0:

$$\chi_0(0) = 0, \chi_0(1) = 1, \chi_0(2) = 1, \chi_0(3) = 1 \text{ and } \chi_0(4) = 1.$$

Another is the Legendre symbol $\left(\frac{n}{5}\right) = \chi_1(n)$ and it is given by

$$\chi_1(0) = 0, \chi_1(1) = 1, \chi_1(2) = -1, \chi_1(3) = -1 \text{ and } \chi_1(4) = 1.$$

If we use the construction of Lemma 7.26.1 and note that 5 has $\varphi(\varphi(5)) = 2$ primitive roots (Theorem 2.22), for example 2 and 3, then we arrive at the following two complex characters modulo 5:

$$\chi_2(0) = 0, \chi_2(1) = 1, \chi_2(2) = i, \chi_2(3) = -i \text{ and } \chi_2(4) = -1.$$

and

$$\chi_3(0) = 0, \chi_3(1) = 1, \chi_3(2) = -i, \chi_3(3) = i \text{ and } \chi_3(4) = -1. \qquad \square$$

One may also construct particular characters modulo k by considering the multiplication table for the reduced residues modulo k and remembering Theorem 7.25.

For example, in the above example, note that $2^2 \equiv 4 \equiv 1 \pmod 3$, and so $\chi(1) = \chi(2^2) = \chi(2)^2$. Thus $\chi(2) = \pm 1$. The different choices give the two characters χ_0 and χ_1 that were listed above since $\chi(0) = 0$ and $\chi(1) = 1$ no matter what. Similar remarks hold for the characters modulo 4.

For the characters modulo 5 things get a bit more complicated. Again we are forced to have $\chi(0) = 0$ and $\chi(1) = 1$. Since

$$1 = \chi(1) = \chi(16) = \chi(4^2) = \chi(4)^2$$

for any character modulo 5 we see that $\chi(4) = \pm 1$. If $\chi(4) = +1$, then since $\chi(2^2) = \chi(3^2) = \chi(4)$ we see that $\chi(2)$ and $\chi(3)$ are either $+1$ or -1. Since

$$1 = \chi(1) = \chi(6) = \chi(3)\chi(2)$$

we see that $\chi(2) = \chi(3)$ and we get the two characters χ_0 and χ_1 listed first. If $\chi(4)$ = -1, then we see that $\chi(2)$ and $\chi(3)$ must be $\pm i$. Now, as above, $\chi(2)\chi(3) = 1$, and so one is $+i$ and $-i$. This gives us the two characters χ_2 and χ_3 listed second.

Similar arguments can be carried out for any modulus.

The following theorem is immediate from Theorem 7.11 and its Corollary, since characters are completely multiplicative.

Theorem 7.27.
   (a) We have $\chi^{*-1} = \mu\chi$.
   (b) We have $\Lambda_\chi = \chi\Lambda$.

We finish this section with the following result which will be useful in proving Dirichlet's theorem on primes in arithmetic progressions (see Section 4 of Chapter 9).

Theorem 7.28. Let χ be any real character and let $A = e*\chi$. Then $A(n) \geq 0$ for all n and $A(n) \geq 1$ if n is a square.

Proof. Since e and χ are multiplicative we see that A is multiplicative, by (a) of Theorem 7.10. Thus we need only evaluate A on prime powers. For prime powers we have

$$A(p^a) = \sum_{t=0}^{a} \chi(p^t) = 1 + \sum_{t=1}^{a} \chi^t(p).$$

Since χ is real valued its only possible values are 0, -1 and +1. If $\chi(p) = 0$, then $A(p^a) = 1$. If $\chi(p) = 1$, then $A(p^a) = a + 1$. If $\chi(p) = -1$, then

$$A(p^a) = \begin{cases} 0 & \text{if } a \text{ is odd} \\ 1 & \text{if } a \text{ is even} \end{cases}.$$

In any case $A(p^a) \geq 1$ if a is even and in all cases $A(p^a) \geq 0$. If $n = p_1^{a_1} \cdots p_r^{a_r}$, then $A(n) = A(p_1^{a_1}) \cdots A(p_r^{a_r}) \geq 0$, since each factor is nonnegative. If n is a square, then all the a_i, $1 \leq i \leq r$, are even. Thus $A(n) = A(p_1^{a_1}) \cdots A(p_r^{a_r}) \geq 1$. ∎

We close this section by noting that the concept of character can be generalized to an arbitrary group instead of just a reduced residue system in the integers.

Problem Set 7.7

1. Construct the characters modulo n, $6 \leq n \leq 10$.

2. Let χ and ψ be two characters modulo k. Show that

$$\sum_{a(k)} \chi(a)\overline{\psi}(a) = \begin{cases} \varphi(k) & \text{if } \chi = \psi \\ 0 & \text{else} \end{cases}.$$

3. An arithmetical function f is said to be **periodic** modulo k if $k > 0$ and $f(m) = f(n)$ whenever $m \equiv n \pmod{k}$. The integer k is called the **period** of f. Prove the following results.

   (a) If f is periodic modulo k, then f has a smallest period k_0 and $k_0 \mid k$.

   (b) Let f be periodic modulo k and completely multiplicative. Let k be the least period of f. Then $f(n) = 0$ if $(n, k) > 1$.

4. Let f be a character modulo k. If k is square-free, then k is the smallest period of f.

5. Let χ be a real character modulo k and let

$$S = \sum_{n=1}^{k} n\chi(n).$$

   (a) Show that if $(a, k) = 1$, then $a\chi(a)S \equiv S \pmod{k}$.

   (b) Let $k = 2^\alpha q$, where q is odd. Show that there exists an integer a, $(a, k) = 1$, such that $a \equiv 3 \pmod{2^\alpha}$ and $a \equiv 2 \pmod{q}$.

   (c) Show that $12S \equiv 0 \pmod{k}$.

6. Let f be an arithmetic function that is periodic modulo k. Show that if f is completely multiplicative, then it is a root of unity.

7. Let f be a periodic completely multiplicative function of period h. Show that there is a period k such that

$$f(n) = 0 \text{ if } (n, k) > 1 \text{ and } f(n) \neq 0 \text{ if } (n, k) = 1.$$

8. If χ is a real character modulo m, show that there is an integer h so that $h \equiv 0$ or $1 \pmod{4}$ and

$$\chi(a) = \left(\tfrac{h}{a}\right),$$

   the Kronecker symbol.

9. Let $k = 2^\alpha p_1^{\alpha_1} \cdots p_r^{\alpha_r}$. Show that all characters modulo k are real if and only if each of $\varphi(2^\alpha), \varphi(p_1^{\alpha_1}), \ldots, \varphi(p_r^{\alpha_r})$ equals 1 or 2. Show that this holds if and only if $k = 1, 2, 3, 4, 6, 8, 12$ or 24.

10. Let N denote the number of solutions of the congruence

$$x^n \equiv a \pmod{p},$$

where $n \mid p - 1$. Show that

$$N = \begin{cases} 1 & \text{if } a \equiv 0 \ (\text{mod } p) \\ \displaystyle\sum_{\chi(p)}{}^{*} \chi(a) & \text{if } a \not\equiv 0 \ (\text{mod } p) \end{cases},$$

where $\displaystyle\sum_{\chi(p)}{}^{*}$ denotes a sum over all those characters χ such that $\chi^{n} = \chi_{0}$.

11. Show that if $\chi \neq \chi_{0}$ is a character modulo k and m and n are positive integers with $n \geq m$, then

$$\left| \sum_{m \leq j \leq r} \chi(j) \right| \leq \tfrac{1}{2} \varphi(k).$$

12. Suppose that j is an integer such that $(k, j) = 1$. Show that

$$\sum_{\chi(k)} \frac{\chi(m)}{\chi(j)} = \begin{cases} \varphi(k) & \text{if } m \equiv j \ (\text{mod } k) \\ 0 & \text{if } m \not\equiv j \ (\text{mod } k) \end{cases}.$$

13. Suppose k_{1} and k_{2} are positive integers with $(k_{1}, k_{2}) = 1$. Let be a character modulo $k_{1}k_{2}$. Show that there are unique characters χ_{1} modulo k_{1} and χ_{2} modulo k_{2} such that $\chi(n) = \chi_{1}(n)\chi_{2}(n)$ for all integers n.

14. Let χ_{1} be a character modulo m_{1} and χ_{2} be a character modulo m_{2}. We say that χ_{1} and χ_{2} are **equivalent** if $(n, m_{1}m_{2}) = 1$ implies that $\chi_{1}(n) = \chi_{2}(n)$. Show that this is an equivalence relation. We denote it by $\chi_{1} \sim \chi_{2}$.

15. Let χ be a character modulo m. If χ is equivalent to a character χ_{1} modulo k, where $k < m$, then we say that χ is an **improper** character. A **proper** character is a nonprincipal character which is not improper. If χ is a nonprincipal character, then the **conductor** f of χ is defined by

$$f = \min \{ k \geq 2 : \chi \sim \chi_{1}, \ \chi_{1} \text{ a character modulo } k \}.$$

Prove the following results.

(a) If χ_{1} is a character modulo m_{1} and χ_{2} is a character modulo m_{2} with $(m_{1}, m_{2}) = 1$ and $\chi_{1} \sim \chi_{2}$, then there exists a character χ modulo k such that $\chi \sim \chi_{1}$.

(b) Any nonprincipal character χ modulo m is equivalent to a unique proper character χ_{1} modulo k with $k \mid m$.

(c) Given any character χ modulo m and an integer $n = dm$, where d is a positive integer, then there exists a character χ_{1} modulo n such that $\chi_{1} \sim \chi$.

(d) Let $m = p_1^{\alpha_1} \cdots p_r^{\alpha_r}$ as the canonical representation. Then χ is a proper character modulo m if and only if the restriction of χ to the reduced residue system of $p_k^{\alpha_k}$ is a proper character for each k, $1 \le k \le r$.

(e) If k is a square-free odd integer, then the only real proper characters modulo k are the Jacobi symbols

$$\chi(n) = \left(\tfrac{n}{k}\right).$$

7.8. TRIGONOMETRICAL SUMS

In this section we discuss a class of arithmetic functions that have wide applications in advanced number theory.

Definition 7.14. Let f be an arithmetic function and let k be a positive integer, $k \ge 2$. If we let

$$S_{f,k} = \sum_{m=1}^{k} f(m) e^{2\pi i m n / k},$$

then $S_{f,k}$ is called a trigonometric (exponential) sum.

Note that if f is periodic of period k, then we have

$$S_{f,k} = \sum_{m(k)} f(m) e^{2\pi i m n / k},$$

where the sum is over any complete residue system modulo k.

We wish to discuss the elementary properties of three trigonometric sums.

Definition 7.15. Let χ be a character modulo k. Then the **Gauss sum** is a sum of the form $G(\chi, n)$

$$G(\chi, n) = \sum_{m(k)} \chi(m) e^{2\pi i m n / k}.$$

In particular the **special Gauss sum** is the sum $G_p(n)$

$$G_p(n) = \sum_{m(k)} \left(\tfrac{m}{p}\right) e^{2\pi i m n / p},$$

where p is an odd prime and $\left(\tfrac{m}{p}\right)$ is the Legendre symbol.

Theorem 7.29. If $p \nmid n$, then

$$G_p(n) = \sum_{m=0}^{p-1} e^{2\pi i n m^2 / p}.$$

Proof. We can write

(7.24)
$$G_p(n) = \sum_{\left(\frac{m}{p}\right)=+1} e^{2\pi i m n / p} - \sum_{\left(\frac{m}{p}\right)=-1} e^{2\pi i m n / p}.$$

Since $e^{2\pi i m n / p}$, $0 \le m \le p - 1$, runs through all the pth roots of unity we have

(7.25)
$$0 = \sum_{m=0}^{p-1} e^{2\pi i m n / p} = 1 + \sum_{\left(\frac{m}{p}\right)=+1} e^{2\pi i m n / p} + \sum_{\left(\frac{m}{p}\right)=-1} e^{2\pi i m n / p}.$$

If we add (7.24) and (7.25), we obtain

$$G_p(n) = 2 \sum_{\left(\frac{m}{p}\right)=+1} e^{2\pi i m n / p} + 1 = 1 + 2 \sum_{m=1}^{(p-1)/2} e^{2\pi i n m^2 / p} = \sum_{m=0}^{p-1} e^{2\pi i n m^2 / p},$$

since $(p - m)^2 \equiv m^2 \pmod{p}$. ∎

Theorem 7.30. Let p be an odd prime and $p \nmid n$. Then

$$G_p(n) = \left(\tfrac{n}{p}\right) G_p(n).$$

Proof. We have, since $\left(\tfrac{n}{p}\right)^2 = 1$ because $p \nmid n$,

$$G_p(n) = \sum_{m(p)} \left(\tfrac{m}{p}\right) e^{2\pi i m n / p} = \left(\tfrac{n}{p}\right) \sum_{m(p)} \left(\tfrac{mn}{p}\right) e^{2\pi i m n / p} = \left(\tfrac{n}{p}\right) \sum_{r(p)} \left(\tfrac{r}{p}\right) e^{2\pi i r / p} = \left(\tfrac{n}{p}\right) G_p(1),$$

by Corollary 2.3.2. ∎

We can explicitly evaluate $G_p(1)$. To do this it is easier to expand the definition of the Gauss sum. Let k be an odd positive integer. We define the **general Gauss sum** by

$$G_k = \sum_{r=0}^{k-1} e^{2\pi i r^2 / k}.$$

Note the when k is a prime, then we just have $G_p(1)$ by Theorem 7.29.

Definition 7.16. Let m and n be positive integers. We define the **Gaussian binomial coefficients**, for $x \ne 1, -1$, by

$$\begin{bmatrix} n \\ m \end{bmatrix} = \frac{(1-x^n)(1-x^{n-1})\cdots(1-x^{n-m+1})}{(1-x)(1-x^2)\cdots(1-x^m)}.$$

We define $\begin{bmatrix} n \\ m \end{bmatrix} = 0$ if $m > n$ and $\begin{bmatrix} n \\ 0 \end{bmatrix} = 1$.

Theorem 7.31. We have, for $n \geq m$,

(a) $\begin{bmatrix} n \\ m \end{bmatrix} = \begin{bmatrix} n \\ n-m \end{bmatrix}$;

(b) $\begin{bmatrix} n \\ m \end{bmatrix} = \begin{bmatrix} n-1 \\ m-1 \end{bmatrix} + x^m \begin{bmatrix} n-1 \\ m \end{bmatrix}$.

Proof of (a). We have

$$\begin{bmatrix} n \\ m \end{bmatrix} = \frac{(1-x^n)(1-x^{n-1})\cdots(1-x^{n-m+1})}{(1-x)(1-x^2)\cdots(1-x^m)}$$

$$= \frac{(1-x^n)\cdots(1-x^{n-m+1})(1-x^{n-m})\cdots(1-x)}{(1-x)\cdots(1-x^m)(1-x)\cdots(1-x^{n-m})} = \frac{(1-x^n)\cdots(1-x^{m+1})}{(1-x)\cdots(1-x^{n-m})} = \begin{bmatrix} n \\ n-m \end{bmatrix}.$$

Proof of (b). We have

$$\begin{bmatrix} n \\ m \end{bmatrix} = \frac{1-x^n}{1-x^m} \cdot \frac{(1-x^{n-1})\cdots(1-x^{n-1-(m-1)+1})}{(1-x)\cdots(1-x^{m-1})}$$

(7.26)
$$= \frac{1-x^n}{1-x^m} \begin{bmatrix} n-1 \\ m-1 \end{bmatrix} = \left(1 + \frac{x^m - x^n}{1-x^m}\right)\begin{bmatrix} n-1 \\ m-1 \end{bmatrix} \qquad ■$$

$$= \left(1 + x^m \frac{1-x^{n-m}}{1-x^m}\right)\begin{bmatrix} n-1 \\ m-1 \end{bmatrix} = \begin{bmatrix} n-1 \\ m-1 \end{bmatrix} + x^m \begin{bmatrix} n-1 \\ m \end{bmatrix}.$$

Corollary 7.31.1. If $n \geq m$, then $\begin{bmatrix} n \\ m \end{bmatrix}$ is a polynomial in x.

Proof. We have

$$\begin{bmatrix} 1 \\ 0 \end{bmatrix} = 1 \text{ and } \begin{bmatrix} 1 \\ 1 \end{bmatrix} = \frac{1-x}{1-x} = 1.$$

Also

$$\begin{bmatrix} 2 \\ 0 \end{bmatrix} = 1, \begin{bmatrix} 2 \\ 1 \end{bmatrix} = \frac{1-x^2}{1-x} = 1 + x \text{ and } \begin{bmatrix} 2 \\ 2 \end{bmatrix} = \frac{(1-x^2)(1-x)}{(1-x)(1-x^2)} = 1.$$

The result follows by induction using (7.26). ■

Theorem 7.32. We have

$$\lim_{x \to 1} \begin{bmatrix} n \\ m \end{bmatrix} = \binom{n}{m},$$

the usual binomial coefficient.

Proof. If $m > n$, then both $\begin{bmatrix} n \\ m \end{bmatrix}$ and $\begin{pmatrix} n \\ m \end{pmatrix}$ are zero. If $m = 0$, then $\begin{bmatrix} n \\ 0 \end{bmatrix} = 1 = \begin{pmatrix} n \\ 0 \end{pmatrix}$. Thus we may assume $n \geq m \geq 1$.

We have

$$\begin{bmatrix} n \\ m \end{bmatrix} = \frac{(1-x)\cdots(1-x)}{(1-x)\cdots(1-x)} = \frac{(1-x)(1+x+\cdots+x^{n-1})\cdots(1-x)(1+x+\cdots+x^{n-m})}{(1-x)\cdots(1-x)(1+x+\cdots+x^{m-1})}$$

$$= \frac{(1+x+\cdots+x^{n-1})\cdots(1+x+\cdots+x^{n-m})}{(1+x)\cdots(1+x+\cdots+x^{m-1})}.$$

Thus

$$\lim_{x \to 1} \begin{bmatrix} n \\ m \end{bmatrix} = \frac{n\cdots(n-m)}{2\cdots m} = \begin{pmatrix} n \\ m \end{pmatrix}.$$ ∎

Thus we may consider the Gaussian binomial coefficient as a generalization of the regular binomial coefficients.

Note that, by Theorem 7.31, we have

(7.27)
$$\begin{bmatrix} n \\ m \end{bmatrix} = \begin{bmatrix} n \\ n-m \end{bmatrix} = \begin{bmatrix} n-1 \\ m \end{bmatrix} + x^{n-m}\begin{bmatrix} n-1 \\ m-1 \end{bmatrix}.$$

Define the polynomial $f(x, m)$ by

$$f(x,m) = \sum_{k=0}^{m} (-1)^k \begin{bmatrix} m \\ k \end{bmatrix}.$$

Then, by (7.27),

$$f(x,m) = \sum_{k=0}^{m} (-1)^k \left\{ \begin{bmatrix} m-1 \\ k \end{bmatrix} + x^{m-k}\begin{bmatrix} m-1 \\ k-1 \end{bmatrix} \right\}$$

(7.28)
$$= \sum_{k=0}^{m} (-1)^k \begin{bmatrix} m-1 \\ k \end{bmatrix} + \sum_{k=0}^{m-1} (-1)^{k+1} x^{m-k-1}\begin{bmatrix} m-1 \\ k \end{bmatrix}$$

$$= \sum_{k=0}^{m} (-1)^k \begin{bmatrix} m-1 \\ k \end{bmatrix}(1 - x^{m-k-1}).$$

Now

(7.29)
$$(1-x^{m-j})\begin{bmatrix} m-1 \\ j-1 \end{bmatrix} = \frac{(1-x^{m-j})\cdots(1-x^{m-j+1})}{(1-x)\cdots(1-x^{j-1})}$$

$$= (1-x^{m-1})\frac{(1-x^{m-2})\cdots(1-x^{m-2-(j-1)+1})}{(1-x)\cdots(1-x^{j-1})} = (1-x^{m-1})\begin{bmatrix} m-2 \\ j-1 \end{bmatrix}.$$

If we combine (7.28) and (7.29), we see that

$$(7.30) \qquad f(x,m) = \sum_{k=0}^{m-2} (-1)^k (1-x^{m-1}) \begin{bmatrix} m-2 \\ k \end{bmatrix} = (1-x^{m-1})f(x,m-2).$$

Theorem 7.33. We have, for $m \geq 1$,

$$\text{(a)} \quad f(x,2m) = \prod_{k=0}^{m-1} (1-x^{2k+1})$$

and

(b) $f(x, 2m + 1) = 0$.

Proof of (a). By definition we have

$$f(x,0) = (-1)^0 \begin{bmatrix} 0 \\ 0 \end{bmatrix} = 1.$$

Thus, by (7.30),

$$f(x,2) = (1-x)f(x,2-2) = (1-x)f(x,0) = 1 - x = \prod_{k=0}^{0} (1-x^{2k+1}).$$

The result then follows by mathematical induction.

Proof of (b). We have

$$f(x,1) = (-1)^0 \begin{bmatrix} 1 \\ 0 \end{bmatrix} + (-1)^1 \begin{bmatrix} 1 \\ 1 \end{bmatrix} = 1 - 1 = 0.$$

Thus, by (7.30) and mathematical induction, the result follows.   ∎

Lemma 7.34.1. We have, for x not an integer and k an odd integer,

$$\frac{\sin(2\pi kx)}{\sin(2\pi x)} = (2i)^{k-1} \prod_{j=1}^{(k-1)/2} \sin(2\pi(x+\tfrac{j}{k}))\sin(2\pi(x-\tfrac{j}{k})).$$

Proof. Instead of providing a purely trigonometric proof of this identity we will use the following fact about polynomials: if two polynomials have the same degree, same leading coefficients and same zeros, then the two polynomials are identical. For a purely trigonometric proof see Problem 4 below.

Recall that $e^{it} - e^{-it} = 2i\sin t$, where t is real (or complex, but we are only interested in real t.) Let, for x real,

$$p(x) = e^{2\pi ix} - e^{-2\pi ix} = (z^2 - 1)/z,$$

where $z = e^{2\pi ix}$. Also note that $p(x) = 2i\sin(2\pi x)$.

We have

(7.31) $$e^{2\pi ikx}p(kx) = e^{4\pi ikx} - 1 = z^{2k} - 1.$$

The roots of this polynomial are $z_j = e^{2\pi ij/2k}$, $0 \le j \le 2k - 1$.

Consider

$$e^{2\pi ikx}p(x) \prod_{j=1}^{(k-1)/2} p(x + j/k)p(x - j/k)$$

$$= z^k((z^2 - 1)/z) \prod_{j=1}^{(k-1)/2} e^{-4\pi ix}(e^{4\pi i(x+j/k)} - 1)(e^{4\pi i(x-j/k)} - 1)$$

(7.32) $$= z^{k-1}(z^2 - 1) \prod_{j=1}^{(k-1)/2} z^{-2}(z^2 e^{4\pi ij/k} - 1)(z^2 e^{-4\pi ij/k} - 1)$$

$$= (z^2 - 1) \prod_{j=1}^{(k-1)/2} (z^2 - e^{-4\pi ij/k})(z^2 - e^{4\pi ij/k}).$$

This is a polynomial of degree $2 + 4((k - 1)/2) = 2k$ with roots at $e^{2\pi ij/2k}$, $0 \le j \le 2k - 1$.

Since the polynomials in (7.31) and (7.32) are both monic with the same roots and degree we see that

$$e^{2\pi ikx}p(kx) = e^{2\pi ikx}p(x) \prod_{j=1}^{(k-1)/2} p(x + j/k)p(x - j/k)$$

or

$$\frac{p(kx)}{p(x)} = \prod_{j=1}^{(k-1)/2} p(x + j/k)p(x - j/k).$$

Since $p(t) = 2i\sin(2\pi t)$ the result follows. ∎

Lemma 7.34.2. If k is an odd positive integer, then

$$k = (2i)^{k-1} \prod_{j=1}^{k-1} \sin(2\pi j/k).$$

Proof. From Lemma 7.34.1, we have

(7.33) $$\frac{\sin(2\pi kx)}{\sin(2\pi x)} = (2i)^{k-1} \prod_{j=1}^{(k-1)/2} \sin(2\pi(x + \tfrac{j}{k}))\sin(2\pi(x - \tfrac{j}{k})).$$

Now

$$\sin(2\pi(x - j/k)) = \sin(2\pi(x - j/k) + 2\pi)$$
$$= \sin(2\pi(x - j/k + 1)) = \sin(2\pi(x + (k - j)/k)).$$

Since $1 \le j \le (k - 1)/2$ if and only if $(k + 1)/2 \le k - j \le k - 1$ we see that

(7.34)
$$\prod_{j=1}^{(k-1)/2} \sin(2\pi(x - j/k)) = \prod_{j=(k+1)/2}^{k-1} \sin(2\pi(x + j/k)).$$

If we combine (7.33) and (7.34), we obtain

(7.35)
$$\frac{\sin(2\pi kx)}{\sin(2\pi x)} = (2i)^{k-1} \prod_{j=1}^{k-1} \sin(2\pi(x + j/k)).$$

Now $\sin t$ is continuous for all t and from calculus we recall that

$$\lim_{u \to 0} \frac{\sin au}{\sin u} = a.$$

Thus, if we take the limit as $x \to 0$ on both sides of (7.35), we obtain

$$k = (2i)^{k-1} \prod_{j=1}^{k-1} \sin(2\pi j/k). \qquad \blacksquare$$

With all of the auxiliary matters out of the way we are now able to evaluate G_k. We follow a proof due to Gauss.

Theorem 7.34. If k is a positive odd integer, say $k = 2m + 1$, $m \ge 0$, we have

$$G_k = i^{m^2} \sqrt{k}.$$

Proof. Let $a = \exp(2\pi i/k)$ be a primitive kth root of unity. From the definition of $f(x, n)$ we have, since k is odd,

$$f(\alpha, k - 1) = 1 - \frac{1 - \alpha^{k-1}}{1 - \alpha} + \cdots + \frac{(1 - \alpha^{k-1})\cdots(1 - \alpha)}{(1 - \alpha)\cdots(1 - \alpha^{k-1})}.$$

Since

$$\frac{1 - \alpha^{k-j}}{1 - \alpha^j} = \frac{1 - \alpha^{-j}}{1 - \alpha^j} = -\alpha^{-j}$$

we have

$$f(\alpha, k - 1) = 1 + \alpha^{-1} + \alpha^{-1}\alpha^{-2} + \cdots + \alpha^{-1}\cdots\alpha^{-k+1} = \sum_{j=0}^{k-1} \alpha^{-j(j+1)/2}.$$

Since k is odd we see that a^{-2} is also a primitive kth root of unity, and so

$$f(\alpha^{-2}, k-1) = \sum_{j=0}^{k-1} \alpha^{j(j+1)}.$$

Since α is a kth root of unity we have

$$\alpha^{(j+(k+1)/2)^2 - ((k+1)/2)^2} = \alpha^{j(j+1)},$$

and so, since a has period k,

$$f(\alpha^{-2}, k-1) = \alpha^{-((k+1)/2)^2} \sum_{j=0}^{k-1} \alpha^{(j+(k+1)/2)^2} = \alpha^{-((k+1)/2)^2} \sum_{j=0}^{k-1} \alpha^{j^2} = \alpha^{-((k+1)/2)^2} G_k.$$

Therefore, by Theorem 7.33,

$$G_k = \alpha^{((k+1)/2)^2} f(\alpha^{-2}, k-1)$$

$$(7.37) \qquad = \alpha^{((k+1)/2)^2} \prod_{j=1}^{k-2} (1 - \alpha^{-2j}) = \alpha^{((k+1)/2)^2} \alpha^{-((k-1)/2)^2} \prod_{j=1}^{m} (\alpha^{2j-1} - \alpha^{1-2j})$$

$$= \prod_{j=1}^{m} (\alpha^{2j-1} - \alpha^{1-2j})$$

$$(7.38) \qquad = (-1)^m \prod_{j=1}^{m} (\alpha^{2j} - \alpha^{-2j})$$

since $a^k = 1$ and $a^{k-j} - a^{j-k} = -(a^j - a^{-j})$. If we combine (7.37) and (7.38) we see that, since $a = \exp(2\pi i/k)$

$$(7.39) \qquad G_k^2 = (-1)^m \prod_{j=1}^{k-1} (\alpha^j - \alpha^{-j}) = (-1)^m (2i)^{k-1} \prod_{j=1}^{k-1} \sin(2\pi j / k) = (-1)^m k,$$

by Lemma 7.34.2. Also, from (7.38), we have

$$(7.40) \qquad G_k = (-1)^m (2i)^m \prod_{j=1}^{m} \sin(4\pi j / k).$$

By (7.40), we have

$$(7.41) \qquad G_k = (-i)^m \sqrt{k} \, \mathrm{sgn} \left(\prod_{j=1}^{m} \sin(4\pi j / k) \right)$$

Now

$$\sin(4\pi j / k) \begin{cases} > 0 & \text{if} \quad 0 < j < k/4 \\ < 0 & \text{if} \ k/4 < j < k/2 \end{cases}.$$

Thus

(7.42) $$\text{sgn}\left(\prod_{j=1}^{m}\sin(4\pi j\,/\,k)\right)=(-1)^{[k/2]-[k/4]}=(-1)^{(k^2-1)/8}.$$

Thus, by (7.41) and (7.42), we have

$$G_k=(-i)^{(k-1)/8}(-1)^{(k^2-1)/8}\sqrt{k}=i^{m^2}\sqrt{k}.\qquad\blacksquare$$

We should remark that we can define G_k for k even and produce a similar evaluation (see [5].)

Corollary 7.34.1. We have

$$G_p(1)=\begin{cases}\sqrt{p} & \text{if }p\equiv1\ (\text{mod }4)\\i\sqrt{p} & \text{if }p\equiv3\ (\text{mod }4)\end{cases}.$$

Proof. If $p\equiv1$ (mod 4), then $p=4j+1$, for some integer j, and so, in the notation of Theorem 7.34, $m=2j$. Thus

$$G_p(1)=i^{(2j)^2}\sqrt{p}=i^{4j^2}\sqrt{p}=\sqrt{p},$$

since $i^4=1$.

If $p\equiv3$ (mod 4), then in the notation of Theorem 7.34, we have $m=2j+1$, for some integer j. Thus

$$G_p(1)=i^{(2j+1)^2}\sqrt{p}=i^{4j^2+4j+1}\sqrt{p}=i\sqrt{p}.\qquad\blacksquare$$

We can apply this corollary to give another proof of Theorem 3.5, the Law of Quadratic Reciprocity. By Theorem 7.30, if p and q are distinct odd primes, then

$$G_p(q)=\left(\tfrac{q}{p}\right)G_p\quad\text{and}\quad G_q(p)=\left(\tfrac{p}{q}\right)G_q.$$

Now, by Corollary 2.6.2,

(7.43)

$$G_p(q)G_q(p)=\sum_{m=0}^{p-1}e^{2\pi iqm^2/p}\sum_{n=0}^{q-1}e^{2\pi ipn^2/q}$$

$$=\sum_{m=0}^{p-1}\sum_{n=0}^{q-1}e^{2\pi i(qm+pn)^2/pq}=\sum_{r(pq)}e^{2\pi ir^2/pq}=G_{pq}.$$

Thus $G_{pq}=G_p(q)G_q(p)=\left(\tfrac{p}{q}\right)\left(\tfrac{q}{p}\right)G_pG_q$. If we consider the four cases

$$p\equiv1,q\equiv1\ (\text{mod }4),p\equiv1,q\equiv3\ (\text{mod }4),$$
$$p\equiv3,q\equiv1\ (\text{mod }4)\text{ and }p\equiv3,q\equiv3\ (\text{mod }4).$$

and apply Theorem 7.34 and Corollary 7.34.1, the result

$$\left(\frac{p}{q}\right)\left(\frac{q}{p}\right) = (-1)^{(p-1)(q-1)/4}$$

follows. For another proof using Gauss sums see [94].

Many of the results stated for special Gauss sums hold in general for Gauss sums (see Problems 6 and 7 below.) To illustrate we prove a multiplicativity property for a more general version of G_k. Let m and n be integers and define $G(m,n)$

$$G(m,n) = \sum_{h=0}^{n-1} e^{2\pi i m h^2 / n}.$$

Since

$$\exp(2\pi i (h + rn)^2 m / n) = \exp((2\pi i h^2 m + 2\pi i h r n m + 2\pi i r^2 n^2)/n) = \exp(2\pi i h^2 m / n)$$

we see that we can also write

$$G(m,n) = \sum_{h(n)} e^{2\pi i m h^2 / n},$$

where the sum is over any complete residue system modulo n.

Theorem 7.35. If $(n_1, n_2) = 1$, then

$$G(m, n_1 n_2) = G(mn_1, n_2)G(mn_2, n_1).$$

Proof. If h_1 and h_2 run through complete residue systems modulo n_1 and n_2, respectively, then, by Corollary 2.6.2, $H = h_1 n_2 + h_2 n_1$ runs through a complete residue system modulo $n_1 n_2$. Also

$$mH^2 = m(h_1 n_2 + h_2 n_1) \equiv m h_1^2 n_2^2 + m h_2^2 n_1^2 \pmod{n_1 n_2}.$$

Thus

$$
\begin{aligned}
G(mn_1, n_2)G(mn_2, n_1) &= \sum_{h_1(n_1)} e^{2\pi i m n_2 h_1^2 / n_1} \sum_{h_2(n_2)} e^{2\pi i m n_1 h_2^2 / n_2} \\
&= \sum_{\substack{h_1(n_1) \\ h_2(n_2)}} e^{2\pi i (mn_1 h_2^2 / n_2 + mn_2 h_1^2 / n_1)} = \sum_{\substack{h_1(n_1) \\ h_2(n_2)}} e^{2\pi i m (n_1^2 h_2^2 + n_2^2 h_1^2)/n_1 n_2} \\
&= \sum_{H(n_1 n_2)} e^{2\pi i m H^2 / n_1 n_2} = G(m, n_1 n_2),
\end{aligned}
$$

which was to be proved.   ■

Note that if $n_1 = p$, $n_2 = q$ and $m = 1$, then this reduces to the result, (7.43), used in the above proof of the Law of Quadratic Reciprocity.

We now turn to another special case of the Gauss sum, namely the **Ramanujan sum**. If we take χ_0 to be the principal character modulo k, then we get

$$G(\chi_0, n) = \sum_{m=0}^{k-1} \chi_0(n) e^{2\pi i mn/k} = \sum_{\substack{m=0 \\ (m,k)=1}}^{k-1} e^{2\pi i mn/k}.$$

This sum is so useful in its own right that it is given its own notation. Since there is no standard notation for the Ramanujan sum we shall pick one. Let

$$c(k,n) = \sum_{\substack{m(k) \\ (m,k)=1}} e^{2\pi i mn/k}.$$

Some of the important properties of the Ramanujan sum are given in the following theorem.

Theorem 7.36.

(a) Suppose $(k_1, k_2) = 1$. Then

$$c(k_1 k_2, n) = c(k_1, n) c(k_2, n).$$

(b) We have

$$c(k,n) = \sum_{d|(n,k)} d\mu(k/d).$$

Proof of (a). By Corollary 2.6.2, we know that if h_1 runs through a reduced residue system modulo k_1 and h_2 runs through a reduced residue system modulo k_2, then $h = h_1 k_2 + h_2 k_1$ runs through a reduced residue system modulo $k_1 k_2$. Thus

$$c(k_1,n)c(k_2,n) = \sum_{\substack{h_1(k_1) \\ (h_1,k_1)=1}} e^{2\pi i h_1 n/k_1} \sum_{\substack{h_2(k_2) \\ (h_2,k_2)=1}} e^{2\pi i h_2 n/k_2} = \sum_{\substack{h_1(k_1),h_2(k_2) \\ (h_1,k_1)=(h_2,k_2)=1}} e^{2\pi i n(h_1 k_2 + h_2 k_1)/k_1 k_2}$$

$$= \sum_{\substack{h(k_1 k_2) \\ (h,k_1 k_2)=1}} e^{2\pi i hn/k_1 k_2} = c(k_1 k_2, n).$$

Proof of (b). By the definition of $\mu(n)$ we have

$$\sum_{d|(n,k)} \mu(d) = \begin{cases} 1 & \text{if } (n,k)=1 \\ 0 & \text{else} \end{cases}.$$

Thus

$$(7.44)$$

$$c(k,n) = \sum_{\substack{h(k) \\ (h,k)=1}} e^{2\pi i h n / k} = \sum_{h(k)} e^{2\pi i h n / k} \sum_{d|(h,k)} \mu(d) = \sum_{d|k} \mu(d) \sum_{\substack{h=1 \\ d|h}}^{k} e^{2\pi i h n / k}$$

$$= \sum_{d|k} \mu(d) \sum_{r=1}^{k/d} e^{2\pi i r d n / k} = \sum_{d|k} \mu(d) \sum_{r=1}^{k/d} e^{2\pi i r n /(k/d)}.$$

Now, in general,

$$(7.45) \sum_{r=1}^{m} e^{2\pi i r n / m} = \sum_{r=1}^{m} (e^{2\pi i n / m})^r = \begin{cases} \dfrac{1-(e^{2\pi i n/m})^m}{1-e^{2\pi i n/m}} e^{2\pi i n/m} & \text{if } m\nmid n \\ m & \text{if } m\mid n \end{cases} = \begin{cases} 0 & \text{if } m\nmid n \\ m & \text{if } m\mid n \end{cases}.$$

If we combine (7.44) and (7.45), we obtain

$$c(k,n) = \sum_{\substack{d|k \\ d|n}} \mu(d)(k/d) = \sum_{\substack{d|k \\ d|n}} d\mu(k/d) = \sum_{d|(k,n)} d\mu(k/d). \qquad \blacksquare$$

Corollary 7.36.1. We have

(a) $c(k, 1) = \mu(k)$;

(b) $c(k, kn) = \varphi(k)$ for any nonnegative integer n;

and

(c) $c(p^k, n) = \begin{cases} p^k - p^{k-1} & \text{if } p^k \mid n \\ -p^{k-1} & \text{if } p^{k-1} \| n. \\ 0 & \text{if } p^{k-1} \nmid n \end{cases}$

Proof of (a). We have, by (b) of Theorem 7.36,

$$c(k,1) = \sum_{d|(k,1)} d\mu(k/d) = \sum_{d|1} d\mu(k/d) = \mu(k).$$

Proof of (b). We have, by (a) of Theorem 7.22 and (b) of Theorem 7.36,

$$c(k,kn) = \sum_{d|(k,kn)} d\mu(k/d) = \sum_{d|k} d\mu(k/d) = \varphi(k).$$

Proof of (c). We have

$$(p^k,n) = \begin{cases} p^k & \text{if } p^k \mid n \\ p^{k-1} & \text{if } p^{k-1} \| n. \\ p^a, 0 \le a \le k-1 & \text{if } p^{k-1} \nmid n \end{cases}$$

Thus, if $p^k \mid n$, we have, by (b) of Theorem 7.36 and Theorem 7.22,

$$c(p^k,n) = \sum_{d \mid (p^k,n)} d\mu(p^k / d) = \sum_{d \mid p^k} d\mu(p^k / d) = \varphi(p^k) = p^k - p^{k-1}.$$

If $p^{k-1} \| n$, then

$$c(p^k,n) = \sum_{d \mid (p^k,n)} d\mu(p^k / d) = \sum_{d \mid p^{k-1}} d\mu(p^k / d) = \sum_{j=0}^{k-1} p^j \mu(p^{k-j}) = p^{k-1}\mu(p) = -p^{k-1},$$

since $\mu(p^m) = 0$ if $m \geq 2$. Finally, if $p^{k-1} \nmid n$, then

$$c(p^k,n) = \sum_{d \mid (p^k,n)} d\mu(p^k / d) = \sum_{d \mid p^a} d\mu(p^k / d) = \sum_{j=0}^{a} p^j \mu(p^{k-j}).$$

Since $0 \leq a \leq k - 2$ we see that $k - j \geq 2$ for all $0 \leq j \leq a$. Thus $\mu(p^{k-j}) = 0$ for $0 \leq j \leq a$. Thus $c(p^k, n) = 0$ in this case. ∎

The Ramanujan sums have many properties and satisfy many identities, but we will content ourselves with just one: an explicit representation for $c(k, n)$ due to Hölder. For more on Ramanujan sums and their generalizations see [75] and [117]. For one generalization see Problem 10 of the Additional Problems below.

Theorem 7.37. Let $(k, n) = d$ and let $k = dK$. Then

$$c(k,n) = \mu(K)\varphi(k) / \varphi(K).$$

Proof. Note that

$$\mu(Kh) = \begin{cases} \mu(K)\mu(h) & \text{if } (K,h) = 1 \\ 0 & \text{if } (K,h) > 1 \end{cases}.$$

Thus, by (b) of Theorem 7.36 and (a) of Theorem 7.12,

$$(7.46) \qquad c(k,n) = d\mu(K) \sum_{\substack{h \mid d \\ (h,K)=1}} \mu(h) / h = d\mu(K) \prod_{\substack{p \mid d \\ p \nmid K}} (1 - 1 / p).$$

By Theorem 7.22, we have

$$(7.47) \qquad \frac{\varphi(k)}{\varphi(K)} = \frac{k}{K} \prod_{\substack{p \mid k \\ p \nmid K}} (1 - \tfrac{1}{p}) = d \prod_{\substack{p \mid k \\ p \nmid K}} (1 - \tfrac{1}{p}) = d \prod_{\substack{p \mid d \\ p \nmid K}} (1 - \tfrac{1}{p}),$$

since $k = dK$. The result follows, by (7.46) and (7.47). ∎

The last trigonometric sum we want to discuss is the **Kloostermann sum**

$$S(u,v,n) = \sum_{\substack{h(n) \\ (h,n)=1}} e^{2\pi i(uh+v\overline{h})/n},$$

where $\overline{h}$ is defined by $h\overline{h} \equiv 1 \pmod{n}$. Note that if $n \mid v$, then

$$S(u, v, n) = c(n, u).$$

Theorem 7.38. We have

    (a) $S(u, v, n) = S(-u, -v, n)$;

    (b) $S(u, v, n) = S(v, u, n)$;

and

    (c) $S(u, v, n) = S(1, uv, n)$ if $(u, n) = 1$ or $(v, n) = 1$.

Proof of (a). Note that if h runs over a reduced residue system modulo n so does $-h$. Also if $h\overline{h} \equiv 1 \pmod{n}$, then $(-h)(-\overline{h}) \equiv 1 \pmod{n}$. Thus $\overline{-h} = -\overline{h}$, by Corollary 2.8.3. Thus

$$S(u,v,n) = \sum_{\substack{h(n) \\ (h,n)=1}} e^{2\pi i(uh+v\overline{h})/n} = \sum_{\substack{h(n) \\ (h,n)=1}} e^{2\pi i(u(-h)+v(-\overline{h}))/n} = S(-u,-v,n).$$

Proof of (b). Note that if h runs over a reduced residue system modulo n, then so does $\overline{h}$. Since $\overline{\overline{h}} = h$, by Corollary 2.8.3, we see that

$$S(u,v,n) = \sum_{\substack{h(n) \\ (h,n)=1}} e^{2\pi i(uh+v\overline{h})/n} = \sum_{\substack{h(n) \\ (h,n)=1}} e^{2\pi i(u\overline{h}+vh)/n} = S(v,u,n).$$

Proof of (c). Suppose $(v, n) = 1$. Then as h runs through a reduced residue system modulo n so does hv, by Theorem 2.5. Thus

$$S(u,v,n) = \sum_{\substack{h(n) \\ (h,n)=1}} e^{2\pi i(uvh+v\overline{hv})/n} = \sum_{\substack{h(n) \\ (h,n)=1}} e^{2\pi i(uvh+\overline{h})/n} = S(uv,1,n) = S(1,uv,n).$$

by (b) since $\overline{hv} = \overline{h}\overline{v}$. Similarly if $(u, n) = 1$, then hu runs over a reduced residue system modulo n. We obtain the analogous result. ∎

We conclude our discussion of Kloostermann sums with their multiplicative property. As might be expected it is a little more involved than those we have seen so far.

Theorem 7.39. Suppose $(n, m) = 1$. If $V = vn^2 + wm^2$, then

$$S(u, v, n)S(u, w, m) = S(u, V, nm).$$

Proof. We have

$$S(u,v,n)S(u,w,m) = \sum_{\substack{h(n),k(m) \\ (h,n)=1,(k,m)=1}} e^{2\pi i((uh+v\overline{h})/n+(uk+w\overline{k})/m)}$$

(7.48)

$$= \sum_{\substack{h(n),k(m) \\ (h,n)=1,(k,m)=1}} e^{2\pi i(u(mh+nk)+v\overline{h}m+w\overline{k}n))/nm} = \sum_{\substack{h(n),k(m) \\ (h,n)=1,(k,m)=1}} e^{2\pi i(uH+K)/nm},$$

where $H = mh + nk$ and $K = v\overline{h}m + w\overline{k}n$. By Corollary 2.6.2, we see that H runs over a reduced residue system modulo mn as h runs over a reduced residue system modulo n and k runs over a reduced residue system modulo m.

Define $\overline{H}$ by $H\overline{H} \equiv 1 \pmod{mn}$. Since $n \mid nm$ and $m \mid nm$ we have

$$\overline{H}mh \equiv \overline{H}(mh + nk) = H\overline{H} \equiv 1 \pmod{n}$$

and

$$\overline{H}nk \equiv \overline{H}(mh + nk) = H\overline{H} \equiv 1 \pmod{m}.$$

Then

$$m\overline{H} \equiv \overline{h} \pmod{n} \quad \text{and} \quad n\overline{H} \equiv \overline{k} \pmod{m},$$

and so

$$m^2\overline{H} \equiv m\overline{h} \pmod{n} \quad \text{and} \quad n^2\overline{H} \equiv n\overline{k} \pmod{m}.$$

Then

$$V\overline{H} = (vm^2 + wn^2)\overline{H} = vm^2\overline{H} + wn^2\overline{H} \equiv v\overline{h}m + w\overline{k}n = K \pmod{n},$$

by Theorem 2.2. Thus, by (7.48),

$$S(u,v,n)S(u,w,m) = \sum_{\substack{h(n),k(m) \\ (h,n)=1,(k,m)=1}} e^{2\pi i(uH+V\overline{H})/nm} = S(u,V,nm). \qquad \blacksquare$$

Problem Set 7.8

1. Show that if p and q are distinct odd primes, then

$$G_p^{q-1}(1) \equiv \left(\tfrac{q}{p}\right) \pmod{q}.$$

2. Show directly that

$$G_p(1) = \begin{cases} \pm\sqrt{p} & \text{if } p \equiv 1 \pmod 4 \\ \pm i\sqrt{p} & \text{if } p \equiv 3 \pmod 4 \end{cases}.$$

(Hint: show that $G_p^2(1) = \left(\frac{-1}{p}\right)p$.)

3. Prove that if $\begin{bmatrix} n \\ m \end{bmatrix}$ is defined with respect to x, then

$$\prod_{j=0}^{n-1}(1+x^j y) = \sum_{k=0}^{n}\begin{bmatrix} n \\ m \end{bmatrix}x^{k(k-1)/2}y^k.$$

Note that if $x = 1$, then this reduces to the binomial theorem. (Hint: use induction.)

4. Provide a purely trigonometric proof of Lemma 7.34.1. (Hint: show that

$$\sin(a+b)\sin(a-b) = \sin^2 a - \sin^2 b$$

and use induction.)

5. Let p be an odd prime, $p \nmid k$. Show that
   (a) $G(k,p^\alpha) = pG(k,p^{\alpha-2})$, for all $a > 2$, and
   (b) $G(k,p^\alpha) = \begin{cases} p^{\alpha/2} & \text{if } \alpha \text{ is even} \\ p^{(\alpha-1)/2}G(k,p) & \text{if } \alpha \text{ is odd} \end{cases}$.
   (c) Evaluate $G(k,2^\alpha)$.

6. (a) Show that if c is a character modulo p and $p \nmid n$, then

$$G(\chi,n) = \overline{\chi}(n)G(\chi,1).$$

   (b) Show that $|G(\chi,1)|^2 = p$, and so $G(\chi,1) = \varepsilon\sqrt{p}$, where $|\varepsilon| = 1$.
   (c) Show that these results hold for arbitrary characters modulo k if χ is restricted to be proper.

7. Let $m = m_1 m_2$, where $(m_1, m_2) = 1$ and let χ be a character modulo m. Suppose that we can write

$$\chi(n) = \chi_1(n)\chi_2(n),$$

where χ_1 is a character modulo m_1 and χ_2 is a character modulo m_2. Show that

$$G(\chi,n) = \chi_1(m_2)\chi_2(m_1)G(\chi_1,n)G(\chi_2,n).$$

Generalize to an arbitrary number of factors. (Hint: use the Chinese Remainder Theorem.)

8. Prove the following inversion formulae, where f is any arithmetic function.

(a)   If $G(n,k) = \sum_{d|n} f(d)c(n/d,k)$, then $nf(n) = \sum_{k=1}^{n} G(n,k)$.

(b)   If $G(n,k) = \sum_{d=1}^{n} f(d)c(d,k)$, then $nf(n) = \sum_{d|n} G(n,d)$.

9. Fix n and define $f(k) = c(k, n)$. Find f^{*-1} and Λ_f.

10. Prove the following identities.

(a) $\displaystyle\sum_{d|k} c(k/d,d) = \begin{cases} \sqrt{k} & \text{if } k \text{ is a square} \\ 0 & \text{else} \end{cases}$

(b) $\displaystyle\sum_{d|(n,k)} c(k/d,n/d) = \sum_{d|(n,k)} \mu(k/d)\sigma(d)$.

(c)   If d_1 and d_2 are divisors of n, then

$$\varphi(d_1)c(d_2,n/d_1) = \varphi(d_2)c(d_1,n/d_2).$$

(d)   Evaluate

$$\sum_{d|n} |c(k,d)|.$$

11.   Prove the following results about Kloostermann sums.

(a) $S(u, v, n)$ is real.

(b)   If $(m, n) = 1$, then $S(u, vm, n) = S(um, v, n)$.

(c)   If $uv \equiv c^2 \pmod{n}$, then $S(u, v, n) = S(c, c, n)$.

(d)   If $m_1, \ldots, m_r$ are pairwise relatively prime, $M = m_1 \cdots m_r$ and, for $1 \leq s \leq r$, $M_s = M/m_s$, then

$$S(u_1,1,m_1)S(u_2,1,m_2)\cdots S(u_r,1,m_r) = S(M_1^2 u_1 + \cdots + M_r^2 u_r, 1, M).$$

12.   We define a modification of the Kloostermann sum called the Salié sum. Let p be an odd prime and a and b integers such that $p \nmid ab$. If $p \nmid h$, define $\bar{h}$ by $h\bar{h} \equiv 1 \pmod{p}$. Then the **Salié sum**, $S_p(a, b)$, is defined to be

$$S_p(a,b) = \sum_{h(p)} \left(\tfrac{h}{p}\right) e^{2\pi i(ah+b\bar{h})/p},$$

where $\left(\tfrac{h}{p}\right)$ denotes the Legendre symbol.

(a)   If $\left(\tfrac{ab}{p}\right) = -1$, then $S_p(a, b) = 0$.

(b)   If $\left(\tfrac{ab}{p}\right) = +1$, then $S_p(a,b) = \left(\tfrac{ac}{p}\right)S_p(c,c)$, where $ab \equiv c^2 \pmod{p}$.

(c) If m and n are integers, $m \neq 2$, show that

$$\left(\tfrac{m-2}{p}\right) \sum_{\substack{n(p) \\ n+\bar{n}\equiv m \ (\mathrm{mod}\ p)}} \left(\tfrac{n}{p}\right) = \sum_{\substack{n(p) \\ n+\bar{n}\equiv m \ (\mathrm{mod}\ p)}} \left(\tfrac{n-1}{p}\right)^2,$$

and so

$$\left(\tfrac{m-2}{p}\right) \sum_{\substack{n(p) \\ n+\bar{n}\equiv m \ (\mathrm{mod}\ p)}} \left(\tfrac{n}{p}\right) = 1 + \left(\tfrac{m^2-4}{p}\right).$$

(d) Show that

$$\sum_{\substack{n(p) \\ n+\bar{n}\equiv m \ (\mathrm{mod}\ p)}} \left(\tfrac{n}{p}\right) = \left(\tfrac{m-2}{p}\right) + \left(\tfrac{m+2}{p}\right).$$

Note that this holds for $m = 2$ as well.

(e) Show that

$$S_p(c,c) = \begin{cases} 2\left(\tfrac{c}{p}\right)\sqrt{p}\,\cos(4\pi c / p) & \text{if } p \equiv 1 \ (\mathrm{mod}\ 4) \\ 2i\left(\tfrac{c}{p}\right)\sqrt{p}\,\cos(4\pi c / p) & \text{if } p \equiv 3 \ (\mathrm{mod}\ 4) \end{cases}.$$

(Hint: sum over fixed values of in the exponent of $S_p(c, c)$ and then use (d) and Theorem 7.30 as well as Corollary 7.34.1.)

7.9. ADDITIVE FUNCTIONS

Definition 7.17. If f is an arithmetic function, then f is said to be **additive** if

(7.49) $f(mn) = f(m) + f(n),$

whenever $(m, n) = 1$. It is said to be **completely additive** if (7.49) holds without any restriction.

The study of additive functions is much more difficult than the study of multiplicative functions in some respects. We mention them here to give the reader a small acquaintance with the subject.

The most well-known additive function is the completely additive function $f(n) = \log(n)$. We define two other additive functions that play a role in much of the study of arithmetic functions.

Definition 7.18. Let n be a positive integer with canonical factorization $n = \prod_{i=1}^{r} p_i^{a_i}$. Then we define the two arithmetic functions Ω and ω by

$\Omega(1) = \omega(1) = 0$ and for $n > 1$ we define

$$\Omega(n) = \sum_{i=1}^{r} a_i$$

and

$$\omega(n) = r.$$

It is clear that both are additive (by unique factorization). We show that Ω is completely additive. Let $n = \prod_{i=1}^{r} p_i^{a_i}$ and $m = \prod_{i=1}^{r} p_i^{b_i}$, where we allow the a_i and b_i, $1 \le i \le r$, to be nonnegative. Then

$$\Omega(mn) = \sum_{i=1}^{r}(a_i + b_i) = \sum_{i=1}^{r} a_i + \sum_{i=1}^{r} b = \Omega(m) + \Omega(n),$$

since if $a_i = 0$ or $b_j = 0$, for some i or j, the sum is unchanged.

Let f be an additive function. Then

$$f(1) = f(1 \cdot 1) = f(1) + f(1) = 2f(1),$$

and so $f(1) = 0$. Thus no additive functions possesses a Dirichlet convolution inverse. This can be gotten around by defining an appropriate convolution for additive functions and then much of the theory of multiplicative functions goes through with the appropriate changes.

There is a very easy connection between additive and multiplicative functions. Let a be a positive number not equal to 1 (indeed one could, by being more careful, let a be complex and $a \ne 1$) and let f be a additive (completely additive) function. Then the function $F(n) = a^{f(n)}$ is multiplicative (completely multiplicative). With regards to Ω and ω, there are two multiplicative functions that can be defined in this way that appear in various guises in number theory. We define $\lambda(n)$

$$\lambda(n) = (-1)^{\Omega(n)}$$

and $\theta(n)$

$$\theta(n) = 2^{\omega(n)}.$$

The function $\lambda(n)$ was first investigated by Liouville and so is called **Liouville's function**. It has deep connections to prime number theory. The second function counts the number of square-free divisors of n (see Example 7.5.) For the number of square-free divisors of n is given by

$$\sum_{d|n} \mu^2(d),$$

which, by (b) of Theorem 7.12, is equal to

$$\prod_{p|n}(1+1) = 2^{\omega(n)} = \theta(n).$$

As for the Liouville function: it has a very useful property if one is trying to identify squares.

Theorem 7.40. We have

$$\sum_{d|n} \lambda(d) = \begin{cases} 1 & \text{if } n \text{ is a perfect square} \\ 0 & \text{else} \end{cases}.$$

Proof. Since λ is multiplicative we see that

$$f(n) = \sum_{d|n} \lambda(d)$$

is multiplicative, by Theorem 7.10. Thus we need only show that the result holds for prime powers. Let $n = p^a$, $a \geq 0$. If $a = 0$, then $n = 1$, and so

$$f(1) = \sum_{d|1} \lambda(d) = \lambda(1) = (-1)^{\Omega(1)} = (-1)^0 = 1.$$

Suppose that $a \geq 1$. Then, since λ is completely multiplicative,

$$f(p^a) = \sum_{d|p^a} \lambda(d) = \sum_{j=0}^{a} \lambda(p^j) = \sum_{j=0}^{a} (\lambda(p))^j = \frac{\lambda(p)^{a+1} - 1}{\lambda(p) - 1}.$$

Now $\lambda(p) = (-1)^{\Omega(p)} = (-1)^1 = -1$. Thus

$$f(p^a) = \frac{(-1)^{a+1} - 1}{-2} = \tfrac{1}{2}\left(1 - (-1)^{a+1}\right) = \begin{cases} 0 & \text{if } a \text{ is odd} \\ 1 & \text{if } a \text{ is even} \end{cases}.$$

The result follows.   ∎

This general connection has been exploited by many writers beginning with [110]. See also [22].

We shall see later (Section 6 of Chapter Nine) how the additive functions Ω and ω behave on the average, which will give some indication of the techniques needed to answer the questions posed in the theory of additive functions.

We close this section with a result, due to Erdös, on a certain class of additive functions. A corollary of this result is that if f is a monotone additive function then it must look like $\log(n)$.

Theorem 7.41. Let $f(n)$ be a real-valued additive function. If

(7.50) $$\liminf_{n \to +\infty}\{f(n+1) - f(n)\} \geq 0,$$

then $f(n) = c\log(n)$ for a suitable constant c.

We omit the proof of this result. However, the proof shows that the constant c can be taken to be $f(2)/\log 2$.

Corollary 7.41.1. Let f be a real-valued additive function. If either

(1) f is monotone increasing on the integers

or

(2) $\lim_{n \to +\infty}\{f(n+1) - f(n)\} = 0,$

then $f(n) = c\log(n)$, for a suitable constant c.

This follows immediately from the theorem.

Problem Set 7.9

1. Show that, for all $n \geq 1$, we have

$$\omega(n) \leq \frac{\log n}{\log 2}.$$

2. (a) Let f and g be arithmetic functions. Show that an arithmetic function h is completely additive if and only if

$$(f*g)h = (f*h)g + f*(gh).$$

   (b) Show that f is completely additive if and only if

$$\sum_{d|n} f(d) = \tfrac{1}{2} f(n)d(n).$$

   (c) Show that

$$\sum_{d|n} \log d = \tfrac{1}{2}(Ld)(n).$$

3. If $n = p_1^{\alpha_1} \cdots p_r^{\alpha_r}$, define $\Omega_k(n) = \alpha_1^k + \cdots + \alpha_r^k$, where k is a nonnegative integer. Show that Ω_k is an additive function. For what values of k, if any, is it completely additive?

4. Show that

$$\sum_{d|n} \omega(d) = \sum_{p|n} d(n/p),$$

where the sum on the right hand side is over all the primes that divide n.

5. Define $e_p(n)$ by $e_p(n) = e$ where $p^e \| n$, where p is a given prime. Show that, for each p, $e_p(n)$ is completely additive.

6. Show that if k is a positive integer, then

$$k^{\omega(n)} \leq d_k(n) \leq k^{\Omega(n)}.$$

7. If $a \neq 1$ is a real number, let $f_a(n) = a^{\omega(n)}$. If $n = p_1^{\alpha_1} \cdots p_r^{\alpha_r}$, show that

$$\sum_{d|n} f_a(d) = \prod_{j=1}^{r} (1 + \alpha_j a).$$

8. Let $a \neq 1$ be real and let $f(n) = (1-a)^{\omega(n)}$ and $g(n) = a^{\Omega(n)}$. Show that $f * g = e$.

9. Show that an arithmetic function f is additive if and only if $(\mu * f)(n) = 0$ when n is not a prime power.

10. Prove the following identities.

(a) $\displaystyle\sum_{d|n} \lambda(d)\mu(d) = \theta(n)$.

(b) $\displaystyle\sum_{d|n} \lambda(d)\mu(n/d) = \theta(n)\lambda(n)$.

(c) $\displaystyle\sum_{d|n} \lambda(d)\theta(n/d) = 1$.

11. Show that $\theta^{*-1} = \lambda\theta$ and $\lambda^{*-1} = \mu^2$. Compute Λ_θ and Λ_λ.

12. Show that

$$\lambda(n) = \sum_{d^2|n} \mu(n/d^2).$$

13. Show that

$$\sum_{k|n} \mu(k)d(k) = (-1)^{\omega(n)}.$$

14. Prove the following identities.

(a) $\displaystyle d(n^2) = \sum_{d|n} \theta(d)$.

(b) $\displaystyle\sum_{k|m} kd((m/k)^2) = \sum_{k|m} \sigma(k)\theta(m/k).$

(c) $\displaystyle\sum_{d|n}(d,n/d) = \sum_{d^2|n} d\theta(n/d^2).$

15. (a) Show that $\sigma(n) - \varphi(n) = \theta(n)$ if and only if n is a prime.

(b) Show that $\sigma(n) + \phi(n) = n\theta(n)$ if and only if $n = 1$ or is a prime.

16. Show that

$$\sum_{dn^2=k} c(d,m) = \sum_{d|(m,k)} d\lambda(k/d).$$

17. Show that

$$\sum_{d|n} \lambda(d)\varphi(d)\sigma(n/d) = \sum_{k^2|n} k^2\theta(n/k^2).$$

7.10. LINEAR RECURSION

In this section we are concerned with a different type of arithmetic function, namely sequences. We are interested in the following special type of sequences.

Definition 7.19. Let k be a positive integer and let $a_1,..., a_k$, and b be given real or complex numbers. The sequence $\{u_n\}$ is defined for $n \geq 0$ by

$$u_{n+k} = a_k u_{n+k-1} + \cdots + a_2 u_{n+1} + a_1 u_n + b,$$

where $u_0,..., u_{k-1}$ are given, is called a **linear recurrence of order k**. If $b = 0$, the recurrence is said to be **homogeneous**.

Our interest is in homogeneous binary recurrences: $k = 2$ and $b = 0$. Thus we take a and b to be given numbers and suppose u_0 and u_1 are known. Then we wish to study sequences satisfying, for $n \geq 0$,

(7.51) $\qquad\qquad\qquad u_{n+2} = au_{n+1} + bu_n.$

These appear in many places, for example, in the study of the Fermat equation (see Corollary 6.34.1) and in the theory of continued fractions (for example the definition of h and k, Equation (4.23)). Probably the most famous binary recurrence is the **Fibonacci series**. Here $u_0 = 0$, $u_1 = 1$ and $a = b = 1$. The first few terms are:

$$u_0 = 0, \ u_0 = 1, \ u_2 = 1, \ u_3 = 2, \ u_4 = 3, \ u_5 = 5 \ \text{and} \ u_6 = 8.$$

There are many simple properties of these sequences that follow from (7.51). In this section we shall prove a few and leave others for the exercises.

We begin with an alternate way of expressing the terms in the sequence. We start with a, b, u_0 and u_1 as given. Note that for any number x

$$u_{n+1} - xu_n = (a-x)(u_n - xu_{n-1}) + (b + ax - x^2)u_{n-1}.$$

Suppose α and β are the zeros of the polynomial $x^2 - ax - b$. First suppose that $\alpha \neq \beta$. Since $\alpha + \beta = a$ we have

$$u_{n+1} - \alpha u_n = \beta(u_n - \alpha u_{n-1})$$

and

$$u_{n+1} - \beta u_n = \alpha(u_n - \beta u_{n-1}).$$

A simple induction argument shows that

$$u_{n+1} - \alpha u_n = \beta^n(u_n - \alpha u_{n-1})$$

and

$$u_{n+1} - \beta u_n = \alpha^n(u_n - \beta u_{n-1}).$$

If we subtract we find that for $n \geq 0$

(7.52)
$$u_n = \frac{(u_1 - u_0)\beta^n - (u_1 - u_0)\alpha^n}{\beta - \alpha},$$

when $\alpha \neq \beta$. (7.52) is often referred to as the **Binet formula.**

Suppose now $\alpha = \beta$. Then $4b + a^2 = 0$. If we let α denote the double zero of $x^2 - ax - b.$, then $\alpha = a/2$ and so the recurrence (7.51) can be written as

(7.53)
$$u_{n+2} = 2\alpha u_{n+1} - \alpha^2 u_n.$$

If we try calculating a few terms of the sequence and rewriting them in terms of u_0 and u_1 we find that for $m \geq 0$

(7.54)
$$u_m = m\alpha^{m-1}u_1 - (m-1)\alpha^m u_0.$$

We shall prove this is the case by induction on m. It is true for $m = 0$ and $m = 1$ and so we suppose (7.54) holds for $m = n \geq 1$. Then for $m = n + 1$, we have, by (7.53),

$$u_{n+1} = 2\alpha u_n - \alpha^2 u_{n-1}$$
$$= 2\alpha(n\alpha^{n-1}u_1 - (n-1)\alpha^n u_0) - \alpha^2((n-1)\alpha^{n-2}u_1 - (n-2)\alpha^{n-1}u_0)$$
$$= (n+1)\alpha^n u_1 - n\alpha^{n+1}u_0,$$

which is (7.54) for $m = n + 1$.

If we combine the above results we have the following theorem.

Theorem 7.42. For any given numbers a, b, u_0 and u_1 define the sequence $\{u_n\}$ by

$$u_{n+2} = au_{n+1} + bu_n,$$

for $n \geq 0$. If $a^2 + 4b = 0$, then (7.54), with $\alpha = a/2$, expresses u_n in terms of the given numbers. If $a^2 + 4b \neq 0$, then the zeros α and β of the polynomial $x^2 - ax - b$ are distinct and (7.52) expresses u_n in terms of the given numbers.

Example 7.14.

    (1) Find a single expression for the Fibonacci series $\{F_n\}$.

    (2) Find a single expression for the Lucas sequence $\{L_n\}$:

$$a = b = 1 \quad \text{and} \quad L_0 = 2, \ L_1 = 1.$$

Solution. For both of these sequences the recurrence relation is the same:

$$u_{n+2} = u_{n+1} + u_n.$$

Thus the associated quadratic equation is

$$x^2 - x - 1 = 0,$$

which has as its roots

$$\alpha = (1 - \sqrt{5})/2 \quad \text{and} \quad \beta = (1 + \sqrt{5})/2.$$

Since $\beta - \alpha = \sqrt{5}$ and $\alpha + \beta = 1$ we see that for the Fibonacci sequence

$$F_n = (\beta^n - \alpha^n)/(\beta - \alpha) = (\beta^n - \alpha^n)/\sqrt{5}.$$

For the Lucas sequence

$$L_n = ((1 - 2\alpha)\beta^n - (1 - 2\beta)\alpha^n)/(\beta - \alpha)$$

and, since $\beta - \alpha = \sqrt{5}, 1 - 2\alpha = \sqrt{5}$ and $1 - 2\beta = -\sqrt{5}$, we see that

$$L_n = \alpha^n + \beta^n. \qquad \qquad \square$$

In what follows we will restrict our attention to the case where the associated quadratic equation has distinct zeros so that we may use (7.52) in our discussion below. Most of the results obtained below hold for both cases and the proofs are only minor modifications of the case considered. (See Problem 1 below.)

In talking about the sequence $\{u_n\}$ it is often useful to introduce the **conjugate sequence** $\{v_n\}$ defined for $n \geq 2$

(7.55) $$v_n = u_{n+1} + bu_{n-1}.$$

If we extend this backward we see that we must have $v_0 = 2u_1 - au_0$ and $v_1 = au_1 + 2bu_0$.

Example 7.15. Show that the Lucas sequence $\{L_n\}$ is conjugate to the Fibonacci sequence $\{F_n\}$.

Solution. We have, since $b = 1$ in this case,

$$F_{n+1} + F_{n-1} = \frac{1}{\sqrt{5}}\left\{\left(\frac{1+\sqrt{5}}{2}\right)^{n+1} - \left(\frac{1-\sqrt{5}}{2}\right)^{n+1}\right\} + \frac{1}{\sqrt{5}}\left\{\left(\frac{1+\sqrt{5}}{2}\right)^{n-1} - \left(\frac{1+\sqrt{5}}{2}\right)^{n-1}\right\}$$

$$= \frac{1}{\sqrt{5}}\left(\left(\frac{1+\sqrt{5}}{2}\right)^{n-1}\left(\left(\frac{1+\sqrt{5}}{2}\right)^2 + 1\right)\right)$$

$$- \frac{1}{\sqrt{5}}\left(\left(\frac{1-\sqrt{5}}{2}\right)^{n-1}\left(\left(\frac{1-\sqrt{5}}{2}\right)^2 + 1\right)\right)$$

$$= \frac{1}{\sqrt{5}}\left\{\left(\frac{1+\sqrt{5}}{2}\right)^{n-1}\left(\frac{5+\sqrt{5}}{2}\right) - \left(\frac{1-\sqrt{5}}{2}\right)^{n-1}\left(\frac{5-\sqrt{5}}{2}\right)\right\}$$

$$= \left(\frac{1+\sqrt{5}}{2}\right)^n + \left(\frac{1-\sqrt{5}}{2}\right)^n = L_n,$$

as desired.   □

This example leads to the following theorem.

Theorem 7.43. If $\{v_n\}$ is the conjugate sequence to $\{u_n\}$, then for $n \geq 0$

(7.56) $$v_n = (u_1 - \alpha u_0)\beta^n + (u_1 - \beta u_0)\alpha^n.$$

Proof. For $n = 0$ and $n = 1$, the result follows by inspection and (7.55). Suppose $n \geq 2$. Then, from (7.52) and the fact that $\alpha\beta = -b$, we have

$$v_n = u_{n+1} + bu_{n-1} = \frac{(u_1 - \alpha u_0)\beta^{n+1} - (u_1 - \beta u_0)\alpha^{n+1}}{\beta - \alpha} + b\frac{(u_1 - \alpha u_0)\beta^{n-1} - (u_1 - \beta u_0)\alpha^{n-1}}{\beta - \alpha}$$

$$= \frac{(u_1 - \alpha u_0)(\beta^{n+1} + b\beta^{n-1}) - (u_1 - \beta u_0)(\alpha^{n+1} + b\alpha^{n-1})}{\beta - \alpha}$$

$$= \frac{(u_1 - \alpha u_0)\beta^n(\beta - \alpha) - (u_1 - \beta u_0)\alpha^n(\alpha - \beta)}{\beta - \alpha} = (u_1 - \alpha u_0)\beta^n + (u_1 - \beta u_0)\alpha^n,$$

which is the result.   ■

Theorem 7.44. The sequence $\{v_n\}$ satisfies the recursion (7.51).

Proof. We have, by (7.55), for $n \geq 1$,

$$av_{n+1} + bv_n = a(u_{n+2} + bu_n) + b(u_{n+1} + bu_{n-1})$$
$$= au_{n+2} + bu_{n+1} + b(au_n + bu_{n-1}) = u_{n+3} + bu_{n+1} = v_{n+2},$$

which is the recursion (7.51). It is easy to show that it also holds for $n = 0$. ∎

Theorem 7.45. We have $v_{n+1} + bv_{n-1} = (a^2 + 4b)u_n$.

Proof. By Theorem 7.43, we have

$$v_{n+1} + bv_{n-1} = \left\{(u_1 - \alpha u_0)\beta^{n+1} + (u_1 - \beta u_0)\alpha^{n+1}\right\} + b\left\{(u_1 - \alpha u_0)\beta^{n-1} + (u_1 - \beta u_0)\alpha^{n-1}\right\}$$
$$= (u_1 - \alpha u_0)\beta^{n-1}(\beta^2 + b) + (u_1 - \beta u_0)\alpha^{n-1}(\alpha^2 + b).$$

We have

$$(7.57) \quad \beta^2 + b = \beta^2 - \alpha\beta = \beta(\beta - \alpha) \quad \text{and} \quad \alpha^2 + b = \alpha(\alpha - \beta) = -\alpha(\beta - \alpha).$$

Thus

$$v_{n+1} + bv_{n-1} = (\beta - \alpha)^2 \left\{ \frac{(u_1 - \alpha u_0)\beta^n - (u_1 - \beta u_0)\alpha^n}{\beta - \alpha} \right\} = (\beta - \alpha)^2 u_n.$$

Since $\beta - \alpha = \sqrt{a^2 + 4b}$ the result follows. ∎

Since the number $(\beta - \alpha)^2 = a^2 + 4b$ will be recurring often in the following we shall give a single symbol, say c. In the case of the Fibonacci sequence $c = 1^2 + 4 \cdot 1 = 5$. Thus Theorem 7.45 says

$$L_{n+1} + L_{n-1} = 5F_n.$$

In order to simplify matters (for example, to make the calculations less tedious and the cases to consider less numerous) we shall now assume that $u_0 = 0$ and $u_1 = 1$. Then (7.52) states that for $n \geq 0$

$$(7.58) \qquad\qquad u_n = \frac{\beta^n - \alpha^n}{\beta - \alpha}$$

and (7.56) states that for $n \geq 0$

$$(7.59) \qquad\qquad v_n = \beta^n + \alpha^n,$$

where $v_0 = 2$ and $v_1 = a$.

Theorem 7.46. We have

$$u_k v_n + u_n v_k = 2u_{n+k}.$$

Proof. By (7.58) and (7.59) we have

$$u_k v_n + u_n v_k = (\beta^k - \alpha^k)(\beta^n + \alpha^n)/(\beta - \alpha) + (\beta^n - \alpha^n)(\beta^k + \alpha^k)/(\beta - \alpha)$$
$$= (2\beta^{k+n} - 2\alpha^{k+n})/(\beta - \alpha) = 2u_{n+k},$$

which is the result. ∎

The following theorem is one of many dual theorems that appear in the study of binary sequences, that is, if one replaces u_n by v_n and v_n by u_n, then one gets a similar expression, except for possible numerical factors.

Theorem 7.47. We have

(a) $v_{n+1}^2 + bv_n^2 = cu_{2n+1}$

and

(b) $u_{n+1}^2 + bu_n^2 = u_{2n+1}.$

Proof of (a). We have, by (7.59),

$$v_{n+1}^2 + bv_n^2 = (\beta^{n+1} + \alpha^{n+1})^2 + b(\beta^n + \alpha^n)^2$$
$$= \beta^{2n}(\beta^2 + b) + \alpha^{2n}(\alpha^2 + b) + 2(\alpha\beta)^n(\alpha\beta + b)$$
$$= \beta^{2n}\beta(\beta - \alpha) + \alpha^{2n}\alpha(\alpha - \beta) = cu_{2n+1},$$

by (7.57) and (7.58).

Proof of (b). We have, by (7.58),

$$u_{n+1}^2 + bu_n^2 = \{(\beta^{n+1} - \alpha^{n+1})^2 + b(\beta^n - \alpha^n)^2\}/c$$
$$= \{\beta^{2n}(\beta^2 + b) + \alpha^{2n}(\alpha^2 + b)\}/c = u_{2n+1},$$

by (7.57) and (7.58). ∎

Theorem 7.48. We have for any nonnegative integer N, if $a + b \neq 1$,

$$\sum_{j=0}^{N} u_j = \frac{1}{b+a-1}(u_{N+1} + bu_N - 1).$$

Proof. By (7.58), we have

$$\sum_{j=0}^{N} u_j = \sum_{j=0}^{N} \frac{\beta^j - \alpha^j}{\beta - \alpha} = \frac{1}{\beta - \alpha}\sum_{j=0}^{N}(\beta^j - \alpha^j) = \frac{1}{\beta - \alpha}\left\{\frac{\beta^{N+1} - 1}{\beta - 1} - \frac{\alpha^{N+1} - 1}{\alpha - 1}\right\}$$
$$= \frac{1}{\beta - \alpha}\left\{\frac{(\alpha\beta)(\beta^N - \alpha^N) - (\beta^{N+1} - \alpha^{N+1}) + (\beta - \alpha)}{(\alpha - 1)(\beta - 1)}\right\},$$

which gives the result, since $(\alpha - 1)(\beta - 1) = \alpha\beta - (\beta + \alpha) + 1 = -b - a + 1$. ∎

For the Fibonacci sequence this reads

$$\sum_{j=0}^{N} F_j = F_{N+2} - 1.$$

We note that we can prove Theorem 7.48 without the use of the Binet formula (7.58). All that is required is the recursion relation (7.51). The idea used here can be applied in many other different contexts involving recursions.

Suppose w_0 and w_1 are given quantities and $w_{n+2} = aw_{n+1} + bw_n$, for $n \geq 0$. Then, for $N \geq 2$,

$$\sum_{j=0}^{N} w_j = \sum_{j=2}^{N} w_j + w_0 + w_1 = \sum_{j=0}^{N-2} w_{j+2} + w_0 + w_1 = \sum_{j=0}^{N-2} (aw_{j+1} + bw_j) + w_0 + w_1$$

$$= a\sum_{j=0}^{N-2} w_{j+1} + b\sum_{j=0}^{N-2} w_j + w_0 + w_1 = a\sum_{j=1}^{N-1} w_j + b\sum_{j=0}^{N-2} w_j + w_0 + w_1$$

$$= a\sum_{j=0}^{N} w_j - aw_N - aw_0 + b\sum_{j=0}^{N} w_j - bw_N - bw_{N-1} + w_0 + w_1$$

$$= (a+b)\sum_{j=0}^{N} w_j - (aw_N + bw_{N-1}) - bw_N + (1-a)w_0 + w_1.$$

Thus

$$(a+b-1)\sum_{j=0}^{N} w_j = w_{N+1} + bw_N - (1-a)w_0 - w_1.$$

Now, if $a + b = 1$, then we have, for all $N \geq 2$,

$$w_{N+1} + bw_N = bw_0 + w_1,$$

and so we are really dealing with a linear recursion of order 1. If we assume $a + b \neq 1$, then we have

$$\sum_{j=0}^{N} w_j = \frac{w_{N+1} + bw_N - (1-a)w_0 - w_1}{a+b-1}.$$

To obtain Theorem 7.48 we recall that for the sequence considered in the theorem we have $u_0 = 0$ and $u_1 = 1$. Thus

$$\sum_{j=0}^{N} u_j = \frac{u_{N+1} + bu_N - 1}{a+b-1}.$$

as desired.

It might also be remarked that many of the results of this section can also be obtained by mathematical induction.

Theorem 7.49. If n and k are positive integers, then

$$u_{k+1}u_{n+1} + bu_nu_k = u_{n+k+1}.$$

Proof. By (7.58), we have, since $\alpha\beta = -b$,

$$u_{k+1}u_{n+1} + bu_nu_k = \left\{(\beta^{n+1} - \alpha^{n+1})(\beta^{k+1} - \alpha^{k+1}) + b(\beta^n - \alpha^n)(\beta^k - \alpha^k)\right\}/c$$

$$= \left\{\beta^{n+k}(\beta^2 + b) + \alpha^{n+k}(\alpha^2 + b) - (\beta^n\alpha^k + \beta^k\alpha^n)(\alpha\beta + b)\right\}/c$$

$$= (\beta^{n+k+1} - \alpha^{n+k+1})/(\beta - \alpha) = u_{n+k+1},$$

by (7.57) and (7.58), which gives the result.   ∎

We now wish to consider some divisibility properties of the sequence $\{u_n\}$. Thus we need the u_n to be integers. Since $u_0 = 0$, $u_1 = 1$ we see that $u_2 = a$ and $u_3 = a^2 + b$. Thus for u_n to be an integer for all n it is necessary and sufficient that a and b be integers.

Theorem 7.50. If k and N are positive integers and $k \mid N$, then $u_k \mid u_N$.

Proof. If $N = 1$, then $k \mid N$ implies that $k = N = 1$ so that $u_k = u_N = 1$, and so the result holds. Now assume the result holds for all M such that $1 \le M \le N$, $N \ge 2$ and suppose $k \mid N$. If $k = N$, the result follows. If $k < N$, let $N = rk$, with $r > 1$. By Theorem 7.49, we have

$$u_N = u_{rk} = u_{(r-1)k+k} = bu_{(r-1)k}u_{k-1} + u_{(r-1)k+1}u_k.$$

Since $u_k \mid u_{(r-1)k}$ by hypothesis, and $u_k \mid u_k$ in any case we see that $u_k \mid u_{rk} = u_N$. ∎

Theorem 7.51. If not both k and j are zero, then

$$(u_k, u_j) = u_{(k,j)}.$$

Proof. If k or j is zero, then the result follows since $u_0 = 0$.

Suppose $kj \ne 0$ and let $d = (k, j)$. Since $d \mid k$ and $d \mid j$ we see, by Theorem 7.50, that $u_d \mid u_k$ and $u_d \mid u_j$, and so

$$u_d \mid (u_k, u_j)$$

Let x and y be integers such that $d = kx + jy$. Then, by Theorem 7.49, we have

$$u_d = bu_{kx}u_{jy-1} + u_{kx+1}u_{jy}.$$

Since $u_k \mid u_{kx}$ and $u_k \mid u_{jy}$, by Theorem 7.50, we see that $(u_k, u_j) \mid u_d$. From this the result follows. ■

There is an analogous result for (v_k, v_j) though not quite as general as Theorem 7.51 (see Problem 9 below).

We now prove some congruences involving the numbers u_n.

Theorem 7.52. For $k \geq 0$ we have

$$2^{k-1} u_k = \sum_{j=0}^{[(k-1)/2]} \binom{k}{2j+1} a^{k-2j-1} c^j.$$

Proof. By (7.58), we have, since $\beta = (a + \sqrt{c})/2$ and $\alpha = (a - \sqrt{c})/2$,

$$2^k u_k = \frac{2^k}{\beta - \alpha}(\beta^k - \alpha^k) = \frac{1}{\sqrt{c}}\left\{(a + \sqrt{c})^k - (a - \sqrt{c})^k\right\}$$

$$= \frac{1}{\sqrt{c}}\left\{\sum_{j=0}^{k}\binom{k}{j}a^{k-j}(\sqrt{c})^j - \sum_{j=0}^{k}\binom{k}{j}a^{k-j}(-\sqrt{c})^j\right\}$$

$$= \frac{1}{\sqrt{c}}\left\{\sum_{j=0}^{k}\binom{k}{j}a^{k-j}(\sqrt{c})^j(1 - (-1)^j)\right\}$$

$$= 2\sum_{j=0}^{[(k-1)/2]}\binom{k}{2j+1}a^{k-2j-1}c^j,$$

which is the result. ■

Theorem 7.53. We have, for any positive integers n and k,

$$u_{nk+1} \equiv u_{k+1}^n \pmod{u_k^2}.$$

Proof. This is clearly true for $n = 1$. Assume true for $n = m - 1 \geq 1$. Then, by Theorem 7.49, we have

$$u_{mk+1} = u_{(m-1)k+k} = bu_{(m-1)k}u_k + u_{(m-1)k+1}u_{k+1}.$$

By Theorem 7.50 we have $u_k \mid u_{(m-1)k}$, and so $u_k^2 \mid u_{(m-1)k}u_k$. Thus, by the induction hypothesis,

$$u_{mk+1} \equiv u_{(m-1)k+1}u_{k+1} \equiv u_{k+1}^{m-1}u_{k+1} \pmod{u_k^2}$$

and the result follows. ■

Theorem 7.54. Let p be an odd prime.

(a) We have $u_p \equiv \left(\frac{c}{p}\right) \pmod{p}$.

(b) We have $2u_{p+1} \equiv \left(1 + \left(\frac{c}{p}\right)\right)a \pmod{p}$.

(c) We have $2bu_{p-1} \equiv a\left(1 - \left(\frac{c}{p}\right)\right) \pmod{p}$.

Proof of (a). By Theorem 7.52 we have

$$2u_p = \sum_{j=0}^{(p-1)/2} \binom{p}{2j+1} a^{p-2j-1} c^j .$$

By Fermat's theorem (Corollary 2.8.1) we have $2^{p-1} \equiv 1 \pmod{p}$. For $0 \leq j < (p-1)/2$ we have $p \Big| \binom{p}{2j+1}$. Thus

$$u_p \equiv \binom{p}{2(p-1)/2+1} a^{p-2(p-1)/2-1} c^{(p-1)/2} \equiv c^{(p-1)/2} \equiv \left(\frac{c}{p}\right) \pmod{p},$$

by Euler's criterion ((a) of Theorem 3.2).

Proof of (b). By Theorem 7.52 we have

$$2^p u_{p+1} = \sum_{j=0}^{(p-1)/2} \binom{p+1}{2j+1} a^{p+1-2j-1} c^j .$$

By Fermat's theorem $2^p \equiv 2 \pmod{p}$ and for $1 \leq j < (p-1)/2$ we have $p \Big| \binom{p+1}{2j+1}$. Thus

$$2u_{p+1} \equiv a^p + ac^{(p-1)/2} \equiv a + a\left(\frac{c}{p}\right) \pmod{p},$$

by Euler's criterion and Fermat's theorem.

Proof of (c). We have, by (a),

$$2bu_{p-1} = 2u_{p+1} - 2au_p \equiv a\left(1 + \left(\frac{c}{p}\right)\right) - 2a\left(\frac{c}{p}\right) = a\left(1 - \left(\frac{c}{p}\right)\right) \pmod{p},$$

by (a) and (b).   ∎

Corollary 7.54.1. If $(a, p) = 1$, then $p \mid u_{p-\left(\frac{c}{p}\right)}$

We can apply the theory of binary recursions we have developed here to give a proof of the Lucas-Lehmer test for the primality of Mersenne primes, $M_p = 2^p - 1$, which were mentioned above in reference to perfect numbers. In Chapter Ten we will outline another proof using algebraic number theory (see Problem 27 of the Additional Problems for Chapter Ten.)

We need some preliminary results regarding a special sequence defined by $u_0 = 0$, $u_1 = 1$ and, for $n \geq 0$,

$$u_{n+2} = 4u_{n+1} - u_n.$$

Let $\{v_n\}$ denote the conjugate sequence. Then

$$v_0 = 2, \quad v_1 = 4 \quad \text{and} \quad v_{n+2} = 4v_{n+1} - v_n, \text{ for } n \geq 0.$$

We will need some properties of these sequences, many of which we have proved above.

Lemma 7.55.1. We have
    (a) $v_n = u_{n+1} - u_{n-1}$, for $n \geq 1$;
    (b) $u_n = \left(\left(2 + \sqrt{3}\right)^n - \left(2 - \sqrt{3}\right)^n \right) / \sqrt{12}$;
    (c) $v_n = \left(2 + \sqrt{3}\right)^n + \left(2 - \sqrt{3}\right)^n$;

and
    (d) $u_{m+n} = u_m u_{n+1} - u_{m-1} u_n$.

Proof of (a). This is just (7.55), the definition of the conjugate sequence, since $b = -1$ in this case.

Proof of (b) and (c). The associated quadratic equation is

$$x^2 - 4x + 1 = 0$$

whose roots are $\alpha, \beta = (4 \pm \sqrt{16 - 4}) / 2 = 2 \pm \sqrt{3}$. The result then follows from (7.58) and (7.59) since $\alpha - \beta = 2\sqrt{3} = \sqrt{12}$.

Proof of (d). This is simply Theorem 7.49 with $b = -1$, $n = n$ and $m = k + 1$. ∎

Lemma 7.55.2. If n and k are positive integers, then

$$u_{kn} \equiv k u_n u_{n+1}^{k+1} \pmod{u_n^2}.$$

Proof. For $k = 1$, this is clear. Assume, for $k = m - 1 \geq 1$, we have

(7.60) $$u_{(m-1)n} \equiv (m - 1) u_n u_{n+1}^{m-2} \pmod{u_n^2}.$$

By Theorem 7.49, we have

(7.61) $$u_{mn} = u_{(m-1)n+1} u_n + b u_{(m-1)n} u_{n-1}.$$

By Theorem 7.53, we have

$$(7.62) \qquad\qquad u_{(m-1)n+1} \equiv u_{n+1}^{m-1} \pmod{u_n^2}.$$

Since $bu_{n-1} = u_{n+1} - au_n$, we have, by (7.60), (7.61) and (7.62),

$$u_{mn} \equiv u_{n+1}^{m-1} u_n + (m-1)u_n u_{n+1}^{m-2}(u_{n+1} - au_n) \pmod{u_n^2}$$
$$\equiv u_{n+1}^{m-1} u_n + (m-1)u_n u_{n+1}^{m-1} - a(m-1)u_n^2 u_{n+1}^{m-2} \pmod{u_n^2}$$
$$\equiv mu_n u_{n+1}^{m-1} \pmod{u_n^2}.$$

The result follows by mathematical induction.   ∎

Corollary 7.55.2.1. Suppose that for some $e \geq 1$ $p^e | u_n$. Then $p^{e+1} | u_{pn}$.

Proof. By Lemma 7.55.2, we have

$$u_{pn} \equiv pu_n u_{n+1}^{p-1} \pmod{u_n^2}.$$

or, for some integer t,

$$u_{pn} = pu_n u_{n+1}^{p-1} + u_n^2 t.$$

If $u_n = sp^e$, then

$$u_{pn} = su_{n+1}^{p-1}p^{e+1} + s^2 t p^{2e} = p^{e+1}(su_{n+1}^{p-1} + s^2 t p^{e-1}).$$

The result follows since $e \geq 1$.   ∎

Lemma 7.55.3. Let N be a positive integer and let $m(N)$ denote the smallest positive integer m such that

$$u_m \equiv 0 \pmod{N}.$$

Then

$$u_n \equiv 0 \pmod{N}. \text{ if and only if } n \equiv 0 \pmod{m(N)}.$$

Proof. By Theorem 7.46, we see that if $m = m(N)$, then

$$u_{m+n} \equiv u_{m+1}u_n \pmod{N}.$$

Since $(u_{m+1}, u_m) = u_{(m,m+1)} = u_1 = 1$, by Theorem 7.51, and $N | u_m$ we see that $(u_{m+1}, N) = 1$. Thus $N | u_{m+n}$ if and only if $N | u_n$, and so if and only if $n \equiv 0 \pmod{m(N)}$.   ∎

The number $m(N)$ is called the **rank of apparition** of N.

Lemma 7.55.4. We have

    (a) $v_{2n} = v_n^2 - 2(-b)^n$;

    (b) $2u_{n+1} = au_n + v_n$;

and

    (c) $(u_n, v_n) \le 2$.

Proof of (a). We have, by (7.59),

$$v_{2n} = \alpha^{2n} + \beta^{2n} = \alpha^{2n} + 2\alpha^n\beta^n + \beta^{2n} - 2\alpha^n\beta^n = (\alpha^n + \beta^n)^2 - 2(\alpha\beta)^n = v_n^2 - 2(-b)^n.$$

since a and b are the roots of $x^2 - ax - b = 0$.

Proof of (b). We have, by (7.55),

$$au_n + v_n = au_n + u_{n+1} + bu_{n-1} = au_n + bu_{n-1} + u_{n+1} = 2u_{n+1}.$$

Proof of (c). Let $d = (u_n, v_n)$. Then, by (b), $d \mid 2u_{n+1}$. Since, as above, $(u_n, u_{n+1}) = 1$ we see that $d \mid 2$. Thus $d = (u_n, v_n) \le 2$. ∎

    For the true state of affairs about (u_n, v_n) see Problem 6 below.

    We now prove the Lucas-Lehmer primality test. It, or modifications of it, are still the mainstay in searching for larger Mersenne primes.

Theorem 7.55. Let p be an odd prime and define the sequence $\{r_n\}$ by

$$r_0 = 4 \text{ and } r_{n+1} = r_n^2 - 2, \text{ for } n \ge 0$$

Then $2^p - 1$ is a prime if and only if $r_{p-2} \equiv 0 \pmod{2^p - 1}$.

Proof. Since $r_0 = 4 = v_1$ and $b = -1$ for this sequence of $\{v_n\}$ we see, by (a) of Lemma 7.55.4 and induction, that $r_n = v_{2^n}$, for $n \ge 0$. By (c) of Lemma 7.55.4, we see that u_n and v_n have no odd prime factor in common. Thus, if $r_{p-2} \equiv 0 \pmod{2^p - 1}$, we must have, by Theorem 7.46,

$$u_{2^{p-1}} = u_{2^{p-2}} v_{2^{p-2}} \equiv 0 \pmod{2^p - 1}$$

and $u_{2^{p-2}} \not\equiv 0 \pmod{2^p - 1}$.

    Let $m = m(2^p - 1)$ be the rank of apparition of $2^p - 1$. By Lemma 7.55.3, we see that $m \mid 2^{p-1}$ and, by the above, $m \nmid 2^{p-2}$. Thus $m = 2^{p-1}$. Suppose $2^p - 1 = q_1^{\alpha_1} \cdots q_r^{\alpha_r}$, where the q_i are odd primes since $2^p - 1$ is odd. Also $2^p - 1 \equiv (-1)^p - 1 = -2 \pmod 3$. Thus all of the q_i must be greater than 3. By Corollary 7.55.2.1, Corollary 7.54.1 and Lemma 7.55.3, we see that $u_n \equiv 0 \pmod{2^p - 1}$, where

$$n = [q_1^{\alpha_1 - 1}(q_1 + \varepsilon_1), \ldots, q_r^{\alpha_r - 1}(q_r + \varepsilon_r)]$$

and each ε_j is 1 or -1. Thus we must have that n is a multiple of $m = 2^{p-1}$.

Let

$$Q = \prod_{j=1}^{r} q_j^{\alpha_j - 1}(q_j + \varepsilon_j).$$

Then $q_i \geq 5$ implies that

$$Q \leq \prod_{j=1}^{r} q_j^{\alpha_j - 1}(q_j + q_j / 5) = (6/5)^r (2^p - 1).$$

Since each $q_j + \varepsilon_j$ is even we also have $n \leq Q/2^{r-1}$ since a factor of 2 is lost each time the lcm of two even numbers is computed. Therefore

$$m \leq n \leq 2(3/5)^r(2^p - 1) < 4(3/5)^r m < 3m.$$

This implies that $r \leq 2$, and so $n = m$ or $n = 2m$. In either case n is a power of 2. Thus $a_1 = 1$ and $a_r = 1$. If 2^p - 1 is not a prime, we must have

$$2^p - 1 = (2^k + 1)(2^l - 1),$$

where $2^k + 1$ and 2^l - 1 are both primes. Since p is odd this can't happen. Thus 2^p - 1 is a prime.

Now suppose that 2^p - 1 is a prime. We have, by Lemma 7.55.1 and the binomial theorem,

(7.63)
$$v_{2^{p-1}} = \left(2 + \sqrt{3}\right)^{2^{p-1}} + \left(2 - \sqrt{3}\right)^{2^{p-1}} = \left(\frac{\sqrt{2} + \sqrt{6}}{6}\right)^{2^p} + \left(\frac{\sqrt{2} - \sqrt{6}}{6}\right)^{2^p}$$

$$= 2^{1-2^p} \sum_{k=0}^{2^p} \binom{2^p}{2k}(\sqrt{2})^{2^p - 2k}(\sqrt{6})^{2k} = 2^{1-2^p} \sum_{k=0}^{2^p} \binom{2^p}{2k}3^k.$$

Since 2^p - 1 is prime and

$$\binom{2^p}{2k} = \binom{2^p - 1}{2k} + \binom{2^p - 1}{2k - 1}$$

we see that the binomial coefficients in (7.63) are all divisible by 2^p - 1 except for $k = 0$ and $k = 2^{p-1}$. Thus

(7.64)
$$2^{2^{p-1}-1} v_{2^{p-1}} \equiv 1 + 3^{2^{p-1}} \pmod{2^p - 1}.$$

Now $2 \equiv \left(2^{(p+1)/2}\right)^2 \pmod{2^p - 1}$. Thus, by Fermat's theorem,

$$2^{2^{p-1}-1} \equiv \left(2^{(p+1)/2}\right)^{2^p - 2} \equiv 1 \pmod{2^p - 1}.$$

Since $2^p - 1 \equiv 1 \pmod 3$ and $2^p - 1 \equiv 3 \pmod 4$ we see, by Euler's criterion and the law of quadratic reciprocity (Theorem 3.5), that

$$3^{2^{p-1}-1} \equiv -1 \pmod{2^p - 1}.$$

Thus, by (7.64), $v_{2^{p-1}} \equiv -2 \pmod{2^p - 1}$. Thus

$$v_{2^{p-2}}^2 = v_{2^{p-1}} + 2 \equiv 0 \pmod{2^p - 1}.$$

and the result follows.   ∎

Example 7.16. Show, by Theorem 7.55, that

   (a) $2^5 - 1$ is prime;

   (b) $2^{13} - 1$ is prime.

Solution of (a) Here $p = 5$. Thus we want to check r_3. We have $r_0 = 4$, $r_1 = 42 - 2 = 14$, $r_2 = 14^2 - 2 = 194$ and $r_3 = 194^2 - 2 = 37634$. Since $2^5 - 1 = 31$ and $37634 = 31 \cdot 1214$ we see, by Theorem 7.55, that 31 is prime.

Solution of (b) As we can see the sequence of r's grows quite rapidly. Here we want to check r_{11}. Clearly this is going to get out of hand quite quickly. Since $2^{13} - 1 = 8191$ we can, with computer or calculator, make this more manageable if we compute r_n modulo 8191. Then we want to show that r_{11} is 0 modulo 8191. We have, modulo 8191,

$$r_0 \equiv 4, r_1 \equiv 14, r_2 \equiv 194, r_3 \equiv 4870, r_4 \equiv 3953, r_5 \equiv 5970,$$

$$r_6 \equiv 1857, r_7 \equiv 36, r_8 \equiv 1294, r_9 \equiv 3470, r_{10} \equiv 128$$

and $r_{11} \equiv 0 \pmod{8191}$. Thus 8181 is prime.   □

These results just barely scratch the surface of what can be done. The general topic of difference equations looks at these results from a more algebraic/analytic point of view. On the other hand, one can just study the relations between these objects as we have done here. Indeed, there is a journal, *The Fibonacci Quarterly*, devoted to the study and application of recursion relations.

Problem Set 7.10

1. Modifying the results if necessary prove the analogs of the results in this section for the case when the quadratic equation of Theorem 7.42 has a double root.

    To make the calculations below simpler assume that $u_0 = 0$ and $u_1 = 1$. It will make a good exercise to extend these results to the general case.

2. Show that $u_{3k} = u_k(v_k^2 - (-b)^k)$.

3. (a) $u_{n-k}u_{n+k} - u_n^2 = -(-b)^{n-k}u_k^2$.

   (b) $v_{n+1}v_{k+1} + bv_nv_k = cu_{n+k+1}$.

   (c) $v_{k+1}u_{n+1} + bv_ku_n = v_{n+k+1}$.

4. Find formulas for each of the following sums.

   (a) $\displaystyle\sum_{j=0}^{N} v_j$

   (d) $\displaystyle\sum_{j=0}^{N} u_ju_{j+1}$

   (b) $\displaystyle\sum_{j=1}^{N} ju_j$

   (e) $\displaystyle\sum_{j=0}^{N} v_jv_{j+1}$

   (c) $\displaystyle\sum_{j=1}^{N} jv_j$

   (f) $\displaystyle\sum_{j=0}^{N} u_j^2$

5. Prove the following identities.

   (a) $v_n^2 - cu_n^2 = 4(-b)^n$.

   (b) $v_nv_{n+2} - cu_{n+1}^2 = a^2(-b)^n$.

   (c) $cu_nu_{n+2} - v_{n+1}^2 = -a^2(-b)^n$.

6. (a) If a is odd and b is even, show that $(u_n, v_n) = 1$.

   (b) If a is even and b is odd, show that

$$(u_n, v_n) = \begin{cases} 2 & \text{if } 2|n \\ 1 & \text{else} \end{cases}.$$

   (c) If a and b are both odd, show that

$$(u_n, v_n) = \begin{cases} 2 & \text{if } 3|n \\ 1 & \text{else} \end{cases}.$$

7. Prove the following identities.

   (a) $\displaystyle\sum_{k=0}^{n} \binom{n}{k} b^{n-k}a^k u_{k+p} = u_{2n+p}$.

   (c) $\displaystyle\sum_{k=0}^{n} \binom{n}{k} a^{n-k}(-1)^k u_{k+p} = -(-b)^p u_{n-p}$.

   (b) $\displaystyle\sum_{k=0}^{n} \binom{n}{k} b^{n-k}a^k v_{k+p} = v_{2n+p}$.

   (d) $\displaystyle\sum_{k=0}^{n} \binom{n}{k} a^{n-k}(-1)^k v_{k+p} = (-b)^p v_{n-p}$.

8. Prove the analog of Theorem 7.52 for the sequence $\{v_n\}$.

9. This exercise calculates (v_k, v_j) in one case.

   (a) Let $k, j \geq 1$. If $k | j$ and j/k is odd, show that $v_k | v_j$.

   (b) Let m and n be odd with $(m, n) = 1$. Show that there exist integers x and y such that $mx + ny = 1$ with x odd and $y \equiv 2(\text{mod } 4)$.

(c) Let $k, j \geq 1$ and $d = (k, j)$. If k/d and j/d are both odd, show that $(v_k, v_j) = v_d$.

(d) Can this result be extended?

10. Show that if n and k are positive integers, then

$$au_{nk} \equiv u_{k+1}^n - b^n u_{k-1}^n \pmod{u_k^3}.$$

11. Let p be an odd prime.

(a) If $p \nmid ac$, show that

$$2av_p \equiv a(a+b)\left(1+\left(\tfrac{c}{p}\right)\right) - b^2\left(\tfrac{c}{p}\right) \pmod{p},$$

where $\left(\tfrac{c}{p}\right)$ is the Legendre symbol.

(b) If $p \mid a$, show that

$$v_p \equiv 0 \pmod{p}.$$

(c) If $p \nmid a$ and $p \mid c$, show that

$$2v_p \equiv (a+b) \pmod{p}.$$

12. In the general setting (that is, not assuming $u_0 = 0$ and $u_1 = 1$) compute

$$\lim_{n \to +\infty} \frac{u_{n+1}}{u_n} \quad \text{and} \quad \lim_{n \to +\infty} \frac{v_{n+1}}{v_n}.$$

13. Define the sequence $\{u_n\}$ by $u_0 = 0$, $u_1 = 1$ and, for $n \geq 0$,

$$u_{n+2} = -u_{n+1} - 2u_n.$$

Show that, for all $n \geq 1$, $2^{n+1} - 7u_{n-1}^2$ is a perfect square.

14. Given the sequence $\{u_n\}$ as defined by (7.58) for $n \geq 0$ define another sequence $w_n = c\alpha^n + d\beta^n$, where c and d are constants.

(a) Show that if x satisfies the equation $x^2 - ax - b = 0$, then, for $n \geq 1$,

$$x^n = u_n x + bu_{n-1}.$$

(Hint: use induction.)

(b) Show that if $j \geq 2$, then

$$\sum_{k=0}^{n} \binom{n}{k} \left(bu_{j-1}\right)^{n-k} u_j^k w_k = w_{jn}.$$

15. Suppose the sequence $\{g_n\}$ satisfies $g_0 = 0$, $g_1 = 4$ and, for $n \geq 0$,

$$g_{n+2} = 3g_{n+1} - g_n - 2.$$

Express g_n in terms of Fibonacci and/or Lucas numbers. (Hint: relate the solutions of the two recursions $u_{n+2} = au_{n+1} + bu_n$ and $w_{n+2} = aw_{n+1} + bw_n + f$, where f is a constant.)

16. Find an analog of Theorem 7.55 for repunits: $R_p = \frac{1}{9}(10^p - 1)$.

Additional Problems For Chapter Seven

General Problems

1. Let $A_1, \ldots, A_n$ be subsets of a finite set B. Show that

$$\#\{B - (A_1 \cup \cdots \cup A_n)\} = \#B + \sum_{j=1}^{n}(-1)^j \sum_{1 \le i_1 < i_2 < \cdots < i_j \le n} \#\{A_{i_1} \cap \cdots \cap A_{i_j}\}.$$

Here $\#$ denotes the number of elements in the set. (Hint: use induction.)

2. Let f be a function defined on the rationals in the interval $[0, 1]$. Define the two arithmetic functions F_1 and F_2 by

$$F_1(n) = \sum_{1 \le k \le n} f(k/n) \text{ and } F_2(n) = \sum_{\substack{1 \le k \le n \\ (k,n)=1}} f(k/n)$$

(a) Show that $F_2 = \mu * F_1$.

(b) Show that

$$\mu(n) = \sum_{\substack{1 \le k \le n \\ (k,n)=1}} e^{2\pi i k/n}.$$

(c) Show that if a and b are integers, then

$$(a,b) = \sum_{m=0}^{a-1}\sum_{n=0}^{a-1} \frac{1}{a} e^{2\pi i bmn/a}.$$

3. Let f be a multiplicative function such that $F = f * e$ is not identically zero. Show that

$$\frac{\displaystyle\sum_{\substack{e|n \\ e \text{ even}}} f(n/e)}{\displaystyle\sum_{\substack{d|n \\ d \text{ odd}}} f(n/d)} = \begin{cases} 0 & \text{if } n \text{ is odd} \\ F(2^{k-1})/f(2^k) & \text{if } 2^k\|n, \text{ if } k \ge 1 \end{cases}.$$

4. Let $\rho(n)$ be the number of $a \le n$ such that $a^m \equiv a \pmod{n}$ for some $m > 1$. Prove that ρ is multiplicative.

5. Prove the following inversion formula due to I. M. Vinogradov. Consider an arbitrary set of n pairs $\{(\alpha_j, d_j)\}_{j=1}^n$, where the α_j may be complex and the d_j are positive integers. If m is any integer, let

$$S_m = \sum_{d_j \equiv 0 \ (\text{mod } m)} \alpha_j$$

and let

$$S = \sum_{d_j = 1} \alpha_j.$$

Show that

$$S = \sum_{m=1}^{+\infty} \mu(m) S_m.$$

6. If $f(1) = 1$ and $f * f = a$, where a is a multiplicative function, show that f is also multiplicative. Show that the other solution to $f * f = a$ is $-f$. Find the multiplicative function f so that $f * f = a$, where a is one of the following multiplicative functions.

   (a) $a = \delta$

   (b) $a = e$

   (c) $a = \mu$

   (d) $a = \mu^2$

   (e) $a = \varphi$

   (f) $a = \sigma$

   (g) $a = \sigma_k, \ k \geq 1$

   (h) $a = d$

   (i) $a = d_k, \ k \geq 3$.

7. Find all real-valued multiplicative functions f such that

   (a) $\sum_{d|n} f(d) = (\mu f)(n);$

   (b) $\sum_{d|n} f(d) = f(n);$

   (c) $\sum_{d|n} \varphi(d) f(n/d) = (df)(n);$

   (d) $\sum_{d|n} \mu^2(d) f(n/d) = f(n^2);$

   (e) $\sum_{d|n} f(d^2) = f^2(n).$

   (Hint: check values on prime powers.)

8. If f is a multiplicative function, define the **norm of** f, N_f, to be the arithmetic function

$$N_f(n) = \sum_{d|n^2} f(n^2/d)\lambda(d) f(d).$$

where λ is Liouville's function. Find all f such that

   (a) $N_f = \delta;$

   (b) $N_f = e;$

   (c) $N_f = \mu;$

   (d) $N_f = \mu^2.$

9. Let f be a multiplicative function and g be a completely multiplicative function. If

$$f(m)f(n) = \sum_{d|(m,n)} g(d)f(mn/d^2)$$

for all positive integers m and n, then f is said to be a **specially multiplicative function** (or **quadratic function**) with associated completely multiplicative function g. For this exercise f and g will have this meaning.

(a) Show that σ_k and $r_2(n)/4$ are specially multiplicative functions.

(b) Show that for all $n \geq 1$ and primes p

$$f(p^{n+1}) = f(p)f(p^n) - g(p)f(p^{n-1}).$$

(c) Show that for all positive integers m and n

$$f(mn) = \sum_{d|(m,n)} f(m/d)f(n/d)g(d)\mu(d).$$

(d) Show that

$$f^{*-1}(p^e) = \begin{cases} -f(p) & \text{if } e = 1 \\ g(p) & \text{if } e = 2. \\ 0 & \text{if } e \geq 3 \end{cases}$$

(e) Show that

$$\Lambda_f(p^e) = \begin{cases} 0 & \text{if } e = 0 \\ f(p)\log p & \text{if } e = 1. \\ (f(p^e) - g(p)f(p^{e-2}))\log p & \text{if } e \geq 2 \end{cases}$$

(f) Show that if e is a positive integer, then

$$f(p^e) = \sum_{j=1}^{[e/2]} (-1)^j \binom{e-j}{j} f^{e-2j}(p)g^j(p).$$

(g) Show that $f^2 = N_f * \mu^2 g * g$, where N_f is the norm of f (see Problem 8 above.) Show that

$$f(n^2) = (f^2 * g^{*-1})(n) = (N_f * \mu^2 g)(n).$$

(h) Define the arithmetic functions G_g and G_1 by

$$G_g(n) = \sum_{d^2|n} g^{*-1}(d) \text{ and } G_1(n) = \sum_{d|n} g^{*-1}(d)G_g(n/d).$$

Show that

(i) $f\sigma_k = f * T_k f * T_{k/2} G_1;$

(ii) $fr_2 = 4f * f\chi_4 * G_{g\chi_4},$

where χ_4 is the nonprincipal character modulo 4. (Hint: see Theorem 6.8.)

(i) Let h be an arithmetic function and define h_k by

$$h_k(n) = \begin{cases} h(n) & \text{if } n|k \\ 0 & \text{else} \end{cases}$$

and let $H_k = \mu * h_k$, with $H = H_0$. Show that

$$\sum_{d|(m,n)} f(m/d)f(n/d)g(d)H_k(d) = \sum_{d|(m,n,k)} h(d)g(d)f(mn/d^2).$$

Show that

(i) $\displaystyle\sum_{d|(m,n)} f(m/d)f(n/d)g(d)c(k,d) = \sum_{d|(m,n,k)} dg(d)f(mn/d^2);$

(ii) $\displaystyle\sum_{d|(m,n)} \sigma_k(m/d)\sigma_k(n/d)d^k c(k,d) = \sum_{d|(m,n,k)} d^{k+1}\sigma_k(mn/d^2).$

(j) Show that

$$\sum_{d|(m,n)} f(m/d)f(n/d)g(d)\varphi(d) = \sum_{d|(m,n)} dg(d)f(mn/d^2).$$

(k) If h is completely multiplicative, show that hf is specially multiplicative with associated completely multiplicative function $h^2 g$.

(l) If F and G are completely multiplicative functions, then $H = F * G$ is specially multiplicative with associated completely multiplicative function FG.

10. Let f and g be multiplicative functions and define $S(n, k)$ by

$$S(n,k) = \sum_{d|(n,k)} f(d)h(n/d).$$

(a) Show that if $(n_1, n_2) = (k_1, k_2) = (k_1, n_2) = (n_1, k_2) = 1$, then

$$S(n_1 n_2, k_1 k_2) = S(n_1, k_1)S(n_2, k_2).$$

(b) Let h be a multiplicative function and let $g = h\mu$. Let f be a completely multiplicative function such that $f(p) \neq 0$ and $f(p) \neq h(p)$ for all primes p. If $F = f * h$, show that

$$S(n,k) = h(n/(n,k))\mu(n/(n,k))F(n)/F(n/(n,k)).$$

(c) Show that the function F of (b) satisfies

$$F(mn) = F(m)F(n)f((m,n))/F((m,n)).$$

(d) Let
$$\alpha_n(m) = \tfrac{1}{n} \sum_{d|(m,n)} dg(d) f(n / d).$$

Show that
$$S(n,k) = \sum_{m(n)} \alpha_n(m) e^{2i\pi k m / n}.$$

(e) Let h be an arithmetic function and define the function H_k by
$$H_k(n) = \sum_{d|(k,n)} \mu(k / d) h(d).$$

(i) Show that $H_1(n) = h(1)$ and that for $a \geq 1$
$$H_{p^a}(n) = \begin{cases} h(p^a) - h(p^{a-1}) & \text{if } p^a | n \\ -h(p^{a-1}) & \text{if } p^{a-1} \| n . \\ 0 & \text{if } p^{a-1} | n \end{cases}$$

(Hint: see Corollary 7.3.6.1.)

(ii) Show that $H_k (n)$ is a multiplicative function of k.

(iii) Let $\gamma(n)$ denote the core of n (see Problem 4 of Problem Set 7.2) and let $n^* = n / \gamma(n)$. Show that
$$\mu(k) H_k(n^*) = \sum_{d|(n,k)} h(d) \mu(d).$$

(iv) If h is a completely multiplicative function, show that
$$H_k(nk^*) = h(k^*) H_{\gamma(k)}(n).$$

(v) If k and n are square-free, show that
$$\mu(k) H_k(n) = \mu(n) H_n(k).$$

(vi) If h is a completely multiplicative function, $h(n) \neq 0$ and $h(k) \neq 0$ for all n and k, show that
$$\frac{\mu(\gamma(k))}{h(k^*)} H_k(nk^*) = \frac{\mu(\gamma(n))}{h(n^*)} H_n(kn^*).$$

11. If h is a nonnegative integer, let
$$\delta_h(n) = \sum_{d|n} \mu(d) \log^h d.$$

(a) Compute δ_0, δ_1 and δ_2.

(b) Show that $\delta_h(n) = 0$ if n is divisible by more than h distinct primes.

12. If $f*e = g$, then show that

$$\sum_{d^2|m} f(d)d(m/d^2) = \sum_{d^2|m} g(d)\theta(m/d^2).$$

13. Let

$$a(n-k,k) = \sum_{\substack{d|n-k \\ d>k}} b(d).$$

Show that

$$\sum_{k=0}^{n-1} a(n-k,k) = \sum_{m=1}^{n} b(m).$$

14. Let

$$\sigma_k^*(n) = \sum_{\substack{d|n \\ (d,n/d)=1}} d^k$$

Find formulas for σ_1^*, σ_2^* and σ_3^*.

15. Show that
    (a) if $3 \mid n$, $5 \nmid n$ and $9 \nmid n$, then $n \in P_3$ implies that $45n \in P_4$;
    (b) if $3 \nmid n$ and $3n \in P_{4k}$, then $n \in P_{3k}$;
    (c) if $n \in P_3$ and 7, 9 and 13 do not divide n, then $273n \in P_4$;
    (d) generalize (a), (b) and (c);
    (e) prove that if $n > 4$, then a P_n has at least $n + 1$ distinct prime factors;
    (f) prove that a P_3 has at least 3 distinct prime factors, a P_4 has at least 4, a P_6 at least 9 and a P_7 at least 14. '

16. Let N be an odd perfect number.
    (a) If m is given, show that there is at most one prime power $p^r, p \equiv r \equiv 1 \pmod 4, p \nmid m$, such that $N = p^r m^2$.
    (b) If $\omega(N) = 3$, show that two of the factors must be 3 and 5.
    (c) Show that there do not exist not exist any N such that $\omega(N) = 3$.

17. (a) Let $p > 3$ be a prime. Show that there does not exist a k such that $3p \mid \sigma_k(3p)$.
    (b) Show that $\sigma(kq - 1) \equiv 0 \pmod k$ for all positive integers q if and only if $k > 1$ and $k \mid 24$.

18. Show that $\#\{(m, n) \in \mathbf{Z}^2: [m, n] = N\} = d(N^2)$.

19. Let, for k a positive integer,

$$\rho_k(n) = \sum_{\substack{1 \le m \le n \\ (m,n)=(m+k,n)=1}} 1 \quad .$$

(a) Show that

$$\rho_1(n) = n\prod_{p|n}(1-2/p).$$

(b) Show that

$$1+\rho_2(n) = \begin{cases} n\displaystyle\prod_{p|n}(1-2/p) & \text{if } (2,n)=1 \\ 2^{\alpha-1}N\displaystyle\prod_{p|N}(1-2/p) & \text{if } n=2^\alpha N, \alpha \geq 1 \text{ and } (2,N)=1 \end{cases}.$$

(c) Show that

$$\sum_{d|(n,r)}\varphi(nr/d^2)d\mu(d) = \varphi(n/u)\varphi(r/u)\rho_1(u)$$

where u is the greatest square-free divisor of n and r such that $(n, n/u) = (r, r/u)$.

20. Let $p_1, \ldots, p_n$ be the distinct primes dividing N. Show that

$$\frac{\sigma(p_1\cdots p_n)}{p_1\cdots p_n} \leq \frac{\sigma(N)}{N} \leq \frac{p_1\cdots p_n}{\varphi(p_1\cdots p_n)}.$$

21. (a) Recall the definition of $\varphi_k(n)$ from Problem 12 of Problem Set 7.6. Find closed form expressions of $\varphi_2(n)$ and $\varphi_3(n)$ in terms of the prime factors of n. (Hint: see Example 7.12.)

(b) Show that, in general, if

$$S_k(m) = 1^k + \cdots + m^k,$$

then

$$\varphi_k(n) = \sum_{d|n}\mu(d)d^k S_k(n/d).$$

22. (a) Let n be an integer, $n > 2$. If $n \equiv r \pmod 4$, with $r = -1, 0, 1$ or 2, then

$$\sum_{\substack{1\leq m\leq n/2 \\ (m,n)=1}} m = \frac{1}{8}\left(n\varphi(n) - |r|\prod_{p|n}(1-p)\right).$$

(b) Find a similar expression for

$$\sum_{\substack{1\leq m\leq n/2 \\ (m,n)=1}} m^2.$$

23. Show that

$$\mu(n)\mu(m) = \sum_{d|(n,m)} \mu(nm/d)\mu(d).$$

24. Let f be an increasing multiplicative function. Show that there is an $\alpha \geq 0$ such that $f = T_\alpha e$.

25. Prove the Brauer-Rademacher identity

$$\varphi(r) \sum_{\substack{d|r \\ (d,n)=1}} d\mu(r/d)\varphi(d) = \mu(r) \sum_{d|(r,n)} d\mu(r/d).$$

26. Let $r \geq 1$ and $n \geq 1$ be integers and let $t = (n, r)$.

   (a) Let f be a strongly multiplicative function (see Problem 4 of Problem Set 7.3.) Let h be a multiplicative function with $h(p) = f(p) - 1$. Show that

   $$\sum_{\substack{d|r \\ (d,n)=1}} f(d)\mu(r/d) = \mu(r)\mu(r/t)h(r/t).$$

   (b) Show that

   $$\sum_{\substack{d|r \\ (d,n)=1}} \theta(d)\mu(r/d) = \mu(r)\mu(r/t).$$

27. Let k be a positive integer and define ψ_k by

$$\psi_k(n) = \sum_{d|n} \mu^2(n/d)d^k.$$

($\psi = \psi_1$ is called **Dedekind's function**.)

   (a) Show that ψ_k is multiplicative.

   (b) Show that

   $$\psi_k(n) = \sum_{d|n} \left(\frac{d}{(d,n/d)}\right)^k J_k\left(\frac{d}{(d,n/d)}\right).$$

   where J_k is the Jordan totient (see Problem 23 of Problem Set 7.6.)

   (c) Show that

   $$\psi_k(n) = n^k \prod_{p|n}\left(1 + 1/p^k\right) = J_{2k}(n)/J_k(n).$$

   (d) Find ψ_k^{*-1} and Λ_{ψ_k}.

   (e) Show that $\psi_k^2(n) \leq 2^{\omega(n)}\sigma_{2k}(n)$. (Hint: use Cauchy's inequality.)

28. Let

$$\alpha_k(n) = \sum_{d|n}\left\{d^k - (d-1)^k\right\}d\mu(n/d).$$

   (a) Show that $\alpha_0 = \mu$ and $\alpha_1 = \varphi$.

   (b) If $k \geq 1$, show that $\varphi(n) | \alpha_k(n)$.

29. (a) Show that $\sigma(n) + \varphi(n) = 2n$ if and only if $n = 1$ or n is a prime.

   (b) If $n = 2^a 3q$, where q is a prime of the form $7 \cdot 2^{a-2} - 1$, show that
        $\sigma(n) + \varphi(n) = 3n$.

   (c) Find all solutions of the form n $= 2^a 3^b pq$, where p and q are distinct odd
        primes, of the equation $\sigma(n) + \varphi(n) = 4n$.

   (d) What can be said in general about the equation $\sigma(n) + \varphi(n) = kn$?

30. Let $f(x)$ be a polynomial with integer coefficients and let $\rho_f(n) = \#\{m:$
    $0 \leq m \leq n - 1$ such that $(f(m), n) = 1\}$.

   (a) Show that if $(m, n) = 1$, then $\rho_f(mn) = \rho_f(m)\rho_f(n)$.

   (b) Show that $\rho_f(p^a) = p^{a-1}(p - \alpha_p)$, where $\alpha_p = \#\{m: 0 \leq m \leq p - 1$
        such that $p | f(m)\}$.

   (c) Show that if $f(x) = (x - e_1)\cdots(x - e_k)$, where the $e_1, \ldots, e_k$ are integers,
        then α_p is the number of e_j that are incongruent modulo p.

   (d) Compute $\rho_f(n)$ for $f(x) = x$, $x(x + 1)$, $x^3 - x$, $x(x + 1)/2$ and
        $x(x + 3)/2$.

31. Let p be an odd prime and let $F(x)$ be a function with period p. Let $\left(\frac{x}{p}\right)$ denote
    the Legendre symbol.

   (a) Show that

$$\sum_{x=1}^{p-1} F(x) + \sum_{x=1}^{p-1}\left(\frac{x}{p}\right)F(x) = \sum_{x=1}^{p-1} F(x^2).$$

   (b) Show that if $p \nmid a$, then

$$\sum_{x=0}^{p-1} F(x) + \sum_{x=0}^{p-1}\left(\frac{x^2-4a}{p}\right)F(x) = \sum_{x=1}^{p-1} F(x + a\bar{x}),$$

   where $x\bar{x} \equiv 1 \pmod{p}$.

   (c) Show that

$$\sum_{x=0}^{p-1}\left(\frac{x-2}{p}\right)F(x) + \sum_{x=0}^{p-1}\left(\frac{x+2}{p}\right)F(x) = \sum_{x=1}^{p-1}\left(\frac{x}{p}\right)F(x + \bar{x}).$$

   (Note that (a) generalizes a property of Gauss sums, Theorem 7.29. A version
   of (b) can be found in Problem 30 of the Additional Problems for Chapter 3.
   The last result is due to K. S. Williams and generalizes identities appearing in
   Problem 12 of Problem Set 7.8.)

32. Let $f(x)$ be a polynomial with integer coefficients and let

$$S_{f(x)}(n) = \sum_{m=1}^{n} e^{2\pi i f(m)/n}$$

where we assume that the coefficients of f have no factors in common with n. Show that if $(n, m) = 1$, then

$$S_{f(x)}(nm) = S_{f(mx)/m}(n)S_{f(nx)/n}(m).$$

33. Let $S(u, v, n)$ denote the Kloostermann sum.

(a) If p is a prime such that $p \mid qrn$ and ρ and μ are positive integers, show that

$$S(rp^\rho, qp^\mu, n) = S(r, qp^{\rho+\mu}, n) + pS(rp^{\rho-1}, qp^{\mu-1}, n/p).$$

(b) Show that if $(n, m) = 1$, $n\overline{n} \equiv 1 \pmod{m}$ and $m\overline{m} \equiv 1 \pmod{n}$, then

$$S(u, v, nm) = S(u, v\overline{m}^2, n)S(u, v\overline{n}^2, m).$$

(c) Show that

$$S(u, v, n) = \sum_{d \mid (u, v, n)} dS(1, uv/d^2, n/d).$$

(d) Show that

$$S(1, uv, n) = \sum_{d \mid (u, v, n)} d\mu(d)S(u/d, v/d, n/d).$$

34. Let $m(n)$ denote the rank of apparition of the number n in the binary recursion sequence $\{u_k\}$: $u_0 = 0$, $u_1 = 1$ and, for $n \geq 0$, $u_{n+2} = au_{n+1} + bu_n$.

(a) If n is a positive integer, $n = n_1n_2$ with $(n_1, n_2) = 1$ and if $m(n_1)$ and $m(n_2)$ both exist, show that $m(n)$ exists and that

$$m(n) = [m(n_1), m(n_2)],$$

where $[,]$ denotes the lcm.

(b) Show that $m(p)$ exists for any prime p.

(c) Suppose that $p^\alpha \| u_{m(p)}$ and $p^\beta \| u_{pm(p)}$. Show that if n is a positive integer and if $p \nmid b$, then

(i) $m(p^n)$ exists

and

(ii) $m(p^n) = \begin{cases} m(p) & \text{if } 1 \leq n \leq \alpha \\ pm(p) & \text{if } \alpha < n \leq \beta. \\ p^{n-\beta+1}m(p) & \text{if } n \geq \beta \end{cases}$

35. If $m, n \geq 1$, show that

    (a) $u_{n-1} \mid u_{nm-1}$

    and

    (b) $(u_{n-1}, u_{m-1}) = u_{(n,m)-1}$.

    (c) Can an analogous statement be made for (v_{m-1}, v_{n-1})?.

36. Show that if n and k are positive integers, then

    $$a^n u_{nk} \equiv f_n u_k u_{k+1}^{n-1} \pmod{u_k^2}.$$

    where $f_1 = a$ and $f_m = bf_{m-1} + a^m$, for $m \geq 2$.

37. Find closed form expressions for

    $$\sum_{k=0}^{N} u_{ak+b} \text{ and } \sum_{k=0}^{N} v_{ak+b}.$$

38. If $x_0 = 1$, $x_1 = 2$, $x_2 = 1$ and $x_{n+3} = x_{n+2} + 4x_{n+1} - 4x_n$, for $n \geq 0$, find a formula for x_n. (Hint: consider the expression $x_{n+1} + kx_n + jx_{n-1}$.) Prove results analogous to Theorems 7.42 to 7.54.

39. (a) If a_0 is given, $a_1 = ma_0 + p$, p a given integer, and $a_{n+2} = 2ma_{n+1} - a_n$, for $n \geq 0$, show that $(m^2 - 1)(a_n^2 - a_0^2) + p^2$ is a perfect square.

    (Hint: write a_n as a function of n, without the recurrence.)

    (b) What relations must hold to obtain an analogous result in the general case?

40. Show that if $n = p_1^{\alpha_1} \cdots p_r^{\alpha_r}$ is a perfect number, then

    $$2 < \prod_{i=1}^{r} \frac{p_i}{p_i - 1} < 4.$$

    Show that if n is odd, then the upper bound of 4 can be reduced to $2\sqrt[3]{2}$.

41. Show that

    $$\sum_{k \mid m} d^2(k) J_l(m/k) = \sum_{k \mid m} (m/k)^l d(k^2).$$

42. Let m be an integer, $m \geq 1$, and let $T(n) = \#\{1 \leq n \leq m : (m,n) = 1$ and $m \not\equiv n \pmod 2\}$. Show that

    $$T(n) = \begin{cases} \varphi(n) & \text{if } n \equiv 0 \pmod 2 \\ \frac{1}{2}\varphi(n) & \text{if } n \equiv 1 \pmod 2 \end{cases}.$$

43. Show that

    $$\sum_{d \mid n} \sum_{e \mid d} J_k(n/d) J_l(n/e) e^k = n^{k+l}.$$

    (This identity is due to G. Métrod.)

44. Show that if $n \geq 1$, then there exists at least one highly composite number m such that $n < m \leq 2n$.

45. Let f be a multiplicative function and define the function $K(m, n)$ by

$$K(m,n) = \begin{cases} (-1)^r & \text{if } \gamma(m) = \gamma(n) \text{ and } \omega(m) = k \\ 0 & \text{else} \end{cases}.$$

Show that Vaidyanathaswamy's identity holds:

$$f(mn) = \sum_{\substack{d_1 \mid m \\ d_2 \mid n}} f(m/d_1)f(n/d_2)f^{*-1}(d_1 d_2)K(d_1, d_2).$$

46. Let f and g be completely multiplicative functions and define $\varphi_{f,g}$ and $\sigma_{f,g}$ by $\varphi_{f,g} = f * \mu g$ and $\sigma_{f,g} = f * g$. Show that

$$\sum_{d \mid n} \lambda(d) \varphi_{f,g}(d) \sigma_{f,g}(n/d) = \sum_{k^2 \mid n} f(k^2)g(n/k^2)\theta(n/k^2).$$

47. If f is a completely multiplicative function, show that

$$\prod_{\substack{m \leq n \\ (m,n)=1}} f(m) = (f(n))^{\varphi(n)} \prod_{d \mid n} \left(f(d)^d / f(d!) \right)^{\mu(n/d)}.$$

(Hint: see Problem 21 of Problem Set 7.6.)

48. Let k be a positive integer. Show that a multiplicative function is the convolution of k completely multiplicative functions if and only if $f^{*-1}(p^a) = 0$ for all primes p and all $a \geq k + 1$. (Hint: see Problem 9(c) above.)

49. We say that an operation D on A is a **derivation** if and only if $Df \in A$ for all $f \in A$ and

    (i) $D(\alpha f + \beta g) = \alpha Df + \beta Dg$, for all complex numbers α and β

and

    (ii) $D(f * g) = f * Dg + g * Df$, for all $f, g \in A$

(a) Let h be an arithmetic function. Show that $Df = hf$ defines a derivation on A if and only if h is completely additive.

(b) Let D be a derivation on A and let $g \in A$. Show that $D_1 f = Df * g$ is also a derivation on A.

50. (a) Let f be an arithmetical function. Show that there exists a multiplicative function g such that

$$\sum_{k=1}^{n} f((k,n)) = \sum_{d \mid n} f(d)g(n/d).$$

(b) Show that

$$\sum_{k=1}^{n}(k,n)\mu((k,n)) = \mu(n).$$

(c) Show that

$$\sum_{k=1}^{n}(k,n)^{r} = \sum_{d|n}d^{r}\varphi(n/d).$$

51. This exercise deals with Jacobi sums which are useful in problems related to equations over finite fields (see [56].) If χ and λ are characters modulo p, then the **Jacobi sum**, $J(\chi,\lambda)$, is defined to be

$$J(\chi,\lambda) = \sum_{a(p)}\chi(a)\lambda(1-a).$$

(a) Show that if χ_0 is the principal character modulo p, then

$$J(\chi_0, \chi_0) = p$$

and if χ is a nonprincipal character modulo p, then

$$J(\chi_0, \chi) = 0.$$

(b) Show that if χ is a nonprincipal character modulo p, then

$$J(\chi,\overline{\chi}) = -\chi(-1).$$

(c) Show that if χ and λ are nonprincipal characters modulo p and $\chi\lambda \neq \chi_0$, then

$$J(\chi,\lambda) = \frac{G(\chi,1)G(\lambda,1)}{G(\chi\lambda,1)}.$$

(d) Let $\psi(n) = \left(\frac{n}{p}\right)$, the Legendre symbol. Show that

$$\sum_{n=0}^{p-1}\chi(1-n^{2}) = J(\chi,\psi).$$

(Hint: see Problem 7 of Problem Set 3.2.)

52. Let $A_m(n) = S(m, m, n)$, where S is the Kloostermann sum. If $a \geq 2$, show that, for any odd prime p,

$$A_m(p^a) = \begin{cases} 2p^{a/2}\cos(4\pi m/p^a) & \text{if } \alpha \text{ is even} \\ 2\left(\frac{m}{p}\right)p^{a/2}\cos(4\pi m/p^a) & \text{if } \alpha \equiv 1 \pmod 4. \\ -2\left(\frac{m}{p}\right)p^{a/2}\sin(4\pi m/p^a) & \text{if } \alpha \equiv 3 \pmod 4 \end{cases}$$

53. Suppose n is not a perfect square and that $n - 1 > \varphi(n) > n - n^{2/3}$. Show that n is a product of two distinct primes.

54. Let a, b, c and d be reals and suppose that the two sequences $\{x_n\}$ and $\{y_n\}$ satisfy the double recurrence

$$x_{n+1} = ax_n + by_n$$
$$y_{n+1} = cx_n + dy_n$$

for $n \geq 0$, where x_0 and y_0 are given. Find linear recurrences for the x_n and y_n involving only x_n and y_n, respectively. (Recall that double recurrences of this type showed up in our work on the equation $x^2 - Dy^2 = N$. See Corollary 6.34.1.)

55. If α and β are the roots of $x^2 - ax + 1 = 0$, where a is an integer, show that, for any positive integer n, we have

(a) $\alpha^n + \beta^n$ is an integer

and

(b) if p is a prime, $p \mid (a - 1)$, then $p \nmid (\alpha^n + \beta^n)$.

Can you generalize this result to the roots of $x^2 - ax + b = 0$, where a and b are integers?

56. Solve the nonhomogeneous binary recurrence

$$y_{n+1} = ay_n + by_{n-1} + cn + d,$$

for $n \geq 1$, where y_1 and y_2 are given. (Hint: let $\{w_n\}$ be a solution to $y_{n+1} = ay_n + by_{n-1}$, $n \geq 1$, and set $y_n = w_n + p(n)$, with $p(n)$ to be determined.)

57. (a) Show that

$$\sum_{\substack{d \mid n \\ \mu(n) = +1}} 1 = 2^{\omega(m)-1}.$$

(Hint: see Problem 1 above.)

(b) Show that

$$\sum_{\substack{d \mid n \\ \mu(n) = -1}} 1 = 2^{\omega(m)-1}.$$

58. Show that

$$\prod_{m=1}^{n} m^{2[n/m]-d(m)} = 1.$$

59. Call a number N **quasiperfect** if $\sigma(N) = 2N + 1$. Prove that every quasiperfect number is the square of an odd integer.

60. If f and g are multiplicative functions, let

$$F(n,r) = \sum_{\substack{d|r \\ (d,n)=1}} f(d)g(r/d).$$

Show that F is a multiplicative function of r. Is it also a multiplicative function of n?

61. Show that n is a product of twin primes if and only if either $\sigma(n) = n + 1 + 2\sqrt{n+1}$ or $\varphi(n) = n + 1 - \sqrt{n+1}$.

62. Let m be a positive integer. Show that

$$\sum_{\substack{abc|m \\ (a,b)=1}} \varphi(abc)\varphi(c) = m^2.$$

63. Prove the following result due to Polya and Vinogradov. Let χ be a nonprincipal character modulo p, p a prime. Then, for any integers N and H, $H > 0$, we have

$$\left| \sum_{n=N+1}^{N+H} \chi(n) \right| < 2\sqrt{p} \log p.$$

(Hint: see Problem 6 of Problem Set 7.8 and recall that $\sin x$ is symmetric about $\pi/2$ and Jordan's lemma: $0 \le \theta \le \pi/2$ implies that $\sin\theta \ge 2/\pi$.)

64. Prove that if p is a prime such that $2p + 1$ is composite, then $\varphi(n) = 2p$ has no solutions.

Computer Problems

1. (a) Write computer programs to calculate $d(n)$, $\sigma(n)$, $\varphi(n)$ and $\mu(n)$.
   (b) Calculate the values of $d(n)$, $\sigma(n)$, $\varphi(n)$ and $\mu(n)$ for $200 \le n \le 300$.

2. Find all values of n, $1 \le n \le 1000$, such that
   (a) $d(n) = d(n + 1)$;
   (b) $d(n) = d(n + 1) = d(n + 2)$;
   (c) $d(n) = d(n + 1) = d(n + 2) = d(n + 3)$.
   (d) Can you make any conjectures? Can you prove any of these conjectures?

3. Find all values of n, $1 \le n \le 1000$, such that
   (a) $\sigma(n) = \sigma(n + 1)$;
   (b) $\sigma(n) = \sigma(n + 1) = \sigma(n + 2)$.
   (c) Can you make any conjectures? Can you prove any of these conjectures?

4. Find all values of n, $1 \le n \le 1000$, such that
   (a) $\varphi(n) = \varphi(n+1)$;
   (b) $\varphi(n) = \varphi(n+1) = \varphi(n+2)$.
   (c) Can you make any conjectures? Can you prove any of these conjectures?

5. Let $s_0(n) = \sigma(n) - n$, for $n \ge 2$, and, for $k \ge 0$, define $s_0(n) = s_0(s_k(n))$. Write a program to calculate $s_k(n)$. Tabulate the values of $s_k(n)$ for $1 \le n \le 100$ and $1 \le k \le 50$. (Meissner, in 1907, conjectured that this sequence always ended in a prime, a perfect number or an amicable pair. Can you prove this?)

6. Find all values of n, $1 \le n \le 1000$, such that $d(n) = \varphi(n)$. Can you make/prove any conjectures?

7. Define a sequence $\{n_k\}$ by $n_0 = n$, $n_1 = \varphi(n)$, and for $k \ge 1$, $n_{k+1} = \varphi(n_k)$, where n is a positive integer.
   (a) Show that there exists an integer k such that $n_k = 1$.
   (b) Write a program to calculate k given n. Calculate k for $1 \le n \le 1000$. Can you make/prove any conjectures?

8. Define a sequence $\{n_k\}$ by $n_0 = n$ and, for $k \ge 0$, $n_{k+1} = d(n_k)$, where n is a positive integer.
   (a) Show that there exists an integer k such that $n_k = 2$ for $k \ge r$.
   (b) Write a program to calculate k given n. Calculate k for $1 \le n \le 1000$. Can you make/prove any conjectures?

9. (a) Write a program to calculate $\sigma_k(n)$, where $k \ge 2$ is an integer.
   (b) Tabulate the values of $\sigma_k(n)$ for $1 \le n \le 100$ and $2 \le k \le 10$.

10. (a) Write a program to calculate $d_k(n)$, where $k \ge 3$ is an integer.
    (b) Tabulate the values of $d_k(n)$ for $1 \le n \le 100$ and $3 \le k \le 10$.

11. (a) Write a program to calculate $J_k(n)$, where $k \ge 2$ is an integer.
    (b) Tabulate the values of $J_k(n)$ for $1 \le n \le 100$ and $2 \le k \le 10$.

12. (a) Write a program to calculate $c(k, n)$, the Ramanujan sum.
    (b) Tabulate the values of $c(k, n)$ for $1 \le k, n \le 50$.

13. (a) Given reals a and b and values for u_0 and u_1 write a program to calculate u_n, for $n \ge 2$, using the recursion relation

$$u_{n+2} = a u_{n+1} + b u_n,$$

   for $n \ge 0$.
    (b) Use (7.58) to calculate $\{u_n\}$.

    (c) Theorem 7.46 implies that $u_{2n} = u_n v_n$ and Theorem 7.47 says that $u_{2n+1} = u_{n+1}^2 + b u_n^2$. Can you use these results to speed up your calculations?

    (d) Try the methods of (a), (b) and (c) to calculate $\{u_n\}$ for $1 \le n \le 50$. Which is fastest?

14. (a) Write a program to implement the Lucas-Lehmer test for the primality of $2^p - 1$.

    (b) Show that $2^{31} - 1$ is prime.

    (c) Show that $2^{23} - 1$ is not a prime.

15. Consider the polynomial

$$f(x, y) = 2xy^4 + x^2 y^3 - 2x^3 y^2 - y^5 - x^4 y + 2y.$$

Evaluate $f(x, y)$ for $-10 \le x, y \le 10$. (This result is due to J. P. Jones.)

CHAPTER EIGHT
The Average Order of Arithmetic Functions

8.1. INTRODUCTION

In the previous chapter we met with a few of the more wellknown arithmetic functions. As can be seen most of these are poorly behaved at first glance. For example, the divisor function, $d(n)$, takes the value 2 infinitely often since there are infinitely many primes, and since $d(p^a) = a + 1$, when p is a prime, we see that it can be as large as we like infinitely often. It turns out that if we consider $d(n)$ on the average, then its values get smoothed out into a move tractable form.

Definition 8.1. An arithmetic function $f(n)$ is said to have **average order** $g(n)$ if

$$\lim_{x \to +\infty} \left\{ \sum_{n \le x} f(n) \bigg/ \sum_{n \le x} g(n) \right\} = 1.$$

The hope is that the function $g(n)$ is somewhat simpler than the arithmetic function. As we shall see in Section 4 below the average value of $d(n)$ is $\log n$. We introduce some notation that will be of use in this and the next chapter. Let a be an extended real number (that is, a may be $+\infty$ or $-\infty$ as well as any finite real number).

1. We say that $f(x)$ is **asymptotic** to $g(x)$, as $x \to a$, written

$$f(x) \sim g(x) \text{ as } x \to a,$$

if the following holds

$$\lim_{x \to +\infty} \left\{ \frac{f(x)}{g(x)} \right\} = 1.$$

2. We say that $f(x)$ is **big O** of $g(x)$, as $x \to a$, written $f(x) = O(g(x))$ as $x \to a$, if there exists a constant K and a neighborhood N of a such that

$$\left| f(x) \right| < K \left| g(x) \right|$$

for all x in N.

3. We say that $f(x)$ is **little o** of $g(x)$, as $x \to a$, written $f(x) = o(g(x))$ as $x \to a$, if the following holds

$$\lim_{x \to a} \left\{ \frac{f(x)}{g(x)} \right\} = 0.$$

Example 8.1. We have $x \sim x + 3/4$ as $x \to +\infty$, $\sin x \sim x$ as $x \to 0$, and $(\sin x)/x \sim 0$ as $x \to -\infty$. We have $\sin x = O(1)$, as $x \to +\infty$, $x = O(x^2)$ as $x \to 0$, and $x^2 = O(x)$ as $x \to 0$. We have $\log^2 x = o(\sqrt{x})$ as $x \to +\infty$, $\sin x = o(x)$ as $x \to -\infty$, and $x^2 = o(x)$ as $x \to 0$. $\qquad\qquad\qquad\qquad\qquad\qquad\qquad$ □

Sometimes we have $f(x) = O(g(x))$ and $g(x) = O(f(x))$, in which case we write

$$f(x) \approx g(x).$$

Problem Set 8.1

1. Prove the following results.
   (a) If $f(x) = O(g(x))$ and $g(x) = O(h(x))$, then $f(x) = O(h(x))$.
   (b) If $f(x) \sim g(x)$, then $f(x) = g(x) + o(g(x))$.
   (c) If $f(x) = O(h(x))$ and $g(x) = O(h(x))$, then $f(x) + g(x) = O(h(x))$.
2. Prove that if $f(x) = O(1)$, then

$$\sum_{n \le x} f(n) = O(x).$$

3. Show that if $f(x) = o(g(x))$, then $f(x) = O(g(x))$.
4. Prove that if $d > 0$, then $\log^m x = o(x^d)$, for any $m > 0$.
5. Show that if $f(x) = O(h(x))$ and $g(x) = O(H(x))$, then $f(x)g(x) = O(h(x)H(x))$.
6. Show that if $f(x) = O(h(x))$ and $g(x) = o(H(x))$, then $f(x)g(x) = o(h(x)H(x))$ and $f(x) + g(x) = O(|h(x)| + |H(x)|) = O(\max(|h(x)|, |H(x)|))$.
7. Show that if $f(x) = O(g(x))$ and $g(x) = o(h(x))$, then $f(x) = o(h(x))$.
8. Show that if $f(x) = o(h(x))$ and $g(x) = o(h(x))$, then $f(x) + g(x) = o(h(x))$.
9. Show that if $f(x) > K > 0$ for all $x > 0$ and $f(x) = O(g(x))$, then $\log f(x) = O(\log g(x))$.
10. Show that if $f(x) = o(g(x))$ and f is increasing, then $e^{f(x)} = o(e^{g(x)})$. Show that this does not hold with o replaced by O.
11. Suppose that $f(x) \sim g(x)$, as $x \to +\infty$.
    (a) If $g(x) \to +\infty$, as $x \to +\infty$, then $\log f(x) \sim \log g(x)$.
    (b) If $g(x) \to 0$, as $x \to +\infty$, then $e^{f(x)} \sim e^{g(x)}$.
12. Show that as $x \to 0+$ we have $e^x = 1 + O(x)$.

8.2. THE GREATEST INTEGER FUNCTION

We begin with the study of a function that will be very useful in the work to follow. We have briefly discussed this function in Section 1.2 (see Definition 1.2.)

Definition 8.2. The **greatest integer function**, denoted $[x]$, is defined on the reals and is the largest integer less than or equal to x. In other words, it is that integer n such that

$$n \le x < n + 1.$$

Example 8.2. We have $[-3] = -3$, $[2.5] = 2$, $[-3.6] = -4$ and $[\pi] = 3$.

The following theorem lists many of the properties of this function.

Theorem 8.1. Let x and y be real numbers . Then we have

(a) $[x] \le x < [x] + 1$, $x - 1 < [x] \le x$, $0 \le x - [x] < 1$;

(b)   if n is an integer, then $[x + n] = [x] + n$;

(c)   if $x \ge 0$, then

$$[x] = \sum_{1 \le n \le x} 1;$$

(d) $[x] + [y] \le [x + y] \le [x] + [y] + 1$;

(e) $[x] + [-x] = \begin{cases} 0 & \text{if } x \text{ is an integer} \\ -1 & \text{else} \end{cases}$.

(f)   if m and n are integers, with m positive, then

$$\left[\frac{n + x}{m} \right] = \left[\frac{n + [x]}{m} \right];$$

and

(g)   if $x \ge 0$ and a is a positive integer, then $[x/a]$ is the number of positive integers $\le x$ that are divisible by a.

Proof of (a). This follows immediately from Definition 8.2.

Proof of (b). This follows immediately from Definition 8.2 and the fact that if M is any integer, then there are no integers strictly between M and $M + 1$.

Proof of (c). If $x < 1$, then the sum is vacuous and if $0 \le x < 1$, then $[x] = 0$, by Definition 8.2.

If $x > 1$, then the sum counts the number of integers less than or equal to x, which is clearly just $[x]$.

Proof of (d). Write $x = n + v$ and $y = m + u$, where $n = [x]$, $m = [y]$, $0 \le v < 1$, and $0 \le u < 1$. Then, by (b),

$$[x] + [y] = n + m \le [n + v + m + u] = [x + y]$$
$$= n + m + [v + u] \le n + m + 1 = [x] + [y] + 1,$$

since $0 < v + u < 2$ implies that $0 \le [v + u] \le 1$.

Proof of (e). Write $x = n + v$, where $n = [x]$ and $0 \le v < 1$. Then

$$[x] + [-x] = n + [-n - v] = n + [-n - 1 + 1 - v] = -1 + [1 - v].$$

If $v = 0$, that is, x is an integer, then $[x] + [-x] = 0$. If x is not an integer, then $0 < v < 1$, and so $0 < 1 - v < 1$. Then $[1 - v] = 0$, and so $[x] + [-x] = -1$.

Proof of (f). Let $x = p + v$, where $p = [x]$ and $0 \le v < 1$, and let $p + n = qm + r$, where $0 \le r \le m - 1$. Then by (b),

$$\left[\frac{n + x}{m} \right] = \left[\frac{p + n + v}{m} \right] = \left[\frac{qm + r + v}{m} \right] = q + \left[\frac{r + v}{m} \right] = q,$$

since $0 \le r + v < m$ implies $0 \le (r + v)/< 1$. Also, by (b),

$$\left[\frac{n + [x]}{m} \right] = \left[\frac{n + p}{m} \right] = \left[\frac{qm + r}{m} \right] = q + \left[\frac{r}{m} \right] = q,$$

since $0 \le r < m$ implies $0 \le r/m < 1$.

Proof of (g). Note that if $0 \le x < 1$, then $[x/a] = 0$, since $x/a < 1/a < 1$ for any positive integer a, and this is the number of positive multiples of a less than or equal to x. Suppose $x \ge 1$ and $a, 2a, ..., ja$ are all the positive multiples of a less than or equal to x. Then we must show that $[x/a] = j$. We have, by definition of j,

$$ja \le x < (j + 1)a \text{ or } j \le x/a < j + 1.$$

Thus, by Definition 8.2, we have $j = [x/a]$.   ∎

Properties (c) and (g) of Theorem 8.1 show the use of $[x]$ as a counting function, something many of the functions we deal with in this chapter do: count the number of objects in sets. We give another application of $[x]$ to the factorization of factorials.

Theorem 8.2. Let p be a prime and n be a positive integer. If $p^e \| n!$, then

$$e = \sum_{m=1}^{+\infty} \left[\frac{n}{p^m} \right].$$

Proof. Note that if $p^m > n$ (that is, $m > (\log n)/(\log p)$), then $[n/p^m] = 0$. Thus the sum is really a finite sum containing $[\log n/\log p]$ nonzero terms.

We begin with a general observation. Let $a_1, ..., a_n$ be a set of nonnegative integers and if k is a positive integer, let $f(k)$ be the number of them that are greater than or equal to k. Then

$$a_1 + \cdots + a_n = f(1) + f(2) + \cdots,$$

since a_i contributes 1 to each of the numbers $f(1), f(2), ..., f(a_i)$.

Now for $1 \le j \le n$, let $p^{a_j} \| j$. Then $e = a_1 + \cdots + a_n$. In this case $f(1)$ counts the number of integers $\le n$ that are divisible by p and, in general, $f(k)$ counts the numbers that are divisible by p^k . Thus, by (g) of Theorem 8.1,

$$f(k) = [n/p^k].$$

Thus, by the observation above,

$$e = a_1 + \cdots + a_n = \sum_{m=1}^{+\infty}\left[\frac{n}{p^m}\right],$$

which is our result.   ∎

Example 8.3.

   (a)   What is the highest power of 7 that divides 1000!?

   (b)   How many zeros does 15000! end in?

   (c)   Show that if $a_1 + \cdots + a_r = n$, then $n!/\,a_1!\cdots a_r!$ is an integer.

Solution of (a). We apply Theorem 8.2. By (f) of Theorem 8.1, we have

$$\left[n/p^{m+1}\right] = \left[\left[n/p^m\right]/p\right].$$

Thus we calculate

$$[1000/7] = 142, \ [142/7] = 20, \ [20/7] = 2 \ \text{and} \ [2/7] = 0.$$

Thus the exponent is $142 + 20 + 2 + 0 = 164$, and so $7^{164} \| 1000!$.

Solution of (b). The number of zeros that 15000! ends in will be the highest power of 10 that divides 15000!, which is the highest power of 5 that divides 15000!. The calculations are

$$[15000/5] = 3000, \ [3000/5] = 600, \ [600/5] = 120,$$
$$[120/5] = 24, \ [24/5] = 4 \ \text{and} \ [4/5] = 0.$$

Thus the number of zeros at the end of 15000! is

$$3000 + 600 + 120 + 24 + 4 + 0 = 3744.$$

Solution of (c). This is, of course, the coefficient of $x_1^{a_1} x_2^{a_2} \cdots x_r^{a_r}$ in the expansion of $(x_1 + \cdots + x_r)^n$, and so must be an integer. We give an alternate proof. To do this it suffices to show that every prime divides the numerator to at least as high a power as it divides the denominator. By Theorem 8.2 we need only show that

$$\sum_m \left[\frac{n}{p^m} \right] \geq \sum_{i=1}^{r} \sum_m \left[\frac{a_i}{p^m} \right],$$

but, by (d) of Theorem 8.1, we have

$$\left[a_1 / p^m \right] + \cdots + \left[a_r / p^m \right] \leq \left[(a_1 + \cdots + a_r) / p^m \right] = \left[n / p^m \right].$$

The result follows.   □

A function related to $[x]$ that is useful in many problems is the so-called "saw tooth function." The name derives from its graph.

Definition 8.3. Let x be a real number. Then the function denoted by $((x))$ is defined as follows:

$$((x)) = \begin{cases} x - [x] - 1/2 & \text{if } x \text{ is not an integer} \\ 0 & \text{else} \end{cases}.$$

The reason for the split definition comes from the theory of Fourier series and need not concern us. Some of its properties are given in the following theorem.

Theorem 8.3. Let x be a real number.

    (a)   The function $y = ((x))$ is periodic of period 1 and is piecewise linear.

    (b)   Let $B_1(x) = x - 1/2$. Then $((x)) = B_1(x - [x])$ for nonintegral x.

    (c)   We have $-1/2 < ((x)) < 1/2$.

    (d)   We have $((-x)) = -((x))$.

    (e)   If h and k are positive integers, then

$$\sum_{m=1}^{k-1} \left(\left(\frac{mh}{k} \right) \right) = 0.$$

    (f)   We have

$$\int_0^1 ((x)) dx = 0.$$

(g) We have

$$\left| \int_a^b ((x))dx \right| \le \tfrac{1}{4},$$

for all real a and b.

(h) If

$$y(x) = \int_0^x ((t))dt,$$

then $|y(x)| \le 1/4$.

Proof of (a). This follows immediately from Definition 8.3 and (b) of Theorem 8.1.

Proof of (b). This is just a restatement of Definition 8.3.

Proof of (c). By (a) of Theorem 8.1, we have, for nonintegral x,

$$0 < x - [x] < 1,$$

and so

$$-1/2 < x - [x] - 1/2 < 1/2.$$

Since $((x)) = 0$, when x is an integer, the result follows.

Proof of (d). If x is an integer, then both sides are zero. If x is not an integer, then, by (e) of Theorem 8.1, we,have

$$((-x)) = -x - [-x] - 1/2 = -x - (-[x] - 1) - 1/2$$
$$= -x + [x] + 1/2 = -(x - [x] - 1/2) = -((x)).$$

Proof of (e). Making the change of summation variable $m = k - n$ gives, by the periodicity of $((x))$,

$$\sum_{m=1}^{k-1}\left(\left(\frac{mh}{k}\right)\right) = \sum_{n=1}^{k-1}\left(\left(\frac{(k-n)h}{k}\right)\right) = \sum_{m=1}^{k-1}\left(\left(\frac{(k-m)h}{k}\right)\right) = \sum_{m=1}^{k-1}\left(\left(\frac{-mh}{k}\right)\right) = -\sum_{m=1}^{k-1}\left(\left(\frac{mh}{k}\right)\right),$$

by (d), and the result follows.

Proof of (f). In the range $0 < x < 1$ we have $((x)) = x - 1/2$ since $[x] = 0$. Thus

$$\int_0^1 ((x))dx = \int_0^1 (x - \tfrac{1}{2})dx = \left(\frac{x^2}{2} - \frac{x}{2}\right)_0^1 = 0.$$

Proof of (g). Since $((x))$ is periodic of period 1, by (a), we have

$$\int_{n}^{n+1}((x))dx = \int_{0}^{1}((y-n))d(y-n) = \int_{0}^{1}((y))dy = 0.$$

Thus, we have

$$(8.1) \int_{a}^{b}((x))dx = \int_{a}^{[a]+1}((x))dx + \sum_{n=[a]+1}^{[b]-1}\int_{n}^{n+1}((x))dx + \int_{[b]}^{b}((x))dx = \int_{a}^{[a]+1}((x))dx + \int_{[b]}^{b}((x))dx.$$

Note that if either a or b is an integer, then the corresponding integral on the right hand side of (8.1) is zero. Also it is clear that we may assume that there are no integers between a and b by adding

$$\int_{[a]+1}^{[b]}((x))dx = 0,$$

if necessary. Finally, by translation, using the periodicity again, we may assume $a,b \in (0,1)$. Then

$$\left|\int_{a}^{b}((x))dx\right| \leq \left|\int_{0}^{1}((x))dx\right| = \int_{0}^{1/2}(\tfrac{1}{2}-x)dx + \int_{1/2}^{1}(x-\tfrac{1}{2})dx = \tfrac{1}{4}.$$

Proof of (h). This follows immediately from (g). ■

The function $((x))$ will show up later in the exercises and is used to define the important Dedekind sum (see Problem 3 of the Additional Problems below).

Problem Set 8.2

1. Show that $[2x] - 2[x] = 0$ or 1. Determine when each case occurs.
2. Show that, for any real number a and any positive integer n,

$$\sum_{k=0}^{n-1}\left[a+\frac{k}{n}\right] = [na].$$

3. Show that

$$[x][y] \leq [xy] \leq [x][y] + [x] + [y].$$

4. (a) Show that

$$\frac{(ab)!}{a!(b!)^{a}}$$

is an integer.

(b) Let $n = a_1 + \cdots + a_r, a_i \geq 0, 1 \leq i \leq r$, and let $d = (a_1, \ldots, a_r)$. Show that

$$\frac{d(n-1)!}{a_1! \cdots a_r!}$$

is an integer.

5. (a) Find the exact powers of 2, 3 and 5 in 5238!.

(b) How many zeros does 17356! end in?

6. (a) Let p be a prime and let

$$m = \sum_{j=0}^{r} a_j p^j,$$

where $0 \leq a_j \leq p-1, 0 \leq j \leq r, 1 \leq a_r \leq p-1$. Show that the highest power of p dividing $m!$ is

$$\frac{m - \displaystyle\sum_{j=0}^{r} a_j}{p-1}.$$

(Hint: see Example 1.4.)

(b) Find the highest power of p in $(p^n - 1)!$.

7. (a) Show that

$$\sum_{\substack{n=1 \\ (n,k)=1}}^{k-1} \left(\left(\frac{hn}{k} \right) \right) = 0.$$

(Hint: use the periodicity and oddness of $((x))$ as well as $(n, k) = (n - k, n)$.)

(b) Show that, if $(h, k) = 1$,

$$\sum_{\substack{n=1 \\ (n,k)=1}}^{k-1} \left[\frac{hn}{k} \right] = \frac{(h-1)\varphi(k)}{2}.$$

(Hint: use (a) or use the facts that $(n, k) = (n - k, n)$ and $[-x] = -1 - [x]$, if x is not an integer.)

8. Show that 2 is the highest power of 2 dividing $\displaystyle\binom{2^{p+1}}{2^p}$.

9. If n is a positive integer, show that $\left[\sqrt{n} + \sqrt{n+1} \right] = \left[\sqrt{4n+2} \right]$. (Hint: use the fact that $\sqrt{x}$ has a negative second derivative.)

10. Show that either x or y must be an integer if both

$$[x + y] = [x] + [y] \text{ and } [-x - y] = [-x] + [-y]$$

   hold.

11. Show that a real number x is rational if and only if there is an integer m such that $[mx] = mx$.

12. Let a, b and n be positive integers. Show that $[ab/n] \geq a[b/n]$.

13. Show that if $m \geq 5$, then

$$\frac{4^m}{m} \leq \frac{(2m)!}{(m!)^2} < 4^{m-1}.$$

14. Show that no integer is closer to the real number x than the integer $[x + 1/2]$. Show that if two integers are equally close, then $[x + 1/2]$ is the larger of the two.

15. Let x be a real number. Compute $\big|[x]\big|$.

16. Show that

$$n \left| \binom{2n-2}{n-1} \right.$$

   for all positive integers n.

17. Show that

$$\lim_{m \to +\infty} \left[\cos^2(m!\pi x) \right] = \begin{cases} 0 & \text{if } x \text{ is irrational} \\ 1 & \text{if } x \text{ is rational} \end{cases}.$$

18. Let a and b be irrational numbers such that $1/a + 1/b = 1$. If x is a real number, let $S(x) = \{[nx]: n = 1, 2, 3, \ldots\}$. Show that every positive integer m is in exactly one of $S(a)$ or $S(b)$.

8.3. PRELIMINARIES

Definition 8.4. If f is an arithmetic function, then the function F, defined by

$$F(x) = \sum_{n \leq x} f(n),$$

is called the **summatory function** of f.

   By (c) of Theorem 8.1, we see that $[x]$ is the summatory function of the arithmetic function $e(n)$. Also the function $F(x) \equiv 1$ is the summatory function of the function $\delta(n)$.

Definition 8.5. If f is an arithmetic function and a is a function defined on the nonnegative real numbers, then we define the operation $\circ$ by

$$(f \circ a)(x) = \sum_{n \le x} f(n)a(x/n).$$

Note that if a is only nonzero on the positive integers and m is a positive integer, then

$$(f \circ a)(m) = (f * a)(m).$$

The properties of this operation are given in the following theorem.

Theorem 8.4.

(a) If f and g are any arithmetic functions and a is a function defined on the nonnegative real numbers, then

$$(f \circ (g \circ a)) = (f * g) \circ a.$$

(b) For any function a defined on the nonnegative real numbers,

$$\delta \circ a = a.$$

(c) If $f \in A_1$ and a and b are functions defined on the nonnegative real numbers, then

$$a = f \circ b \text{ if and only if } b = f^{*-1} \circ a.$$

(d) If f is a completely multiplicative function and a and b are functions defined on the nonnegative reals, then

$$a = f \circ b \text{ if and only if } b = \mu f \circ a.$$

Proof of (a). For $x \ge 0$, we have

$$\{f \circ (g \circ a)\}(x) = \sum_{n \le x} f(n) \sum_{m \le x/n} g(m)a(x/mn) = \sum_{mn \le x} f(n)g(m)a(x/mn)$$

$$= \sum_{k \le x} \left\{ \sum_{n|k} f(n)g(n/k) \right\} a(x/k) = \sum_{k \le x} (f * g)(k)a(x/k)$$

$$= \{(f * g) \circ a\}(x).$$

Proof of (b). We have, for $x \ge 0$,

$$(\delta \circ a)(x) = \sum_{n \le x} \delta(n)a(x/n) = \delta(1)a(x) = a(x).$$

Proof of (c). If $a = f \circ b$, then, by (a) and (b),

$$f^{*-1} \circ a = f^{*-1} \circ (f \circ b) = (f^{*-1} * f) \circ b = \delta \circ b = b.$$

The converse is proved similarly.

Proof of (d). This follows from (c) and Theorem 7.13.   ∎

Corollary 8.4.1. Let f and g be functions defined on the nonnegative reals. Then

$$f(x) = \sum_{n \le x} g(x / n) \text{ if and only if } g(x) = \sum_{n \le x} \mu(n) f(x / n).$$

Many of the sums we deal with are of the form

$$\sum_{n \le x} (f * g)(n),$$

where f and g are arithmetic functions. The idea is to use information about the summatory functions of f and g to gain knowledge about the summatory function of $f * g$. The following theorem gives two ways of dealing with this problem.

Theorem 8.5. Let f and g be arithmetic functions with summatory functions F and G, respectively, and let $h = f * g$ with summatory function H. Then

(a) $H(x) = \sum_{n \le x} f(n) G(x / n) = \sum_{n \le x} g(n) F(x / n);$

(b) $\sum_{n \le x} \sum_{d \mid n} f(d) = \sum_{n \le x} f(n)[x / n] = \sum_{n \le x} F(x / n);$

and

(c) if a and b are positive real numbers so that $ab = x$, then

$$H(x) = \sum_{n \le a} f(n) G(x / n) + \sum_{n \le b} g(n) F(x / n) - F(a) G(b).$$

Proof of (a). Define the function U by

$$U(x) = \begin{cases} 0 & \text{if } 0 < x < 1 \\ 1 & \text{if } x \ge 1 \end{cases}.$$

Then $F = f \circ U$ and $G = g \circ U$ and we have, by (a) of Theorem 8.4,

$$f \circ G = f \circ (g \circ U) = (f * g) \circ U = H$$

and

$$g \circ F = g \circ (f \circ U) = (g * f) \circ U = H.$$

Proof of (b). This follows immediately from (a) upon taking $g = e$, since, in that case,

$$G(x) = \sum_{n \leq x} g(n) = \sum_{n \leq x} 1 = [x].$$

Proof of (c). Consider the figure below.

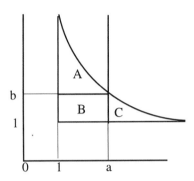

Now

$$H(x) = \sum_{n \leq x} (f*g)(n) = \sum_{qd \leq x} f(d)g(q).$$

Thus $H(x)$ is summed over all the points under the hyperbola that have positive integer coordinates. We then split this region into two parts: one over those points in $A \cup B$ and one over those points in $B \cup C$. The points in B are covered twice so that we have

$$H(x) = \sum_{d \leq a} \sum_{q \leq x/d} f(d)g(q) + \sum_{q \leq b} \sum_{d \leq x/q} f(d)g(q) - \sum_{d \leq a} \sum_{q \leq b} f(d)g(q)$$

$$= \sum_{d \leq a} f(d)G(x/d) + \sum_{q \leq b} g(q)F(x/q) - F(a)G(b),$$

which is our result.   ∎

The question might be raised as to why have two different formulas for the same sum. The answer lies in the relative sizes of F and G. If F is large and G small, say, then the first formula is the better one to use, since little will be gained from accounting for both sums. On the other hand, if both F and G are approximately the same size, then the third formula is the better one to use. We shall illustrate this in the sections below.

One might consider the above to be the elementary work on summatory functions. One further tool that is quite useful in this work is the Riemann-Stieltjes integral. We shall now briefly describe this in as much generality as we need. For those who do not wish to wade through the theory that follows (or those who know this material already) may wish only to check the results and take them on faith since the details are much like that for the ordinary Riemann integral. For a more thorough study see [6], for example.

Let $I = [a, b]$ be an interval and let f and g be two functions that are bounded on I. A **partition** of I is a finite collection of nonoverlapping subintervals whose union is I. We usually describe such partitions by describing the set P of endpoints of the partitioning intervals. Then $P = \{x_0, x_1, ..., x_n\}$, say, where

$$a = x_0 \leq x_1 \leq \cdots \leq x_{n-1} \leq x_n = b.$$

If P and Q are partitions of I, then Q is said to be a **refinement** of P if every subinterval in Q is contained in some subinterval in P. Note that this equivalent to saying that every partition point of P is a partition point of Q.

Definition 8.6. If P is a partition of I, then a **Riemann-Stieltjes sum** of f with respect to g and corresponding to the partition $P = \{x_0, x_1, ..., x_n\}$ is the real number $S(P; f, g)$ defined by

$$S(P; f, g) = \sum_{k=1}^{n} f(\xi_k)\{g(x_k) - g(x_{k-1})\},$$

where the ξ_k satisfy $x_{k-1} \leq \xi_k \leq x_k, 1 \leq k \leq n$.

If we take $g(x) = x$, then we just have the ordinary Riemann sums.

Definition 8.7. We say that f is **integrable** with respect to g on I if there exists a real number A such that for every positive number ε there is a partition P_ε of I such that if P is any refinement of P_ε and $S(P; f, g)$ is any Riemann-Stieltjes sum corresponding to P, then

$$|S(P; f, g) - A| < \varepsilon.$$

In this case the number A is uniquely determined and is denoted by

$$A = \int_a^b f dg = \int_a^b f(t) dg(t).$$

It is called the **Riemann-Stieltjes integral** of f with respect to g over $I = [a, b]$. We call f the **integrand** and g the **integrator**.

Before starting the theorems we recall some definitions and notation. If F is a function defined on $[a, b]$ and x is in $[a, b]$, then we define

$$F(x+) = \lim_{h \to 0^+} F(x + h)$$

and

$$F(x-) = \lim_{h \to 0^+} F(x - h),$$

providing the limits exist. If $F(x) = F(x+)$, then F is said to be **continuous from the right** at x and if $F(x) = F(x-)$, then F is said to be **continuous from the left** at x. Finally, if $F(x-) = F(x) = F(x+)$, then F is said to be **continuous** at x.

Example 8.4.

(a) Let $g(x)$ be constant on I. Then any function f is integrable with respect to g and the value of the integral is always zero.

(b) Let

$$g(x) = \begin{cases} \alpha & \text{if } a \le x < c \\ \beta & \text{if } c \le x \le b \end{cases}.$$

and suppose f is left continuous at c . Then it is an easy calculation to show that f is integrable with respect to g and that

$$\int_a^b f dg = (\beta - \alpha) f(c).$$

The second example is a special case of one that will be quite important to us in our later work: that is, when the integrator is piecewise constant.

Theorem 8.6. The function f is integrable with respect to g over $I = [a, b]$ if and only if for each positive real number ε there is a partition Q_ε of I such that if P and Q are refinements of Q_ε and $S(P; f, g)$ and $S(Q; f, g)$ are any corresponding Riemann-Stieltjes sums, then

(8.2) $\left| S(P; f, g) - S(Q; f, g) \right| < \varepsilon .$

Proof. If f is integrable, then there is a partition P_ε such that if P and Q are refinements of P_ε , then any corresponding Riemann-Stieltjes sums satisfy

$$\left| S(P; f, g) - A \right| < \varepsilon / 2 \text{ and } \left| S(Q; f, g) - A \right| < \varepsilon / 2.$$

Then (8.2) follows by the triangle inequality.

Conversely, suppose (8.2) is satisfied. Let Q_1 be a partition of I such that if P

and Q are refinements of Q_1, then

$$|S(P; f, g) - S(Q; f, g)| < 1.$$

Inductively, we define Q_n to be a refinement of Q_{n-1} such that if P and Q are refinements of Q_n then

$$|S(P; f, g) - S(Q; f, g)| < 1/n.$$

Consider the sequence $\{S(Q_n; f, g)\}_{n=1}^{+\infty}$ obtained in this way. Since Q_n is a refinement of Q_m, whenever $n \geq m$, this sequence is a Cauchy sequence of real numbers regardless of how the intermediate points are chosen. Thus the sequence converges to some real number A. Thus if $\varepsilon > 0$, there exists an integer N such that $2/N < \varepsilon$ and

$$|S(Q_N; f, g) - A| < \varepsilon / 2.$$

If P is a refinement of Q_N then by the construction of Q_N we have

$$|S(P; f, g) - S(Q_N; f, g)| < 1/N < \varepsilon / 2.$$

Thus for any refinement P of Q_N and any corresponding Riemann-Stieltjes sum we have

$$|S(P; f, g) - A| < \varepsilon .$$

Thus f is integrable over I with respect to g and the value of the integral is A.   ∎

Theorem 8.7.

    (a) If f_1 and f_2 are integrable over I with respect to g and α and β are any real numbers, then $\alpha f_1 + \beta f_2$ is integrable over I with respect to g and

$$\int_a^b (\alpha f_1 + \beta f_2)dg = \alpha \int_a^b f_1 dg + \beta \int_a^b f_2 dg.$$

    (b) If f is integrable over I with respect to both g_1 and g_2 and α and β are any real numbers, then f is integrable over I with respect to $\alpha g_1 + \beta g_2$ and

$$\int_a^b f d(\alpha g_1 + \beta g_2) = \alpha \int_a^b f dg_1 + \beta \int_a^b f dg_2.$$

Proof of (a). Let $\varepsilon > 0$ and let $P_1 = \{x_0, x_1, ..., x_n\}$ and $P_2 = \{y_0, y_1, ..., y_n\}$ be partitions of I such that if Q is a refinement of both P_1 and P_2, then for any corresponding Riemann-Stieltjes sums we have

(8.3) $|I_1 - S(Q; f_1, g)| < \varepsilon$ and $|I_2 - S(Q; f_2, g)| < \varepsilon$.

Let P_ε be a partition of I which is a refinement of both P_1 and P_2 (for example, $P_1 \cup P_2$. If Q is a refinement of P_ε, then (8.3) still holds. When the same intermediate points are used we have

$$S(Q; \alpha f_1 + \beta f_2, g) = \alpha S(Q; f_1, g) + \beta S(Q; f_2, g).$$

Thus

$$|\alpha I_1 + \beta I_2 - S(Q; \alpha f_1 + \beta f_2, g)| = |\alpha(I_1 - S(Q; f_1, g)) + \beta(I_2 - S(Q; f_2, g))| \le (|\alpha| + |\beta|)\varepsilon.$$

Thus $\alpha I_1 + \beta I_2$ is integral of $\alpha f_1 + \beta f_2$ with respect to g.

Proof of (b). The proof is similar so we omit it. ∎

Theorem 8.8.

   (a) Suppose $a < c < b$ and that f is integrable with respect to g over both $[a, c]$ and $[c, b]$. Then f is integrable with respect to g over $[a, b]$ and

(8.4)
$$\int_a^b f \, dg = \int_a^c f \, dg + \int_c^b f \, dg.$$

   (b) Let f be integrable with respect to g on $[a, b]$ and let c satisfy $a < c < b$. Then f is integrable with respect to g over $[a, c]$ and $[c, b]$ and (8.4) holds.

Proof of (a). Let $\varepsilon > 0$ and let P'_ε be a partition of $[a, c]$ such that if P' is a refinement of P'_ε, then for any Riemann-Stieltjes sum we have

$$|S(P'; f, g) - A'| < \varepsilon / 2.$$

Let P''_ε be a corresponding partition of $[c, b]$. If $P_\varepsilon = P'_\varepsilon \cup P''_\varepsilon$ is the corresponding partition of $[a, b]$ and if P is a refinement of P_ε, then

$$S(P; f, g) = S(P'; f, g) + S(P''; f, g),$$

where P' and P'' denote the partitions of $[a, c]$ and $[c, b]$ induced by P. Thus

$$\left| \int_a^c f \, dg + \int_c^b f \, dg - S(P; f, g) \right| < \left| \int_a^c f \, dg - S(P'; f, g) \right| + \left| \int_c^b f \, dg - S(P''; f, g) \right| < \varepsilon.$$

Thus f is integrable with respect to g on $[a, b]$ and (8.4) holds.

Proof of (b). Since f is integrable over $[a, b]$, if $\varepsilon > 0$, then there exists a partition Q_ε of $[a, b]$ such that if P and Q are refinements of Q_ε, then

$$|S(P; f, g) - S(Q; f, g)| < \varepsilon$$

for any corresponding Riemann-Stieltjes sums. It is clear that we may assume that $c \in Q_\varepsilon$ and let Q'_ε be the induced partition of $[a, c]$. Let P' and Q' be partitions of $[a, c]$ which are refinements of Q'_ε and extend them to partitions P and Q of $[a, b]$ by using the points of Q_ε in $[c, b]$. Since P and Q are refinements of Q_ε and are identical on $[c, b]$ we see that if we use the same intermediate points, then

$$|S(P';f, g) - S(Q';f, g)| = |S(P;f, g) - S(Q;f, g)| < \varepsilon.$$

Thus, by Theorem 8.5, we see that f is integrable over $[a, c]$ with respect to g. A similar argument applies to the interval $[c, b]$ and the result then follows from (a). ∎

We now prove one of the key theorems in our work: the formula for integration by parts.

Theorem 8.9. A function f is integrable with respect to g over $[a, b]$ if and only if g is integrable with respect to f over $[a, b]$. In this case

(8.5)
$$\int_a^b f dg + \int_a^b g df = f(b)g(b) - f(a)g(a).$$

Proof. Suppose f is integrable with respect to g. Let $\varepsilon > 0$ and let P_ε be a partition of $[a, b]$ such that if Q is a refinement of P_ε and $S(Q;f, g)$ is any corresponding Riemann-Stieltjes sum, then

(8.6)
$$\left| S(Q;f,g) - \int_a^b f dg \right| < \varepsilon.$$

Let P be a refinement of P_ε and consider a Riemann-Stieltjes sum $S(P; g, f)$ given by

$$S(P;g,f) = \sum_{k=1}^n g(\xi_k)\{f(x_k) - f(x_{k-1})\}.$$

where $x_{k-1} \le \xi_k \le x_k$. Let $Q = \{y_0, y_1, \ldots, y_{2n}\}$ of $[a, b]$ obtained by using both the ξ_k and the x_k as partition points. Then $y_{2k} = x_k$ and $y_{2k-1} = \xi_k$. Add and subtract the terms $f(y_{2k})g(y_{2k})$, $0 \le k \le n$, to $S(P; g, f)$ and rearrange to obtain

$$S(P;g,f) = f(b)g(b) - f(a)g(a) - \sum_{k=1}^{2n} f(\eta_k)\{g(y_k) - g(y_{k-1})\},$$

where the points η_k are selected to be the points x_j Thus

$$S(P;g,f) = f(b)g(b) - f(a)g(a) - S(Q;f,g),$$

where $Q = \{y_0, y_1, \ldots, y_{2n}\}$ is a refinement of P_ε. Thus, by (8.6),

$$\left|S(P;g,f)-\{f(b)g(b)-f(a)g(a)\}-\int_a^b fdg\right| < \varepsilon,$$

provided P is a refinement of P_ε. Thus g is integrable with respect to f over $[a, b]$ and (8.5) holds. The other direction is similar. ∎

So far we have yet to prove that any functions are integrable with respect to any integrators. That is the subject of the next theorem. First we need some definitions.

Definition 8.8. We say that a function F is of **bounded variation** on $[a, b]$ if there exists a constant M such that for any partition $P = \{x_0, x_1, ..., x_n\}$ of $[a, b]$ we have

(8.7)
$$\sum_{k=1}^{n}|F(x_k) - F(x_{k-1})| < M.$$

If F is of bounded variation on $[a, b]$ we denote by $V(F)$ the **total variation** of F on $[a, b]$ and it is defined to be

$$\sup_P \sum_{k=1}^{n}|F(x_k) - F(x_{k-1})|,$$

where the suprema is taken over all partitions, P, of $[a, b]$. If F is defined on the nonnegative reals we define the extended real valued function $V(F)(x)$ to be the total variation of F on $[0, x]$.

Suppose F is monotone increasing on $[a, b]$, that is, if $b \geq x_1 \geq x_2 \geq a$, then $F(x_1) \geq F(x_2)$. Then $|F(x_k) - F(x_{k-1})| = F(x_k) - F(x_{k-1})$. Thus the sum in (8.7) telescopes and we can take $M = F(b) - F(a)$. A similar result obtains if F is monotone decreasing on $[a, b]$, that is, if $b \geq x_1 \geq x_2 \geq a$, then $F(x_1) \leq F(x_2)$, by considering the function $-F$ which is then monotone increasing.

Another class of examples that are of interest are the summatory functions:

$$F(x) = \sum_{n \leq x} f(n)$$

Then F is constant between integers and has a jump of $f(n)$ at $x = n$. Let $P = \{x_0, x_1, ..., x_n\}$ be a partition of $[1, x]$. Then

$$|F(x_k) - F(x_{k-1})| = \begin{cases} 0 & \text{if there are no integers in } [x_{k-1}, x_k] \\ \sum_{j=m}^{l}|f(j+1) - f(j)| & \text{if the integers } m, m+1, ..., l \text{ are in } [x_{k-1}, x_k] \end{cases}$$

Thus

$$\sum_{k=1}^{n}|F(x_k) - F(x_{k-1})| < |f(1)| + |f(2)| + \cdots + |f([x])|,$$

independent of the partition P. With regard to the following theorem summatory functions are continuous from the right.

Theorem 8.10. Let f and g be of bounded variation on $[a, b]$ and assume that f is continuous from the right on $[a, b]$ and g is continuous from the left on $[a, b]$. Then

$$\int_a^b f dg$$

exists.

Proof. Let $\varepsilon > 0$. We shall find a partition $P = \{t_0, ..., t_n\}$ of $[a, b]$ such that for any refinements P' and P'' and any choices of intermediate points we have

(8.8) $|S(P'; f, g) - S(P''; f, g)| < \varepsilon.$

The result then follows from Theorem 8.6.

We shall take $\{t_1, ..., t_n\}$ as intermediate points for P and show that if P' is any refinement of P and $S(P'; f, g)$ is the corresponding Riemann-Stieltjes sum with any choice of intermediate points, then

(8.9) $|S(P'; f, g) - S(P; f, g)| < \varepsilon / 2.$

This clearly implies (8.8).

The bounded variation condition implies that there exist a finite number of points x in $[a, b]$ at which either of the inequalities

(8.10) $|f(x) - f(x+)| \geq \varepsilon / \{4 + 4V(g)(b)\}$

or

(8.11) $|g(x) - g(x-)| \geq \varepsilon / \{4 + 4V(f)(b)\}$

holds. Include all such points in P.

Next insert more points into P so that it never happens that t_{i+1} satisfies (8.11) while t_i satisfies (8.10) for the same index i.

Finally, insert enough points into P so that if $t_i < t_{i+1}$ are consecutive points of P, then

$$V(f)(t_{i+1}) - V(f)(t_i) < \varepsilon / \{4 + 4V(g)(b)\}$$

provided t_i does not satisfy (8.10) and

$$V(g)(t_{i+1}) - V(g)(t_i) < \varepsilon / \{4 + 4V(f)(b)\}$$

provided t_{i+1} does not satisfy (8.11).

To estimate (8.9) we consider consecutive partition points t_i and t_{i+1} of P. If P' has points in (t_i, t_{i+1}) we let

$$P' \cup [t_i, t_{i+1}] = \{t_i = x_0 < x_1 < \cdots < x_m = t_{i+1}\}$$

and let $\xi_{i_0}, \dots, \xi_{i_{m-1}}$ be the corresponding intermediate points. Let S_i denote the contribution to (8.9) arising from the interval $[t_i, t_{i+1}]$. Then

(8.14)
$$\begin{aligned}
|S_i| &= \left| \sum_{j=0}^{m-1} f(\xi_{i_j})\{g(x_{j+1}) - g(x_j)\} - f(t_{i+1})\{g(t_{i+1}) - g(t_i)\} \right| \\
&= \left| \sum_{j=0}^{m-1} \{f(\xi_{i_j}) - f(t_{i+1})\}\{g(x_{j+1}) - g(x_j)\} \right| \\
&\leq \sum_{j=0}^{m-1} \{V(f)(t_{i+1}) - V(f)(t_i)\}\left| g(x_{j+1}) - g(x_j) \right| \\
&\leq \sum_{j=0}^{m-1} \{V(f)(t_{i+1}) - V(f)(t_i)\}\{V(g)(t_{j+1}) - V(g)(t_j)\}
\end{aligned}$$

Now suppose t_i satisfies (8.10). Then t_{i+1} does not satisfy (8.11) and we can apply (8.13). In this case

$$|S_i| < \{V(f)(t_{i+1}) - V(f)(t_i)\}\varepsilon / \{4 + 4V(f)(b)\}.$$

If t_i does not satisfy (8.10), then (8.12) can be applied, and so

$$|S_i| < \{V(g)(t_{i+1}) - V(g)(t_i)\}\varepsilon / \{4 + 4V(g)(b)\}.$$

In either case $|S_i|$ is certainly less than the sum of the last two estimates. If we now sum on i we obtain

$$|S(P'; f, g) - S(P; f, g)| < \left\{ \frac{V(f)(b) - V(f)(a)}{4 + 4V(f)(b)} + \frac{V(g)(b) - V(g)(a)}{4 + 4V(g)(b)} \right\} \varepsilon < \varepsilon / 2,$$

which finishes the proof.   ∎

Lemma 8.11.1. Let f and g satisfy the hypotheses of Theorem 8.10 with g increasing.

(a) We have

$$\left|\int_a^b f dg\right| \le \int_a^b |f| dg \le \sup_{x\in[a,b]} \{V(g)(b) - V(g)(a)\}.$$

(b) If $m \le f(x) \le M$ on $[a, b]$, then

$$m\{V(g)(b) - V(g)(a)\} \le \int_a^b f dg \le M\{V(g)(b) - V(g)(a)\}$$

Proof. If $P = \{x_0, ..., x_n\}$ is a partition of $[a, b]$ and $\{\xi_k\}$ is a set of intermediate points, then, for $k = 1, 2,..., n$, we have

(8.15) $-\sup|f(x)| \le -|f(\xi_k)| \le f(\xi_k) \le |f(\xi_k)| \le \sup|f(x)|.$

By Theorem 8.10, we know that $|f|$ is integrable with respect to g if f is, since absolute value preserves continuity. Then, by (8.15),

$$-\sup|f(x)|\{V(g)(b) - V(g)(a)\} \le -S(P;|f|,g)$$
$$\le S(P;f,g) \le S(P;|f|,g)$$
$$\le \sup|f(x)|\{V(g)(b) - V(g)(a)\},$$

and so we have

$$|S(P;f,g)| \le S(P;|f|,g) \le \sup|f(x)|\{V(g)(b) - V(g)(a)\}$$

which gives the result of part (a).

Since the proof of (b) is similar we omit it (see Problem 1 below).   ■

Theorem 8.11. Let f and g be as in Theorem 3.10. If $a < x < b$, let

$$\phi(x) = \int_a^b f dg.$$

Then ϕ is continuous from the right and $\phi(x) - \phi(x-) = f(x)\{g(x) - g(x-)\}$.

Proof. It suffices to show that if $c \in (a, b)$, then both

$$\int_c^{c+\varepsilon} f dg \text{ and } \int_{c-\varepsilon}^c f dg$$

have limits as $\varepsilon \downarrow 0$. The first integral is zero since, by Lemma 8.11.1,

$$\left|\int_c^{c+\varepsilon} f dg\right| \le \sup_{c\le u\le c+\varepsilon} |f(u)|\{V(g)(c + \varepsilon) - V(g)(c)\}$$

and the last factor tends to zero, since g is of bounded variation and continuous from the right. To treat the variable lower limit set

$$\int_{c-\varepsilon}^{c} f dg = f(c)\{g(c) - g(c - \varepsilon)\} + \int_{c-\varepsilon}^{c} \{f(t) - f(c)\} dg(t).$$

Now the last integral tends to zero as $\varepsilon \downarrow 0$ since f is continuous from the left. ∎

This theorem shows that a discontinuity of g at the right limit b is reflected in the value of the integral, but a discontinuity of g at the left limit a does not affect the value of the integral at all.

If g is as in Theorem 8.10, then we have

$$\int_{1}^{x} 1 dg = g(x) - g(1).$$

Usually we want to include any contribution arising from a possible discontinuity of g at 1. For example, if F is the summatory function of an arithmetic function f, then F has a jump of $f(1)$ at the point 1. Thus if f and g satisfy the hypotheses of Theorem 8.10 on the interval $[a - \varepsilon, b]$ for some $\varepsilon > 0$, then we define

$$\int_{a-}^{b} f dg = \lim_{\delta \to 0^+} \int_{a-\delta}^{b} f dg.$$

We are now in position to give our two main examples of the Riemann-Stieltjes integral that are most important for our further work.

Theorem 8.12.

(a) Let f be an arithmetic function with summatory function F. Let $0 < a < b$ be given and assume that g is continuous from the left at all integers in $[a, b]$. Then

$$\sum_{a < n \leq b} f(n)g(n) = \int_{a}^{b} g(t) dF(t)$$

and

$$\sum_{a \leq n \leq b} f(n)g(n) = \int_{a-}^{b} g(t) dF(t).$$

(b) Let f and g be Riemann integrable functions on $[1, x]$ for any finite x. Let, for all $x \geq 1$,

$$F(x) = \int_{1}^{x} f(t) dt.$$

If $1 \le a < b < +\infty$, then

(8.16)
$$\int_a^b g \, dF = \int_a^b g(x) F'(x) \, dx = \int_a^b g(x) f(x) \, dx.$$

Proof of (a). This follows immediately from Theorem 8.8 and (b) of Example 8.4.

Proof of (b). We will assume that f is continuous on $[a, b]$. If it is only of bounded variation, then we can adapt the proof in a way analogous to that of Theorem 8.10.

We see that f is uniformly continuous on $[a, b]$. If $\varepsilon > 0$, let $P = \{x_0, \ldots, x_n\}$ be a partition of $[a, b]$ such that if ξ_k and ζ_k belong to $[x_{k-1}, x_k]$, then $|f(\xi_k) - f(\zeta_k)| < \varepsilon$. We consider the difference between the Riemann-Stieltjes sums $S(P; g, F)$ and the Riemann sums $S(P; fg)$ using the same intermediate points ξ_k. This gives us a sum of terms of the form

$$g(\xi_k)\{F(x_k) - F(x_{k-1})\} - g(\xi_k) f(\xi_k)\{x_k - x_{k-1}\}.$$

We apply the mean value theorem to F, and so the differences are then of the form

$$g(\xi_k)\{f(\zeta_k) - f(\xi_k)\}\{x_k - x_{k-1}\},$$

where ζ_k is some point in the interval $[x_{k-1}, x_k]$. Each of these terms is dominated by $\varepsilon \sup|g(x)|(x_k - x_{k-1})$ so that

$$|S(P; g, F) - S(P; gf)| < \varepsilon \sup|g(u)|(b - a),$$

where the sup is over $[a, b]$, provided the partition is sufficiently fine. Since the integral on the left hand side of (8.16) exists and is the limit of the Riemann-Stieltjes sums $S(P; g, f)$ we infer that the integral on the right hand side of (8.16) exists and the equality holds. ∎

Example 8.5. We have

$$\sum_{1 \le n \le x} \log n = \int_{1-}^{x} \log t \, d[t],$$

since, by (c) of Theorem 8.1,

$$\sum_{1 \le n \le x} 1 = [x].$$

In general, we have

$$\sum_{a \le n \le b} f(n) = \int_{a-}^{b} f(t) \, d[t],$$

provided f is continuous from the left at the integers in $[a, b]$.

Example 8.6. Find upper and lower estimates for the sum

$$\sum_{n=1}^{N} \sqrt{n},$$

where N is a positive integer.

Solution. By Example 8.5, we have

$$\sum_{n=1}^{N} \sqrt{n} = \int_{1-}^{N} \sqrt{t}\, d[t] = \int_{1}^{N} \sqrt{t}\, dt + \int_{1-}^{N} \sqrt{t}\, d([t] - t),$$

since t is a continuous integrator and by (b) of Theorem 8.7. If we integrate the second integral by parts (Theorem 8.9) and evaluate the first integral we have

$$\sum_{n=1}^{N} \sqrt{n} = \tfrac{2}{3} t^{3/2}\Big|_1^N + ([t] - t)\sqrt{t}\Big|_{1-}^N + \int_1^N (t - [t]) \frac{dt}{2\sqrt{t}} = \tfrac{2}{3} N^{3/2} - \tfrac{2}{3} + 1 + \int_1^N (t - [t]) \frac{dt}{2\sqrt{t}},$$

since $\lim_{t \to 1^-}([t] - t)\sqrt{t} = -1$. Since $(t - [t]) \geq 0$ we see that the integral on the right hand side is nonnegative, and so

$$\sum_{n=1}^{N} \sqrt{n} \geq \tfrac{2}{3} N^{3/2} + \tfrac{1}{3}.$$

On the other hand $(t - [t]) \leq 1$, and so

$$\sum_{n=1}^{N} \sqrt{n} \leq \tfrac{2}{3} N^{3/2} + \tfrac{1}{3} + \int_1^N \frac{dt}{2\sqrt{t}} = \tfrac{2}{3} N^{3/2} + \sqrt{N} - \tfrac{2}{3} < \tfrac{2}{3} N^{3/2} + \sqrt{N}.$$

Thus

$$\tfrac{2}{3} N^{3/2} + \tfrac{1}{3} \leq \sum_{n=1}^{N} \sqrt{n} < \tfrac{2}{3} N^{3/2} + \sqrt{N}. \qquad \square$$

This example illustrates a standard technique that will be used quite often in what follows. By Theorem 8.12, we have

$$\sum_{a \leq n \leq b} f(n)g(n) = \int_{a-}^{b} g(t)\, dF(t).$$

Suppose $F(t) = A(t) + E(t)$, where $A(t)$, say, is continuous and $E(t)$ is at least of bounded variation. Then we may write, by (b) of Theorem 8.7,

$$\sum_{a \leq n \leq b} f(n)g(n) = \int_{a-}^{b} g(t)\, d(A(t) + E(t)) = \int_{a}^{b} g(t)\, dA(t) + \int_{a-}^{b} g(t)\, dE(t).$$

The hope is that the first integral on the right will be easy to evaluate and that we can somehow estimate the second integral. Sometimes the estimation is easy, as in the above example, and sometimes it is not.

Note that we could write the result of Example 8.6 as

$$\sum_{n=1}^{N}\sqrt{n} = \tfrac{2}{3}N^{3/2} + O(\sqrt{N})$$

as $N \to +\infty$.

Theorem 8.13. Let f be an arithmetic function with summatory function F. Suppose g has a derivative on the interval $[a, b]$, where $0 < a < b$. Then

$$\sum_{a<n\le b}f(n)g(n) = F(b)g(b) - F(a)g(a) - \int_a^b F(t)g'(t)dt.$$

Proof. By Theorem 8.9 and (a) of Theorem 8.12, we have

$$\sum_{a\le n\le b}f(n)g(n) = \int_{a-}^b g(t)dF(t) = g(b)F(b) - g(a)F(a) - \int_a^b F(t)dg(t).$$

Since g has a derivative on $[a, b]$, the result follows by (b) of Theorem 8.12. ∎

We now give one of the major applications of the previous theorem: the Euler-Maclaurin summation formula.

Theorem 8.14.

(a) Let f be a continuous function on $[a, b]$ with a piecewise continuous derivative there. Then

$$\sum_{a\le n\le b}f(n) = \int_a^b f(t)dt - (t - [t] - \tfrac{1}{2})f(t)\big|_{a-}^b + \int_a^b (t - [t] - \tfrac{1}{2})f'(t)dt.$$

(b) Suppose f also possesses a piecewise continuous second derivative on $[a, b]$ and let

$$\phi(x) = \int_0^x (t - [t] - \tfrac{1}{2})dt.$$

Then

$$\sum_{a\le n\le b}f(n) = \int_a^b f(t)dt - (t - [t] - \tfrac{1}{2})f(t)\big|_{a-}^b - \phi(t)f'(t)\big|_a^b + \int_a^b \phi(t)f''(t)dt.$$

Proof of (a). By Theorems 8.7, 8.9 and Example 8.5 we have

$$\sum_{a \le n \le b} f(n) = \int_{a-}^{b} f(t)d[t] = \int_{a}^{b} f(t)dt - \int_{a-}^{b} f(t)d(t - [t] - \tfrac{1}{2})$$

$$= \int_{a}^{b} f(t)dt - (t - [t] - \tfrac{1}{2})f(t)\big|_{a-}^{b} + \int_{a}^{b} (t - [t] - \tfrac{1}{2})df(t)$$

$$= \int_{a}^{b} f(t)dt - (t - [t] - \tfrac{1}{2})f(t)\big|_{a-}^{b} + \int_{a}^{b} (t - [t] - \tfrac{1}{2})f'(t)dt,$$

by (b) of Theorem 8.12.

Proof of (b). This follows from (a) by another integration by parts and (b) of Theorem 8.12.   ■

Example 8.7. If $x \ge 1$, compute

(a) $\displaystyle\sum_{1 \le n \le x} n.$

(b) $\displaystyle\sum_{1 \le n \le x} \frac{1}{n}.$

Solution of (a). The function $f(x) = x$ is infinitely differentiable so we may apply (b) of Theorem 8.14. This gives, if $x \notin \mathbf{Z}$,

$$\sum_{1 \le n \le x} n = \int_{1}^{x} t\,dt - (t - [t] - \tfrac{1}{2})\big|_{1-}^{x} - \phi(t) \cdot 1\big|_{1}^{x} + \int_{1}^{x} \phi(t) \cdot 0\,dt$$

$$= \left(\frac{x^2}{2} - \tfrac{1}{2}\right) - \{x(x - [x] - \tfrac{1}{2}) - 1(1 - \tfrac{1}{2})\} - \{\phi(x) - \phi(1)\}$$

$$= \tfrac{1}{2}(x^2 + x) + \{-x^2 + x[x] - \phi(x)\} = \tfrac{1}{2}x^2 + O(x),$$

since

$$\left|-x^2 + x[x] - \phi(x)\right| < |x|(x - [x]) + |\phi(x)| \le |x| + \tfrac{1}{4} = O(x).$$

Note that if x is a positive integer, say $x = N$, then

$$\sum_{n=1}^{N} n = \frac{N^2}{2} - \tfrac{1}{2} - \left(N(-\tfrac{1}{2}) - \tfrac{1}{2}\right) - \phi(N) = \tfrac{1}{2}(N^2 + N),$$

the usual formula, since, as in the proof of (f) of Theorem 8.3, $\phi(N) = 0$.

Solution of (b). We have, by (a) of Theorem 8.12,

$$\sum_{1 \le n \le x} \frac{1}{n} = \int_{1}^{x} \frac{dt}{t} - \frac{(t - [t] - \tfrac{1}{2})}{t}\Big|_{1-}^{x} - \int_{1}^{x} \frac{(t - [t] - \tfrac{1}{2})}{t^2}dt = \log x - \frac{x - [x]}{x} + 1 - \int_{1}^{x} \frac{t - [t]}{t^2}dt.$$

Note that, since $0 \le t - [t] < 1$, we have

$$0 \le \int_1^x \frac{t-[t]}{t^2}dt \le \int_1^x \frac{1}{t^2}dt,$$

and so, by the comparison test,

$$\int_1^{+\infty} \frac{t-[t]}{t^2}dt$$

exists and defines some real number. Thus

$$\sum_{1 \le n \le x} \frac{1}{n} = \log x + 1 - \int_1^{+\infty} \frac{t-[t]}{t^2}dt - \frac{x-[x]}{x} + \int_x^{+\infty} \frac{t-[t]}{t^2}dt.$$

Now

$$\left| -\frac{x-[x]}{x} + \int_x^{+\infty} \frac{t-[t]}{t^2}dt \right| \le \frac{1}{x},$$

and so, if we let

$$\gamma = 1 - \int_1^{+\infty} \frac{t-[t]}{t^2}dt,$$

we have

$$\sum_{1 \le n \le x} \frac{1}{n} = \log x + \gamma + \frac{\theta}{x},$$

where $|\theta| \le 1$. The number γ is called Euler's constant (or sometimes the Euler-Mascheroni constant) and is approximately equal to .5772+.   □

Note that the same argument as above shows that if $\alpha > 0$, then

$$\int_1^{+\infty} t^{-\alpha-1}(t-[t])dt$$

converges. The **Riemann zeta function**, $\zeta(\alpha)$, is defined by

(8.17)
$$\zeta(\alpha) = \frac{-\alpha}{1-\alpha} - \alpha \int_1^{+\infty} t^{-\alpha-1}(t-[t])dt,$$

for $\alpha > 0$, $\alpha \ne 1$. If $\alpha > 1$, then integrating by parts shows that

(8.18)
$$\zeta(\alpha) = \lim_{N \to +\infty} \left\{ \frac{-\alpha}{1-\alpha} - \alpha \int_1^N t^{-\alpha-1}(t-[t])dt \right\}$$

$$= \lim_{N \to +\infty} \int_1^N t^{-\alpha}d[t] = \lim_{N \to +\infty} \sum_{n=1}^N n^{-\alpha} = \sum_{n=1}^{+\infty} n^{-\alpha}.$$

In connection with the last example one should note that the following statement is false

$$" \int_1^x (t - [t]) t^{-2} dt = \int_1^x O(t^{-2}) dt = O(x^{-1}), "$$

since the first integral is actually convergent and the O inequalities hold only for large values, not small values. This shows that one must be careful when trying to integrate O-statements. Similar remarks apply to o-statements and $\sim$-statements.

This concludes the survey of tools needed for the material to follow. We close with one last piece of notation. We will write

$$\sum_{n \leq x} f(n) \text{ to mean } \sum_{1 \leq n \leq x} f(n),$$

unless otherwise specified.

Problem Set 8.3

1. Provide the details of the proof of (b) of Lemma 8.11.1.

2. (a) Estimate the sum

$$\sum_{n \leq N} \log n.$$

Find both upper and lower bounds.

(b) Show that if $k \geq 1$ is an integer, then

$$\sum_{n \leq x} \log^k n = \int_1^x \log^k t \, dt + O(\log^k x).$$

3. Use the Euler-Maclaurin summation formula to improve the error estimate in

$$\sum_{n \leq N} \sqrt{n}.$$

4. If k is a positive integer, let

$$\gamma_k = \lim_{N \to +\infty} \left\{ \sum_{n=1}^N \frac{\log^k n}{n} - \frac{\log^{k+1} N}{k+1} \right\}.$$

(a) Show that γ_k exists.

(b) Show that $\gamma_k = O(2^{-k} k!)$, as $k \to +\infty$.

5. (a) Show that

$$\sum_{n \leq x} \lambda(n) \left[\frac{x}{n} \right] = \left[\sqrt{x} \right].$$

(b) Show that

$$\sum_{n\le x}\varphi(n)\left[\frac{x}{n}\right] = \tfrac{1}{2}[x][x+1].$$

(c) Show that

$$\sum_{n\le x}\mu(n)\left[\frac{x}{n}\right] = 1.$$

(d) Show that

$$\left|\sum_{n\le x}\frac{\mu(n)}{n}\right|\le 1 \text{ and } \left|\sum_{n\le x}\frac{\lambda(n)}{n}\right|< 2.$$

(e) Show that

$$\sum_{m=1}^{n}\sigma(m) = \sum_{m=1}^{n}m[n/m].$$

6. Suppose $g(x)$ is a monotone function with continuous derivative on the interval $[1, +\infty)$.

(a) Show that

$$\sum_{m\le x}g(m) = \int_{1}^{x}g(t)dt + O(|g(1)|+|g(x)|).$$

(b) Suppose, in addition, that g tends monotonically to 0 as $x \to +\infty$. Show that

$$\sum_{m\le x}g(m) = \int_{1}^{x}g(t)dt + c + O(|g(x)|),$$

where

$$c = \int_{1}^{+\infty}(t-[t])g'(t)dt.$$

7. Show that

$$\sum_{\substack{n\le N \\ d|n}}n = \frac{N^2}{2d} + O(N).$$

8. Let F and G be complex-valued functions on $(0, +\infty)$ such that $F(x) = G(x) = 0$, for $0 < x < 1$, and $F(1) \ne 0$. Suppose that for all $x \ge 1$ we have

$$G(x) = \sum_{n\le x}F(x/n).$$

Show that if h is an arithmetic function such that, for all $x \ge 1$, $F(x) = (h \circ G)(x)$, then $h = \mu$. (Compare with Problem 2 of Problem Set 7.2.)

9. Let λ denote Lioville's function. Show that if $0 < C < 1/4$, then the number of n, $n \le x$, such that $\lambda(n) = \lambda(n+1)$ is greater than Cx.

10. Suppose f is continously differentiable on $[0, +\infty)$. Show that

(a) $\displaystyle\sum_{0 < n \le N} f(a + nd) = \frac{1}{d} \int_a^{a+dN} f(x)dx + \int_a^{a+dN} f'(x)\left(\frac{x-a}{d} - \left[\frac{x-a}{d}\right]\right)dx$

and

(b) $\displaystyle\sum_{0 \le n < N} f(a + nd) = \frac{1}{d} \int_a^{a+dN} f(x)dx + \int_a^{a+dN} f'(x)\left(\frac{x-a}{d} + \left[\frac{x-a}{d}\right]\right)dx$.

11. Let $a < b$ and let f be continously differentiable on $[a, b]$. Show that

$$\left| f\left(\frac{a+b}{2}\right)\right| \le \frac{1}{b-a}\int_a^b |f(x)|dx + \frac{1}{2}\int_a^b |f'(x)|dx.$$

(Hint: consider the intervals $[a, (a + b)/2]$ and $[(a + b)/2, b]$ separately.) This identity is due to Sololev and Gallagher and has wide uses in advanced number theory.

12. Evaluate

$$\sum_{n \le x} \log \frac{x}{n}.$$

13. Evaluate

$$\sum_{2 \le n \le x} \frac{1}{n \log n}.$$

14. The following results are useful when summing over an arithmetic progression.

(a) Show that

$$\sum_{\substack{n \le x \\ n \equiv a \,(\mathrm{mod}\, m)}} f(n) = \frac{1}{\varphi(m)} \sum_{\chi(m)} \overline{\chi}(n) \sum_{n \le x} \chi(n) f(n).$$

(b) Show that if $f = g*h$, then

$$\sum_{n \le x} \chi(n) f(n) = \sum_{d \le x} \chi(d)g(d) \sum_{n \le x/d} \chi(n)h(n).$$

(c) Show that

$$\sum_{\substack{n \le x \\ (n,m)=1}} f(n) = \sum_{d|m} \mu(d) \sum_{n \le x/d} f(nd).$$

15. Show that if $\alpha > -1$, $\alpha \ne 1$ then we have

$$\zeta(\alpha) = \frac{1}{\alpha - 1} + \frac{1}{2} - \alpha \int_1^{+\infty} \frac{((x))}{x^{\alpha+1}}dx.$$

(Hint: see (8.17) and use (g) of Theorem 8.3.)

16. Suppose $a(n)$ is an arithmetic function satisfying

$$A(x) = \sum_{n \le x} a(n) = Ax^\alpha \log^\beta x + O(x^\gamma)$$

where $\gamma < \alpha$. Give estimates for the following sums.

(a) $\displaystyle\sum_{n \le x} \frac{a(n)}{n}$.

(b) $\displaystyle\sum_{n \le x} \frac{a(n)}{n^\alpha}$.

(c) $\displaystyle\sum_{n \le x} a(n) \log n$.

(d) $\displaystyle\sum_{n \le x} a(n) \log \frac{x}{n}$.

8.4. SUM OF DIVISORS FUNCTION

Lemma 8.15.1.

(a) If $\alpha \ge 0$, then, as $x \to +\infty$,

$$\sum_{n \le x} n^\alpha = \frac{x^{1+\alpha}}{1+\alpha} + O(x^\alpha).$$

(b) If $\alpha > 0$, $\alpha \ne 1$, then, as $x \to +\infty$,

$$\sum_{n \le x} n^{-\alpha} = \frac{x^{1-\alpha}}{1-\alpha} + \zeta(\alpha) + O(x^{-\alpha}).$$

Proof of (a). By Theorem 8.14, we have

$$\sum_{n \le x} n^\alpha = \int_1^x t^\alpha dt + \alpha \int_1^x (t - [t] - \tfrac{1}{2}) t^{\alpha-1} dt - (t - [t] - \tfrac{1}{2}) t^{\alpha-1} \Big|_{1-}^x$$

$$= \frac{x^{\alpha+1} - 1}{\alpha+1} + \alpha \int_1^x (t - [t]) t^{\alpha-1} dt + 1 - (x - [x]) x^\alpha$$

$$= \frac{x^{\alpha+1}}{\alpha+1} + O\left(\alpha \int_1^x t^{\alpha-1} dt + x^\alpha \right) = \frac{x^{\alpha+1}}{\alpha+1} + O(x^\alpha),$$

as $x \to +\infty$.

Proof of (b). By Theorem 8.14, we have

$$\sum_{n \le x} n^{-\alpha} = \int_1^x t^{-\alpha} dt + \alpha \int_1^x (t - [t]) t^{-\alpha-1} dt + 1 - (x - [x]) x^{-\alpha}$$

$$= \frac{x^{1-\alpha}}{1-\alpha} - \frac{1}{1-\alpha} + 1 - \alpha \int_1^{+\infty} (t - [t]) t^{-\alpha-1} dt + O(x^{-\alpha}) = \frac{x^{1-\alpha}}{1-\alpha} + \zeta(\alpha) + O(x^{-\alpha}),$$

as $x \to +\infty$, by (8.17).   ■

Lemma 8.15.2. We have

$$\sum_{k=1}^{m} \cot^2\left(\frac{k\pi}{2m+1}\right) = \frac{m(2m-1)}{3},$$

where m is any positive integer.

Proof. By DeMoivre's formula we have

$$\cos(n\theta) + i\sin(n\theta) = (\cos\theta + i\sin\theta)^n = \sin^n(\theta)(\cot\theta + i)^n = \sin^n(\theta)\sum_{k=0}^{n}\binom{n}{k}i^k \cot^{n-k}\theta.$$

Equating imaginary parts gives

$$\sin(n\theta) = \sin^n(\theta)\left\{\binom{n}{1}\cot^{n-1}(\theta) - \binom{n}{3}\cot^{n-3}(\theta) + \cdots + \binom{n}{\mu}(-1)^\mu\cot^{n-\mu}(\theta)\right\}$$

where μ is the largest odd number $\leq n$, that is, $\mu = 1 + 2[(n-1)/2]$. Take $n = 2m + 1$ and write this as

$$\sin\{(2m+1)\theta\} = \sin^{2m+1}(\theta)P_m(\cot^2\theta),$$

with $0 < \theta < \pi/2$, where P_m is a polynomial of degree m given by

$$P_m(x) = \binom{2m+1}{1}x^m - \binom{2m+1}{3}x^{m-1} + \cdots + (-1)^m.$$

Since $\sin\theta \neq 0$ for $0 < \theta < \pi/2$ we see that $P_m(\cot^2\theta) = 0$ if and only if $(2m+1)\theta = k\pi$ for some integer k. Thus $P_m(x)$ vanishes at the m distinct points $x_k = \cot^2\left(\frac{k\pi}{2m+1}\right)$, for $1 \leq k \leq m$. These are all the zeros of $P_m(x)$, and so

$$\sum_{k=1}^{m}\cot^2\left(\frac{k\pi}{2m+1}\right) = \frac{\binom{2m+1}{3}}{\binom{2m+1}{1}} = \frac{m(2m-1)}{3},$$

as claimed.   ■

Lemma 8.15.3. We have

$$\zeta(2) = \sum_{k=1}^{+\infty}k^{-2} = \frac{\pi^2}{6}.$$

Proof. If $0 < x < \pi/2$ we have $\sin x < x < \tan x$. If we take reciprocals and square we get

$$\cot^2 x < x^{-2} < 1 + \cot^2 x.$$

Put $x = k\pi/(2m + 1)$, $1 \le k \le m$. If we sum on k we get

$$\sum_{k=1}^{m} \cot^2\left(\frac{k\pi}{2m+1}\right) < \frac{(2m+1)^2}{\pi^2} \sum_{k=1}^{m} k^{-2} < m + \sum_{k=1}^{m} \cot^2\left(\frac{k\pi}{2m+1}\right),$$

which, by Lemma 8.15.2, is equivalent to

$$\frac{m(2m-1)}{3} < \frac{(2m+1)^2}{\pi^2} \sum_{k=1}^{m} k^{-2} < m + \frac{m(2m-1)}{3}.$$

If we multiply this inequality by $\pi^2/(4m^2)$ and let $m \to +\infty$, we see that

$$\zeta(2) = \sum_{k=1}^{+\infty} k^{-2} = \lim_{m \to +\infty} \sum_{k=1}^{m} k^{-2} = \frac{\pi^2}{6}.$$

as claimed.   ∎

This proof is due to Apostol.

We now show that if $\alpha > 0$, then the average order of $\sigma_\alpha(n)$ is $\zeta(\alpha + 1)n^\alpha$ and if $\alpha < 0$, then the average order of $\sigma_\alpha(n)$ is $\zeta(1 - \alpha)$.

Theorem 8.15.

(a) If $\alpha > 0$, $\alpha \ne 1$, then, as $x \to +\infty$,

$$\sum_{n \le x} \sigma_\alpha(n) = \frac{\zeta(\alpha+1)}{\alpha+1} x^{\alpha+1} + O(x^\beta),$$

where $\beta = \max(1, \alpha)$.

(b) We have, as $x \to +\infty$,

$$\sum_{n \le x} \sigma(n) = \frac{\pi^2}{12} x + O(x \log x).$$

Proof of (a). Since $\sigma_\alpha = e * T_\alpha e$, we have, by (b) of Theorem 8.5,

$$\sum_{n \le x} \sigma_\alpha(n) = \sum_{n \le x} \sum_{q \mid n} q^\alpha = \sum_{d \le x} \sum_{q \le x/d} q^\alpha$$

(8.19)

$$= \sum_{d \le x} \left\{ \frac{1}{\alpha+1} \left(\frac{x}{d}\right)^{\alpha+1} + O\left(\left(\frac{x}{d}\right)^\alpha\right) \right\},$$

by Lemma 8.15.1. Now, by (8.18),

$$\sum_{d \leq x} \frac{1}{\alpha+1} \left(\frac{x}{d}\right)^{\alpha+1} = \frac{x^{\alpha+1}}{\alpha+1} \sum_{d \leq x} d^{-\alpha-1} = \frac{x^{\alpha+1}}{\alpha+1} \left\{ \sum_{d=1}^{+\infty} d^{-\alpha-1} - \sum_{d > x} d^{-\alpha-1} \right\}$$

(8.20)

$$= \frac{x^{\alpha+1}}{\alpha+1} \zeta(\alpha+1) - \frac{x^{\alpha+1}}{\alpha+1} \sum_{d > x} d^{-\alpha-1}.$$

Since $d^{-\alpha-1} > 0$, we have

$$\sum_{d > x} d^{-\alpha-1} < \int_x^{+\infty} t^{-\alpha-1} dt = \frac{1}{\alpha} x^{-\alpha},$$

and so

(8.21)
$$\frac{x^{\alpha+1}}{\alpha+1} \sum_{d > x} d^{-\alpha-1} = O(x).$$

By Lemma 8.15.1, we have

$$\sum_{d \leq x} d^{-\alpha} = \frac{x^{1-\alpha}}{1-\alpha} + \zeta(\alpha) + O(x^{-\alpha}),$$

and so

(8.22) $$O\left(\sum_{d \leq x} \left(\frac{x}{d}\right)^{\alpha}\right) = O\left\{ x^{\alpha} \frac{x^{1-\alpha}}{1-\alpha} + \zeta(\alpha) + O(x^{-\alpha}) \right\} = O\left(x + x^{\alpha} + 1\right).$$

If we combine (8.19), (8.20), (8.21) and (8.22), we get

$$\sum_{n \leq x} \sigma_\alpha(n) = \frac{\zeta(\alpha+1)}{\alpha+1} x^{1+\alpha} + O(1 + x + x^{\alpha}) = \frac{\zeta(\alpha+1)}{\alpha+1} x^{1+\alpha} + O(x^{\beta}).$$

Proof of (b). With $\alpha = 1$ in (8.19) we have

$$\sum_{n \leq x} \sigma(n) = \frac{x^2}{2} \sum_{d \leq x} d^{-2} + O\left(x \sum_{d \leq x} d^{-1} \right).$$

By (8.20)

$$\sum_{d \leq x} d^{-2} = \zeta(2) + O(x^{-1})$$

and, by (b) of Example 8.7, we have

$$\sum_{d \leq x} d^{-1} = \log x + \gamma + O(x^{-1})$$

Thus

$$\sum_{n \leq x} \sigma(n) = \tfrac{1}{2} \zeta(2) x^2 + O(x \log x)$$

and the result follows from Lemma 8.15.3. ∎

If we tried to apply (c) of Theorem 8.5 to this problem we would have (taking just the case $\alpha = 1$)

$$\sum_{n \leq x} \sigma(n) = \sum_{n \leq x} (e * T_1 e)(n)$$

$$= \sum_{n \leq z} e(n) \sum_{m \leq x/n} (T_1 e)(m) + \sum_{n \leq y} (T_1 e)(n) \sum_{m \leq x/n} e(m) - \sum_{n \leq y} (T_1 e)(n) \sum_{n \leq z} e(n),$$

where $yz = x$. Using the estimates above this becomes

$$\sum_{n \leq x} \sigma(n) = \tfrac{1}{2}\zeta(2)x^2 + O(x \log z + y^2 + x + xy).$$

The problem is now to choose y and z to make the error term as small as possible. Usually one takes $y = x^\alpha$, $z = x^{1-\alpha}$ where $0 < \alpha < 1$, and tries to choose α to equalize, as nearly as possible, the error terms. But with any $\alpha > 0$ we have $x\log z + y^2 + x + xy = O(xy)$ and no choice of α will make this as small as the previously obtained error term of the theorem. The only choice is really $y = \log x$ and $z = x/\log x$. The problem is that by doing the estimate this way we gave undue weight to the summatory function of e and not enough to $T_1 e$. As is seen from the final choice of y and z that we really are giving most of the weight to the summatory function of $T_1 e$. As we shall see in the next section if the two summatory functions are very nearly equal in size, then (c) of Theorem 8.5 works quite well and gives results better than obtainable from parts (a) or (b) of Theorem 8.5.

We now deal with the case $\alpha < 0$. (The case $\alpha = 0$ will be dealt with in the next section since $\sigma_0 = d$.)

Theorem 8.16. Let $\beta > 0$ and $\delta = \max(0, \ 1 - \beta)$. If $x > 1$, then, as $x \to +\infty$

$$\sum_{n \leq x} \sigma_{-\beta}(n) = \begin{cases} \zeta(\beta+1)x + O(x^\delta) & \text{if } \beta \neq 1 \\ \pi^2 x / 6 + O(\log x) & \text{if } \beta = 1 \end{cases}.$$

Proof. We have, by (b) of Theorem 8.5,

$$\sum_{n \leq x} \sigma_{-1}(n) = \sum_{d \leq x} d^{-1} \sum_{q \leq x/d} 1 = \sum_{d \leq x} d^{-1} \left\{ \frac{x}{d} + O(1) \right\} = x \sum_{d \leq x} d^{-2} + O\left(\sum_{d \leq x} d^{-1} \right)$$

$$= x\zeta(2) + O(1 + \log x) = \zeta(2)x + O(\log x).$$

We could do the other part the same way, but we prefer to illustrate the use of Riemann-Stieltjes integration. We have

$$\sigma_{-\beta}(n) = \sum_{d \mid n} d^{-\beta} = n^{-\beta} \sum_{d \mid n} (n/d)^{-\beta} = n^{-\beta} \sigma_\beta(n),$$

since as d runs through the divisors of n so does n/d. Thus, for $\beta \neq 1$,

$$\sum_{n \leq x} \sigma_{-\beta}(n) = \sum_{n \leq x} n^{-\beta} \sigma_{\beta}(n) = \int_{1-}^{x} t^{-\beta} dS_{\beta}(t),$$

where

$$S_{\beta}(t) = \sum_{n \leq x} \sigma_{\beta}(n).$$

Now we imitate the proof of the Euler–Maclaurin summation formula (Theorem 8.14) and get, by (a) of Theorem 8.15,

$$\sum_{n \leq x} \sigma_{-\beta}(n) = \int_{1-}^{x} t^{-\beta} d\left\{ S_{\beta}(t) - \frac{\zeta(\beta+1)}{\beta+1} t^{\beta+1} \right\} + \zeta(\beta+1) \int_{1}^{x} t^{-\beta} t^{\beta} dt$$

$$= \zeta(\beta+1)x - \zeta(\beta+1) + \left\{ S_{\beta}(t) - \frac{\zeta(\beta+1)}{\beta+1} t^{\beta+1} \right\} t^{-\beta} \Big|_{1-}^{x}$$

$$+ \beta \int_{1}^{x} \left\{ S_{\beta}(t) - \frac{\zeta(\beta+1)}{\beta+1} t^{\beta+1} \right\} t^{-\beta-1} dt$$

$$= \zeta(\beta+1)x - \zeta(\beta+1) + \left\{ S_{\beta}(t) - \frac{\zeta(\beta+1)}{\beta+1} \right\} x^{-\beta}$$

$$+ \frac{\zeta(\beta+1)}{\beta+1} + \beta \int_{1}^{x} O(t^{\max(-\beta,-1)}) dt$$

$$= \zeta(\beta+1)x - \zeta(\beta+1) + \frac{\zeta(\beta+1)}{\beta+1} + O(x^{\max(1-\beta,0)})$$

$$= \zeta(\beta+1)x - \zeta(\beta+1) + \frac{\zeta(\beta+1)}{\beta+1} + O(x^{\max(1-\beta,0)})$$

$$+ \beta \int_{1}^{x} O(t^{\max(-\beta,-1)}) dt$$

$$= \zeta(\beta+1)x + O(x^{\max(1-\beta,0)}),$$

which proves the first part.   ∎

Note that in this case one could simply integrate the O-term because the integral

$$\int_{1}^{x} O(t^{\max(-\beta,-1)}) dt$$

diverges.

Problem Set 8.4

1. Evaluate

$$\sum_{n \le x} \sigma(n) \log \frac{x}{n}.$$

2. Let g be a bounded arithmetic function and let a be a real number. Define the arithmetic function f_a by

$$f_a(n) = \sum_{d \mid n} d^a g(n / d).$$

Show that there is a constant A such that if $a > 0$, then

$$\sum_{n \le x} f_a(n) = \frac{A}{a+1} x^{a+1} + E_a(x),$$

where

$$E_a(x) = \begin{cases} O(x^a) & \text{if } a > 1 \\ O(x \log x) & \text{if } a = 1. \\ O(x) & \text{if } a < 1 \end{cases}$$

What is the value of A?

3. (a) Show that if k is a positive integer, then

$$\sum_{\substack{n \le x \\ (n,k)=1}} 1 = \frac{\varphi(k)}{k} x + O(1).$$

(b) Show that if k is a positive integer, then

$$\sum_{\substack{n=1 \\ (n,k)=1}}^{+\infty} n^{-2} = \frac{\pi^2}{6} \prod_{p \mid k} \left(1 - \frac{1}{p^2} \right).$$

(c) If h and k are positive integers such that $(h, k) = 1$, show that

$$\sum_{\substack{n \le x \\ n \equiv h \pmod{k}}} \sigma_{-1}(n) = \frac{\pi^2 x}{6k} \prod_{p \mid k} \left(1 - \frac{1}{p^2} \right) + O(\log x).$$

(d) Show that if k is a positive integer, then

$$\sum_{\substack{n \le x \\ (n,k)=1}} \sigma_{-1}(n) = \frac{\pi^2}{6} \cdot \frac{\varphi(k) J_2(k)}{k^3} x + O(f(k) \log x)$$

uniformly in h and x, where $f(k)$ is a certain arithmetic function of k alone.
What is the value of $f(k)$?

4. Evaluate the following sums.

   (a) $\displaystyle\sum_{n\le x} n\sigma(n)$.

   (b) $\displaystyle\sum_{n\le x} \frac{\sigma(n)}{n^2}$.

8.5. THE NUMBER OF DIVISORS FUNCTIONS

In this section we shall deal with case $\alpha = 0$ of $\sigma_\alpha(n)$ as well as $d_k(n)$, at least for $k = 3$. As we shall see these problems give us a perfect problem for (c) of Theorem 8.5 to handle. Indeed, this summation formula was devised by Dirichlet to handle the problem of determining the average order of $d(n)$.

Theorem 8.17. The average order of $d(n)$ is $\log n$. More precisely,

$$\sum_{n\le x} d(n) = x\log x + (2\gamma - 1)x + O(\sqrt{x}),$$

as $x \to +\infty$.

Proof. Recall that $d = e*e$. Thus, by (c) of Theorem 8.5, we have, with $yz = x$,

$$\sum_{n\le x} d(n) = \sum_{n\le y} e(n) \sum_{m\le x/n} e(m) + \sum_{n\le z} e(n) \sum_{m\le x/m} e(m) - \sum_{m\le y} e(m)\sum_{n\le z} e(n)$$

$$= \sum_{n\le y}[\frac{x}{n}] + \sum_{m\le z}[\frac{x}{n}] - [y][z].$$

Now, since the sums are the same, it would seem reasonable to take $y = z = \sqrt{x}$, but we will continue on in our generality and at the end we shall see that $y = z = \sqrt{x}$ is indeed the optimal choice. Then there exist $|\theta_i| < 1$, $i = 1, 2, 3, 4$, such that

$$\sum_{n\le x} d(n) = \sum_{n\le y}\left(\frac{x}{n} + \theta_1\right) + \sum_{n\le z}\left(\frac{x}{n} + \theta_2\right) - (y + \theta_3)(z + \theta_4)$$

$$= x\sum_{n\le y}\frac{1}{n} + O(y) + x\sum_{n\le z}\frac{1}{n} + O(z) - yz + O(y+z+1)$$

$$= x(\log y + \gamma + O(1/y)) + x(\log z + \gamma + O(1/z)) - x + O(y+z+1)$$

$$= x\log x + (2\gamma - 1)x + O(y+z),$$

where we have used (b) of Example 8.7. In order to equalize our error terms we see that indeed we want to take $y = z = \sqrt{x}$. This gives the result. ∎

One of the unsolved problems in number theory is to find out exactly what the order of magnitude of the error term is. If we define θ by

$$\theta = \inf\{\alpha\colon \sum_{n\leq x} d(n) - x\log x - (2\gamma - 1)x = O(x^\alpha)\},$$

then it is known that $1/4 \leq \theta \leq 12/37$.

One might ask: why not use (b) of Theorem 8.5? The reason is that it does not give as good an error term. For, by (b) of Theorem 8.5,

$$\sum_{n\leq x} d(n) = \sum_{n\leq x}(e*e)(n) = \sum_{n\leq x}[\frac{x}{n}] = x\sum_{n\leq x}\frac{1}{n} + O(x) = x\log x + O(x),$$

which is not as good as Theorem 8.17. The reason is that we failed to take advantage of the symmetry involved in this problem; the order of magnitude of the two summatory functions involved were the same.

The next theorem will again use (c) of Theorem 8.5, even though in this case the summatory functions do not quite have the same order of magnitude. We shall estimate the summatory function for $d_3(n)$ (for $d_k(n)$, in general, see Problem 3 below). We need two preliminary estimates which we collect in the following lemma.

Lemma 8.18.1. As $z \to +\infty$, we have

(a) $\displaystyle\sum_{n\leq z}\frac{\log n}{n} = \frac{1}{2}\log^2 z + A + O\left(\frac{\log z}{z}\right),$

where A is a certain constant defined in the proof, and

(b) $\displaystyle\sum_{n\leq z}\frac{d(n)}{n} = \frac{1}{2}\log^2 z + 2\gamma\log z + A_1 + O(z^{-1/2}),$

where A_1 is a certain constant defined below.

Proof of (a). We have

$$\sum_{n\leq z}\frac{\log n}{n} = \int_{1-}^{z}\frac{\log t}{t}d[t] = \int_{1}^{z}\frac{\log t}{t}dt - \int_{1-}^{z}\frac{\log t}{t}d(t-[t])$$

$$= \frac{1}{2}\log^2 z - (t-[t])\frac{\log t}{t}\Big|_{1-}^{z} + \int_{1}^{z}(t-[t])\frac{1-\log t}{t^2}dt$$

$$= \frac{1}{2}\log^2 z + O\left(\frac{\log z}{z}\right) + \int_{1}^{+\infty}(t-[t])\frac{1-\log t}{t^2}dt - \int_{z}^{+\infty}(t-[t])\frac{1-\log t}{t^2}dt$$

$$= \frac{1}{2}\log^2 z + A + O\left(\frac{\log z}{z}\right),$$

where

$$A = \int_{1}^{+\infty}(t-[t])\frac{1-\log t}{t^2}dt.$$

Proof of (b). We have, if $D(t)$ is the summatory function of $d(n)$,

$$\sum_{n \le z} \frac{d(n)}{n} = \int_{1-}^{z} \frac{1}{t} dD(t) = \int_{1-}^{z} \frac{1}{t} d(t \log t + (2\gamma - 1)t) + \int_{1-}^{z} \frac{1}{t} d\{D(t) - t \log t - (2\gamma - 1)t\}$$

$$= \tfrac{1}{2} \log^2 z + 2\gamma \log z + \frac{D(t) - t \log t - (2\gamma - 1)t}{t} \Big|_{1-}^{z} + \int_{1}^{z} \frac{D(t) - t \log t - (2\gamma - 1)t}{t^2} dt$$

$$= \tfrac{1}{2} \log^2 z + 2\gamma \log z + \int_{1}^{+\infty} \frac{D(t) - t \log t - (2\gamma - 1)t}{t^2} dt + O(z^{-1/2})$$

$$= \tfrac{1}{2} \log^2 z + 2\gamma \log z + A_1 + O(z^{-1/2}),$$

where

$$A_1 = \int_{1}^{+\infty} \frac{D(t) - t \log t - (2\gamma - 1)t}{t^2} dt,$$

which converges since the integrand is $O(t^{-3/2})$. $\blacksquare$

Theorem 8.18. There exist constants a and b such that

$$\sum_{n \le x} d_3(n) = \tfrac{1}{2} x \log^2 x + ax \log x + bx + O(x^{2/3} \log x),$$

as $x \to +\infty$

Proof. We use (c) of Theorem 8.5 and take $y = x^\alpha$ and $z = x^{1-\alpha}$, $0 < \alpha < 1$. Then, by Lemma 8.18.1,

$$\sum_{n \le x} d_3(n) = \sum_{n \le x^\alpha} d(n) \left[\frac{x}{n}\right] + \sum_{n \le x^{1-\alpha}} D\left(\frac{x}{n}\right) - D(x^\alpha)[x^{1-\alpha}]$$

$$= x \sum_{n \le x^\alpha} \frac{d(n)}{n} + O(x^\alpha \log x) + \sum_{n \le x^{1-\alpha}} \left\{ \frac{x}{n} \log \frac{x}{n} + (2\gamma - 1) \frac{x}{n} + O\left(\sqrt{\frac{x}{n}}\right) \right\}$$

$$- \left\{ x^\alpha \log x^\alpha + (2\gamma - 1)x^\alpha + O(x^{\alpha/2}) \right\} \left\{ x^{1-\alpha} + O(1) \right\}$$

$$= x \left\{ \tfrac{1}{2} \log^2 x^\alpha + 2\gamma x^\alpha + A_1 + O(x^{-\alpha/2}) \right\} + O(x^\alpha \log x)$$

$$+ \left\{ x \log x + (2\gamma - 1)x \right\} \left\{ \log x^{1-\alpha} + \gamma + O(x^{\alpha-1}) \right\}$$

$$- x \left\{ \tfrac{1}{2} \log^2 x^{1-\alpha} + A + O(x^{\alpha-1} \log^{1-\alpha} x) \right\} + O(x^{1-\alpha/2})$$

$$- \left\{ x \log x^\alpha - (2\gamma - 1)x + O(x^{1-\alpha/2}) + O(x^\alpha \log x) \right\}$$

$$= \left\{ \tfrac{1}{2} \alpha^2 - \tfrac{1}{2}(1-\alpha)^2 + 1 - \alpha \right\} x \log^2 x$$

$$+ \left\{ 2\gamma\alpha + \gamma + (2\gamma - 1)(1 - \alpha) - \alpha \right\} x \log x$$

$$+ \left\{ A_1 + (2\gamma - 1)\gamma - A + (2\gamma - 1) \right\} x + O(x^{1-\alpha/2} + x^\alpha \log x)$$

$$= \tfrac{1}{2} x \log^2 x + (3\gamma - 1)x \log x + Cx + O(x^{1-\alpha/2} + x^\alpha \log x),$$

where $C = A_1 - A + 2\gamma^2 + \gamma - 1$. To equalize the error terms we take $\alpha = 2/3$, which gives the result. ∎

Problem Set 8.5

1. Use induction to show that, for any positive integer k and all $x \geq 1$, we have

   (a) $\displaystyle\sum_{n \leq x} d^k(n) \leq x \log^{2^k - 1}(2ex)$,

   (b) $\displaystyle\sum_{n \leq x} \frac{d^k(n)}{n} \leq \log^{2^k}(2ex)$

   and

   (c) $\displaystyle\sum_{n \leq x} \frac{d_k(n)}{n} \leq \log^k(2ex)$.

   (Hint: see Problem 22 of Problem Set 7.5.)

2. Use induction to show that if k is a positive integer and $x \geq 1$, then

$$\sum_{n \leq x} \frac{d_k^2(n)}{n} \leq \log^{k^2}(2ex) \quad \text{and} \quad \sum_{n \leq x} d_k^2(n) \leq x \log^{k^2 - 1}(2ex).$$

   (Hint: see Problem 16 of Problem Set 7.5.)

3. Prove that if $k \geq 2$ is an integer, then

$$\sum_{n \leq x} d_k(n) = x P_{k-1}(\log x) + O(x^{1-1/k} \log^{k-2} x),$$

   where $P_{k-1}(z)$ is a polynomial of degree $k - 1$ in z. (Hint: use induction on k.)

4. Show that if $x \geq 2$, $\alpha > 0$, $\alpha \neq 1$, then

$$\sum_{n \leq x} \frac{d(n)}{n^\alpha} = \frac{x^{1-\alpha} \log x}{1 - \alpha} + \zeta^2(\alpha) + O(x^{1-\alpha}).$$

5. Compute

$$\sum_{n \leq x} d(n) \log \frac{x}{n}.$$

6. Show that for certain constants A, B, C and D we have

$$\sum_{n \leq x} d^2(n) = Ax \log^3 x + Bx \log^2 x + Cx \log x + Dx + O(x^\theta),$$

   where $0 < \theta < 1$. (Hint: see Problem 15 of Problem Set 7.5.)

7. Show that

$$\sum_{k=1}^{2n} d(k) - \sum_{k=1}^{n} \left[\frac{2n}{k}\right] = n.$$

8. Show that the constant A of Lemma 8.18.1 is equal to the constant γ_1 of Problem 4 of Problem Set 8.3.

8.6. EULER'S FUNCTION

Theorem 8.19. Suppose $F(s)$ and $G(s)$ are two functions which can be represented by the Dirichlet series

$$F(s) = \sum_{n=1}^{+\infty} f(n)n^{-s}$$

and

$$G(s) = \sum_{n=1}^{+\infty} g(n)n^{-s},$$

which converge absolutely for $s > 1$. Then, for $s > 1$,

$$F(s)G(s) = \sum_{n=1}^{+\infty} (f*g)(n)n^{-s}.$$

Proof. If $s > 1$, then we have

$$F(s)G(s) = \sum_{n=1}^{+\infty} f(n)n^{-s} \sum_{n=1}^{+\infty} g(n)n^{-s} = \sum_{n=1}^{+\infty}\sum_{m=1}^{+\infty} f(n)g(m)(nm)^{-s}.$$

Since we have absolute convergence we can rearrange this last series any way we wish, say by summing over the products $k = mn$. Thus

$$F(s)G(s) = \sum_{k=1}^{+\infty} \left\{ \sum_{mn=k} f(n)g(m) \right\} k^{-s} = \sum_{k=1}^{+\infty} (f*g)(k)k^{-s},$$

which is our result.   ∎

Corollary 8.19.1. We have

$$\sum_{n=1}^{+\infty} \mu(n)n^{-2} = 6/\pi^2.$$

Proof. We have, by Definition 7.5, $\mu*e = \delta$. Thus

$$\sum_{n=1}^{+\infty} \mu(n)n^{-2} \sum_{n=1}^{+\infty} e(n)n^{-2} = \sum_{n=1}^{+\infty} (\mu*e)(n)n^{-2} = 1.$$

By Lemma 8.15.3, we have

$$\sum_{n=1}^{+\infty} e(n)n^{-2} = \pi^2 / 6$$

and the result follows. ∎

Note that, by Corollary 8.19.1,

$$(8.23) \quad \sum_{n \le x} \mu(n)n^{-2} = \sum_{n=1}^{+\infty} \mu(n)n^{-2} - \sum_{n>x} \mu(n)n^{-2} = \frac{6}{\pi^2} + O\left(\sum_{n \ge x} n^{-2}\right) = \frac{6}{\pi^2} + O(1/x).$$

Theorem 8.20. If $x > 1$, then

$$\sum_{n \le x} \varphi(n) = \frac{3}{\pi^2}x^2 + O(x \log x),$$

as $x \to +\infty$

Proof. Recall that $\varphi = \mu * T_1 e$. Thus, by (a) of Theorem 8.5,

$$\sum_{n \le x} \varphi(n) = \sum_{n \le x} \sum_{d|n} \mu(d)(n/d) = \sum_{d \le x} \mu(d) \sum_{n \le x/d} n = \sum_{d \le x} \mu(d)\{\tfrac{1}{2}(x/d)^2 + O(x/d)\}$$

$$= \tfrac{1}{2}x^2 \sum_{d \le x} \frac{\mu(d)}{d^2} + O\left(x \sum_{d \le x} \frac{1}{d}\right) = \frac{3x^2}{\pi^2} + O(x) + O(x \log x) = \frac{3x^2}{\pi^2} + O(x \log x),$$

by (8.23) and (b) of Example 8.7. ∎

We now give an application of Theorem 8.20.

Theorem 8.21. There is a quadratic nonresidue of p between 1 and $\sqrt{p}$ for all large primes p.

Proof. Corresponding to each pair of integers x and y, with $(x, y) = 1$, $0 < x < \sqrt{p}$ and $0 < y < \sqrt{p}$, there is an integer z, unique modulo p, such that

$$(8.24) \qquad\qquad x \equiv yz \ (\mathrm{mod}\ p).$$

Note that different pairs x_1, x_2 and y_1, y_2 yield different z's. For, if

$$x_1 \equiv y_1 z \ (\mathrm{mod}\ p) \quad \text{and} \quad x_2 \equiv y_2 z \ (\mathrm{mod}\ p),$$

then

$$x_1 y_2 \equiv x_2 y_1 \ (\mathrm{mod}\ p).$$

and so, since $0 < x_i, y_i < \sqrt{p}$, $i = 1, 2$, $x_1 y_2 = x_2 y_1$. Since $(x_1, y_1) = (x_2, y_2) = 1$ we see that $x_1 = x_2$ and $y_1 = y_2$.

Now there are $2\varphi(m)$ ordered pairs (x, y) of relatively prime positive integers whose largest element is m, if $1 < m < \sqrt{p}$, and one pair both of whose elements are 1. Thus the total number of pairs is

$$1 + 2 \sum_{2 \le m < \sqrt{p}} \varphi(m) = 2 \sum_{m < \sqrt{p}} \varphi(m) - 1.$$

If this number is larger than $p/2$, then there are more than $(p - 1)/2$ different residue classes z, and, since there are only $(p - 1)/2$ quadratic residues modulo p, at least one z must be a quadratic nonresidue. But then it follows from (8.24) that one of x and y is a quadratic nonresidue and each of these is less than $\sqrt{p}$. Thus the proof will be complete when we show that

(8.25) $$2 \sum_{m < \sqrt{p}} \varphi(m) - 1 > p / 2$$

for $p > p_0$ some fixed $p_0 > 0$. By Theorem 8.20, we have

$$2 \sum_{m < \sqrt{p}} \varphi(m) - 1 = \frac{6}{\pi^2} \left(\sqrt{p} \right)^2 + O(\sqrt{p} \log p) = \frac{6p}{\pi^2} + O(\sqrt{p} \log p).$$

Since $6 / \pi^2 > 1/2$ we see that (8.25) holds with some $p_0 > 0$. ∎

By refining the argument slightly one can show that $p_0 = 10^4$ will do. Then by checking tables one sees that there is a quadratic nonresidue in the interval $[1, \sqrt{p}]$ for all primes except $p = 2, 3, 7,$ or 23.

Problem Set 8.6

1. Prove the following Dirichlet series identities.

(a) $\sum_{n=1}^{+\infty} d_k(n) n^{-s} = \zeta^k(s)$, for $s > 1$ and k is a positive integer.

(b) $\sum_{n=1}^{+\infty} \sigma_k(n) n^{-s} = \zeta(s)\zeta(s-k)$, for $s > \max(1, \ k + 1)$.

(c) $\sum_{n=1}^{+\infty} \varphi(n) n^{-s} = \zeta(s-1) / \zeta(s)$, for $s > 2$.

(d) $\sum_{n=1}^{+\infty} c(n,m) n^{-s} = \sigma_{s-1}(m) m^{1-s} / \zeta(s)$, for $s > 1$.

(e) $\sum_{n=1}^{+\infty} \lambda(n) n^{-s} = \zeta(2s) / \zeta(s)$, for $s > 1$.

(f) $\sum_{n=1}^{+\infty} 2^{\omega(n)} n^{-s} = \zeta^2(s) / \zeta(2s)$, for $s > 1$.

(g) $\sum_{n=1}^{+\infty} d^2(n) n^{-s} = \zeta^4(s) / \zeta(2s)$, for $s > 1$.

(h) $\displaystyle\sum_{n=1}^{+\infty} r_2(n)n^{-s} = 4\zeta(s)L(s)$, for $s > 1$, where

$$L(s) = \sum_{n=1}^{+\infty} \chi(n)n^{-s}$$

and χ is the nonprincipal character modulo 4.

(i) Find a relation for the Dirichlet series

$$\sum_{n=1}^{+\infty} r_4(n)n^{-s}.$$

2. Estimate the sum

$$\sum_{n\le x} \varphi(n)\log\frac{x}{n}.$$

3. Let $Q(x)$ denote the number of square-free integers less than x. As an application of Corollary 8.19.1 show that

$$Q(x) = \frac{6x}{\pi^2} + O(\sqrt{x}).$$

(Hint: see Problem 12 of Problem Set 7.2.)

4. (a) Show that if $x \ge 2$, then

$$\sum_{n\le x} \frac{\varphi(n)}{n^2} = \frac{6}{\pi^2}\log x + \frac{6x}{\pi^2} - A + O\left(\frac{\log x}{x}\right),$$

where

$$A = \sum_{n=1}^{+\infty} \mu(n)n^{-2}\log n.$$

(b) If $a > 1$, $a \ne 2$, $x \ge 2$, show that

$$\sum_{n\le x} \frac{\varphi(n)}{n^a} = \frac{6}{\pi^2}\cdot\frac{x^{2-a}}{2-a} + \frac{\zeta(a-1)}{\zeta(a)} + O(x^{1-a}\log x).$$

(c) If $a \le 1$, show that

$$\sum_{n\le x} \frac{\varphi(n)}{n^a} = \frac{6}{\pi^2}\cdot\frac{x^{2-a}}{2-a} + O(x^{1-a}\log x).$$

(Hint: $T_{-a}\varphi = T_{-a}\mu * T_{1-a}e$.)

5. Let $\varphi_1 = \mu^2 * T_1 e$.

(a) Show that

$$\varphi_1(n) = \sum_{d^2|n} \mu(d)\sigma(n/d^2).$$

(Hint: see Problem 12 of Problem Set 7.2.)

(b) Show that

$$\sum_{n\le x}\varphi_1(n) = \sum_{d\le\sqrt{x}}\mu(d)S(x/d^2),$$

where

$$S(z) = \sum_{n\le z}\sigma(n).$$

(c) Show that, as $x \to +\infty$,

$$\sum_{n\le x}\varphi_1(n) = \frac{\zeta(2)}{2\zeta(4)}x^2 + O(x\log x).$$

6. Let

$$\Phi(t) = \sum_{n\le t}\varphi(n).$$

Let f be an arithmetical function.

(a) Show that

$$\sum_{\substack{a\le x\\ b\le x}} f((a,b)) = 2\sum_{n\le x} f(n)\Phi(x/n) - \sum_{n\le x} f(n).$$

(b) Show that if $k > 1$ is a positive integer, then

$$\sum_{a,b\le x}\sigma_k((a,b)) = \frac{x^{k+1}}{k+1}\left(2\zeta(k) - \zeta(k+1)\right) + O(E_k(x))$$

and

$$\sum_{a,b\le x}\sigma((a,b)) = x^2\left(\log x + 2\gamma - \tfrac{1}{2} - \frac{\zeta(2)}{2}\right) + O(x^{3/2}\log x),$$

where

$$E_k(x) = \begin{cases} x^a & \text{if } k > 2 \\ x^2\log x & \text{if } k = 2 \\ x^2 & \text{if } 1 < k < 2 \end{cases}.$$

(c) Show that

$$\sum_{a,b\le x}\varphi((a,b)) = \frac{x^2}{\zeta^2(2)}\left(\log x + 2\gamma - \tfrac{1}{2} - \frac{\zeta(2)}{2} - \frac{12}{\pi^2}A\right) + O(x^{3/2}\log x),$$

where A is the same constant as in 4(a) above.

7. Let J_k denote the Jordan totient function. Show that, for $k \ge 2$, we have

$$\sum_{n\le x}J_k(n) = \frac{x^{k+1}}{(k+1)\zeta(k+1)} + O(x^k).$$

8. Let $S(n) = \{k \in \mathbf{Z}^+: n/k - [n/k] \geq 1/2\}$. Show that

$$\sum_{k \in S(n)} \varphi(k) = n^2.$$

(Hint: see Problem 1 of Problem Set 8.2 and Problem 5 of Problem Set 8.3.)

8.7. LATTICE POINT PROBLEMS

Definition 8.9. In a coordinate system a **lattice point** is a point with integer coordinates.

We will only deal with the usual rectangular coordinate. The key result is contained in the following theorem.

Theorem 8.22. Let $f(x)$ be continuous and nonnegative on the interval $[Q, R]$. Then the number of lattice points in the plane region $Q \leq x \leq R, 0 < y \leq f(x)$ is equal to

$$\sum_{Q \leq n \leq R} [f(n)].$$

Proof. On the ordinate of the point of the curve $y = f(x)$ with abcissa n there are $[f(n)]$ lattice points in our region. Summing on n gives the result. ∎

The easiest regions in the plane to consider are circles. Let

$$R_2(x) = \sum_{0 \leq a^2 + b^2 \leq x} 1,$$

where the sum is over integer values of both a and b. This is the number of lattice points in and on the circle centered at the origin of radius $\sqrt{x}$. If we recall that $r_2(n)$ is the number of ways of writing n as a sum of two squares, then, for x not an integer,

$$R_2(x) = \sum_{n \leq x} r_2(n),$$

so that we are also calculating the average value of the arithmetic function $r_2(n)$.

Theorem 8.23. As $x \to +\infty$ we have

$$R_2(x) = \pi x + O(\sqrt{x}).$$

Proof. By symmetry, the number of lattice points in $a^2 + b^2 \leq x$ is 4 times the number contained in the first quadrant. The equation of the quarter circle is $s = \sqrt{x - r^2}$. If we use Theorem 8.22 with $Q = 0$ and $R = \sqrt{x}$, then

(8.26) $$R_2(x) = 4 \sum_{0 \leq n \leq \sqrt{x}} \left[\sqrt{x - n^2}\right] = 4 \sum_{0 \leq n \leq \sqrt{x}} \sqrt{x - n^2} + 4c_1\sqrt{x}$$

where $0 \le c_1 < 2$, since there are at most $\sqrt{x} + 1$ terms in the sum. Now

(8.27)
$$\sum_{0 \le n \le \sqrt{x}} \sqrt{x - n^2} = \int_{0-}^{\sqrt{x}} \left(x - t^2\right)^{1/2} d[t] = \int_0^{\sqrt{x}} \left(x - t^2\right)^{1/2} dt + \int_{0-}^{\sqrt{x}} \left(x - t^2\right)^{1/2} d([t] - t)$$

$$= \int_0^{\sqrt{x}} \left(x - t^2\right)^{1/2} dt + \left(x - t^2\right)^{1/2} ([t] - t) \Big|_{0-}^{\sqrt{x}} + \int_0^{\sqrt{x}} \frac{t([t] - t)}{\sqrt{x - t^2}} dt.$$

From elementary calculus we know that

$$\int_0^{\sqrt{x}} \left(x - t^2\right)^{1/2} dt = \frac{\pi x}{4},$$

the area of the quarter circle of radius $\sqrt{x}$. The second term in (8.27) is

$$(x - x)^{1/2}(\sqrt{x} - [\sqrt{x}]) - (x - 0)^{1/2}(-1 - 0) = \sqrt{x}$$

and the second integral in (8.27) is

$$\le \int_0^{\sqrt{x}} \frac{t}{\sqrt{x - t^2}} dt = \lim_{\varepsilon \to 0^+} \int_0^{\sqrt{x} - \varepsilon} \frac{t}{\sqrt{x - t^2}} dt = \frac{-1}{2} \lim_{\varepsilon \to 0^+} 2(x - t^2)^{1/2} \Big|_0^{\sqrt{x} - \varepsilon} = \sqrt{x}.$$

Thus

(8.28)
$$\sum_{0 \le n \le \sqrt{x}} \sqrt{x - n^2} = \frac{\pi x}{4} + c_2 \sqrt{x},$$

where $0 \le c_2 \le 2$. Thus, by (8.26) and (8.28), we have

$$R_2(x) = 4\left(\pi x / 4 + c_2 \sqrt{x}\right) + 4c_1 \sqrt{x} = \pi x + c_3 \sqrt{x},$$

where $0 \le c_3 \le 16$. ∎

We can now proceed by induction on k and calculate the number of lattice points in and on a k-dimensional sphere of radius $\sqrt{x}$ which we denote by $R_k(x)$. The number of lattice points in a k-dimensional sphere of radius $\sqrt{x}$ is the number of k-tuples of integers $(a_1, \ldots, a_k)$ such that $a_1^2 + \cdots + a_k^2 \le x$. Let $r_k(n)$ be the number of ways of writing n as a sum of k squares. Then

(8.29)
$$R_k(x) = \sum_{a_1^2 + \cdots + a_k^2 \le x} 1 = \sum_{n \le \sqrt{x}} \sum_{|m| \le x - n^2} r_{k-1}(|m|)$$

$$= 2 \sum_{n \le \sqrt{x}} \sum_{0 \le m \le x - n^2} r_{k-1}(m) + O\left(\sqrt{x}\right).$$

Now we need an expression for

$$\sum_{N \le z} r_{k-1}(N).$$

We illustrate the use of this reduction formula by computing the number of lattice inside a sphere of radius $\sqrt{x}$. Here we have, by (8.29) and Theorem 8.23,

$$R_3(x) = 2 \sum_{n \le \sqrt{x}} \sum_{0 < m \le x - n^2} r_2(m) + O(\sqrt{x})$$

$$= 2 \sum_{n \le \sqrt{x}} \left\{ \pi(x - n^2) + O\left(\sqrt{x - n^2}\right) \right\} + O(\sqrt{x})$$

$$= 2\pi \sum_{n \le \sqrt{x}} (x - n^2) + O\left(\sqrt{x} + \sum_{n \le \sqrt{x}} \sqrt{x - n^2}\right).$$

The error term is, by (8.28), $O(x)$. The first sum in (8.30) is easy to compute by elementary principles. We have

$$(8.31) \quad \sum_{n \le \sqrt{x}} (x - n^2) = x \sum_{n \le \sqrt{x}} 1 - \sum_{n \le \sqrt{x}} n^2 = x[\sqrt{x}] - [\sqrt{x}][\sqrt{x} + 1][2\sqrt{x} + 1]/6$$

$$= x^{3/2} + O(x) - \tfrac{1}{6}\{\sqrt{x}(\sqrt{x} + 1)(2\sqrt{x} + 1) + O(x)\} = \tfrac{2}{3}x^{3/2} + O(x).$$

Thus, by (8.30) and (8.31),

$$R_3(x) = 2\pi(\tfrac{2}{3}x^{3/2} + O(x)) + O(x) = \tfrac{4}{3}x^{3/2} + O(x).$$

Note that $\frac{4\pi}{3}x^{3/2}$ is the volume of a sphere of radius $\sqrt{x}$, just as πx is the area of the circle of radius $\sqrt{x}$. One might guess that $R_k(x)$ is asymptotic to the volume of the k-dimensional sphere of radius $\sqrt{x}$. This is indeed true (see Problem 4 below).

To get better results we do not want to estimate

$$\sum_{n \le x} [f(n)] = \sum_{n \le x} f(n) - \sum_{n \le x} (f(n) - [f(n)]) = \sum_{n \le x} f(n) + E(x),$$

by always taking $E(x) = O(x)$ since $0 \le f(n) - [f(n)] \le 1$. For example, it can be shown that

$$\sum_{n \le x} ((An + B) - [An + B]) = \tfrac{1}{2}x + O(\log x)$$

instead of just $O(x)$. (See [125].) The details are too intricate to go into here, but see [53, pp 129-138] where it is shown, for example, that

$$\sum_{n \le x} r_2(n) = \pi x + O(x^{1/3} \log x).$$

For a general result see Problem 10 below.

Problem Set 8.7

1. (a) Show that

$$\sum_{k=1}^{n}\left[\frac{k}{2}\right]=\left[\frac{n^2}{4}\right].$$

(b) Show that

$$\sum_{k=1}^{n}\left[\frac{k}{3}\right]=\left[\frac{n(n-1)}{6}\right].$$

(c) Can you generalize this result to

$$\sum_{k=1}^{n}\left[\frac{k}{m}\right],$$

where $m \geq 4$ is an integer?

2. (a) Show that if $f(x, y)$ is defined on the positive integers, then

$$\sum_{\substack{m,n\leq x \\ (m,n)=1}} f(m,n) = \sum_{d\leq x} \mu(d) \sum_{m,n\leq x/d} f(md,nd).$$

(b) Show that

$$\sum_{\substack{0\leq u^2+v^2\leq x \\ (u,v)=1}} 1 = \frac{6}{\pi}x + O\left(\sqrt{x}\log x\right).$$

3. (a) Show that if $(m, n) = 1$, then

$$\sum_{k=1}^{n-1}\left[\frac{mk}{n}\right]=\tfrac{1}{2}(m-1)(n-1).$$

(b) If $(m, n) = d$, show that

$$\sum_{k=0}^{n-1}\left[\frac{mk+b}{n}\right]=\tfrac{1}{2}(m-1)(n-1)+\frac{d-1}{2}.$$

(Hint: count lattice points under an appropiate line.)

4. Show that if

$$R_k(x) =\#\left\{(n_1,\ldots,n_k):0 \leq n_1^2+\cdots+n_k^2 \leq x\right\}$$

then

$$R_k(x) = \alpha_k x^{k/2} + O\left(x^{(k-1)/2}\right),$$

where

$$\alpha_k = \begin{cases} \pi^{(k/2)} / (k/2)! & \text{if } k \text{ is even} \\ \pi^{(k-1)/2} 2^k (\frac{k-1}{2})! / k! & \text{if } k \text{ is odd} \end{cases}.$$

5. (a) For $k \geq 2$, let

$$R_{k,2}(x) = \sum_{0 \leq n_1^k + n_2^k \leq x} 1.$$

Show that

$$R_{k,2}(x) = \frac{2\Gamma^2(1/k)}{k\Gamma(2/k)} x^{k/2} + O\left(x^{\theta_k}\right),$$

where $2/k - 1 \leq \theta_k \leq 1/k$. You can use the fact that

$$\frac{\Gamma(\alpha)\Gamma(\beta)}{\Gamma(\alpha+\beta)} = \int_0^1 (1-t)^{\alpha-1} t^{\beta-1} dt.$$

for $\alpha, \beta > 0$.

(b) For $k, l \geq 2$, let

$$R_{k,l}(x) = \sum_{0 \leq n_1^k + \cdots + n_l^k \leq x} 1.$$

Show that

$$R_{k,l}(x) = \frac{\Gamma^l(1/k)}{(k/2)^l \Gamma(l/k+1)} x^{l/k} + O\left(x^{\theta_{k,l}}\right),$$

where $l/k \leq \theta_{k,l} \leq (l-1)/k$. (Hint: use induction on l.)

6. Find the number of lattice points inside the ellipse

$$\frac{x^2}{a^2} + \frac{y^2}{b^2} = 1.$$

7. Find the number of lattice points in the region

$$1 \leq x \leq R, 0 \leq y \leq \frac{1}{a}\sqrt{x^2 - 1}.$$

8. Let p be a prime, $p \equiv 1 \pmod 4$. Show that

$$\sum_{n=1}^{(p-1)/4} \left[\sqrt{pn}\right] = \tfrac{1}{12}(p^2 - 1).$$

This result is due to Buniakovsky. What happens if $p \equiv 3 \pmod 4$? (Hint: count the lattice points above the parabola $y = \sqrt{px}$ in an appropiate rectangle.)

9. Let $(a, b) = d$. Show that

$$\sum_{n=1}^{a-1}\left[\frac{bn}{a}\right] + \sum_{n=1}^{b-1}\left[\frac{an}{b}\right] = (a-1)(b-1) + (d-1).$$

10. This problem details a result of Jarnik regarding lattice points inside a closed curve. A curve is said to be **rectifiable** if it has an arclength.

   (a) Let C be a rectifiable curve inside a unit square with two end points on the boundary of the square. If C closes the two diagonals of the square, then its length must be at least 1. (Hint: a picture and the triangle inequality may be helpful.)

   (b) Let C be a rectifiable curve inside a unit square with two endpoints on the boundary of the square so that the square is partitioned into two regions. Suppose that C does not pass through the center of the square, and denote by D the region which does not contain the center. Then the area of D must be less than the length of C. (Hint: consider the various cases depending on which sides the ends are on.)

   (c) (Jarnik's Theorem) Let $l \geq 1$ be the length of a rectifiable simple closed curve and let A be the area of the region bounded by the curve. If N is the number of lattice points inside the curve, then

$$|A - N| < l.$$

(Hint: cover the the cartesian plane with unit squares and consider those squares that intersect the curve. Count the number of lattice points inside these squares using parts (a) and (b). Then count the rest of the lattice points.)

Additional Problems For Chapter Eight

General Problems

1. (a) If a, b, c and d are positive real numbers such that

$$[na] + [nb] = [nc] + [nd]$$

for all positive integers n, show that $a + b = c + d$.

   (b) Suppose a, b, c and d are positive irrational numbers such that $a + b = c + d$. Show that

$$[na] + [nb] = [nc] + [nd]$$

for all positive integers n.

2. Let $\alpha = (1 + \sqrt{5}) / 2$ and let F_n denote the nth Fibonacci number and L_n denote the nth Lucas number. Show that

   (a) $F_n = \left[\alpha^n / \sqrt{5} + 1/2 \right]$

   for $n \geq 0$,

   (b) $L_n = \left[\alpha^n + 1/2 \right]$

   for $n \geq 2$,

   (c) $F_{n+1} = \left[\alpha F_n + 1/2 \right]$

   for $n \geq 2$,

   and

   (d) $L_{n+1} = \left[\alpha L_n + 1/2 \right]$

   for $n \geq 4$.

   Can you generalize these results to the more general resurrence $u_0 = 0$, $u_1 = 1$ and, for $n \geq 0$, $u_{n+2} = a u_{n+1} + b u_n$?

3. Define the **Dedekind sum** $s(h, k)$ by

$$s(h,k) = \sum_{m(k)} \left(\left(\frac{m}{k} \right) \right) \left(\left(\frac{mh}{k} \right) \right).$$

   Prove the following properties of $s(h, k)$.

   (a) If $h_1 \equiv h_2 \pmod{k}$, then $s(h_1, k) = s(h_2, k)$.

   (b) If $h\bar{h} \equiv 1 \pmod{k}$, then $s(\bar{h}, k) = s(h, k)$.

   (c) $s(-h, k) = -s(h, k)$.

   (d) If $hh^* \equiv -1 \pmod{k}$, then $s(h^*, k) = -s(h, k)$.

   (e) If $h^2 \equiv -1 \pmod{k}$, then $s(h, k) = 0$.

   (f) $s(1, k) = -1/4 + 1/6k + k/2$.

   (g) If $(h, k) = 1$ and $h, k > 0$, then

   $$s(h, k) + s(k, h) = -1/4 + (h/k + k/h + 1/hk)/12.$$

   (h) If g is an integer, then $s(gh, gk) = s(h, k)$.

   (i) Let F_n be as in Problem 2 above. If $h = F_{2n}$ and $k = F_{2n+1}$, then $s(h, k) = 0$. If $h = F_{2n-1}$ and $k = F_{2n}$, then $12hks(h, k) = h^2 + k^2 - 3hk + 1$.

   (j) If $(h, k) = 1$ and $k \equiv 1 \pmod{h}$, then

   $$12hks(h, k) = (k - 1)(k - h^2 - 1).$$

   (k) Assume that $0 < h < k$ and let $r_0, r_1, \ldots, r_{n+1}$ denote the sequence of remainders in the Euclidean algorithm for calculating (h, k), so that

$$r_0 = k, r_1 = h, r_{j+1} \equiv r_{j-1} \pmod{r_j}, 1 \le r_{j+1} < r_j \text{ and } r_{n+1} = 1.$$

Show that

$$s(h,k) = \tfrac{1}{12} \sum_{j=1}^{n+1} \left\{ (-1)^{j+1} \frac{r_j^2 + r_{j-1}^2 + 1}{r_j r_{j-1}} \right\} - \frac{(-1)^n + 1}{8}.$$

For more on Dedekind sums see [95].

4. Show that $\zeta(2) = \pi^2/6$ by evaluating the double integral

$$I = \int_0^1 \int_0^1 \frac{dx\,dy}{1 - xy}$$

in two ways. (Hint: obtain $I = \zeta(2)$ from the expansion

$$\frac{1}{1 - xy} = 1 + xy + (xy)^2 + \cdots,$$

which is valid for $|xy| < 1$ and then evaluate the integral directly by rotating the axes $45°$.)

5. Let

$$\varphi(x,n) = \sum_{\substack{m \le x \\ (m,n)=1}} 1.$$

Show that

(a) $\displaystyle\sum_{d|n} \varphi(x/d, n/d) = [x]$

and

(b) $\varphi(x,n) = \displaystyle\sum_{d|n} \mu(d)[x/d].$

(c) Show that

$$|\varphi(x,n) - x\varphi(n)/n| < \theta(n).$$

6. Let a be a positive integer. Show that there is a constant $\alpha(a)$ such that

$$\sum_{\substack{n \le x \\ (n,a)=1}} \frac{1}{n} = \frac{\varphi(a)}{a} \log x + \alpha(a) + O\!\left(\frac{d(a)}{x} \right),$$

where the implied O-constant is independent of x and a. (Hint: see Problem 14 of Problem Set 8.3.)

7. Let $k \ge 2$ be an integer. We say that n is a **k-free integer** if $p^\alpha \| n$ implies

$\alpha < k$ for all primes $p \mid n$. Let $Q_k(x)$ be the number of k-free integers less than x. Show that

$$|Q_k(x) - x / \zeta(k)| < 2x^{1/k} + 1.$$

(Hint: find an analog to Problem 12 of Problem Set 7.2 for k-free integers or show that

$$[x] = \sum_{n \leq x^{1/k}} Q_k(x / n^k).$$

8. Let $(m, n) = 1$ and let

$$S = \sum_{k=0}^{m-1} \sum_{j=0}^{n-1} a(k)b(j)e^{2\pi i k j n / m}.$$

Show that

$$|S| \leq \left(m \sum_{k=0}^{m-1} |a(k)|^2 \sum_{l=0}^{n-1} |b(l)|^2 \right)^{1/2}.$$

9. Suppose $(a, m) = 1$. Show that

$$\sum_{k=1}^{m-1} k \left[\frac{ak}{m} \right] = \tfrac{1}{4}(a-1)(m-1)(2m-1) - \tfrac{1}{2} \sum_{k=1}^{a-1} \left[\frac{mk}{a} \right]^2.$$

(Hint: count the lattice points under the line $y = ax/m$ in two ways.)

10. Let $\delta(n)$ be the greatest odd divisor of the positive integer n.

  (a) Show that

$$\left| \sum_{n \leq N} \frac{\delta(n)}{n} - \frac{2N}{3} \right| < 1$$

  for all positive integers N. (Hint: break up the sum into even and odd terms and exploit the values of $\delta(2m+1)$ and $\delta(2m)$.)

  (b) Show that

$$\left| \sum_{n \leq N} \delta(n) - \frac{N^2}{3} \right| < \frac{4N-1}{3}.$$

11. Estimate the sum

$$\sum_{n \leq x} \frac{1}{\varphi(n)}.$$

(Hint: see Problem 15 of Problem Set 7.6.)

12. Estimate the sum

$$\sum_{n \le x} \psi_k(n).$$

(Hint: see Problem 28 of Additional Problems for Chapter Seven.)

13. (a) Show that if a is a positive integer, then

$$\sum_{n \le x} d(an) = \alpha(a)x \log x + \beta(a)x + O\!\left(d_3(a)\sqrt{x}\right),$$

for certain functions $\alpha(a)$ and $\beta(a)$. Identify these functions. (Hint: see Problem 9, parts (c) and (l), of Additional Problems for Chapter Seven.)

(b) Estimate

$$\sum_{n \le x} d(n)\sigma(n).$$

14. Show that there is a constant α so that

$$\sum_{n \le x} \theta(n) = \frac{6}{\pi^2} x \log x + \alpha x + O\!\left(\sqrt{x} \log x\right).$$

(Hint: see Problem 14 of Problem Set 7.9 and Problem 12 of Additional Problems of Chapter Seven. These will show that

$$\theta(m) = \sum_{k^2 \mid m} \mu(k) d(m/k^2).$$

15. (a) If $k \ge 1$ and $s > 0$, find a formula for

$$\sum_{\substack{n \le x \\ (n,k)=1}} n^{-s}.$$

(b) If $k \ge 1$ and $s > 1$, find a formula for

$$\sum_{\substack{n=1 \\ (n,k)=1}}^{+\infty} n^{-s}.$$

16. (a) If f is an arithmetic function, show that

$$\sum_{n \le x} f(n) \log n = \sum_{n \le x} \Lambda_f(n) \sum_{m \le x/n} f(m).$$

(b) If g is completely multiplicative, show that

$$\sum_{n \le x} g(n) \log n = \sum_{p^\alpha \le x} g^\alpha(p) \log p \sum_{m \le x/p^\alpha} g(m).$$

17. Let $\alpha \neq \beta$ be positive numbers. Show that

$$\sum_{m^\alpha n^\beta \leq x} 1 = \zeta(\beta / \alpha)x^{1/\alpha} + \zeta(\alpha / \beta)x^{1/\beta} + O\left(x^{1/(\alpha+\beta)}\right).$$

(Hint: use Theorem 8.5.)

18. Show that

$$\sum_{n \leq x} \sigma^2(n) = \tfrac{5}{6}\zeta(3)x^3 + O\left(x^2 \log^2 x\right).$$

(Hint: see Problem 18 of Problem Set 7.4.)

19. Let

$$\varphi_k(n) = n \sum_{d^k \mid n} \mu(d) / d.$$

Estimate the sum

$$\sum_{n \leq x} \varphi_k(n).$$

20. Show that

$$\left[\sqrt{an+d\pm1} + \sqrt{an+d}\right] = \left[\sqrt{4an+2(2d\pm1)}\right].$$

Can you generalize this to results of the form

$$\left[\sqrt{an+b} + \sqrt{cn+d}\right] = \left[\sqrt{An+B}\right]?$$

21. Prove the following result of Landau. Let $m_1, \ldots, m_n$ be nonnegative integers and let, for $1 \leq r \leq h$ and $1 \leq s \leq k$,

$$u_r = \sum_{j=1}^{n} a_{rj}m_j \text{ and } v_s = \sum_{j=1}^{n} b_{sj}m_j,$$

where a_{rj} and b_{sj} are also nonnegative integers. Then

$$\frac{u_1! \cdots u_h!}{v_1! \cdots v_k!}$$

is integral for all choices of $m_1, \ldots, m_n$ if and only if

$$\sum_{r=1}^{h}\sum_{j=1}^{n}\left[a_{rj}x_j\right] = \sum_{s=1}^{k}\sum_{j=1}^{n}\left[b_{sj}x_j\right]$$

for all real $x_1, \ldots, x_n$ in [0, 1].

22. Let $(a, b) = 1$. Show that if a and b are odd, then

$$\sum_{n=1}^{(b-1)/2}\left[\frac{an}{b}\right] + \sum_{n=1}^{(a-1)/2}\left[\frac{bn}{a}\right] = \tfrac{1}{4}(a-1)(b-1).$$

(Hint: see Problem 9 of Problem Set 8.6.)

23. Let f be an arithmetic function and let $S(n)$ be the set defined in Problem 8 of Problem Set 8.6. Define the arithmetic function g by

$$g(n) = \sum_{k=1}^{n} f(k)[n / k].$$

Show that

$$\sum_{k \in S(n)} f(k) = g(2n) - 2g(n).$$

This result is due to Apostol.

24. If f is an arithmetic function, show that

$$\sum_{n \le x} \mu^2(n) f(n) = \sum_{d \le \sqrt{x}} \mu(d) \sum_{n \le x/d^2} f(nd^2).$$

(Hint: see Problem 12 of Problem Set 7.2.)

25. Let $k \ge 2$ be an integer. Show that

$$\sum_{a,b=1}^{+\infty} [a,b]^{-k} = \zeta^3(k) / \zeta(2k),$$

where [,] denotes the least common multiple. (Hint: recall that $[a, b] = (a, b)/ab$ and sum over fixed values of (a, b).)

26. This problem gives a two-dimensional Euler-Maclaurin formula. Let f be a function of two variables whose second order mixed partial, f_{xy}, is continuous on $R = \{(x, y): a \le x \le b, c \le y \le d\}$, with a, b, c and d integers. Show that

$$\sum_{\substack{c < n \le d \\ a < m \le b}} f(m,n) = \int_c^d \int_a^b f(x,y)dxdy + \int_c^d \int_a^b f_x(x,y)(x - [x])dxdy$$

$$+ \int_c^d \int_a^b f_y(x,y)(y - [y])dxdy + \int_c^d \int_a^b f_{xy}(x,y)(x - [x])(y - [y])dxdy.$$

(Hint: sum over one variable holding the other fixed and use Theorem 8.14. Then use Theorem 8.14 for the sum on the other variable.)

27. Show that every positive integer greater than 1 can be written as a sum of two square-free integers. (Hint: show that $Q_2(x) > (x + 1)/2$, in the terminology of Problem 7 above.) Can this be extended to k-free representations?

28. Estimate the sum

$$\sum_{\substack{n \le x \\ (n,k)=1}} \mu^2(n),$$

where k is a positive integer.

29. (a) Evaluate

$$\sum_{\substack{n_1,\ldots,n_r=1 \\ (n_1,\ldots,n_r)=1}}^{+\infty} n_1^{-k_1} n_2^{-k_2} \cdots n_r^{-k_r} .$$

(Hint: in the sum $\sum_{n_1,\ldots,n_r=1}^{+\infty}$ sum on the value of $(n_1, \ldots, n_r)$ from 1 to $+\infty$.)

(b) Evaluate

$$\sum_{n=2}^{+\infty} \left(\frac{1}{n} \sum_{m=2}^{+\infty} \left(\frac{1}{m} \right)^n \right).$$

(Hint: invert the order of summation and recall the series for $\log(1 - x)$.)

30. If a and b are positive integers, show that

$$\sum_{n \le x} d(an+b) = \left(\sum_{d|(a,b)} \frac{\varphi(a/d)}{a/d} \right) x \log x + \beta(a,b)x + O_{a,b}(\sqrt{x}),$$

where $\beta(a, b)$ is a constant depending on a and b and $O_{a,b}$ indicates that the O-constant depends on a and b. Identify $\beta(a, b)$ and make explicit the O-constant's dependence on a and b.

31. If n is a positive integer, then, by unique factorization, we may write $n = mN^2$, where m is square-free. Denote by $q(n)$, the square-free component of n, that is, $q(n) = m$. Estimate

$$\sum_{n \le x} q(n).$$

(Hint: organize the integers up to x by their largest square divisor, N^2 is the above notation, and determine how many integers have the same square divisor.)

32. Show that there exists a constant A so that for all real numbers $x \ge 1$ there is a square-free integer n such that

$$x < n \le x + Ax^{1/3}.$$

Can you prove an analogous result for k-free integers? (Hint: we wish to show that there is a function $f(x)$ so that

$$Q_k(x + f(x)) - Q_k(x) \ge 1$$

for all $x \ge 1$.)

33. Compute, for $k \geq 0$,

$$\sum_{n \leq x} \frac{\mu(n)}{n} \log^k \frac{x}{n}.$$

(Hint: use induction on k.)

34. Show that, for $k \geq 1$, we have

$$\sum_{n \leq x} \log^k \frac{x}{n} = kx + O(\log^{k-1} x).$$

(Hint: use induction on k.)

35. Prove the following result due to Heilbronn. Suppose f is a real nonnegative function defined on the interval $[a, b]$ and that f'' is continuous on $[a, b]$. Suppose also that

(1) $|f(x)| \leq B_1$

(2) $|f'(x)| \leq B_2$

(3) $f'(x)f''(x)$ is nonzero on $[a, b]$,

where B_1 and B_2 are constants. Then

$$\sum_{a \leq n \leq b} e^{2\pi i f(n)} = \int_a^b e^{2\pi i f(x)} dx + R,$$

where $|R| \leq 2 + 8B_2 + 40B_1 B_2$. (Hint: use Theorem 8.14 and the mean value theorem for integrals.)

36. The following identity due to Vaughan and is useful in the theory of prime numbers. Show that

$$\Lambda(n) = -\sum_{\substack{kl=n \\ k>v \\ l>u}} \Lambda(k) \sum_{\substack{d|l \\ d \leq u}} \mu(d) + \sum_{\substack{kl=n \\ l \leq u}} \mu(k) \log l - \sum_{\substack{klm=n \\ l \leq u \\ m \leq v}} \Lambda(k)\mu(m)$$

Can you prove the analog of this result for the generalized van-Mangoldt function Λ_f?

37. Prove the following result due to Axer. Suppose $B(x)$ is defined for all real x and $a(n)$ is an arithmetic function. Suppose a and B satisfy the following conditions.

(a) $B(x)$ is of bounded variation on every finite interval.

(b) We have, as $x \to +\infty$,

$$\sum_{n \leq x} a(n) = o(x).$$

(c) Either $B(x) = O(1)$ and $\displaystyle\sum_{n \le x} |a(n)| = O(x)$. or $B(x) = O(x^\alpha)$, $0 \le \alpha$

   < 1, and $a(n) = o(1)$.

Then

$$\sum_{n \le x} a(n) B\left(\frac{x}{n}\right) = o(x).$$

This is also useful in the discussion of some results in the theory of prime numbers. (Hint: write

$$\sum_{n \le x} a(n) B\left(\frac{x}{n}\right) = \sum_{n \le \sqrt{x}} a(n) B\left(\frac{x}{n}\right) + \sum_{\sqrt{x} < n \le x} a(n) B\left(\frac{x}{n}\right).$$

Now the estimates on B and a can be applied as appropiate in each of the summations.)

Computer Problems

1. Compute $s(h, k)$ for $1 \le h, k \le 30$.

2. One can sometimes use the computer to find asymptotic formulas, or at least constants associated with them. For example, write a program to compute the sum

$$\sum_{n=1}^{N} d(n)d(n+1).$$

Include in your program the ratios of this number compared to $N\log^2 N$, $N^2 \log N$ and $N^2 \log^2 N$, say. If you do the computations up to $N = 1000$ or perhaps 5000 or 10000, then you may notice one of these ratios tending to a constant. Can you identify this constant? (Hint: there is a π^2 involved.) Finally, can you prove your asymptotic formula?

CHAPTER NINE
Prime Number Theory

9.1. INTRODUCTION

In this chapter we wish to discuss the behavior of the prime numbers. In the first section we prove some elementary results about the distribution of the prime numbers. The next section is devoted to the proof of the prime number theorem. We then prove Dirichlet's theorem on primes in arithmetic progressions. The last two sections are devoted to the applications of the prime number theorem to the study of arithmetic functions.

The interest in primes stems from two results of Euclid:

1. every integer greater than one can be written as a product of primes

and

2. there are infinitely many primes.

After consulting a table of primes (which can be computed using the Sieve of Eratosthenes (see Section 1.3)) one notices that the primes seem to be rather randomly distributed. If one considers the sequence of composite numbers,

$$n! + 2, \ n! + 3, \ ..., \ n! + n - 1, \ n! + n,$$

for $n \geq 2$, one sees that there could be rather long strings of composite numbers. Both of these observations lead to the following questions.

1. How far do we have to go from one prime before we hit another prime?
2. How many primes are there in any given interval $[1, x]$?

An answer to the first question was conjectured by J. Bertrand and proved by P. Chebychev and is one of the main results in the next section, namely, for $n \geq 2$, there exists a prime p such that

$$n < p < 2n.$$

If p_k denotes the $k\underline{th}$ prime, then this result states that

$$0 < p_{k+1} - p_k < p_k.$$

One can attempt to sharpen this result and replace p_k by p_k^θ where $\theta < 1$. This has been done, though the final answer is not in, and one knows that we can take any θ that satisfies

$$1/2 - \varepsilon < \theta < 11/20,$$

for every $\varepsilon > 0$.

As to the second problem, C. F. Gauss, looking at a table of primes, guessed that if $\pi(x)$ is the number of primes in $[1, x]$, then

$$\pi(x) \sim x/\log x.$$

Chebychev tried to prove this, but could not. He did show that if $\pi(x)/(x/\log x)$ had a limit, then that limit had to be 1. Part of what he proved is the second major result of the next section. After Chebychev tried, and failed, G. B. Riemann attacked the problem and nearly succeeded. In the process he bequeathed to number theory many problems and new techniques that have led to much of what is modern day number theory. Finally, in 1896, J. Hadamard and C. de la Vallee-Poussin proved the prime number theorem, the proof of which is rightly called the crowning achievement of 19th century analysis. The proof given in 1896 was based on the theory of functions of a complex variable. The proof given in Section 3 is a variation of the proof obtained, independently in 1948, by A. Selberg and P. Erdös. Their original proof included virtually no analysis at all, being, what is technically called, elementary, though it is quite complicated.

If we ignore 2, then all primes are odd, that is, they lie in the arithmetic progression $2n + 1$. One might ask about other arithmetic progressions $an + b$: are there infinitely many primes? Clearly if there is to be more than one prime, then we must have $(a, b) = 1$. It turns out that this condition is also sufficient, as we shall see in Section 5. This result is due to P.G.L. Dirichlet, who was the first to use analysis to solve a strictly number theoretic problem, though both L. Euler and C. G. Jacobi obtained arithmetic results from analytic researches.

The next two sections deal with various applications of the prime number theorem to the study of multiplicative and additive functions. In the case of multiplicative functions we are interested in their order of magnitude, that is their local size, and in the case of additive functions we are interested in the average order.

9.2. BERTRAND'S POSTULATE AND CHEBYCHEV'S THEOREM

In this section we are concerned with two problems. The first is how far apart the primes can be and the second is how many primes there are.

We begin with some preliminary results, the first of which is of interest in its own right.

Theorem 9.1. If $x \geq 2$, then

$$\prod_{p \leq x} p < 4^x.$$

Proof. The theorem is clearly true for $2 \leq x < 3$. If it is true when x is odd, say $x = n \geq 3$, then it is true for all real numbers y such that $n \leq y < n + 2$ since

$$\prod_{p \leq y} p = \prod_{p \leq n} p < 4^n < 4^y.$$

Thus we need only prove the result for x an odd integer ≥ 3. We proceed by induction on n.

Note that the result holds for $n = 3$ since

$$\prod_{p \leq 3} p = 2 \cdot 3 < 4^3.$$

Assume the result holds for all odd integers m with $1 \leq m \leq n, n \geq 5$. Let $k = (n \pm 1)/2$, where the sign is chosen so that k is odd. Then $k \geq 3$. Also $n - k = 2k \mp 1 - k \leq k + 1$ and so if p is a prime such that $k < p < n$, then p is odd and $p \mid n!$, $p \nmid k!$ and $p \nmid (n - k)!$. Thus the product of all such primes divides $\binom{n}{k}$. Thus

$$\prod_{k < p \leq n} p \leq \binom{n}{k}$$

But $\binom{n}{k} = \binom{n}{n-k}$ and both occur in the expansion of $(1 + 1)^n$. Thus $\binom{n}{k} < 2^{n-1}$.

Thus, by the induction hypothesis, we have

$$\prod_{p \leq n} p = \prod_{p \leq k} p \cdot \prod_{k < p \leq n} p < 4^k 2^{n-1} = 2^{n+2k-1} \leq 2^{2n} = 4^n,$$

since $n \geq 2k - 1$. ∎

Better estimates are known. For example, it is known that

$$2^x < \prod_{p \leq x} p < (2.83)^x.$$

The lower bound is trivial since $p \geq 3$ implies $p > 2$. The upper bound is much harder and is a consequence of the work in [105].

We now give two lemmas on the binomial coefficient $\binom{2n}{n}$.

Lemma 9.2.1. If $n > 1$, then

$$\binom{2n}{n} > 4^n \left(2\sqrt{n}\right)^{-1}.$$

Proof. For $n = 2$, we have

$$4^2\left(2\sqrt{2}\right)^{-1} = 4\sqrt{2} < 6 = \binom{4}{2}.$$

Thus the result holds for $n = 2$ and we proceed by induction on n. Assume the result holds for all $m \leq n$, $n \geq 2$. Since $(2n + 1)^2 > 4n(n + 1)$ we have

$$\binom{2(n+1)}{n+1} = 2\frac{2n+1}{n+1}\binom{2n}{n} > \frac{2(2n+1)4^n}{2\sqrt{n(n+1)}\sqrt{n+1}} > \frac{4^{n+1}}{2\sqrt{n+1}},$$

which is our result.   ∎

The following lemma concerns the canonical factorization of $\binom{2n}{n}$.

Lemma 9.2.2. Let

$$\binom{2n}{n} = \prod p^{e_p},$$

be the canonical factorization. Then

(a) if $p \geq \sqrt{2n}$, then $e_p \leq 1$;

(b) if $p^a \leq 2n < p^{a+1}$, then $e_p \leq a$;

(c) if $2n/3 < p \leq n$, $n > 2$, then $e_p = 0$; and

(d) if $n < p < 2n$, then $e_p = 1$.

Proof. We know that if $m! = \prod p^{v_p}$, then, by Theorem 8.2,

$$v_p = \sum_{a=1}^{+\infty}\left[\frac{m}{p^a}\right].$$

Thus

(9.1) $$e_p = \sum_{a=1}^{+\infty}\left\{\left[\frac{2n}{p^a}\right] - 2\left[\frac{n}{p^a}\right]\right\}$$

Now

$$\left[\frac{2n}{p^a}\right] - 2\left[\frac{n}{p^a}\right] = \begin{cases} 0 & \text{if } [2n/p^a] \text{ is even} \\ 1 & \text{if } [2n/p^a] \text{ is odd} \end{cases},$$

and so e_p is the number of odd terms in the sequence $\{[2n/p^a]\}$. This gives the following results.

    (a) If $p \geq \sqrt{2n}$, then $p^2 \geq 2n$. If $p^2 = 2n$, then $p = 2 = n$ and the result follows. If $p^2 > 2n$, then all the $a > 1$ terms in (9.1) are zero, and so $e_p < 1$.

    (b) If $p^a \leq 2n < p^{a+1}$, then, as in (a), the first a terms are the only possible nonzero terms, and so $e_p \leq a$.

    (c) If $2n/3 < p \leq n$ and $n > 2$, then $2n/p < 3$. Also we must have $p \geq 3$, since $p = 2$ implies that $n < 3$. Now $p \geq 3$ implies $p^2 \geq 2np/3 \geq 2n$, and so $1 \leq n/p < 3/2$. Thus $2 \leq 2n/p < 3$. Thus $[2n/p] - 2[n/p] = 0$, and so $e_p = 0$.

    (d) If $n < p < 2n$, then $[2n/p] = 1$, since $1 < 2n/p < 2$. Also $n < p$, that is, $n/p < 1$, so that $[n/p] = 0$. Thus $e_p = 1$.   ■

Theorem 9.2 (Bertrand-Chebychev). If $n \geq 2$, then there exists a prime p such that

$$n < p < 2n.$$

Proof. It suffices to show that the product

$$P_n = \prod_{n<p<2n} p$$

is not empty, that is, we need only show that $P_n > 1$. Let

$$\binom{2n}{n} = Q_n P_n.$$

By (c) of Lemma 9.2.2 we see that all the prime factors of Q_n are less than $2n/3$. Thus

$$Q_n = \prod_{p \leq 2n/3} p^{e_p},$$

of (a) of Lemma 9.2.2. Now $e_p > 1$ for at most those $p < \sqrt{2n}$ and their number is at most $\sqrt{2n}/2 = \sqrt{n/2}$. Thus the product of the corresponding primes is $\leq (2n)^{\sqrt{n/2}}$. Thus, by Theorem 9.1,

$$Q_n < \prod_{p \le 2n/3} p \cdot \prod_{p < \sqrt{2n}} p^{e_p} < 4^{2n/3}(2n)^{\sqrt{n/2}}$$

so that, by Lemma 9.2.1,

(9.2) $$P_n > \binom{2n}{n} Q_n^{-1} > 4^n \left(2\sqrt{n}\right)^{-1} Q_n^{-1} > 4^{n/3}\left(2\sqrt{n}\right)^{-1}(2n)^{-\sqrt{n/2}}.$$

By squaring we see that $P_n > 1$ for those n such that

$$4^{2n/3} > 2(2n)^{\sqrt{2n}+1}.$$

If we let $2n = x^2$, then we must have

$$4^{x^2/3} > 2x^{2(x+1)}.$$

For $x = 12$ this is valid and by taking derivatives we see that $4^{x^2/3} - 2x^{2(x+1)}$ is increasing for $x \ge 7$. Thus for all n such that $n > x^2/2 \ge 72$ we have $P_n > 1$. For $2 \le n \le 72$ the primes 2, 3, 5, 7, 13, 23, 43, and 83 satisfy the theorem.   ∎

Corollary 9.2.1. If p_k denotes the kth prime, then

$$0 < p_{k+1} - p_k < p_k$$

for $k \ge 1$.

By refining the method used in the proof of Theorem 9.2 one can do better. It can be shown, in an elementary way, that there exists a prime p such that $a_n \le x < p < (1 + 1/n)x$, where $a_1 = 2$, $a_2 = 8$, $a_3 = 9$, $a_4 = 24$ and $a_5 = 25$. See [79].

We now give a quantitative version of Theorem 9.2 which shows that there are many primes between n and $2n$ for n large enough.

Theorem 9.3. If $n > 1$, then

$$\frac{n}{3\log(2n)} < \pi(2n) - \pi(n) < \frac{7n}{5\log n}.$$

Proof. In the proof of Theorem 9.2 we had, by (9.2),

(9.3) $$P_n > 4^{n/3}\left(2\sqrt{n}(2n)^{\sqrt{n/2}}\right)^{-1}.$$

Let the right hand side of (9.3) be $(2n)^x$. Then, by (d) of Lemma 9.2.1, we have

$$x < \pi(2n) - \pi(n).$$

For $n \geq 2500$ we have

$$x \log(2n) = (n/3)(\log 4 - (3\log(4n))/2n) - (3\log(2n)/\sqrt{2n}) > n/3.$$

For $n < 2500$ we check tables of primes and find

$$\pi(2n) - \pi(n) \geq 2,\ 3,\ 9,\ 25,\ 50,\ \text{and}\ 100$$

for

$$n \geq 6,\ 9,\ 36,\ 135,\ 321;\ \text{and}\ 720,$$

respectively. Since $(n/3)/\log(2n) < 1,\ 3,\ 9,\ 25,\ 50,$ and 100 for $n \leq 8,\ 39,\ 150,$ $500,\ 1000,$ and 2500, respectively, we have

(9.4) $$\pi(2n) - \pi(n) > x > n/(3\log(2n)).$$

For the other side we use $\dbinom{2n}{n} < 4^n$. For $n = 1$ we have $\dbinom{2}{1} = 2 < 4$. The

induction steps follows from $\dbinom{2n}{n} / \dbinom{2n-2}{n-1} = \dfrac{2(2n-1)}{n} = 4 - \dfrac{2}{n} < 4$. Thus

(9.5) $$(\pi(2n) - \pi(n))\log n \leq \sum_{n < p \leq 2n} \log p = \log P_n \leq \log\dbinom{2n}{n} \leq n\log 4 < 7n/5.$$

The result follows from (9.4) and (9.5).   ■

The following immediate corollary relates to Chebychev's theorem which follows.

Corollary 9.3.1. If $n > 1$, then

$$\frac{n}{6\log n} < \pi(2n) - \pi(n) < \frac{7n}{5\log n}.$$

Theorem 9.4 (Chebychev). There exist positive constants A and B such that

$$\frac{Ax}{\log x} \leq \pi(x) \leq \frac{Bx}{\log x}.$$

Proof. If $n \geq 2$ is a positive integer, then let $p^{e_p} \left\| \dbinom{2n}{n} \right.$ and define f_p by

$p^{f_p} \leq 2n < p^{f_p + 1}$. If $j > f_p$ then

$$\left[\frac{2n}{p^j}\right] - 2\left[\frac{n}{p^j}\right] = 0 - 0 = 0.$$

Also

$$\left[\frac{2n}{p^j}\right] - 2\left[\frac{n}{p^j}\right] < \frac{2n}{p^j} - 2\frac{n}{p^{j-1}} = 2.$$

Thus, by (9.1),

$$e_p \le \sum_{j=1}^{f_p} 1 = f_p.$$

Thus

$$\binom{2n}{n} \Bigg| \prod_{p \le 2n} p^{f_p}.$$

If $n < p \le 2n$, then $p \mid (2n)!$, but $p \nmid n!$. Thus

$$\prod_{n < p \le 2n} p \Bigg| \binom{2n}{n}.$$

Thus

$$\prod_{n < p \le 2n} n \le \prod_{n < p \le 2n} p \le \binom{2n}{n} \le \prod_{p \le 2n} p^{f_p} \le \prod_{p \le 2n} 2n,$$

that is,

(9.6) $$n^{\pi(2n) - \pi(n)} \le \binom{2n}{n} \le (2n)^{\pi(2n)}.$$

Now $\binom{2n}{n} \le 2^{2n}$, as above, and

$$\binom{2n}{n} = \frac{(2n)(2n-1)\cdots(n+1)}{n(n-1)\cdots 1} = \prod_{j=1}^{n}\left(\frac{n+j}{j}\right) \ge \prod_{j=1}^{n} 2 = 2^n.$$

If we take logarithms of the inequalities (9.6) we obtain

(9.7) $$\pi(2n) - \pi(n) \le \frac{2n\log 2}{\log n} \quad \text{and} \quad \pi(2n) \ge \frac{n\log 2}{\log(2n)}.$$

If $x \ge 2$, let $2n$ be the largest even number $\le x$. Then

$$\pi(x) \ge \pi(2n) \ge \frac{n\log 2}{\log(2n)} \ge \frac{n\log 2}{\log x} \ge \frac{(2n+2)\log 2}{4\log x} \ge \left(\tfrac{1}{4}\log 2\right)\frac{x}{\log x}.$$

Thus we may take $A = (\log 2)/4 = .1732868^+$.

To begin the estimate for B note that

(9.8) $$\pi(64) < \frac{128}{6\log 2} < 14.78.$$

Suppose that for $n \geq 6$ we have

(9.9)
$$\pi(2^n) < \frac{2^{n+1}}{n \log 2} = \frac{2(2^n)}{\log 2^n}.$$

Then, by Theorem 9.3, we have

(9.10)
$$\pi(2^{n+1}) < \pi(2^n) + \frac{7 \cdot 2^n}{5 \log 2^n} < \frac{2^{n+1}}{n \log 2} + \frac{7 \cdot 2^n}{5n \log 2}$$
$$= \frac{2^{n+2}}{(n+1) \log 2} \cdot \frac{17(n+1)}{20n} \leq \frac{2^{n+2}}{(n+1) \log 2}.$$

since $n \geq 6$. Note that (9.9) is true for $n = 1, 2, 3, 4$ and 5. Thus, from (9.10), by mathematical induction, we see that (9.9) is true for all $n \geq 1$. Let x be a real number such that $2^{n-1} \leq x < 2^n$. Then, by (9.9) and the fact that $\pi(x)$ is increasing, we see that

(9.11)
$$\pi(x) \leq \pi(2^n) < \frac{2^{n+1}}{n \log 2} = \frac{4 \cdot 2^{n-1}}{\log 2^n} < 4 \frac{x}{\log x}$$

for $x \geq 2$. From (9.11) we see that we can take $B = 4$. ∎

     Our proof is somewhat wasteful. Chebychev, in his original proof, obtained

$$A = .92129^+ \text{ and } B = 1.10555^+.$$

See [17]. In Problem 1 below it is shown how to get better results straight from Corollary 9.3.1.

Corollary 9.4.1. The sum of the reciprocals of the primes diverges.

Proof. We have, if N is a positive integer

(9.12)
$$\sum_{p \leq N} \frac{1}{p} = \int_{3/2}^N \frac{d\pi(t)}{t} = \frac{\pi(N)}{N} + \int_2^N \frac{\pi(t)}{t^2} dt.$$

By Theorem 9.4, there is a positive constant A such that

$$\pi(x) \geq \frac{Ax}{\log x}.$$

If we put this inequality into (9.12), we obtain

(9.13)
$$\sum_{p \leq N} \frac{1}{p} \geq \frac{A}{\log N} + A \int_2^N \frac{A \, dt}{t \log t} = \frac{A}{\log N} + A \log \log N - A \log \log 2.$$

     Since the right hand side of (9.13) goes to infinity as $N \to +\infty$ we see that

$$\sum_{p=2}^{+\infty} \frac{1}{p}$$

diverges. ∎

One can actually show that

$$\sum_{p \le x} \frac{1}{p} = \log \log x + B + O\left(\frac{1}{\log x}\right),$$

where B is a certain constant. (See Problem 1 of Problem Set 9.3.)

We close this section with an exact formula, due to Legendre, for $\pi(x)$. For improvements see [103].

Theorem 9.5. If N is a positive integer, then

$$\pi(N) - \pi\left(\sqrt{N}\right) + 1 = N - \sum_{p \le \sqrt{N}} \left[\frac{N}{p}\right] + \sum_{p < q \le \sqrt{N}} \left[\frac{N}{pq}\right] - \cdots.$$

Proof. For each prime p $[N/p]$ is the number of integers less than or equal to N which are divisible by p ((g) of Theorem 8.1.) Then

$$N - \sum_{p \le \sqrt{N}} \left[\frac{N}{p}\right]$$

gives the exact number of integers between 1 and N which are not divisible by any primes less than $\sqrt{N}$. Unfortunately any number divisible by two or more such primes has been counted twice. The term

$$\sum_{p < q \le \sqrt{N}} \left[\frac{N}{pq}\right]$$

adds these back, but now we have counted twice any number divisible by three or more such primes. This continual compensation then gives the alternating sum. This is a finite sum since there are $2^{\pi(\sqrt{N})}$ summands. ■

Example 9.1. Calculate $\pi(50)$ by Theorem 9.5.

Solution. We have

$$\pi(50) = \pi\left(\sqrt{50}\right) - 1 + 50 - \sum_{p \le \sqrt{50}} \left[\frac{50}{p}\right] + \sum_{p < q \le \sqrt{50}} \left[\frac{50}{pq}\right] - \cdots$$

$$= 4 + 49 - [50/2] - [50/3] - [50/5] - [50/7]$$
$$+ [50/2 \cdot 3] + [50/2 \cdot 5] + [50/2 \cdot 7] + [50/3 \cdot 5] + [50/3 \cdot 7] + [50/5 \cdot 7]$$
$$- [50/2 \cdot 3 \cdot 5] - [50/2 \cdot 3 \cdot 7] - [50/2 \cdot 5 \cdot 7] - [50/3 \cdot 5 \cdot 7] + [50/2 \cdot 3 \cdot 5 \cdot 7]$$
$$= 53 - 25 - 16 - 10 - 7 + 8 + 3 + 5 + 3 + 2 + 1 - 1 - 1 - 0 - 0 + 0 = 15,$$

which is the correct value.   □

Problem Set 9.2

In the problems that follow p_n will denote the nth prime, that is, $p_1 = 2$, $p_2 = 3$, $p_3 = 5$, etc.

1. Prove Theorem 9.4 as a direct corollary of Corollary 9.3.1. (Hint: note that if

$$\frac{\alpha x}{\log x} < \pi(2x) - \pi(x) < \frac{\beta x}{\log x},$$

then

$$\frac{\alpha(x/2)}{\log(x/2)} < \pi(x) - \pi(x/2) < \frac{\beta(x/2)}{\log(x/2)}$$

etc., where this series of inequalities is finite, and so if we add them up we have a telescoping series.)

2. Use Bertrand's postulate to prove that if $n \geq n_0(s)$, then between n and $2n$ there exists at least one number which is a product of s different primes. Show that we can take $n_0(2) = 4$, $n_0(3) = 5$ and, for $s \geq 4$, $n_0(s) = p_1 \cdots p_s$, the product of the first s primes.

3. Let p_n denote the nth prime number. Show that if

$$\frac{\alpha x}{\log x} < \pi(x) < \frac{\beta x}{\log x},$$

then there exist constants a, b and c such that

$$an \log n < p_n < bn(\log n + c).$$

What values does Theorem 9.4 give for a, b and c?

4. Use Bertrand's postulate to show that if m and k are positive integers, then

$$\frac{1}{m} + \frac{1}{m+1} + \cdots + \frac{1}{m+k}$$

is never an integer unless $k = 0$ and $m = 1$.

5. (a) Prove that in every solution set in positive integers of $x^n = y!$ we have $n = 1$ or $x = 1$.

   (b) Show that if x, y, m and n are positive integers, then $(x!)^m = (y!)^n$ implies that $x = y$ and $m = n$.

6. Suppose that for some constants $\alpha < 1 < \beta$ we have, for $x \geq 2$,

$$\frac{\alpha x}{\log x} < \pi(x) < \frac{\beta x}{\log x}.$$

Show that there exists a number ρ such that $\pi(xy) > \pi(x) + \pi(y)$, for $x \geq 3$, $y \geq \rho$ and $y \leq x$. What value of ρ does Theorem 9.4 imply? (Hint: consider a lower bound for $\pi(xy) - \pi(y)$ and show that $f(x) = x/\log x$ is increasing for $x \geq e$.)

7. (a) Show that if $n \geq 5$, then

$$p_{n+1}^2 < \prod_{k=2}^{n} p_k.$$

   (b) What is the largest integer N such that if k is odd, $1 < k < N$ and $(k, N) = 1$, then k is a prime?

8. Show that $p_n < (n^2 + 3n + 4)/4$. (Hint: see problem 3 above.)

9. Show that if $n \geq 6$, then

$$p_{n+1}^3 < \prod_{k=2}^{n} p_k.$$

10. (a) Show that 30 is the largest integer n such that if $1 < k < n$ and $(k, n) = 1$, then k is a prime.

   (b) Show that 42 is the largest integer n such that if $1 < k < n$ and $(k, n) = 1$, then k is a prime or the power of a prime.

   (c) Can you generalize these results? Suppose r is a given positive integer. For (a) we want to find an integer N such that it is the largest integer n such that if $1 < k < n$ and $(k, n) = 1$, then k is square-free and $\omega(k) = r$. For (b) we want to find N so that it is the largest integer n such that $1 < k < n$ and $(k, n) = 1$ implies that $\omega(k) = r$.

11. Show that the ratio

$$\frac{\pi(2x) - \pi(x)}{\pi(x)}$$

   is bounded.

12. Show that for all $n \geq 1$ we have

$$\binom{2n}{n} < \frac{2^{2n}}{\sqrt{n+1}}.$$

13. In this exercise we sharpen some of the results obtained earlier on highly composite integers. (See Theorem 7.21.)

   (a) Show that for $n \geq 1$ we have $p_{n+1} < p_n^2$.

(b) Show that for $n \geq 1$ we have $p_n^3 > p_{n+3}$.

(c) Show that if $5 \leq p_{n+2} \leq 19$, then $p_n p_{n+3} < p_{n+1} p_{n+2}$.

(d) Show that if $p_{n+2} \geq 11$, then $p_n p_{n+1} p_{n+2} > p_{n+3} p_{n+4}$.

In what follows let $N = 2^{a_1} 3^{a_2} \cdots p_n^{a_n}$ denote a highly composite number.

Theorem 7.21 says that $a_1 \geq a_2 \geq \ldots \geq a_n \geq 1$.

(e) Show that $a_n \leq 2$. (Hint: use (a).)

(f) Suppose the last three primes in the factorization of N are p, q and r, $p < q < r$, and write

$$N = 2^{a_1} 3^{a_2} \cdots p^{a_p} q^{a_q} r^{a_r}.$$

Show that $a_p \leq 4$. (Hint: use (b).)

(g) Suppose N has at least three prime factors. In the notation of (f) assume that $r \leq 19$. Show that $a_r = 1$. (Hint: use (c).)

(h) Suppose N has at least three prime factors and $r \geq 11$. Show that $a_r = 1$. (Hint: use (d).)

(i) Show that if N is a highly composite number with at most two prime factors, then N must have the form

$$N = 2^a 3^b$$

with $a \geq b$ and $0 \leq b \leq 2$.

(j) Show that if $b = 0$, then $a = 2$.

(k) Show that if $b = 2$, then $a = 2$.

(l) Conclude that if N is a highly composite integer, then the final exponent $a_n = 1$ except when $N = 4$ or $N = 36$.

9.3. THE PRIME NUMBER THEOREM

After Theorem 9.4 one might reasonably guess that $\pi(x)$ and $x/\log x$ are asymptotic to each other as $x \to +\infty$. This is indeed the case as we shall show in this section. The proof that follows is a modification of a proof due to Wright.

Lemma 9.6.1. As $x \to +\infty$ we have

(a) $\displaystyle\sum_{n \leq x} \log n = x \log x - x + O(\log x)$

and

(b) $\displaystyle\sum_{n \leq x} \log^2 n = x \log^2 x - 2x \log x + 2x + O(\log^2 x)$.

Proof. We have, for $k \geq 1$,

$$\sum_{n \leq x} \log^k n = \int_1^x \log^k t \, d[t] = \int_1^x \log^k t \, dt + \int_1^x \log^k t \, d([t] - t)$$

$$= \int_1^x \log^k t \, dt + (\log^k t)([t] - t)\Big|_1^x - k \int_1^x ([t] - t) \log^{k-1} t \, \frac{dt}{t}$$

$$= \int_1^x \log^k t \, dt + O\left(\log^k t + \int_1^x \frac{\log^{k-1} t}{t} \, dt \right) = \int_1^x \log^k t \, dt + O(\log^k x).$$

Now

$$\int_1^x \log t \, dt = x \log x - x + 1$$

and

$$\int_1^x \log^2 t \, dt = x \log^2 x - 2x \log x + 2x - 2$$

and the results follow.   ∎

Lemma 9.6.2. As $x \to +\infty$, we have

(a) $\displaystyle\sum_{n \leq x} \frac{\mu(n)}{n} = O(1)$,

(b) $\displaystyle\sum_{n \leq x} \frac{\mu(n)}{n} \log \frac{x}{n} = O(1)$,

and

(c) $\displaystyle\sum_{n \leq x} \frac{\mu(n)}{n} \log^2 \frac{x}{n} = 2 \log x + O(1)$.

Proof. We use Corollary 8.4.1 with different choices of g and (b) of Example 8.6.

(a) Take $g(x) = 1$. Then

$$f(x) = \sum_{n \leq x} g(x/n) = [x] = x + O(1).$$

Thus

$$1 = \sum_{n \leq x} \mu(n)\left\{ \frac{x}{n} + O(1) \right\} = x \sum_{n \leq x} \frac{\mu(n)}{n} + O(x).$$

If we divide by x the result follows.

(b) Take $g(x) = x$. Then $f(x) = x \log x + \gamma x + O(1)$. Thus

$$x = \sum_{n \leq x} \mu(n)\left\{ \frac{x}{n} \log \frac{x}{n} + \gamma \frac{x}{n} + O(1) \right\}$$

$$= x\left\{ \sum_{n \leq x} \frac{\mu(n)}{n} \log \frac{x}{n} + \gamma \sum_{n \leq x} \frac{\mu(n)}{n} \right\} + O(x)$$

$$= x \sum_{n \leq x} \frac{\mu(n)}{n} \log \frac{x}{n} + O(x),$$

by (a). If we divide by x, the result follows.

(c) Take $g(x) = x\log x$. Then $f(x) = \frac{1}{2}x\log^2 x + \gamma x\log x + Ax + O(\log x)$, by (a) of Lemma 8.18.1. Thus

$$
\begin{aligned}
x\log x &= \sum_{n\le x}\mu(n)\left\{\frac{x}{2n}\log^2\frac{x}{n} + \gamma\frac{x}{n}\log\frac{x}{n} + A\frac{x}{n} + O\left(\log\frac{x}{n}\right)\right\} \\
&= \frac{1}{2}x\sum_{n\le x}\frac{\mu(n)}{n}\log^2\frac{x}{n} + \gamma x\sum_{n\le x}\frac{\mu(n)}{n}\log\frac{x}{n} + Ax\sum_{n\le x}\frac{\mu(n)}{n} + O\left(\sum_{n\le x}\log\frac{x}{n}\right).
\end{aligned}
$$
(9.14)

Now, by (a) of Lemma 9.5.1,

$$
\begin{aligned}
\sum_{n\le x}\log\frac{x}{n} &= \log x\sum_{n\le x}1 - \sum_{n\le x}\log n \\
&= x\log x + O(\log x) - (x\log x - x + O(\log x)) = x + O(\log x).
\end{aligned}
$$
(9.15)

Thus, by (9.14) and (9.15) and (a) and (b) above,

$$
x\log x = \frac{1}{2}x\sum_{n\le x}\frac{\mu(n)}{n}\log^2\frac{x}{n} + O(x).
$$

If we divide by $x/2$ the result follows. ∎

Theorem 9.6 (Selberg). As $x \to +\infty$, we have

$$
\sum_{n\le x}\left\{(L\Lambda)(n) + (\Lambda*\Lambda)(n)\right\} = 2x\log x + O(x).
$$

Proof. Let $f = e$ in Theorem 7.10. Then

$$
(L\Lambda)(n) + (\Lambda*\Lambda)(n) = (\mu*L^2e)(n).
$$

Thus, by (a) of Theorem 8.5 and Lemma 9.5.2, we have

$$
\begin{aligned}
\sum_{n\le x}\left\{(L\Lambda)(n) + (\Lambda*\Lambda)(n)\right\} &= \sum_{n\le x}(\mu*L^2e)(n) \\
&= \sum_{n\le x}\mu(n)\sum_{d\le x/d}\log^2 d = \sum_{n\le x}\mu(n)\left\{\frac{x}{n}\log^2\frac{x}{n} - 2\frac{x}{n}\log\frac{x}{n} + 2\frac{x}{n}O\left(\log^2\frac{x}{n}\right)\right\} \\
&= x\sum_{n\le x}\frac{\mu(n)}{n}\log^2\frac{x}{n} - 2x\sum_{n\le x}\frac{\mu(n)}{n}\log\frac{x}{n} + 2x\sum_{n\le x}\frac{\mu(n)}{n} + O\left(\sum_{n\le x}\log^2\frac{x}{n}\right) \\
&= 2x\log x + O(x) + O(x) + O(x) + O\left(\sum_{n\le x}\log^2\frac{x}{n}\right).
\end{aligned}
$$
(9.16)

Now, by Lemma 9.5.1,

$$\sum_{n \leq x} \log^2 \frac{x}{n} = \log^2 x \sum_{n \leq x} 1 - 2 \log x \sum_{n \leq x} \log n + \sum_{n \leq x} \log^2 n$$

(9.17)
$$= \log^2 x (x + O(1)) - 2 \log x (x \log x - x + O(\log x))$$
$$+ x \log^2 x - 2x \log x + 2x + O(\log^2 x)$$
$$= 2x + O(\log^2 x),$$

and so, by (9.16) and (9.17),

$$\sum_{n \leq x} \{ (L\Lambda)(n) + (\Lambda * \Lambda)(n) \} = 2x \log x + O(x),$$

which is our result. ∎

We are interested in getting an estimate for the function $\pi(x)$. Unfortunately $\pi(x)$ does not behave well. A function that behaves better is

$$\psi(x) = \sum_{n \leq x} \Lambda(n).$$

Since e is multiplicative and $\Lambda = \Lambda_e$ we know that Λ lives only on prime powers. We have, for $a \geq 1$

$$\Lambda(p^a) = (\mu * Le)(p^a) = \sum_{j=0}^{a} \mu(p^j)(Le)(p^{a-j})$$

$$= \mu(1)(Le)(p^a) + \mu(p)(Le)(p^{a-1}) = \log(p^a) - \log(p^{a-1}) = \log p.$$

Thus we could rewrite ψ as

$$\psi(x) = \sum_{p^a \leq x} \log p.$$

The relationship between $\pi(x)$ and $\psi(x)$ is given by the following theorem.

Theorem 9.7. For all x we have

$$\frac{\psi(x)}{\log x} \leq \pi(x) \leq \frac{\psi(x)}{\log x} + O\left(\frac{x}{\log^2 x} \right).$$

Proof. Note that $p^a \leq x$ if and only if $a \leq \log x / \log p$. Thus

$$\psi(x) = \sum_{p^a \leq x} \log p = \sum_{p \leq x} \sum_{a \leq \log x / \log p} \log p = \sum_{p \leq x} \left[\frac{\log x}{\log p} \right] \log p \leq \sum_{p \leq x} \log x = \pi(x) \log x,$$

which proves the left hand inequality.

We have

$$\pi(x) = \sum_{p \leq x} 1 \leq \sum_{p^a \leq x} \frac{\log p}{\log(p^a)} = \sum_{n \leq x} \frac{\Lambda(n)}{\log n} = \int_{2-}^{x} \frac{d\psi(t)}{\log t} = \frac{\psi(x)}{\log x} + \int_{2}^{x} \frac{\psi(t) dt}{t \log^2 t}$$

By Theorem 9.4 we know that

$$\pi(x) = O(x / \log x),$$

and so, by the first inequality,

(9.18) $$\psi(x) = O(x).$$

Thus

$$\pi(x) \le \frac{\psi(x)}{\log x} + O\left(\int_2^x \frac{dt}{\log^2 t}\right) = \frac{\psi(x)}{\log x} + O\left(\frac{x}{\log^2 x}\right).$$

by l'Hospital's rule.   ∎

As an immediate consequence we have the following result.

Corollary 9.7.1. As $x \to +\infty$ we have

$$\pi(x) \sim \frac{x}{\log x} \text{ if and only if } \psi(x) \sim x.$$

Thus to prove the prime number theorem we need only show that $\psi(x) \sim x$ as $x \to +\infty$. Recall (see Section 9.1) that it is equivalent to showing that, as $x \to +\infty$,

$$r(x) = \psi(x) - x = o(x).$$

This is our goal for the rest of this section.

Lemma 9.8.1. As $x \to +\infty$, we have

$$\sum_{n \le x} (L\Lambda)(n) = \psi(x) \log x + O(x).$$

Proof. We have

$$\sum_{n \le x} (L\Lambda)(n) = \int_1^x \log t \, d\psi(t) = \psi(x) \log x - \int_1^x \frac{\psi(t)}{t} dt = \psi(x) \log x + O(x),$$

by (9.18).   ∎

An immediate consequence of this lemma and Theorem 9.5 is the following result.

Corollary 9.8.1.1. As $x \to +\infty$, we have

$$\psi(x) \log x + \sum_{n \le x} (\Lambda * \Lambda)(n) = 2x \log x + O(x).$$

Lemma 9.8.2. As $x \to +\infty$, we have

(a) $\displaystyle\sum_{n \le x} \psi(x/n) = x \log x - x + O(x)$

and

(b) $\displaystyle\sum_{n \le x} \frac{\Lambda(n)}{n} = \log x + O(1).$

Proof. We use (a) of Theorem 8.5 and a consequence of the definition of Λ, namely: $Le = e * \Lambda$.

(a)   We have, by (a) of Lemma 9.6.1, that as $x \to +\infty$,

$$x \log x - x + O(\log x) = \sum_{n \le x} \log n = \sum_{n \le x} (e * \Lambda)(n)$$

(9.19)

$$= \sum_{n \le x} e(n) \sum_{m \le x/n} \Lambda(m) = \sum_{n \le x} \psi(x/n).$$

(b)   We return to (9.19) and split it the other way. Thus, as $x \to +\infty$

$$x \log x - x + O(\log x) = \sum_{n \le x} (e * \Lambda)(n)$$

$$= \sum_{n \le x} \Lambda(n)[x/n] = \sum_{n \le x} \Lambda(n)\left(\frac{x}{n} + O(1)\right)$$

$$= \sum_{n \le x} \frac{\Lambda(n)}{n} + O\left(\sum_{n \le x} \Lambda(n)\right) = x \sum_{n \le x} \frac{\Lambda(n)}{n} + O(x),$$

by (9.18). The result follows upon dividing by x.   ∎

Let $a(n) = (L\Lambda + \Lambda * \Lambda)$. Then Theorem 9.5 states that

$$A(x) = \sum_{n \le x} a(n) = 2x \log x + O(x).$$

Lemma 9.8.3. We have, as $x \to +\infty$,

$$|r(x)|\log^2 x \le \sum_{n \le x} a(n)|r(x/n)| + O(x \log x).$$

Proof. By Corollary 9.8.1.1 and (a) of Theorem 8.5, we have

(9.20) $\displaystyle (x + r(x))\log x + \sum_{n \le x} \Lambda(n)(x/n + r(x/n)) = 2x \log x + O(x).$

By (b) of Lemma 9.8.2, we have, from (9.20),

(9.21) $\displaystyle r(x)\log x + \sum_{n \le x} \Lambda(n)r(x/n) = O(x).$

Thus

$$(9.22) \qquad r(x/n)\log(x/n) + \sum_{m \le x/n} \Lambda(m)r(x/nm) = O(x/n).$$

Now multiply (9.21) by $\log x$ and (9.22) by $\Lambda(n)$ and sum on $n \le x$. If we subtract these from each other we get, by Lemma 9.8.2,

$$\log x \left\{ r(x)\log x + \sum_{n \le x} \Lambda(n)r(x/n) \right\} - \sum_{n \le x} \Lambda(n) \left\{ r(x/n)\log(x/n) + \sum_{m \le x/n} \Lambda(m)r(x/mn) \right\}$$

$$= O(x\log x) + O\left(x \sum_{n \le x} \frac{\Lambda(n)}{n} \right) = O(x\log x).$$

Thus

$$r(x)\log^2 x = \sum_{n \le x} (L\Lambda)(n)r(x/n) + \sum_{n \le x} (\Lambda * \Lambda)(n)r(x/n) + O(x\log x)$$

$$= \sum_{n \le x} a(n)r(x/n) + O(x\log x).$$

If we take absolute values and note that $a(n) \ge 0$, the result follows. ∎

Lemma 9.8.4. As $x \to +\infty$, we have

$$\sum_{n \le x} a(n)|r(x/n)| = 2\int_1^x |r(x/t)|\log t\,dt + O(x\log x).$$

Proof. Let $\rho(x) = |r(x)| = |\psi(x) - x|$. Then $d\rho(x) = dx + d\psi(x)$ and we have

$$\sum_{n \le x} a(n)|r(x/n)| = \int_1^x |r(x/t)|dA(t)$$

$$= \int_1^x |r(x/t)|d(2t\log t) + \int_1^x |r(x/t)|d(A(t) - 2t\log t)$$

$$= 2\int_1^x |r(x/t)|(\log t + 1)dt + |r(x/t)|(A(t) - 2t\log t)\Big|_1^x$$

$$(9.23) \qquad\qquad - \int_1^x (A(x/t) - 2(x/t)\log(x/t))d\rho(t)$$

$$= 2\int_1^x |r(x/t)|\log t\,dt + O(x\log x)$$

$$- \int_1^x (A(x/t) - 2(x/t)\log(x/t))d\rho(t).$$

since $r(x) = O(x)$, by (9.18). The last integral in (9.23) is

(9.24) $\quad = O\left(\int_1^x \frac{x}{t}d(\psi(t)+t)\right) = O\left(x\int_1^x \frac{d\psi(t)}{t} + x\int_1^x \frac{dt}{t}\right) = O(x\log x),$

by Lemma 9.8.2, since

$$\int_1^x \frac{d\psi(t)}{t} = \sum_{n\le x} \frac{\Lambda(n)}{n}$$

If we combine (9.23) and (9.24) we obtain the result.   ∎

If we combine the results of Lemmas 9.8.3 and 9.8.4 we have the following result.

Corollary 9.8.4.1. As $x \to +\infty$, we have

(9.25) $\quad |r(x)|\log^2 x \le 2\int_1^x |r(x/t)|\log t\, dt + O(x\log x).$

Now let $V(x) = e^{-x}r(e^x) = e^{-x}\psi(e^x) - 1$. Then, by (9.18), we have

$$V(x) = O(1).$$

We need to show that $V(x) = o(1)$ as $x \to +\infty$.

If we let $x = e^\zeta$ and $t = xe^{-\eta}$, then

$$\int_1^x |r(x/t)|\log t\, dt = x\int_0^\zeta |V(\eta)|(\zeta - \eta)d\eta = x\int_0^\zeta |V(\eta)|\left\{\int_\eta^\zeta d\xi\right\}d\eta = x\int_0^\zeta \int_0^\xi |V(\eta)|d\eta d\xi$$

Thus (9.25) reads

$$\left|e^\zeta V(\zeta)\zeta^2\right| \le 2e^\zeta \int_0^\zeta \int_0^\xi |V(\eta)|d\eta d\xi + O(\zeta e^\zeta)$$

or, dividing out the e^ζ,

(9.26) $\quad \zeta^2|V(\zeta)| \le 2\int_0^\zeta \int_0^\xi |V(\eta)|d\eta d\xi + O(\zeta).$

Let

$$\alpha = \limsup_{\zeta\to+\infty}|V(\zeta)| \quad \text{and} \quad \beta = \liminf_{\zeta\to+\infty} \zeta^{-1}\int_0^\zeta |V(\eta)|d\eta.$$

Since, by (9.18), $V(\eta) = O(1)$ we see that both α and β exist. Thus

$$|V(\zeta)| \le \alpha + o(1),$$

as $\zeta \to +\infty$. Also

$$\int_0^\zeta |V(\eta)|d\eta \le \beta\zeta + o(\zeta),$$

as $\zeta \to +\infty$. Thus, by (9.26),

$$\zeta^2 |V(\zeta)| \leq 2 \int_0^\zeta \{\beta\xi + o(\xi)\} d\xi + O(\zeta) = \beta\zeta^2 + o(\zeta^2),$$

that is,

$$|V(\zeta)| \leq \beta + o(1),$$

as $\zeta \to +\infty$. Thus

$$\alpha \leq \beta.$$

Lemma 9.8.5. As $x \to +\infty$, we have

$$\int_1^x \frac{\psi(t)}{t^2} dt = \log x + O(1).$$

Proof. We have, by Lemma 9.7.2 and (9.18),

$$\log x + O(1) = \sum_{n \leq x} \frac{\Lambda(n)}{n} = \int_1^x \frac{d\psi(t)}{t} = \frac{\psi(x)}{x} + \int_1^x \frac{\psi(t)}{t^2} dt = O(1) + \int_1^x \frac{\psi(t)}{t^2} dt$$

and the result follows. ∎

Lemma 9.8.6. There exists a positive constant A such that for every positive ζ_1 and ζ_2 we have

$$\left| \int_{\zeta_1}^{\zeta_2} V(\eta) d\eta \right| < A.$$

Proof. Let $\zeta = \log x$ and $\eta = \log t$. Then, by Lemma 9.8.5,

$$\int_0^\zeta V(\eta) d\eta = \int_1^x \left\{ \frac{\psi(t)}{t^2} - \frac{1}{t} \right\} dt = O(1).$$

Thus

$$\int_{\zeta_1}^{\zeta_2} V(\eta) d\eta = \int_0^{\zeta_2} V(\eta) d\eta - \int_0^{\zeta_1} V(\eta) d\eta = O(1),$$

which gives the result. ∎

Lemma 9.8.7. If $\eta_0 > 0$ and $V(\eta_0) = 0$, then

$$\int_0^\alpha |V(\eta_0 + t)| dt \leq \tfrac{1}{2}\alpha^2 + O(\eta_0^{-1}).$$

Proof. By Corollary 9.8.1.1, we have

$$\sum_{n \le x} (\Lambda * \Lambda)(n) + \psi(x)\log x = 2x\log x + O(x).$$

If $x > x_0 \ge 1$, then this holds for x_0 with the same error term. Thus

$$\psi(x)\log x - \psi(x_0)\log x_0 + \sum_{x_0 < n \le x}(\Lambda * \Lambda)(n) = 2\left(x\log x - x_0\log x_0\right) + O(x).$$

Since $\Lambda(n) \ge 0$ for all n and $\psi(t)\log t$ is increasing in t for $t > 0$ we have

$$0 \le \psi(x)\log x - \psi(x_0)\log x_0 \le 2\left(x\log x - x_0\log x_0\right) + O(x)$$

or

(9.27) $$\left|r(x)\log x - r(x_0)\log x_0\right| \le x\log x - x_0\log x_0 + O(x).$$

If $\alpha \ge t \ge 0$, $x = e^{\eta_0 + t}$ and $x = e^{\eta_0}$, then $r(x_0) = 0$ and, by (9.27),

$$\left|V(\eta_0 + t)e^{\eta_0 + t}(\eta_0 + t)\right| \le e^{\eta_0 + t}(\eta_0 + t) - e^{\eta_0}\eta_0 + O(e^{\eta_0 + t}).$$

Thus

$$\left|V(\eta_0 + t)\right| \le 1 - \frac{\eta_0}{\eta_0 + t}e^{-t} + O((\eta_0 + t)^{-1}) = 1 - e^{-t} + O(\eta_0^{-1}) \le t + O(\eta_0^{-1}),$$

since $$\frac{te^{-t}}{\eta_0 + t} \le \frac{t}{\eta_0 + t} < \frac{t}{\eta_0} = O(\eta_0^{-1}) \text{ since } t \le \alpha. \text{ Thus}$$

$$\int_0^\alpha \left|V(\eta_0 + t)\right|dt \le \int_0^\alpha \left(t + O(\eta_0^{-1})\right)dt + \tfrac{1}{2}\alpha^2 + O(\eta_0^{-1}),$$

since α is fixed.   ■

Theorem 9.8. As $x \to +\infty$ we have

$$\psi(x) \sim x.$$

Proof. To prove this theorem it suffices to show that $\alpha = 0$. We shall assume $\alpha > 0$ and derive a contradiction. If $\alpha > 0$, let

(9.28) $$\delta = (3\alpha^2 + 4A)/2\alpha.$$

Then $\delta > \alpha$.

By definition we see that $V(\eta)$ is decreasing except at points of discontinuity where it increases. Let $\zeta > 0$ and $\eta \in [\zeta, \zeta + \delta - \alpha]$. Then V is either zero somewhere or changes sign at most once on this interval. If $V(\eta_0) = 0$ and $\eta_0 \in [\zeta, \zeta + \delta - \alpha]$. we have, by Lemma 9.8.7,

$$\int_{\zeta}^{\zeta+\delta}|V(\eta)|d\eta = \left\{\int_{\zeta}^{\eta_0} + \int_{\eta_0}^{\eta_0+\alpha} + \int_{\eta_0+\alpha}^{\zeta+\delta}\right\}|V(\eta)|d\eta$$

$$\leq \alpha(\eta_0 - \zeta) + \tfrac{1}{2}\alpha^2 + o(1) + \alpha(\zeta + \delta - \eta_0 - \alpha)$$

$$= \alpha(\delta - \tfrac{1}{2}\alpha) + o(1) = \alpha'\delta + o(1),$$

for ζ sufficiently large, where $\alpha' = \alpha(1 - \alpha/2\delta) < \alpha$. If V changes sign exactly once, say at $\eta = \eta_1 \in [\zeta, \zeta + \delta - \alpha]$, then

$$\int_{\zeta}^{\zeta+\delta-\alpha}|V(\eta)|d\eta = \left|\int V(\eta)d\eta\right| + \left|\int V(\eta)d\eta\right| < 2A,$$

by Lemma 9.8.6. Finally, if V does not change sign at all on $[\zeta, \zeta + \delta - \alpha]$, then

$$\int_{\zeta}^{\zeta+\delta-\alpha}|V(\eta)|d\eta = \left|\int_{\zeta}^{\zeta+\delta-\alpha} V(\eta)d\eta\right| < A,$$

by Lemma 9.8.6. In the last two cases we have

$$\int_{\zeta}^{\zeta+\delta}|V(\eta)|d\eta = \int_{\zeta}^{\zeta+\delta-\alpha}|V(\eta)|d\eta + \int_{\zeta+\delta-\alpha}^{\zeta+\delta}|V(\eta)|d\eta < 2A + \alpha^2 + o(1) = \alpha''\delta + o(1),$$

by Lemma 9.8.7 and the assumption that $\alpha > 0$, where, by (9.28),

$$\alpha'' = (2A + \alpha^2)/\delta = \alpha(4A + 2\alpha^2)/(4A + 3\alpha^2) = \alpha(1 - \alpha/2\delta) = \alpha'.$$

Thus, in all three cases, we have

$$\int_{\zeta}^{\zeta+\delta}|V(\eta)|d\eta \leq \alpha'\delta + o(1).$$

If $M = [\lambda/\delta]$, then

$$\int_{0}^{\lambda}|V(\eta)|d\eta = \left(\sum_{m=0}^{M-1}\int_{m\delta}^{(m+1)\delta} + \int_{M\delta}^{\lambda}\right)|V(\eta)|d\eta \leq \alpha'\delta M + o(M) + O(1) = \alpha'\lambda + o(\lambda).$$

Thus

$$\beta = \limsup_{\lambda\to+\infty}\lambda^{-1}\int_{0}^{\lambda}|V(\eta)|d\eta \leq \alpha' < \alpha,$$

which is our contradiction. Thus $\alpha = 0$.   ■

Corollary 9.8.1. As $x \to +\infty$,

$$\pi(x) \sim x/\log x.$$

This follows immediately from Corollary 9.7.1 and Theorem 9.8.

There are many other ways to prove this theorem. For other elementary proofs see [109], [93], [73] and [21]. The proof in [109] is Selberg's original proof and in which almost all analysis has been banned. (For example, there is no integration.) The proof in [93] is based on properties of the summatory function of $\mu(n)$. The proof in [21] takes a totally different approach, though, as is to be expected, there are certain similarities with the proof given above. Finally, the proof in [73] is presented in such a way as to explain why each step goes the way it does.

There are several papers in the literature which prove more than just

$$\pi(x) = x/\log x + o(x/\log x).$$

What has been proved are results of the form

$$\pi(x) = x/\log x + O(xR(x)),$$

where $R(x) = o(x/\log x)$. For an elementary proof of

$$\pi(x) = x/\log x + O(x/\log^m x),$$

for any $m > 0$, see [131] and [8]. There have been further improvements, see [24], for example. For a general survey of elementary methods see [23]. The method of Wirsing, in [131], is roughly to smooth out the process over and over again, whereas the method of Bombieri, in [8], uses an approach somewhat similar to that used in the proof of Theorem 9.8.

If one is willing to employ the theory of functions of a complex variable (and so cease to have an elementary proof), then one can usually get shorter proofs. For a proof of Corollary 9.8.1, directly, see [41, ch. 8 and 9] and for a simpler proof of Theorem 9.8 see [2, ch. 12 and 13].

The following theorem gives us some information about the summatory function of the Möbius function.

Theorem 9.9. As $x \to +\infty$, we have

(9.29)
$$\sum_{n \le x} \mu(n) = o(x).$$

Proof. Let

$$M(x) = \sum_{n \le x} \mu(n).$$

By Theorem 7.9, we have

$$e * L\mu = \Lambda_\mu = -\Lambda.$$

Thus

$$L\mu = -\mu * \Lambda$$

or, recalling that $\mu * e = \delta$, we have

$$L\mu = -\mu * \Lambda + \mu * e - \delta = -\mu * (\Lambda - e) - \delta.$$

Thus

$$\sum_{n \leq x}(L\mu)(n) = -\sum_{n \leq x}(\mu * (\Lambda - e))(n) - \sum_{n \leq x}\delta(n)$$

$$= -\sum_{n \leq x}\mu(n)\sum_{m \leq x/n}\{\Lambda(m) - e(m)\} - 1$$

(9.30)

$$= -\sum_{n \leq x}\mu(n)(\psi(x/n) - [x/n]) - 1$$

$$= -\sum_{n \leq x}\mu(n)o(x/n) - 1 = o\left(\sum_{n \leq x}(x/n)\right) - 1 = o(x \log x).$$

On the left hand side we have

$$(9.31)\sum_{n \leq x}(L\mu)(n) = \int_1^x \log t \, dM(t) = M(x)\log x - \int_1^x \frac{M(t)}{t}dt = M(x)\log x + O(x),$$

since $|M(t)| \leq x$. If we combine (9.30) and (9.31) and divide by $\log x$, we obtain the result (9.29). ∎

One can show the converse holds, but it is more difficult. The converse implication is the basis of the Postnikov-Romanov proof [93].

Problem Set 9.3

1. (a) Show that

$$\sum_{p \leq x}\frac{\log p}{p} = \log x + O(1).$$

(Hint: use (b) of Lemma 9.8.2 as well as the proof of Theorem 9.7 to show that

$$\log x + O(1) = \sum_{n \leq x}\frac{\Lambda(n)}{n} = \sum_{p \leq x}\frac{\log p}{p} + \sum_{\substack{p^a \leq x \\ a \geq 2}}\frac{\log p}{p^a} = \sum_{p \leq x}\frac{\log p}{p} + O(1).)$$

(b) Show that there is a constant B such that

$$\sum_{p \leq x}\frac{1}{p} = \log \log x + B + O\left(\frac{1}{\log x}\right).$$

(Hint: we have

$$\sum_{p \leq x}\frac{1}{p} = \sum_{p \leq x}\frac{\log p}{p} \cdot \frac{1}{\log p}.)$$

2. Let

$$\theta(x) = \sum_{p \le x} \log p.$$

This function was also introduced by Chebychev.

(a) Show that

$$\psi(x) = \theta(x) + O\left(\sqrt{x} \log x\right).$$

(Hint: treat the higher powers of the primes as in 1(a) above.)

(b) Show that $\psi(x) \sim x$ if and only if $\theta(x) \sim x$. This is why many proofs of the prime number theorem use $\theta(x)$ instead of $\psi(x)$.

(c) Show that

$$\theta(x) = \sum_{n=1}^{+\infty} \mu(n) \psi(x^{1/n}).$$

(Hint: show that the sum, for fixed x, is really a finite sum and then use Corollary 8.4.1.)

(d) Show that

$$\sum_{n \le x} \theta(x/n) = x \log x + O(x).$$

3. Define $a(n)$ by

$$a(n) = \begin{cases} \frac{1}{k} & \text{if } n = p^k, \ p \text{ a prime} \\ 0 & \text{else} \end{cases}.$$

(a) Show that

$$a(n) = \frac{\Lambda(n)}{\log n}.$$

(b) Show that

$$\sum_{n \le x} a(n) = \pi(x) + O\left(\sqrt{x} \log\log x\right).$$

4. Find all positive integers n such that n is the sum of all the primes less than n.

5. If p_n denotes the nth prime, show that $p_n \sim n \log n$, as $n \to +\infty$.

6. (a) Show that

$$\psi(x) \log^2 x + \log x \sum_{n \le x} (\Lambda * \Lambda)(n) = 2x \log^2 x + O(x \log x).$$

(b) Show that

$$\psi(x) \log^2 x = 2 \sum_{n \le x} (\Lambda * \Lambda * \Lambda)(n) + O(x \log x).$$

(c) Show that

$$\psi(x)\log^2 x + 2\sum_{n\le x}(\Lambda * L\Lambda)(n) = 2x\log^2 x + O(x\log x).$$

7. (a) Show that

$$\sum_{n\le x}\Lambda(n)\frac{\log n}{n} = \tfrac{1}{2}\log^2 x + O(\log x).$$

(b) Show that

$$\sum_{n\le x}(T_{-1}\Lambda * T_{-1}\Lambda)(n) = \tfrac{1}{2}\log^2 x + O(\log x).$$

(c) Show that

$$\sum_{n\le x}\frac{\Lambda(n)}{n}\log\frac{x}{n} = \tfrac{1}{2}\log^2 x + O(\log x).$$

8. (a) Show that

$$\sum_{n\le x}(\Lambda * L\Lambda)(n) = \tfrac{1}{2}\log x\sum_{n\le x}(\Lambda * \Lambda)(n) + O(x\log x).$$

(b) Show that

$$\sum_{n\le x}(\Lambda * L\Lambda)(n) + \sum_{n\le x}(\Lambda * \Lambda * \Lambda)(n) = x\log^2 x + O(x\log x).$$

(c) Show that

$$\log x\sum_{n\le x}(\Lambda * \Lambda)(n) + 2\sum_{n\le x}(\Lambda * \Lambda * \Lambda)(n) = 2x\log^2 x + O(x\log x).$$

(d) Compute

$$\sum_{n\le x}\Lambda^{*k}(n).$$

(Hint:

$$\Lambda^{*k} = \Lambda * \Lambda^{*(k-1)}$$

and use induction.)

9. Show that

$$\int_1^x \frac{\psi(t)}{t^2}\,dt = \log x + O(1).$$

(Hint: see the end of the proof of Lemma 9.8.4.)

10. (a) Show that if $0 \le x \le 1/2$, then $\log(1 - x) + x = O(x^2)$.

(b) Show that there is a constant c such that

$$\prod_{p \le x}\left(1 - \frac{1}{p}\right) = \frac{c}{\log x}\left(1 + O\left(\frac{1}{\log x}\right)\right).$$

(Hint: take logarithms and also recall Problem 12 of Problem Set 8.1.)

Relate the constant c to the constant B of Problem 1(b) above.

11. Show that

$$\theta(x)\log x + 2 \sum_{p \le \sqrt{x}} \theta(x / p)\log p = 2x\log x + o(x \log x).$$

12. (a) Let F and G be functions defined for $x \ge 1$ and related by

$$G(x) = \log x \sum_{n \le x} F(x / n).$$

Show that

$$F(x)\log x + \sum_{n \le x} F(x / n)\Lambda(n) = \sum_{n \le x} \mu(n)G(x / n).$$

(b) Deduce Corollary 9.8.1.1 from (a).

13. Show that

$$M(x)\log x + \sum_{n \le x} M(x / n)\Lambda(n) = O(x).$$

9.4. PRIMES IN ARITHMETIC PROGRESSIONS

In this section we prove a quantitative form of Dirichlet's theorem on the infinitude at primes in arithmetic progressions. The idea behind this proof goes back to A. Selberg [108].

Lemma 9.10.1. Let $f(x)$ be nonnegative and monotone decreasing such that

(9.33) $\lim_{x \to +\infty} f(x) = 0.$

If χ is a nonprincipal character modulo m, then

$$\sum_{n \ge z} \chi(n)f(n) = O(f(z)).$$

Proof. Let

$$S(x) = \sum_{n \le x} \chi(n).$$

Since $\chi \ne \chi_0$ we have, by Theorem 7.25, that

(9.34)
$$|S(x)| = \left| \sum_{n \le x} \chi(n) \right| \le \sum_{n \le x} |\chi(n)| = m.$$

Then

$$\sum_{n \ge z} \chi(n) f(n) = \int_{z-}^{+\infty} f(t) dS(t) = f(t) S(t) \Big|_{z-}^{+\infty} - \int_{z}^{+\infty} S(t) df(t)$$

$$= O(f(z)) + O\left(\int_{z}^{+\infty} df(t) \right) = O(f(z)),$$

by (9.33) and (9.34) since f is monotone and nonnegative.   ∎

If we let $f(x) = x^{-1}$, $x^{-1} \log x$ and $x^{1/2}$, then Lemma 9.10.1 applies and we see that the following three infinite series converge:

$$L_0(\chi) = \sum_{n=1}^{+\infty} \frac{\chi(n)}{n}, L_1(\chi) = \sum_{n=1}^{+\infty} \frac{\chi(n) \log n}{n} \text{ and } L_2(\chi) = \sum_{n=1}^{+\infty} \frac{\chi(n)}{\sqrt{n}}.$$

Also, by Lemma 9.10.1, we have, as $x \to +\infty$,

(9.35)
$$\sum_{n \le x} \frac{\chi(n)}{n} = L_0(\chi) + O(x^{-1}),$$

(9.36)
$$\sum_{n \le x} \frac{\chi(n) \log n}{n} = L_1(\chi) + O(x^{-1} \log x)$$

and

(9.37)
$$\sum_{n \le x} \frac{\chi(n)}{\sqrt{n}} = L_2(\chi) + O(x^{-1/2}).$$

Lemma 9.10.2. If χ is a nonprincipal real character modulo m, then

$$L_0(\chi) \ne 0.$$

Proof. Define the arithmetic function

$$A = e * \chi.$$

Then, by Theorem 7.28 we see that for all n we have $A(n) \ge 0$ and $A(n^2) \ge 1$. Thus the series

$$G = \sum_{n=1}^{+\infty} \frac{A(n)}{\sqrt{n}}$$

diverges since it is minorized by the harmonic series.

By Theorem 7.5 we have

$$T_{-1/2}A = T_{-1/2}e * T_{-1/2}\chi.$$

Thus, if we let

$$G(x) = \sum_{n \le x} A(n)n^{-1/2}$$

then, by Theorem 8.5,

$$G(x) = \sum_{d \le x} \frac{\chi(d)}{\sqrt{d}} \sum_{m \le x/d} \frac{1}{\sqrt{m}} = \sum_{d \le x} \frac{\chi(d)}{\sqrt{d}} \left\{ 2\sqrt{\frac{x}{d}} + \zeta(1/2) + O\left(\sqrt{\frac{d}{x}}\right) \right\}$$

$$= 2\sqrt{x} \sum_{d \le x} \frac{\chi(d)}{d} + \zeta(1/2) \sum_{d \le x} \frac{\chi(d)}{\sqrt{d}} + O\left(\frac{1}{\sqrt{x}}\right) = 2\sqrt{x} \sum_{d \le x} \frac{\chi(d)}{d} + O(1).$$

using the estimates (9.24) and (9.37) and (b) of Lemma 8.15.1 and the fact that χ is nonprincipal. If we apply the estimate (9.35), we have

$$G(x) = 2\sqrt{x} L_0(\chi) + O(1).$$

Thus, if $L_0(\chi) = 0$, then $G(x) = O(1)$, which contradicts the fact that the series G diverges. ∎

Lemma 9.10.3. As $x \to +\infty$, we have, for any character χ

$$L_1(\chi) \sum_{n \le x} \frac{\mu(n)\chi(n)}{n} = \begin{cases} -\log x + O(1) & \text{if } L_0(\chi) = 0 \\ O(1) & \text{if } L_0(\chi) \ne 0 \end{cases}$$

Proof. Define $g_1(x)$ by

$$g_1(x) = \sum_{n \le x} \chi(n) \frac{x}{n}.$$

Then, by (9.35),

$$g_1(x) = x \sum_{n \le x} \frac{\chi(n)}{n} = x L_0(\chi) + O(1).$$

Since χ is completely multiplicative, by Definition 7.14, we have, by (d) of Theorem 8.4, that

$$x = \sum_{n \le x} \mu(n)\chi(n) \left\{ \frac{x}{n} L_0(\chi) + O(1) \right\} = x L_0(\chi) \sum_{n \le x} \frac{\mu(n)\chi(n)}{n} + O(x),$$

and so, if we divide both sides by x,

(9.38) $$L_0(\chi) \sum_{n \le x} \frac{\mu(n)\chi(n)}{n} = O(1).$$

Define $g_2(x)$ by

$$g_2(x) = \sum_{n \le x} \chi(n) \frac{x}{n} \log \frac{x}{n}.$$

Then, by (9.35) and (9.36),

$$g_2(x) = x \log x \sum_{n \le x} \frac{\chi(n)}{n} - x \sum_{n \le x} \frac{\chi(n) \log n}{n} = L_0(\chi) x \log x - x L_1(\chi) + O(\log x).$$

If $L_0(\chi) = 0$, then

$$g_2(x) = -x L_1(\chi) + O(\log x),$$

and so, by (d) of Theorem 8.4, we have, as above using Lemma 9.5.1,

$$(9.39) \qquad\qquad x \log x = -x L_1(\chi) \sum_{n \le x} \frac{\mu(n) \chi(n)}{n} + O(x).$$

Thus, if $L_0(\chi) = 0$, then $L_1(\chi) \ne 0$. Thus, since $L_0(\chi)$ and $L_1(\chi)$ are finite numbers, the results follow from (9.38) and (9.39). ∎

Lemma 9.10.4. If χ is a nonprincipal character, then

$$\sum_{p \le x} \frac{\chi(p) \log p}{p} = \begin{cases} -\log x + O(1) & \text{if } L_0(\chi) = 0 \\ O(1) & \text{if } L_0(\chi) \ne 0 \end{cases}.$$

Proof. We have

$$\sum_{p \le x} \frac{\chi(p) \log p}{p} = \sum_{p^s \le x} \frac{\chi(p^s) \log p}{p^s} - O(1) = \sum_{n \le x} \frac{\chi(n) \Lambda(n)}{n} - O(1),$$

since

$$\left| \sum_{\substack{p^s \le x \\ s \ge 2}} \frac{\chi(p^s) \log p}{p^s} \right| < \sum_{n \le x} \frac{\log n}{n^2} = \sum_{n=1}^{+\infty} \frac{\log n}{n^2} + O\left(\frac{\log x}{x} \right)$$

and the infinite series converges by the integral test. Now, by Definition 7.8,

$$\sum_{n \le x} \frac{\chi(n) \Lambda(n)}{n} = \sum_{n \le x} \frac{\chi(n)}{n} \sum_{d|n} \mu(d) \log(n/d)$$

$$= \sum_{mn \le x} \frac{\chi(mn) \mu(m) \log n}{mn} = \sum_{m \le x} \frac{\chi(m) \mu(m)}{m} \sum_{n \le x/m} \frac{\chi(n) \log n}{n}$$

$$= L_1(\chi) \sum_{m \le x} \frac{\chi(m) \mu(m)}{m} + O(1),$$

by (9.36). The result follows from Lemma 9.10.3. ∎

Lemma 9.10.5. If χ is a nonprincipal character, then $L_0(\chi) \neq 0$.

Proof. If χ is real, the result is just Lemma 9.10.2. Let N be the number of nonprincipal characters χ, modulo m, such that $L_0(\chi) = 0$. Then, by Theorem 7.25, we have

$$Q(x) = \varphi(m) \sum_{\substack{p \leq x \\ p \equiv 1 \ (\mathrm{mod}\ m)}} \frac{\log p}{p} = \sum_{\chi(m)} \sum_{p \leq x} \frac{\chi(p) \log p}{p}$$

$$= \sum_{\substack{p \leq x \\ (p,m)=1}} \frac{\log p}{p} + \sum_{\chi \neq \chi_0} \sum_{p \leq x} \frac{\chi(p) \log p}{p}.$$

Now

$$(9.40) \qquad \sum_{p \leq x} \frac{\log p}{p} = \sum_{p^s \leq x} \frac{\log p}{p^s} - O(1) = \sum_{n \leq x} \frac{\Lambda(n)}{n} - O(1) = \log x + O(1),$$

by Lemma 9.8.2. Thus, by Lemma 9.10.4

$$Q(x) = (1 - N)\log x + O(1).$$

Since $Q(x) > 0$ we see that $0 \leq N \leq 1$. If χ is a complex character such that $L_0(\chi) = 0$, then

$$0 = \overline{0} = \overline{L_0(\chi)} = \overline{\sum_{n=1}^{+\infty} \frac{\chi(n)}{n}} = \sum_{n=1}^{+\infty} \frac{\overline{\chi(n)}}{n} = L_0(\overline{\chi}).$$

Thus, if $L_0(\chi) = 0$ for a complex character, we have $L_0(\overline{\chi}) = 0$, and so $N \geq 2$ in this case. Thus $N = 0$ and the result follows. ∎

Theorem 9.10. If $(a, m) = 1$, then

$$(9.41) \qquad \sum_{\substack{p \leq x \\ p \equiv a \ (\mathrm{mod}\ m)}} \frac{\log p}{p} = \frac{1}{\varphi(m)} \log x + O_m(1).$$

Proof. By Lemmas 9.10.4 and 9.10.5 we see that if χ is a nonprincipal character, then

$$\sum_{p \leq x} \frac{\chi(p) \log p}{p} = O(1),$$

as $x \to +\infty$. By (9.40) and Theorem 7.26, we have, if $(a, m) = 1$,

$$\varphi(m) \sum_{\substack{p \leq x \\ p \equiv a \ (\mathrm{mod}\ m)}} \frac{\log p}{p} = \sum_{\chi(m)} \overline{\chi(a)} \sum_{p \leq x} \frac{\chi(p) \log p}{p}$$

$$= \sum_{\substack{p \leq x \\ (p,m)=1}} \frac{\log p}{p} + \sum_{\chi \neq \chi_0} \overline{\chi(a)} \sum_{p \leq x} \frac{\chi(p) \log p}{p} = \log x + O_m(1),$$

since

$$\sum_{\substack{p \le x \\ p \mid m}} \frac{\log p}{p} = O_m(1),$$

which gives the result.   ∎

Corollary 9.10.1 (Dirichlet). If $(a, m) = 1$, then there are infinitely many primes p such that $p \equiv a \pmod m$.

Proof. This follows immediately from Theorem 9.10 since the right hand side of (9.41) tends to $+\infty$ as $x \to +\infty$.   ∎

One can rearrange the proof of Theorem 9.8 to prove a prime number theorem for arithmetic progressions. If

$$\pi(x;m,a) = \sum_{\substack{p \le x \\ p \equiv a \ (\mathrm{mod}\ m)}} 1 \ ,$$

then one can show

$$\pi(x;m,a) \sim \frac{1}{\varphi(m)} \cdot \frac{x}{\log x}.$$

For a version along the lines of the Postnikov-Romanov proof of the prime number theorem see [40, ch. 3]. See also Problem 2 below.

Problem Set 9.4

1. If $(a, m) = 1$ with $m > 0$, show that there exists a constant $A(m, a)$ such that

$$\sum_{\substack{p \le x \\ p \equiv a \ (\mathrm{mod}\ m)}} \frac{1}{p} = \frac{1}{\varphi(m)} \log\log x + A(m,a) + O_m\left(\frac{1}{\log x}\right),$$

   where O_m means that the O-constant depends on m.

2. This exercise outlines a proof of the prime number theorem for arithmetic progressions.

   (a) Let $(a, m) = 1$ and

$$\psi(x;m,a) = \sum_{\substack{n \le x \\ n \equiv a \ (\mathrm{mod}\ m)}} \Lambda(n).$$

   Show that

$$\pi(x;m,a)\log x = \psi(x;m,a) + O(x / \log x).$$

(b) Let, for $(a, m) = 1$,

$$M(x;m,a) = \sum_{\substack{n \leq x \\ n \equiv a \ (\text{mod } m)}} \mu(n).$$

Show that if $M(x) = o(x)$ and $M(x; m, a) = o(x)$, then

$$\psi(x;m,a) \sim x / \varphi(m).$$

(Hint: show that for any arithmetic function f we have

$$\psi(x;m,a) = \sum_{\substack{n \leq x \\ n \equiv a \ (\text{mod } m)}} (\mu*(Le - f*e))(n) + \sum_{\substack{n \leq x \\ n \equiv a \ (\text{mod } m)}} f(n)$$

and choose an appropriate f after showing that

$$\sum_{n \leq x} \left(\log n - \frac{x}{n} + (\gamma + 1)e(n) \right) = O(\log x).)$$

(c) Let

$$R(x;m,a) = \sum_{\substack{n \leq x \\ n \equiv a \ (\text{mod } m)}} \sum_{d \mid n} \mu(d) \log^2 \frac{n}{d}.$$

Show that

$$R(x;m,a) = \frac{2}{\varphi(m)} x \log x + O(x).$$

(Hint: show first that

$$\sum_{d \mid n} \mu(d) \log^2 \frac{n}{d} = (L\Lambda + \Lambda*\Lambda)(n)$$

and then show that

$$R(x;m,a) = \psi(x;m,a) \log x + \sum_{\substack{n \leq x \\ n \equiv a \ (\text{mod } m)}} (\Lambda*\Lambda)(n) + O(x).)$$

(d) Show that

$$\sum_{\substack{n \leq x \\ n \equiv a \ (\text{mod } m)}} \frac{\mu(n)}{n} = O(1).$$

(e) Show that

$$|M(x;m,a)| \log x \leq \sum_{n \leq x} \Lambda(n) \psi(x;m,\overline{n}a) + O(x),$$

where $n\bar{n} \equiv 1 \pmod{m}$. (Hint: begin with

$$\sum_{\substack{n \le x \\ n \equiv a \pmod{m}}} \mu(n) \log \frac{x}{n}.)$$

(f) Show that if χ_0 is the principal character modulo m, then

$$|M(x;m,a)| \log x \le \frac{1}{\varphi(m)} \sum_{q=1}^{m} \chi_0(q) \sum_{n \le x} \left| M\left(\frac{x}{n};m,q\right) \right| + O(x \log \log x).$$

(g) If $x > t > 1$, show that there exists a q, $x \le q \le tx$, and a constant C such that

$$|M(x;m,a)| < Cq/\log t.$$

(h) Show that if $m > 1$, $(a, m) = 1$, then $M(x; m, a) = o(x)$.

(i) Conclude that

$$\pi(x;m,a) \sim \frac{x}{\varphi(m) \log x}.$$

3. Let $a_1, \ldots, a_m$ be pairwise relatively prime integers such that no a_k, $1 \le k \le m$, is a square. Let $\varepsilon_1, \ldots, \varepsilon_m$ be either +1 or -1. Show that there are infinitely many primes p such that

$$\left(\frac{a_k}{p}\right) = \varepsilon_k, \ 1 \le k \le m.$$

(Hint: use the Chinese Remainder Theorem (Theorem 3.7) as well as Problem 12 of Problem Set 3.3.)

4. Show that if $(a, m) = 1$, then there is a constant $A(m, a)$ such that

$$\prod_{\substack{p \le x \\ p \equiv a \pmod{m}}} \left(1 - \frac{1}{p}\right) = \frac{A(m,a)}{(\log x)^{1/\varphi(m)}} \left(1 + O\left(\frac{1}{\log x}\right)\right).$$

5. Let $(a, m) = 1$. Let k be a positive integer and let

$$N = \prod_{n=1}^{k} (a + mn).$$

Show that $m(N + 1) + a$, $m(N + 2) + a$, $\ldots$, $m(N + k) + a$ are all composite.

6. Show that for any integers a, b and c the constant quadratic polynomial $an^2 + bn + c$ must, for infinitely many n, have prime divisors of the form

$4k + 1$. (The analogous result for the arithmetic progression $4k + 3$ is false as the polynomial $4n^2 + 1$ shows.)

7. Show that there exists an infinite set of primes P such that if p_1, p_2 are in P, then $((p_1 - 1)/2, (p_2 - 1)/2) = (p_1, p_2 - 1) = (p_1 - 1, p_2) = 1$. (Hint: given $p_1, \ldots, p_n$ let

$$d_n = \prod_{k=1}^{n} \tfrac{1}{2} p_k (p_k - 1)$$

and apply Dirichlet's theorem appropriately.)

8. (a) Show that if A is a positive integer, then there exist infinitely many primes p such that

$$\left(\frac{-1}{p} \right) = \left(\frac{q_k}{p} \right) = 1, \ 1 \le k \le n$$

where $q_1, \ldots, q_n$ are the primes less than or equal to A.

   (b) If p is an odd prime, let g be the least positive primitive root modulo p and let G be the greatest negative primitive root modulo p. Show that if A is a positive integer, then there exist infinitely primes p such that

$$A < g < p - A \ \text{ and } \ -p + A < G < -A .$$

9. This exercise gives a "Dirichlet's Theorem" for deficient and abundant numbers. Let h and k be integers.

   (a) If (h, k) is deficient, show that there exist infinitely many deficient numbers n such that $n \equiv h \pmod{k}$. (Hint: let $d = (h, k)$ and consider the arithmetic progression $(h/d) + (k/d)m$.)

   (b) Show that there exist infinitely many abundant integers n such that $n \equiv h \pmod{k}$. (Hint: recall that $\sum 1 / p$ diverges.)

10. (a) If D is a positive nonsquare integer, show that there are infinitely many primes p such that

$$p \equiv 3 \pmod{4} \text{ and } \left(\frac{D}{p} \right) = -1.$$

   (b) Let d be an odd positive integer and let $b_1, b_2, b_3, \ldots$ be the sequence of numbers representable as a sum of two squares listed in their natural order. Prove that there are infinitely many n such that

$$b_{n+1} - b_n = d$$

is solvable. (Hint: find integers k such that $k^2 + (d - 1)/2 + j, j = 1, \ldots,$ $d - 1$, cannot be represented as a sum of two squares (Corollary 6.8.2).)

(c) Can you prove an analogous result for d even?

9.5. ORDER OF MAGNITUDE OF MULTIPLICATIVE FUNCTIONS

Theorem 9.11. Let $f(n)$ be a multiplicative function. If $f(p) \to 0$ as $p^m \to +\infty$, then $f(n) \to 0$ as $n \to +\infty$.

Proof. Let $\varepsilon > 0$. Then there exists a positive constant A, independent of p and m, such that $|f(p^m)| < A$ for all p and m. Also there exists a positive number $N(\varepsilon)$, depending only on ε, such that

(9.42) $|f(p^m)| < \varepsilon$, whenever $p^m > N(\varepsilon)$.

In particular, there exists an absolute positive constant B such that

$$|f(p^m)| < 1, \text{ whenever } p^m > B.$$

Now, if

$$n = p_1^{a_1} \cdots p_r^{a_r},$$

then not more than C of the prime powers can satisfy, where C is a positive constant depending only on B. Those factors contribute to $f(n)$ a term that is in absolute value $< A^C$ and the rest of the factors are all less than 1.

The number of integers which can be formed by multiplication of factors $p^a \leq N(\varepsilon)$ is $M(\varepsilon)$ and every such number is less than $P(\varepsilon)$, where $M(\varepsilon)$ and $P(\varepsilon)$ depend only on ε. Thus if $n > P(\varepsilon)$, there is at least one prime factor p of n such that $p^a > N(\varepsilon)$, and so, by (9.42)

$$|f(p^a)| < \varepsilon.$$

Thus

$$|f(n)| < A^C \varepsilon,$$

whenever $n > P(\varepsilon)$. The result follows.   ∎

Corollary 9.11.1. For any $\eta > 0$ we have $d(n) = O(n^\eta)$.

Proof. Let $f(n) = n^{-\eta} d(n)$. Then f is multiplicative and

$$f(p^m) = (m+1)p^{-m\eta} \leq 2mp^{-m\eta} = 2p^{-m\eta}\frac{\log(p^m)}{\log p} \leq \frac{2}{\log 2} \cdot \frac{\log(p^m)}{(p^m)^\eta} \to 0$$

as $p^m \to +\infty$. Thus $f(n) \to 0$ as $n \to +\infty$. Thus

$$d(n) = o(n^\eta) = o(n^{\eta'})$$

for any $\eta' \geq \eta$ and the result follows.   ∎

Corollary 9.11.2. For any $\eta > 0$ we have $\sigma_k(n) = O(n^{k+\eta})$.

Proof. We have

$$\sigma_k(n) = \sum_{d|n} d^k \leq n^k \sum_{d|n} 1 = d(n)n^k$$

and the result follows from Corollary 9.11.1.   ∎

Corollary 9.11.3. For any $\eta > 0$ we have, for $k \geq 2$, $d_k(n) = O(n^\eta)$.

Proof. The result is true for $k = 2$ by Corollary 9.11.1. Assume $d_k(n) = O(n^\eta)$ for some $k \geq 2$. Then

$$d_{k+1}(n) = \sum_{t|n} d_k(t) = O\left(\sum_{t|n} t^\eta\right) = O(\sigma_\eta(n)) = O(n^{2\eta}).$$

The result follows by mathematical induction.   ∎

Corollary 9.11.4. For any $\eta > 0$ we have $\varphi(n)/n^{1-\eta} \to +\infty$ as $n \to +\infty$.

Proof. Let $f(n) = n^{1-\eta}/\varphi(n)$, which is multiplicative. Then

$$\frac{1}{f(p^m)} = \frac{\varphi(p^m)}{p^{m(1-\eta)}} = p^{m\eta}\left(1 - \frac{1}{p}\right) \geq \tfrac{1}{2}p^{m\eta} \to +\infty.$$

The result follows from Theorem 9.11.   ∎

Lemma 9.12.1. Let x and y be any positive real numbers and let $0 < \theta < 1$. Then

$$(1 - \theta)x + \theta y \geq x^{1-\theta}y^\theta.$$

Proof. We have that if x is a positive real number and

$$f(x) = x^\theta - 1 - \theta(x - 1),$$

then

$$f'(x) = \theta x^{\theta-1} - \theta,$$

so that $x = 1$ is the critical point for $f(x)$. Since

$$f''(x) = \theta(\theta - 1)x^{\theta-2}$$

we see that $x = 1$ gives a local maximum, which, since $0 < \theta < 1$ implies that

$f'(x) > 0$ and $x > 1$ implies that $f'(x) < 0$, implies that $x = 1$ is also a global maximum. Thus

$$f(x) \geq f(1) = 0,$$

in fact we have equality if and only if $x = 1$.

Now let x and y be positive real numbers. Then by the above result we have

$$(y / x)^\theta - 1 \leq \theta((y / x) - 1)$$

or

$$y^\theta - x^\theta \leq \theta(y - x)x^{\theta - 1}$$

or

$$y^\theta x^{1-\theta} \leq \theta y + (1 - \theta)x,$$

which is our result.   ∎

Lemma 9.12.2. Let $A > 0$, $0 < \theta < 1$. Then there exist constants B and λ such that

$$B\delta^{-\lambda} + a\delta \log 2 \geq Aa^\theta$$

for all $a \geq 1$ and all $\delta > 0$.

Proof. Let

$$\lambda = \theta / (1 - \theta) \quad B = (A\theta / \lambda)^{\lambda/\theta}(\lambda / \log 2)^\lambda$$
$$x = B / \delta^\lambda \qquad y = (a\delta \log 2) / \lambda \qquad .$$

The result then follows from the inequality of Lemma 9.12.1.   ∎

Theorem 9.12 (Shiu). Let $f(n)$ be a multiplicative function satisfying the following conditions.

(1)   There exist constants A and θ, $0 < \theta < 1$, such that

$$f(2^a) \leq \exp(Aa^\theta)$$

 for all $a \geq 1$.

(b)   For all primes p and all $a \geq 1$ we have

$$f(p^a) = f(2^a) \geq 1.$$

Then we have

$$\limsup_{n \to +\infty} \frac{\log(f(n))\log\log n}{\log n} = \log M,$$

where

$$M = \max_{a \geq 1}\left(f(2^a)\right)^{1/a}.$$

Proof. M exists since

$$1 \le \left(f(2^a)\right)^{1/a} \le \exp(Aa^{\theta-1}) \to 1$$

as $a \to +\infty$. By the definition of M we can choose b such that

$$M^b = f(2^b).$$

Let $p_1, \ldots, p_r$ be the first r primes and let

$$n = (p_1{}^a \cdots \cdot p_r{}^{})^b,$$

so that

$$f(n) = (f(2^b))^r = M^{br} = M^{\pi(pr)}.$$

By Theorem 9.8, we have, as $r \to +\infty$,

$$\pi(p_r) \sim p_r / \log p_r,$$

and so

$$\sum_{p \le p_r} \log p \sim p_r.$$

Thus, as $r \to +\infty$,

$$\log n = b \sum_{p \le p_r} \log p \sim b p_r$$

and

$$\log \log n = \log p_r + O(1).$$

Thus, as $r \to +\infty$,

$$\log(f(n)) = (b \log M)\pi(p_r) \sim (b \log M)p_r / \log p_r \sim (\log M)(\log n) / \log \log n,$$

and so, by the definition of lim sup, we have

(9.43)
$$\limsup_{n \to +\infty} \frac{(\log f(n)) \log \log n}{\log n} \ge \log M.$$

Let $n = \prod p^a$ and $\delta > 0$. Then

(9.44)
$$\frac{f(n)}{n^\delta} = \prod \frac{f(p^a)}{p^{a\delta}} = \prod \frac{f(2^a)}{p^{a\delta}}.$$

By (a), we have, for all p and a, that

$$f(2^a)p^{-a\delta} \le \exp(Aa^\theta - a\delta \log 2) \le \exp(B\delta^{-\lambda}),$$

by Lemma 9.12.2. For $p \ge M^{1/\delta}$ we also have

$$f(2^a)p^{-a\delta} \le f(2^a)M^{-a} \le 1.$$

Thus, by (9.44),

$$f(n)n^{-\delta} \le \prod_{p \le M^{1/\delta}} \exp(B\delta^{-\lambda}) \le \exp(BM^{1/\delta}\delta^{-\lambda}),$$

and so

$$\log f(n) \le \delta \log n = BM^{1/\delta}\delta^{-\lambda}.$$

Let $\varepsilon > 0$ and set

$$\delta = (1 + \varepsilon/2)\log M / \log\log n.$$

Then

$$\log f(n) \le \frac{(1+\varepsilon/2)\log M \log n}{\log\log n} + \frac{B(\log n)^{1/(1+\varepsilon/2)}(\log\log n)^\lambda}{(1+\varepsilon/2)^\lambda \log^\lambda M} \le \frac{(1+\varepsilon)\log M \log n}{\log\log n}$$

for n large enough. Thus

(9.45)
$$\limsup_{n \to +\infty} \frac{(\log f(n))\log\log n}{\log n} \le \log M.$$

The result follows from (9.43) and (9.45). ∎

Note that if we only knew that

$$f(p^a) \le \exp(Aa^\theta)$$

for every prime and all $a \ge 1$, then from the proof it is clear that we could show

$$\limsup_{n \to +\infty} \frac{(\log f(n))\log\log n}{\log n} \le \log M^*,$$

where

$$M^* = \sup_p \left\{ \max_{a \ge 1} \left(f(p^a) \right)^{1/a} \right\}.$$

Corollary 9.12.1. Let k be an integer, $k \ge 2$. Then

$$\limsup_{n \to +\infty} \frac{(\log d_k(n))\log\log n}{\log n} = \log k.$$

Proof. We have, by Theorem 7.19, that

$$d_k(p^a) = d_k(2^q) = \binom{a+k-1}{k-1}.$$

Since $\binom{a+k-1}{k-1} < a^k$ we see that $d_k(n)$ satisfies conditions (a) and (b) of Theorem 9.12. Now

(9.46)
$$M = \max_{a \ge 1}\binom{a+k-1}{k-1}^{1/a} = \max_{a \ge 1}\left\{\frac{(a+k-1)\cdots(a+1)}{(k-1)!}\right\}^{1/a}$$

$$= \max_{a \ge 1} a^{(k-1)/a}\left\{\frac{(1+(k-1)/a)\cdots(1+1/a)}{(k-1)!}\right\}^{1/a}.$$

Now the quantity in braces reaches its maximum of k at $a = 1$. The function $f(x) = x^{(k-1)/x}$ has its global maximum at $x = e$. However,

$$\left(\frac{k+1}{k-1}\right)^{1/2} = \sqrt{k(k+1)/2} < k,$$

since $k \ge 2$. Thus the maximum occurs at $a = 1$ and has the value k. Thus $M = k$ and the result follows from Theorem 9.12.   ∎

In a similar way one can show the following result.

Corollary 9.12.2. Let $\theta(n) = 2^{\omega(n)}$. Then

$$\limsup_{n \to +\infty}\frac{(\log \theta(n))\log\log n}{\log n} = \log 2.$$

Unfortunately this theorem does not apply to Euler's function or the sum of divisors functions. These two functions are not as well behaved as $d_k(n)$ since their definitions do involve the prime divisors of the integers. The following theorem does show that the order of magnitude of $\sigma(n)$ and $\varphi(n)$ are closely related.

Theorem 9.13. There exists a positive constant A such that

$$A < \frac{\sigma(n)\varphi(n)}{n^2} < 1.$$

Proof. Let $n = \prod_{p^a \| n} p^a$. Then

$$\sigma(n) = \prod_{p^a \| n}\frac{p^{a+1}-1}{p-1} = n\prod_{p^a \| n}\frac{1-p^{-a-1}}{1-p^{-1}}$$

and

$$\varphi(n) = n\prod_{p|n}\left(1-p^{-1}\right).$$

Thus

$$\prod_{p|n}\left(1-p^{-2}\right) < \frac{\sigma(n)\varphi(n)}{n^2} < \prod_{p^a\|n}\left(1-p^{-a-1}\right) < 1.$$

Now

$$\prod_{p|n}\left(1-p^{-2}\right) > \prod_{p}\left(1-p^{-2}\right)$$

and the infinite product converges to some nonzero finite number since

$$0 < \sum_{p}\frac{1}{p^2} < \sum_{n=1}^{+\infty}\frac{1}{n^2} = \zeta(2) = \frac{\pi^2}{6}.$$

The result follows.   ■

Lemma 9.14.1. As $x \to +\infty$, we have

$$\sum_{p\leq x}\frac{1}{p} = \log\log x + B + O\left(\frac{1}{\log x}\right).$$

where B is a certain constant.

Proof. Let

$$C(x) = \sum_{p\leq x}\frac{\log p}{p},$$

which, by (9.40), satisfies

(9.47) $C(x) = \log x + O(1),$

as $x \to +\infty$. Then

$$\sum_{p\leq x}\frac{1}{p} = \int_{2-}^{x}\frac{1}{\log t}dC(t) = \int_{2}^{x}\frac{1}{\log t}dt + \int_{2-}^{x}\frac{1}{\log t}d(C(t)-\log t)$$

$$= \log\log x - \log\log 2 + O\left(\frac{1}{\log x}\right) + \int_{2}^{+\infty}\frac{C(t)-\log t}{t\log^2 t}dt - \int_{x}^{+\infty}\frac{C(t)-\log t}{t\log^2 t}dt$$

$$= \log\log x + B + O\left(\frac{1}{\log x}\right),$$

by (9.47), since $C(t)$ - $\log t = O(1)$ implies that

$$\int_{2}^{+\infty}\frac{C(t)-\log t}{t\log^2 t}dt$$

converges.   ■

Theorem 9.14. There exists a positive constant A such that, for $n \geq 3$,

$$\varphi(n) > An / \log \log n.$$

Proof. By Theorem 7.22, we have

$$\frac{\varphi(n)}{n} = \prod_{p|n}\left(1 - p^{-1}\right).$$

Thus

$$(9.48) \qquad \log\left(\frac{\varphi(n)}{n}\right) = \sum_{p|n}\log\left(1 - p^{-1}\right) = -\sum_{p|n}\frac{1}{p} + \sum_{p|n}\left\{\log\left(1 - p^{-1}\right) + \frac{1}{p}\right\}.$$

Now

$$\sum_{p|n}\left\{\log\left(1 - p^{-1}\right) + \frac{1}{p}\right\} = -\sum_{p|n}\left\{\log\frac{1}{\left(1 - p^{-1}\right)} - \frac{1}{p}\right\}$$

$$(9.49)$$

$$= -\sum_{p|n}\sum_{k=2}^{+\infty}\frac{1}{kp^k} > \frac{-1}{2}\sum_{p}\sum_{k=2}^{+\infty}\frac{1}{p^k} = \sum_{p}\frac{1}{p(p-1)} = -A_1,$$

where A_1 is a positive constant. Thus, by (9.48) and (9.49), we have

$$(9.50) \qquad \log\left(\frac{\varphi(n)}{n}\right) \geq -\sum_{p|n}\frac{1}{p} - A_1 = -\sum_{\substack{p|n \\ p < \log n}}\frac{1}{p} - \sum_{\substack{p|n \\ p \geq \log n}}\frac{1}{p} - A_1.$$

Let N be the number of terms in the second sum on the right hand side of (9.50). Then

$$n \geq \prod_{\substack{p|n \\ p \geq \log n}}p \geq (\log n)^N,$$

so that

$$N(\log \log n) \leq \log n.$$

Thus

$$(9.51) \qquad \sum_{\substack{p|n \\ p \geq \log n}}\frac{1}{p} \leq \frac{N}{\log n} \leq \frac{\log n}{\log \log n} \cdot \frac{1}{\log n} = \frac{1}{\log \log n} \leq \frac{1}{\log \log 3},$$

since $n \geq 3$. Finally, by Lemma 9.14.1, we have

$$(9.52) \qquad \sum_{\substack{p|n \\ p < \log n}}\frac{1}{p} \leq \sum_{p < \log n}\frac{1}{p} \leq \log \log n + A_2,$$

for some positive constant A_2. Thus, by (9.50), (9.51) and (9.52), we have

$$\log\left(\frac{\varphi(n)}{n}\right) > -\log\log\log n - A_2 - \frac{1}{\log\log 3} - A_1$$

(9.53)

$$= -\log\log\log n - A_3,$$

where A_3 is a positive constant. Exponenting (9.53) gives

$$\frac{\varphi(n)}{n} > \exp\left(-\log\log\log n - A_3\right) = A(\log\log n)^{-1}$$

and the result follows. ∎

The following result follows immediately from Theorems 9.13 and 9.14.

Corollary 9.14.1. If $n \geq 3$, then there is a positive constant C such that

$$\sigma(n) < Cn\log\log n.$$

Problem Set 9.5

1. Show that the constant A, in Theorem 9.13, can be taken to be $6/\pi^2$, that is, show that

$$\prod_p \left(1 - p^{-2}\right) = 1/\zeta(2) = 6/\pi^2.$$

(Hint: if we write

$$\prod_{p \leq N}\left(1 - p^{-2}\right) = \sum_{n \leq N} f(n) + F(N),$$

what can be said about f and F (think unique factorization)?)

2. Let f be a multiplicative function such that

$$f(p^k) = P_k(p),$$

where $P_k(x)$ is a monic polynomial whose coefficients all lie in the interval $[-1, K]$, where K is a positive absolute constant.

(a) Show that there is an integer N and a constant C_1 such that for $n \geq N$

$$f(n) \leq C_1 n(\log\log n)^K.$$

(b) Show that there is a constant C_2 such that

$$f(n) \geq C_2 m / \log\log m,$$

where $n = 2^k m$, $m > 1$ is odd.

(Hint: adapt the proof of Theorem 9.14.)

3. Show that $d_k(n) = O(\log^m n)$ is false for every constant m, if $k \geq 2$.

4. Show that for all $n \geq 2$ we have $\sigma(n) = O(n \log\log n)$. Do this from scratch by adapting the proof of Theorem 9.14, but not its result.

5. Let $k \geq 2$ be an integer and let $\varepsilon > 0$.

   (a) Show that there is an $N = N(\varepsilon)$ such that if $n \geq N$, then

$$d_k(n) < k^{(1+\varepsilon)\log n/\log\log n}.$$

   (b) Show that there are infinitely many n such that

$$d_k(n) > k^{(1-\varepsilon)\log n/\log\log n}.$$

   (Hint: Corollary 9.12.1.)

   (c) Prove the analogs of (a) and (b) for $\theta(n)$.

6. (a) Show that for every m $r_2(n) = O(\log^m n)$ is false.

   (b) Show that for every $\eta > 0$ we have $r_2(n) = O(n^\eta)$.

   (c) Show that

$$\limsup_{n \to +\infty} \frac{(\log r_2(n))\log\log n}{\log n} = \tfrac{1}{2}\log 2.$$

   (Hint: you will have to modify the proof of Theorem 9.11 since $r_2(p) = 0$ if $p \equiv 3 \pmod 4$.)

7. (a) Show that

$$\limsup_{n \to +\infty} \frac{\varphi(n+1)}{\varphi(n)} = +\infty \text{ and } \liminf_{n \to +\infty} \frac{\varphi(n+1)}{\varphi(n)} = 0.$$

   (b) Show that

$$\limsup_{n \to +\infty} \frac{\sigma(n+1)}{\sigma(n)} = +\infty \text{ and } \liminf_{n \to +\infty} \frac{\sigma(n+1)}{\sigma(n)} = 0.$$

8. (a) Let $f(n) = \sigma(n)\varphi(n)n^{-2}$. Show that f is multiplicative and that, for $n \geq 4$,

$$f(n) \geq \tfrac{1}{2}\left(1 + \left[\sqrt{n}\right]^{-1}\right).$$

   (Hint: show that

$$f(n) = \prod_{p^\alpha \| n}(1 - p^{-\alpha-1})$$

   and find a lower bound for this product (remember that $\alpha \geq 1$).)

   (b) Show that, for $n \geq 3$, $\sigma(n)\varphi(n)n^{-2} \geq \tfrac{1}{2}\left(1 + n^{-1/2}\right)$.

   (c) Show that

$$\varphi(n) \geq \frac{n}{2}\left(1 + n^{-1/2}\right)(1 + \log n)^{-1}.$$

   (Hint: see Problem 1 of Problem Set 7.4.)

(d) Show that, for $n > 1$,

$$\varphi(n) > \frac{n}{4\log n}.$$

(e) Can you improve the constant 1/4?

9. Show that

$$\limsup_{n\to+\infty}\frac{\varphi(n)}{n} = 1 \text{ and } \liminf_{n\to+\infty}\frac{\varphi(n)}{n} = 0.$$

9.6. SUMMATORY FUNCTIONS OF ADDITIVE FUNCTIONS

In this section we derive summatory functions for classes of additive functions and apply this to proving a result of Turán on the average order of additive functions.

Theorem 9.15. If h is an additive function such that $h(p^k)$ does not depend on p, but only on k, and $|h(p^k)| < c_1 2^{k/2}$, for all positive integers k, then

$$\sum_{n\le x} h(n) = h(p)x\log\log x + c_2 x + O\!\left(\frac{x}{\log x}\right).$$

Proof. Since h is an additive function we have

$$h(n) = \sum_{p^m \| n} h(p^m) = \sum_{\substack{p^m \| n \\ m\ge 1}} \{h(p^m) - h(p^{m-1})\},$$

since the sum telescopes and $h(1) = 0$. Thus

$$\sum_{n\le x} h(n) = \sum_{n\le x}\sum_{\substack{p^m \| n \\ m\ge 1}} \{h(p^m) - h(p^{m-1})\} = \sum_{\substack{p^m \le x \\ m\ge 1}} \{h(p^m) - h(p^{m-1})\}[x/p^m]$$

(9.54)

$$= \sum_{p\le x} h(p)[x/p] + \sum_{\substack{p^m \le x \\ m\ge 2}} \frac{\{h(p^m) - h(p^{m-1})\}}{p^m} + O\!\left(\sum_{\substack{p^m \le x \\ m\ge 2}} |h(p^m)|\right).$$

Now $h(p)$ has the same value for any prime p. Also, by Lemma 9.14.1, we have

(9.55) $$\sum_{p\le x}\left[\frac{x}{p}\right] = x\sum_{p\le x}\frac{1}{p} + O(\pi(x)) = x\log\log x + Bx + O\!\left(\frac{x}{\log x}\right).$$

Thus, by (9.55),

(9.56) $$\sum_{p\le x} h(p)[x/p] = h(p)x\log\log x + Bh(p)x + O\!\left(\frac{x}{\log x}\right).$$

Also, by Theorem 9.4,

$$(9.57) \quad \sum_{\substack{p^m \le x \\ m \ge 2}} |h(p^m)| = O\left\{ \sum_{2 \le m \le \frac{\log x}{\log 2}} \sum_{p \le x^{1/m}} 2^{m/2} \right\} = O\left\{ \pi(\sqrt{x}) \sum_{m \le \frac{\log x}{\log 2}} 2^{m/2} \right\}$$

$$= O\left(\frac{\sqrt{x}}{\log x} 2^{\log x / 2 \log 2} \right) = O\left(\frac{x}{\log x} \right).$$

Since $h(p^m)$ depends only on m and for $k \ge 2$, by Theorem 9.4,

$$\sum_{p \ge y} p^{-k} = O\left(y^{1-k} \log^{-1} y \right),$$

as $y \to +\infty$, we have

$$\sum_{\substack{p^m > x \\ 2 \le m \le \frac{\log x}{\log 2}}} \frac{h(p^m) - h(p^{m-1})}{p^m}$$

$$= \sum_{m=2}^{[\log x / \log 2]} \sum_{p > x^{1/m}} \frac{h(p^m) - h(p^{m-1})}{p^m}$$

$$= \sum_{m=3}^{[\log x / \log 2]} \sum_{p > x^{1/m}} \frac{h(p^m) - h(p^{m-1})}{p^m} + O\left(\sum_{p > \sqrt{x}} p^{-2} \right)$$

$$(9.58) \quad = O\left\{ \sum_{m=3}^{\lceil \log x / \log 2 \rceil} 2^{m/2} \frac{mx^{1/m}}{x \log x} + \frac{1}{\sqrt{x} \log x} \right\}$$

$$= O\left\{ \frac{1}{x \log x} x^{1/3} (\sqrt{2})^{\log x / \log 2} \log^2 x + \frac{1}{\sqrt{x} \log x} \right\}$$

$$= O\left(x^{-1/6} \log x \right) = O\left(\log^{-1} x \right).$$

If we let

$$c_2 = \sum_{m=2}^{+\infty} \sum_{p} \frac{h(p^m) - h(p^{m-1})}{p^m} + Bh(p),$$

then the result follows from (9.55), (9.56), (9.57), and (9.58). ∎

Corollary 9.15.1. We have, as $x \to +\infty$,

$$\sum_{n \le x} \omega(n) = x \log \log x + Bx + O\left(\frac{x}{\log x} \right)$$

and

$$\sum_{n\le x}\Omega(n) = x\log\log x + c_3 x + O\!\left(\frac{x}{\log x}\right),$$

where

$$c_3 = B + \sum_p \frac{1}{p(p-1)}.$$

Proof. Note that we have

$$\omega(p) = \Omega(p) = 1.$$

Also

$$\sum_{m=2}^{+\infty}\sum_p \frac{\omega(p^m) - \omega(p^{m-1})}{p^m} + B\omega(p) = B$$

since $\omega(p^k) = 1$ for all $k \ge 1$, and

$$\sum_{m=2}^{+\infty}\sum_p \frac{\Omega(p^m) - \Omega(p^{m-1})}{p^m} + B\Omega(p) = B + \sum_{m=2}^{+\infty}\sum_p \frac{m - (m-1)}{p^m} = B + \sum_p \frac{1}{p(p-1)}.$$

The result then follows from Theorem 9.15. ∎

   The next theorem gives a result on the average deviation from the average value given in Theorem 9.15 for a subclass of the additive functions considered above. Corollary 9.16.2 is the main result on average deviation.

Lemma 9.16.1. As $x \to +\infty$, we have

$$\sum_{n\le x}\omega^2(n) = x(\log\log x)^2 + O(x\log\log x).$$

Proof. Let us consider the number of pairs of different prime factors p and q of n, that is, $p \ne q$, where we count the pair q, p as being distinct from the pair p, q. There are $\omega(n)$ possibilities for p and so $\omega(n)$ - 1 choices for q. Thus

$$\omega(n)(\omega(n)-1) = \sum_{\substack{pq\mid n \\ p\ne q}}1 = \sum_{pq\mid n}1 - \sum_{p^2\mid n}1.$$

Thus

$$\sum_{n\le x}\omega^2(n) - \sum_{n\le x}\omega(n) = \sum_{n\le x}\left(\sum_{pq\mid n}1 - \sum_{p^2\mid n}1\right) = \sum_{pq\le x}\left\lfloor\frac{x}{pq}\right\rfloor - \sum_{p^2\le x}\left\lfloor\frac{x}{p^2}\right\rfloor.$$

Now

$$\sum_{p^2 \le x}\left[\frac{x}{p^2}\right] \le \sum_{p^2 \le x}\frac{x}{p^2} \le x\sum_{p}\frac{1}{p^2} = O(x).$$

Thus, by Corollary 9.15.1, we have

(9.59)
$$\sum_{n \le x}\omega^2(n) = x\sum_{pq \le x}\frac{1}{pq} + x\log\log x + O(x)$$

We have

(9.60)
$$\left(\sum_{p \le \sqrt{x}}\frac{1}{p}\right)^2 \le \sum_{pq \le x}\frac{1}{pq} \le \left(\sum_{p \le x}\frac{1}{p}\right)^2,$$

since if $pq \le x$, then $p \le x$ and $q \le x$, while if $p \le \sqrt{x}$ and $q \le \sqrt{x}$, then $pq \le x$. Now the outside sums in (9.60) are both, by Lemma 9.14.1,

$$(\log\log x + O(1))^2 = (\log\log x)^2 + O(\log\log x),$$

and so

$$\sum_{n \le x}\omega^2(n) = x(\log\log x)^2 = O(x\log\log x),$$

by (9.59).   ∎

Theorem 9.16. Let g be a real valued arithmetic function such that $g(0) = 0$ and $g(1) \ne 0$. Let $f(1) = 0$ and, for $n > 1$,

(9.61)
$$f(n) = g(a_1)+\cdots+g(a_r),$$

where

$$n = \prod_{j=1}^{r}p_j^{a_j}.$$

If $|g(n)| \le c_4 2^{n/2}$, then

$$\sum_{n \le x}\{f(n) - g(1)\log\log x\}^2 \le c_5 x\log\log x.$$

Proof. We see, from (9.61), that f is an additive function such that $f(p^k)$ depends only on k and not on p. If we let

$$f_1(n) = \sum_{\substack{p^m|n \\ m \ge 2}}\{g(m) - g(m-1)\},$$

then

$$f(n) = \sum_{\substack{p^m|n \\ m \ge 1}}\{g(m) - g(m-1)\} = g(1)\omega(n) + f_1(n),$$

so that

(9.62) $$f^2(n) = g^2(1)\omega^2(n) + 2g(1)\omega(n)f_1(n) + f_1^2(n).$$

Since

$$1 \le \omega(kp^m) \le \omega(k) + \omega(p^m) = \omega(k) + 1 \le 2\omega(k)$$

and

$$|g(m)| \le c_4 2^{m/2}$$

we have

$$\sum_{n \le x} \omega(n)f_1(n) = \sum_{n \le x}\sum_{\substack{p^m|n \\ m \ge 2}} \omega(n)\{g(m) - g(m-1)\}$$

$$= \sum_{\substack{p^m \le x \, k \le xp^{-m} \\ m \ge 2}} \omega(kp^m)\{g(m) - g(m-1)\}$$

(9.63)

$$= O\left(\sum_{\substack{p^m \le x \, k \le xp^{-m} \\ m \ge 2}} \omega(k)2^{m/2}\right).$$

By Corollary 9.15.1, we have

(9.64) $$\sum_{k \le xp^{-m}} \omega(k) = O(xp^{-m}\log\log(xp^{-m})) = O(xp^{-m}\log\log x).$$

Also, as above,

(9.65) $$\sum_{\substack{p^m \le x \\ m \ge 2}} \left(\sqrt{2}/p\right)^m = O(1).$$

Thus, by (9.63), (9.64) and (9.65), we have

(9.66) $$\sum_{n \le x} \omega(n)f_1(n) = O(x\log\log x).$$

We have

$$\sum_{n \le x} f_1^2(n) = \sum_{n \le x}\sum_{\substack{p^m|n \\ m \ge 2}} f(n)\{g(m) - g(m-1\}$$

(9.67)

$$= O\left(\sum_{\substack{p^m \le x \, k \le xp^{-m} \\ m \ge 2}} |f_1(n)|2^{m/2}\right).$$

Now

$$\left| f_1(kp^m) \right| = \left| \sum_{\substack{q^i | kp^m \\ i \geq 2}} \{g(i) - g(i-1)\} \right| \leq \sum_{\substack{q^i | kp^m \\ i \geq 2}} \left| g(i) - g(i-1) \right|$$

$$= O\left(\sum_{\substack{q^i | kp^m \\ i \geq 2}} 2^{i/2} \right) = O\left(2^{m/2} + \sum_{\substack{q^i | k \\ i \geq 2}} 2^{(i+m)/2} \right) = O\left(2^{m/2} \left\{ 1 + \sum_{\substack{q^i | k \\ i \geq 2}} 2^{i/2} \right\} \right),$$

so that

(9.68)

$$\sum_{k \leq t} \left| f_1(kp^m) \right| = O\left(2^{m/2} t + 2^{m/2} \sum_{\substack{q^i \leq t \\ i \geq 2}} \sum_{j \leq t/q} 2^{i/2} \right)$$

$$= O\left(2^{m/2} t + 2^{m/2} t \sum_{\substack{q^i \leq t \\ i \geq 2}} \left(\sqrt{2} / p \right)^i \right) = O\left(2^{m/2} t \right).$$

Thus, by (9.67) and (9.68), we have

(9.69)
$$\sum_{n \leq x} f_1^2(n) = O\left(x \sum_{\substack{p^m \leq x \\ m \geq 2}} \left(\sqrt{2} / p \right)^m \right) = O(x).$$

Thus, by (9.62), (9.66) and (9.69), we have

(9.70)
$$\sum_{n \leq x} f^2(n) = g^2(1) \sum_{n \leq x} \omega^2(n) + O(x \log \log x).$$

Now

(9.71)
$$\sum_{n \leq x} \left(f(n) - g(1) \log \log x \right)^2 = \sum_{n \leq x} f^2(n) - 2g(1) \log \log x \sum_{n \leq x} f(n)$$
$$+ g^2(1) x (\log \log x)^2 + O(\log \log x).$$

By Theorem 9.15, we have

(9.72)
$$\sum_{n \leq x} f(n) = g(1) x \log \log x + O(x),$$

and by Lemma 9.16.1, we have

(9.73)
$$\sum_{n \leq x} \omega^2(n) = x(\log \log x)^2 + O(x \log \log x).$$

The result follows from (9.70), (9.71), (9.72), and (9.73). ∎

Corollary 9.16.1. As $x \to +\infty$, we have

$$\sum_{n \le x} (\omega(n) - \log \log x)^2 = O(x \log \log x)$$

and

$$\sum_{n \le x} (\Omega(n) - \log \log x)^2 = O(x \log \log x).$$

Corollary 9.16.2. Let f be as in Theorem 9.16. Let $\delta > 0$ and

$$N_{f,\delta}(x) = \#\{n \le x : |f(n) - g(1) \log \log n| > (\log \log n)^{1/2+\delta}\}.$$

Then, for any $\delta > 0$, we have, as $x \to +\infty$,

$$N_{f,\delta}(x) = o(x).$$

Proof. It suffices to prove that if

$$M_{f,\delta}(x) = \#\{n \le x : |f(n) - g(1) \log \log x| > (\log \log x)^{1/2+\delta}\},$$

then, as $x \to +\infty$, we have, for any $\delta > 0$,

(9.74) $$M_{f,\delta}(x) = o(x).$$

For if $x^{1/e} \le n \le x$, then

$$\log \log x - 1 \le \log \log n \le \log \log x,$$

so that

(9.75) $$N_{f,\delta}(x) = M_{f,\delta}(x) + O(x^{1/e}) = M_{f,\delta}(x) + o(x).$$

Now, if there exist more than ηx numbers $n \le x$ such that

$$|f(n) - g(1) \log \log x| > (\log \log x)^{1/2+\delta},$$

then

$$\sum_{n \le x} (f(n) - g(1) \log \log x)^2 \ge \eta x (\log \log x)^{1+2\delta}.$$

For x large enough this contradicts Theorem 9.16, since it would hold for any $\eta > 0$. Thus (9.74) holds and the result follows by (9.75). ∎

This corollary says that the value of $f(n)$ is around $g(1)\log\log n$ for most n, in the sense that the number of n for which $f(n)$ is far from $g(1)\log\log n$ is small in comparison with the number of n for which it is nearer $g(1)\log\log n$. The result itself is due to Turán and generalizes a similar result of Hardy and Ramanujan for the functions $\omega(n)$ and $\Omega(n)$.

Problem Set 9.6

1. Give an estimate for

$$\sum_{n\le x}\log(d_k(n)).$$

(Hint: see Theorem 7.19.)

2. (a) Show that if $n\ge 1$, then

$$\Omega(n)\le\log n\,/\log 2.$$

(b) Show that if $n\ge 3$, then for any $\varepsilon>0$,

$$\omega(n)\le\frac{((1+\varepsilon)\log 2)\log n}{\log\log n}.$$

(Hint: see Problem 5 of Problem Set 9.5 and Problem 6 of Problem Set 7.9.)

3. Let $f(n)$ be the additive function defined by

$$f(n)=\sum_{p\mid n}p^m,$$

where m is a positive integer. Estimate the sum

$$\sum_{n\le x}f(n).$$

(Hint: it will be useful to take the prime number theorem in the form

$$\pi(x)=\frac{x}{\log x}+O\!\left(\frac{x}{\log^2 x}\right)$$

and to use Theorem 8.5.)

4. (a) Show that there exist constants a and b so that

$$\sum_{pq\le x}\frac{1}{pq}=(\log\log x)^2+a\log\log x+b+O\!\left(\frac{\log\log x}{\log x}\right).$$

(b) Show that there exist constants c and d so that

$$\sum_{n\le x}\omega^2(n)=x(\log\log x)^2+cx\log\log x+dx+O\!\left(\frac{x\log\log x}{\log x}\right).$$

(c) Show that there exist constants e and f so that

$$\sum_{n\le x}(\omega(n)-\log\log x)^2=ex\log\log x+fx+O\!\left(\frac{x\log\log x}{\log x}\right).$$

(d) Relate the constants a, b, c, d, e and f.

(e) Prove analogs of (b), (c) and (d) for $\Omega(n)$.

5. (a) Let $f(n)$ be an arithmetic function. Show that

$$\sum_{n \leq x} f(n)\omega(n) = \sum_{p \leq x} \sum_{n \leq x/p} f(np).$$

(b) Show that

$$\sum_{n \leq x} d(n)\omega(n) = 2x \log x \log \log x + O(x \log x).$$

(Hint: show that if p is a prime, then $d(np) = 2d(n) - d(n/p)$, where $d(x) = 0$ if x is not an integer.)

(c) Show that

$$\sum_{n \leq x} d(n)\omega^2(n) = 4x \log x (\log \log x)^2 + O(x \log x \log \log x).$$

(Hint: show that if p and q are distinct primes, then $d(npq) = 4d(n) - 2d(n/p) - 2d(n/q) + d(n/pq)$.)

(d) Can you generalize the results of (b) and (c)? (See Problem 24 of the Additional Problems below for related results.)

Additional Problems For Chapter Nine

General Problems

1. Let $f(n) \geq 0$ for n a nonnegative integer and for j a nonnegative integer, let

$$F_j(x) = \sum_{n \leq x} \frac{1}{j!} f(n) \log^j \left(\frac{x}{n} \right).$$

For each $j \geq 1$, show that $F_j(x) \sim x$ if and only if $F_0(x) \sim x$. (Hint: try to write the sum on the right hand side as an integral.)

2. Prove the following generalization of Bertrand's postulate. If $\varepsilon > 0$, then, for $x > x_0(\varepsilon)$, there exists a prime p such that

$$x < p < (1 + \varepsilon)x.$$

3. Prove that

$$\sum_{n \leq x} \mu(n) = o(x) \text{ implies that } \psi(x) = x + o(x).$$

(Hint: modify the hint to Problem 2(b) of Problem Set 9.4.)

4. This exercise outlines another proof of the prime number theorem. Let $r(x) = \psi(x) - x$.

(a) Show that

$$|r(x)|\log^2 x \le 2\sum_{n\le x}(\Lambda*\Lambda)(n)|r(x/n)| + O(x\log x).$$

(b) Show that

$$|r(x)|\log^2 x \le \sum_{n\le x}(\log n)|r(x/n)| + O(x\log x).$$

(c) Let $k > 1$ be an arbitrary, but fixed, real number. Show that there exists a positive constant C, which is independent of k and a number x_0, depending only on k, such that for every $x \ge x_0$ there is an integer N in $[x, kx]$ such that

$$|r(N)/N| < C/\log k.$$

(d) Let $\delta > 0$ be such that $\delta < 8\log 2$. Then there exists an x_0, depending only on δ, such that if $x \ge x_0$, then the interval $\left[x, xe^{C/\delta}\right]$ contains a subinterval $I = \left[N, Ne^{\delta/8}\right]$ such that for all $y \in I$ we have

$$|r(y)/y| < \delta.$$

(e) Conclude that $r(x) = o(x)$.

5. Let $\{a(n)\}$ be a sequence of nonnegative real numbers such that

$$\sum_{n\le x}a(n)[x/n] = x\log x + O(x).$$

Show that

(a) for $x \ge 1$, we have

$$\sum_{n\le x}\frac{a(n)}{n} = \log x + O(1);$$

(b) there exists a constant $B > 0$ so that, for $x \ge 1$,

$$\sum_{n\le x}a(n) \le Bx;$$

(c) there exist constants $A > 0$ and $x_0 > 0$ such that, for $x \ge x_0$,

$$\sum_{n\le x}a(n) \ge Ax.$$

6. (a) Let f be an arithmetic function such that

$$\sum_{p\le x}f(p)\log p = (ax+b)\log x + cx + O(1).$$

Show that there is a constant A, which depends on f, such that

$$\sum_{p \leq x} f(p) = ax + (a+c)\left(\frac{x}{\log x} + \int_2^x \frac{dt}{\log^2 t}\right) + b \log\log x + A + O\left(\frac{1}{\log x}\right).$$

(b) Deduce Lemma 9.14.1 from (a)

7. Let $S(x)$ and $T(x)$ be real valued functions such that

$$T(x) = \sum_{n \leq x} S(x/n).$$

If $S(x) = O(x)$ and there exists a positive constant c such that, as $x \to +\infty$, $S(x) \sim cx$, then, as $x \to +\infty$,

$$T(x) \sim cx\log x.$$

8. This exercise gives a generalization of Corollary 9.8.1.1. Let $A(x)$ be defined on $(0, +\infty)$ and assume that

$$\sum_{n \leq x} A(x/n) = ax\log x + bx + o\left(\frac{x}{\log x}\right),$$

where a and b are constants. Show that, as $x \to +\infty$,

$$A(x)\log x + \sum_{n \leq x} A(x/n)\Lambda(n) = 2ax\log x + o(x\log x).$$

(Hint: see Problem 13 of Problem Set 9.3.)

9. Suppose $a(n) \geq 0$ and

$$A(x) = \sum_{n \leq x} a(n)$$

satisfies the hypotheses of Problem 8.

(a) Show that

$$\sum_{n \leq x} A(x/n)\Lambda(n) = \sum_{n \leq \sqrt{x}} A(x/n) + \sum_{n \leq \sqrt{x}} \psi(x/n)a(n) + O(x).$$

(b) Show that

$$\frac{A(x)}{x} + \frac{1}{x\log x}\sum_{n \leq \sqrt{x}} A(x/n)\Lambda(n) + \frac{1}{x\log x}\sum_{n \leq \sqrt{x}} \psi(x/n)a(n) = 2a + o(1).$$

(c) Let

$$\alpha = \liminf_{n \to +\infty} \frac{A(x)}{x} \quad \text{and} \quad \beta = \limsup_{n \to +\infty} \frac{A(x)}{x}.$$

If $\varepsilon > 0$, show that $\alpha + \beta/2 + a/2 + \varepsilon/2 + a\varepsilon/2 > 2a$, and so (since is arbitrary)

$$\alpha + \beta/2 \geq 3a/2.$$

(Hint: if x/t is large, then $A(x/t) < (\beta + \varepsilon)x/t$ and $\psi(x/t) < (1+\varepsilon)x/t$.)

(d) Similarly show that

$$\beta + \alpha/2 \leq 3a/2.$$

(e) Show that $\alpha = \beta = a$, and so $A(x) \sim ax$, as $x \to +\infty$.

10. In Problem 9 the prime number theorem was used in a disguised form. Suppose we only have Chebychev-type estimates:

$$\gamma = \liminf_{x \to +\infty} \frac{\psi(x)}{x} \text{ and } \delta = \limsup_{x \to +\infty} \frac{\psi(x)}{x},$$

with $\gamma \leq \delta$.

(a) Show that

$$\alpha + \beta/2 + a\delta/2 \geq 2a \text{ and } \beta + \alpha/2 + a\gamma/2 \leq 2a.$$

(b) Show that

$$a\gamma \leq \alpha \leq \beta \leq a\delta.$$

11. (a) Let

$$A(x) = \sum_{n \leq x} a(n) \text{ and } A_1(x) = \int_1^x A(t)dt.$$

Assume that $a(n) \geq 0$ for all n. Show that if $A_1(x) \sim Lx^c$, as $x \to +\infty$, for some $c > 0$ and $L > 0$, then we also have

$$A(x) \sim cLx^{c-1}, \text{ as } x \to +\infty.$$

(b) Let $A(u)$ be a nonnegative increasing function of u, for $u \geq 1$, and let

$$A_2(x) = \int_1^x \frac{A(t)}{t}dt.$$

If $A_2(x) \sim Lx^c$, as $x \to +\infty$, where $c, L > 0$, show that

$$A(x) \sim cLx^c, \text{ as } x \to +\infty.$$

(Hint: let $\varepsilon > 0$ and consider what happens to $A_2(x + \varepsilon x) - A_2(x)$ as $\varepsilon \to 0$ (remember that A is increasing).)

12. Show that

$$\int_1^{+\infty} \frac{\psi(x) - x}{x^2} dx = -(1 + \gamma).$$

13. (a) Show that

$$\sum_{p \leq x} \frac{\log^2 p}{p} = \tfrac{1}{2} \log^2 x + O(\log x).$$

(b) Show that

$$\sum_{pq \leq x} \frac{\log p \log q}{pq} = \tfrac{1}{2} \log^2 x + O(\log x).$$

14. Define a sequence of primes $\{p_n\}$ by letting p_1 be given let p_{n+1} be determined by

$$2^{p_n} < p_{n+1} \leq 2^{p_n+1}.$$

(a) Show that $p_{n+1} + 1 < 2^{p_n+1}$.

(b) If $L(x) = \log_2 x$, define the sequence of iterates by $L^{(0)}(x) = x$ and, for $n \geq 1$,

$$L^{(n)}(x) = L^{(n-1)}(L(x)).$$

If $u_n = L^{(n)}(p_n)$ and $v_n = L^{(n)}(p_n + 1)$, show that

$$u_n < u_{n+1} < v_{n+1} < v_n.$$

(c) Show that there exists a real number α such that

$$\alpha = \lim_{n \to +\infty} u_n.$$

Also show that

$$u_n < \alpha < v_n.$$

(d) (Miller) If $\alpha_0 = \alpha$ and, for $n \geq 0$, $\alpha_{n+1} = 2^{\alpha_n}$, then $[\alpha_n]$ is always prime.

15. In this exercise we will improve Theorem 8.21. Let p be a prime.

(a) If $\left(\frac{n}{p}\right)$ denotes the Legendre symbol modulo p, show that

$$\sum_{n \leq x} \tfrac{1}{2}\left(1 - \left(\tfrac{n}{p}\right)\right) > \frac{x}{2} - \tfrac{1}{2}\sqrt{p} \log p.$$

(Hint: see Problem 62 of the Additional Problems for Chapter Seven.)

(b) Show that if $c > 3$ is such that $1 + \varepsilon(2\sqrt{e}) > e^{1/c}$ and $x = \left[c\sqrt{p} \log p\right]$, then there are at least $\frac{1}{2}(c-2)\sqrt{p}\log p$ quadratic nonresidues modulo p in $[1, x]$.

(c) Suppose $\delta \in (0,1/2)$ is arbitrary. Show that if all primes less than p^δ are quadratic residues of p, then we must have

$$\frac{c-2}{2c} = -\log(2\delta).$$

(Hint: show that if n is a nonresidue modulo p, then there is a prime $q < p$ such that $q \mid n$ and $\left(\frac{q}{p}\right) = -1$. Now estimate the number of such primes with the aid of Lemma 9.14.1.)

(d) Show that if $\delta > e^{1/c}/2\sqrt{e}$, then we have a contradiction.

(e) Conclude that if $\varepsilon > 0$, then, for p sufficiently large, there is a quadratic nonresidue modulo p less than $p^{1/2\sqrt{e}+\varepsilon}$.

16. Let f be an arithmetic function such that

$$\sum \frac{f(n)}{n} = 0.$$

(a) Show that

$$\sum_{m \le x} (Le * f)(m) = O(x).$$

(b) Let

$$E_f(x) = \sum_{m \le x} (e * f)(n).$$

Show that

$$\sum_{m \le x} (\Lambda * e * f)(n) = \sum_{n \le x} \Lambda(n) E_f(x/n).$$

(c) If

$$\delta_r(n) = \begin{cases} 1 & \text{if } n = r \\ 0 & \text{else} \end{cases},$$

let

$$f = \delta_1 - \delta_2 - \delta_3 - \delta_5 + \delta_{30}.$$

Show that

$$E_f(x + 30) = E_f(x) \text{ and } E_f(x) + E_f(30 - x) = 1.$$

(d) Show that for this f we have

$$\sum_{m \leq x} (Le* f)(m) = Ax + O(\log ex)$$

for

$$A = \tfrac{1}{2}\log 2 + \tfrac{1}{3}\log 3 + \tfrac{1}{5}\log 5 - \tfrac{1}{30}\log 30.$$

(e) Show that

$$\sum_{n \leq x} \Lambda(n) E_f(x/n) = \begin{cases} \leq \displaystyle\sum_{n \leq x} \Lambda(n) = \psi(x) \\[2ex] \geq \displaystyle\sum_{x/6 < n \leq x} \Lambda(n) = \psi(x) - \psi(x/6) \end{cases}$$

(f) Show that

$$Ax + O(\log ex) \leq \psi(x) \leq 1.2Ax + O(\log ex).$$

This is Chebychev's estimate.

17. (a) Show that the prime number theorem implies that

$$\lim_{x \to +\infty} \frac{\pi(ax)}{\pi(bx)} = \frac{a}{b}.$$

(b) Show that if $a < b$, then there is at least one prime p such that $ax < p < bx$ if $x > x_0$.

18. Show that

$$\lim_{x \to +\infty} \frac{\pi(2x) - \pi(x)}{\pi(x)} = 1.$$

19. (a) Let m be a positive integer and let $M = \{1 \leq n \leq m : (n, m) = 1\}$. If $m > 2$, show that

$$\sum_{n \in M} \frac{1}{n}$$

is not an integer.

(b) If m and n are positive integers, then

$$\frac{1}{m} + \frac{1}{m+n} + \cdots + \frac{1}{m+xn}$$

is never an integer unless $x = 0$ and $m = 1$.

20. Let k and M be positive integers with $k \geq 2$. Show that there are infinitely many positive integers n such that between n^k and $(n+1)^k$ there are at least M primes.

21. Let P be the set of odd primes and let $f(k)$ be the number of ways that $2k$ can be written as a sum of two elements out of P.

   (a) Show that

   $$\left(\sum_{p \in P} x^p\right)^2 = \sum_{k=2}^{+\infty} f(k) x^{2k}.$$

   (b) Suppose $f(k) \le M$ for all $k \ge 1$. Show that

   $$\sum_{p \in P} x^{p-1} \le \frac{\sqrt{M} x}{\sqrt{1 - x^2}}.$$

   (c) Show that for each positive integer N there exist infinitely many even integers which can be written in more than N ways as a sum of two odd primes. (Hint: note that

   $$\sum_{3 \le p \le u} \frac{1}{p} = \sum_{3 \le p \le u} \int_0^1 x^{p-1} dx.)$$

22. (a) Show that if $x \ge x_0(\varepsilon)$, then

   $$\frac{bx}{\log x} - 1 \ge \frac{(b - \varepsilon)x}{\log x}.$$

   Give an estimate for $x_0(\varepsilon)$.

   (b) Show that if $x \ge x_0$ and $a > b$, then

   $$\frac{ax}{\log(ax)} \ge \frac{bx}{\log x}.$$

   Give an estimate for x_0.

   (c) Find a value for B so that if $A > 1$ and $\alpha \le 1 \le \beta$, then

   $$\frac{\alpha A x}{\log(A x)} - \frac{\beta x}{\log x} - 1 \ge \frac{B x}{\log x}$$

   for $x \ge x_0$. Give an estimate for x_0.

   (d) Using a Chebychev type estimate

   $$\frac{\alpha x}{\log x} < \pi(x) < \frac{\beta x}{\log x}$$

   show that there is a λ such that

   $$p_{n+1} - p_n < \lambda \log p_n$$

   for infinitely many n. Give a lower bound for λ.

   (e) Show that there are infinitely many values of n such that

   $$p_{n+1} - p_n > \log p_n.$$

(Hint: use Chebychev in the form $\alpha\, n\log n < p_n < \beta\, n\log n$, for some $\alpha \le 1 \le \beta$ and use the Chinese Remainder Theorem.)

23. Let $\{a_n\}$ be a sequence of distinct positive integers. Let

$$A(x) = \sum_{a_n \le x} 1 = \#\{n : a_n \le x\}.$$

(a) If

$$\limsup_{x \to +\infty} \frac{A(x)}{x} = +\infty,$$

then for every integer N there exists a positive integer $c = c(N)$ such that $a_n + c$ represents more than N primes.

(b) If

$$\limsup_{x \to +\infty} \frac{A(2x) - A(x)}{\log x} = +\infty,$$

then for every integer N there exists a positive integer $c = c(N)$ such that $a_n - c$ represents more than N primes.

(c) Let $P(x) = b_m x^m + b_{m-1} x^{m-1} + \cdots + b_1 x + b_0$ be a polynomial with integer coefficients and $b_m > 0$. Show that for every positive integer N there exists an integer $c = c(N)$ such that $P(x) + c$ is prime for more than N integers x. (Hint: note that $b_m > 0$ implies that $P(x)$ is eventually increasing, and so consider the sequence $a_n = P(n + n_0) - b_0$, where n_0 is an integer such that $P(x)$ is increasing for $x \ge n_0$.)

These results are due to Abel and Siebert.

24. Let $f(n)$ be a specially multiplicative function with associated completely multiplicative function $g(n)$. (See Problem 9 of Additional Problems of Chapter Seven.)

(a) Show that

$$\sum_{n \le x} f(n)\omega(n) = \sum_{p \le x}\left\{ f(p) \sum_{n \le x/p} f(n) - g(p) \sum_{n \le x/p^2} f(n) \right\}.$$

(b) Show that

$$\sum_{n \le x} f(n)\omega^2(n) = \sum_{\substack{pq \le x \\ p \ne q}}\left\{ f(p)f(q) \sum_{n \le x/pq} f(n) - f(p)g(q) \right.$$

$$\left. - f(q)g(p) \sum_{n \le x/p^2 q} f(n) + g(pq) \sum_{n \le x/p^2 q^2} f(n) \right\} \sum_{n \le x/pq^2} f(n).$$

(c) Show that

$$\sum_{n\le x}\sigma(n)\omega(n)=\frac{\pi^2 x^3}{12\log x}+O\!\left(\frac{x^3}{\log^2 x}\right).$$

(Hint: use the prime number theorem in the form

$$\pi(x)=\frac{x}{\log x}+O\!\left(\frac{x}{\log^2 x}\right).)$$

(d) Show that

$$\sum_{n\le x}r_2(n)\omega(n)=2\pi x\log\log x+O\!\left(\frac{x}{\log x}\right).$$

(e) Give an estimate for

$$\sum_{n\le x}\sigma(n)\omega^2(n).$$

(f) Give an estimate for

$$\sum_{n\le x}r_2(n)\omega^2(n).$$

(g) Extend (c) and (e) for $\sigma_k(n)$ for any $k>0$.

25. Show that there is a constant C_k so that

$$\sum_{\substack{n\le x\\ \Omega(n)-\omega(n)=k}}1\ =C_k x+O\!\left(\frac{x}{\log x}\right)$$

for k a nonnegative integer. (Hint: note, for example, that $\Omega(n)=\omega(n)$ if and only if n is square-free and $\Omega(n)=\omega(n)+1$ if and only if $n=p^2q$, where q is square-free and p is a prime with $p\nmid q$. Proceed by induction on k.)

26. Let f be a multiplicative function such that $|f(n)|\le 1$ for all positive integers n. Suppose that

$$\sum_{n\le x}\frac{f(n)}{n}=O(1).$$

Let

$$F(x)=\sum_{n\le x}f(n).$$

This exercise generalizes Theorem 9.9.

(a) Show that

$$F(x)\log x - \sum_{n \le x} f(n)\Lambda(n)F(x/n) = O(x).$$

(Hint: note that $\log(x/n) = \log x - \log n$ and use the Euler-Maclaurin Summation Formula (Theorem 8.14).)

(b) Let

$$A(x) = \int_2^x \frac{F(t)}{t}dt.$$

Show that

$$A(x)\log x - \sum_{n \le x} f(n)\Lambda(n)A(x/n) = O(x).$$

(c) Show that there is a constant c so that

$$|A(x)|\log^2 x \le \sum_{n \le x}(L\Lambda + \Lambda * \Lambda)(n)|A(x/n)| + cx\log x.$$

(Hint: this is the analog of Lemma 9.8.3.)

(d) Show that if $\varepsilon \ge 0$, then $|A(x)| \le \varepsilon^2 x/3$ for $x \ge x_0$.

(e) Show that $x + F(x)$ is nondecreasing.

(f) Show that $F(x) = o(x)$, as $x \to +\infty$. (Hint: argue as in Problem 13(b) above.)

27. Let

$$\psi(x,y) = \#\{n \le x: p|n \Rightarrow p \le y\}.$$

(a) Let p be a prime. Show that the number of integers n, $1 \le n \le x$, whose largest prime factor is p is $\psi(x/p, p)$.

(b) Show that

$$\psi(x,y) = \sum_{p \le y} \psi(x/p, p).$$

(c) Show that if $y \ge x$, then $\psi(x,y) = [x]$.

(d) Show that if $\sqrt{x} \le y \le x$, then

$$\psi(x,y) = [x] - \sum_{y < p \le x}\left[\frac{x}{p}\right].$$

(e) Show that if $1 \le u \le 2$, then

$$\psi(x, x^{1/u}) = (1 - \log u)x + O\left(\frac{x}{\log x}\right).$$

28. In this exercise we determine the value of the constant in Problem 11(b) of Problem Set 9.3.

(a) Show that if $x \geq 2$, then

$$\sum_{\substack{p,k \\ p \leq x \\ p^k > x}} \frac{1}{kp^k} = O\left(\frac{1}{\log x}\right).$$

(b) Take logarithms of the result of 11(b) to show that for $x \geq 2$

$$\sum_{n \leq x} \frac{\Lambda(n)}{n \log n} = \log \log x - \log c + O\left(\frac{1}{\log x}\right).$$

(c) Show that, for $x \geq 1$,

$$\sum_{n \leq x} \frac{\Lambda(n)}{n \log n} = \sum_{n \leq \log x} \frac{1}{n} - (\gamma + \log c) + O\left(\frac{1}{\log 2x}\right).$$

(d) If we write the result in (c) as $S_1 = S_2 + S_3 + S_4$, let, for $k = 1, 2, 3$ and 4,

$$I_k(x) = x \int_1^{+\infty} S_k(t) t^{-1-k} dt.$$

Show that

(i) $I_1(x) = \log(1/x) + O(x)$,

(ii) $I_2(x) = -\log(1 - e^{-x}) = \log(1/x) + O(x)$,

(iii) $I_3(x) = -\gamma - \log c$,

and

(iv) $I_4(x) = O(x \log(1/x))$.

(e) Combine the results in (d) to show that $c = e^{-\gamma}$.

29. Let $P(x)$ denote the number of odd perfect numbers less than x. Show that

$$P(x) = O\left(x\sqrt{\log \log x / \log x}\right).$$

(Hint: use Theorems 7.16 and 9.4 and the Cauchy-Schwarz inequality.)

30. Let $E(x)$ denote the number of even perfect numbers less than x. Show that

$$E(x) = O\left(\frac{\log x}{\log \log x}\right).$$

(Hint: use Theorems 7.15 and 9.4 and determine how $n \leq x$, n an even perfect number, relates to the exponent involved in n.)

Computer Problems

1. (a) One might from a probabilistic point of view, see the prime number theorem as saying that the probability that an integer n in $[1, x]$ is prime is $1/\log x$. Thus one might argue that the probability that p and $p + 2$ are both primes is $1/\log^2 x$. This doesn't work (see [47, pp. 371-372]), but it does give a ball park figure. It is conjectured that

$$\#\{p \le x: p+2 \text{ is prime}\} \sim \frac{cx}{\log^2 x}.$$

Write a program to compute the number of twin primes less than a preassigned number N and compare this with $N/\log^2 N$. Does this ratio approach a limit? (The value of c is conjectured to be 1.32032^+.) Do this for $N = 1000$, 5000 and 10000.

   (b) Do analogous computations for the triples $p, p + 2$ and $p + 6$ and p, $p + 4$ and $p + 6$. Why can't we do $p, p + 2$ and $p + 4$?

2. (a) Write a program to implement Legendre's formula (Theorem 9.5).

   (b) Use this to compute $\pi(1000)$, $\pi(10000)$ and $\pi(25000)$.

3. As we have seen above (Problem 32 of the Additional Problems for Chapter Eight) that for every $n \ge 1$ there is a square-free number in the interval $[n, n + \sqrt{n}]$. Legendre conjectured that there is an $N > 0$ such that $[n, n + \sqrt{n}]$ always contains a prime, if $n \ge N$. Explore this on the computer to see if you can find such an N. (Legendre's conjecture is still open.)

4. In Problem 24 we discussed upper and lower bounds for

$$\delta_n = \frac{p_{n+1} - p_n}{\log p_n}.$$

   (a) Show that

$$\liminf_{n \to +\infty} \delta_n \le 1.$$

   (b) Calculate δ_n for $n = 1, \ldots, 1000$ to see if any patterns emerge. (The exact nature of δ_n is still unknown.)

CHAPTER TEN
An Introduction To Algebraic Number Theory

10.1. INTRODUCTION

In this chapter we will briefly discuss some of the properties of algebraic number fields, that is, extensions of the field of rational numbers. In Section 2 we discuss some of the general properties of algebraic number fields and then in Section 3 specialize to the case of quadratic number fields, about which most is known. In Section 4 we apply some of this information to the study of Diophantine equations.

Probably the first person to seriously use algebraic numbers in connection with number theory was L. Euler in his proof of the impossibility of solving the equation

$$x^3 + y^3 = z^3$$

in nonzero integers, which is the case $n = 3$ of the great Fermat conjecture. C. F. Gauss then used algebraic numbers in his researches on biquadratic reciprocity followed by G. Eisenstein for cubic reciprocity. Even so there was no real foundation of algebraic number theory. It wasn't until E. E. Kummer tried to prove Fermat's conjecture in general that the study of algebraic number fields became more structured with the aid of R. Dedekind and L. Kronecker. For a more thorough history see [27].

10.2. GENERAL CONSIDERATIONS

Definition 10.1. A complex number α is called an **algebraic number** if it is a root of a polynomial equation with integer coefficients, that is, there exist integers a_0, $a_1,..., a_n$, not all zero, such that

(10.1)
$$a_n\alpha^n + a_{n-1}\alpha^{n-1}+\cdots+a_1\alpha + a_0 = 0.$$

If α is not an algebraic number it is said to be **transcendental**.

Due to lack of space we will not be dealing with transcendental numbers.

Example 10.1. Show that if r is a rational number, then it is algebraic. Show that if a, b, c, and m are integers, $c \neq 0$, then $\left(a + b\sqrt{m}\right)/c$ is an algebraic number.

Solution. Let $r = a/b$, where a and b are integers and $b > 0$. Then

$$br - a = 0,$$

and so r is an algebraic number.

Let $\alpha = \left(a + b\sqrt{m}\right)/c$. Then $c\alpha - a = b\sqrt{m}$. Thus

$$c^2\alpha^2 - 2ac\alpha + a^2 = (c\alpha - a)^2 = (b\sqrt{m})^2 = b^2 m$$

or

$$c^2\alpha^2 - 2ac\alpha + a^2 - b^2 m = 0,$$

and so is an algebraic number.   □

Definition 10.2. An algebraic number α is said to be an **algebraic integer** if it is the root of an equation of the type (10.1) with $a_n = 1$.

Example 10.2. Show that the only rational algebraic integers are the ordinary integers.

Solution. Let $r = a/b$ be a rational number with $(a, b) = 1$ and $b > 0$. If r is an algebraic integer, then it satisfies an equation of the form

$$r^n + a_{n-1}r^{n-1} + \cdots + a_1 r + a_0 = 0,$$

where the a_i are integers. Thus

$$a^n + a_{n-1}ba^{n-1} + \cdots + a_1 ab^{n-1} + a_0 b^n = 0.$$

or

$$b(a_{n-1}a^{n-1} + \cdots + a_1 ab^{n-2} + a_0 b^{n-1}) = -a^n.$$

Thus $b \mid a$ and since $(a, b) = 1$ we see that $b = 1$. Thus r is an integer.   □

In general determining the form of the algebraic integers is not an easy problem. In the future when we mean the integers in $\mathbf{Z}$ we speak of the rational integers.

Definition 10.3. We say that an algebraic number α is of **degree n** if it satisfies a polynomial equation, with integer coefficients, of degree n, but no such polynomial equation of lower degree.

Thus all nonzero algebraic numbers have degree at least one.

Example 10.3. Determine the degree of a rational number and the degree of the numbers of the form $\left(a+b\sqrt{m}\right)/c$, where a, b, c, and m are integers and m is not the square of a rational integer.

Solution. If $r = a/b$, then $br - a = 0$, that is, r is a root of the first degree polynomial $bx - a$. Thus rational numbers have degree 1.

In Example 10.1 we saw that $\left(a+b\sqrt{m}\right)/c$ satisfies a quadratic equation. Since m is not the square of a rational integer we know that $\sqrt{m}$ is irrational. Thus $(a+b\sqrt{m})/c$ cannot be the root of a linear polynomial. Thus the degree of $(a+b\sqrt{m})/c$ is 2.   ☐

Before going too much further we must develop some results on polynomials. We introduce some notation. If $\mathbf{Q}$ denotes the set of rational numbers, then $\mathbf{Q}[x]$ denotes the set of all polynomials with rational coefficients and $\mathbf{Z}[x]$ is the set of all polynomials with integer coefficients. If $f(x)$ is a polynomial, which is not identically zero, then $\deg f$ denotes its degree.

Definition 10.4. We say that a polynomial $f(x)$ over $\mathbf{Q}$ (over $\mathbf{Z}$), which is not identically zero, **divides** another polynomial $g(x)$ over $\mathbf{Q}$ (over $\mathbf{Z}$), if there exists a polynomial $q(x)$ in $\mathbf{Q}[x]$ ($\mathbf{Z}[x]$) such that

$$g(x) = f(x)q(x).$$

We also say that $f(x)$ is a **factor** of $g(x)$ and write $f(x) \,|\, g(x)$.

The divisibility theory for polynomials over $\mathbf{Q}$ or $\mathbf{Z}$ over parallels the divisibility theory for the rational integers in many respects. Many of the results we will prove below are true in more generality.

In what follows we shall assume that all polynomials under discussion are in $\mathbf{Q}[x]$ (which, of course, contains $\mathbf{Z}[x]$).

Theorem 10.1 (Division algorithm). Let $f(x)$ and $g(x)$ be polynomials with $f(x)$ not identically zero. Then there exist unique polynomials $q(x)$ and $r(x)$ such that

$$g(x) = f(x)q(x) + r(x),$$

where either $r(x)$ is identically zero or $\deg r < \deg f$.

Proof. If either $g(x)$ is identically zero or $\deg g < \deg f$ let $q(x) \equiv 0$ and $r(x) = g(x)$. Otherwise divide $f(x)$ into $g(x)$ to get a quotient $q(x)$ and a remainder $r(x)$. It is clear that $q(x)$ and $r(x)$ are in $\mathbf{Q}[x]$ and either $r(x) \equiv 0$ or $\deg r < \deg f$, if the division is carried out to completion. If there were another pair $q_1(x)$ and $r_1(x)$, then we would have

$$g(x) = f(x)q_1(x) + r_1(x),$$

and so

$$r(x) - r_1(x) = f(x)\{q_1(x) - q(x)\}.$$

Thus $f(x)$ is a divisor of the polynomial $r(x) - r_1(x)$, which, unless it is identically zero, has degree less than deg f. Thus $r(x) - r_1(x) \equiv 0$ and it follows that $q(x) = q_1(x)$.   ∎

Theorem 10.2. Any polynomials $f(x)$ and $g(x)$, not both identically zero, have a common divisor $h(x)$ which is a linear combination of $f(x)$ and $g(x)$. Thus $h(x) \mid f(x)$ and $h(x) \mid g(x)$ and

(10.2) $$h(x) = f(x)F(x) + g(x)G(x),$$

for some polynomials $F(x)$ and $G(x)$ in $\mathbf{Q}[x]$.

Proof. From all the polynomials of the form (10.2) that are not identically zero choose any one of least degree and call it $h(x)$. If $h(x)$ were not a divisor of $f(x)$, then, by Theorem 101, we would have

$$f(x) = h(x)q(x) + r(x),$$

where deg $r <$ deg h and $r(x)$ is not identically zero. But then

$$r(x) = f(x) - h(x)q(x) = f(x)\{1 - F(x)q(x)\} - q(x)\{G(x)q(x)\},$$

which is of the form (10.2), in contradiction to the minimality of the degree of h. Thus $r(x)$ is identically zero, and so $h(x) \mid f(x)$. Similarly, $h(x) \mid g(x)$.   ∎

Theorem 10.3. To any polynomials $f(x)$ and $g(x)$, not both identically zero, there corresponds a unique monic polynomial $d(x)$ such that

(a) $d(x) \mid f(x)$ and $d(x) \mid g(x)$,

(b) $d(x)$ is a linear combination of $f(x)$ and $g(x)$, as in (10.2)

and

(c)   any common divisor of $f(x)$ and $g(x)$ is a divisor of $d(x)$ and thus there is no common divisor having higher degree than that of $d(x)$.

Proof. Let $h(x)$ be as in Theorem 10.2 and let c be the leading coefficient of $h(x)$. Let $d(x) = c^{-1}h(x)$. Then (a) and (b) follow for $d(x)$ from the corresponding properties for $h(x)$. From (10.2) we have

$$d(x) = c^{-1}f(x)F(x) + c^{-1}g(x)G(x),$$

and so if $m(x)$ is a common divisor of $f(x)$ and $g(x)$ we see that $m(x) \mid d(x)$. Suppose $d(x)$ and $d_1(x)$ have properties (a), (b) and (c). Then we have $d(x) \mid d_1(x)$ and $d_1(x) \mid d(x)$, that is,

$$d_1(x) = q(x)d(x) \text{ and } d(x) = q_1(x)d_1(x),$$

for some $q(x)$, $q_1(x)$ in $\mathbf{Q}[x]$. Thus

$$q(x)q_1(x) = 1,$$

and so both $q(x)$ and $q_1(x)$ must have degree 0. Since both $d(x)$ and $d_1(x)$ are monic we must have $q(x) = q_1(x) = 1$ and $d(x) = d_1(x)$.　■

Definition 10.5. The polynomial $d(x)$ determined in Theorem 10.3 is called the **greatest common divisor** of $f(x)$ and $g(x)$. We write

$$d(x) = (f(x), g(x)).$$

Definition 10.6. A polynomial $f(x)$, not identically zero, is **irreducible**, or **prime**, over $\mathbf{Q}$ (over $\mathbf{Z}$) if there is no factoring $f(x) = h(x)g(x)$ of $f(x)$ into two polynomials $g(x)$ and $h(x)$ in $\mathbf{Q}[x]$ (or $\mathbf{Z}[x]$) which both have positive degrees over $\mathbf{Q}$ (over $\mathbf{Z}$).

Example 10.4. The polynomial $x^2 - 2$ is irreducible over $\mathbf{Q}$ and since $\sqrt{2}$ is irrational, but over the real numbers we have the factorization $x^2 - 2 = (x - \sqrt{2})(x + \sqrt{2})$.

Theorem 10.4. If an irreducible polynomial $p(x)$ divides a product $f(x)g(x)$, then $p(x)$ divides at least one of the polynomials $f(x)$ or $g(x)$.

Proof. If either $f(x) \equiv 0$ or $g(x) \equiv 0$, then the result is obvious.

If neither is identically zero, let us assume that $p(x) \nmid f(x)$. Then we must have $(p(x), f(x)) = 1$, and so, by Theorem 10.3, there exist polynomials $F(x)$ and $G(x)$ in $\mathbf{Q}[x]$ such that

$$1 = p(x)F(x) + f(x)G(x)$$

Thus

$$g(x) = p(x)F(x)g(x) + f(x)g(x)G(x).$$

Since $p(x) \mid f(x)g(x)$ and $p(x) \mid p(x)F(x)g(x)$ we see that $p(x) \mid g(x)$, which is the result.　■

This is the same result that leads up to unique factorization of the rational integers (Theorem 1.9). It does the same here.

Theorem 10.5 (Unique factorization for $\mathbf{Q}[x]$). Any polynomial $f(x)$ in $\mathbf{Q}[x]$ of positive degree can be factored into a product of the form

$$f(x) = cp_1(x)p_2(x)\cdots p_k(x),$$

where the $p_j(x)$ are irreducible monic polynomials over $\mathbf{Q}$, $1 \leq j \leq k$, and c is a rational number. This factoring is unique apart from order.

Proof. Clearly $f(x)$ can be factored repeatedly until it becomes a product of irreducible polynomials and the constant c can be adjusted to make all factors monic.

Let us consider another factoring, say

$$f(x) = dq_1(x)q_2(x)\cdots q_m(x)$$

into irreducible monic polynomials. Clearly, $c = d$, since it is the leading coefficient of $f(x)$. By Theorem 10.4, we know that $p_1(x)$ divides some $q_i(x)$ and we can reorder the $q_i(x)$, if necessary, so that $p_1(x)\,|\,q_1(x)$. Since both are irreducible and monic we see that $p_1(x) = q_1(x)$. A repetition of this argument yields

$$p_2(x) = q_2(x), \ \ldots, \ p_k(x) = q_k(x) \text{ and } k = m,$$

which gives uniqueness. ∎

Definition 10.7. A polynomial $f(x) = a_0x^n + \cdots + a_n$, in $\mathbf{Z}[x]$, is said to be **primitive** if $(a_0,\ldots, a_n) = 1$.

A key result for primitive polynomials is the following.

Theorem 10.6. The product of two primitive polynomials is primitive.

Proof. Let $a_0x^n + \cdots + a_n$ and $b_0x^n + \cdots + b_n$ be primitive polynomials and let $c_0x^{n+m} + \cdots + c_{n+m}$ denote their product. Suppose that the product is not primitive, so that there is a prime p that divides every c_k, $0 \leq k \leq n + m$. Since $a_0x^n + \cdots + a_n$ is primitive there exists at least one coefficient not divisible by p. Let a_i denote the first such coefficient and let b_j denote the first coefficient of $b_0x^n + \cdots + b_n$ not divisible by p. Then the coefficient of $x^{n+m-i-j}$ in the product polynomial is

$$c_{i+j} = \sum_{\substack{0 \leq k \leq n \\ 0 \leq i+j-k \leq m}} a_k b_{i+j-k}.$$

Note that in this sum any a_k with $k < i$ is a multiple of p and any b_{i+j-k} with $k > i$

will have a factor of p. The term $a_i b_j$, for $k = i$, is not divisible by p. Thus

$$a_i b_j \equiv c_{i+j} \equiv 0 \ (\mathrm{mod} \ p),$$

which is a contradiction since $p \nmid a_i$ and $p \nmid b_i$. Thus the product must be primitive. ∎

Theorem 10.7 (Gauss). If a monic polynomial $f(x)$ in $\mathbf{Z}[x]$ factors into two monic polynomials in $\mathbf{Q}[x]$, say

$$f(x) = g(x)h(x),$$

then $g(x)$ and $h(x)$ are in $\mathbf{Z}[x]$.

Proof. Let c be the least positive integer such that $cg(x) \in \mathbf{Z}[x]$; if $g(x) \in \mathbf{Z}[x]$ we take $c = 1$. Then $cg(x)$ is a primitive polynomial, since if p is a divisor of its coefficients and $p \mid c$, because c is the leading coefficient, and so $(c/p)g(x)$ would be in $\mathbf{Z}[x]$, contrary to the minimality of c. Similarly, let c_1 be the least positive integer such that $c_1 h(x) \in \mathbf{Z}[x]$. Then $c_1 h(x)$ is also primitive. Thus, by Theorem 10.6, the polynomial $cg(x)c_1 h(x) = cc_1 f(x)$ is primitive. Since $f(x)$ has integral coefficients and is monic we see that $cc_1 = 1$, that is $c = c_1 = 1$, which gives the result. ∎

We now return to our study of algebraic numbers. Recall that an algebraic number is a complex number satisfying a polynomial equation over $\mathbf{Q}$. Clearly, there are lots of polynomials that any one algebraic number can satisfy. The following theorem establishes a uniqueness result for at least one of these polynomials.

Theorem 10.8. An algebraic number α satisfies a unique irreducible monic polynomial equation $f(x) = 0$ over $\mathbf{Q}$. Moreover, every polynomial equation over $\mathbf{Q}$ satisfied by α is divisible by $f(x)$.

Proof. From all polynomial equations over $\mathbf{Q}$ satisfied by α choose one of lowest degree, say $F(x) = 0$. If the leading coefficient of $F(x)$ is c, define $f(x) = c^{-1}F(x)$, so that $f(\alpha) = 0$ and $f(x)$ is monic. The polynomial $f(x)$ is irreducible since $f(x) = g(x)h(x)$ implies that either $g(\alpha) = 0$ or $h(\alpha) = 0$ contrary to the assumption that $F(x) = 0$ and $f(x) = 0$ are polynomial equations over $\mathbf{Q}$ of least degree satisfied by α.

Next let $G(x) = 0$ be any polynomial equation over $\mathbf{Q}$ having α as a root. By Theorem 10.1, we have

$$G(x) = f(x)q(x) + r(x),$$

where $r(x) \equiv 0$ or $\deg r < \deg f$. The latter can't happen since $G(\alpha) = f(\alpha) = 0$

imply $r(\alpha) = 0$ and f has minimal degree among such polynomials over $\mathbf{Q}$. Thus $f(x)$ is a divisor of $G(x)$.

Finally, to show that $f(x)$ is unique suppose that $H(x)$ is an irreducible monic polynomial over $\mathbf{Q}$ such that $H(\alpha) = 0$. Then $f(x) \mid H(x)$, by the argument above, say $H(x) = f(x)q(x)$. But the irreducibility of $H(x)$ implies that $q(x)$ is a constant and since both H and f are monic we must have $q(x) \equiv 1$. Thus $H(x) = f(x)$. ∎

Definition 10.8. The **minimal equation** of a algebraic number α is the equation $f(x) = 0$ described in Theorem 10.9. The **minimal polynomial** of α is $f(x)$.

Thus the degree of α (Definition 10.3) is the degree of the minimal polynomial.

Theorem 10.9. The minimal polynomial of an algebraic integer is a monic polynomial in $\mathbf{Z}[x]$.

Proof. The polynomial is monic by definition, so we need only show that it must have integer coefficients. Let α satisfy $f(x) = 0$ as in Definition 10.2 and let its minimal polynomial be $g(x)$, monic and irreducible over $\mathbf{Q}$. By Theorem 10.8 we know that $g(x) \mid f(x)$, say $f(x) = g(x)h(x)$, where $h(x)$ is monic and $\mathbf{Q}[x]$. By Theorem 10.7, we see that $g(x)$ has integer coefficients. ∎

Theorem 10.10. Let n be a positive rational integer and α a complex number. Suppose that the complex numbers $\theta_1, \theta_2, \ldots, \theta_n$ not all zero, satisfy the equations

(10.3) $$\alpha\theta_j = a_{j1}\theta_1 + a_{j2}\theta_2 + \cdots + a_{jn}\theta_n,$$

for $j = 1, 2, \ldots, n$, where the n^2 coefficients a_{ji} are rational. Then α is an algebraic number. Moreover, if the a_{ji} are rational integers, then α is an algebraic integer.

Proof. The equations (10.3) can be thought of as a system of homogeneous linear equations in $\theta_1, \theta_2, \ldots, \theta_n$. Since the θ_j are not all zero the determinant of the coefficients must vanish, that is,

$$\begin{vmatrix} \alpha - a_{11} & -a_{12} & \cdots & -a_{1n} \\ -a_{21} & \alpha - a_{22} & \cdots & -a_{2n} \\ & & \vdots & \\ -a_{n1} & -a_{n2} & \cdots & \alpha - a_{nn} \end{vmatrix} = 0.$$

Expansion of this determinant gives an equation

$$\alpha^n + b_1\alpha^{n-1} + \cdots + b_n = 0,$$

where the b_i are polynomials in the a_{jk}. Thus the b_i are rational since the a_{jk} are. If the a_{jk} are integers, so are the b_i. ∎

Theorem 10.11. If α and β are algebraic numbers, so are $\alpha + \beta$ and $\alpha\beta$. If α and β are algebraic integers, so are $\alpha + \beta$ and $\alpha\beta$.

Proof. Suppose α and β satisfy

$$\alpha^m + a_1\alpha^{m-1} + \cdots + a_m = 0$$

and

$$\beta^r + b_1\beta^{r-1} + \cdots + b_r = 0$$

over $\mathbf{Q}$. Let $n = mr$ and define the complex numbers $\theta_1, \ldots, \theta_n$ by

$$\left\{\theta_j\right\}_{j=1}^n = \left\{\alpha^i\beta^j\right\}_{i=0,j=0}^{m-1,r-1}.$$

Thus for any θ_j we have

$$\alpha\theta_j = \alpha^{s+1}\beta^t = \begin{cases} \text{some } \theta_k & \text{if } s+1 \le m-1 \\ (-a_1\alpha^{m-1} - \cdots - a_m)\beta^t & \text{if } s+1 = m \end{cases}$$

In either case we see that there are rational numbers $h_{j1}, \ldots, h_{jn}$ such that

(10.4)
$$\alpha\theta_j = h_{j1}\theta_1 + \cdots + h_{jn}\theta_n, \quad j = 1, \ldots, n.$$

Similarly there are rational constants $k_{j1}, \ldots, k_{jn}$ such that

$$\beta\theta_j = k_{j1}\theta_1 + \cdots + k_{jn}\theta_n, \quad j = 1, \ldots, n.$$

Thus

$$(\alpha + \beta)\theta_j = (h_{j1} + k_{j1})\theta_1 + \cdots + (h_{jn} + k_{jn})\theta_n, \quad j = 1, \ldots, n$$

and so, by Theorem 10.10, we see that $\alpha + \beta$ is an algebraic number. Moreover, if they are algebraic integers, then the a_i and b_i are rational integers. Thus the h_{ji} and k_{ji} are rational integers, and so, by Theorem 10.10, we see that $\alpha + \beta$ is an algebraic integer.

We have, for $j = 1, \ldots, n$,

$$\alpha\beta\theta_j = \alpha(k_{j1}\theta_1 + \cdots k_{jn}\theta_n)$$
$$= k_{j1}\alpha\theta_1 + \cdots + k_{jn}\alpha\theta_n$$
$$= c_{j1}\theta_1 + \cdots + c_{jn}\theta_n,$$

where, by (10.4), we have

$$c_{ji} = k_{j1}h_{1i} + \cdots + k_{jn}h_{ni}.$$

We again use Theorem 10.10 to see that $\alpha\beta$ is an algebraic number and is an algebraic integer if both α and β are. ∎

The following theorem states a little more.

Theorem 10.12. The set of all algebraic numbers form a field and the set of all algebraic integers form a ring.

Proof. The rational integers 0 and 1 serve as the zero and unit in either system. Theorem 10.11 gives the closure property for addition and multiplication. The other properties for fields and rings are easily seen to hold, except possibly for inverses, since we are dealing with a subset of the complex numbers.

Let $\alpha \neq 0$ be a solution of the equation

$$a_0 x^n + a_1 x^{n-1} + \cdots + a_{n-1} x + a_n = 0.$$

Then $-\alpha$ and α^{-1} are solutions of

$$a_0 x^n - a_1 x^{n-1} + a_2 x^{n-2} + \cdots + (-1)^n a_n = 0$$

and

$$a_0 + a_1 x + a_2 x^2 + \cdots + a_{n-1} x^{n-1} + a_n x^n = 0,$$

respectively. Therefore $-\alpha$ and α^{-1} are algebraic numbers if α is: If α is an algebraic integer, then so is $-\alpha$, but not necessarily α^{-1} since $a_0 = 1$ does not imply that $a_n = 1$. Thus the algebraic numbers form a field, while the algebraic integers form a ring. ∎

Definition 10.9. A subfield of the field of all algebraic numbers is called an **algebraic number field**.

If α is an algebraic number, then it is easy to see that all numbers of the form $f(\alpha)/g(\alpha)$, $g(\alpha) \neq 0$, where f and g are in $\mathbf{Q}[x]$, constitutes a field. This field is denoted by $\mathbf{Q}(\alpha)$ and is called the **extension** of $\mathbf{Q}$ by a formed by adjoining α to $\mathbf{Q}$.

Theorem 10.13. If α is an algebraic number of degree n, then every number in $\mathbf{Q}(\alpha)$ can be written uniquely in the form

(10.5) $$a_0 + a_1 \alpha + a_2 \alpha^2 + \cdots + a_{n-1} \alpha^{n-1},$$

where the a_i are rational numbers.

Proof. Consider any number $f(\alpha)/g(\alpha)$ of $\mathbf{Q}(\alpha)$. If the minimal polynomial of α is $h(x)$, then $h(x) \nmid g(x)$, since $g(\alpha) \neq 0$. Since $h(x)$ is irreducible we see that the greatest common divisor of $g(x)$ and $h(x)$ is 1. Thus, by Theorem 10.3, there exist polynomials $G(x)$ and $H(x)$ such that

$$g(x)G(x) + h(x)H(x) = 1.$$

Thus, since $h(\alpha) = 0$, we see that $G(\alpha) = 1/g(\alpha)$, and so $f(\alpha)/g(\alpha) = f(\alpha)G(\alpha)$. Let $k(x) = f(x)G(x)$, so that $f(\alpha)/g(\alpha) = k(\alpha)$. Then, dividing $k(x)$ by $h(x)$, we obtain

$$k(x) = h(x)g(x) + r(x),$$

where $r(x) \equiv 0$ or deg $r <$ deg h. Then

$$f(\alpha)/g(\alpha) = k(\alpha) = r(\alpha),$$

where $r(\alpha)$ is of the form (10.5).

To prove the form (10.5) is unique we suppose that $r(\alpha)$ and $r_1(\alpha)$ are expressions of the form (10.5). If $r(x) - r_1(x)$ is not identically zero, then it is a polynomial of degree less than n. Since the minimal polynomial of α has degree n we have that $r(\alpha) - r_1(\alpha) \neq 0$, that is $r(\alpha) \neq r_1(\alpha)$, unless $r(x)$ and $r_1(x)$ are the same polynomial. ∎

We state a theorem which is useful for the general study we are engaged in. In the problems below we shall indicate another way of proceeding that does not require this result. (See Problem 1 of the Additional Problems below.)

Theorem 10.14 (Fundamental Theorem of Algebra). If $f(z) = a_n z^n + \cdots + a_0$, with $n > 0$ and $a_n \neq 0$, is a polynomial that has complex coefficients, then f has a complex root.

Proof. We proceed by induction on the degree of f. In the case $n = 1$, the result is trivial. Suppose $n = 2$. If the coefficients are real, then the quadratic formula provides the answer. If the coefficients are complex, then we might have to extract square roots of complex numbers. Once we do this the quadratic formula will again yield the result. Suppose we need a square root of $a + bi$, where a and b are real. Let $a + bi = (x + yi)^2$. Then

$$a = x^2 - y^2 \text{ and } b = 2xy$$

or

$$4x^4 - 4ax^2 - b^2 = 0.$$

Thus, if $b \neq 0$,

$$x = \pm\sqrt{\frac{a + \sqrt{a^2 + b^2}}{2}} \text{ and } y = \frac{b}{2x}.$$

If $b = 0$, then from the fact that $\sqrt{a^2} = |a|$ we see that if $a > 0$, then $x = \pm\sqrt{a}$ and $y = 0$ and if $a < 0$, then $x = 0$ and $y = \pm\sqrt{-a}$ Finally, if $a = b = 0$, then $x = y = 0$.

Before proceeding to the general case note that we may write

$$f(x + yi) = G(x, y) + iH(x, y),$$

where G and H are polynomials in the real variables x and y with real coefficients. Now G and H are continuous throughout the xy-plane, and so $|f(z)|$ is continuous throughout the complex z-plane $z = x + yi$, as well. Moreover, if $n > 0$ and $a_n \neq 0$ (as we have assumed), then

$$\lim_{|z| \to \infty} |f(z)| = +\infty,$$

since if $A = \max\{|a_1|, \ldots, |a_n|\}$, then, for $|z| > \max(2nA / |a_n|, 1)$

$$|f(z)| \geq |a_n z^n| - \left(|a_0| + |a_1 z| + \cdots + |a_{n-1} z^{n-1}|\right)$$
$$\geq |a_n z^n| \left(1 - nA / |a_n z|\right)$$
$$> |a_n z^n| / 2,$$

and the result follows.

Since $|f(z)|$ is continuous it assumes a minimum value at some point in any closed circular disk with center at the origin. Since $|f(z)|$ becomes infinite with $|z|$, the disk can be chosen so huge that this minimum occurs at an interior point α. We must show that $|f(\alpha)| = 0$.

Now suppose that every polynomial of degree less than n, $n > 3$, with complex coefficients, has a complex zero, and that f has degree n with its minimum assumed at $z = \alpha$. Suppose $f(\alpha) = M \neq 0$. Let

$$g(z) = f(z + \alpha) / M = 1 + b_1 z + \cdots + b_n z^n.$$

Then $|g(z)| > 1$ for all z. Let k be the smallest index such that $b_k \neq 0$ so that

$$g(z) = 1 + b_k z^k + \cdots + b_n z^n, \quad k \leq n.$$

Suppose $k < n$. Then, by the induction hypothesis, the equation

$$1 + b_k z^k = 0$$

has a root, say η. Let $z = \delta \eta$, where $0 < \delta < 1$. Then

$$g(\delta \eta) = 1 + b_k \delta^k \eta^k + b_{k+1} \delta^{k+1} \eta^{k+1} + \cdots + b_n \delta^n \eta^n$$
$$= 1 - \delta^k + \left(b_{k+1} \eta^{k+1} + \cdots + b_n \delta^{n-k+1} \eta^n\right) \delta^{k+1}.$$

If $|b_j| < B$ for $k < j \leq n$, then

$$\delta^{k+1} \left|b_{k+1} \eta^{k+1} + \cdots + b_n \eta^n \delta^{n-k+1}\right| \leq B(1 + |\eta|)^n \delta^{k+1} \left(1 + \delta + \cdots + \delta^{n-k+1}\right)$$
$$\leq Bn(1 + |\eta|)^n \delta^{k+1} = C\delta^{k+1}.$$

Thus

$$|g(\delta\eta)| < 1 - \delta^k + C\delta^{k+1} = 1 - \delta^k(1 - C\delta),$$

and so, for $0 < \delta < 1/C$, we have $|g(\delta\eta)| < 1$, which contradicts the assumption that 1 is the minimum of $|g(z)|$.

If $k = n$, then $g(z) = 1 + b_n z^n$. If n is even, then the equation

$$z^{n/2} - \sqrt{-1/b_n} = 0$$

is solvable, by the induction hypothesis, and any root of it is also a root of $g(z) = 0$. Thus we may suppose that n is odd. Let $b_n = c + di$. If $c \neq 0$, we let $z = -\delta \operatorname{sgn} c$ and obtain

$$\left|1 + (c + di)z^n\right|^2 = \left|1 - |c|\delta^n - \delta^n di \operatorname{sgn} c\right|^2$$
$$= 1 - 2|c|\delta^n + (c^2 + d^2)\delta^{2n}.$$

Now the last expression is again less than 1 if δ is sufficiently small yielding the same contradiction as before. If $c = 0$, then $d \neq 0$. Moreover, a sign can be chosen so that $(\pm i)^n = i$. If $z = +i\delta \operatorname{sgn} d$, then

$$\left|1 + id(\pm i\delta \operatorname{sgn} d)^n\right| = \left|1 - |d|\delta^n\right|$$

and this is less than 1 for δ sufficiently small, which gives the same contradiction as before.

Thus in all cases we must have $M = 0$, that is, α is a root of $f(z)$. ■

This proof is a modification of a proof due to Gauss, the first to give a complete proof of this theorem.

An immediate corollary is the following result.

Corollary 10.14.1. Let $f(z)$ be as in Theorem 10.14. Then there exist complex numbers $\alpha_1, \ldots, \alpha_n$ such that

$$f(z) = a_n(z - \alpha_1) \cdots (z - \alpha_n).$$

Definition 10.10. Let α be an algebraic number with minimal polynomial $f(x)$. If $f(x)$ is of degree n and has as its n zeros $\alpha, \alpha_1, \ldots, \alpha_{n-1}$, then the algebraic numbers $\alpha_1, \ldots, \alpha_{n-1}$ are called the **conjugate** of α. If $\xi \in \mathbf{Q}(\alpha)$ and $\xi = \phi(\alpha)$, as in Theorem 10.13, then the numbers $\xi_i = \phi(\alpha_i)$, $i = 1, \ldots, n-1$, are called the **field conjugates** of ξ.

Note that the field conjugates of ξ are just its conjugates. For if g is the minimal polynomial of ξ, then $g(\phi(x))$ vanishes for $x = \alpha$. Thus $f(x) \mid g(\phi(x))$, and so $g(\phi(\alpha_i))) = g(\xi_i)$ that if α is an algebraic integer, then so are its conjugates.

To prove the converse of this result we need a result on symmetric polynomials.

Definition 10.11. A polynomial in the variables $x_1, x_2,..., x_n$ is called a **symmetric polynomial** if it is unchanged when the variables are changed in any order whatsoever. The **elementary symmetric polynomials**, in the variables x_1, $x_2,..., x_n$, are the following polynomials:

$$S_1 = x_1 + \cdots + x_n$$
$$S_2 = x_1 x_2 + x_1 x_3 + \cdots + x_{n-1} x_n$$
$$\vdots$$
$$S_n = x_1 \cdots x_n.$$

The basic result is contained in the following theorem.

Theorem 10.15. Every symmetric polynomial of the variables $x_1, x_2,..., x_n$ can be expressed as a polynomial in the elementary symmetric polynomials. Moreover, the coefficients of this polynomial are built up by addition and subtraction of the coefficients of the symmetric polynomial. In particular, if the latter are integers, the former will be integers also.

Proof. We proceed by induction on n, the number of variables. For $n = 2$, we have the two symmetric functions

$$S_1 = x_1 + x_2 \quad \text{and} \quad S_2 = x_1 x_2.$$

Let $F(x_1, x_2)$ be a symmetric polynomial. Arranged in powers of x_1, it can be written as

$$F(x_1, x_2) = A_m x_1^m + A_{m-1} x_1^{m-1} + \cdots + A_0,$$

where the A_i, $0 \leq i \leq m$, are polynomials in x_2. Replace x_2 by $S_1 - x_1$ and arrange the resulting polynomial in powers x_1 to give

$$F(x_1, x_2) = B_l x_1^l + B_{l-1} x_1^{l-1} + \cdots + B_0,$$

where the B_i, $0 \leq i \leq l$, are polynomials in S_1. Let

$$\phi(t) = B_l t^l + \cdots + B_0.$$

and divide $\phi(t)$ by

$$f(t) = t^2 - S_1 t + S_2.$$

Then, identically in t, we have, by Theorem 10.1,

(10.6) $$\phi(t) = f(t)Q(t) + Ct + D,$$

where C and D are polynomials in S_1 and S_2 with coefficients built up as in the statement of the theorem. In (10.6) set $t = x_1$. Since $f(x_1) = 0$ and $\phi(x_1) = F(x_1, x_2)$ we have

(10.7) $$F(x_1, x_2) = Cx_1 + D.$$

Now interchange x_1 and x_2. Since F is symmetric, as are C and D, being polynomials in S_1 and S_2, we have, from (10.7), that

$$F(x_1, x_2) = Cx_2 + D.$$

Thus

$$Cx_1 + D = Cx_2 + D$$

or

$$C(x_1 - x_2) = 0,$$

identically in x_1 and x_2. Thus $C \equiv 0$, and so, by (10.7),

$$F(x_1, x_2) = D,$$

where D is a polynomial in S_1 and S_2. This proves the theorem in the case $n = 2$.

Now assume the theorem to be true for any symmetric polynomial in $n - 1$ variables. Let $\sigma_1, \ldots, \sigma_{n-1}$ be the elementary symmetric polynomials of the $n - 1$ variables $x_2, x_3, \ldots, x_n$. Then

$$S_1 = x_1 + \sigma_1,$$
$$S_2 = x_1\sigma_1 + \sigma_2$$
$$\vdots$$
$$S_{n-1} = x_1\sigma_{n-2} + \sigma_{n-2},$$

and conversely,

$$\sigma_1 = -x_1 + S_1$$
$$\sigma_2 = x_1^2 - x_1 S_1 + S_2$$
$$\vdots$$
$$\sigma_{n-1} = (-1)^{n-1}\left[x_1^{n-1} - x_1^{n-2}S_1 + \cdots + (-1)^{n-1}S_{n-1}\right].$$

Let $F = F(x_1,..., x_n)$ be any symmetric polynomial in n variables. We arrange it in powers of x_1 and write

$$F = A_m x_1^m + A_{m-1} x_1^{m-1} + \cdots + A_0,$$

where the A_i, $0 \leq i \leq m$, are symmetric functions in $x_2, x_3,..., x_n$. Then, by the induction hypothesis, we may write the A_i, $0 \leq i \leq m$, as polynomials in the σ_j, $1 \leq j \leq n - 1$. If we substitute for the σ_j, $1 \leq j \leq n -1$, their expressions in terms of x_1 and the S_k, $1 \leq k \leq n -1$, then the coefficients A_i, $0 \leq i \leq m$, will be expressed as polynomials in x_1 and the S_k, $1 \leq k \leq n -1$. If we substitute these expressions into F and rearrange the resulting expression again in powers of x_1, we obtain

$$F = B_l x_1^l + B_{l-1} x_1^{l-1} + \cdots + B_0,$$

where the B_i, $0 \leq i \leq l$, are polynomials in the S_k, $1 \leq k \leq n -1$. Let

$$\phi(t) = B_l t^l + B_{l-1} t^{l-1} + \cdots + B_0$$

and divide $\phi(t)$ by

$$f(t) = t^n - S_1 t^{n-1} + S_2 t^{n-2} - \cdots + (-1)^n S_n.$$

Then, identically in t, we have, by Theorem 10.1,

(10.8) $$\phi(t) = f(t)Q(t) + C_0 t^{n-1} + C_1 t^{n-2} + \cdots + C_{n-1},$$

where the C_i, $0 \leq i \leq n -1$, are polynomials in the S_k, $1 \leq k \leq n$, with coefficients built up as stated in the theorem. In (10.8) we let $t = x_1$ and, since $\phi(x_1) = F(x_1,...,x_n)$ and $f(x_1) = 0$, we see that

(10.9) $$F(x_1,...,x_n) = C_0 x_1^{n-1} + C_1 x_1^{n-2} + \cdots + C_{n-1}.$$

Now, since F is symmetric and the C_i, $0 \leq i \leq n -1$, are polynomials in the symmetric functions S_k, $1 \leq k \leq n$, we see that if we successively replace x_1 by x_2, x_1 by x_3, and so on, we obtain the following

$$C_0 x_1^{n-1} + C_1 x_1^{n-2} + \cdots + C_{n-1} - F = 0$$

$$\vdots$$

$$C_0 x_n^{n-1} + C_1 x_n^{n-2} + \cdots + C_{n-1} - F = 0.$$

This means that the polynomial

$$C_0 t^{n-1} + C_1 t^{n-2} + \cdots + C_{n-1} - F$$

vanishes for $t = x_1, x_2,..., x_n$, which can only happen if

$$C_0 = C_1 = \cdots = C_{n-1} - F = 0.$$

Thus

$$F = C_{n-1},$$

which is a polynomial in the S_k, $1 \le k \le n$. ∎

Theorem 10.16. The set of field conjugates of an element ξ in $Q(\alpha)$ is either identical with the set of conjugates of ξ or consists of several copies of the set of conjugates of ξ. The polynomial whose zeros are the field conjugates of ξ is a power of the minimal polynomial of ξ. If it is equal to the minimal polynomial, then $Q(\xi) = Q(\alpha)$.

Proof. Form the field polynomial for :

$$f(x) = (x - \xi_1)(x - \xi_2) \cdots (x - \xi_n).$$

Its coefficients are symmetric polynomials in the ξ_i, and so are symmetric polynomials in the α_1, ..., α_n. Thus the coefficients are rational numbers, by Theorem 10.15. Factor it into its irreducible monic factors in $Q[x]$, say

$$f(x) = f_1(x)f_2(x) \cdots f_m(x),$$

and let $f_1(x)$ be a factor which vanishes for $x = \xi$. Then $f_1(\phi(\alpha)) = 0$, and so, if $p(x)$ is the minimal polynomial for α, we have

$$p(x) \big| f_1(\phi(x))$$

and $f_1(x)$ vanishes at $\xi_1, \ldots, \xi_{n-1}$ as well. If these are distinct, then $f_1(x)$ has degree n, and so $f(x)$ is irreducible. If they are not, let $\xi, \xi_1, \ldots, \xi_t$ be a maximal distinct set of ξ's. Then $f_2(x)$ vanishes for some ξ_k, so that $f_1(x) \big| f_2(x)$. Since f_2 is irreducible we must have

$$f_2(x) = c f_1(x)$$

and since f_1 and f_2 are monic we must have $c = 1$. If there are other factors of $f(x)$, the argument can be repeated. Eventually, we find that

$$f(x) = (f_1(x))^{n/t}.$$

Since the zeros of $f_1(x)$, which is the minimal polynomial of ξ, are the conjugates of ξ, those of f, which are the field conjugates of ξ, consist of n/t copies of the set of conjugates of ξ.

Suppose $f_1(x) = f(x)$. Let

$$\phi(x) = f(x)\left[\frac{\alpha}{x - \xi} + \frac{\alpha_1}{x - \xi_1} + \cdots + \frac{\alpha_{n-1}}{x - \xi_{n-1}}\right].$$

Then ϕ is a polynomial of degree $n - 1$ with rational coefficients. Since

$$\phi(\xi) = \alpha(\xi - \xi_1)\cdots(\xi - \xi_{n-1}) = \alpha f'(\xi)$$

we see that

$$\alpha = \phi(\xi) / f'(\xi)$$

is in $Q(\xi)$, so that $Q(\alpha)$ is a subfield of $Q(\xi)$. Thus $Q(\alpha) = Q(\xi)$. ■

Thus if one field $Q(\xi)$ is a proper subfield of a second field $Q(\alpha)$, then we must have $\deg\xi < \deg\alpha$. For if $\deg\xi = \deg\alpha$, then the field polynomial of ξ with respect to $Q(\alpha)$ is irreducible, so that $f_1(x) = f(x)$, and $Q(\alpha) = Q(\xi)$.

Definition 10.12. If $\xi \in Q(\alpha)$ is such that $Q(\alpha) = Q(\xi)$, then is called a **primitive element** of $Q(\alpha)$.

Given the algebraic number field $Q(\alpha_1)$ and α_2 any other algebraic number we denote by $Q(\alpha_1, \alpha_2)$ the field formed by taking all rational functions of α_2 with coefficients in $Q(\alpha_1)$. Thus if α_2 is in $Q(\alpha_1)$, then $Q(\alpha_1, \alpha_2) = Q(\alpha_1)$.

Theorem 10.17. If θ and η are algebraic numbers, then the field formed by adjoining η to $Q(\theta)$ is the same field, $Q(\theta, \eta)$, as the field formed by adjoining θ to $Q(\eta)$. Moreover, there exists an algebraic number ζ such that $Q(\theta, \eta) = Q(\zeta)$.

Proof. The first part is clear, since both $Q(\theta, \eta)$ and $Q(\eta, \theta)$ are identical with the field consisting of all numbers of the form

$$q_1(\theta, \eta)/q_2(\theta, \eta),$$

where $q_2(\theta, \eta) \neq 0$ and $q_1(x, y)$ and $q_2(x, y)$ are polynomials in two variables with rational coefficients.

If $\eta \in Q(\theta)$, then $Q(\theta, \eta) = Q(\theta)$, as remarked above. Assume that θ and η do not lie in the fields $Q(\eta)$ and $Q(\theta)$, respectively. Let their minimal polynomials be $p_1(x)$ and $p_2(x)$, respectively, and their conjugates $\theta_1, \ldots, \theta_n$ and $\eta_1, \ldots, \eta_n$, respectively. Let a and b be rational numbers and let $\zeta = \zeta_1, \ldots, \zeta_{nm}$ be all expressions of the form $a\theta_j + b\eta_k$. Since the conjugates of θ are distinct, as are the conjugates of η, there are only a finite set of ratios a/b for which some two of the ζ's are equal. Choose a and b so that a/b is not in this set and order the ζ_i so that $\zeta = \zeta_1 = a\theta_1 + b\eta_1 = a\theta + b\eta$. Let

$$f(x) = (x - \zeta_1)(x - \zeta_2)\cdots(x - \zeta_{nm}).$$

This polynomial has no multiple zeros and its coefficients, being symmetric in the θ's and η's, are rational numbers, by Theorem 10.15.

We shall show that $Q(\theta, \eta) = Q(\zeta)$. Clearly, every element of $Q(\zeta)$ is in $Q(\theta, \eta)$, since ζ is a linear combination, over Q, of θ and η. Suppose $\rho \in Q(\theta, \eta)$ and that

$$\rho = q_1(\theta, \eta) / q_2(\theta, \eta),$$

where $q_2(\theta, \eta) \neq 0$. Define $\rho = \rho_1,\ldots,\rho_{nm}$, by

$$\rho_i = q_1(\theta_j, \eta_k) / q_2(\theta_j, \eta_k),$$

where the same subscripts appear on θ and η in the definition of ρ_i as in the definition of ζ_i for $1 \leq i \leq nm$. Let

$$F(x) = f(x)\left[\frac{\rho_1}{x - \zeta_1} + \frac{\rho_2}{x - \zeta_2} + \cdots + \frac{\rho_{nm}}{x - \zeta_{nm}}\right].$$

Then, by Theorem 10.15, the coefficients of F are rational. If $i > 1$, then the polynomial

$$f(x)\frac{\rho_i}{(x - \zeta_i)} = \rho_i(x - \zeta_1)\cdots(x - \zeta_{i-1})(x - \zeta_{i+1})\cdots(x - \zeta_{nm})$$

vanishes for $x = \zeta_1 = \zeta$. From the representation

$$f(x)\frac{\rho}{(x - \zeta)} = \rho(x - \zeta_2)\cdots(x - \zeta_{nm})$$

we have

$$F(\zeta) = \rho(\zeta - \zeta_2)\cdots(\zeta - \zeta_{nm}) = \rho f'(\zeta),$$

and so

$$\rho = F(\zeta) / f'(\zeta),$$

that is $\rho \in Q(\zeta)$. ∎

Example 10.4. Show that $Q\left(\sqrt{n}, \sqrt{m}\right) = Q\left(\sqrt{n} + \sqrt{m}\right)$.

Solution. It is clear that

$$Q\left(\sqrt{n} + \sqrt{m}\right) \subset Q\left(\sqrt{n}, \sqrt{m}\right)$$

since $\sqrt{n} + \sqrt{m} \in Q\left(\sqrt{n}, \sqrt{m}\right)$. To prove the inclusion the opposite way it suffices to

show that

$$\sqrt{n}, \sqrt{m} \in \mathbf{Q}\left(\sqrt{n} + \sqrt{m}\right).$$

To do this we follow along the lines of the proof of Theorem 10.17.

We take $\theta = \sqrt{n}$ and $\eta = \sqrt{m}$, whose conjugates are $-\sqrt{n}$ and $-\sqrt{m}$, respectively. Then we see that the numbers ζ_i are

$$\zeta_1 = a\sqrt{n} + b\sqrt{m}, \quad \zeta_2 = a\sqrt{n} - b\sqrt{m},$$
$$\zeta_3 = -a\sqrt{n} + b\sqrt{m} \text{ and } \zeta_4 = -a\sqrt{n} - b\sqrt{m}.$$

If $ab \neq 0$, then it is easy to see that all four of these numbers are distinct. Thus we can take $\zeta = \sqrt{n} + \sqrt{m}$. Then

$$f(x) = \left(x - \sqrt{n} - \sqrt{m}\right)\left(x - \sqrt{n} + \sqrt{m}\right)\left(x + \sqrt{n} - \sqrt{m}\right)\left(x + \sqrt{n} + \sqrt{m}\right)$$
$$= x^4 - 2(n+m)x^2 + (n+m)^2.$$

It suffices to show that $\sqrt{n} \in \mathbf{Q}\left(\sqrt{n} + \sqrt{m}\right)$, since the proof for $\sqrt{m}$ is almost identical. Then we may take

$$\sqrt{n} = \sqrt{n} / 1,$$

that is $q_1(x, y) = x$ and $q_2(x, y) = 1$. The numbers ρ_1, ρ_2, ρ_3 and ρ_4 are then given by

$$\rho_1 = \sqrt{n}, \quad \rho_2 = \sqrt{n}, \quad \rho_3 = -\sqrt{n} \text{ and } \rho_4 = -\sqrt{n}.$$

Let

$$F(x) = f(x)\left[\frac{\sqrt{n}}{x - \sqrt{n} - \sqrt{m}} + \frac{\sqrt{n}}{x - \sqrt{n} + \sqrt{m}} - \frac{\sqrt{n}}{x + \sqrt{n} - \sqrt{m}} - \frac{\sqrt{n}}{x + \sqrt{n} + \sqrt{m}}\right]$$
$$= 4n(x^2 - (n-m)).$$

Then

$$\sqrt{n} = \frac{4n\left((\sqrt{n} + \sqrt{m})^2 - (n-m)\right)}{4(\sqrt{n} + \sqrt{m})^3 - 4(n+m)(\sqrt{n} + \sqrt{m})}.$$

In a similar way, one can show

$$\sqrt{m} = \frac{4m\left((\sqrt{m} + \sqrt{n})^2 - (m-n)\right)}{4(\sqrt{m} + \sqrt{n})^3 - 4(m+n)(\sqrt{m} + \sqrt{n})}.$$

Thus the two algebraic number fields are the same. $\quad\square$

We return to the subject of algebraic integers with the following result.

Theorem 10.18. If α is a root of an equation of the form

$$f(x) = x^n + \beta_1 x^{n-1} + \cdots + \beta_n = 0,$$

where the $\beta_1, \ldots, \beta_n$ are algebraic integers, then α is an algebraic integer.

Proof. By Theorem 5.17 we know that $\beta_1, \ldots, \beta_n$ all lie in some extension field $\mathbf{Q}(\theta)$, of degree m, say. Let

$$\{\beta_1, \ldots, \beta_n\}, \; \{\beta_1^{(2)}, \ldots, \beta_n^{(2)}\}, \ldots, \{\beta_1^{(m)}, \ldots, \beta_n^{(m)}\}$$

be the sets of field conjugates, in this field for these algebraic integers and form the polynomials

$$f_2(x) = x^n + \beta_1^{(2)} x^{n-1} + \cdots + \beta_n^{(2)},$$

$$\vdots$$

$$f_m(x) = x^n + \beta_1^{(m)} x^{n-1} + \cdots + \beta_n^{(m)}.$$

Then the product $f(x)f_2(x)\cdots f_m(x)$ has rational coefficients and is monic. Thus, by Definition 10.2, α is an algebraic integer. ∎

We let O_α denote the algebraic integers of the field $\mathbf{Q}(\alpha)$. We know that the sum and product of two algebraic integers are both algebraic integers and they lie in $\mathbf{Q}(\alpha)$, since it is a field, and so they lie in O_α. Thus O_α is actually a ring (additive inverses belong by Theorem 10.12). The determination, in general, of the integers in a field is no easy task. We shall content ourselves with doing this only in one case: the quadratic case. However, the following result shows that the integers aren't too far away from the algebraic numbers.

Theorem 10.19. If α is an algebraic number, then there exists a nonzero rational integer a such that $a\alpha$ is an algebraic integer. If α satisfies an equation

$$\beta_n x^n + \beta_{n-1} x^{n-1} + \cdots + \beta_0 = 0,$$

where the β_i, $0 \leq i \leq n$, are algebraic integers, then α is an algebraic integer.

Proof. Let the minimal polynomial for α be

$$p(x) = x^n + r_1 x^{n-1} + \cdots + r_n,$$

where the r_i, $1 \leq i \leq n$, are rational numbers. Let the least common multiple of the denominators of the fractions $r_1, \ldots, r_n$ be a. Then the polynomial

$$a^n p(x/a) = x^n + ar_1 x^{n-1} + \cdots + a^n r_n$$

has rational integral coefficients and has as zeros $a\alpha, a\alpha_2,..., a\alpha_{n-1}$ each of which is therefore an algebraic integer.

The proof of second part is similar and uses Theorem 10.18. ∎

Since $\mathbf{Q}(\alpha)$ and $\mathbf{Q}(a\alpha)$ are identical if a is a nonzero rational integer, we see that any algebraic number field can be considered as an extension of $\mathbf{Q}$ by adjoining an algebraic integer.

If α is an algebraic integer, then so are its conjugates, and hence so are its field conjugates.

Definition 10.13. Let α be an algebraic number in $\mathbf{Q}(\theta)$ of degree n and let $\alpha_2,..., \alpha_n$ be its field conjugates. Then the product $\alpha\alpha_2\cdots\alpha_n$ is called the **norm** of α over $\mathbf{Q}(\theta)$, which we denote by $N(\alpha)$.

Since the norm depends on the algebraic number field $\mathbf{Q}(\theta)$ one should really denote the norm by $N_{\mathbf{Q}(\theta)}(\alpha)$, but we will suppress the dependence on θ if it is clear which number field we are considering.

Theorem 10.20. The norm of an algebraic number is a rational number and the norm of an algebraic integer is a rational integer.

Proof. Let α have minimal polynomial

$$x^m + s_1 x^{m-1} + \cdots + s_m.$$

Then the norm of α (over any $\mathbf{Q}(\theta)$ containing α) is plus or minus a power of s_m, by Theorem 10.16. Thus $N(\alpha)$ is a rational number if α is an algebraic number and a rational integer if α is an algebraic integer. ∎

The key property of the norm is given in the following theorem.

Theorem 10.21. If α and β are algebraic numbers in $\mathbf{Q}(\theta)$, then

$$N(\alpha\beta) = N(\alpha)N(\beta).$$

Moreover, $N(\alpha) = 0$ if and only if $\alpha = 0$. Finally, if $\beta \neq 0$, then

$$N(\alpha/\beta) = N(\alpha)/N(\beta).$$

Proof. As in Theorem 10.13, if n is degree of θ, let

$$\alpha = a_0 + a_1\theta + \cdots + a_{n-1}\theta^{n-1}$$

and

$$\beta = b_0 + b_1\theta + \cdots + b_{n-1}\theta^{n-1}.$$

Then in the product $\alpha\beta$ powers of θ higher than the $(n-1)$st can be reduced using the equation

(10.10) $$\theta^{n+j} = -\theta^j(r_1\theta^{n-1}+\cdots+r_n),$$

where $p(x) = x^n + r_1x^{n-1}+\cdots+r_n$ is the minimal polynomial for θ. Also $\alpha^{(k)}$ and $\beta^{(k)}$, $2 \leq k \leq n$, the field conjugates of α and β, can be obtained from (10.9), by replacing θ by θ_k and in the product $\alpha^{(k)}\beta^{(k)}$ higher powers of θ_k can be reduced using (10.10) with θ replaced by θ_k. Hence the field conjugates $(\alpha\beta)^{(1)}$, $(\alpha\beta)^{(2)}$, ..., $(\alpha\beta)^{(n)}$ of $\alpha\beta$ are simply $\alpha\beta$, $\alpha^{(2)}\beta^{(2)}$, ..., $\alpha^{(n)}\beta^{(n)}$. Thus

$$N(\alpha\beta) = (\alpha\beta)^{(1)}(\alpha\beta)^{(2)}\cdots(\alpha\beta)^{(n)}$$
$$= \alpha^{(1)}\beta^{(1)}\alpha^{(2)}\beta^{(2)}\cdots\alpha^{(n)}\beta^{(n)}$$
$$= (\alpha^{(1)}\alpha^{(2)}\cdots\alpha^{(n)})(\beta^{(1)}\beta^{(2)}\cdots\beta^{(n)})$$
$$= N(\alpha)N(\beta).$$

Clearly $\alpha = 0$ implies $N(\alpha) = 0$. If $\alpha \neq 0$, then none of its field conjugates can be zero either since the minimal polynomial of α is irreducible over $\mathbf{Q}$. Thus, if $\alpha \neq 0$, then $N(\alpha) = \alpha^{(1)}\cdots\alpha^{(n)} \neq 0$.

If $\beta \neq 0$, then α/β is defined and $N(\beta) \neq 0$. Then

$$N(\alpha/\beta)N(\beta) = N((\alpha/\beta)\beta) = N(\alpha),$$

since α/β is an algebraic number in $\mathbf{Q}(\theta)$, and the result follows. ∎

Note that $N(1) = 1$ since $\alpha \neq 0$ implies that $N(1) = N(\alpha/\alpha) = N(\alpha)/N(\alpha) = 1$.

Definition 10.14. Let α and β be in O_θ. If $\beta \neq 0$, we say that β **divides** α, written $\beta \mid \alpha$, if there exists a $\gamma \in O_\theta$ such that $\alpha = \beta\gamma$. An integer ε such that $\varepsilon \mid 1$ is called a **unit** of O_θ. We say that α is an **associate** of β, if there exists a unit, ε, of O_θ such that $\alpha = \varepsilon\beta$. An integer π in O_θ is said to be **prime** if $\pi = \alpha\beta$ implies that one of α and β is an associate of π and the other is a unit in O_θ.

The following theorem gives a characterization of units.

Theorem 10.22. An element of O_θ is a unit if and only if its norm (over $\mathbf{Q}(\theta)$) is ± 1.

Proof. If ε is a unit, then there is an integer δ, in O_θ, such that $\varepsilon\delta = 1$. Then, by Theorem 10.21,

(10.11) $$1 = N(\varepsilon\delta) = N(\varepsilon)N(\delta).$$

By Theorem 10.20, we know that $N(\varepsilon)$ is a rational integer and hence $N(\varepsilon) = \pm 1$.

Conversely, suppose $N(\varepsilon) = \pm 1$. Then we have

$$\varepsilon \varepsilon^{(2)} \cdots \varepsilon^{(n)} = \pm 1,$$

where $\varepsilon^{(2)}, \ldots, \varepsilon^{(n)}$ are the field conjugates of ε. Since they are integers if ε is, we see that $\varepsilon \mid 1$, and so is a unit. ∎

Corollary 10.22.1. The set of units in O_θ forms a multiplicative group.

Proof. To show that the units form a group it suffices to show that if ε_1, and ε_2 are units, then so are $\varepsilon_1 \varepsilon_2$ and ε_1^{-1}.

If ε_1 and ε_2 are units, then $N(\varepsilon_1) = \pm 1$ and $N(\varepsilon_2) = \pm 1$, by Theorem 10.22. Thus $N(\varepsilon_1 \varepsilon_2) = N(\varepsilon_1)N(\varepsilon_2) = \pm 1$, by Theorem 10.21, and so, by Theorem 10.22, $\varepsilon_1 \varepsilon_2$ is a unit. Similarly, $N(\varepsilon_1^{-1}) = N(1)/N(\varepsilon_1) = 1/(\pm 1) = \pm 1$, and so, by Theorem 5.22, ε_1^{-1} is a unit. ∎

Theorem 10.23. (a) If ε is a unit of O_θ, then so is ε_1, where ε_1 is any of the field conjugates of ε.

(b) If ε is a unit of O_θ and α is an integer of O_θ, then $\varepsilon \mid \alpha$.

Proof of (a). Let ε_1 be a field conjugate of ε. Then the set of field conjugates of ε_1 is the set of field conjugates of ε, by Theorem 10.16. Thus $N(\varepsilon_1) = N(\varepsilon) = \pm 1$. Thus ε_1 is a unit by Theorem 10.22.

Proof of (b) We have $1 \mid \alpha$ for any a in O_θ. Since ε is a unit of O_θ there exists a δ in O_θ such that $\varepsilon \delta = 1$. Thus

$$\alpha = \alpha \cdot 1 = \alpha(\varepsilon \delta) = \varepsilon(\alpha \delta),$$

that is, $\varepsilon \mid \alpha$, since α is in O_θ. ∎

Theorem 10.24. Let α, and β be in O_θ.

(a) If $\alpha \mid \beta$ and $\alpha \mid \gamma$, then $\alpha \mid (\beta \delta + \gamma \varepsilon)$ for any δ and ε in O_θ.
(b) If $\alpha \mid \beta$ and $\beta \mid \gamma$, then $\alpha \mid \gamma$.

Proof of (a) If $\alpha \mid \beta$ and $\alpha \mid \gamma$, then there exist η and ζ in O_θ such that $\beta = \alpha \eta$ and $\gamma = \alpha \zeta$. Thus

$$\beta \delta + \gamma \varepsilon = \alpha \eta \delta + \alpha \zeta \varepsilon = \alpha(\eta \delta + \zeta \varepsilon).$$

Since $\eta \delta + \zeta \varepsilon$ is in O_θ we have $\alpha \mid (\beta \delta + \gamma \varepsilon)$.

Proof of (b) If $\alpha \mid \beta$ and $\beta \mid \gamma$, then there exist η and ζ in O_θ such that $\beta = \alpha\eta$ and $\gamma = \beta\zeta$. Thus $\gamma = \beta\zeta = \alpha(\eta\zeta)$. Since $\eta\zeta$ is in O_θ we see that $\alpha \mid \gamma$. ∎

Theorem 10.25. Let α, and β be in O_θ. Then

  (a) α is an associate of β if and only if β is an associate of α;

  (b) α and β are associates if and only if $\alpha \mid \beta$ and $\beta \mid \alpha$;

  (c) if α and β are associates and $\gamma \mid \alpha$, then $\gamma \mid \beta$

  (d) if α and β are associates and $\alpha \mid \delta$, then, $\beta \mid \delta$; and

  (e) α is a prime if and only if every associate of α is a prime.

Proof of (a) If α is an associate of β, then there is a unit, ε, of O_θ such that $\alpha = \varepsilon\beta$. Since ε^{-1} is also a unit of O_θ, by Corollary 5.22.1, and $\alpha\varepsilon^{-1} = \beta$, we see that β is an associate of α.

Proof of (b) If α and β are associates, then we have $\alpha = \varepsilon_1\beta$ and $\beta = \varepsilon_2\alpha$ where ε_1 and ε_2 are units of O_θ. Thus $\beta \mid \alpha$ and $\alpha \mid \beta$. Suppose $\alpha \mid \beta$ and $\beta \mid \alpha$. Then α/β and β/α are integers in O_θ. Since

$$1 = (\alpha/\beta)(\beta/\alpha)$$

we see that α/β, and β/α are units in O_θ. Since $\alpha = (\alpha/\beta)\beta$ and $\beta = (\beta/\alpha)\alpha$ we see that α and β are associates.

Proof of (c) If α and β are associates, then there is a unit ε of O_θ such that $\beta = \varepsilon\alpha$. If $\gamma \mid \alpha$, then there exists a δ in O_θ such that $\alpha = \gamma\delta$. Thus $\beta = \varepsilon\alpha = \varepsilon(\gamma\delta) = \gamma(\varepsilon\delta)$. Since $\varepsilon\delta$ is in O_θ we see that $\gamma \mid \beta$.

Proof of (d) If α and β are associates, then there exists a unit ε of O_θ such that $\alpha = \varepsilon\beta$. If $\alpha \mid \delta$, then there exists a γ in O_θ such that $\delta = \alpha\gamma$. Thus $\delta = (\beta\varepsilon)\gamma = \beta(\varepsilon\gamma)$. Since $\varepsilon\gamma$ is in O_θ we see that $\beta \mid \delta$.

Proof of (e) If α and β are associates, then there is a unit ε of O_θ such that $\alpha = \varepsilon\beta$. If $\beta = \gamma\delta$, where γ and δ are in O_θ, then $\alpha = (\varepsilon\gamma)\delta$. If α is a prime, then either δ is a unit or $\varepsilon\gamma$ is a unit. If $\varepsilon\gamma$ is a unit, then so is γ, since ε is a unit. Thus α a prime implies that γ or δ is a unit, that is, β is a prime. If β is composite, then neither γ nor δ need be units, and so neither $\varepsilon\gamma$ nor δ need be units. Since γ and δ can be chosen to be nonunits, when β is composite, we see that α is not a prime. ∎

Theorem 10.26. If α is in O_θ and $\mid N(\alpha) \mid$ is a rational prime, then α is a prime in O_θ.

Proof. Suppose $\alpha = \beta\delta$, where β and δ are in O_θ. Then, by Theorem 10.21, we have $N(\alpha) = N(\beta)N(\delta)$. Also $N(\beta)$ and $N(\delta)$ are rational integers, by Theorem 10.20. Thus if p is a rational prime such that $p = |N(\alpha)|$, then $p = |N(\beta)N(\delta)|$, as a product of rational integers. Thus either $|N(\beta)| = 1$ or $|N(\delta)| = 1$, that is, either β or δ is a unit, by Theorem 10.22, and hence α is a prime in O_θ.   ■

We close this section of general considerations with the following result.

Theorem 10.27. Every nonzero nonunit in O_θ can be written as a product of primes.

Proof. If α is in O_θ is not a unit or not zero, then we must have $|N(\alpha)| > 1$. If α is a prime, we have the trivial representation $\alpha = \alpha$. If α is not a prime, then we have a factorization

$$\alpha = \beta\gamma,$$

where neither β nor γ is a unit. Also, by Theorem 10.21, we have $N(\alpha) = N(\beta)N(\gamma)$, where

$$1 < |N(\beta)| < |N(\alpha)| \text{ and } 1 < |N(\gamma)| < |N(\alpha)|.$$

If either β or γ is not a prime, then it may be factored. This process must terminate since the rational integer $N(\alpha)$ has only a finite number of divisors of absolute value greater than 1.   ■

Thus we can write every nonzero integer α in O_θ in the form

$$\alpha = \varepsilon\pi_1\cdots\pi_r,$$

where ε is a unit and $\pi_1, \ldots, \pi_r$ are primes. If α is a unit we interpret this to mean $r = 0$ and $\alpha = \varepsilon$. If α is not a unit, then $r > 1$ and we can take $\varepsilon = 1$ if we wish.

Corollary 10.27.1. If α is a nonzero nonunit in O_θ, then it has at most $(\log|N(\alpha)|)/\log 2$ prime factors.

Proof. We write α as a product of primes

$$\alpha = \pi_1\cdots\pi_r.$$

Then

$$|N(\alpha)| = |N(\pi_1)|\cdots|N(\pi_r)|.$$

Since primes are not units we have $|N(\pi_r)| \geq 2$, for $1 \leq i \leq r$. Thus

$$|N(\alpha)| \geq 2^r$$

and taking logarithms gives the result.   ■

This is as far as we can go. It is not true, in general, that in O_θ an arbitrary algebraic integer can be uniquely factored into primes. This fact was rediscovered several times in the 1800's mostly in connection with attempts to prove Fermat's conjecture on the equation $x^n + y^n = z^n$. In the next section we will give an example of nonunique factorization (Example 10.8.) Some sort of unique factorization can be restored by introducing the concept of ideals. (See Problem 39 of the Additional Problems below), but this still does not completely solve the problem.

Problem Set 10.2

1. Use the fundamental theorem of algebra to show that sums and products of algebraic numbers are algebraic numbers.

2. Let α be an algebraic number and m the least positive integer such that $m\alpha$ is an algebraic integer. If b is a rational integer such that $b\alpha$ is an algebraic integer, show that $m \mid b$.

3. Supply the details to the proof of the second part of Theorem 10.19.

4. Prove that there are infinitely many primes in. (Hint: look at $\left| N(\pi_1 \cdots \pi_r) \right| + 1$.)

5. (a) If a is a rational integer in $\mathbf{Q}(\theta)$, a field of degree n, then $N(a) = a^n$.

   (b) If $\mathbf{Q}(\theta)$ is a field of degree n, then any prime of O_θ has at most n prime factors.

6. (a) Let x be a root of $x^3 + 2x + 6 = 0$. Compute $N(3x - 2)$ in $\mathbf{Q}(x)$. (Hint: find the equation satisfied by $3x - 2$.)

   (b) Let x be a root of $x^3 - 2x + 5 = 0$. Compute $N(2x - 1)$ in $\mathbf{Q}(x)$.

   (c) Let x_1, x_2 and x_3 be the roots of $x^3 - x^2 - x - 2 = 0$. Compute $N(3x_i - 2)$ in $\mathbf{Q}(x_i)$ for $i = 1, 2$ and 3. (Hint: is the polynomial irreducible?)

7. (a) Show that the integers of $\mathbf{Q}\left(\sqrt{2} + \sqrt{3}\right)$ are given by

$$\xi = a + b\sqrt{3} + (c\sqrt{2} + d\sqrt{6})/2,$$

   where a, b, c and d are rational integers with c and d of the same parity.

   (b) Find the conjugates of ξ in this ring.

   (c) What are the units in this ring?

8. Let m and l be relatively prime rational integers, with $m, l > 1$. Let ζ be a primitive mth root of unity and ξ a primitive lth root of unity. Show that $1 - \zeta\xi$ is a unit in the ring of integers of $\mathbf{Q}(\zeta, \xi)$. (An nth root of unity is a **primitive** nth **root of unity** if it is not a pth root of unity for any $p < n$.)

9. Prove the following result due to Eisenstein. Let

$$f(x) = a_n x^n + a_{n-1} x^{n-1} + \cdots + a_1 x + a_0$$

be a polynomial in $\mathbf{Z}[x]$. suppose there is a prime p such that
(a) $p \nmid a_n$,
(b) $p \mid a_k, \ 0 \leq k \leq n - 1$
and
(c) $p^2 \nmid a_0$.

Show that, apart from constant factors, $f(x)$ is irreducible over $\mathbf{Z}$, and so irreducible over $\mathbf{Q}$. (Hint: assume the contrary, that is, $f = gh$, where g and h are polynomials in $\mathbf{Z}[x]$. Show that p can divide only one of the constant terms of g and h. Show that if $p \mid g_0$, then there is an a_k, $0 \leq k \leq n - 1$, such that $p \nmid a_k p \nmid a_k$.)

10. In this exercise we discuss the discriminant of a number field. This is useful for, among other things, helping top determine an integral basis for the integers of an algebraic number field. Let $\mathbf{Q}(\theta)$ be an algebraic number field of degree n and let $\alpha_1, \ldots, \alpha_n$ be n elements in $\mathbf{Q}(\theta)$ with field conjugates $\alpha_i^{(k)}$, $1 \leq k \leq n$, $1 \leq i \leq n$. Define

$$\Delta(\alpha_1, \ldots, \alpha_n) = \begin{vmatrix} \alpha_1^{(1)} & \alpha_1^{(2)} & \cdots & \alpha_1^{(n)} \\ & & \vdots & \\ \alpha_n^{(1)} & \alpha_n^{(2)} & \cdots & \alpha_n^{(n)} \end{vmatrix}^2 .$$

This is called the **discriminant** of $\alpha_1, \ldots, \alpha_n$.
(a) Show that if $\alpha_1, \ldots, \alpha_n$ are in O_θ, then $\Delta(\alpha_1, \ldots, \alpha_n)$ is a rational integer.
(b) If $f(x) = x^3 + px + q$ and θ is a root, show that

$$\Delta(1, \theta, \theta^2) = -27q^2 - 4p^3.$$

(c) Show that, in general, if $\theta, \theta^{(2)}, \ldots, \theta^{(3)}$ are roots of

$$f(x) = a_n x^n + a_{n-1} x^{n-1} + \cdots + a_1 x + a_0,$$

which is irreducible over $\mathbf{Q}$, then

$$a_n^n \Delta(1, \theta, \ldots, \theta^{n-1}) = (-1)^{n(n-1)/2} \prod_{i=1}^{n} f'(\theta^{(i)}).$$

Here $f'(x) = n a_n x^{n-1} + \cdots + a_1$. (Hint: recall Vandermonde determinants.)
(d) We say that $\omega_1, \ldots, \omega_n$ form an **integral basis** for O_θ if any element ρ of

O_θ can be written in the form

$$\rho = x_1\omega_1 + \cdots + x_n\omega_n,$$

where x_k, $1 \le k \le n$, are rational integers. Show that if ω_1, ..., ω_n are elements of O_θ for which $|\Delta(\omega_1,...,\omega_n)|$ is of minimal nonzero value, then ω_1, ..., ω_n form an integral basis for O_θ.

(e) Show that if α_1, ..., α_n are elements of O_θ such that $\Delta(\alpha_1,...,\alpha_n)$ is squarefree, then α_1, ..., α_n forms an integral basis for O_θ.

(f) Let $\zeta = \exp(2\pi i/p)$, where p is an odd prime. Show that

$$\Delta(1,\zeta,...,\zeta^{p-2}) = (-1)^{(p-1)/2} p^{p-2}.$$

(g) Show that 1, ζ, ..., ζ^{p-2} form an integral basis for O_ζ.

11. Prove that **Q** is a subfield of any algebraic number field.

12. Let N be a positive integer and let m be a positive integer, $m > 1$, which is not a perfect kth power for any $k \mid N$, $k > 1$. Show that the number $\alpha = m^{1/N}$ is an algebraic integer of degree N. (Hint: we have $\alpha^N - m = 0$. Show that $x^N - m$ cannot be factored over **Z**.)

10.3. QUADRATIC NUMBER FIELDS

In this section we mostly restrict our attention to the more concrete case of the algebraic number field of degree 2: the quadratic number fields. At the end of the last section we showed that every algebraic integer in a given algebraic number field can be written as a product of primes. We also stated that unique factorization does not hold in general. Although we shall give an example of a quadratic field where unique factorization fails, we shall determine a number of quadratic fields where unique factorization does hold and apply this, in the next section, to the study of some Diophantine equations. Many of the properties that hold for the quadratic case also hold in general, and so some of our results will be proved in the general case of O_θ, though the applications will be to quadratic number fields.

We are concerned with algebraic numbers of the second degree, that is, those algebraic numbers that are zeros of polynomials of the type

$$ax^2 + bx + c,$$

where a, b and c are rational integers such that $a \ne 0$. By the quadratic formula we know that the zeros are

$$(-b \pm \sqrt{b^2 - 4ac}) / 2a,$$

which may be written in the form

(10.12) $(A + B\sqrt{m})/C,$

where the A, B, C, and m are rational integers such that $C > 0$, $|m|$ is a squarefree integer and $(A, B, C) = 1$. We saw in Example 10.3 that the numbers of the form (10.12) are indeed algebraic numbers of degree 2 so that any quadratic polynomial which is irreducible over $\mathbf{Q}$ is the minimal polynomial for its zeros. Let $\mathbf{Q}(\sqrt{m})$ denote the algebraic number field generated by the numbers of the form (10. 12). If α is in $\mathbf{Q}(\sqrt{m})$, then there are polynomials $q_1(x)$ and $q_2(x)$ in $\mathbf{Z}[x]$ and a number of the form (10.12) such that

$$\alpha = \frac{q_1\left((A + B\sqrt{m})/C\right)}{q_2\left((A + B\sqrt{m})/C\right)} = \frac{A_1 + B_1\sqrt{m}}{A_2 + B_2\sqrt{m}} = A_3 + B_3\sqrt{m},$$

where A_3 and B_3 are rational numbers. Thus all numbers in the algebraic number field $\mathbf{Q}(\sqrt{m})$ are rational linear combinations of 1 and $\sqrt{m}$.

Definition 10.15. If $m > 0$, then $\mathbf{Q}(\sqrt{m})$ is called a **real quadratic field** and if $m < 0$, then $\mathbf{Q}(\sqrt{m})$ is called a **complex quadratic field**.

Theorem 10.28. If $m_1 \neq m_2$, then $\mathbf{Q}(\sqrt{m_1}) \cap \mathbf{Q}(\sqrt{m_2}) = \mathbf{Q}$.

Proof. Suppose that for some rational numbers a, b, c and d we have

$$a + b\sqrt{m_1} = c + d\sqrt{m_2}.$$

Then subtracting and squaring gives, if $bd \neq 0$,

(10.13) $\dfrac{(d^2 m_2 + b^2 m_1) - (a - c)^2}{2bd} = \sqrt{m_1 m_2}.$

Now the number on the left hand side of (10.13) is a rational number. Since $m_1 \neq m_2$ and both are squarefree we see that $\sqrt{m_1 m_2}$ is irrational. Thus we must have $bd = 0$, which, since both $\sqrt{m_1}$ and $\sqrt{m_2}$ are irrational, implies that $b = d = 0$ and $a = c$, that is, the only elements common to $\mathbf{Q}(\sqrt{m_1})$ and $\mathbf{Q}(\sqrt{m_2})$ are the rational numbers.   ∎

Note that if $\alpha = a + b\sqrt{m}$, where a and b are rational numbers, is in $\mathbf{Q}(\sqrt{m})$, then α satisfies the equation

$$x^2 - 2ax + a^2 - b^2 m = 0,$$

whose other root, the conjugate of α, is $a - b\sqrt{m}$, which we denote by $\overline{\alpha}$ is also in

$\mathbf{Q}\left(\sqrt{m}\right)$. This is not always the case as is illustrated by the field $\mathbf{Q}\left(\sqrt[3]{2}\right)$, whose conjugates are $\sqrt[3]{2}, \omega\sqrt[3]{2}$ and $\overline{\omega}\sqrt[3]{2}$, where $\omega = (1+\sqrt{-3})/2$ is a cube root of unity. This is one of the things that makes the study of quadratic fields a little easier.

Theorem 10.29. Let α and β be in $\mathbf{Q}\left(\sqrt{m}\right)$. Then

   (a) $\overline{(\overline{\alpha})} = \alpha$;
   (b) $\overline{(\alpha+\beta)} = \overline{\alpha} + \overline{\beta}$;
   (c) $\overline{(\alpha\beta)} = \overline{\alpha}\overline{\beta}$;
   (d) $\alpha = \overline{\alpha}$ if and only if α is a rational number; and
   (e) if $\beta \neq 0$, then $\overline{(\alpha/\beta)} = \overline{\alpha}/\overline{\beta}$.

Proof. Most of these results follow immediately from the general considerations above. However, it is useful to demonstrate them from scratch. To this end let $\alpha = a + b\sqrt{m}$ and $\beta = c + d\sqrt{m}$, where a, b, c and d are rational numbers.

Proof of (a) We have

$$\overline{(\overline{\alpha})} = \overline{(a - b\sqrt{m})} = a + b\sqrt{m} = \alpha.$$

Proof of (b) We have

$$\overline{(\alpha+\beta)} = \overline{((a+c) + (b+d)\sqrt{m})} = (a+c) - (b+d)\sqrt{m} = \overline{\alpha} + \overline{\beta}.$$

Proof of (c) We have

$$\overline{(\alpha\beta)} = \overline{((ac+mbd) + (ad+bc)\sqrt{m})} = (ac+mbd) - (ad+bc)\sqrt{m} = \overline{\alpha}\overline{\beta}.$$

Proof of (d) If $\alpha = \overline{\alpha}$, then $a + b\sqrt{m} = a - b\sqrt{m}$ or $b\sqrt{m} = -b\sqrt{m}$. Since $\sqrt{m}$ is irrational (and so nonzero) we see that $b = -b$, that is, $b = 0$. If $b = 0$, then $\alpha = a + 0\sqrt{m} = a - 0\sqrt{m} = \overline{\alpha}$.

Proof of (e) If $\beta \neq 0$, then not both c and d are zero. Thus $\overline{\beta} = c - d\sqrt{m} \neq 0$. Then, since

$$1/(\beta\overline{\beta}) = 1/(c^2 - d^2m),$$

which is a rational number, we have

$$\overline{(\alpha/\beta)} = \overline{(1/(\beta\overline{\beta})\alpha\overline{\beta})} = \overline{(1/(\beta\overline{\beta}))}\overline{\alpha(\overline{\beta})} = 1/(\beta\overline{\beta})\overline{\alpha}\beta = \overline{\alpha}/\overline{\beta},$$

by (c) and (d). ∎

Note that in $\mathbf{Q}\left(\sqrt{m}\right)$ we have, if $\alpha = a + b\sqrt{m}$,

$$N(a+b\sqrt{m}) = \alpha\bar{\alpha} = a^2 - b^2 m.$$

Theorem 10.30. If a is a rational number, then $N(a) = a^2$. If $m < 0$, then $N(\alpha)$ ≥ 0 for all α in $\mathbf{Q}\left(\sqrt{m}\right)$.

Proof. If a is rational, then $a = \bar{a}$, by (d) of Theorem 10.29. Thus

$$N(a) = a\bar{a} = a^2.$$

If $m < 0$, then

$$N(\alpha) = a^2 - b^2 m = a^2 + b^2(-m) \geq 0,$$

since $-m > 0$. ∎

We now determine the form of the algebraic integers in $\mathbf{Q}\left(\sqrt{m}\right)$.

Theorem 10.31. The elements of $O_{\sqrt{m}}$ are

    (a)   the rational integers,

    (b)   if $m \equiv 1 \pmod 4$, then all elements of $\mathbf{Q}\left(\sqrt{m}\right)$ of the form

$$\frac{a+b\sqrt{m}}{2},$$

        where a and b are rational integers of the same parity

and

    (c)   if $m \equiv 2$ or $3 \pmod 4$, then all elements of $\mathbf{Q}\left(\sqrt{m}\right)$ of the form

$$a+b\sqrt{m},$$

        where a and b are rational integers.

Proof. Let $a+b\sqrt{m}$, be in $\mathbf{Q}\left(\sqrt{m}\right)$.

If $b = 0$, then $a+b\sqrt{m}$, is an integer if and only if a is a rational integer. Suppose $b \neq 0$. Then the minimal polynomial for $a+b\sqrt{m}$, is

$$x^2 - 2ax + a^2 - b^2 m = 0.$$

Thus if $a+b\sqrt{m}$, is in $O_{\sqrt{m}}$ we must have that $2a$ and $a^2 - mb^2$ are both rational integers. Thus $(2a)^2 - 4(a^2 - mb^2) = 4mb^2$ must also be a rational integer and, since m is squarefree, we see that $2b$ must be a rational integer.

Suppose $a = k + 1/2$, where k is a rational integer. Then

$$0 \equiv 4a^2 - 4mb^2 \equiv 4k^2 + 4k + 1 - 4mb^2 \equiv 1 - 4mb^2 \pmod 4,$$

and so $2b \equiv 1$ (mod 2) and $m \equiv 1$ (mod 4). Conversely, if a and b were halves of odd integers and $m \equiv 1$ (mod 4), then the minimal polynomial for $a + b\sqrt{m}$, has rational integer coefficients. Thus the algebraic integers, in the case $m \equiv 1$ (mod 4), are either of the form

$$a + b\sqrt{m} = (2a + 2b\sqrt{m}) / 2$$

or

$$(c + d\sqrt{m}) / 2,$$

where c and d are both odd.

If $m \equiv 2$ or 3 (mod 4), then a must be a rational integer, by the argument above. If $b = k + 1/2$, where k is a rational integer, then we would have

$$0 \equiv 4a^2 - 4mb^2 \equiv -(4k^2 + 4k + 1)m \equiv -m \text{ (mod 4)},$$

which cannot hold since m is squarefree. Thus, in this case, the integers are of the form

$$a + b\sqrt{m},$$

where a and b are rational integers. ∎

Corollary 10.31.1. If $m \equiv 1$ (mod 4), then we may also write the elements of $O_{\sqrt{m}}$ in the form

$$a + \tfrac{1}{2}b(\sqrt{m} \pm 1),$$

where a and b are any rational integers.

Proof. We have that if a and b are rational integers, then

$$a + \tfrac{1}{2}b(\sqrt{m} \pm 1) = ((2a \pm b) + b\sqrt{m}) / 2.$$

Since $2a \pm b \equiv b$ (mod 2) we see that $2a \pm b$ and b have the same parity. Thus $a + \tfrac{1}{2}b(\sqrt{m} \pm 1)$ is an element of $O_{\sqrt{m}}$. Conversely, if a and b have the same parity, then

$$\frac{a + b\sqrt{m}}{2} = \frac{a \mp b}{2} + \tfrac{1}{2}b(\sqrt{m} \pm 1),$$

Since $a \equiv \pm b$ (mod 2), we see that $(a \mp b) / 2$ and b are rational integers. ∎

Example 10.6. What form do the integers in $\mathbf{Q}(i)$, $\mathbf{Q}((\sqrt{-3}))$ and $\mathbf{Q}(\sqrt{-5})$ have?

Solution. We have $i = \sqrt{-1}$, so we take $m = -1 \equiv 3 \pmod 4$. Thus

$$O_i = \{a + bi : a, b \in \mathbf{Z}\}.$$

We have $-3 \equiv 1 \pmod 4$. Thus

$$O_{\sqrt{-3}} = \left\{a + \tfrac{1}{2}b\left(\frac{\sqrt{-3}-1}{2}\right) : a, b \in \mathbf{Z}\right\}.$$

We have $-5 \equiv 3 \pmod 4$. Thus

$$O_{\sqrt{-5}} = \{a + b\sqrt{-5} : a, b \in \mathbf{Z}\}. \qquad \square$$

Example 10.7. Show that the converse of Theorem 10.26 is not true.

Solution. Consider O_i. Here $N(3) = 9$. Suppose $3 = (a + bi)(c + di)$. Then

$$9 = (a^2 + b^2)(c^2 + d^2).$$

A simple congruence argument shows that we can' t have $3 = a^2 + b^2 = c^2 + d^2$. Thus we must have $a^2 + b^2 = 1$ or $c^2 + d^2 = 1$, that is $a + bi$ or $c + di$ is, by Theorem 10.22, a unit. Thus 3 is a prime of O_i. $\qquad \square$

Theorem 10.32. (a) If α is in $O_{\sqrt{m}}$, then $\overline{\alpha}$ is in $O_{\sqrt{m}}$

(b) If α and β are in $O_{\sqrt{m}}$ and $\alpha \mid \beta$, then $\overline{\alpha} \mid \overline{\beta}$.

Proof of (a) If α is in $O_{\sqrt{m}}$ then $\alpha = a + b\sqrt{m}$, where a and b are rational integers, and so $\overline{\alpha} = a - b\sqrt{m}$ is also in $O_{\sqrt{m}}$, or $\alpha = (a + b\sqrt{m})/2$ where a and b are rational integers of the same parity, and so $\overline{\alpha} = (a - b\sqrt{m})/2$ is also in $O_{\sqrt{m}}$, since a and $-b$ are also of the same parity.

Proof of (b) If $\alpha \mid \beta$, then there exists a γ in $O_{\sqrt{m}}$ such that Then $\beta = \alpha\gamma$, by (c) of Theorem 10.29. Since $\overline{\alpha}, \overline{\beta}$ and $\overline{\gamma}$ are all in $O_{\sqrt{m}}$ we see that $\overline{\alpha} \mid \overline{\beta}$. ■

Theorem 10.33. Let $\mathbf{Q}(\sqrt{m})$ be a complex quadratic field. Then $O_{\sqrt{m}}$ has only a finite number of units. If $m = -1$, then the units are 1, -1, i and $-i$. If $m = -3$, then the units are 1, -1, $(1+\sqrt{-3})/2$, $(1-\sqrt{-3})/2$, $(-1+\sqrt{-3})/2$ and $(-1-\sqrt{-3})/2$. In all other cases with $m < 0$, then $O_{\sqrt{m}}$ possesses only the units 1 and -1.

Proof. Let $m = -\mu$. Then the norm of an element of $O_{\sqrt{m}}$ is given by

$$\begin{cases} a^2 + \mu b^2 & \text{if } m \equiv 2 \text{ or } 3 \pmod 4 \\ (a - \tfrac{1}{2}b)^2 + \tfrac{1}{4}\mu b^2 & \text{if } m \equiv 1 \pmod 4 \end{cases}.$$

Since, in either case, the norm is a sum of squares it is clear that there can only be a finite number of units.

If $\mu \neq 1$ or 3, then the only solutions of

$$\begin{cases} a^2 + \mu b^2 = 1 & \text{if } \mu \equiv 1 \text{ or } 2 \pmod 4 \\ (a - \tfrac{1}{2}b)^2 + \tfrac{1}{4}\mu b^2 = 1 & \text{if } \mu \equiv 3 \pmod 4 \end{cases}$$

are given by $a = \pm 1$ and $b = 0$, since in the first case $\mu \geq 2$, and so $a^2 + \mu b^2 \geq a^2 + 2b^2$, and in the second case $\mu \geq 7$, and so $(a - \tfrac{1}{2}b)^2 + \tfrac{1}{4}\mu b^2 \geq a^2 - ab + 2b^2$.

If $\mu = 1$, then we must solve

$$a^2 + b^2 = 1,$$

whose only solutions, in rational integers, are given by $a = \pm 1$, $b = 0$ or $a = 0$, $b = \pm 1$.

If $\mu = 3$, then we must solve

$$a^2 - ab + b^2 = 1$$

or

$$(2a - b)^2 + 3b^2 = 4$$

whose only solutions, in rational integers, are given by $a = \pm 1$, $b = 0$, $a = 0$, $b = \pm 1$, $a = 1$, $b = 0$, or $a = -1$, $b = -1$. ∎

Definition 10.16. We say that the ring of integers in an algebraic number field $\mathbf{Q}(\theta)$ is a **unique factorization domain** if, whenever is a nonzero nonunit of O_θ and we have two factorizations of the form

$$\alpha = \varepsilon \pi_1 \cdots \pi_r \quad \text{and} \quad \alpha = \varepsilon' \pi_1' \cdots \pi_s',$$

where ε and ε' are units and $\pi_1, \ldots, \pi_r, \pi_1', \ldots, \pi_s'$ are primes (not necessarily distinct), then $r = s$ and the primes $\pi_1', \ldots, \pi_s'$ can be given new subscripts in such a way that π_j and π_j' are associates, $j = 1, \ldots, r$. We denote this by UFD.

Example 10.7. Show that not all rings $O_{\sqrt{m}}$ are UFD's.

Solution. Consider $O_{\sqrt{-5}}$. The integers have, by Example 10.6, the form $a + b\sqrt{-5}$, where a and b are rational integers. It is not difficult to show that 2, 3, $1 + \sqrt{-5}$ and $1 - \sqrt{-5}$ are primes in $O_{\sqrt{-5}}$. For example, if

$$1 + \sqrt{-5} = (a + b\sqrt{-5})(c + d\sqrt{-5})$$

then

$$6 = (a^2 + 5b^2)(c^2 + 5d^2)$$

Thus, if neither $a + b\sqrt{-5}$ nor $c + d\sqrt{-5}$ is a unit, we must have $a^2 + 5b^2 = 2$ or 3.

A simple congruence argument shows that this can't happen. By Theorem 10.33, we see that 2 is not an associate of either $1+\sqrt{-5}$ or $1-\sqrt{-5}$. Note that

$$6 = 2 \cdot 3 = (1+\sqrt{-5})(1-\sqrt{-5}),$$

which gives two distinct prime factorizations. Thus, by Definition 10.16, we see that $O_{\sqrt{-5}}$ is not a UFD.   □

Thus we do not have unique factorization in general. Recall the steps of many of the proofs of unique factorization in the rational integers:

   I.   Euclidean algorithm

   II. $p \mid ab$ implies $p \mid a$ or $p \mid b$

   III. unique factorization.

If such a proof is examined closely we see that I is really the key step, since it immediately implies II and, as the next theorem shows, II and III are equivalent.

Theorem 10.34. O_θ is a unique factorization domain if and only if O_θ has the following property: if $\pi \mid \alpha\beta$, where π is a prime and α and β are integers in O_θ, then $\pi \mid \alpha$ or $\pi \mid \beta$.

Proof. Suppose O_θ is a UFD and suppose $\pi \mid \alpha\beta$, where π is a prime and α and β are integers in O_θ. Then, by definition, there exists an integer γ in O_θ such that $\alpha\beta = \pi\gamma$. By Theorem 10.27, we know there exist primes $\pi_1, \pi_2, \ldots, \pi_n$ and a unit ε in O_θ such that

$$\gamma = \varepsilon\pi_1 \cdots \pi_n$$

and so

$$\alpha\beta = \varepsilon\pi\pi_1 \cdots \pi_n$$

is a factorization of into primes. Likewise, there are primes $\pi_1', \ldots, \pi_r'$ and $\pi_1'', \ldots, \pi_s''$ and units ε_1 and ε_2 such that

$$\alpha = \varepsilon_1\pi_1' \cdots \pi_r' \quad \text{and} \quad \beta = \varepsilon_2\pi_1'' \cdots \pi_s'',$$

so that

$$\alpha\beta = (\varepsilon_1\varepsilon_2)\pi_1' \cdots \pi_r'\pi_1'' \cdots \pi_s''.$$

The number $\alpha\beta$ is not a unit since it is divisible by the prime π. It may be that α, β or γ is a unit, but not both α and β can be units. If α, β or γ is a unit, then we take $r = 0$, $s = 0$, or $n = 0$. This gives us two factorizations into primes

(10.14)
$$\alpha\beta = \varepsilon\pi\pi_1\cdots\pi_n = (\varepsilon_1\varepsilon_2)\pi_1'\cdots\pi_r'\pi_1''\cdots\pi_s''.$$

Since O_θ is a UFD we know that one of the primes on the right hand side of (10.14) is an associate of π, and so, by (d) of Theorem 10.25, we see that π divides either α or β, depending on whether this associate is a π' or a π''.

Now suppose O_θ has the property that if $\pi \mid \alpha\beta$, where π is a prime and α and β are integers in O_θ, then $\pi \mid \alpha$ or $\pi \mid \beta$. Suppose α is a nonunit nonzero integer in O_θ and

(10.15)
$$\alpha = \varepsilon\pi_1\cdots\pi_r = \varepsilon'\pi_1'\cdots\pi_s',$$

where ε and ε' are units and $\pi_1,\ldots,\pi_r,\pi_1',\ldots,\pi_s'$ are primes in O_θ. Either $r \leq s$ or $s \leq r$, so we may assume $r \leq s$, say. Since $\pi_1 \mid \alpha$ and, by (c) of Theorem 10.25, all associates of α, we see that

$$\pi_1 \mid (\pi_1'\cdots\pi_{s-1}')\pi_s'.$$

Thus, by hypothesis, $\pi_1 \mid \pi_1'\cdots\pi_{s-1}'$ or $\pi_1 \mid \pi_s'$. If $\pi_1 \mid \pi_s'$, π_s' must be an associate of π_1 since π_s' is a prime. If π_s' is not an associate of π_1 then we must have

$$\pi_1 \mid (\pi_1'\cdots\pi_{s-2}')\pi_{s-1}'.$$

If we continue in this way, we see that one of the π' must be an associate of π_1. By renumbering we may assume that π_1 and π_1' are associates and that ε_1 is a unit such that $\pi_1' = \varepsilon_1\pi_1$. Thus we have, by (10.15),

(10.16)
$$\varepsilon\pi_2\cdots\pi_r = (\varepsilon'\varepsilon_1)\pi_2'\cdots\pi_s',$$

where $\varepsilon'\varepsilon_1$ is a unit.

We repeat the above process with π_2 instead of π_1 and we find that one of the π''s is an associate of π_2 say π_2'. This gives, from (10.16),

$$\varepsilon\pi_3\cdots\pi_r = (\varepsilon'\varepsilon_1\varepsilon_2)\pi_3'\cdots\pi_s',$$

where ε_2 is a unit such that $\pi_2' = \varepsilon_2\pi_2$, and so $\varepsilon'\varepsilon_1\varepsilon_2$ is a unit. We repeat this whole process with π_3 and then π_4 and find that each π_j' is an associate of π_j for $1 \leq j \leq r$ -1. This gives

(10.17)
$$\varepsilon\pi_r = (\varepsilon'\varepsilon_1\cdots\varepsilon_{r-1})\pi_r'\cdots\pi_s',$$

where $(\varepsilon'\varepsilon_1\cdots\varepsilon_{r-1})$ is a unit. Now the integer of the left hand side of (10.17) is a prime and therefore so is the number on the right hand side. If $s > r$, then the number on the right hand side of (10.17) is composite. Thus $s = r$, and so $\varepsilon\pi_r = \varepsilon''\pi_r'$ or $\pi_r' = \varepsilon'''\pi_r$, that is, π_r and π_r' are associates. Thus O_θ is a UFD. ∎

We now give a definition that gives a generalization of 1, which we shall show also implies unique factorization.

Definition 10.17. A ring of integers O_θ is said to be **Norm-Euclidean** (or just **Euclidean**) if given any integers α and β and in O_θ, with $\beta \neq 0$, there exist integers γ and ρ in O_θ such that

$$\alpha = \beta\gamma + \rho,$$

with $|N(\rho)| < |N(\beta)|$.

Theorem 10.35. If O_θ is an Euclidean ring and α and β are any integers in O_θ, not both zero, then there exists an integer δ in O_θ such that

   (a) $\delta \mid \alpha$ and $\delta \mid \beta$

and

   (b) if $\gamma \mid \alpha$ and $\gamma \mid \beta$, then $\gamma \mid \delta$.

Any integer σ in O_θ possessing properties (a) and (b) is an associate of δ. If δ has properties (a) and (b), then there exist integers η and ζ in O_θ such that

   (c) $\delta = \alpha\eta + \beta\zeta$.

Proof. Since α and β are not both zero we may assume $\beta \neq 0$, say. By Definition 10.17, there exist integers in O_θ such that

$$\alpha = \beta\gamma_1 + \beta_1,$$

where $|N(\beta_1)| < |N(\beta)|$. If $\beta_1 = 0$ we stop. If $\beta_1 \neq 0$ then there exist integers γ_2 and β_2 such that

$$\alpha = \beta_1\gamma_2 + \beta_2,$$

where $|N(\beta_2)| < |N(\beta_1)|$. If $\beta_2 = 0$ we stop. If $\beta_2 \neq 0$ then we continue this process as long as we do not get some $\beta_n = 0$. In this way we get a sequence of integers $\beta, \beta_1, \beta_2, \ldots$ in O_θ such that

$$|N(\beta)| > |N(\beta_1)| > |N(\beta_2)| > \cdots.$$

Since each $|N(\beta_j)|$ is a positive rational integer (since we are assuming each $\beta_j \neq 0$), this sequence must terminate. As a result, there is a last β_j in the sequence, say β_{n-1}. Thus we must have $\beta_n = 0$. Thus we have a sequence of equations

$$\alpha = \beta\gamma_1 + \beta_1$$
$$\beta = \beta_1\gamma_2 + \beta_2$$

(10.18) $$\vdots$$

$$\beta_{n-3} = \beta_{n-2}\gamma_{n-1} + \beta_{n-1}$$
$$\beta_{n-2} = \beta_{n-1}\gamma_n + 0.$$

From the equations (10.18) we see that $\beta_{n-1} \mid \beta_{n-2}$, and so, by (b) of Theorem 10.24, we have that $\beta_{n-1} \mid \beta_{n-3}$. By moving up the equations and applying (b) of Theorem 10.24 repeatedly we see that in the second equation $\beta_{n-1} \mid \beta_1$ and $\beta_{n-1} \mid \beta_2$ and hence β and, from the first equation $\beta_{n-1} \mid \alpha$. This proves property (a).

If we work the equation from the next to the last one up, solving for each β_j in sequence, we obtain

$$\beta_{n-1} = \beta_{n-3} - \beta_{n-2}\gamma_{n-1}$$
$$= \beta_{n-3} - \gamma_{n-1}(\beta_{n-4} - \gamma_{n-2}\beta_{n-3})$$
$$= -\gamma_{n-1}\beta_{n-4} + (1 + \gamma_{n-1}\gamma_{n-2})\beta_{n-3}$$
$$\vdots$$
$$= \alpha\eta + \beta\zeta,$$

for some integers η and ζ in O_θ. Thus if $\gamma \mid \alpha$ and $\gamma \mid \beta$, then, by (a) of Theorem 10.24, we have that $\gamma \mid \alpha\eta + \beta\zeta$, that is, $\gamma \mid \beta_{n-1}$ which proves (b).

By Theorem 10.25, we see that any associate of β_{n-1} has properties (a) and (b). If δ has these properties, then, by (b), we see that $\delta \mid \beta_{n-1}$ and $\beta_{n-1} \mid \delta$. Hence, by (b) of Theorem 10.25, we see that δ and β_{n-1} are associates.

Let ε be a unit of O_θ. Then

$$\varepsilon\beta_{n-1} = \alpha(\varepsilon\eta) + \beta(\varepsilon\zeta),$$

that is any associate of β_{n-1} can be written in the form required by (c). Thus, any with properties (a) and (b), must have (c).   ■

Definition 10.18. Let α and β be integers of O_θ and let δ be any integer in O_θ that possesses properties (a) and (b) of Theorem 10.35. Then δ is called a **greatest common divisor** and is denoted by (α, β). This notation is unique, by Theorem 10.35, up to associates. If (α, β) is a unit, then and are said to be **relatively prime**.

Theorem 10.36. Any Euclidean ring O_θ is a unique factorization domain.

Proof. By Theorem 5.34 it suffices to show that if $\pi \mid \alpha\beta$, then $\pi \mid \alpha$ or $\pi \mid \beta$, where π is a prime and α and β are integers in O_θ.

Suppose $\pi \nmid \alpha$. Then no associate of π divides α, by (d) of Theorem 10.24. Also any divisor of π must be a unit or an associate of π. Thus (α, π) is a unit. Since any unit divides 1, we see 1 possesses the properties (a) and (b) of Theorem 10.35. Hence, by (c) of Theorem 10.35, there exist integers η and ζ in O_θ such that

$$\alpha\eta + \pi\zeta = 1.$$

Thus

$$(\alpha\beta)\zeta + \pi\beta\eta = \beta.$$

Since $\pi \mid \alpha\beta$, we have, by (a) of Theorem 10.24, that $\pi \mid (\alpha\beta\zeta + \pi\beta\eta)$, that is, $\pi \mid \beta$. Thus, by Theorem 10.34, O_θ is a UFD. ∎

We now proceed to produce some Euclidean rings among the $O_{\sqrt{m}}$. These rings of integers will then be UFD's. However, as we shall mention, there are quadratic number fields that are UFD's, but are not Euclidean rings. The proof of unique factorization in these cases is more difficult.

Theorem 10.37. O_θ is an Euclidean ring if and only if for every δ in $\mathbf{Q}(\theta)$, there exists a κ in O_θ such that

$$|N(\delta - \kappa)| < 1.$$

Proof. Suppose α and β are any integers in O_θ, with $\beta \neq 0$. Then α/β is in $\mathbf{Q}(\theta)$, by Theorem 10.13. Thus there exists an integer κ in O_θ such that

$$|N(\alpha/\beta - \kappa)| < 1$$

or

$$|N(\alpha - \beta\kappa)| < |N(\beta)|,$$

by Theorem 10.21. If we let $\rho = \alpha - \beta\kappa$, then we have

$$\alpha = \beta\kappa + \rho,$$

where $|N(\rho)| < |N(\beta)|$. Thus O_θ is Euclidean.

Suppose O_θ is Euclidean and δ is in $\mathbf{Q}(\theta)$. If δ is in O_θ, then we take $\kappa = \delta$. If δ is not in O_θ, then, by Theorem 10.19, there exists a nonzero rational integer c such that $c\delta$ is an integer in O_θ. Thus there exist integers κ and ρ in O_θ such that

$$c\delta = c\kappa + \rho,$$

where $|N(\rho)| < |N(c)|$. Thus

$$|N(\delta - \kappa)| = |N(\rho/c)| = |N(\rho)| / |N(c)| < 1,$$

which is the result. ∎

First we tackle the complex Euclidean quadratic fields.

Theorem 10.38. The only negative values of m for which $O_{\sqrt{m}}$ is a Euclidean ring are given by

integer and suppose $z = z_1 z_2$. Then $\pi \mid z$ implies, by Theorem 10.34, that $\pi \mid z_1$ or $\pi \mid z_2$. This is a contradiction unless one of z_1 or z_2 is 1, that is, z is a rational prime. Thus π divides at least one rational prime.

Suppose π divides two distinct rational primes p_1 and p_2. Then there exist rational integers x and y such that

$$p_1 x + p_2 y = 1.$$

Then, by (b) of Theorem 10.24, we have $\pi \mid 1$, which is a contradiction, since π is a prime. The result follows.   ∎

From the second paragraph of the above proof the following result is immediate.

Corollary 10.40.1. Let π be a prime in O_θ, a UFD or not. Then divides at most one rational prime.

Theorem 10.41. Let $O_{\sqrt{m}}$ be a UFD.

(a)   Any rational prime p is either a prime π of $O_{\sqrt{m}}$ or the product $\pi_1 \pi_2$ of two not necessarily distinct primes of $O_{\sqrt{m}}$.

(b)   The totality of primes π, π_1, π_2 obtained by applying (a) to all rational primes, together with their associates, constitutes the set of all primes of $O_{\sqrt{m}}$.

(c)   An odd prime p satisfying $(p, m) = 1$ is a product $\pi_1 \pi_2$ of two primes in $O_{\sqrt{m}}$ if and only if $\left(\dfrac{m}{p}\right) = +1$. Moreover, if $p = \pi_1 \pi_2$ the product of two primes, then π_1 and π_2 are not associates, but π_1 and $\overline{\pi_2}$ and π_2 and $\overline{\pi_1}$.

(d)   If $(2, m) = 1$, then 2 is the associate of a square of a prime if $m \equiv 3 \pmod 4$; 2 is prime if $m \equiv 5 \pmod 8$; and 2 is the product of two distinct primes if $m \equiv 1 \pmod 8$.

(e)   Any rational prime p that divides m is the associate of the square of a prime in $O_{\sqrt{m}}$.

Proof of (a) If p is a rational prime, then $p = \pi\beta$, for some prime π and some integer in $O_{\sqrt{m}}$. Then we have $N(\pi)N(\beta) = N(p) = p^2$. Since $N(\pi) \neq \pm 1$, we must have either $N(\beta) = \pm 1$ or $N(\beta) = \pm p$. If $N(\beta) = \pm 1$, then β is a unit, by Theorem 10.22, and so π is an associate of p, which then must be a prime in $O_{\sqrt{m}}$, by (e) of Theorem 10.25. If $N(\beta) = \pm p$, then β is a prime, by Theorem 10.26, and so p is a product $\pi\beta$ of two primes in $O_{\sqrt{m}}$.

Proof of (b) This follows directly from (a) and Theorem 10.40.

Proof of (c) If p is an odd rational prime such that (p, m) and $\left(\dfrac{m}{p}\right) = +1$, then there exists a rational integer x such that

$$x^2 \equiv m \ (\text{mod } p)$$

or

$$p \mid (x^2 - m) = (x - \sqrt{m})(x + \sqrt{m}).$$

Now, if p were a prime in $O_{\sqrt{m}}$, then it would, by Theorem 10.34, divide one of the factors $x - \sqrt{m}$ or $x + \sqrt{m}$, so that one of

$$(x / p) - (\sqrt{m} / p) \text{ and } (x / p) + (\sqrt{m} / p)$$

would be an integer in $O_{\sqrt{m}}$. This is impossible, by Theorem 10.31, and so p is not a prime in $O_{\sqrt{m}}$. Thus, by (a) $p = \pi_1 \pi_2$ if $\left(\dfrac{m}{p}\right) = +1$.

Suppose now that p is an odd rational prime such that $(m, p) = 1$ and p is not a prime in $O_{\sqrt{m}}$. From the proof of (a) we see that $p = \pi\beta$, where $N(\pi) = N(\beta) = \pm p$. We can write $p = a + b\sqrt{m}$, where a and b are rational integers, or, if $m \equiv 1$ (mod 4), possibly halves of odd rational integers. Then

(10.24) $$a^2 - mb^2 = N(\pi) = \pm p$$

or

(10.25) $$(2a)^2 - m(2b)^2 = \pm 4p.$$

Thus

$$(2a)^2 \equiv m(2b)^2 \ (\text{mod } p).$$

Now $2a$ and $2b$ are rational integers and neither is a multiple of p, for if p divided either it would, by (10.24), divide the other, and so we would have, by (10.25), $p \mid 4p$, which is impossible if p is odd. Thus $(2b, p) = 1$, and so there is a rational integer c such that $(2b)c \equiv 1$ (mod p). Thus

$$(2ac)^2 \equiv m(2bc)^2 \equiv m \ (\text{mod } p),$$

that is, $\left(\dfrac{m}{p}\right) = +1$.

Also, with the notation of the preceding paragraph we prove that π and β are not associates, but that π and $\overline{\beta}$ and $\overline{\pi}$ and β are. From $p = \pi\beta$ and $N(\pi) = a^2 - mb^2 = \pm p$ we have

$$\beta = p / \pi = p / (a + b\sqrt{m}) = \pm(a - b\sqrt{m}),$$

and so $\overline{\beta} = \pm(a - b\sqrt{m})$. Thus π and $\overline{\beta}$ are associates. On the other hand, we have

$$\pi / \beta = \pm(a + b\sqrt{m}) / (a - b\sqrt{m}) = \{(2a)^2 + m(2b)^2\} / 4p + 8ab\sqrt{m} / 4p,$$

which is not an integer of $O_{\sqrt{m}}$, and so not a unit, since $p \nmid 8ab$. Thus π and β are not associates.

Proof of (d) If $m \equiv 3 \pmod 4$, then

$$(m + \sqrt{m})(m - \sqrt{m}) = m^2 - m = 2\frac{m^2 - m}{2}$$

and $2 \nmid (m \pm \sqrt{m})$. Thus 2 is not a prime of $O_{\sqrt{m}}$. Thus 2 is divisible by a prime $x + y\sqrt{m}$ and this prime must have norm ± 2. Now

$$\pm\frac{x - y\sqrt{m}}{x + y\sqrt{m}} = \frac{x^2 + my^2}{2} - xy\sqrt{m}$$

and similarly

$$\pm\frac{x + y\sqrt{m}}{x - y\sqrt{m}} = \frac{x^2 + my^2}{2} + xy\sqrt{m}.$$

Thus $(x - y\sqrt{m})(x + y\sqrt{m})^{-1}$ and its inverse are both integers of $O_{\sqrt{m}}$, and hence must be units. Thus $x - y\sqrt{m}$ and $x + y\sqrt{m}$ are associates.

If $m \equiv 1 \pmod 4$ and 2 is not a prime in $O_{\sqrt{m}}$, then 2 is divisible by a prime $\frac{1}{2}(x + y\sqrt{m})$ having norm ± 2. This would mean that there are rational integers x and y, of the same parity, such that

(10.26) $\qquad\qquad\qquad x^2 - my^2 = \pm 8.$

If x and y are even, say $x = 2x_0$ and $y = 2y_0$, then (10.26) would require that

$$x_0^2 - my_0^2 = \pm 2,$$

which is impossible since $m \equiv 1 \pmod 4$ implies that $x_0^2 - my_0^2$ is either odd or a multiple of 4. Thus (10.26) can have solutions only if x and y are odd. Then $x^2 \equiv y^2 \equiv 1 \pmod 8$ and (10.26) implies that

$$x^2 - my^2 \equiv 1 - m \equiv 0 \pmod 8, \text{ if } m \equiv 1 \pmod 8.$$

It follows that 2 is a prime in $O_{\sqrt{m}}$ if $m \equiv 5 \mod 8$).

If $m \equiv 1 \pmod 8$, note that

$$\tfrac{1}{2}(1-\sqrt{m})\tfrac{1}{2}(1+\sqrt{m}) = \tfrac{1}{4}(1-m) = 2\frac{1-m}{8}$$

and, since $2\!\!\!/\,(1\pm\sqrt{m})/2$, we see that 2 is not a prime in $O_{\sqrt{m}}$. Thus (10.26) has solutions in odd rational integers x and y. Now the primes $\tfrac{1}{2}(x+y\sqrt{m})$ and $\tfrac{1}{2}(x-y\sqrt{m})$ are not associates in $O_{\sqrt{m}}$ because their quotient is not a unit. In fact, their quotient is

$$\frac{x+y\sqrt{m}}{x-y\sqrt{m}} = \pm\frac{x^2+my^2}{8} \pm \frac{xy}{4}\sqrt{m},$$

which is not even an integer in $O_{\sqrt{m}}$.

Proof of (e) Let p be a rational prime divisor of m. If $p = |m|$, then $p = \pm\sqrt{m}\sqrt{m}$, and so p is the associate of the square of a prime of $O_{\sqrt{m}}$, by Theorem 10.26. If $p < |m|$, note that,

(10.27) $\sqrt{m}\sqrt{m} = m = p(m/p).$

Now p is not a divisor of $\sqrt{m}$ in $O_{\sqrt{m}}$, by Theorem 10.31, and so p is not a prime of $O_{\sqrt{m}}$. Thus p is divisible by a prime π, with $N(\pi) = \pm p$, and so is not a divisor of m/p. By (10.27), π is also a divisor of $\sqrt{m}$. Thus π^2 divides m, and so π^2 is a divisor of p. The result follows from (a).   ∎

The following corollary is a restatement of parts of the above theorem.

Corollary 10.41.1. Let $O_{\sqrt{m}}$ be a UFD.

(a) If p is a rational prime such that $(p, 2m) = 1$ and $\left(\dfrac{m}{p}\right) = -1$, then p is a prime of $O_{\sqrt{m}}$.

(b) If p is a rational prime such that $(p, 2m) = 1$ and $\left(\dfrac{m}{p}\right) = +1$, $p = \pi_1\pi_2$ a product of two distinct primes of $O_{\sqrt{m}}$.

This result is really a special case of a more general result due to Kummer. Note that what really determines whether or not p remains prime in $O_{\sqrt{m}}$ is whether or not

(10.28) $x^2 - m \equiv 0 \pmod p$

is solvable or not, providing $(p, 2m) = 1$. (This latter condition only removes a few special cases where p is the square of a prime, usually.) If (10.28) is solvable, then $\left(\dfrac{m}{p}\right) = +1$, and so $p = \pi_1 \pi_2$ as a product of distinct primes. If (10.28) is not solvable, then $\left(\dfrac{m}{p}\right) = -1$, and so p remains a prime. The polynomial $x^2 - m$ is the minimal polynomial for $\sqrt{m}$ and this is the key relationship. We give here the result of Kummer that is analogous to Theorem 10.41. The result really holds in a more general setting, but the beauty and simplicity of Kummer's theorem can still be seen here. We will not be able to prove this theorem here.

Theorem 10.42. Let O_θ be a UFD with $F(x)$ the minimal polynomial for θ. Let p be a rational prime and suppose that

$$F(x) \equiv f_1(x)^{a_1} \cdots f_r(x)^{a_r} \pmod{p},$$

where $f_1, ..., f_r$ are polynomials that are irreducible modulo p. Then the p is decomposable into a product of r distinct primes of O_θ. Moreover, these primes can be so chosen so that if $\pi_1, ..., \pi_r$ are the appropriate choice, then

(10.29) $$p = \pi_1^{a_1} \cdots \pi_r^{a_r}.$$

The question of how to arrange the primes so that (10.29) holds requires some work, but it is not too difficult to achieve. For a proof of this theorem (in its more general setting) see [66, sec. 8 of ch. 1] or [129, ch. 3].

We now give some examples of our theorem.

Example 10.9. Determine the primes of O_i, $O_{\sqrt{-3}}$ and $O_{\sqrt{2}}$.

Solution. O_i. Here $m = -1$ and we have $2m = -2$, $1^2 + 1^2 = 2$ and $\overline{1+i} = 1-i$. Here

$$\left(\frac{m}{p}\right) = \begin{cases} +1 & \text{if } p = 4k+1 \\ -1 & \text{if } p = 4k+3 \end{cases}.$$

For each rational prime of the form $4k + 1$ the equation $p = x^2 + y^2$ has a solution since $x^2 + y^2 = -p$ is clearly impossible. For each such p choose a solution $x = a_p$ and $y = b_p$. The primes of O_i are then $1 + i$, the rational primes of the form $4k + 3$ and all $a_p + ib_p$ and $a_p - ib_p$, corresponding to the rational primes of the form $4k + 1$, and the associates of all these primes. Note that $1 - i = \overline{1+i}$ has not been included since $1 - i = (-i)(1 + i)$ and $-i$ is a unit of O_i, that is $1 - i$ and $1 + i$ are associates.

$O_{\sqrt{-3}}$. Here $m = -3$ and we have $2m = -6$, $3^2 + 3 \cdot 1^2 = 4 \cdot 3$ and

$\overline{(3+\sqrt{-3})/2} = (3-\sqrt{-3})/2$. Here

$$\left(\frac{m}{p}\right) = \begin{cases} +1 & \text{if } p = 3k+1, \ (p,6) = 1 \\ -1 & \text{if } p = 3k+2, \ (p,6) = 1 \end{cases}.$$

Note that $x^2 + 3y^2 = \pm 4 \cdot 2$ has no solution. For each rational prime p of the form $3k + 1$ choose a_p and b_p such that $a_p^2 + 3b_p^2 = 4p$ (which can be done by Theorem 3.6.) Then the primes of $O_{\sqrt{-3}}$ are 2, $(3+\sqrt{-3})/2$, all odd rational primes of the form $3k + 2$, and all $(a_p + b_p\sqrt{-3})/2$ and $(a_p - b_p\sqrt{-3})/2$, corresponding to the rational primes of the form $3k + 1$, and the associates of all these primes. Again we omit $(3-\sqrt{-3})/2$ since it is easily seen to be an associate of $(3+\sqrt{-3})/2$.

$O_{\sqrt{2}}$. Here $m = 2$ and we have $2m = 4$, $3^2 - 2 \cdot 2^2 = 2$. Here

$$\left(\frac{m}{p}\right) = \begin{cases} +1 & \text{if } p = 8k \pm 1 \\ -1 & \text{if } p = 8k \pm 3 \end{cases}.$$

For each rational prime of the form $8k \pm 1$ choose a_p and b_p such that $a_p^2 - 2b_p^2 = p$. (Since $x^2 - 2y^2 = -1$ is solvable, this will suffice.) Then the primes of $O_{\sqrt{2}}$ are $\sqrt{2}$ and its associates, all rational primes of the form $8k \pm 3$ and their associates, and all $a_p + b_p\sqrt{2}$ and $a_p - b_p\sqrt{2}$, corresponding to the rational primes $8k \pm 1$, and their associates. In this case there are an infinite number of units since both $x^2 - 2y^2 = \pm 1$ are solvable.   □

Example 10.10. Show that the ring $O_{\sqrt{-14}}$ cannot be a UFD.

Solution. By (e) of Theorem 10.41, if $O_{\sqrt{-14}}$ were a UFD, then 2 would factor into two primes. Thus we shall show that 2 is a prime. If 2 is not a prime of $O_{\sqrt{-14}}$, then we have

$$2 = \pm(a + b\sqrt{-14})(a - b\sqrt{-14})$$

for some rational integers a and b. This gives

$$2 = \pm(a^2 + 14b^2),$$

which is clearly impossible in rational integers a and b. Thus $O_{\sqrt{-14}}$ is not a UFD. □

This concludes our general survey of quadratic number fields. What we have covered is a small amount of what can be done with quadratic number fields, but we hope that it indicates what more can be done from the general case if more information is added. For more on quadratic number fields see, for example, [1, ch. 7-11] or [19]. In the latter text there is more of a mixture of the general and quadratic case, as was done here. Both texts give applications to Diophantine equations. For a mixture of algebraic and analytic number theory see [40, ch. 4 and 12]. In Chapter 4 of this text they prove a prime number theorem for the primes of O_i with norm less than x.

We now turn to some of the applications of the theory of quadratic number fields to the study of Diophantine equations.

Problem Set 10.3

1. Let γ and γ_1, with $\gamma_1 \neq 0$, be two Gaussian integers, that is, elements of O_i. Show that there exist Gaussian integers κ and γ_2 such that

$$\gamma = \kappa \gamma_1 + \gamma_2,$$

   where $N(\gamma_2) \leq \frac{1}{2} N(\gamma_1)$.

2. Find the greatest common divisors of the following pairs of numbers in the given ring.

   (a) $15 + 12i$, $3 - 9i$ (in O_i)   (b) $3 + 8i$, $12 + i$ (in O_i)

   (c) $1 + \sqrt{-2}$, $5 + 3\sqrt{-2}$ (in $O_{\sqrt{-2}}$)   (e) $\dfrac{1 - \sqrt{-3}}{2}$, $\dfrac{5 + 7\sqrt{-3}}{2}$ (in $O_{\sqrt{-3}}$)

   (d) $3 - \sqrt{2}$, $17 + 5\sqrt{2}$ (in $O_{\sqrt{2}}$)   (f) $17 + 34\sqrt{2}$, $1 - \sqrt{2}$ (in $O_{\sqrt{2}}$).

3. If π is a prime in O_i, define the function φ by $\varphi(\pi) = N(\pi) - 1$.

   (a) Show that if π is a prime dividing the rational prime p, which is of the form $4k + 1$, then $\varphi(\pi) = p - 1$.

   (b) Show that if $\pi = q$, a rational prime of the form $4k + 3$, then $\varphi(\pi) = q^2 - 1$.

   (c) Show that if $N(\pi)$ is odd and, then

   $$\alpha^{\varphi(\pi)} \equiv 1 \ (\text{mod } \pi).$$

   (This is the analog of Fermat's Little Theorem in O_i.)

4. Show that $O_{\sqrt{10}}$ is not a UFD.

5. (a) Show that the primes of $O_{\sqrt{3}}$ are $\sqrt{3} - 1$, $\sqrt{3}$, all rational primes of the form $12k \pm 5$ and all $a + b\sqrt{3}$ of rational primes of the $12k \pm 1$ and the associates of all these numbers.

   (b) Explain why the equation $2 \cdot 11 = (5 + \sqrt{3})(5 - \sqrt{3})$ does not contradict the fact that $O_{\sqrt{3}}$ is a UFD.

6. Factor each of the following numbers in the indicated ring.

(a) $7 + 70i$, $1 - 7i$, $17 + i$, 86961 (in O_i);

(b) $7 + 70\sqrt{3}$, $1 + \sqrt{3}$, 86961 (in $O_{\sqrt{3}}$);

(c) $6 - 7\sqrt{2}$, $5 + \sqrt{2}$, 86961 (in $O_{\sqrt{2}}$);

(d) $9 + 3\sqrt{-3}$, $\dfrac{5 - \sqrt{-3}}{2}$, 86961 (in $O_{\sqrt{-3}}$);

(Hint: sometimes it helps to factor the norms.)

7. (a) Show that the units in $O_{\sqrt{5}}$ are $\pm\tau^n$, where $\tau = (1 + \sqrt{5})/2$ and n runs through all rational integers.

(b) Show that the primes of $O_{\sqrt{3}}$ are $\sqrt{5}$, the rational primes of the form $5k \pm 2$ and the factors $a + b\tau$ of the rational primes of the rational primes of the form $5k \pm 1$ and the associates of these numbers.

(c) Let p and q be rational primes of the form $5k \pm 1$ and $5k \pm 2$, respectively. If π is a prime in $O_{\sqrt{5}}$ let $\varphi(\pi) = \left| N(\pi) \right| - 1$. Then

$$\varphi(\pi) = \begin{cases} p - 1 & \text{if } \pi | p \\ q^2 - 1 & \text{if } \pi = q \end{cases}.$$

(d) If $(\alpha, \pi) = 1$, then

$$\alpha^{\varphi(\pi)} \equiv 1 \ (\mathrm{mod}\ \pi)$$
$$\alpha^{p-1} \equiv 1 \ (\mathrm{mod}\ \pi), \ \text{if } \pi | p$$

and

$$\alpha^{q+1} \equiv N(\alpha) \ (\mathrm{mod}\ q).$$

Furthermore, if $\pi | p$ and $(\alpha, \pi) = (\alpha, \overline{\pi}) = 1$, then

$$\alpha^{p-1} \equiv 1 \ (\mathrm{mod}\ p).$$

8. Let α and β be integers of $O_{\sqrt{m}}$. Show that if $\alpha | \beta$, then $N(\alpha) | N(\beta)$.

9. If m is squarefree, $m < -1$ and $|m|$ is not a prime, show that $O_{\sqrt{m}}$ is not a UFD. (Hint: use (e) of Theorem 10.41.)

10. If α is an integer of $O_{\sqrt{m}}$, show that $\alpha^2 + (\overline{\alpha})^2$ is a rational integer.

11. Let α and β be two integers in $O_{\sqrt{m}}$. If g is a rational integer such that $g | \alpha\overline{\alpha}$, $g | \beta\overline{\beta}$ and $g | (\alpha\overline{\beta} + \beta\overline{\alpha})$, show that $g | \alpha\overline{\beta}$ and $g | \overline{\alpha}\beta$. (Hint: consider the equations satisfied by $\alpha\overline{\beta}$ and $\beta\overline{\alpha}$.)

12. Show that if ε is a unit of $O_{\sqrt{m}}$ such that $\sqrt{\varepsilon}$ is an integer of $O_{\sqrt{m}}$, then $\sqrt{\varepsilon}$ is a unit of $O_{\sqrt{m}}$.

13. Use $\alpha = 1 + \sqrt{m}$, if $m \leq -5$, $m \not\equiv 1$ (mod 4) or $\alpha = (1 + \sqrt{m})/2$, $m \leq -15$, $m \equiv 1$ (mod 4) and $\beta = 2$ to show that $O_{\sqrt{m}}$ is not Euclidean.

14. (a) If $5 \mid m$, show that 2 is a prime of $O_{\sqrt{m}}$.

    (b) If $m \not\equiv 1$ (mod 4) and $5 \mid m$, show that $O_{\sqrt{m}}$ is not a UFD.

15. Show that the integers of norm $n > 0$ in a complex quadratic field satisfy an equation of the type $\alpha^2 - A\alpha + n = 0$, where $A \leq 2\sqrt{n}$.

16. Consider the ring $R = \{a + b\sqrt{-3} : a, b \in \mathbb{Z}\}$.

    (a) Show that 1 and -1 are the only units of R.

    (b) Show that 2, $1 + \sqrt{-3}$ and $1 - \sqrt{-3}$ are nonassociates and prime in R.

    (c) Use the fact that to show that R is not a UFD.

(Notice the contrast with $O_{\sqrt{-3}}$.)

17. The following result is due to E. Gethner and provides a "gap" result for Gaussian primes somewhat like Theorem 1.17. Let k be a positive rational integer and let $p_1, \ldots, p_{k^2}$ be the first k^2 rational primes. For $j = 1, \ldots, k$, let $m_j = p_{k+j-1} \cdots p_{jk}$ and $M_j = p_j p_{k+j} \cdots p_{k^2-k+j}$. Let a be a solution of the system of congruences

$$x \equiv -1 \ (\text{mod } m_1), \ldots, x \equiv -k \ (\text{mod } m_k)$$

and let b be a solution to the system of congruences

$$y \equiv -1 \ (\text{mod } M_1), \ldots, y \equiv -k \ (\text{mod } M_k).$$

Show that all the Gaussian integers $(a + r) + (b + s)i$, $r, s = 1, \ldots, k - 1$, are composite. Can you extend this result to other quadratic number fields?

10.4. APPLICATIONS TO DIOPHANTINE EQUATIONS

We begin with a result that is really a corollary of Theorem 10.41.

Theorem 10.43. Let $O_{\sqrt{m}}$ be a UFD and let p be a rational prime such that $(p, 2m) = 1$ and $\left(\dfrac{m}{p}\right) = +1$.

    (a) If $m \not\equiv 1$ (mod 4), then at least one of the equations

$$x^2 - my^2 = \pm p$$

has a solution.

    (b) If $m \equiv 1$ (mod 4), then at least one of the equations

$$x^2 - my^2 = \pm 4p$$

has a solution with x and y of the same parity.

This is a consequence of the fact that if p is as in the statement of the theorem, then p is a product of two distinct primes of $O_{\sqrt{m}}$.

A result of more consequence is the following one.

Theorem 10.44. Let O_θ be a UFD. If α, β and γ are integers and ε is a unit in O_θ such that $(\alpha, \beta) = 1$ and

$$\alpha\beta = \varepsilon\gamma^n,$$

where n is a positive rational integer, then there are units ε' and ε'' and integers δ and ζ in O_θ such that

$$\alpha = \varepsilon'\delta^n \text{ and } \beta = \varepsilon''\zeta^n.$$

Proof. If γ is a unit, then $\alpha\beta$ is a unit, and so both α and β are units. In this case the result is trivial: put $\varepsilon' = \alpha$, $\varepsilon'' = \beta$ and $\delta = \zeta = 1$. If $\gamma = 0$, then one of α or β is 0. Since anything divides 0, the only way we can have $(\alpha, \beta) = 1$ is that the other integer is a unit. The result is trivial in this case also: put one of α or β to 0 and the other 1.

Suppose γ is neither 0 nor a unit. Then, by Theorem 10.27, we may write

$$(10.30) \qquad\qquad \gamma = \pi_1 \cdots \pi_r,$$

where the π_i are primes, some of which may be associates. It is sufficient to show that α is a unit times an nth power as the proof for β is identical.

If α is a unit, then set $\varepsilon' = \alpha$ and $\delta = 1$. If α is not a unit, then it is also not 0 since γ is not 0. Thus we may write

$$(10.31) \qquad\qquad \alpha = \sigma_1 \cdots \sigma_s,$$

where the σ_j are primes. Thus, by (10.30) and (10.31), we have

$$(10.32) \qquad\qquad \sigma_1 \cdots \sigma_s \beta = \varepsilon\pi_1^n \cdots \pi_r^n.$$

Unique factorization says that σ_1 is an associate of one of the π_j and we may renumber so that it is π_1. Now, if any associate of π_1 divides β, then $\pi_1 \mid \beta$, and so $\sigma_1 \mid \beta$, by Theorem 10.25. This can't happen since $(\alpha, \beta) = 1$. Thus π_1 or its associates, must show up n times among the primes $\sigma_1, \ldots, \sigma_s$, which can be renumbered so that $\sigma_1, \ldots, \sigma_n$ are the associates of π_1. This means that $s \geq n$. Hence there are units $\varepsilon_1, \ldots, \varepsilon_n$ such that

$$\sigma_1 = \varepsilon_1\pi_1, \ \sigma_2 = \varepsilon_2\pi_1, \ \ldots, \ \sigma_n = \varepsilon_n\pi_1,$$

and so

$$\sigma_1 \cdots \sigma_n = (\varepsilon_1 \cdots \varepsilon_n) \, \pi_1^n.$$

If $s = n$, then we are finished, since $\sigma_1 \cdots \sigma_s = \alpha$. If $s > n$, then we divide both sides of (10.32) by π_1^n and get

(10.33) $(\varepsilon_1 \cdots \varepsilon_n) \sigma_{n+1} \cdots \sigma_s \beta = \varepsilon \pi_2^n \cdots \pi_r^n.$

We repeat the process. By unique factorization one of the π_j's is an associate of σ_{n+1} and they can be renumbered so that it is π_2. No associate of π_2 can divide β, since $(\alpha, \beta) = 1$. Thus π_2 or its associates, show up n times among the primes σ_{n+1}, $\ldots$, σ_s and these can be renumbered so that they are $\sigma_{n+1}, \ldots, \sigma_{2n}$. It follows from this that $s \geq 2n$ and it also follows that there are units $\varepsilon_{n+1}, \ldots, \varepsilon_{2n}$ such that

$$\sigma_{n+1} = \varepsilon_{n+1}\pi_2, \ \sigma_2 = \varepsilon_2\pi_2, \ \ldots, \ \sigma_{2n} = \varepsilon_{2n}\pi_2,$$

and so

$$\sigma_{n+1} \cdots \sigma_{2n} = (\varepsilon_{n+1} \cdots \varepsilon_{2n}) \, \pi_2^n.$$

If $s = 2n$, then $\alpha = (\varepsilon_1 \cdots \varepsilon_{2n})(\pi_1 \pi_2)^n$ and we're done. If $s > 2n$, then divide both sides of (10.33) by π_2^n and repeat the process a third time.

Since there are a finite number of primes in the factorization of α, by Corollary 10.27.1, the repetition of this process must eventually come to an end. When we have finally gone through this process for the last time, say on the $\underline{k}$th repetition, we will have $s = kn$. Also we will have renumbered the σ's and π's and found units $\varepsilon_1, \ldots, \varepsilon_{kn}$ such that

$$\sigma = \sigma_1 \cdots \sigma_{kn} = (\varepsilon_1 \cdots \varepsilon_{kn}) \, (\pi_1 \cdots \pi_k)^n.$$

If we take

$$\varepsilon' = \varepsilon_1 \cdots \varepsilon_{kn} \ \text{and} \ \delta = \pi_1 \cdots \pi_k,$$

we will have α written in the required form.   ■

A direct application of this theorem is to the following result.

Theorem 10.45. Let x, y and z be rational integers so that

(10.34) $x^2 + y^2 = z^l,$

where $l > 1$, $(x, y) = 1$ and $z > 0$. Then the solutions are given by

$$x + iy = i^r(a + ib)^l, \ x - iy = (-i)^r(a - ib)^l \ \text{and} \ z = a^2 + b^2,$$

where a and b are relatively prime rational integers of opposite parity and $r = 0, 1, 2$ or 3.

Proof. Note that $(x, y) = 1$ implies that z must be odd. For if z even, then $z^l \equiv 0$ (mod 4) implies $x^2 + y^2 \equiv 0$ (mod 4), which is impossible.

We rewrite (10.34) as

$$(x + iy)(x - iy) = z^l$$

and let $d = (x + iy, x - iy)$. Then $d \mid (2x, 2y) = 2(x, y) = 2$. Thus $d = 1, 1 + i$ or 2. Since $2 = -i(1 + i)^2$, to show that $d = 1$, that is a unit, it suffices to show that $1 + i$ cannot divide $x + yi$ and $x - yi$. If $1 + i$ divides $x + yi$ and $x - iy$, then, by (10.34), we have $1 + i \mid z^l$. Now

$$z^l / (1 + i) = (z^l/2) - (z^l/2)i,$$

which is not in O_i, since z^l is odd. Thus $x + iy$ and $x - iy$ are relatively prime.

Since O_i is a UFD, we have, by Theorem 10.44, that

$$x + iy = \varepsilon \alpha^l$$

where ε is a unit and α is an integer of O_i. Now the units of O_i are of the form i^r, $r = 0, 1, 2$, or 3, by Theorem 10.33. If we let $\alpha = a + ib$ and recall that, by Theorem 10.29, $\overline{\alpha^l} = (\overline{\alpha})^l$, the result follows. ∎

The case $l = 2$ is classical and is given by the following corollary, which is also Theorem 5.1.

Corollary 10.45.1. The solution, in rational integers, of

$$x^2 + y^2 = z^2$$

with $(x, y) = 1$, $z > 0$, are given by

$$x = \pm(p^2 - q^2), \; y = \pm 2pq \text{ and } z = p^2 + q^2,$$

where p and q are relatively prime rational integers of opposite parity.

Corollary 10.45.2. The solution in rational integers of the equation

(10.35) $x^2 + y^2 = 2z^2,$

where $(x, y, z) = 1$ and $z > 0$, are given by

$$x = p^2 - q^2 - 2pq, \; y = p^2 - q^2 + 2pq \text{ and } z = p^2 + q^2,$$

where p and q are relatively prime rational integers of opposite parity.

Proof. Since $(x, y, z) = 1$ we see that we must have $(x, y) = 1$. Since $x^2 + y^2$ is even we see that both x and y must be odd. Thus, as in the proof of Theorem 10.45, we see that if $d = (x + iy, x - iy)$, then $d = 1$ or $1 + i$. We have

$$(10.36) \qquad \frac{x + iy}{1 + i} = \frac{x + y}{2} - \frac{x - y}{2} i,$$

which is an integer of O_i if and only if x and y have the same parity. Since $(1 + i)^2 \,|\, 2z^2$ we see that $x + iy$ is divisible by $1 + i$. Let $x + iy = (1 + i)(u + vi)$, where u and v are rational integers of opposite parity, by (10.36). This gives

$$x = u - v \text{ and } y = u + v.$$

If we substitute these into (10.35) we obtain the equation

$$u^2 + v^2 = z^2.$$

The result then follows from Corollary 10.45.1.   ∎

We can also tackle the problem of the sum of two squares, which was done in Corollary 6.8.2.

Theorem 10.46. Let n be a fixed positive rational integer. Then the equation

$$(10.37) \qquad x^2 + y^2 = n$$

has a solution in rational integers x and y if and only if n can be written in the form $n = m^2 k$, where m and k are positive rational integers and k has no positive rational prime divisors of the form $4l + 3$.

Proof. Suppose $n = m^2 k$, where m and k are as in the statement of the theorem. If $k = 1$, then $n = m^2 + 0^2$. If $k > 1$ we can write

$$k = p_1 \cdots p_r,$$

where $p_i \equiv 1 \pmod 4$ or $p_i = 2$, $1 \le i \le r$. Then there are primes $\pi_1, \ldots, \pi_r$ of O_i such that, for $1 \le i \le r$,

$$N(\pi_i) = p_i.$$

Let

$$a + bi = m \pi_1 \cdots \pi_r.$$

Then

$$a^2 + b^2 = N(a + bi) = N(m)N(\pi_1) \cdots N(\pi_r) = m^2 p_1 \, p_r = m^2 k = n,$$

and so (10.37) has solutions.

Suppose (10.37) is solvable and that $a^2 + b^2 = n$, where a and b are rational integers. Then

$$N(a + bi) = n.$$

If $a + bi$ is a unit, then $n = 1$, which can be put in the form $1^2 \cdot 1$ as is required by the theorem. If $a + bi$ is not a unit, then we can factor it into a product of primes, say

$$a + bi = \pi_1 \cdots \pi_r.$$

By Theorem 10.41, we can, by renumbering if necessary, assume that $\pi_1, \ldots, \pi_s$ are associates of rational primes $p_1, \ldots, p_s$, all congruent to 3 modulo 4, while $\pi_{s+1}, \ldots,$ π_r have norms $p_{s+1}, \ldots, p_r$ which are rational primes either equal to 2 or congruent to 1 modulo 4. Let $m = p_1 \cdots p_s$ and $k = p_{s+1} \cdots p_r$. Then

$$n = N(a + bi) = N(\pi_1) \cdots N(\pi_s) N(\pi_{s+1}) \cdots N(\pi_r)$$
$$= p_1^2 \cdots p_s^2 p_{s+1} \cdots p_r = m^2 k.$$

Since the prime divisors of k are not congruent to 3 modulo 4 the result follows. ■

The above results deal with Diophantine equations that always have solutions, perhaps under an extra condition. The next Diophantine equation we shall consider does not always have solutions. Also the above results were obtained basically by use of factorization theory. The following result is obtained from factorization theory and some elementary number theory.

We consider the so-called Mordell equation

$$x^3 = y^2 + k,$$

where k is a nonzero rational integer. We have tackled this problem from various points of view previously. (See the Problem Sets in Chapters 3 and 6.) L. J. Mordell has shown that this equation has only a finite number of solutions in rational integers x and y for any value of k. We shall consider the case $k = 2$.

Theorem 10.47. The equation

(10.38) $y^2 + 2 = x^3$

has as its only rational integer solutions $x = 3$ and $y = \pm 5$.

Proof. We see that x and y must be odd, since if y is even, then x is even and the equation is impossible modulo 4.

By Corollary 10.38.1, $O_{\sqrt{-2}}$ is a UFD. We write (10.38) as

$$(y + \sqrt{-2})(y - \sqrt{-2}) = x^3.$$

Since x is odd it is not divisible by the prime $\sqrt{-2}$, and so $\sqrt{-2}$ is not a divisor of y + $\sqrt{-2}$ or y - $\sqrt{-2}$. Let $\alpha = (y + \sqrt{-2}, y - \sqrt{-2})$. Then $\alpha \mid 2y$ and $\alpha \mid 2\sqrt{-2} = -\left(\sqrt{-2}\right)^3$. If α is not a unit, let π be a prime of $O_{\sqrt{-2}}$ that divides α. Then $\pi = \sqrt{-2}$ or $-\sqrt{-2}$ (the associate of $\sqrt{-2}$ in $O_{\sqrt{-2}}$). But $\sqrt{-2}$ does not divide $y + \sqrt{-2}$ or $y -$ $\sqrt{-2}$. Thus α must be a unit, that is, $y + \sqrt{-2}$ and $y - \sqrt{-2}$ are relatively prime. Thus, by Theorem 10.44, we have

$$y + \sqrt{-2} = \varepsilon_1\beta^3 \text{ and } y - \sqrt{-2} = \varepsilon_2(\bar{\beta})^3,$$

where ε_1 and ε_2 are units and β is an integer of $O_{\sqrt{-2}}$. By Theorem 10.33, the only units of $O_{\sqrt{-2}}$ are ± 1. If we let $\alpha = a + b\sqrt{-2}$, where a and b are rational integers, and equate coefficients of $\sqrt{-2}$ we obtain

$$1 = b(3a^2 - 2b^2).$$

Thus $b = 1$ and $a = \pm 1$. This gives $x = 3$ and $y = \pm 5$. ∎

So far we have been factorizing equations over $\mathbf{Z}$ into factors in appropriate quadratic number fields. In our last example we work almost exclusively in the quadratic number field. We shall deal with the case $n = 3$ of the Great Fermat Conjecture, which states that if $n \geq 3$, then there are no rational integer solutions of

(10.39) $x^n + y^n = z^n$

that satisfy $xyz \neq 0$. We will be doing our work in the UFD $O_{\sqrt{-3}}$.

Let $\omega = (-1 + \sqrt{-3})/2$ and $\lambda = 1 - \omega$, which is a prime of $O_{\sqrt{-3}}$, by Theorem 10.41. If α, β and γ are integers of $O_{\sqrt{-3}}$, we write $\alpha \equiv \beta \pmod{\gamma}$ if there exists an integer δ in $O_{\sqrt{-3}}$ such that $\alpha - \beta = \gamma\delta$.

We shall do more than show that (10.39) does not hold when $n = 3$ and x, y and z are rational integers such that $xyz \neq 0$. We shall show that

(10.40) $\xi^3 + \eta^3 + \zeta^3 = 0$

has no solutions in integers ξ, η and ζ in $O_{\sqrt{-3}}$ such that

(10.41) $(\xi, \eta) = (\eta, \zeta) = (\xi, \zeta) = 1$.

This clearly includes (10.39), when $n = 3$. To do this we shall have to prove some results about the arithmetic of $O_{\sqrt{-3}}$.

Theorem 10.48. All integers of $O_{\sqrt{-3}}$ fall into three classes modulo λ typified by 0, 1 or -1.

Proof. If γ is an integer of $O_{\sqrt{-3}}$, then, by Corollary 10.31.1, we have, for some rational integers a and b

$$\gamma = a + b\omega = a + b - b\lambda \equiv a + b \pmod{\lambda}.$$

Since $3 = (1 - \omega)(1 - \omega^2)$ we have $\lambda \mid 3$. Since $a + b$ has one of the three residues 0, 1 or -1 modulo 3, γ has one of the same residues modulo λ. These residues are incongruent since neither $N(1) = 1$ nor $N(2) = 4$ are divisible by $N(\lambda) = 3$. ∎

Theorem 10.49. (a) 3 is an associate of λ^2.

(b)   The numbers $\pm(1 - \omega)$, $\pm(1 - \omega^2)$, $\pm\omega(1- \omega)$ are the associates of λ.

Proof of (a). We have

$$\lambda^2 = 1 - 2\omega + \omega^2 = -3\omega$$

since $1 + \omega + \omega^2 = 0$, and the result follows since $-\omega$ is a unit, by Theorem 10.33.

Proof of (b). We have

$$\pm(1 - \omega) = \pm\lambda, \ \pm(1 - \omega^2) = \mp\lambda\omega^2 \text{ and } \pm\omega(1 - \omega) = \pm\lambda\omega$$

and the result follows, by Theorem 10.33. ∎

Theorem 10.50. If ρ is not divisible by λ, then $\rho^3 \equiv \pm1 \pmod{\lambda^4}$.

Proof. By Theorem 10.48, we know that is congruent to 1, -1 or 0 modulo λ. Since $\lambda \nmid \rho$ we know

$$\rho \equiv \pm1 \pmod{\lambda}.$$

Choose $\alpha = \pm\rho$ such that

$$\alpha \equiv 1 \pmod{\lambda}$$

and define β by

$$\alpha = 1 + \beta\lambda.$$

Then

$$\pm(\rho^3 \mp 1) = \alpha^3 - 1 = (\alpha - 1)(\alpha - \omega)(\alpha - \omega^2)$$
$$= \beta\lambda(\beta\lambda + 1 - \omega)(\beta\lambda + 1 - \omega^2)$$
$$= \lambda^3\beta(\beta + 1)(\beta - \omega^2),$$

by Theorem 10.49. Also

$$\omega^2 \equiv 1 \pmod{\lambda},$$

and so

$$\beta(\beta + 1)(\beta - \omega^2) \equiv \beta(\beta + 1)(\beta - 1) \pmod{\lambda}.$$

Since one of β, $\beta + 1$ or $\beta - 1$ must be divisible by λ, by Theorem 10.48, the result follows. ∎

Theorem 10.51. If (10.40) holds, then one of ξ, η or ζ is divisible by λ.

Proof. Suppose $\lambda \nmid \xi$, $\lambda \nmid \eta$ and $\lambda \nmid \zeta$. Then

$$0 = \xi^3 + \eta^3 + \zeta^3 \equiv \pm 1 \pm 1 \pm 1 \pmod{\lambda^4},$$

and so $\pm 1 \equiv 0$ or $\pm 3 \equiv 0 \pmod{\lambda^4}$. Thus $\lambda^4 \mid 1$ or $\lambda^4 \mid 3$. The first hypothesis is untenable since λ is a prime, and so not a unit. The second cannot hold since 3 is an associate of λ^2, by (a) of Theorem 10.49, and so not divisible by λ^4. Thus one of ξ, η or ζ must be divisible by λ. ∎

Suppose (10.40) and (10.41) hold. By Theorem 10.51 we know that λ divides one of ξ, η or ζ, say $\lambda \mid \zeta$. We write

$$\zeta = \lambda^n \gamma,$$

where n is a positive rational integer and $\lambda \nmid \gamma$. Then, by (10.41), we have $\lambda \nmid \eta$ and $\lambda \nmid \xi$. Thus (10.40) is equivalent to

$$\xi^3 + \eta^3 + \lambda^{3n} \gamma^3 = 0,$$

where

(10.42) $(\xi, \eta) = 1$, $n \geq 1$, $\lambda \nmid \xi$, $\lambda \nmid \eta$ and $\lambda \nmid \gamma$.

Theorem 10.52. Suppose ξ, η and γ satisfy (10.42) and

(10.43) $\xi^3 + \eta^3 + \varepsilon \lambda^{3n} \gamma^3 = 0$,

where ε is a unit. Then $n \geq 2$.

Proof. By Theorem 10.50, we have

$$-\varepsilon \lambda^{3n} \gamma^3 = \xi^3 + \eta^3 \equiv \pm 1 \pm 1 \pmod{\lambda^4}.$$

If the signs are the same, then

$$-\varepsilon \lambda^{3n} \gamma^3 \equiv \pm 2 \pmod{\lambda^4},$$

which cannot hold since $\lambda \nmid 2$ (both λ and 2 are primes in $O_{\sqrt{-3}}$, by Theorem 10.41). Thus the signs are opposite and we have

$$-\varepsilon \lambda^{3n} \gamma^3 \equiv 0 \pmod{\lambda^4}.$$

Since $\lambda \nmid \gamma$, by (10.42), and $\lambda \nmid (-\varepsilon)$, since λ is a prime, we see that $3n \geq 4$, that is, $n \geq 2$. ∎

Theorem 10.53. If (10.43) holds for $n = m > 1$, then it holds for $n = m - 1$.

Proof. From (10.43) we have

$$(10.44) \qquad -\varepsilon\lambda^{3m}\gamma^3 = (\xi + \eta)(\xi + \omega\eta)(\xi + \omega^2\eta).$$

The differences of the factors on the right hand side of (10.44) are $\eta\lambda$, $\omega\eta\lambda$, and $\omega^2\eta\lambda$, which are all associates of $\eta\lambda$. Each of them is divisible by λ, but not λ^2, since $\lambda \nmid \eta$ by (10.42).

Since $m \geq 2$, we have $3m > 3$ and one of the factors must be divisible by λ^2. The other two factors must be divisible by λ (since the differences are divisible by λ), but not by λ^2 (since the differences are not so divisible). We may suppose that the factor divisible by λ^2 is $\xi + \eta$, for if not we could replace η by one of its associates. Thus

$$(10.45) \qquad \xi + \eta = \lambda^{3m-2}\kappa_1, \ \xi + \omega\eta = \lambda\kappa_2 \text{ and } \xi + \omega^2\eta = \lambda\kappa_3,$$

where $\lambda \nmid \kappa_i$, $i = 1, 2, 3$.

If $\delta \mid \kappa_2$ and $\delta \mid \kappa_3$, then δ also divides

$$\kappa_2 - \kappa_3 = \omega\eta$$

and

$$\omega\kappa_3 - \omega^2\kappa_2 = \omega\xi$$

Thus, since ω is a unit, δ divides both η and ξ. Thus, by (10.42), we see that δ is a unit and $(\kappa_2, \kappa_3) = 1$. Similarly, $(\kappa_3, \kappa_1) = (\kappa_2, \kappa_1) = 1$.

If we substitute (10.45) into (10.44) we obtain

$$-\varepsilon\gamma^3 = \kappa_1\kappa_2\kappa_3.$$

Since $O_{\sqrt{-3}}$ is a UFD, by Corollary 10.38.1, we see that, by Theorem 10.44, each of the κ's is the associate of a cube. Thus we have

$$\xi + \eta = \lambda^{3m-2}\kappa_1 = \varepsilon_1\lambda^{3m-2}\theta^3$$
$$\xi + \omega\eta = \varepsilon_2\lambda\phi^3$$

and

$$\xi + \omega^2\eta = \varepsilon_3\lambda\psi^3,$$

where θ, ϕ and ψ have no common factor and are not divisible by λ and ε_1, ε_2 and ε_3 are units. Thus

$$0 = (1 + \omega + \omega^2)(\xi + \eta) = \xi + \eta + \omega(\xi + \omega\eta) + \omega^2(\xi + \omega^2\eta)$$
$$= \varepsilon_1\lambda^{3m-2}\theta^3 + \varepsilon_2\omega\lambda\phi^3 + \varepsilon_3\omega^2\lambda\psi^3,$$

so that

$$(10.46) \qquad \phi^3 + \varepsilon_4 \psi^3 + \varepsilon_5 \lambda^{3(m-1)} \theta^3 = 0,$$

where $\varepsilon_4 = \varepsilon_3 \omega / \varepsilon_2$ and $\varepsilon_5 = \varepsilon_1 / \varepsilon_2 \omega$ are units, by Corollary 10.22.1. Since $m \geq 2$, we have

$$\phi^3 + \varepsilon_4 \psi^3 \equiv 0 \pmod{\lambda^2}$$

(in fact, modulo λ^3). But $\lambda \nmid \phi$ and $\lambda \nmid \psi$, and so, by Theorem 10.48, we have

$$\phi^3 \equiv \pm 1 \pmod{\lambda^2} \text{ and } \psi^3 \equiv \pm 1 \pmod{\lambda^2}$$

(in fact, modulo λ^4). Thus

$$\pm 1 \pm \varepsilon_4 \equiv 0 \pmod{\lambda^2},$$

where ε_4 is one of ± 1, $\pm \omega$ or $\pm \omega^2$, by Theorem 10.33. Now none of $\pm 1 \pm \omega$ and $+1 \pm \omega^2$ is divisible by λ^2, since each is an associate of 1 or λ, by Theorem 10.47, and so we have $\varepsilon_4 = \pm 1$. If $\varepsilon_4 = 1$, then (10.46) is an equation of required type. If $\varepsilon_4 = -1$, then replacing ψ by $-\psi$ gives an equation of required type. In either case the result is proven. ■

Corollary 10.53.1. There do not exist nontrivial solutions of (10.40).

Proof. This follows immediately from Theorem 10.53. For, if (10.43) holds with any n, then it holds with $n = 1$, which contradicts Theorem 10.52 and the result follows from Theorem 10.51. ■

Corollary 10.53.2. There do not exist nontrivial solutions of (10.39), when $n = 3$

Proof. This follows from Corollary 10.53.1, since $\mathbf{Z} \subset O_{\sqrt{-3}}$. ■

To show that (10.39) does not have any nontrivial solution for $n > 4$ requires one to investigate the algebraic number fields $\mathbf{Q}(\zeta_p)$, where $\zeta_p = \exp(2\pi i / p)$ is a primitive pth root of unity. The minimal equation here is

$$x^{p-1} + \cdots + x + 1 = 0.$$

To do this in general requires the concept of ideals since O_{ζ_p} is usually not a UFD. For more details, see, for example, [27]. However, the proof by Wiles goes beyond such easy concepts as algebraic number fields.

For further applications of algebraic number theory to the study of Diophantine equations see various chapters of [9] or [77].

Problem Set 10.4

1. Solve the Diophantine equation $x^3 + y^3 = z^2$. (Hint: factor over $O_{\sqrt{-3}}$.)

2. Solve the Diophantine equation $x^2 + 2y^2 = z^2$. (Hint: see Theorem 5.4 for the solution.)

3. (a) Let n be a rational integer. Give a formula for the number of representations of $x^2 + y^2 = n$, in rational integers x and y. (Hint: write

$$n = 2^\alpha \prod_{p \equiv 1 \ (\text{mod } 4)} p^{\alpha_p} \prod_{q \equiv 3 \ (\text{mod } 4)} q^{\beta_q}$$

and then factor this and $x^2 + y^2$ over O_i. Equating the results should give Theorem or Corollary 6.8.1 depending on how you do it.)

   (b) Give criteria for the solvability of $x^2 + dy^2 = n$, for $d = \pm 2$ and ± 3. Give the number of solutions in each case.

4. Show that $x^2 - y^3 = (2c)^3 - 1$ has no rational integer solutions.

5. Find all solutions of $y^2 + k = x^3$, for $k = 1, 4$ and 11.

6. Solve $x^2 + 3 = y^5$.

7. Show that $\xi^4 + \eta^4 = \zeta^4$, $\xi\eta\zeta \neq 0$, has no solutions in O_i. (Hint: mimic the proof in the rational integer case (Theorem 5.2).)

8. Show that $x^3 + y^3 + 3z^3 = 0$, $xyz \neq 0$, has no solutions in rational integers.

9. Show that $x^2 + 7 = 2^n$ has only the solutions $(x, n) = (\pm 1, 3), (\pm 3, 4), (\pm 5, 5), (\pm 11, 7)$ and $(\pm 181, 15)$. This equation is known as the Ramanujan-Nagell equation. (Hint: factor over $O_{\sqrt{-7}}$.)

10. Let $x = 1 + \sqrt{-3}$, $y = 1 - \sqrt{-3}$ and $z = 2$. Show that if $p > 3$ is a prime, then

$$x^p + y^p = z^p.$$

11. (a) Solve $x^2 + 3y^2 = z^3$.

    (b) Solve $x^2 + 3y^2 = z^l$, where $l \geq 2$ is a rational integer.
    (Hint: factor over $O_{\sqrt{-3}}$.)

12. Find all solutions of $x^2 + x + 2 = y^3$. (Hint: factor over $O_{\sqrt{-7}}$ after completing the square.)

13. Solve the following variations on Mordell's equation.

    (a) $x^3 + 1 = -y^2$ (work in O_i.)
    (b) $x^3 + 1 = -2y^2$ (work in $O_{\sqrt{-2}}$.)
    (c) $x^3 + 1 = y^2$ (work in $O_{\sqrt{-3}}$.)

14. Solve the equation $x^6 + 1 = 2y^2$. (Hint: work in $O_{\sqrt{-3}}$ and use the fact that
    $3 \nmid (y^2 + 1)$.)

15. Solve $x^2 + 2y^2 = z^3$. (Hint: work in $O_{\sqrt{-2}}$.)

16. Let b be a fixed rational integer. Consider the equation

    $$x^2 + b^2 = y^5$$

    with $(x, y) = 1$.

    (a) Show that if $|b| < 38$, then the only solution of this equation is when b
        $= \pm 1$ and $x = 0$, $y = 1$.

    (b) Show that if $38 \le |b| < 122$, then the only solution of this equation is
        $41^2 + 38^2 = 5^5$.

    (c) Show that there is an increasing sequence of b's such that the equation is
        solvable.

17. Show that if $y^2 + k = x^3$ has (x_0, y_0) as one rational solution, then others may
    arise as follows. Let $x_1 = x_0 - u$ and $y_1 = y_0 - v$, where u and v are to be
    chosen later. Then

    $$y_1^2 + k = x_1^3 \text{ if and only if } -2y_0v + v^2 = -3x_0^2u + 3x_0u^2 - u^3.$$

    Now choose u and v so that $2y_0v = 3x_0^2u$ and find x by solving the resulting
    cubic. Does this process produce an infinite number of rational solutions? Why
    or why not?

18. Suppose $O_{\sqrt{m}}$ is a UFD. Determine the form of Pythagorean triples in this ring
    of algebraic integers.

10.5. CONCLUDING REMARKS

The material presented in the preceding sections of this chapter just barely
scratches the surface of the study of algebraic number theory. As mentioned at the
end of the last section just to do many more cases of the Great Fermat Conjecture
requires the new concept of ideals. Ideals are a construct that allows one to return
unique factorization to algebraic number fields by embedding them in a larger
structure. Ideal theory allows one to prove many results that are beyond proof in the
algebraic number field alone. In Problem 39 of the Additional Problems below we
give a brief induction to the theory of ideals.

For an approach to algebraic number theory that is nearest to its founders,
Kummer, Dedekind and Kronecker, see [46]. An intermediate approach between that

in Hancock and the more modern treatments can be found in [91] or [129]. Three different modern approaches can be found in [59], [66] or [127].

Additional Problems for Chapter Ten

General Problems

1. This problem shows a way of dealing with algebraic number theory without using the Fundamental Theorem of Algebra (Theorem 10.14.) We say that two polynomials over $\mathbf{Q}$, $f_1(x)$ and $f_2(x)$, are **congruent** modulo a third polynomial $G(x)$, whose degree is at least one, if $G(x) \mid (f_1(x) - f_2(x))$. We write this as $f_1(x) \equiv f_2(x) \pmod{G(x)}$.

  (a) Show that the usual properties of numerical congruences hold for these polynomial congruences. (See Theorem 2.1.)

  The division algorithm then maps division by $G(x)$ onto a unique polynomial $r(x)$, modulo $G(x)$ by

$$f(x) = G(x)q(x) + r(x) \text{ and } f(x) \equiv r(x) \pmod{G(x)}.$$

  (b) Let $G(x)$ be a polynomial over $\mathbf{Q}$ with $n = \deg G \geq 1$. Show that the totality of polynomials

$$r(x) = a_0 + a_1 x + \cdots + a_{n-1} x^{n-1},$$

with coefficients in $\mathbf{Q}$, and with addition and multiplication modulo $G(x)$, forms a ring.

  (c) Show that the ring of polynomials modulo $G(x)$ of (b) is a field if and only if $G(x)$ is irreducible over $\mathbf{Q}$. If $G(x)$ is the minimal polynomial of the algebraic number α, then this field is isomorphic to $\mathbf{Q}(\alpha)$.

  This is basically how Kronecker approached the subject.

  (d) Prove that the field of Gaussian numbers $\mathbf{Q}(i)$ is isomorphic to the field of all polynomials $a + bx$, with a and b in $\mathbf{Q}$, taken modulo $x^2 + 1$.

2. (a) Let c be any positive real number and let $\mathbf{Q}(\theta)$ be an algebraic number field. Show that there are only finitely many algebraic integers x in O_θ such that $|x_j| \leq c$ for all conjugates x_j of x.

  (b) Show that x is a root of unity in $\mathbf{Q}(\theta)$ if and only if x is an algebraic integer in O_θ such that $|x_j| = 1$ for all conjugates x_j of x.

  (c) Show that the set of roots of unity in O_θ is a finite multiplicative cyclic group.

3. Let α be an algebraic integer with conjugates $\alpha_1, \ldots, \alpha_n$. Define the **height** of α, denoted by $\lceil \alpha \rceil$, to be

Let $m > 0$. If $\varepsilon = a + b\sqrt{m}$ is a unit of $O_{\sqrt{m}}$ (where a and b may be halves of odd integers.) and $\varepsilon > 1$, then show that $a, b > 0$. (Hint: look at $|\varepsilon\bar{\varepsilon}|$.)

Use (a) to show that there is a smallest unit greater than 1 in $O_{\sqrt{m}}$. This unit is called the **fundamental unit** of $O_{\sqrt{m}}$ and is denoted by ε_0. (Hint: compare with Theorem 6.34.)

Show that every unit of $O_{\sqrt{m}}$ is of the form $\pm\varepsilon_0^n$, where n is a rational integer.

Of all the units ε in $O_{\sqrt{m}}$, $m > 0$, show that the fundamental unit minimizes $|\varepsilon + \bar{\varepsilon}|$.

Consider the equation

$$x^2 - my^2 = \pm 4,$$

with $x, y > 0$ and $m \not\equiv 1 \pmod 4$. Show that the fundamental unit of $O_{\sqrt{m}}$ is $(x + y\sqrt{m})/2$, where x is the least positive value so that $(x + y\sqrt{m})/2$ is a solution. (Hint: see Problem 39 of the Additional Problems for Chapter Six.)

If $\eta = (a + b\sqrt{m})/2$ is a unit, with a and b rational integers, show that there exist rational integers A and B such that

$$\eta^3 = A + B\sqrt{m}$$

(Hint: show that $\eta^3 = \eta(a^2 \pm 1) \mp a$.)

Let $\alpha = x + y(1 + \sqrt{m})/2$, where x and y are in $\mathbf{Q}(\sqrt{m})$. Show that

$$N(\alpha) = x^2 + xy + [(1 - m)/4]y^2.$$

Suppose $m < 0$ and $m \equiv 1 \pmod 4$. Prove that if $\alpha \in O_{\sqrt{m}}$ and α is not rational, then $N(\alpha) \geq (1 - m)/4$.

If $m < 0$, $m \equiv 1 \pmod 4$ and $O_{\sqrt{m}}$ is a UFD, show that $-m$ is a positive rational prime. (Hint: let $m = -ab$, where $a, b > 1$ and show that $\sqrt{m} \mid ab$, but $\sqrt{m} \nmid a$ and $\sqrt{m} \nmid b$. Use (a) to show that $\sqrt{m}$ is a prime in $O_{\sqrt{m}}$. Note that 3, 7 and 11 are primes and if $m \leq -15$, then $((1 - m)/4)^2 > -m$.)

Let $m < 0$, $m \equiv 1 \pmod 4$ and $O_{\sqrt{m}}$ be a UFD. Let $\alpha = x + y(1 + \sqrt{m})/2$ be an integer in $O_{\sqrt{m}}$, with $(x, y) = 1$ and $y \neq 0$. Show that if $N(\alpha) < ((1 - m)/4)^2$, then $N(\alpha)$ is a rational prime. (Hint: use (a).) The special case $m \leq -15$, $x = -1$ and $y = 2$ is (b). The special case $y = 1$ says

$$\lceil \alpha \rceil = \max\{|\alpha|, |\alpha_1|, \ldots, |\alpha_n|\}.$$

7. (a)

(a) If α is a nonzero algebraic integer which is n
$\lceil \alpha \rceil > 1$.

(b

(b) An algebraic integer is said to be **totally real**
If the algebraic integer is totally real, nonzero ar
where r is rational, then $\lceil \alpha \rceil > 2$.

(c

4. (a) Show that a Gaussian integer $a + bi$ is a sum o
integers if and only if b is even and if $a \equiv 2$ (m

(c

Can you generalize this to other quadratic fields?

(b) Show that a Gaussian integer $a + bi$ is the squai
and only if there are rational integers c, x and y,

$$a^2 + b^2 = c^2, \quad c + a = 2x^2 \text{ and } c - a =$$

In this case

$$a + bi = (\pm x \pm yi)^2,$$

with the signs the same if $b > 0$ and opposite if b

5. (a) Suppose there are nonzero integers α, β and γ in (
ε_3 such that

$$e_1 a^3 + e_2 b^3 + e_3 g^3 = 0.$$

Prove that $\{\varepsilon_1, \varepsilon_2, \varepsilon_3\} = \{1, \omega, \omega^2\}$. (Hint: $\varepsilon_1 \alpha$
may assume ε_1, ε_2 and ε_3 are some of 1, ω or ω^2.)

8.

(b) Prove that there are nonzero integers and units as in

$$e_1 a^3 + e_2 b^3 + e_3 g^3 = 0.$$

6. An integral domain E is said to be **Euclidean** if there
integer valued function g defined on the nonzero eleme
every x and y in E we have

    (a) $g(xy) \geq g(x)$

    (b) if $x \nmid y$, then there exists an element q in E, depei
that

$$g(y - qx) < g(x).$$

Such a function g is called a **Euclidean function**.
imaginary quadratic field. Show that if $O_{\sqrt{m}}$ is Euclidea
function must be the norm function, that is, it must be norm

that $x^2 + x + (1 - m)/4$ is a prime for $0 \le x \le (1 - m)/4 - 1$, a result due to Rabinowitz.

9. (a) Let m and n be squarefree rational integers other than 0 and 1. Suppose there is an $\alpha \in O_{\sqrt{n}}$ such that $N(\alpha) = m$. Show that if $O_{\sqrt{m}}$ is a UFD, then there exists a $\beta \in O_{\sqrt{m}}$ such that $N(\beta) = n$. Show that if $O_{\sqrt{n}}$ and $O_{\sqrt{m}}$ are both UFDs, then either both of the equations

$$N(\alpha) = m, \ \alpha \in O_{\sqrt{n}} \ \text{and} \ N(\beta) = n, \ \beta \in O_{\sqrt{m}}$$

in the unknowns α and β have solutions or neither of them does.

(b) Note that in $\mathbf{Q}(\sqrt{5})$ $N(18 + 7\sqrt{5}) = 79$. Show that there is no β in $O_{\sqrt{79}}$ such that $N(\beta) = 5$. Hence show that $O_{\sqrt{79}}$ is not a UFD.

10. (a) Suppose $m > 0$ and ε_0 is the fundamental unit of $O_{\sqrt{m}}$. If m has no prime divisors of the form $4k + 3$ and $O_{\sqrt{m}}$ is a UFD, show that $N(\varepsilon_0) = -1$.

(b) Let ε_0 be the fundamental unit of $O_{\sqrt{34}}$. Show that $N(\varepsilon_0) = 1$ and hence show that $O_{\sqrt{34}}$ is not a UFD.

11. Prove that the number of Euclidean rings $O_{\sqrt{m}}$, where $m \equiv 2$ or $3 \pmod 4$, is finite.

12. Let D be a rational integer and let

$$d = \begin{cases} D & \text{if } D \equiv 1 \pmod 4 \\ 4D & \text{if } D \not\equiv 1 \pmod 4 \end{cases}.$$

Show that if $d < 0$, then no number α of norm g exists in $\mathbf{Q}(\sqrt{d})$ if $g < |d|/4$, unless g is a perfect square and $\alpha = \pm\sqrt{g}$ is a rational integer.

13. Let ε_0 be the fundamental unit of $O_{\sqrt{m}}$, $m > 0$.

(a) Show that if $m \equiv 2$ or $3 \pmod 4$, then

$$\varepsilon_0 \ge 1 + \sqrt{m}.$$

(b) Show that if $m \equiv 1 \pmod 4$, then

$$\varepsilon_0 \ge (3 + \sqrt{m})/2.$$

14. (a) Show that if $\varepsilon = x + y\sqrt{m}$ is a unit of $O_{\sqrt{m}}$, $m > 0$, and $x > (y^2/2) - 1$, then is the fundamental unit of $O_{\sqrt{m}}$.

(b) Show that the fundamental unit of $O_{\sqrt{m^2-1}}$, $m \ge 2$, is

$$\varepsilon = m + \sqrt{m^2 - 1}.$$

(c) Let $m = a(ay^2 + 2)$, where a and y are positive rational integers. Prove that $1 + ay^2 + y\sqrt{m}$ is the fundamental unit of $O_{\sqrt{m}}$.

(d) Suppose $m > 0$ and that d is a positive rational integer. Show that $O_{\sqrt{m}}$ has infinitely units of the form $a + b\sqrt{m}$, where $d \mid b$.

(Hint: see Section 6.7.)

15. (a) Show that the integers of $Q(\sqrt{2} + i)$ are given by
$$\xi = a + bi + (c\sqrt{2} + di\sqrt{2})/2,$$
where a, b, c and d are rational integers such that $c \equiv d \pmod 2$.

(b) Show that the conjugates of ξ are found by changing the signs of i, $\sqrt{2}$ or both.

(c) With $N(\xi) = \xi\xi_1\xi_2\xi_3$, ξ_i a conjugate of ξ, show that $O_{\sqrt{2}+i}$ is Euclidean.

16. Consider the algebraic number field $Q(\sqrt{2} + \sqrt{3})$ (see Problem 7 of Problem Set 10.2.)

(a) If ξ_i, $i = 1$, 2 and 3, are the conjugates of ξ, let $N(\xi) = \xi\xi_1\xi_2\xi_3$. Show that $O_{\sqrt{2}+\sqrt{3}}$ is Euclidean.

(b) What are the primes of $O_{\sqrt{2}+\sqrt{3}}$?

17. Let m and n be squarefree rational integers. Determine the integers of $Q(\sqrt{m}, \sqrt{n})$.

18. Suppose m and n are squarefree and that the integers of $Q(\sqrt{m}, \sqrt{n})$ form a UFD. Determine the decomposition of the rational primes. (Hint: it will be useful to consider the squarefree rational integer $k = mn/(m, n)^2$.)

19. (a) Let m be a cubefree rational integer. Determine the integers of $Q(\sqrt[3]{m})$.

(Hint: if m is cubefree, then it can be written in the form $m = ab^2$, where ab is squarefree. Break up your considerations into the cases of $9 \mid (a^2 - b^2)$ or $9 \nmid (a^2 - b^2)$.)

(b) Suppose m is cubefree and $O_{\sqrt[3]{m}}$ is a UFD. Determine the decomposition of the rational primes. (Hint: consider the cases $p = 3$, $p \mid m$, $p \nmid m$ and $p \equiv -1 \pmod 3$ and $p \nmid m$ and $p \equiv 1 \pmod 3$.)

20. This problem gives an analog of Fermat's little theorem in the case where $O_{\sqrt{m}}$ is a UFD. We let p and q denote rational primes and let π denote a prime of $O_{\sqrt{m}}$. Let μ be an integer of $O_{\sqrt{m}}$.

(a) If $p = \pi\bar{\pi}$, show that
$$\mu^{p-1} \equiv 1 \pmod{\pi},$$
when $(\mu, \pi) = 1$.

(b) If, in addition to the requirements of (a), we have $(\mu, \overline{\pi}) = 1$, show that

$$\mu^{p-1} \equiv 1 \ (\text{mod } p).$$

(c) If $q = \pi$, show that

$$\mu^q \equiv \overline{\mu} \ (\text{mod } q),$$

when $(\mu, q) = 1$.

(d) If $q = \pi$, show that

$$\mu^{q+1} \equiv N(\mu) \ (\text{mod } q).$$

21. If $\alpha \in O_{\sqrt{m}}$, a UFD, calculate $(\alpha, \overline{\alpha})$. Determine when $(\alpha, \overline{\alpha}) = 1$.

22. (a) Let k be negative and squarefree, $k \equiv 2, 3 \ (\text{mod } 4)$ and $k \neq 1$. Show that if $O_{\sqrt{k}}$ is a UFD, then $y^2 = x^3 + k$ can be solved if and only if there exists a rational integer a such that

$$k = \pm 1 - 3a^2,$$

in which case the solution is $(a^2 - k, \pm a(a^2 + 3k))$.

(b) What happens if $k > 0$?

23. Consider the equation

$$x + y + z = 1 = xyz.$$

(a) Solve this equation in O_i.

(b) Solve this equation in $O_{\sqrt{-3}}$.

(c) Solve this equation in $O_{\sqrt{m}}$, $m < 0$, $m \neq -1, -3$.

(d) Solve this equation in $O_{\sqrt{2}}$.

(e) Generalize the result of (d).

24. Let p be a rational prime with $p \equiv 1 \ (\text{mod } 3)$. Show that the equation

$$4p = x^2 + 27y^2$$

has a unique solution with $x \equiv 1 \ (\text{mod } 3)$. (Hint: work in $O_{\sqrt{-3}}$.)

25. (a) Solve the Diophantine equation $y^2 + 19 = x^3$.

(b) Solve the Diophantine equation $y^2 + 3 = x^3$.

(c) Suppose k is squarefree. Show that if $y^2 - k = x^3$ is soluable in rational integers, then we must have $(k, x) = 1$.

26. Let $p \equiv 1 \ (\text{mod } 4)$ be a prime and suppose $p = r^2 + s^2$. Show that all solutions of

$$x^2 + y^2 = pz^2,$$

with $z > 0$, are given by

$$x + iy = (\pm r \pm is)(a + ib)^2 \text{ and } z = a^2 + b^2,$$

where the signs are independent of each other.

27. In this problem we outline a proof of Theorem 7.55, the Lucas-Lehmer test for the primality of a Mersenne prime. Recall that the result states that if p is a prime, then $M_p = 2^p - 1$ is a prime if and only if M_p divides r_{p-1}, where $r_1 = 4$ and, for $n \geq 2$, $r_n = r_{n-1}^2 - 2$. This proof takes place in $\mathbf{Q}(\sqrt{3})$.

(a) Let $\sigma = 1 + \sqrt{3}$. Show that $\tau = \sigma / \bar{\sigma}$ is a unit in $O_{\sqrt{3}}$.

(b) Show that $r_m = \tau^{2^{m-1}} + \bar{\tau}^{2^{m-1}}$. (Hint: if $t_m = \tau^{2^{m-1}} + \bar{\tau}^{2^{m-1}}$, show that $t_m = t_{m-1}^2 - 2$ and $t_1 = 4$.)

Assume that M_p is a prime.

(c) Show that

$$3^{(M_p - 1)/2} \equiv -1 \pmod{M_p}.$$

(d) Show that

$$\sigma^{M_p} \equiv \bar{\sigma} \pmod{M_p}.$$

(e) Show that

$$\sigma^{2^{p-1}} \equiv -1 \pmod{M_p}.$$

(f) Combine (b) and (e) to show that $r_{p-1}^2 \equiv 0 \pmod{M_p}$ and hence $M_p | r_{p-1}$.

(g) Assume that M_p is composite. Show that there exists a rational prime q such that $q | M_p$ and $q \equiv \pm 5 \pmod{12}$. (Hint: to what residue class modulo 12 does M_p belong?) Show that there is a prime π in $O_{\sqrt{3}}$ such that $\pi | M_p$.

(Hint: see Problem 5 of Problem Set 9.3.)

Assume M_p is composite and $M_p | r_{p-1}$.

(h) Show that

$$r_p \equiv -2 \pmod{M_p}.$$

(i) Show that

$$\tau^{2^{p-2}} + \bar{\tau}^{2^{p-2}} \equiv 0 \pmod{M_p}.$$

(j) Show that

$$\tau^{2^{p-1}} \equiv -1 \pmod{M_p}$$

and hence that

$$\tau^{2^{p-1}} \equiv -1 \;(\text{mod } q) \text{ and } \tau^{2^{p}} \equiv 1 \;(\text{mod } q).$$

(k) Let $d = (q + 1, 2^p)$. Show that if $q \equiv 7 \;(\text{mod } 12)$.

(l) Conclude that if $q \equiv 7 \;(\text{mod } 12)$, then $q = M_p$.

(m) If $q \equiv 5 \;(\text{mod } 12)$, show that $q + 1 \equiv 0 \;(\text{mod } 2^{p+1})$.

(n) Show that if M_p is composite, then $M_p | r_{p-1}$.

(o) Conclude the validity of the Lucas-Lehmer test.

28. If θ is an algebraic number, let $\theta^{(1)}, \ldots, \theta^{(n)}$ be its field conjugates. Define the **trace** of by

$$tr(\theta) = \theta + \theta^{(1)} + \cdots + \theta^{(n)}.$$

(a) If θ_1 and θ_2 are elements of the field $\mathbf{Q}(\alpha)$ and $a \in \mathbf{Q}$, show that

  (i) $tr(\theta_1 + \theta_2) = tr(\theta_1) + tr(\theta_2)$

  and

  (ii) $tr(a\theta_1) = atr(\theta_1)$.

(b) Show that if θ is an algebraic integer in $\mathbf{Q}(\alpha)$, then $tr(\theta)$ is a rational integer.

(c) Show that if $\theta_1, \ldots, \theta_n$ are in $\mathbf{Q}(\alpha)$, then

$$\Delta(\theta_1, \ldots, \theta_n) = \begin{vmatrix} tr(\theta_1\theta_1) & \cdots & tr(\theta_1\theta_n) \\ & \vdots & \\ tr(\theta_n\theta_1) & \cdots & tr(\theta_n\theta_n) \end{vmatrix}.$$

(d) Show that if $a \in \mathbf{Q}$ and $\mathbf{Q}(\alpha)$ has degree n, then $tr(a) = na$.

(e) In $\mathbf{Q}(\zeta_p)$, where ζ_p is a pth root of unity, show that

$$tr\left(\zeta_p^k\right) = \begin{cases} -1 & \text{if } k \not\equiv 0 \;(\text{mod } p) \\ p-1 & \text{if } k \equiv 0 \;(\text{mod } p) \end{cases}.$$

(Hint: what are the conjugates of ζ_p?)

29. Let $\mathbf{Q}(\theta)$ be an algebraic number field of degree n and let $\alpha_1, \ldots, \alpha_n$ be a basis for $\mathbf{Q}(\theta)$, where $\alpha_1, \ldots, \alpha_n \in O_\theta$. Let $\Delta = \Delta(\alpha_1, \ldots, \alpha_n)$ be the discriminant of $\alpha_1, \ldots, \alpha_n$. Show that if $\beta \in O_\theta$, then there are rational integers r_k, $1 \le k \le n$, such that

$$\beta = \frac{1}{\Delta}(r_1\alpha_1 + \cdots + r_n\alpha_n).$$

30. Find all rational integers b and m such that $\left(1 + b\sqrt{m}\right) / \left(1 - b\sqrt{m}\right)$ is a unit of $O_{\sqrt{m}}$.

31. Let π be a prime in $O_{\sqrt{-3}}$ which is not associated to $1 - \omega$. (Recall that $\omega = \left(-1+\sqrt{-3}\right)/2$, see Example 10.9.)

   (a) Show that $3 \mid N(\pi) - 1$.

   (b) Show that if any two of 1, ω, ω^2 are congruent modulo π, then $1 \equiv \omega$ (mod π). Show that 1, ω and ω^2 are distinct modulo π.

32. If π is a prime in O_i not associated to $1 + i$ show that $4 \mid N(\pi) - 1$ and hence 1, -1, i and $-i$ are all distinct modulo π.

33. Let θ be an algebraic number. Show that $\mathbf{Q}(\theta) = \mathbf{Q}[\theta]$.

34. Let a and b be rational integers and suppose $(a, b) = d$. Let α be an integer in $\mathbf{Q}\left(\sqrt{m}\right)$ such that $\alpha \mid a$ and $\alpha \mid b$. Show that $\alpha \mid d$. In particular, show that in $O_{\sqrt{m}}$ the only divisors of a and b, if a and b are relatively prime, are units.

35. Let $m = 2p$, where $p \equiv 1$ (mod 4) is a prime. Let $q \equiv 5$ (mod 8) be a prime such that $\left(\frac{q}{p}\right) = -1$. (Such a prime exists by Problem 3 of Problem Set 9.4.)

   (a) Show that $x^2 \equiv 2p$ (mod q) is solvable.

   (b) If $O_{\sqrt{m}}$ is a UFD, show that there exist rational integers a and b such that
   $$a^2 - 2pb^2 = \pm q.$$

   (c) Conclude that $O_{\sqrt{m}}$ is not a UFD by examining the relation in (b) modulo p.

   (d) Can you extend this result to the case where $m = p_1 \cdots p_n$, $n \geq 2$, $p_1 \equiv 1$ (mod 4) and p_k odd, $1 \leq k \leq n$?

36. Let $\theta = \exp(2\pi i/5)$.

   (a) Shop that the integers of O_θ are given by
   $$\xi = a + b\theta + c\theta^2 + d\theta^3,$$
   where a, b, c and d are rational integers.

   (b) Show that the conjugates of ξ are obtained by changing θ into θ^2, θ^3 and θ^4, respectively.

   (c) If we define $N(\xi) = \xi\xi_1\xi_2\xi_3$, where ξ_i, $i = 1,2$ and 3, is a conjugate of ξ, show that O_θ is Euclidean.

   (d) Show that $(1 + \theta)^n$ is a unit for any rational integer n.

   (e) What are the primes of O_θ?

   (f) Prove that
   $$x^5 + y^5 = z^5,$$
   with $(x, y, z) = 1$, has no solutions in rational integers x, y and z with $xyz \neq 0$.

37. (a) Let $\theta = \exp(2\pi i/n)$, where $n \geq 3$. How many primitive nth roots of unity are there?

   (b) If $F_n(x)$ is the minimal polynomial of θ, show that

$$F_n(x) = \prod_{d\mid n} \left(x^{n/d} - 1\right)^{\mu(d)}.$$

   (c) If p is a rational prime, show that $\tau_p = \exp(2\pi i/p^2)$ is not an element of $\mathbf{Q}(\zeta_p)$, where $\zeta_p = \exp(2\pi i/p)$.

   (d) Show that the integers of $\mathbf{Q}(\zeta_p)$ are of the form

$$a_0 + a_1\zeta_p + \cdots + a_{p-2}\zeta_p^{p-2},$$

   where a_k, $0 \leq k \leq p - 2$, is a rational integer.

   (e) If $\lambda = 1 - \zeta_p$, show that λ is a prime in O_{ζ_p}. (Hint: compute $N(\lambda)$.)

   (f) Show that p is an associate of λ^{p-1}.

   (g) Show that if $\alpha, \beta \in O_{\zeta_p}$ and $\alpha \equiv \beta \pmod{\lambda}$, then $\alpha^p \equiv \beta^p \pmod{\lambda^p}$.

   (h) Show that $(1 - \zeta_p^k)/(1 - \zeta_p)$ is a unit for $1 \leq k \leq p - 1$.

   (i) Show that if $\alpha \in O_{\zeta_p}$ is a root of unity, then $\alpha = \pm\zeta_p^s$, for some s, $0 \leq s < p$

   (j) If $\alpha \in O_{\zeta_p}$, show that there exists a rational integer a such that

$$\alpha^p \equiv a \pmod{\lambda^p}.$$

   (k) Let ε be a unit of O_{ζ_p}. Show that $\varepsilon = \zeta_p^g \eta$, where η is a real unit and g is a rational integer.

38. In this problem we work in $\mathbf{Q}\left(\sqrt{-3}\right)$ and prove a law of cubic reciprocity.

   (a) Suppose π is a prime in $O_{\sqrt{-3}}$ such that $N(\pi) \neq 3$ and that $\pi \nmid \alpha$, where $\alpha \in O_{\sqrt{-3}}$. Show that there is a unique integer $m = 0$, 1 or 2 such that

$$\alpha^{(N(\pi)-1)/3} \equiv \omega^m \pmod{\pi}.$$

   (Hint: see Problem 20 above.)

   If $N(\pi) \neq 3$, then we define the **cubic residue character** of α modulo π, denoted by $\left(\frac{\alpha}{\pi}\right)_3$, to be

   (i) $\left(\dfrac{\alpha}{\pi}\right)_3 = 0$, if $\pi \mid \alpha$

and

   (ii) $\alpha^{(N(\pi)-1)/3} \equiv \left(\dfrac{\alpha}{\pi}\right)_3 \pmod{\pi}$, where $\left(\dfrac{\alpha}{\pi}\right)_3$ is 1, ω, ω^2.

(b) Prove the following properties of the cubic residue symbols.

(i) $\left(\dfrac{\alpha}{\pi}\right)_3 = 1$ if and only if $x^3 \equiv \alpha \pmod{\pi}$ is solvable.

(ii) $\alpha^{(N(\pi)-1)/3} \equiv \left(\dfrac{\alpha}{\pi}\right)_3 \pmod{\pi}$.

(iii) $\left(\dfrac{\alpha\beta}{\pi}\right)_3 = \left(\dfrac{\alpha}{\pi}\right)_3 \left(\dfrac{\beta}{\pi}\right)_3$.

(iv) If $\alpha \equiv \beta \pmod{\pi}$, then $\left(\dfrac{\alpha}{\pi}\right)_3 = \left(\dfrac{\beta}{\pi}\right)_3$.

(Note the similarity to the properties of the Legendre symbol.)

(c) Show that

(i) $\overline{\left(\dfrac{\alpha}{\pi}\right)_3} = \left(\dfrac{\alpha^2}{\pi}\right)_3$

and

(ii) $\overline{\left(\dfrac{\alpha}{q}\right)_3} = \left(\dfrac{\alpha^2}{q}\right)_3$.

(d) Let q be a rational prime. Show that $\left(\dfrac{\overline{\alpha}}{q}\right)_3 = \left(\dfrac{\alpha^2}{q}\right)_3$ and $\left(\dfrac{n}{q}\right)_3 = 1$ if n is a rational integer prime to q.

(e) If π is a prime, we say that π is **primary** if $\pi \equiv 2 \pmod 3$. Show that if $\pi = \alpha + \beta\omega$ is a complex prime, then π is primary if and only if $a \equiv 2 \pmod 3$ and $b \equiv 0 \pmod 3$. Show that if q is a rational prime that is prime in $O_{\sqrt{-3}}$, then q is primary.

(f) Suppose $N(\pi) = p \equiv 1 \pmod 3$. Show that among the associates of π exactly one is primary.

(g) Show that if π is a complex prime such that $N(\pi) = p \equiv 1 \pmod 3$, then the following results hold.

(i) We have
$$pJ\left(\left(\dfrac{\alpha}{\pi}\right)_3, \left(\dfrac{\alpha}{\pi}\right)_3\right) = G^3\left(\left(\dfrac{\alpha}{\pi}\right)_3, 1\right).$$

(Hint: recall Problem 6 of Problem 7.8 and Problem 52 of the Additional Problems for Chapter Seven.)

(ii) If
$$J\left(\left(\dfrac{\alpha}{\pi}\right)_3, \left(\dfrac{\alpha}{\pi}\right)_3\right) = a + b\omega,$$

then $a \equiv -1 \pmod 3$ and $b \equiv 0 \pmod 3$. (Hint: use (i) and the definition of $G(\chi, 1)$.)

(iii) If π is a primary prime, then

$$J\left(\left(\frac{\alpha}{\pi}\right)_3, \left(\frac{\alpha}{\pi}\right)_3\right) = \pi.$$

(Hint: first show that

$$J\left(\left(\frac{\alpha}{\pi}\right)_3, \left(\frac{\alpha}{\pi}\right)_3\right) \bar{J}\left(\left(\frac{\alpha}{\pi}\right)_3, \left(\frac{\alpha}{\pi}\right)_3\right) = p$$

and then use the definition of $J(\chi, \chi)$ and Problem 14 of Problem Set 2.3.)

(iv) We have

$$G^3\left(\left(\frac{\alpha}{\pi}\right)_3, 1\right) = p\pi.$$

(h) (Law of Cubic Reciprocity) Let π_1 and π_2 be primary, with $N(\pi_1)$, $N(\pi_2)$ $\neq 3$ and $N(\pi_1) \neq N(\pi_2)$. Show that

$$\left(\frac{\pi_2}{\pi_1}\right)_3 = \left(\frac{\pi_1}{\pi_2}\right)_3.$$

(i) Suppose $N(\pi) \neq 3$. If $\pi = q$ is rational, write $q = 3m - 1$ and if $\pi = a + b\omega$ is a primary complex prime, write $a = 3m - 1$. Show that

$$\left(\frac{1-\omega}{\pi}\right)_3 = \omega^{2m}.$$

There is a related biquadratic reciprocity law that operates in $\mathbf{Q}(i)$. See [56, ch. 9].

The remaining problems discuss the theory of ideals. Consider the algebraic number field $\mathbf{Q}(\theta)$ of degree n over $\mathbf{Q}$ with ring of integers O_θ. Let $\alpha_1, \ldots, \alpha_n \in O_\theta$. Then a set of the form

$$A = \left\{\lambda_1\alpha_1 + \cdots + \lambda_k\alpha_k : \lambda_1, \ldots, \lambda_k \in O_\theta\right\}$$

is called an **ideal** and the integers $\alpha_1, \ldots, \alpha_n$ are called the **generators** of A.

In some ways this may be thought of as a generalization of the greatest common divisor. Indeed we write $A = (\alpha_1, \ldots, \alpha_n)$

39. Prove the following properties of ideals.
  (a) If $\alpha, \beta \in O_\theta$ and $\alpha, \beta \in A$, A an ideal and $\lambda \in O_\theta$, then $\lambda\alpha \in A$ and $\alpha \pm \beta \in A$.

(b) We say that an ideal A is **principal** if we have $A = (\alpha)$ for some $\alpha \in O_\theta$. If $A = (\alpha)$ and $B = (\beta)$ are principal, then $A = B$ if and only if α and β are associates.

(c) If $\alpha_1, \ldots, \alpha_m, \beta_1, \ldots, \beta_k \in O_\theta$ and $A = (\alpha_1, \ldots, \alpha_m)$ and $B = (\beta_1, \ldots, \beta_k)$, then we define the **product** of A and B, denoted by AB, to be

$$AB = (\alpha_1\beta_1, \alpha_1\beta_2, \ldots, \alpha_1\beta_k, \alpha_2\beta_1, \ldots, \alpha_m\beta_k)$$

Show that multiplication of ideals is commutative and associative.

(d) We say that if A and B are ideals, then A **divides** B if there is an ideal C such that $B = AC$ and we say that A and C are **factors** of B. We write $A \mid B, C \mid B$.

(i) If $A = (\alpha)$ and $B = (\beta)$ are principal ideals, $B \mid A$ if and only if $\beta \mid \alpha$.

(ii) If $A \mid B$ and $B \mid C$, then $A \mid C$.

(iii) If D is any ideal, then $A \mid B$ implies $AD \mid BD$.

(iv) $(1) \mid A$ for all ideals A.

(v) $A \mid A$ for all ideals A.

(e) Recall that A and B are also sets. Then $A \mid B$ implies $A \subset B$.

(f) If A is an ideal and $A \mid (1)$, then $A = (1)$.

40. Let A be an ideal and suppose $\alpha_1, \ldots, \alpha_n \in A$. We say that $\alpha_1, \ldots, \alpha_n$ is a **basis** for A if $\alpha \in A$ implies that $\alpha = a_1\alpha_1 + \cdots + a_n\alpha_n$, for some rational integers a_k, $1 \le k \le n$. Show that if A is an ideal, $A \ne (0)$, then A has a basis.

41. Let n be a rational integer. Then there are only finitely many ideals $A_1, \ldots, A_m$ such that $n \in A_k$, $1 \le k \le m$.

42. Show that an ideal A has only a finite number of factors. (Hint: show that if $\alpha \in A$, then $N(\alpha) \in A$ and apply Problem 41 above.)

43. (a) Let $A = (\alpha_0, \alpha_1, \ldots, \alpha_m)$ be an ideal of the field $\mathbf{Q}(\theta)$ of degree n. Show that there is a polynomial $p_i(x)$ in $\mathbf{Z}[x]$ such that $\alpha_i = p_i(\theta)$.

(b) Let

$$g(x) = \prod_{j=1}^{n} \{p_0(\theta^{(j)})x^m + \cdots + p_m(\theta^{(j)})\} = \sum_{j=0}^{N} c_j x^{N-j}.$$

Show that the coefficients c_j are symmetric polynomials, with rational integer coefficients, of $\theta = \theta^{(1)}, \ldots, \theta^{(n)}$. Hence show that they are polynomials, with rational integer coefficients, in the coefficients $a_1, \ldots, a_n$ of the irreducible polynomial for θ.

(c) Show that the c_k, $0 \le k \le N$, are rational numbers. Indeed, show that they are rational integers.

(d) Let

$$f(x) = \alpha_0 x^m + \alpha_1 x^{m-1} + \cdots + \alpha_{m-1} x + \alpha_m.$$

Show that $f(x) \mid g(x)$ in $O_\theta[x]$.

(e) Let $g(x)/f(x) = h(x) \in O_\theta[x]$ (by (d)) and suppose

$$h(x) = \beta_0 x^k + \cdots + \beta_{k-1} x + \beta_k.$$

Let $B = (\beta_0, \ldots, \beta_k)$ be an ideal and let $c = \gcd(c_0, c_1, \ldots, c_N)$. Show that $A \cdot B = (c)$. (Hint: use Gauss' Lemma (Theorem 10.7) to show that $(c) \supset (\alpha_0\beta_0, \alpha_0\beta_1, \ldots, \alpha_m\beta_k) = A \cdot B$. Show that there exist rational integers $a_0, \ldots, a_N$ such that

$$a_0 c_0 + \cdots + a_N c_N = c.$$

to show that $(c) \subset A \cdot B$.

44. (a) If (γ) is a principal ideal and $(\gamma)A = (\gamma)B$, show that $A = B$.

(b) If A, B and C are ideals such that $AC = BC$, then show that $A = B$. (Hint: use (a) and 43(e).)

45. Show that if $A \supset B$, then $A \mid B$.

46. Show that if $A \mid B$ and $A \ne B$, then A has fewer factors than B.

47. We say that an ideal P is a **prime ideal** if $P \ne (1)$ and it has no factors except P and (1). Show that every ideal can be factored into prime ideals.

48. (a) If p and q are distinct rational primes and P is a prime ideal such that $P \mid (p)$, show that $P \nmid (q)$.

(b) Show that there exist infinitely many prime ideals in $Q(\theta)$.

49. (a) Let A and B be ideals of $Q(\theta)$. We say that an ideal D is the **greatest common ideal divisor** of A and B if

   (i) $D \mid A$ and $D \mid B$ and

   (ii) if $C \mid A$ and $C \mid B$, then $C \mid D$.

We denote this by $D = (A, B)$. Show that if A and B are any two ideals, then they have a gcd. (Hint: if $A = (\alpha_1, \ldots, \alpha_r)$ and $B = (\beta_1, \ldots, \beta_s)$, consider the ideal $(\alpha_1, \ldots, \alpha_r, \beta_1, \ldots, \beta_s)$.)

(b) Show that the elements of (A, B) are of the form $\alpha + \beta$, where $\alpha \in A$ and $\beta \in B$.

(c) We say that two ideals are coprime if $(A, B) = (1)$. Show that if $(A, B) = (1)$, then there exist $\alpha \in A$ and $\beta \in B$ such that $\alpha + \beta = 1$.

50. (a) Let P be a prime ideal and suppose $P \mid AB$. Show that $P \mid A$ or $P \mid B$.

(b) Let $\alpha, \beta \in O_\theta$. and P be a prime ideal such that $P \mid (\alpha)(\beta)$. Show that if $P \nmid (\alpha)$, then $P \mid (\beta)$.

(c) Let $A_1, \ldots, A_r$ be ideals and let P be a prime ideal. Show that if $P \mid A_1 \cdots A_r$, then $P \mid A_k$, for some k, $1 \le k \le r$.

51. Show that the ideals of $\mathbf{Q}(\theta)$ factor into prime ideals and that this factorization is unique up to order.

52. Let α and β be in O_θ. We say that α and β are **congruent modulo an ideal** A, written $\alpha \equiv \beta \pmod{A}$, if and only if $A \mid (\alpha - \beta)$.

(a) Show that congruence is an equivalence relation.

(b) Show that if $\alpha \equiv \beta \pmod{A}$ and $B \mid A$, then $\alpha \equiv \beta \pmod{A}$.

(c) Let α, β, γ and δ be in O_θ. Show that if $\alpha \equiv \beta \pmod{A}$ and $\gamma \equiv \delta \pmod{A}$, then $\alpha \pm \gamma \equiv \beta \pm \delta \pmod{A}$ and $\alpha\gamma \equiv \beta\delta \pmod{A}$.

(d) Show that the set of residue classes modulo classes modulo an ideal is finite. (Hint: let A be an ideal and let B be an ideal such that $A \cdot B = (c)$, with $c > 0$. If $\alpha \in O_\theta$ and $\omega_1, \ldots, \omega_n$ is an integral basis for O_θ, write

$$\alpha = a_1\omega_1 + \cdots + a_n\omega_n.$$

Now work modulo c.)

53. If A is an ideal we define the **norm** of A, written $N(A)$, to be the number of residue classes modulo A. Show that if $\alpha \in O_\theta$, then $N((\alpha)) = |N(\alpha)|$, where $N((\alpha))$ is the norm of the ideal (α) and $N(\alpha)$ is the ordinary norm.

54. (a) Let A be an ideal and let P be a prime ideal. Show that there is an $\alpha \in O_\theta$ such that $AP \nmid (\alpha)$, but $A \mid (\alpha)$.

(b) Let $\alpha_1, \ldots, \alpha_{N(A)}$ be representations of the residue classes modulo A and let $\pi_1, \ldots, \pi_{N(P)}$ be representatives of the residue classes modulo P. Consider the set $C = \{\alpha\pi_j + \alpha_k \colon 1 \le j \le N(P), 1 \le k \le N(A)\}$. Show that

(i) no two of the elements in C are congruent modulo AP and

(ii) every integer in O_θ is congruent to some element of C.

(c) Show that $N(A)N(P) = N(AP)$.

55. Let A, B and C be ideals such that $AB = C$. Show that $N(A)N(B) = N(C)$.

56. If A is an ideal, show that $A \mid (N(A))$

57. Let m be a positive rational integer. Show that there exist only finitely many ideals A such that $N(A) = m$.

58. Two ideals A and B are said to be **equivalent**, written $A \sim B$, if there exist integers $\alpha, \beta \in O_\theta$ such that $(\alpha)A = (\beta)B$. Show that the equivalence of ideals is an equivalence relation. The classes induced by $\sim$ are called the **ideal classes** of $\mathbf{Q}(\theta)$ and the class equivalent to (1) is called the **principal class**.

59. Let A, B, C and D be ideals. If $A \sim B$ and $C \sim D$, then $AC \sim BD$.

60. Show that there exists a positive rational integer m, depending on θ, so that if A is an ideal in $\mathbf{Q}(\theta)$, then there is an integer $\alpha \in O_\theta$, with $\alpha \in A$, such that $|N(\alpha)| \leq mN(A)$. (Hint: let $\omega_1, \ldots, \omega_n$ be an integral basis for O_θ and let

$$M = \prod_{k=1}^{n} \sum_{j=1}^{n} |\omega_j^{(k)}|.$$

Show that $m = [M] + 1$ will do. Let r be a positive rational integer such that

$$r^n \leq N(A) < (r + 1)^n$$

and consider the α's with $\alpha = a_1\omega_1 + \cdots + a_n\omega_n$, with $0 \leq a_j \leq r$. Now apply the pigeon hole principle.)

61. With m as in Problem 60, show that in each class of ideals there is an ideal A such that $N(A) \leq m$.

62. Show that for $\mathbf{Q}(\theta)$ the number of ideal classes is finite. The number of these ideal classes is denoted by h and is called the **class number** of $\mathbf{Q}(\theta)$.

63. Show that if h is the class number of $\mathbf{Q}(\theta)$ and A is an ideal, then A^h is a principal ideal.

64. Show that a necessary and sufficient condition for O_θ to be a UFD is that $h = 1$.

Computer Problems

1. (a) Let $\mathbf{Q}(\sqrt{m})$ be a Euclidean domain. Write a program that will compute the remainder in the division algorithm for two integers in $O_{\sqrt{m}}$.

   (b) Write a program to compute $\gcd(\alpha, \beta)$ if $\alpha, \beta \in O_{\sqrt{m}}$, a Euclidean domain.

   (c) Modify the program in (b) to write $\gcd(\alpha, \beta)$ as a linear combination of α and β.

   (d) Find the gcds of the following quadratic integers.

   (i) $1 + 2\sqrt{3}$, $5 + 4\sqrt{3}$ (iii) $7 + 5i$, $16 + 2i$

   (ii) $2 - 5i$, $6 + 7i$ (iv) $15 + 6\sqrt{7}$, $5 + 40\sqrt{7}$

(v) $5+\sqrt{2}$, $1-\sqrt{2}$ (vii) $5-\sqrt{-2}$, $6+19\sqrt{-2}$

(vi) $6-\sqrt{-3}$, $\left(5+3\sqrt{-3}\right)/2$ (viii) $\left(5+\sqrt{29}\right)/2$, $\left(7+3\sqrt{29}\right)/2$.

(e) Write the gcds found in (d) as linear combinations of the given pair of integers.

2. (a) Suppose $\mathbf{Q}\left(\sqrt{m}\right)$ is a UFD. Write a program to factorize integers in $O_{\sqrt{m}}$ by trial division. (Hint: recall that if $\beta\mid\alpha$, then $N(\beta)\le N(\alpha)$ so that we have an upper limit on the search.)

(b) Find the factors of

(i) 71 (in $\mathbf{Q}(i)$)   (iv) $84 + 49i$

(ii) 71 (in $\mathbf{Q}\left(\sqrt{-3}\right)$)   (v) $\left(17+5\sqrt{-3}\right)/2$

(iii) 71 (in $\mathbf{Q}\left(\sqrt{2}\right)$)   (vi) $25+3\sqrt{2}$.

(c) Modify the program in (a) to list all possible factorizations.

(d) Find all factorizations of the numbers given in (b).

3. (a) Suppose $\mathbf{Q}\left(\sqrt{m}\right)$ is a UFD. Modify the program of 2(a) so that it first factors $N(\alpha)$. Then factor the rational integer factors of $N(\alpha)$ over $\mathbf{Q}\left(\sqrt{m}\right)$ using Theorem 10.41.

(b) Find the factors of the numbers in 2(b) using the program in 3(a). (Hint: use Example 10.9.) Which program is faster?

4. (a) In Problem 15 of the Additional Problems above it is shown that $O_{\sqrt{2}+i}$ is Euclidean. Write a program to compute gcds in this set of integers.

(b) Find the greatest common divisor of the following pairs of integers of $O_{\sqrt{2}+i}$.

(i) $3+5i+\sqrt{2}-4\sqrt{-2}$, $1+i+\left(3\sqrt{2}-\sqrt{-2}\right)/2$

(ii) $i+\sqrt{-2}$, $1+\left(\sqrt{2}+\sqrt{-2}\right)/2$

(iii) $\sqrt{2}-\sqrt{-2}$, $3-i+\left(5\sqrt{2}-\sqrt{-2}\right)/2$

(iv) 4, 7

(c) Modify the program in (a) to write the gcd as a linear combination.

(d) Write the gcds found in (b) as linear combinations of the given pair of integers.

5. Repeat Problem 4 for the rings of integers $O_{\sqrt{2}+\sqrt{3}}$ (see Problem 16 of the Additional Problems above) and O_θ, $\theta = \exp(2\pi i/5)$ (see Problem 36 of the Additional Problems above) to the pairs of integers:

(a) $3 + 5\sqrt{3} + (\sqrt{2} + 5\sqrt{6})/2$, $1 + \sqrt{3} - 4\sqrt{2} + 7\sqrt{6}$

(b) $2 - \sqrt{3} + (7\sqrt{2} - \sqrt{6})/2$, $7 + 8\sqrt{3} - (-\sqrt{2} + 19\sqrt{6})/2$

(c) $1 + 2\theta + 5\theta^2 + \theta^3$, $3 + 7\theta + \theta^2 - 4\theta^3$

(d) $2\theta + 3\theta^3$, $1 + 5\theta^2$

6. (a) Write a program to implement the algorithm of Problem 17 of the Problem Set 10.4 to find rational solutions of $y^2 + k = x^3$.

(b) Given that (3, 5) and (3, -5) are solutions to $y^2 + 2 = x^3$ apply the program of (a) to find further rational solutions, if any.

(c) Given that (-1, 0), (0, 1), (0, -1), (2, 3) and (2, -3) are solutions to $y^2 - 1 = x^3$ apply the program of (a) to find further rational solutions. Do you find any? Can you explain this result?

(d) What happens with the equations $y^2 + 1 = x^3$ and $y^2 + 4 = x^3$?

7. Read Jeremiah 51:63.

TABLES

TABLE I

This table gives the least prime factor of all odd integers not divisible by 5 between 3 and 1000. When the odd integer is prime we have indicated this by putting the number in boldface.

	1	3	7	9		1	3	7	9
000	—	3	**7**	3	001	**11**	**13**	**17**	**19**
002	3	**23**	3	**29**	003	**31**	3	**37**	3
004	**41**	**43**	**47**	7	005	3	**53**	3	**59**
006	**61**	3	**67**	3	007	**71**	**73**	7	**79**
008	3	**83**	3	**89**	009	7	3	**97**	3
010	**101**	**103**	**107**	**109**	011	3	**113**	3	7
012	11	3	**127**	3	013	**131**	7	**137**	**139**
014	3	11	3	**149**	015	**151**	3	**157**	3
016	7	**163**	**167**	13	017	3	**173**	3	**179**
018	**181**	3	11	3	019	**191**	**193**	**197**	**199**
020	3	7	3	11	021	**211**	3	7	3
022	13	**223**	**227**	**229**	023	3	**233**	3	**239**
024	**241**	3	13	3	025	**251**	11	**257**	7
026	3	**263**	3	**269**	027	**271**	3	**277**	3
028	**281**	**283**	7	17	029	3	**293**	3	13
030	7	3	**307**	3	031	**311**	**313**	**317**	11
032	3	17	3	7	033	**331**	3	**337**	3
034	11	7	**347**	**349**	035	3	**353**	3	**359**
036	19	3	**367**	3	037	7	**373**	13	**379**
038	3	**383**	3	**389**	039	17	3	**397**	3
040	**401**	13	11	**409**	041	3	7	3	**419**

	1	3	7	9		1	3	7	9
042	421	3	7	3	043	431	433	19	439
044	3	443	3	449	045	11	3	457	3
046	461	463	467	7	047	3	11	3	479
048	13	3	487	3	049	491	17	7	499
050	3	503	3	509	051	7	3	11	3
052	521	523	17	23	053	3	13	3	7
054	541	3	547	3	055	19	7	557	13
056	3	563	3	569	057	571	3	577	3
058	7	11	587	19	059	3	593	3	599
060	601	3	607	3	061	13	613	617	619
062	3	7	3	17	063	631	3	7	3
064	641	643	647	11	065	3	653	3	659
066	661	3	23	3	067	11	673	677	7
068	3	683	3	13	069	691	3	17	3
070	701	19	7	709	071	3	23	3	719
072	7	3	727	3	073	17	733	11	739
074	3	743	3	7	075	751	3	757	3
076	761	7	13	769	077	3	773	3	19
078	11	3	787	3	079	7	13	797	17
080	3	11	3	809	081	811	3	19	3
082	821	823	827	829	083	3	7	3	839
084	29	3	7	3	085	23	853	857	859
086	3	863	3	11	087	13	3	877	3
088	881	883	887	7	089	3	19	3	29
090	17	3	907	3	091	911	11	7	919
092	3	13	3	929	093	7	3	937	3
094	941	23	947	13	095	3	953	3	7
096	31	3	967	3	097	971	7	977	11
098	3	983	3	23	099	991	3	997	3

TABLE II

This table gives, for the primes p, $3 \le p \le 500$, the least positive primitive root g modulo p. We have also given the prime factorization of $p - 1$ and indicated whether or not 10 is a primitive root modulo p with an asterisk.

p	$p-1$	g	p	$p-1$	g	p	$p-1$	g
3	2	2	109*	$2^2 \cdot 3^3$	6	263*	$2 \cdot 131$	5
5	2^2	2	113*	$2^4 \cdot 7$	3	269*	$2^2 \cdot 67$	2
7*	$2 \cdot 3$	3	127	$2 \cdot 3^2 \cdot 7$	3	271	$2 \cdot 3^3 \cdot 5$	6
11	$2 \cdot 5$	2	131*	$2 \cdot 5 \cdot 13$	2	277	$2^2 \cdot 3 \cdot 23$	5
13	$2^2 \cdot 3$	2	137	$2^3 \cdot 17$	3	281	$2^3 \cdot 5 \cdot 7$	3
17*	2^4	3	139	$2 \cdot 3 \cdot 23$	2	283	$2 \cdot 3 \cdot 47$	3
19*	$2 \cdot 3^2$	2	149*	$2^2 \cdot 37$	2	293	$2^2 \cdot 73$	2
23	$2 \cdot 11$	5	151	$2 \cdot 3 \cdot 5^2$	6	307	$2 \cdot 3^2 \cdot 17$	5
29*	$2^2 \cdot 7$	2	157	$2^2 \cdot 3 \cdot 13$	5	311	$2 \cdot 5 \cdot 31$	17
31	$2 \cdot 3 \cdot 5$	3	163	$2 \cdot 3^4$	2	313*	$2^3 \cdot 3 \cdot 13$	10
37	$2^2 \cdot 3^2$	2	167*	$2 \cdot 83$	5	317	$2^2 \cdot 79$	2
41	$2^3 \cdot 5$	6	173	$2^2 \cdot 43$	2	331	$2 \cdot 3 \cdot 5 \cdot 11$	3
43	$2 \cdot 3 \cdot 7$	3	179*	$2 \cdot 89$	2	337*	$2^4 \cdot 3 \cdot 7$	10
47*	$2 \cdot 23$	5	181*	$2^2 \cdot 3^2 \cdot 5$	2	347	$2 \cdot 173$	2
53	$2^2 \cdot 13$	2	191	$2 \cdot 5 \cdot 19$	19	349	$2^2 \cdot 3 \cdot 29$	2
59*	$2 \cdot 29$	2	193*	$2^6 \cdot 3$	5	353	$2^5 \cdot 11$	3
61*	$2^2 \cdot 3 \cdot 5$	2	197	$2^2 \cdot 7^2$	2	359	$2 \cdot 179$	7
67	$2 \cdot 3 \cdot 11$	2	199	$2 \cdot 3^2 \cdot 11$	3	367*	$2 \cdot 3 \cdot 61$	6
71	$2 \cdot 5 \cdot 7$	7	211	$2 \cdot 3 \cdot 5 \cdot 7$	2	373	$2^2 \cdot 3 \cdot 31$	2
73	$2^3 \cdot 3^2$	5	223*	$2 \cdot 3 \cdot 37$	3	379*	$2 \cdot 3^3 \cdot 7$	2
79	$2 \cdot 3 \cdot 13$	3	227	$2 \cdot 113$	2	383*	$2 \cdot 191$	5
83	$2 \cdot 41$	2	229*	$2^2 \cdot 3 \cdot 19$	6	389*	$2^2 \cdot 97$	2
89	$2^3 \cdot 11$	3	233*	$2^3 \cdot 29$	3	397	$2^2 \cdot 3^2 \cdot 11$	5
97*	$2^5 \cdot 3$	5	239	$2 \cdot 7 \cdot 17$	7	401	$2^4 \cdot 5^2$	3
101	$2^2 \cdot 5^2$	2	241	$2^4 \cdot 3 \cdot 5$	7	409	$2^3 \cdot 3 \cdot 17$	21
103	$2 \cdot 3 \cdot 17$	5	251	$2 \cdot 5^3$	6	419*	$2 \cdot 11 \cdot 19$	2
107	$2 \cdot 53$	2	257*	2^3	3	421	$2^2 \cdot 3 \cdot 5 \cdot 7$	2

p	$p-1$	g	p	$p-1$	g	p	$p-1$	g
431	$2 \cdot 5 \cdot 43$	7	457	$2^3 \cdot 3 \cdot 19$	13	479	$2 \cdot 239$	13
433*	$2^4 \cdot 3^3$	5	461*	$2^2 \cdot 5 \cdot 23$	2	487*	$2 \cdot 3^5$	3
439	$2 \cdot 3 \cdot 73$	15	463	$2 \cdot 3 \cdot 7 \cdot 11$	3	491*	$2 \cdot 5 \cdot 7^2$	2
443	$2 \cdot 13 \cdot 17$	2	467	$2 \cdot 233$	2	499*	$2 \cdot 3 \cdot 83$	7
449	$2^6 \cdot 7$	3						

TABLE III

This table gives the values of the arithmetic functions $d(n)$, $\sigma(n)$, $\varphi(n)$ and $\mu(n)$ for all n, $1 \le n \le 200$.

n	$d(n)$	$\sigma(n)$	$\varphi(n)$	$\mu(n)$	n	$d(n)$	$\sigma(n)$	$\varphi(n)$	$\mu(n)$
1	1	1	1	1	29	2	30	28	-1
2	2	3	1	-1	30	8	72	8	-1
3	2	4	2	-1	31	2	32	30	-1
4	3	7	2	0	32	6	63	16	0
5	2	6	4	-1	33	4	48	20	1
6	4	12	2	1	34	4	54	16	1
7	2	8	6	-1	35	4	48	24	1
8	4	15	4	0	36	9	91	12	0
9	3	13	6	0	37	2	38	36	-1
10	4	18	4	1	38	4	60	18	1
11	2	12	10	-1	39	4	56	24	1
12	6	28	4	0	40	8	90	16	0
13	2	14	12	-1	41	2	42	40	-1
14	4	24	6	1	42	8	96	12	-1
15	4	24	8	1	43	2	44	42	-1
16	5	31	8	0	44	6	84	20	0
17	2	18	16	-1	45	6	78	24	0
18	6	39	6	0	46	4	72	22	1
19	2	20	18	-1	47	2	48	46	-1
20	6	42	8	0	48	10	124	16	0
21	4	32	12	1	49	3	57	42	0
22	4	36	10	1	50	6	93	20	0
23	2	24	22	-1	51	4	72	51	1
24	8	60	12	0	52	6	98	24	0
25	3	31	20	0	53	2	54	52	-1
26	4	42	12	1	54	8	120	18	0
27	4	40	18	0	55	4	72	40	1
28	6	56	12	0	56	8	120	24	0

n	$d(n)$	$\sigma(n)$	$\varphi(n)$	$\mu(n)$	n	$d(n)$	$\sigma(n)$	$\varphi(n)$	$\mu(n)$
57	4	80	36	1	89	2	90	88	-1
58	4	90	28	1	90	12	234	24	0
59	2	60	58	-1	91	4	112	72	1
60	12	168	16	0	92	6	168	44	0
61	2	62	60	-1	93	4	128	60	1
62	4	96	30	1	94	4	144	46	1
63	6	104	36	0	95	4	120	72	1
64	7	127	32	0	96	12	252	32	0
65	4	84	48	1	97	2	98	96	-1
66	8	144	20	-1	98	6	171	42	0
67	2	68	66	-1	99	6	156	60	0
68	6	126	32	0	100	9	217	40	0
69	4	96	44	1	101	2	103	100	-1
70	8	144	24	-1	102	8	216	32	-1
71	2	72	70	-1	103	2	104	102	-1
72	12	195	24	0	104	8	210	48	0
73	2	74	72	-1	105	8	192	48	-1
74	4	114	36	1	106	4	162	52	1
75	6	124	40	0	107	2	108	106	-1
76	6	140	36	0	108	12	280	36	0
77	4	96	60	1	109	2	110	108	-1
78	8	168	24	-1	110	8	216	40	-1
79	2	80	78	-1	111	4	152	72	1
80	10	186	32	0	112	10	248	48	0
81	5	121	54	0	113	2	114	112	-1
82	4	126	40	1	114	8	240	36	-1
83	2	84	82	-1	115	4	144	88	1
84	12	224	24	0	116	6	210	56	0
85	4	108	64	1	117	6	182	72	0
86	4	132	42	1	118	4	180	58	1
87	4	120	56	1	119	4	144	96	1
88	8	180	40	0	120	16	360	32	0

n	$d(n)$	$\sigma(n)$	$\varphi(n)$	$\mu(n)$	n	$d(n)$	$\sigma(n)$	$\varphi(n)$	$\mu(n)$
121	3	133	110	0	153	6	234	96	0
122	4	186	60	1	154	8	288	60	-1
123	4	168	80	1	155	4	192	120	1
124	6	224	60	0	156	12	392	48	0
125	4	156	100	0	157	2	158	156	-1
126	12	312	36	0	158	4	240	78	1
127	2	128	126	-1	159	4	216	104	1
128	8	255	64	0	160	12	378	64	0
129	4	176	84	1	161	4	192	132	1
130	8	252	48	-1	162	10	363	54	0
131	2	132	130	-1	163	2	164	162	-1
132	12	336	40	0	164	6	294	80	0
133	4	160	108	1	165	8	288	80	-1
134	4	204	66	1	166	4	252	82	1
135	8	240	72	0	167	2	168	166	-1
136	8	270	64	0	168	16	480	48	0
137	2	138	136	-1	169	3	183	156	0
138	8	288	44	-1	170	8	324	64	-1
139	2	140	138	-1	171	6	260	108	0
140	12	336	48	0	172	6	308	84	0
141	4	192	92	1	173	2	174	172	-1
142	4	216	70	1	174	8	360	56	-1
143	4	168	120	1	175	6	248	120	0
144	15	403	48	0	176	10	372	80	0
145	4	180	112	1	177	4	240	116	1
146	4	222	72	1	178	4	270	88	1
147	6	228	84	0	179	2	180	178	-1
148	6	266	72	0	180	18	546	48	0
149	2	150	148	-1	181	2	182	180	-1
150	12	372	40	0	182	8	336	72	-1
151	2	152	150	-1	183	4	248	120	1
152	8	300	72	0	184	8	360	88	0

n	$d(n)$	$\sigma(n)$	$\varphi(n)$	$\mu(n)$	n	$d(n)$	$\sigma(n)$	$\varphi(n)$	$\mu(n)$
185	4	228	144	1	193	2	194	192	-1
186	8	384	60	-1	194	4	294	96	1
187	4	216	160	1	195	8	336	96	-1
188	6	336	92	0	196	9	399	84	0
189	8	320	108	0	197	2	198	196	-1
190	8	360	72	-1	198	12	468	60	0
191	2	192	190	-1	199	2	200	198	-1
192	14	508	64	0	200	12	465	80	0

TABLE IV

This table gives, for n a nonsquare and $2 \le n \le 99$, the continued fraction expansion of $\sqrt{n}$ and the fundamental solution to $x^2 - ny^2 = \pm 1$ with the last column indicating which equation.

n	$\sqrt{n} = [a_0; \overline{a_1, \ldots, a_r}]$	(x, y)	$(\pm)$
2	$\sqrt{2} = [1; \overline{2}]$	(1, 1)	$-$
3	$\sqrt{3} = [1; \overline{1, 2}]$	(2, 1)	$+$
5	$\sqrt{5} = [2; \overline{4}]$	(2, 1)	$-$
6	$\sqrt{6} = [2; \overline{2, 4}]$	(5, 2)	$+$
7	$\sqrt{7} = [2; \overline{1, 1, 1, 4}]$	(8, 3)	$+$
8	$\sqrt{8} = [2; \overline{1, 4}]$	(3, 1)	$+$
10	$\sqrt{10} = [3; \overline{6}]$	(3, 1)	$-$
11	$\sqrt{11} = [3; \overline{3, 6}]$	(10, 3)	$+$
12	$\sqrt{12} = [3; \overline{2, 6}]$	(7, 2)	$+$
13	$\sqrt{13} = [3; \overline{1, 1, 1, 1, 6}]$	(18, 5)	$-$
14	$\sqrt{14} = [3; \overline{1, 2, 1, 6}]$	(15, 4)	$+$
15	$\sqrt{15} = [3; \overline{1, 6}]$	(4, 1)	$+$
17	$\sqrt{17} = [4; \overline{8}]$	(4, 1)	$-$
18	$\sqrt{18} = [4; \overline{4, 8}]$	(17, 4)	$+$
19	$\sqrt{19} = [4; \overline{2, 1, 3, 1, 2, 8}]$	(170, 39)	$+$
20	$\sqrt{20} = [4; \overline{2, 8}]$	(9, 2)	$+$
21	$\sqrt{21} = [4; \overline{1, 1, 2, 1, 1, 8}]$	(55, 12)	$+$
22	$\sqrt{22} = [4; \overline{1, 2, 4, 2, 1, 8}]$	(197, 43)	$+$
23	$\sqrt{23} = [4; \overline{1, 3, 1, 8}]$	(24, 5)	$+$
24	$\sqrt{24} = [4; \overline{1, 8}]$	(5, 1)	$+$
26	$\sqrt{26} = [5; \overline{10}]$	(5, 1)	$-$
27	$\sqrt{27} = [5; \overline{5, 10}]$	(26, 5)	$+$
28	$\sqrt{28} = [5; \overline{3, 2, 3, 10}]$	(127, 24)	$+$
29	$\sqrt{29} = [5; \overline{2, 1, 1, 2, 10}]$	(70, 13)	$-$
30	$\sqrt{30} = [5; \overline{2, 10}]$	(11, 2)	$+$
31	$\sqrt{31} = [5; \overline{1, 1, 3, 5, 3, 1, 1, 10}]$	(1520, 273)	$+$
32	$\sqrt{32} = [5; \overline{1, 1, 1, 10}]$	(17, 3)	$+$

n	$\sqrt{n}=[a_0;\overline{a_1,\dots,a_r}]$	(x,y)	$(\pm)$
33	$\sqrt{33}=[5;\overline{1,2,1,10}]$	$(23,4)$	$+$
34	$\sqrt{34}=[5;\overline{1,4,1,10}]$	$(35,6)$	$+$
35	$\sqrt{35}=[5;\overline{1,10}]$	$(6,1)$	$+$
37	$\sqrt{37}=[6;\overline{12}]$	$(6,1)$	$-$
38	$\sqrt{38}=[6;\overline{6,12}]$	$(37,6)$	$+$
39	$\sqrt{39}=[6;\overline{4,12}]$	$(25,4)$	$+$
40	$\sqrt{40}=[6;\overline{3,12}]$	$(19,3)$	$+$
41	$\sqrt{41}=[6;\overline{2,2,12}]$	$(32,5)$	$-$
42	$\sqrt{42}=[6;\overline{2,12}]$	$(13,2)$	$+$
43	$\sqrt{43}=[6;\overline{1,1,3,1,5,1,3,1,1,12}]$	$(3482,531)$	$+$
44	$\sqrt{44}=[6;\overline{1,1,1,2,1,1,1,12}]$	$(199,30)$	$+$
45	$\sqrt{45}=[6;\overline{1,2,2,2,1,12}]$	$(161,24)$	$+$
46	$\sqrt{46}=[6;\overline{1,3,1,1,2,6,2,1,1,3,1,12}]$	$(24335,3588)$	$+$
47	$\sqrt{47}=[6;\overline{1,5,1,12}]$	$(48,7)$	$+$
48	$\sqrt{48}=[6;\overline{1,12}]$	$(7,1)$	$+$
50	$\sqrt{50}=[7;\overline{14}]$	$(7,1)$	$-$
51	$\sqrt{51}=[7;\overline{7,14}]$	$(50,7)$	$+$
52	$\sqrt{52}=[7;\overline{4,1,2,1,4,14}]$	$(649,90)$	$+$
53	$\sqrt{53}=[7;\overline{3,1,1,3,14}]$	$(182,25)$	$-$
54	$\sqrt{54}=[7;\overline{2,1,6,1,2,14}]$	$(485,66)$	$+$
55	$\sqrt{55}=[7;\overline{2,2,2,14}]$	$(89,12)$	$+$
56	$\sqrt{56}=[7;\overline{2,14}]$	$(15,2)$	$+$
57	$\sqrt{57}=[7;\overline{1,1,4,1,1,14}]$	$(151,20)$	$+$
58	$\sqrt{58}=[7;\overline{1,1,1,1,1,1,14}]$	$(99,13)$	$-$
59	$\sqrt{59}=[7;\overline{1,2,7,2,1,14}]$	$(530,69)$	$+$
60	$\sqrt{60}=[7;\overline{1,2,1,14}]$	$(31,4)$	$+$
61	$\sqrt{61}=[7;\overline{1,4,3,1,2,2,1,3,4,1,14}]$	$(29{,}718,3{,}805)$	$-$
62	$\sqrt{62}=[7;\overline{1,6,1,14}]$	$(63,8)$	$+$
63	$\sqrt{63}=[7;\overline{1,14}]$	$(8,1)$	$+$
65	$\sqrt{65}=[8;\overline{16}]$	$(8,1)$	$-$
66	$\sqrt{66}=[8;\overline{8,16}]$	$(65,8)$	$+$
67	$\sqrt{67}=[8;\overline{5,2,1,1,7,1,1,2,5,16}]$	$(48{,}842,5{,}967)$	$+$
68	$\sqrt{68}=[8;\overline{4,16}]$	$(33,4)$	$+$

n	$\sqrt{n} = [a_0; \overline{a_1, \ldots, a_r}]$	(x, y)	$(\pm)$
69	$\sqrt{69} = [8; \overline{3,3,1,4,1,3,3,16}]$	$(7{,}775, 935)$	$+$
70	$\sqrt{70} = [8; \overline{2,1,2,1,2,16}]$	$(251, 30)$	$+$
71	$\sqrt{71} = [8; \overline{2,2,1,7,2,2,16}]$	$(3{,}480, 413)$	$+$
72	$\sqrt{72} = [8; \overline{2,16}]$	$(17, 2)$	$+$
73	$\sqrt{73} = [8; \overline{1,1,5,5,1,1,16}]$	$(1{,}068,\ 125)$	$-$
74	$\sqrt{74} = [8; \overline{1,1,1,1,16}]$	$(43, 5)$	$-$
75	$\sqrt{75} = [8; \overline{1,1,1,16}]$	$(26, 3)$	$+$
76	$\sqrt{76} = [8; \overline{1,2,1,1,5,4,5,1,1,2,1,16}]$	$(57{,}799,\ 6{,}630)$	$+$
77	$\sqrt{77} = [8; \overline{1,3,2,3,1,16}]$	$(351, 40)$	$+$
78	$\sqrt{78} = [8; \overline{1,4,1,16}]$	$(53, 6)$	$+$
79	$\sqrt{79} = [8; \overline{1,7,1,16}]$	$(80, 9)$	$+$
80	$\sqrt{80} = [8; \overline{1,16}]$	$(9, 1)$	$+$
82	$\sqrt{82} = [9; \overline{1,18}]$	$(9, 1)$	$-$
83	$\sqrt{83} = [9; \overline{9,18}]$	$(82, 9)$	$+$
84	$\sqrt{84} = [9; \overline{6,18}]$	$(55, 6)$	$+$
85	$\sqrt{85} = [9; \overline{4,1,1,4,18}]$	$(378, 41)$	$-$
86	$\sqrt{86} = [9; \overline{3,1,1,1,8,1,1,1,3,18}]$	$(10{,}405,\ 1{,}122)$	$+$
87	$\sqrt{87} = [9; \overline{3,18}]$	$(28, 3)$	$+$
88	$\sqrt{88} = [9; \overline{2,1,1,1,2,18}]$	$(197, 21)$	$+$
89	$\sqrt{89} = [9; \overline{2,3,3,2,18}]$	$(500, 53)$	$-$
90	$\sqrt{90} = [9; \overline{2,18}]$	$(19, 2)$	$+$
91	$\sqrt{91} = [9; \overline{1,1,5,1,5,1,1,18}]$	$(1{,}574,\ 165)$	$+$
92	$\sqrt{92} = [9; \overline{1,1,2,4,2,1,1,18}]$	$(1{,}151,\ 120)$	$+$
93	$\sqrt{93} = [9; \overline{1,1,1,4,6,4,1,1,1,18}]$	$(12{,}151,\ 1{,}260)$	$+$
94	$\sqrt{94} = [9; \overline{1,2,3,1,1,5,1,8,1,5,1,1,3,2,1,18}]$	$(2{,}143{,}295,\ 221{,}064)$	$+$
95	$\sqrt{95} = [9; \overline{1,2,1,8}]$	$(39, 4)$	$+$
96	$\sqrt{96} = [9; \overline{1,3,1,18}]$	$(49, 5)$	$+$
97	$\sqrt{97} = [9; \overline{1,5,1,1,1,1,1,1,5,1,18}]$	$(5{,}604,\ 569)$	$-$
98	$\sqrt{98} = [9; \overline{1,8,1,18}]$	$(99, 10)$	$+$
99	$\sqrt{99} = [9; \overline{1,18}]$	$(10, 1)$	$+$

Bibliography

1. W. Adams, L. J. Goldstein, *Introduction to the Theory of Numbers*, Prentice-Hall, 1976.

2. T. M. Apostol, *Introduction to Analytic Number Theory*, Springer-Verlag, 1976.

3. —, Some properties of completely multiplicative arithmetical functions, Amer. Math. Monthly 78(1971), 266-271.

4. R. G. Archibald, *An Introduction to the Theory of Numbers*, Merrill, 1970.

5. R. Ayoub, *An Introduction to the Analytic Theory of Numbers,* Amer. Math. Soc., 1963.

6. R. G. Bartle, *The Elements of Real Analysis*,. John Wiley & Sons, 1964.

7. P. T. Bateman, On the representation of a number as the sum of three squares, Trans. Amer. Math. Soc. 71(1951), 70-101 .

8. E. Bombieri, Sulle formule di A. Selberg generalizzate per classe di funzioni arithmetiche e la applicazioni a problema del resto net "Primzahlsatz," Rev. Mat. Univ. Parma 3(1962), 293-340.

9. Z. I. Borevich, I. R. Shaferevich, *Number Theory*, Academic Press, 1966.

10. B. Bosworth, *Codes Ciphers and Computers*, Hayden, 1982.

11. C. Boyer, *A History of Mathematics*, John Wiley & Sons, 1990.

12. D. M. Bressoud, *Factorization and Primality Testing*, Springer-Verlag, 1989.

13. R. D. Carmichael, *Diophantine Analysis*, Dover, 1964.

14. J. W. S. Cassels, *.An Introduction to Diophantine Approximation* , Cambridge Univ. Press, 1957.

15. A. B. Chace, *The Rhind Mathematical Papyrus*, National Council of Teachers of Mathematics, 1979.

16. H. Chatland, H. Davenport, Euclid's algorithm in real quadratic fields, Can. J. Math. 2(1950), 289-296.

17. P. L. Chebychev, Memoire sur les nombres premiers, J. Math. pure et appl. 17(1852), 366-390.

18. G. Chrystal, *Algebra*, Chelsea, 1964.

19. H. Cohn, *Advanced Number Theory*, Dover, 1980.

20. D. A. Cox, *Primes of the Form x² +ny²,* Wiley Interscience, 1989.

21. H. Daboussi, Sur le theoreme de nombres premiers, C. R. Acad. Sci. Paris 298(1984), 161-164.

22. J-M. DeKoninck, A. Ivić, *Topics in Arithmetical Functions*, North Holland, 1980.

23. H. Diamond, Elementary methods in the study of the distribution of prime numbers, Bull. Amer. Math. Soc. 7(1982), 553-589.

24. —, J. Steinig, An elementary proof of the prime number theorem with remainder term, Invent. Math. 11(1970), 199-258.

25. L. E. Dickson, *The History of the Theory of Numbers*, Chelsea, 1950.

26. P. G. L. Dirichlet, *Vorlesungen über Zahlentheorie*, Chelsea, 1968.

27. H. E. Edwards, *Fermat's Last Theorem* , Springer-Verlag, 1978.

28. —, *Riemann's Zeta Function* , Academic Press, 1974.

29. R. B. Eggleton, C. B. Lacampagne, J. L. Selfridge, Euclidean quadratic fields, Amer. Math. Mon. 99(1992), 829-837.

30. W. Ellison, Waring's problem, Amer. Math. Mon. 78(1971),10-36.

31. —, F. Ellison, *Prime Numbers* , Wiley Interscience, 1985.

32. P. Erdös, On a new method in elementary number theory which leads to an elementary proof of the prime number theorem, Proc. Nat. Acad. Sci. USA 35(1949), 374-384.

33. E. B. Escott, Amicable numbers, Scripta Math. 12(1946), 61-72.

34. T. Estermann, On the representation of a number as the sum of three squares, Proc. London Math. Soc. 9(1959), 575-594.

35. Euclid, *The Elements*, Dover, 1956.

36. L. Euler, *Elements of Algebra*, Springer-Verlag, 1984.

37. G. Frei, Leonhard Euler's convenient numbers, Math. Intelligencer, 7(1985), 55-58, 64.

38. G. W. Fung, H. C. Williams, Quadratic polynomials which have a high density of prime values, Math. Comp. 55(1990), 345-353.

39. C. F. Gauss, *Disquistiones Arithmeticae*, Yale Univ. Press, 1966.

40. A. O. Gelfond, Yu. U. Linnik, *Elementary Methods in the Analytic Theory of Numbers*, MIT Press. 1966.

41. E. Grosswald, *Topics from the Theory of Numbers*, Birkhauser, 1984.

42. —, *Representations of Integers as Sums of Squares*, Springer-Verlag, 1985.

43. R. Guy, *Reviews in Number Theory*, Amer. Math. Soc., 1988.

44. J. Hadamard, Sur la distribution des zeros de la fonction $\zeta(s)$ et ses consequences arithmetique, Bull. Soc. Math. France 24(1896), 199-220.

45. H. Halberstam, H. E. Richert, *Sieve Methods*, Academic Press, 1974.

46. H. Hancock, *Foundations of the Theory of Algebraic Numbers*, Dover, 1964.

47. G. H. Hardy, E. M. Wright, *An Introduction to the Theory of Numbers*, Claredon Press, 1979.

48. T. L. Heath, *Diophantus of Alexandria*, Dover, 1964.

49. E. Hecke, *Lectures on of Algebraic Numbers*, Springer-Verlag, 1981.

50. E. Heppner, Die maximale Ordnung primzahl-unabhangiger multiplikativer Funktionen, Arch. Math. 24(1973), 63-66.

51. A. Hildebrand, Characterization of the logarithm as an additive arithmetic function, *Number Theory*, Walter de Gruyter, 1990.

52. A. P. Hillman, G. L. Alexanderson, *A First Undergraduate Course in Abstract Algebra*, Wadsworth, 1988.

53. L. K. Hua, *Introduction to Number Theory*, Springer-Verlag, 1982.

54. A. Hurwitz, *Lectures on Number Theory*, Springer-Verlag, 1986.

55. A. E. Ingham, *Distribution of Prime Numbers*, Hafner, 1971 .

56. K. Ireland, M. Rosen, *A Classical Introduction to Modern Number Theory*, Springer-Verlag, 1982.

57. A. Ivić, A property of Ramanujan's sum concerning totally multiplicative functions, Univ. Beograd Publ. Elektrotehn. Fak. Ser. Mat.-Fiz. 577-598(1977), 74-78.

58. —, *The Riemann Zeta-Function*, John Wiley & Sons, 1985.

59. J. Janusz, *Algebraic Number Fields*, Academic Press, 1973.

60. A. Y. Khinchin, *Continued Fractions*, Univ. Chicago Press, 1964.

61. N. Koblitz, *A Course in Number Theory and Cryptography*, Springer-Verlag, 1987.

62. R. Kortum, G. McNiel, *A Table of Periodic Continued Fractions*, Lockheed, 1961.

63. E. Lanczi, Unique prime factorization in imaginary number fields, Acta Math. Acad. Sci. Hung. 26(1965), 453-466.

64. E. Landau, *Elementary Number Theory*, Chelsea, 1966.

65. —, *Handbuch von der Lehre von der Verteilung der Primzahlen*, Chelsea, 1953.

66. S. Lang, *Algebraic Number Theory*, Addison-Wesley, 1970.

67. E. Lee, J. Madachy, The history and discovery of amicable numbers, J. Rec. Math. 5(1972), 77-93, 153-173, 231-249.

68. A. M. Legendre, *Essai sur la theorie des nombres*, Courcier, 1808.

69. Leonardo, *Book of Squares*, Academic Press, 1987.

70. W. J. LeVeque, *Reviews in Number Theory*, Amer. Math. Soc., 1974.

71. —, *Topics in Number Theory*, vol. I and II, Addison-Wesley, 1956.

72. B. V. Levin, A. S. Fainleib, Application of some integral equations to problems of number theory, Russian Math. Surveys 22(1967), 119-204.

73. N. Levinson, A motivated account of an elementary proof of the prime number theorem, Amer. Math. Monthly 76(1969), 225-244.

74. S. Louboutin, Prime producing quadratic polynomials and class-numbers of real quadratic fields, Can. J. Math. 42(1990), 315-341.

75. P. J. McCarthy, *Introduction to Arithmetic Functions*, Springer-Verlag, 1986.

76. W. Mills, A prime representing function, Bull. Amer. Math. Soc. 29(1954), 63-71.

77. L. J. Mordell, *Diophantine Equations*, Academic Press, 1969.

78. T. Nagell, *Introduction to Number Theory*, Chelsea, 1964.

79. J. Nagura, On the interval containing at least one prime number, Proc. Japan. Acad. 28(1952), 177-181.

80. W. Narkiewicz, *Elementary and Analytic Theory of Algebraic Numbers*, PWN, 1974.

81. —, *Number Theory*, World Scientific, 1983.

82. Nicomachus, *Introduction to Arithmetic*, Macmillan, 1926.

83. I. Niven, *Diophantine Approximation*, John Wiley & Sons, 1963.

84. —, H. S. Zuckerman, H. L. Montgomery, *Introduction to the Theory of Numbers*, John Wiley & Sons, 1980.

85. C. D. Olds, *Continued Fractions*, Random House, 1963.

86. O. Ore, *Number Theory and its History*, McGraw-Hill, 1949.

87. D. Parent, *Exercises in Number Theory*, Springer-Verlag, 1984.

88. S. J. Patterson, *An Introduction to the Theory of the Riemann Zeta-Function*, Cambridge Univ. Press, 1988.

89. O. Perron, *Die Lehre von der Kettenbrüchen*, Chelsea, 1950.

90. C. de la Vallee Poisson. Recherches analytique sur la theorie des nombres; premier partie: la fonction $\zeta(s)$ de Riemann et les nombres premiers en general, Ann. Soc. Sci. Bruxelles 20(1896), 183-256.

91. H. Pollard, H. G. Diamond. *The Theory of Algebraic Numbers* , Math. Assn. Amer., 1975.

92. A. G. Postnikov, *Introduction to Analytic Number Theory*, Amer. Math. Soc., 1988.

93. —, N. P. Romanov, A simplification of A. Selberg's proof of the asymptotic law of the distribution of prime numbers, Amer. Math. Soc. Trans. 113(1979)),75-87.

94. H. Rademacher, *Lectures on Elementary Number Theory*, Blaisdell, 1964.

95. —, E. Grosswald, *Dedekind Sums*, Math. Assn. Amer., 1972.

96. S. Ramanujan, Highly composite numbers, Proc. London Math. Soc. 14(1915), 347-409.

97. D. Redmond, Some remarks on a paper of A. Ivić, Univ. Beograd Publ. Elektrotehn. Fak. Ser. Mat.-Fiz. 656(1979), 137-142.

98. —, R. Sivaramakrishnan, Specially multiplicative functions, J. Number Theory 13(1981), 210-227.

99. P. Ribenboim, *Algebraic Numbers*, John Wiley & Sons, 1972.

100. —, *The Book of Prime Number Records*, Springer-Verlag, 1989.

101. —, *13 Lectures on Fermat's Last Theorem*, Springer-Verlag, 1979.

102. B. Riemann, Über die Anzahl der Primzahlen unter einer gegebener Grösse, *Gesammelte Werke*, Dover, 1953. (Translated in Edwards [28].)

103. H. Riesel, *Prime Numbers and Computer Methods for Factorization*, Birkhauser, 1985.

104. R. M. Robinson, Unsymmetrical approximation of irrational numbers, Bull. Amer. Math. Soc. 53(1947), 351-361.

105. J. B. Rosser, L. Schoenfeld, Approximate formulas for some functions of prime numbers, Ill. J. Math. 6(1962), 64-94.

106. W. Scharlau, H. Opolka, *From Fermat to Minkowski*, Springer-Verlag, 1985.

107. M. Schroeder, *Number Theory in Science and Communication*, Springer-Verlag, 1987.

108. A. Selberg, An elementary proof of Dirichlet's theorem about primes in arithmetic progressions, Ann. Math. 50(1949), 297-304.

109. —, An elementary proof of the prime number theorem, Ann Math. 50(1949), 305-313.

110. —, Notes on a paper by L. G. Sathe, J. Indian Math. Soc. 18(1954), 83-87.

111. J. Sesiano, *Books IV to VII of Diophantus' Arithmetica*, Springer-Verlag, 1982.

112. H. N. Shapiro, G. H. Sparer, Power quadratic Diophantine equations and descent, Comm. Pure Appl. Math. 3(1978),185-203.

113. P. Shiu, The maximum orders of multiplicative functions, Quart. J. Math. 31(1980), 247-252.

114. W. Sierpinski, *Pythagorean Triangles*, Scripta Math., 1962.

115. —, *Theory of Numbers*, PWN, 1964.

116. A. Sinkov, *Elementary Cryptanalysis*, Math. Assn. Amer., 1966.

117. R. Sivaramakrishnan, *Classical Theory of Arithmetical Functions*, Marcel Dekker, 1989.

118. D. D. Spence, *Computers in Number Theory*, Computer Science Press, 1982.

119. H. M. Stark, A complete determination of complex quadratic fields of class number one, Mich. Math. J. 14(1967),1-27.

120. J. Steinig, On Euler's idoneal numbers, Elem. Math. 21 (1966), 73-88.

121. B. Stolt, On a Diophantine equation of the second degree, Ark Mat. 3(1956), 381-390.

122. E. Trost, Primzahlen, Birkhauser, 1968.

123. P. Turán, Über einige verallgemeinerungen eines Satzes von Hardy und Ramanujan, J. London Math. Soc. 11(1936), 125-133.

124. J. Uspensky, M. A. Heaslet, *Elementary Number Theory*, McGraw-Hill, 1939.

125. I. M. Vinogradov, *Elements of Number Theory*, Dover, 1954.

126. A. Weil, *Number Theory: An Approach Through History*, Birkhauser, 1984.

127. E. Weiss, *Algebraic Number Theory*, McGraw-Hill, 1963.

128. A. E. Western, J. C. P. Miller, *Indices and Primitive Roots*, Cambridge Univ. Press, 1968.

129. H. Weyl, *Algebraic Number Theory*, Princeton Univ. Press, 1940.

130. E. E. Whitford, *Pell's Equation*

131. E. Wirsing, Elementare Beweise des Primzahlsatz mit Restglied, II, J. reine angew. Math. 214/215(1964), 1-18.

132. E. M. Wright, A class of representing functions, J. London Math. Soc. 29(1954), 63-71 .

Index